Student Solutions Guide for

Elementary and Intermediate Algebra: A Combined Course

Third Edition

Larson/Hostetler

Gerry C. Fitch
Carolyn F. Neptune

Houghton Mifflin Company

Boston   New York

Editor in Chief, Mathematics: Jack Shira
Managing Editor: Cathy Cantin
Senior Associate Editor: Maureen Ross
Associate Editor: Laura Wheel
Assistant Editor: Carolyn Johnson
Supervising Editor: Karen Carter
Project Editor: Patty Bergin
Editorial Assistant: Christine E. Lee
Art Supervisor: Gary Crespo
Marketing Manager: Ros Kane, Michael Busnach
Marketing Associate: Erin Dionne
Composition and Art: Meridian Creative Group
Senior Manufacturing Coordinator: Sally Culler

Printed in the U.S.A.

ISBN: 0-395-97646-4

123456789-HES-03 02 01 00 99

Preface

This *Student Solutions Guide* is a supplement to *Elementary and Intermediate Algebra: A Combined Course*, Third Edition, by Ron Larson and Robert P. Hostetler. This guide includes solutions for the odd-numbered exercises in the text including the integrated reviews, mid-chapter quizzes, chapter reviews, chapter tests and cumulative tests.

These solutions give step-by-step details of each exercise. The algebraic steps are clearly shown and explanatory comments are included where appropriate. There are usually several "correct" ways to arrive at a solution to a problem in mathematics. Therefore, you should not be concerned if you have approached problems differently than we have.

We have made every effort to see that the solutions are correct. However, we would appreciate hearing about any errors or other suggestions for improvement.

We hope you find the *Student Solutions Guide* to be a helpful supplement as you use the textbook, and we wish you well in your study of algebra.

Gerry C. Fitch
Carolyn F. Neptune

Contents

Part I
Solutions to Integrated Review Exercises

CHAPTER 2   Fundamentals of Algebra

SECTION 2.1   Writing and Evaluating Algebraic Expressions

1. Commutative Property of Multiplication

2. Additive Inverse Property

3. Distributive Property

4. Associative Property of Addition

5. $10 - |-7| = 10 - 7 = 3$

6. $6 - (10 - 12) = 6 - (-2)$
$$= 6 + 2$$
$$= 8$$

7. $\dfrac{3 - (5 - 20)}{4} = \dfrac{3 - (-15)}{4} = \dfrac{18}{4} = \dfrac{9}{2}$

8. $\dfrac{6}{7} - \dfrac{4}{7} = \dfrac{6 - 4}{7} = \dfrac{2}{7}$

9. $-\dfrac{3}{4}\left(\dfrac{28}{33}\right) = -\dfrac{\cancel{3}(\cancel{4})(7)}{\cancel{4}(\cancel{3})(11)} = -\dfrac{7}{11}$

10. $\dfrac{5}{8} \div \dfrac{3}{16} = \dfrac{5}{8} \cdot \dfrac{16}{3} = \dfrac{5(\cancel{8})(2)}{\cancel{8}(3)} = \dfrac{10}{3}$

11. $50(12)(10) = 6000$

You will set aside \$6000 during the 10 years.

12. $\frac{120}{8} = 15$

Each piece of rope will be 15 feet long.

SECTION 2.2   Simplifying Algebraic Expressions

1. $a^m \cdot a^n = a^{m+n}$

2. Distributive Property

3. $0 - (-12) = 0 + 12 = 12$

4. $60 - (-60) = 60 + 60 = 120$

5. $-12 - 2 + |-3| = -12 - 2 + 3 = -11$

6. $-730 + 1820 + 3150 + (-10,000) = -5760$

7. $72 + (-37) = 35$

8. $250 - 600 = -350$

9. $\dfrac{5}{16} - \dfrac{3}{10} = \dfrac{5(5)}{16(5)} - \dfrac{3(8)}{10(8)}$
$$= \dfrac{25}{80} - \dfrac{24}{80}$$
$$= \dfrac{1}{80}$$

10. $\dfrac{9}{16} + 2\dfrac{3}{12} = \dfrac{9}{16} + \dfrac{27}{12}$
$$= \dfrac{9(3)}{16(3)} + \dfrac{27(4)}{12(4)}$$
$$= \dfrac{27}{48} + \dfrac{108}{48}$$
$$= \dfrac{135}{48}$$
$$= \dfrac{45}{16}$$

11. $832,000 - (-1,530,000) = 832,000 + 1,530,000$

$$= 2,362,000$$

The profit during the last two quarters was \$2,362,000.

12. $\frac{676}{13} = 52$

The average speed was 52 miles per hour.

SECTION 2.3   Algebra and Problem Solving

1. The other factor is negative.

2. $7 + 4 + 4 = 15$

Because 15 is divisible by 3, 744 is divisible by 3.

3. False

$-4^2 = -16$

4. True

$(-4)^2 = 16$

5. $(-6)(-13) = 78$

6. $|4(-6)(5)| = |-120| = 120$

7. $\left(-\frac{4}{3}\right)\left(-\frac{9}{16}\right) = \frac{\cancel{4}(\cancel{3})(3)}{\cancel{3}(\cancel{4})(4)} = \frac{3}{4}$

8. $\frac{7}{8} \div \frac{3}{16} = \frac{7}{8} \cdot \frac{16}{3} = \frac{7(\cancel{8})(2)}{\cancel{8}(3)} = \frac{14}{3}$

9. $\left|-\frac{5}{9}\right| + 2 = \frac{5}{9} + \frac{18}{9} = \frac{23}{9}$

10. $-7\frac{3}{5} - 3\frac{1}{2} = -\frac{38}{5} - \frac{7}{2} = -\frac{76}{10} - \frac{35}{10} = -\frac{111}{10}$

11. $30(4) = 120$

After saving for 4 weeks, you would not have enough for the coat.

$$30(5) = 150$$

$150 - 133.50 = 16.50$

You must save for 5 weeks to buy the coat, and you will have \$16.50 left.

12. Width = 8 meters

Length = $1.5(8) = 12$ meters

Perimeter = 2(Length) + 2(Width)

$$= 2(12) + 2(8)$$

$$= 24 + 16$$

$$= 40$$

The perimeter of the rectangle is 40 meters.

SECTION 2.4   Introduction to Equations

1. Negative

2. Positive

The product of an even number of negative factors is positive.

3. $6 + 10 = 10 + 6$

4. Multiplicative Inverse Property

5. $t^2 \cdot t^5 = t^{2+5} = t^7$

6. $(-3y^3)y^2 = -3y^{3+2} = -3y^5$

7. $(u^3)^2 = u^{3(2)} = u^6$

8. $2(ab)^5 = 2a^5b^5$

9. $(3a^2)(4ab) = 12a^{2+1}b = 12a^3b$

10. $2(x+3)^2(x+3)^3 = 2(x+3)^{2+3} = 2(x+3)^5$

11. Perimeter: $4\left(\dfrac{3x}{2}\right) = \dfrac{12x}{2} = 6x$

Area: $\left(\dfrac{3x}{2}\right)^2 = \dfrac{(3x)^2}{2^2} = \dfrac{9x^2}{4}$

12. Perimeter: $(2x + 1) + (2x + 1) + (5x - 4) = 2x + 1 + 2x + 1 + 5x - 4 = 9x - 2$

Area: $\frac{1}{2}(2x)(5x - 4) = x(5x - 4) = 5x^2 - 4x$

CHAPTER 3   Linear Equations and Problem Solving

SECTION 3.1   Solving Linear Equations

1. (a) $(ab)^n = a^n b^n$

(b) $(a^m)^n = a^{mn}$

2. Associative Property of Addition

3. $(u^2)^4 = u^{2 \cdot 4} = u^8$

4. $(-3a^3)^2 = (-3)^2(a^3)^2 = 9a^6$

5. $-3(x - 5)^2(x - 5)^3 = -3(x - 5)^{2+3}$
$$= -3(x - 5)^5$$

6. $(4rs)(-5r^2)(2s^3) = -40r^{1+2}s^{1+3}$
$$= -40r^3 s^4$$

7. $\dfrac{2m^2}{3n} \cdot \dfrac{3m}{5n^3} = \dfrac{6m^{2+1}}{15n^{1+3}}$
$$= \dfrac{2m^3}{5n^4}$$

8. $\dfrac{5(x + 3)^2}{10(x + 8)} = \dfrac{\cancel{5}(x + 3)^2}{\cancel{5}(2)(x + 8)}$
$$= \dfrac{(x + 3)^2}{2(x + 8)}$$

9. $-3(3x - 2y) + 5y = -9x + 6y + 5y$
$$= -9x + 11y$$

10. $3v - (4 - 5v) = 3v - 4 + 5v$
$$= 8v - 4$$

11. $\frac{3}{4} - \frac{2}{3} = \frac{9}{12} - \frac{8}{12} = \frac{1}{12}$

The last person runs $\frac{1}{12}$ mile.

12. $10\frac{1}{3} + 7\frac{3}{5} + 12\frac{5}{6} = \frac{31}{3} + \frac{38}{5} + \frac{77}{6}$
$$= \frac{310}{30} + \frac{228}{30} + \frac{385}{30}$$
$$= \frac{923}{30}$$
$$= 30\frac{23}{30}$$

A total of $30\frac{23}{30}$ tons of soybeans were purchased during the first quarter.

SECTION 3.2   Equations That Reduce to Linear Form

1. (a) Add the numerators and write the sum over the common denominator.
$$\frac{1}{5} + \frac{7}{5} = \frac{1 + 7}{5} = \frac{8}{5}$$

(b) Determine the least common multiple of the denominators. Rewrite each fraction with that denominator. Then add the numerators and write the sum over the common denominator.
$$\frac{1}{5} + \frac{7}{3} = \frac{3}{15} + \frac{35}{15} = \frac{38}{15}$$

2. Answers will vary. Here are two examples:

$3x^2 + 2\sqrt{x}$ and $\dfrac{4}{x^2 + 1}$

3. $(-2x)^2 x^4 = (-2)^2 x^2 \cdot x^4 = 4x^6$

4. $-y^2(-2y)^3 = -y^2(-2)^3y^3$

$\qquad = -y^{2+3}(-8)$

$\qquad = 8y^5$

5. $5z^3(z^2)^2 = 5z^3 \cdot z^4 = 5z^7$

6. $(a+3)^2(a+3)^5 = (a+3)^{2+5}$

$\qquad = (a+3)^7$

7. $\dfrac{5x}{3} - \dfrac{2x}{3} - 4 = \dfrac{5x - 2x}{3} - 4$

$\qquad = \dfrac{3x}{3} - 4$

$\qquad = x - 4$

8. $2x^2 - 4 + 5 - 3x^2 = (2x^2 - 3x^2) + (-4 + 5)$

$\qquad = -x^2 + 1$

9. $-y^2(y^2 + 4) + 6y^2 = -y^4 - 4y^2 + 6y^2$

$\qquad = -y^4 + 2y^2$

10. $5t(2 - t) + t^2 = 10t - 5t^2 + t^2$

$\qquad = -4t^2 + 10t$

11. $1 - \frac{5}{8} = \frac{8}{8} - \frac{5}{8} = \frac{3}{8}$

$\frac{3}{8}(20) = \frac{60}{8} = 7.5$

Thus, 7.5 gallons of gasoline were used.

12. (a) Total = Down payment + 36(Monthly payments)

$\qquad = 1800 + 36(625)$

$\qquad = 1800 + 22,500$

$\qquad = 24,300$

You will pay a total of $24,300.

(b) $24,300 - 19,999 = 4301$

The finance charges and other fees cost $4301.

SECTION 3.3 Problem Solving with Percents

1.

When the two numbers are graphed on a number line, -28 is to the left of 63; therefore $-28 < 63$.

2. 0

3. $8 - |-7 + 11| + (-4) = 8 - |4| - 4$

$\qquad = 8 - 4 - 4$

$\qquad = 0$

4. $34 - [54 - (-16 + 4) + 6] = 34 - [54 - (-12) + 6]$

$\qquad = 34 - [54 + 12 + 6]$

$\qquad = 34 - 72$

$\qquad = -38$

5. $-300 - 230 = -530$

6. $|17 - (-12)| = |17 + 12| = 29$

7. $4(2x - 5) = 4(2x) - 4(5) = 8x - 20$

8. $-z(xz - 2y^2) = -z(xz) - (-z)(2y^2)$

$\qquad = -xz^2 + 2y^2z$

9. (a) $4^2 - 3^2 = 16 - 9 = 7$

(b) $(-5)^2 - 3^2 = 25 - 9 = 16$

10. (a) $\dfrac{1^2 + 2}{1^2 - 1} = \dfrac{1 + 2}{1 - 1} = \dfrac{3}{0}$; undefined

Division by 0 is undefined.

(b) $\dfrac{2^2 + 2}{2^2 - 1} = \dfrac{4 + 2}{4 - 1} = \dfrac{6}{3} = 2$

11. Cost of call = Cost of 1st minute + Cost of additional minutes

$$= 1.37 + 0.95(15 - 1)$$

$$= 1.37 + 0.95(14)$$

$$= 1.37 + 13.30$$

$$= 14.67$$

The cost of the call is $14.67.

12. Distance = (Rate)(Time)

$$= r(5)$$

$$= 5r$$

SECTION 3.4   Ratios and Proportions

1. Divide out any common factors from the numerator and denominator.

$$\frac{15}{12} = \frac{3(5)}{3(4)} = \frac{5}{4}$$

2. Multiply the first fraction by the reciprocal of the divisor.

$$\frac{3}{5} \div \frac{x}{2} = \frac{3}{5} \cdot \frac{2}{x} = \frac{6}{5x}$$

3. $(3x)y = 3(xy)$

4. Additive Identity Property

5. $3^2 - (-4) = 9 + 4 = 13$

6. $(-5)^3 + 3 = -125 + 3 = -122$

7. $9.3 \times 10^6 = 9,300,000$

8. $\dfrac{-|7 + 3^2|}{4} = \dfrac{-|7 + 9|}{4}$

$$= \frac{-|16|}{4}$$

$$= \frac{-16}{4}$$

$$= -4$$

9. $(-4)^2 - (30 \div 50) = 16 - \dfrac{30}{50}$

$$= \frac{80}{5} - \frac{3}{5}$$

$$= \frac{77}{5}$$

10. $(8 \cdot 9) + (-4)^3 = 72 - 64 = 8$

11. $2(n - 10)$

12. Area = $\frac{1}{2}$(Base)(Height)

$$= \tfrac{1}{2}b\left[\tfrac{1}{2}(b + 6)\right]$$

$$= \tfrac{1}{4}b(b + 6)$$

SECTION 3.5   Geometric and Scientific Applications

1. If n is an integer, $2n$ is an even integer and $2n + 1$ is an odd integer.

2. $\quad 2x - 3 = 10$

$$2x - 3 + 3 = 10 + 3$$

$$2x = 13$$

3. $(-3.5y^2)(8y) = -28y^{2+1}$

$$= -28y^3$$

4. $(-3x^2)^4 = (-3)^4 x^{2 \cdot 4} = 81x^8$

5. $\dfrac{24u}{15} \cdot \dfrac{25u^2}{6} = \dfrac{4(6)(5)(5)u^{1+2}}{3(5)(6)}$

$$= \frac{20u^3}{3}$$

6. $12\left(\dfrac{3y}{18}\right) = \dfrac{2(6)(3)y}{6(3)} = 2y$

7. $5x(2 - x) + 3x = 10x - 5x^2 + 3x$
$$= -5x^2 + 13x$$

8. $3t - 4(2t - 8) = 3t - 8t + 32$
$$= -5t + 32$$

9. $3(v - 4) + 7(v - 4) = 3v - 12 + 7v - 28$
$$= (3v + 7v) + (-12 - 28)$$
$$= 10v - 40$$

Alternate method:

$3(v - 4) + 7(v - 4) = 10(v - 4)$
$$= 10v - 40$$

10. $5[6 - 2(x - 3)] = 5[6 - 2x + 6]$
$$= 5[-2x + 12]$$
$$= -10x + 60$$

11. Total cost = Computer cost + Tax

Total cost = Computer cost + (Computer cost)(Tax rate)

$2915 = 2750 + 2750p$

$165 = 2750p$

$\frac{165}{2750} = p$

$0.06 = p$

The sales tax rate is 6%.

12. Catalog: $109.95 + $14.25 = $124.20

Store: $139.99(100\% - 20\%) = $139.99(0.80) \approx 111.99$

The store's 20% off sale is the better bargain.

SECTION 3.6   Linear Inequalities

1. Distributive Property

2. $(x + 2) - 4 = x + (2 - 4)$

3. If $a < 0$, then $|a| = -a$.

4. If $a < 0$ and $b > 0$, then $a \cdot b < 0$.

5. $-\frac{1}{2} > -7$

6. $-\frac{1}{3} < -\frac{1}{6}$

7. $-\pi < -3$

8. $-6 > -\frac{13}{2}$

9. Perimeter $= (x^2 + 2x) + x^2 + (5x - 3)$
$$= x^2 + 2x + x^2 + 5x - 3$$
$$= 2x^2 + 7x - 3$$
Area $= \frac{1}{2}x^2(5x - 3)$
$$= \frac{5}{2}x^3 - \frac{3}{2}x^2$$

10. Perimeter $= (2y - 1) + (2y - 1) + (2y^2 - 4y + 2)$
$$= 2y^2$$
Area $= \frac{1}{2}(2y^2 - 4y + 2)(y)$
$$= (y^2 - 2y + 1)y$$
$$= y^3 - 2y^2 + y$$

11. $-312,500 + 275,500 + 297,750 + 71,300 = 332,050$

The profit for the year was $332,050.

12. (Rate)(Time) = Distance

$r(7) = 371$

$r = \frac{371}{7}$

$r = 53$

Their average speed was 53 miles per hour.

SECTION 3.7   Absolute Value Equations and Inequalities

1. If n is an integer, then $2n$ is an even integer and $2n - 1$ is an odd integer.

2. No, they are not equal.

$(-3x)^2 = (-3)^2x^2$

$\qquad = 9x^2 \neq -3x^2$

3. Divide numerator and denominator by the common factor of 3.

$\dfrac{27}{12} = \dfrac{3(9)}{3(4)} = \dfrac{9}{4}$

4. Multiply the first fraction by the reciprocal of the divisor.

$\dfrac{2}{3} \div \dfrac{5}{3} = \dfrac{2}{3} \cdot \dfrac{3}{5} = \dfrac{2(3)}{3(5)} = \dfrac{2}{5}$

5. $3 > -2$

6. $-3 < -2$

7. $-\frac{1}{2} > -3$

8. $-\frac{1}{3} > -\frac{2}{3}$

9. $\frac{1}{2} > \frac{5}{16}$

10. $4 < \frac{45}{11}$

11. $76{,}300 - 75{,}926 = 374$ and $374 < 500$

No, the \$374 difference is less than \$500.

12. $39{,}632 - 37{,}800 = 1832$ and $1832 > 500$

Yes, the difference is more than \$500.

CHAPTER 4   Graphs and Functions

SECTION 4.1   Ordered Pairs and Graphs

1. Yes, $3x = 7$ is a linear equation because it has the form $ax + b = c$. No, $x^2 + 3x = 2$ is not a linear equation because it cannot be written in the form $ax + b = c$; it has an x^2-term.

2. Substitute 3 for x in the equation to verify that it satisfies the equation.

$5(3) - 4 = 11$

3. $-y = 10$

$\quad y = -10$

4. $10 - t = 6$

$\quad -t = -10 + 6$

$\quad -t = -4$

$\quad\ \ t = 4$

5. $3x = 42$

$\quad x = \frac{42}{3}$

$\quad x = 14$

6. $64 - 16x = 0$

$\quad -16x = -64$

$\qquad x = \dfrac{-64}{-16}$

$\qquad x = 4$

7. $125(r - 1) = 625$

$\quad 125r - 125 = 625$

$\qquad\ \ 125r = 750$

$\qquad\quad r = \dfrac{750}{125}$

$\qquad\quad r = 6$

8. $2(3 - y) = 7y + 5$

$\quad 6 - 2y = 7y + 5$

$\quad 6 - 9y = 5$

$\qquad -9y = -1$

$\qquad\quad y = \dfrac{-1}{-9}$

$\qquad\quad y = \dfrac{1}{9}$

9. $\quad 20 - \dfrac{1}{9}x = 4$

$9\left(20 - \dfrac{1}{9}x\right) = 9(4)$

$\quad 180 - x = 36$

$\qquad -x = -144$

$\qquad\ \ x = 144$

10. $0.35x = 70$

$\quad x = \dfrac{70}{0.35}$

$\quad x = 200$

11. Cost of lot + Cost of house = 154,000

$$x + 7x = 154,000$$

$$8x = 154,000$$

$$x = 19,250$$

The cost of the lot is $19,250.

12. Amount earned on 1st job + Amount earned on 2nd job = 450

$$9.50(40) + 8t = 450$$

$$380 + 8t = 450$$

$$8t = 70$$

$$t = \frac{70}{8}$$

$$t = \frac{35}{4}$$

You should work $8\frac{3}{4}$ hours, or 8 hours and 45 minutes, at the second job.

SECTION 4.2   Graphs of Equations in Two Variables

1. $x - 2 + c > 5 + c$

2. $(x - 2)c > 5c$

3. $x\left(\dfrac{1}{x}\right) = 1$

4. Commutative Property of Addition

5. $-3(3x - 2y) + 5y = -9x + 6y + 5y = -9x + 11y$

6. $3z - (4 - 5z) = 3z - 4 + 5z = 8z - 4$

7. $-y^2(y^2 + 4) + 6y^2 = -y^4 - 4y^2 + 6y^2 = -y^4 + 2y^2$

8. $5t(2 - t) + t^2 = 10t - 5t^2 + t^2$

$$= -4t^2 + 10t$$

9. $3[6x - 5(x - 2)] = 3[6x - 5x + 10]$

$$= 3[x + 10]$$

$$= 3x + 30$$

10. $5(t - 2) - 5(t - 2) = 5t - 10 - 5t + 10 = 0$

11. (a) Cost per day + Mileage cost = 50.80

$$30 + 0.32x = 50.80$$

$$0.32x = 20.80$$

$$x = \frac{20.80}{0.32}$$

$$x = 65$$

She drives 65 miles.

(b) The minimum cost for 2 days is 2($30) = $60. Her bill was less than $60, so it was for 1 day.

(c) There are three possibilities:

1 day: $30 + 0.32x = 96.80$

$$0.32x = 66.80$$

$$x = \frac{66.80}{0.32}$$

$$x = 208.75 \text{ miles}$$

2 days: $2(30) + 0.32x = 96.80$

$$60 + 0.32x = 96.80$$

$$0.32x = 36.80$$

$$x = \frac{36.80}{0.32}$$

$$x = 115 \text{ miles}$$

3 days: $2(30) + 0.32x = 96.80$

$$90 + 0.32x = 96.80$$

$$0.32x = 6.80$$

$$x = \frac{6.80}{0.32}$$

$$x = 21.25 \text{ miles}$$

12. 2(Length) + 2(Width) = Perimeter

$$2(x) + 2\left(\frac{3}{5}x\right) = 80$$

$$2x + \frac{6x}{5} = 80$$

$$5\left(2x + \frac{6x}{5}\right) = 5(80)$$

$$10x + 6x = 400$$

$$16x = 400$$

$$x = \frac{400}{16}$$

$$x = 25 \text{ and } \frac{3}{5}(25) = 15$$

The dimensions of the mirror are 25 inches by 15 inches.

SECTION 4.3   Relations, Functions, and Graphs

1. $a < c$

Transitive Property

2. $7x = 21$

$$\frac{7x}{7} = \frac{21}{7}$$

$$x = 3$$

3. $4s - 6t + 7s + t = (4s + 7s) + (-6t + t) = 11s - 5t$

4. $2x^2 - 4 + 5 - 3x^2 = (2x^2 - 3x^2) + (-4 + 5) = -x^2 + 1$

5. $\dfrac{5}{3}x - \dfrac{2}{3}x - 4 = \dfrac{5 - 2}{3}x - 4$

$$= x - 4$$

6. $3x^2y + xy - xy^2 - 6xy = 3x^2y - x^2y + (xy - 6xy)$

$$= 3x^2y - x^2y - 5xy$$

7. $3x + 9 = 0$

$$3x = -9$$

$$x = -3$$

8. $\dfrac{x}{4} + \dfrac{x}{3} = \dfrac{1}{3}$

$$12\left(\frac{x}{4} + \frac{x}{3}\right) = 12\left(\frac{1}{3}\right)$$

$$3x + 4x = 4$$

$$7x = 4$$

$$x = \frac{4}{7}$$

9. $\dfrac{2x - 3}{4} = \dfrac{3}{2}$

$$2(2x - 3) = 4(3)$$

$$4x - 6 = 12$$

$$4x = 18$$

$$x = \frac{18}{4}$$

$$x = \frac{9}{2}$$

10. $-(4 - 3x) = 2(x - 1)$

$$-4 + 3x = 2x - 2$$

$$-4 + x = -2$$

$$x = 2$$

11. Interest = 8190 − 7500 = 690

Interest = Prt

$$690 = 7500(r)(1)$$

$$690 = 7500r$$

$$\frac{690}{7500} = r$$

$$0.092 = r$$

The interest rate is 9.2%.

12. Distance = (Rate)(Time)

$$2500 = r(3)$$

$$\frac{2500}{3} = r$$

$$833\frac{1}{3} = r$$

The average speed is $833\frac{1}{3}$ miles per hour.

SECTION 4.4   Slope and Graphs of Linear Equations

1. Equivalent equations

2. $5x = 6 + 2$

3. $(x^2)^3 \cdot x^3 = x^6 \cdot x^3 = x^9$

4. $(y^2z^3)(z^2) = y^2z^{3+2} = y^2z^5$

5. $(u^4v^2)^2 = u^{4\cdot2}v^{2\cdot2} = u^8v^4$

6. $(ab)^4 = a^4b^4$

7. $(25x^3)(2x^2) = 50x^{3+2} = 50x^5$

8. $(3yz)^2(6yz^3) = 9y^2z^2(6yz^3) = 54y^3z^5$

9. $x^2 - 2x - x^2 + 3x + 2 = (x^2 - x^2) + (-2x + 3x) + 2$
$$= x + 2$$

10. $x^2 - 5x - 2 + x = x^2 + (-5x + x) - 2$
$$= x^2 - 4x - 2$$

11. $x + x + 3(x) = 10$
$$5x = 10$$
$$x = 2 \text{ and } 3(2) = 6$$
The length of each of the two shorter pieces is 2 feet and the length of the longer piece is 6 feet.

12. Cost of parts + Cost of labor = Total bill
$$65 + 32t = 113$$
$$32t = 48$$
$$t = \frac{48}{32}$$
$$t = \frac{3}{2}$$
The repair work took 1.5 hours.

SECTION 4.5   Equations of Lines

1. Find the prime factorization of each number. The greatest common factor is the product of all the common prime factors.
$$180 = 2^2 \cdot 3^2 \cdot 5$$
$$300 = 2^2 \cdot 3 \cdot 5^2$$
The greatest common factor is $2^2 \cdot 3 \cdot 5 = 60$.

2. Find the prime factorization of each number. The least common multiple is the product of the highest powers of the prime factors of the numbers.
$$180 = 2^2 \cdot 3^2 \cdot 5$$
$$300 = 2^2 \cdot 3 \cdot 5^2$$
The least common multiple is $2^2 \cdot 3^2 \cdot 5^2 = 900$.

3. $4(3 - 2x) = 12 - 8x$

4. $x^2(xy^3) = x^{2+1}y^3 = x^3y^3$

5. $3x - 2(x - 5) = 3x - 2x + 10 = x + 10$

6. $u - [3 + (u - 4)] = u - [3 + u - 4]$
$$= u - [u - 1]$$
$$= u - u + 1$$
$$= 1$$

7. $3x + y = 4$
$$y = -3x + 4$$

8. $4 - y + x = 0$
$$-y = -x - 4$$
$$y = x + 4$$

9. $4x - 5y = -2$
$$-5y = -4x - 2$$
$$y = \frac{-4x}{-5} + \frac{-2}{-5}$$
$$y = \frac{4}{5}x + \frac{2}{5}$$

10. $3x + 4y - 5 = 0$
$$4y = -3x + 5$$
$$y = -\frac{3}{4}x + \frac{5}{4}$$

SECTION 4.6   Graphs of Linear Inequalities

1. $a + 5 < b + 5$

2. $2a < 2b$

3. $-3a > -3b$

4. $a < c$

5. $x + 3 > 0$

$x > -3$

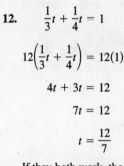

6. $2 - x \geq 0$

$-x \geq -2$

$x \leq 2$

7. $2t - 11 \leq 5$

$2t \leq 16$

$t \leq 8$

8. $\dfrac{3}{2}y + 8 < 20$

$2\left(\dfrac{3}{2}y + 8\right) < 2(20)$

$3y + 16 < 40$

$3y < 24$

$y < 8$

9. $5 < 2x + 3 < 15$

$5 - 3 < 2x + 3 - 3 < 15 - 3$

$2 < 2x < 12$

$\dfrac{2}{2} < \dfrac{2x}{2} < \dfrac{12}{2}$

$1 < x < 6$

10. $-2 < -\dfrac{x}{4} < 1$

$-2(4) < -\dfrac{x}{4}(4) < 1(4)$

$-8 < -x < 4$

$8 > x > -4$

11. $a = pb$

$544.50 = 0.045b$

$\dfrac{544.50}{0.045} = b$

$12{,}100 = b$

The sales were \$12,100.

12. $\dfrac{1}{3}t + \dfrac{1}{4}t = 1$

$12\left(\dfrac{1}{3}t + \dfrac{1}{4}t\right) = 12(1)$

$4t + 3t = 12$

$7t = 12$

$t = \dfrac{12}{7}$

If they both work, the project can be completed in $1\frac{5}{7}$ hours.

CHAPTER 5   Exponents and Polynomials

SECTION 5.1   Adding and Subtracting Polynomials

1. An algebraic expression is a collection of letters (called variables) and real numbers (called constants) combined by using addition, subtraction, multiplication, or division.

2. The terms of an algebraic expression are those parts separated by addition or subtraction.

3. $10(x - 1) = 10x - 10$

4. $4(3 - 2z) = 12 - 8z$

5. $-\frac{1}{2}(4 - 6x) = -2 + 3x$

6. $-25(2x - 3) = -50x + 75$

7. $8y - 2x + 7x - 10y = (-2x + 7x) + (8y - 10y)$

$= 5x - 2y$

8. $\frac{5}{6}x - \frac{2}{3}x + 8 = \frac{5}{6}x - \frac{4}{6}x + 8$

$= \frac{1}{6}x + 8$

9. $10(x - 1) - 3(x + 2) = 10x - 10 - 3x - 6$

$= (10x - 3x) + (-10 - 6)$

$= 7x - 16$

10. $-3[x + (2 + 3x)] = -3[x + 2 + 3x]$

$= -3[4x + 2]$

$= -12x - 6$

11.

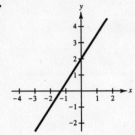

12.

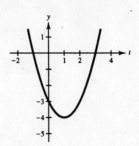

SECTION 5.2   Multiplying Polynomials: Special Products

1. The point $(3, -2)$ is located 3 units to the right of the y-axis and 2 units below the x-axis.

2. Quadrant I: $(3, 4)$

Quadrant II: $(-3, 4)$

Quadrant III: $(-3, -4)$

Quadrant IV: $(3, -4)$

3. $\frac{3}{4}x - \frac{5}{2} + \frac{3}{2}x = \left(\frac{3}{4}x + \frac{6}{4}x\right) - \frac{5}{2} = \frac{9}{4}x - \frac{5}{2}$

4. $4 - 2(3 - x) = 4 - 6 + 2x = 2x - 2$

5. $2(x - 4) + 5x = 2x - 8 + 5x$

$= (2x + 5x) - 8$

$= 7x - 8$

6. $4(3 - y) + 2(y + 1) = 12 - 4y + 2y + 2$

$= -2y + 14$

7. $-3(z - 2) - (z - 6) = -3z + 6 - z + 6$

$= -4z + 12$

8. $(u - 2) - 3(2u + 1) = u - 2 - 6u - 3$

$= (u - 6u) + (-2 - 3)$

$= -5u - 5$

9. $a = pb$

$1600 = 0.055b$

$\dfrac{1600}{0.055} = b$

$29,090.91 \approx b$

The sales were \$29,090.91.

10. Distance of first runner = Distance of second runner

$$4\left(t + \frac{1}{4}\right) = 5t$$

$4t + 1 = 5t$

$1 = t$

It will take the second runner 1 hour to overtake the first runner. Each runner will have run 5 miles at that point.

11.

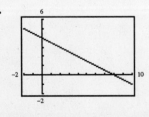

$(0, 4), (8, 0)$

12.

$(0, 0), (4, 0)$

SECTION 5.3   Negative Exponents and Scientific Notation

1. The graph of the function is the set of all solution points of the function.

2. Construct a table of solution points, plot these solution points on a rectangular coordinate system, and use the pattern to connect the points with a smooth curve or line.

3. There are infinitely many answers. Here are some examples:

   $(0, 0), (9, 3), (25, 5)$

4. To find the x-intercept, solve the equation $f(x) = 0$ for x. To find the y-intercept, find $f(0)$.

$$f(x) = 0 \qquad\qquad f(0) = 3(0 - 2)$$
$$3(x - 2) = 0 \qquad\qquad = 3(-2)$$
$$3x - 6 = 0 \qquad\qquad = -6$$
$$3x = 6$$
$$x = 2$$

The x-intercept is $(2, 0)$ and the y-intercept is $(0, -6)$.

5. $x^2 \cdot x^3 = x^{2+3} = x^5$

6. $(y^2z^3)(z^2)^4 = (y^2z^3)(z^8) = y^2z^{11}$

7. $\left(\dfrac{x^2}{y}\right)^3 = \dfrac{x^{2\cdot3}}{y^3} = \dfrac{x^6}{y^3}$

8. $\dfrac{a^2b^3}{c} \cdot \dfrac{2a}{3} = \dfrac{2a^{2+1}b^3}{3c} = \dfrac{2a^3b^3}{3c}$

9.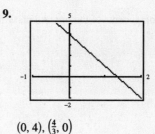

   $(0, 4), \left(\frac{4}{3}, 0\right)$

10.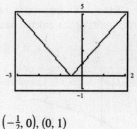

    $\left(-\frac{1}{2}, 0\right), (0, 1)$

11.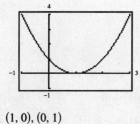

    $(1, 0), (0, 1)$

12.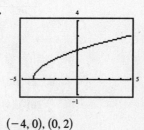

    $(-4, 0), (0, 2)$

SECTION 5.4   Dividing Polynomials

1. Divide out the common factor of 6 from the numerator and denominator.

$$\dfrac{24x}{18} = \dfrac{\cancel{6}(4)x}{\cancel{6}(3)} = \dfrac{4x}{3}$$

2. Quadrant II

   Since the x-coordinate is negative, the point lies to the left of the y-axis. Since the y-coordinate is positive, the point lies above the x-axis.

3. $\dfrac{8}{12} = \dfrac{\cancel{4}(2)}{\cancel{4}(3)} = \dfrac{2}{3}$

4. $\dfrac{18}{144} = \dfrac{1\cancel{(18)}}{8\cancel{(18)}} = \dfrac{1}{8}$

5. $\dfrac{60}{150} = \dfrac{2\cancel{(30)}}{5\cancel{(30)}} = \dfrac{2}{5}$

6. $\dfrac{175}{42} = \dfrac{7(25)}{7(6)} = \dfrac{25}{6}$

7. $-2x^2(5x^3) = -10x^{2+3}$
$$= -10x^5$$

8. $(2z + 1)(2z - 1) = (2z)^2 - 1^2$
$$= 4z^2 - 1$$

9. $(x + 7)^2 = x^2 + 2(x)(7) + 7^2$

$\qquad = x^2 + 14x + 49$

10. $(x + 4)(2x - 5) = 2x^2 - 5x + 8x - 20$

$\qquad = 2x^2 + 3x - 20$

11. $(2n + 1)(2n + 3) = 4n^2 + 6n + 2n + 3$

$\qquad = 4n^2 + 8n + 3$

12. Distance = (Rate)(Time)

$\qquad 180 = r(3.5)$

$\qquad \dfrac{180}{3.5} = r$

$\qquad 51.4 \approx r$

The average speed is approximately 51.4 miles per hour.

CHAPTER 6 Factoring and Solving Equations

SECTION 6.1 Factoring Polynomials with Common Factors

1. A function is a set of ordered pairs in which no two ordered pairs have the same first component and different second components.

2. The domain of a function is the set of all first components of the ordered pairs. The range of a function is the set of all second components of the ordered pairs.

3.

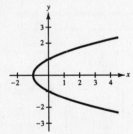

4.

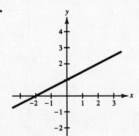

5. (a) $f(0) = \frac{1}{2}(0) + 1 = 0 + 1 = 1$

(b) $f(4) = \frac{1}{2}(4) + 1 = 2 + 1 = 3$

(c) $f(-3) = \frac{1}{2}(-3) + 1 = -\frac{3}{2} + \frac{2}{2} = -\frac{1}{2}$

(d) $f\left(-\frac{3}{2}\right) = \frac{1}{2}\left(-\frac{3}{2}\right) + 1 = -\frac{3}{4} + \frac{4}{4} = \frac{1}{4}$

6. (a) $g(0) = 0(0 - 4) = 0(-4) = 0$

(b) $g(4) = 4(4 - 4) = 4(0) = 0$

(c) $g(-2) = -2(-2 - 4) = -2(-6) = 12$

(d) $g\left(-\frac{5}{2}\right) = -\frac{5}{2}\left(-\frac{5}{2} - \frac{8}{2}\right) = -\frac{5}{2}\left(-\frac{13}{2}\right) = \frac{65}{4}$

7. (a) $F(0) = \sqrt{2(0) + 1} = \sqrt{0 + 1} = \sqrt{1} = 1$

(b) $F(4) = \sqrt{2(4) + 1} = \sqrt{8 + 1} = \sqrt{9} = 3$

(c) $F\left(-\frac{1}{2}\right) = \sqrt{2\left(-\frac{1}{2}\right) + 1} = \sqrt{-1 + 1} = \sqrt{0} = 0$

(d) $F(10) = \sqrt{2(10) + 1} = \sqrt{20 + 1} = \sqrt{21}$

8. (a) $h(0) = |0 - 3| = |-3| = 3$

(b) $h(4) = |4 - 3| = |1| = 1$

(c) $h(2) = |2 - 3| = |-1| = 1$

(d) $h(-3) = |-3 - 3| = |-6| = 6$

9. $\qquad a = pb$

$\qquad 1620 = p(54,000)$

$\qquad \frac{1620}{54,000} = p$

$\qquad 0.03 = p$

The commission rate is 3%.

10. $\frac{1}{10}t + \frac{1}{6}t = 1$

$30\left(\frac{1}{10}t + \frac{1}{6}t\right) = 30(1)$

$3t + 5t = 30$

$8t = 30$

$t = \frac{30}{8}$

$t = \frac{15}{4}$

It would take $3\frac{3}{4}$ hours, or 3 hours and 45 minutes for them to complete the project working together.

11.

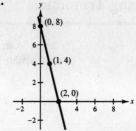

12.

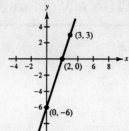

13.

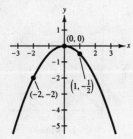

14.

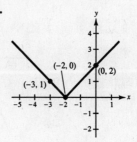

SECTION 6.2   Factoring Trinomials

1. If there are two y-intercepts, then there are two values of y that correspond to an x-value of 0. Such a relation is not a function.

2. In standard form, the polynomial is $4x^3 - 7x^2 + 3x - 4$. The leading coefficient is 4.

3. $y(y + 2) = y^2 + 2y$

4. $-a^2(a - 1) = -a^3 + a^2$

5. $(x - 2)(x - 5) = x^2 - 5x - 2x + 10$

$= x^2 - 7x + 10$

6. $(v - 4)(v + 7) = v^2 + 7v - 4v - 28$

$= v^2 + 3v - 28$

7. $(2x + 5)(2x - 5) = (2x)^2 - 5^2$

$= 4x^2 - 25$

8. $x^2(x + 1) - 5(x^2 - 2) = x^3 + x^2 - 5x^2 + 10$

$= x^3 - 4x^2 + 10$

9. $1{,}475{,}000 - (-2{,}500{,}000) = 1{,}475{,}000 + 2{,}500{,}000$

$= 3{,}975{,}000$

The profit for the second 6 months was $3,975,000.

10. $5(12)(11.95) = 717$

The total cost of the order is $717.

11. $R > C$

$75x > 62.5x + 570$

$12.5x > 570$

$x > \dfrac{570}{12.5}$

$x > 45.6$

To produce a profit, the number of units must be at least 46; $x \geq 46$.

12. $40\left(3\frac{1}{2}\right) \leq \text{Distance} \leq 65\left(3\frac{1}{2}\right)$

$140 \leq \text{Distance} \leq 227.5$

The distance x in miles would satisfy the inequality $140 \leq x \leq 227.5$.

SECTION 6.3   More About Factoring Trinomials

1. Prime

2. The sum of the digits, $2 + 5 + 5$, is 12 and 12 is divisible by 3. Therefore, 255 is divisible by 3.

3. $500 = 2^2 \cdot 5^3$

4. $315 = 3^2 \cdot 5 \cdot 7$

5. $792 = 2^3 \cdot 3^2 \cdot 11$

6. $2275 = 5^2 \cdot 7 \cdot 13$

7. $(2x - 5)(x + 7) = 2x^2 + 14x - 5x - 35$
$$= 2x^2 + 9x - 35$$

8. $(3x - 2)^2 = (3x)^2 - 2(3x)(2) + 2^2$
$$= 9x^2 - 12x + 4$$

9.

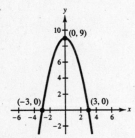

10.

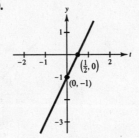

11. (a)

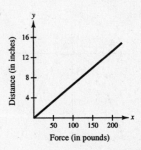

(b) $y = 0.066(100) = 6.6$ inches

SECTION 6.4   Factoring Polynomials with Special Forms

1. Quadrant II

2. Quadrant I or II

3. $(-4, 0)$

4. $(9, -6)$

5. $7 + 5x = 7x - 1$
$$7 - 2x = -1$$
$$-2x = -8$$
$$x = 4$$

6. $2 - 5(x - 1) = 2[x + 10(x - 1)]$
$$2 - 5x + 5 = 2[x + 10x - 10]$$
$$-5x + 7 = 2[11x - 10]$$
$$-5x + 7 = 22x - 20$$
$$-27x + 7 = -20$$
$$-27x = -27$$
$$x = 1$$

7. $2(x + 1) = 0$
$$2x + 2 = 0$$
$$2x = -2$$
$$x = -1$$

8. $\dfrac{3}{4}(12x - 8) = 10$
$$9x - 6 = 10$$
$$9x = 16$$
$$x = \dfrac{16}{9}$$

9. $\dfrac{x}{5} + \dfrac{1}{5} = \dfrac{7}{10}$
$$10\left(\dfrac{x}{5} + \dfrac{1}{5}\right) = 10\left(\dfrac{7}{10}\right)$$
$$2x + 2 = 7$$
$$2x = 5$$
$$x = \dfrac{5}{2}$$

10. $\dfrac{3x}{4} + \dfrac{1}{2} = 8$

$4\left(\dfrac{3x}{4} + \dfrac{1}{2}\right) = 4(8)$

$3x + 2 = 32$

$3x = 30$

$x = 10$

11. $a = pb$

$8345 = 1.20b$

$\dfrac{8345}{1.20} = b$

$6954 \approx b$

The station had approximately 6954 members last year.

12. $a = pb$

$a = 0.26(46{,}750)$

$a = 12{,}155$

You can spend \$12,155 on housing.

SECTION 6.5   Polynomial Equations and Applications

1. Additive Inverse Property

2. Multiplicative Identity Property

3. Distributive Property

4. Associative Property of Addition

5. (a) $2(-3) + 9 = -6 + 9 = 3$

(b) $(-5)^2 + 3 = 25 + 3 = 28$

6. (a) $4 - \dfrac{5}{2} = \dfrac{8}{2} - \dfrac{5}{2} = \dfrac{3}{2}$

(b) $\dfrac{|18 - 25|}{6} = \dfrac{|-7|}{6} = \dfrac{7}{6}$

7. $\left(-\dfrac{7}{12}\right)\left(\dfrac{3}{28}\right) = -\dfrac{7(3)}{3(4)(7)(4)} = -\dfrac{1}{16}$

8. $\dfrac{4}{3} \div \dfrac{5}{6} = \dfrac{4}{3} \cdot \dfrac{6}{5} = \dfrac{4(3)(2)}{3(5)} = \dfrac{8}{5}$

9. $2t(t - 3) + 4t + 1 = 2t^2 - 6t + 4t + 1$

$= 2t^2 - 2t + 1$

10. $2u - 5(2u - 3) = 2u - 10u + 15$

$= -8u + 15$

11. $I = Prt$

$I = 1000(0.075)(10)$

$I = 750$

The interest is \$750.

12. Distance of car = Distance of truck

$r(2.5) = 50(2.5 + 1)$

$2.5r = 50(3.5)$

$2.5r = 175$

$r = \dfrac{175}{2.5}$

$r = 70$

The speed of the car is 70 miles per hour.

CHAPTER 7   Systems of Equations

SECTION 7.1   Solving Systems of Equations by Graphing and Substitution

1. Answers vary.

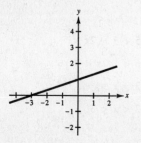

2. Answers vary.

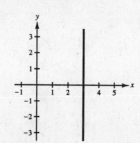

3. $\dfrac{3}{2}$

$m_1 \cdot m_2 = -1$

$-\dfrac{2}{3} \cdot \dfrac{3}{2} = -1$

4. The line with $m = -3$ is steeper because this line's slope is the greater absolute value.

5. $y - 3(4y - 2) = 1$

$y - 12y + 6 = 1$

$-11y = -5$

$y = \frac{5}{11}$

6. $x + 6(3 - 2x) = 4$

$x + 18 - 12x = 4$

$-11x = -14$

$x = \frac{14}{11}$

7. $\frac{1}{2}x + \frac{1}{5}x = 15$

$5x + 2x = 150$

$7x = 150$

$x = \frac{150}{7}$

8. $\frac{1}{10}(x - 4) = 6$

$x - 4 = 60$

$x = 64$

9. $3x + 4y - 5 = 0$

$4y = -3x + 5$

$y = -\frac{3}{4}x + \frac{5}{4}$

10. $-2x - 3y + 6 = 0$

$-3y = 2x - 6$

$y = \frac{2}{-3}x + 2$

11. $y = -3x + 2$

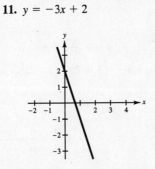

12. $4x - 2y = -4$

$-2y = -4x - 4$

$y = 2x + 2$

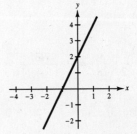

13. $3x + 2y = 8$

$2y = -3x + 8$

$y = -\frac{3}{2}x + 4$

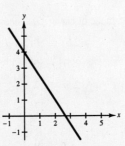

14. $x + 3 = 0$

$x = -3$

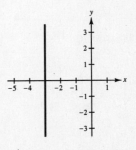

SECTION 7.2   Solving Systems of Equations by Elimination

1. $2(x + y) = 2x + 2y$

Distributive Property

2. $x - 4 = 7$

$x - 4 + 4 = 7 + 4$

$x = 11$

Addition Property of Equality

3. $1 < 2x + 5 < 9$

$-4 < 2x < 4$

$-2 < x < 2$

4. $0 \le \dfrac{x - 4}{2} < 6$

$0 \le x - 4 < 12$

$4 \le x < 16$

5. $|6x| > 12$

$6x > 12$ or $6x < -12$

$x > 2$ $x < -2$

6. $|1 - 2x| < 5$

$-5 < 1 - 2x < 5$

$-6 < -2x < 4$

$3 > x > -2$

$-2 < x < 3$

7. $4x - 12 < 0$

$\qquad 4x < 12$

$\qquad x < 3$

8. $4x + 4 \geq 9$

$\qquad 4x \geq 5$

$\qquad x \geq \frac{5}{4}$

9. *Verbal Model:* $\boxed{\text{Total cost}} = \boxed{\text{Number of miles}} \cdot \boxed{\text{Cost per mile}} + \boxed{\text{Initial cost}}$

$\qquad\qquad C = 0.45m + 6200$

Equation: $\qquad\qquad C < 15{,}000$

$\qquad\qquad 0.45m + 6200 < 15{,}000$

$\qquad\qquad\qquad 0.45m < 8800$

$\qquad\qquad\qquad m < \dfrac{8800}{0.45}$

$\qquad\qquad\qquad m19{,}555.56$

10. *Verbal Model:* $\boxed{\begin{array}{c}\text{Payment}\\\text{Plan 1}\end{array}} = 2500$

$\boxed{\begin{array}{c}\text{Payment}\\\text{Plan 2}\end{array}} = 4\% \cdot \boxed{\begin{array}{c}\text{Gross}\\\text{sales}\end{array}} + 1500$

$\boxed{\begin{array}{c}\text{Payment}\\\text{Plan 2}\end{array}} > \boxed{\begin{array}{c}\text{Payment}\\\text{Plan 1}\end{array}}$

Labels: Gross sales $= x$

Equation: $0.04x + 1500 > 2500$

$\qquad\qquad 0.04x > 1000$

$\qquad\qquad x > \$25{,}000$

SECTION 7.3   Linear Systems in Three Variables

1. No, $2x + 8 = 7$ has only one solution.

$\qquad 2x + 8 = 7$

$\qquad\quad 2x = -1$

$\qquad\quad\ x = -\dfrac{1}{2}$

2. $\dfrac{t}{6} + \dfrac{5}{8} = \dfrac{7}{4}$

Multiply both sides of the equation by the lowest common denominator, 24.

3. $4x^2(x^3)^2 = 4x^2 \cdot x^6 = 4x^8$

4. $(2x^2y)^3(xy^3)^4 = 8x^6y^3 \cdot x^4y^{12}$

$\qquad\qquad\qquad = 8x^{10}y^{15}$

5. $\dfrac{8x^{-4}}{2x^7} = 4x^{-4-(7)} = 4x^{-11} = \dfrac{4}{x^{11}}$

6. $\left(\dfrac{t^4}{3}\right)^{-1} = \dfrac{3}{t^4}$

7. $|2x - 4| = 6$

$\quad 2x - 4 = 6 \quad$ or $\quad 2x - 4 = -6$

$\qquad 2x = 10 \qquad\qquad 2x = -2$

$\qquad\ x = 5 \qquad\qquad\ x = -1$

8. $\frac{1}{4}(5 - 2x) = 9x - 7x$

$\quad \frac{1}{4}(5 - 2x) = 2x$

$\qquad 5 - 2x = 8x$

$\qquad\qquad 5 = 10x$

$\qquad\qquad \frac{5}{10} = x$

$\qquad\qquad \frac{1}{2} = x$

9. *Verbal Model:* $\boxed{\text{Distance}} = \boxed{\text{Rate}} \cdot \boxed{\text{Time}}$

$\qquad\qquad\qquad d = 15t$

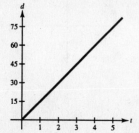

10. *Verbal Model:* $\boxed{\text{Volume}} = (\text{side})^3$

Labels: Volume $= V$

$\qquad\quad$ Side $= s$

Equation: $V = s^3$

11. Area $= \pi \cdot (\text{radius})^2$ Circumference $= 2 \cdot \pi \cdot \text{radius}$

$$A = \pi r^2$$

$$A = \pi\left(\frac{C}{2\pi}\right)^2$$

$$A = \pi \cdot \frac{C^2}{4\pi^2}$$

$$A = \frac{C^2}{4\pi}$$

$$C = 2\pi r$$

$$\frac{C}{2\pi} = r$$

SECTION 7.4   Matrices and Linear Systems

1. $2ab - 2ab = 0$

    Additive Inverse Property

2. $8t \cdot 1 = 8t$

    Multiplicative Identity Property

3. $b + 3a = 3a + b$

    Commutative Property of Addition

4. $3(2x) = (3 \cdot 2)x$

    Associative Property of Multiplication

5. $(-3, 2), \left(-\frac{3}{2}, -2\right)$

$$m = \frac{y_2 - y_1}{x_2 - x_1}$$

$$= \frac{-2 - 2}{-\frac{3}{2} - (-3)}$$

$$= \frac{-4}{-\frac{3}{2} + 3}$$

$$= \frac{-4}{\frac{3}{2}}$$

$$= -4 \cdot \frac{2}{3} = -\frac{8}{3}$$

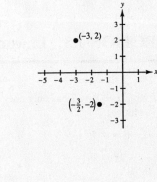

6. $(0, -6), (8, 0)$

$$m = \frac{y_2 - y_1}{x_2 - x_1}$$

$$= \frac{0 - (-6)}{8 - 0}$$

$$= \frac{6}{8}$$

$$= \frac{3}{4}$$

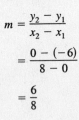

7. $\left(\frac{5}{2}, \frac{7}{2}\right), \left(\frac{5}{2}, 4\right)$

$$m = \frac{y_2 - y_1}{x_2 - x_1}$$

$$= \frac{4 - \frac{7}{2}}{\frac{5}{2} - \frac{5}{2}}$$

$$= \frac{\frac{1}{2}}{0} = \text{undefined}$$

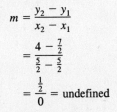

8. $\left(-\frac{5}{8}, -\frac{3}{4}\right), \left(1, -\frac{9}{2}\right)$

$$m = \frac{y_2 - y_1}{x_2 - x_1}$$

$$= \frac{-\frac{9}{2} - \left(-\frac{3}{4}\right)}{1 - \left(-\frac{5}{8}\right)}$$

$$= \frac{-\frac{18}{4} + \frac{3}{4}}{\frac{8}{8} + \frac{5}{8}} = \frac{-\frac{15}{4}}{\frac{13}{8}}$$

$$= -\frac{15}{4} \cdot \frac{8}{13}$$

$$= -\frac{30}{13}$$

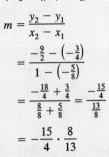

9. $(3, 1.2), (-3, 2.1)$

$$m = \frac{y_2 - y_1}{x_2 - x_1}$$

$$= \frac{2.1 - 1.2}{-3 - 3}$$

$$= \frac{0.9}{-6}$$

$$= -0.15$$

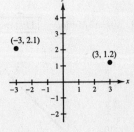

10. $(12, 8), (6, 8)$

$$m = \frac{y_2 - y_1}{x_2 - x_1}$$

$$= \frac{8 - 8}{12 - 6}$$

$$= \frac{0}{6}$$

$$= 0$$

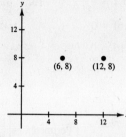

11. *Verbal Model:*

| Current number members | = | Number members before drive | + 10% · | Number members before drive |

Labels: Number members before drive $= x$

Equation: $8415 = x + 0.10x$

$8415 = 1.10x$

$7650 = x$

12. *Verbal Model:*

| Amount increase | = 4% · | Price |

Labels: Amount increase $= x$

Price $= \$23{,}500$

Equation: $x = 0.04(23{,}500)$

$x = \$940$

SECTION 7.5   Determinants and Linear Systems

1. $(px + m)(qx + n) = ax^2 + bx + c$

$pqx^2 + pnx + mqx + mn$

$(pq)x^2 + (pn + mq)x + mn$

So $a = pq$.

2. $(px + m)(qx + n) = ax^2 + bx + c$

$pqx^2 + pnx + mqx + mn$

$(pq)x^2 + (pn + mq)x + mn$

So $b = pn + mq$.

3. $(px + m)(qx + n) = ax^2 + bx + c$

$pqx^2 + pnx + mqx + mn$

$(pq)x^2 + (pn + mq)x + mn$

So $c = mn$.

4. If $a = 1$ then $p = 1$ and $q = 1$ or $p = -1$ and $q = -1$.

5. $3x^2 + 9x - 12 = 0$

$3(x^2 + 3x - 4) = 0$

$3(x + 4)(x - 1) = 0$

$x = -4 \quad x = 1$

6. $x^2 - x - 6 = 0$

$(x - 3)(x + 2) = 0$

$x = 3 \quad x = -2$

7. $4x^2 - 20x + 25 = 0$

$(2x - 5)(2x - 5) = 0$

$x = \frac{5}{2} \quad x = \frac{5}{2}$

8. $x^2 - 16 = 0$

$(x - 4)(x + 4) = 0$

$x = 4 \qquad x = -4$

9. $x^3 + 64 = 0$

$(x + 4)(x^2 - 4x + 16) = 0$

$x = -4 \qquad x^2 - 4x + 16 = 0$

Not real

10. $3x^3 - 6x^2 + 4x - 8 = 0$

$3x^2(x - 2) + 4(x - 2) = 0$

$(x - 2)(3x^2 + 4) = 0$

$x = 2 \qquad 3x^2 + 4 = 0$

Not real

11. *Verbal Model:* $\boxed{\text{Distance}} = \boxed{\text{Rate}} \cdot \boxed{\text{Time}}$

Equation $\quad 320 = r \cdot t$

$\dfrac{320}{r} = t$

12. *Verbal Model:* $\text{Perimeter} = \boxed{\begin{array}{c}\text{Length}\\ \text{side 1}\end{array}} + \boxed{\begin{array}{c}\text{Length}\\ \text{side 2}\end{array}} + \boxed{\begin{array}{c}\text{Length}\\ \text{side 3}\end{array}}$

Equation: $\quad P = (x + 1) + \left(\dfrac{1}{2}x + 5\right) + (3x + 1)$

$= \left(x + \dfrac{1}{2}x + 3x\right) + (1 + 5 + 1)$

$= \dfrac{9}{2}x + 7$

CHAPTER 8   Rational Expressions, Equations, and Functions

SECTION 8.1   Rational Expressions and Functions

1. Slope $= m = \dfrac{y_2 - y_1}{x_2 - x_1}$

2. (a) $m > 0$ \qquad (b) $m < 0$

(c) $m = 0$ \qquad (d) m is undefined.

3. $2(x + 5) - 3 - (2x - 3) = 2x + 10 - 2x$

$= 10$

4. $3(y + 4) + 5 - (3y + 5) = 3y + 12 + 5 - 3y - 5$

$= 12$

5. $4 - 2[3 + 4(x + 1)] = 4 - 2[3 + 4x + 4]$

$= 4 - 2[7 + 4x]$

$= 4 - 14 - 8x$

$= -10 - 8x$

6. $5x + x[3 - 2(x - 3)] = 5x + x[3 - 2x + 6]$

$= 5x + x[9 - 2x]$

$= 5x + 9x - 2x^2$

$= -2x^2 + 14x$

7. $\left(\dfrac{5}{x^2}\right)^2 = \dfrac{25}{x^4}$

8. $-\dfrac{(2u^2v)^2}{-3uv^2} = -\dfrac{4u^4v^2}{-3uv^2} = \dfrac{4u^3}{3}$

9. *Verbal Model:* $\boxed{\begin{array}{c}\text{Gallons}\\ \text{solution 1}\end{array}} \cdot 30\% + \boxed{\begin{array}{c}\text{Gallons}\\ \text{solution 2}\end{array}} \cdot 60\% = \boxed{\begin{array}{c}\text{Total}\\ \text{gallons}\end{array}} \cdot 40\%$

Labels: Gallons solution 1 $= x$

Gallons solution 2 $= 20 - x$

Total gallons $= 20$

Equation: $0.30x + 0.60(20 - x) = 20(0.40)$

$0.30x + 12 - 0.60x = 8$

$-0.30x = -4$

$x = 13\dfrac{1}{3}$ gallons at 30%

$20 - x = 6\dfrac{2}{3}$ gallons at 60%

10. *Verbal Model:* $\boxed{\begin{array}{c}\text{Original}\\\text{price}\end{array}} \cdot 75\% = \boxed{\begin{array}{c}\text{Sale}\\\text{price}\end{array}}$

Labels: Original price $= x$

Sale price $= \$375$

Equation: $x \cdot 0.75 = 375$

$x = \$500$

SECTION 8.2 Multiplying and Dividing Rational Expressions

1. $9t^2 - 4 = (3t - 2)(3t + 2)$

2. $4x^2 - 12x + 9 = (2x - 3)^2$

3. $8x^3 + 64 = (2x + 4)(4x^2 - 8x + 16)$

4. $3x^2 + 13x - 10 = (3x - 2)(x + 5)$

5. $5x - 20x^2 = 5x(1 - 4x)$

6. $64 - (x - 6)^2 = [8 - (x - 6)][8 + (x - 6)]$

$= (8 - x + 6)(8 + x - 6)$

$= (14 - x)(2 + x)$

7. $15x^2 - 16x - 15 = (5x + 3)(3x - 5)$

8. $16t^2 + 8t + 1 = (4t + 1)^2$

9. $y^3 - 64 = (y - 4)(y^2 + 4y + 16)$

10. $8x^3 + 1 = (2x + 1)(4x^2 - 2x + 1)$

11.

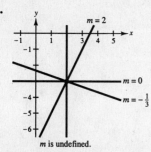

m is undefined.

12.

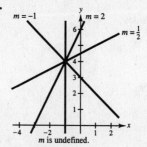

m is undefined.

SECTION 8.3 Adding and Subtracting Rational Expressions

1. (a) $5y - 3x - 4 = 0$

$5y = 3x + 4$

$y = \frac{3}{5}x + \frac{4}{5}$

(b) $5y - 3x - 4 = 0$

$5y = 3x + 4$

$y = \frac{3}{5}x + \frac{4}{5}$

$y - y_1 = \frac{3}{5}(x - x_1) + \frac{4}{5}$ Let $x_1 = 1$.

$y - y_1 = \frac{3}{5}(x - 1) + \frac{4}{5}$

$y - \frac{7}{5} = \frac{3}{5}(x - 1)$ $y_1 = \frac{7}{5}$

(Many answers)

2. If $m > 0$, the line rises from left to right.

If $m < 0$, the line falls from left to right.

3. $-6x(10 - 7x) = -60x + 42x^2$

$\qquad = 42x^2 - 60x$

4. $(2 - y)(3 + 2y) = 6 + 4y - 3y - 2y^2$

$\qquad = 6 + y - 2y^2$

5. $(11 - x)(11 + x) = 121 - x^2$

6. $(4 - 5z)(4 + 5z) = 16 + 20z - 20z - 25z^2$

$\qquad = 16 - 25z^2$

7. $(x + 1)^2 = (x + 1)(x + 1)$

$\qquad = x^2 + 2x + 1$

8. $t(t^2 + 1) - t(t^2 - 1) = t^3 + t - t^3 + t$

$\qquad = 2t$

9. $(x - 2)(x^2 + 2x + 4) = x^3 + 2x^2 + 4x - 2x^2 - 4x - 8$

$\qquad = x^3 - 8$

10. $t(t - 4)(2t + 3) = t(2t^2 + 3t - 8t - 12)$

$\qquad = t(2t^2 - 5t - 12)$

$\qquad = 2t^3 - 5t^2 - 12t$

11. Perimeter = Sum of all sides

$\qquad = 7(x) + (x + 3) + (2x) + (2x + 3)$

$\qquad = 12x + 6$

Area = Area rectangle 1 + Area rectangle 2 + Area rectangle 3

$\qquad = (x \cdot x) + (x + 3)(3x) + (x \cdot x)$

$\qquad = x^2 + 3x^2 + 9x + x^2$

$\qquad = 5x^2 + 9x$

12. Perimeter = Sum of all sides

$\qquad = 3x + 4x + 5x$

$\qquad = 12x$

Area $= \frac{1}{2} \cdot$ Base $\cdot$ Height

$\qquad = \frac{1}{2} \cdot 3x \cdot 4x$

$\qquad = 6x^2$

SECTION 8.4 Solving Rational Equations

1. $(-2, y)$ can be located in quadrants II or III.

2. $(x, 3)$ can be located in quadrants I or II.

3. Points whose y-coordinates are 0 are located on the x-axis.

4. $(9, -6)$

5. $7 - 3x > 4 - x$

$\qquad -2x > -3$

$\qquad x < \frac{3}{2}$

6. $2(x + 6) - 20 < 2$

$\qquad 2x + 12 - 20 < 2$

$\qquad 2x - 8 < 2$

$\qquad 2x < 10$

$\qquad x < 5$

7. $|x - 3| < 2$

$\quad -2 < x - 3 < 2$

$\quad\quad 1 < x < 5$

8. $|x - 5| > 3$

$\quad x - 5 > 3 \quad \text{or} \quad x - 5 < -3$

$\quad\quad x > 8 \quad \text{or} \quad\quad x < 2$

9. $\left|\frac{1}{4}x - 1\right| \geq 3$

$\quad \frac{1}{4}x - 1 \geq 3 \quad \text{or} \quad \frac{1}{4}x - 1 \leq -3$

$\quad\quad \frac{1}{4}x \geq 4 \quad\quad\quad\quad \frac{1}{4}x \leq -2$

$\quad\quad\quad x \geq 16 \quad \text{or} \quad\quad x \leq -8$

10. $\left|2 - \frac{1}{3}x\right| \leq 10$

$\quad -10 \leq 2 - \frac{1}{3}x \leq 10$

$\quad -12 \leq -\frac{1}{3}x \leq 8$

$\quad\quad 36 \geq x \geq -24$

$\quad -24 \leq x \leq 36$

11. *Verbal Model:* $\boxed{\text{Distance}} = \boxed{\text{Rate}} \cdot \boxed{\text{Time}}$

Labels: Distance $= d$

1st jogger's rate $= 6$; 1st jogger's time $= x + \dfrac{5}{60}$

2nd jogger's rate $= 8$; 2nd jogger's time $= x$

Equation: $d = 6\left(x + \dfrac{1}{12}\right)$

$d = 8x$

$6\left(x + \dfrac{1}{12}\right) = 8x$

$6x + \dfrac{1}{2} = 8x$

$\dfrac{1}{2} = 2x$

$\dfrac{1}{4} = x$ hours, or 15 minutes

$d = 8\left(\tfrac{1}{4}\text{ hour}\right) = 2$ miles

12. *Verbal Model:* $\boxed{\begin{array}{c}\text{Amount} \\ \text{at } 7.5\%\end{array}} + \boxed{\begin{array}{c}\text{Amount} \\ \text{at } 9\%\end{array}} = 24{,}000$

$7.5\% \cdot \boxed{\begin{array}{c}\text{Amount} \\ \text{at } 7.5\%\end{array}} + 9\% \cdot \boxed{\begin{array}{c}\text{Amount} \\ \text{at } 9\%\end{array}} = 1935$

Labels: Amount at 7.5% $= x$

Amount at 9% $= y$

System:

$\quad\quad x + y = 24{,}000$

$\quad 0.075x + 0.09y = 1935$

$\quad\quad y = 24{,}000 - x$

$0.075x + 0.09(24{,}000 - x) = 1935$

$0.075x + 2160 - 0.09x = 1935$

$\quad\quad -0.015x = -225$

$\quad\quad x = \$15{,}000$ at 7.5%

$\quad\quad y = \$9000$ at 9%

SECTION 8.5 Graphs of Rational Functions

1. Leading coefficient in $7x^2 + 3x - 4$ is 7. It is the coefficient of the ax^2-term.

2. Degree is 5.

$\quad (x^4 + 3)(x - 4) = x^5 - 4x^4 + 3x - 12$

3. Many answers

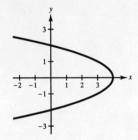

For some x there corresponds more than one value of y.

4. Many answers

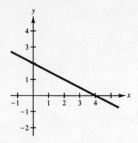

For each x there corresponds exactly one value of y.

5. $-2x^5(5x^3) = -10x^8$

6. $3x(5 - 2x) = 15x - 6x^2$

7. $(2x - 15)^2 = (2x - 15)(2x - 15)$
$$= 4x^2 - 60x + 225$$

8. $(3x + 2)(7x - 10) = 21x^2 + 14x - 30x - 20$
$$= 21x^2 - 16x - 20$$

9. $[(x + 1) - y][(x + 1) + y] = (x + 1)^2 - y^2$
$$= x^2 + 2x + 1 - y^2$$
$$= x^2 - y^2 + 2x + 1$$

10. $(x + 3)(x^2 - 3x + 9) = x^3 - 3x^2 + 9x + 3x^2 - 9x + 27$
$$= x^3 + 27$$

11. *Verbal Model:* $\boxed{\text{Area}} = \dfrac{1}{2} \cdot \boxed{\text{Base}} \cdot \boxed{\text{Height}}$

Labels: Area $= A = 80$
Base $= x$
Height $= x - 12$

Equation: $A = \dfrac{1}{2} \cdot x \cdot (x - 12)$

$$80 = \dfrac{1}{2}x^2 - 6x$$

$$0 = x^2 - 12x - 160$$

$$0 = (x - 20)(x + 8)$$

$$x = 20 \text{ meters} \quad x = -8$$

$x - 12 = 8 \text{ meters}$

Base $= 20$ meters

Height $= 8$ meters

12. *Verbal Model:* $\boxed{\begin{array}{c}\text{Surface}\\\text{area}\end{array}} = \boxed{\begin{array}{c}\text{Area of}\\\text{bottom}\end{array}} + 4 \cdot \boxed{\begin{array}{c}\text{Area of}\\\text{one side}\end{array}}$

Labels: Surface area $= 825$
Area of bottom $= x \cdot x$
Area of one side $= 10 \cdot x$

Equation: $825 = x \cdot x + 4(10 \cdot x)$

$$825 = x^2 + 40x$$

$$0 = x^2 + 40x - 825$$

$$0 = (x + 55)(x - 15)$$

$$x + 55 = 0 \qquad x - 15 = 0$$

$$x = -55 \qquad x = 15 \text{ inches}$$

15 inches $\times$ 15 inches

SECTION 8.6   Variation

1.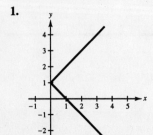

For some x there corresponds more than one value of y.

2.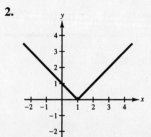

For each x there corresponds exactly one value of y.

3. $f(x) = x^2 - 4x + 9$

Domain: $(-\infty, \infty)$

4. $h(x) = \dfrac{x - 1}{x^2(x^2 + 1)}$

Domain: $x^2(x^2 + 1) \neq 0$

$\qquad x^2 \neq 0 \qquad x^2 + 1 \neq 0$

$\qquad x \neq 0$

$(-\infty, 0) \cup (0, \infty)$

5. $f(x) = 2x^3 - 3x^2 - 18x + 27$

$\qquad = (2x - 3)(x + 3)(x - 3)$

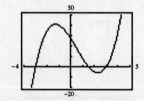

Yes, graphs are the same.

6. $(2x - 3)(x + 3)(x - 3) = (2x - 3)(x^2 - 9)$

$\qquad = 2x^3 - 18x - 3x^2 + 27$

$\qquad = 2x^3 - 3x^2 - 18x + 27$

7. $\dfrac{2x^3 - 3x^2 - 18x + 27}{2x - 3}$

$$
\begin{array}{r}
x^2 \qquad\quad - 9 \\
2x - 3 \overline{\smash{)}\ 2x^3 - 3x^2 - 18x + 27} \\
\underline{2x^3 - 3x^2 \qquad\qquad\quad} \\
-18x + 27 \\
\underline{-18x + 27} \\
0
\end{array}
$$

$x^2 - 9 = (x - 3)(x + 3)$

8. $\dfrac{2x^3 - 3x^2 - 18x + 27}{x^2 - 9} = x^2 - 9 \overline{\smash{)}\ 2x^3 - 3x^2 - 18x + 27}$

$$
\begin{array}{r}
2x - 3 \\
\underline{2x^3 \qquad\quad - 18x} \\
-3x^2 + 27 \\
\underline{-3x^2 + 27} \\
0
\end{array}
$$

9. $f(x) = x^2 - 3$

$\dfrac{f(2 + h) - f(2)}{h} = \dfrac{(2 + h)^2 - 3 - (2^2 - 3)}{h}$

$\qquad = \dfrac{4 + 4h + h^2 - 3 - 4 + 3}{h}$

$\qquad = \dfrac{4h + h^2}{h}$

$\qquad = \dfrac{h(4 + h)}{h}$

$\qquad = 4 + h$

10. $f(x) = \dfrac{3}{x + 5}$

$\dfrac{f(2 + h) - f(2)}{h} = \dfrac{\dfrac{3}{(2 + h) + 5} - \dfrac{3}{2 + 5}}{h}$

$\qquad = \dfrac{\dfrac{3}{7 + h} - \dfrac{3}{7}}{h} \cdot \dfrac{7(7 + h)}{7(7 + h)}$

$\qquad = \dfrac{21 - 3(7 + h)}{7h(7 + h)}$

$\qquad = \dfrac{21 - 21 - 3h}{7h(7 + h)}$

$\qquad = \dfrac{-3h}{7h(7 + h)}$

$\qquad = \dfrac{-3}{7(7 + h)}$

11. *Verbal Model:* $\boxed{\text{Total cost}} = \boxed{\text{Cost per unit}} \cdot \boxed{\text{Number of units}} + \boxed{\text{Fixed costs}}$

Labels: Total cost $= C$

Cost per unit $= \$5.75$

Number of units $= x$

Fixed costs $= \$12,000$

Equation: $C = 5.75x + 12,000$

12. *Verbal Model:* $\boxed{\text{Perimeter}} = 2 \cdot \boxed{\text{Length}} + 2 \cdot \boxed{\text{Width}}$

Labels: Perimeter $= P$

Length $= \frac{3}{2}w$

Width $= w$

Equation: $P = 2 \cdot \frac{3}{2}w + 2 \cdot w$

$= 3w + 2w$

$= 5w$

CHAPTER 9   Radicals and Complex Numbers

SECTION 9.1   Radicals and Rational Exponents

1. $a^m \cdot a^n = a^{m+n}$

2. $(ab)^m = a^m b^m$

3. $(a^m)^n = a^{mn}$

4. $\dfrac{a^m}{a^n} = a^{m-n}$, if $m > n$

5. $3x + y = 4$

$y = -3x + 4$

6. $2x + 3y = 2$

$3y = -2x + 2$

$y = -\dfrac{2}{3}x + \dfrac{2}{3}$

7. $x^2 + 3y = 4$

$3y = 4 - x^2$

$y = \dfrac{4 - x^2}{3} = \dfrac{1}{3}(4 - x^2)$

8. $x^2 + y - 4 = 0$

$y = -x^2 + 4$

9. $2\sqrt{x} - 3y = 15$

$-3y = -2\sqrt{x} + 15$

$y = \dfrac{2}{3}\sqrt{x} - 5$

10. $6|x| - 5y + 10 = 0$

$-5y = -6|x| - 10$

$y = \dfrac{6}{5}|x| + 2$

11. *Verbal Model:* $\boxed{\text{Rate person 1}} + \boxed{\text{Rate person 2}} = \boxed{\text{Rate together}}$

Labels: Your time $= 4$ hours

Friend's time $= 6$ hours

Time together $= x$ hours

Equation: $\dfrac{1}{4} + \dfrac{1}{6} = \dfrac{1}{x}$

$12x\left(\dfrac{1}{4} + \dfrac{1}{6}\right) = \left(\dfrac{1}{x}\right)12x$

$3x + 2x = 12$

$5x = 12$

$x = \dfrac{12}{5}$ hours

12. *Verbal Model:* $\boxed{\text{Total time}} = \boxed{\text{Time 1}} + \boxed{\text{Time 2}}$

$\boxed{\text{Rate}} = \boxed{\dfrac{\text{Distance}}{\text{Time}}}$

Labels: Time 1 $= \dfrac{90}{54}$

Time 2 $= \dfrac{90}{42}$

Equation: Rate $= \dfrac{180}{\frac{90}{54} + \frac{90}{42}}$

Rate $= 47.25$ mph

SECTION 9.2   Simplifying Radical Expressions

1. Graph $x - y = -3$ with a dotted line since the inequality is $>$. Test one point in each half-plane formed by the line. Shade the half-plane that satisfies the inequality.

2. $3x + 4y \le 4$ and $3x + 4y < 4$

The first inequality includes the points on the line $3x + 4y = 4$ and the second does not.

3.
$$-x^3 + 3x^2 - x + 3 = -x^2(x - 3) - 1(x - 3)$$
$$= (x - 3)(-x^2 - 1)$$
$$= -(x - 3)(x^2 + 1)$$

4. $4t^2 - 169 = (2t - 13)(2t + 13)$

5. $x^2 - 3x + 2 = (x - 2)(x - 1)$

6. $2x^2 + 5x - 7 = (2x + 7)(x - 1)$

7. $11x^2 + 6x - 5 = (11x - 5)(x + 1)$

8.
$$4x^2 - 28x + 49 = (2x - 7)(2x - 7)$$
$$= (2x - 7)^2$$

9.
Verbal Model:

$$\boxed{\text{Adult tickets}} + \boxed{\text{Student tickets}} = 1200$$

$$\boxed{\text{Price adult tickets}} \cdot \boxed{\text{Adult tickets}} + \boxed{\text{Price student tickets}} \cdot \boxed{\text{Student tickets}} = 21{,}120$$

Labels: Adult tickets $= x$
Student tickets $= y$

System:
$$x + y = 1200$$
$$20x + 12.50y = 21{,}120$$

Solve by substitution:

$$y = 1200 - x$$
$$20x + 12.50(1200 - x) = 21{,}120$$
$$20x + 15{,}000 - 12.50x = 21{,}120$$
$$7.5x = 6120$$
$$x = 816 \text{ adults}$$
$$y = 1200 - 816 = 384 \text{ students}$$

10.
Verbal Model:

$$\boxed{\frac{\text{Number defective units 1}}{\text{Total number units 1}}} = \boxed{\frac{\text{Number defective units 2}}{\text{Total number units 2}}}$$

Labels: Number defective units (2) $= x$

Equation:
$$\frac{2}{75} = \frac{x}{10{,}000}$$
$$x = \frac{2(10{,}000)}{75}$$
$$x \approx 267 \text{ units}$$

SECTION 9.3   Multiplying and Dividing Radical Expressions

1. $x^2 + bx + c = (x + m)(x + n)$

$mn = c$

2. $x^2 + bx + c = (x + m)(x + n)$

If $c > 0$, the signs of m and n must be the same.

3. If $c < 0$, the signs of m and n must be different.

4. If m and n have like signs, then $m + n = b$.

5. $(-1, -2), (3, 6)$

$$m = \frac{y_2 - y_1}{x_2 - x_1} = \frac{6 - (-2)}{3 - (-1)} = \frac{6 + 2}{3 + 1} = \frac{8}{4} = 2$$

$$y - y_1 = m(x - x_1)$$

$$y - 6 = 2(x - 3)$$

$$y - 6 = 2x - 6$$

$$y = 2x$$

$$0 = 2x - y$$

6. $(1, 5), (6, 0)$

$$m = \frac{y_2 - y_1}{x_2 - x_1} = \frac{0 - 5}{6 - 1} = \frac{-5}{5} = -1$$

$$y - y_1 = m(x - x_1)$$

$$y - 0 = -1(x - 6)$$

$$y = -x + 6$$

$$x + y - 6 = 0$$

7. $(6, 3), (10, 3)$

$$m = \frac{y_2 - y_1}{x_2 - x_1} = \frac{3 - 3}{10 - 6} = \frac{0}{4} = 0$$

$$y - y_1 = m(x - x_1)$$

$$y - 3 = 0(x - 10)$$

$$y - 3 = 0$$

8. $(4, -2), (4, 5)$

$$m = \frac{y_2 - y_1}{x_2 - x_1} = \frac{5 - (-2)}{4 - 4} = \frac{7}{0} = \text{undefined}$$

$$x = 4$$

$$x - 4 = 0$$

9. $\left(\frac{4}{3}, 8\right), (5, 6)$

$$m = \frac{y_2 - y_1}{x_2 - x_1} = \frac{6 - 8}{5 - \frac{4}{3}} = \frac{-2}{\frac{11}{3}} = -\frac{6}{11}$$

$$y - y_1 = m(x - x_1)$$

$$y - 6 = -\frac{6}{11}(x - 5)$$

$$11y - 66 = -6x + 30$$

$$0 = 6x + 11y - 96$$

10. $(7, 4), (10, 1)$

$$m = \frac{y_2 - y_1}{x_2 - x_1} = \frac{1 - 4}{10 - 7} = \frac{-3}{3} = -1$$

$$y - y_1 = m(x - x_1)$$

$$y - 1 = -1(x - 10)$$

$$y - 1 = -x + 10$$

$$x + y - 11 = 0$$

11. *Verbal Model:* $\boxed{\text{Distance}} = \boxed{\text{Rate}} \cdot \boxed{\text{Time}}$

$\dfrac{\boxed{\text{Distance}}}{\boxed{\text{Rate}}} = \boxed{\text{Time}}$

Labels: Time $= t$

Distance $= 360$

Rate $= r$

Equation: $\dfrac{360}{r} = t$

12. *Verbal Model:* $\boxed{\text{Perimeter}} = 2 \cdot \boxed{\text{Length}} + 2 \cdot \boxed{\text{Width}}$

Labels: Perimeter $= P$

Length $= L$

Width $= \dfrac{L}{3}$

Equation: $P = 2L + 2\left(\dfrac{L}{3}\right)$

$$P = \frac{8}{3}L$$

SECTION 9.4   Solving Radical Equations

1. $f(x) = \dfrac{4}{(x+2)(x-3)}$

The function is undefined when the denominator is zero.
Set the denominator equal to zero and solve for x.

$(x+2)(x-3) = 0$

$x + 2 = 0 \qquad x - 3 = 0$

$\quad x = -2 \qquad\quad x = 3$

The domain is all real numbers x such that $x \neq -2$ and $x \neq 3$.

2. $\dfrac{2x^2 + 5x - 3}{x^2 - 9} = \dfrac{2x-1}{x-3}, \quad x \neq -3$

$\dfrac{2(-3)^2 + 5(-3) - 3}{(-3)^2 - 9} = \dfrac{2(-3) - 1}{-3-3}$

$\dfrac{18 - 15 - 3}{9 - 9} = \dfrac{-7}{-6}$

$\dfrac{0}{0} = \dfrac{7}{6}$

Undefined

3. $(-3x^2y^3)^2 \cdot (4xy^2) = 9x^4y^6 \cdot 4xy^2 = 36x^5y^8$

4. $(x^2 - 3xy)^0 = 1$

5. $\dfrac{64r^2s^4}{16rs^2} = 4r^{2-1}s^{4-2} = 4rs^2$

6. $\left(\dfrac{3x}{4y^3}\right)^2 = \left(\dfrac{3x}{4y^3}\right)\left(\dfrac{3x}{4y^3}\right) = \dfrac{9x^2}{16y^6}$

7. $\dfrac{x+13}{x^3(3-x)} \cdot \dfrac{x(x-3)}{5} = \dfrac{x+13}{x^3(3-x)} \cdot \dfrac{-x(3-x)}{5}$

$\qquad = \dfrac{-(x+13)}{5x^2}$

8. $\dfrac{x+2}{5x+15} \cdot \dfrac{x-2}{5(x-3)} = \dfrac{(x+2)(x-2)}{5(x+3)5(x-3)}$

$\qquad = \dfrac{x^2 - 4}{25(x^2 - 9)}$

9. $\dfrac{2x}{x-5} - \dfrac{5}{5-x} = \dfrac{2x}{x-5} + \dfrac{5}{x-5} = \dfrac{2x+5}{x-5}$

10. $\dfrac{3}{x-1} - 5 = \dfrac{3}{x-1} - 5\left(\dfrac{x-1}{x-1}\right)$

$\qquad = \dfrac{3}{x-1} - \dfrac{5(x-1)}{x-1}$

$\qquad = \dfrac{3 - 5x + 5}{x-1}$

$\qquad = \dfrac{8 - 5x}{x-1}$

$\qquad = -\dfrac{5x - 8}{x-1}$

11. $y = 2x - 3$

y-intercept:

$y = 2(0) - 3$

$y = -3$

x-intercept:

$0 = 2x - 3$

$3 = 2x$

$\dfrac{3}{2} = x$

12. $y = -\dfrac{3}{4}x + 2$

y-intercept:

$y = -\dfrac{3}{4}(0) + 2 = 2$

x-intercept:

$0 = -\dfrac{3}{4}x + 2$

$\dfrac{3}{4}x = 2$

$x = 2 \cdot \dfrac{4}{3}$

$x = \dfrac{8}{3}$

SECTION 9.5   Complex Numbers

1. $\dfrac{3t}{5} \cdot \dfrac{8t^2}{15} = \dfrac{(3t)(8t^2)}{(5)(15)} = \dfrac{24t^3}{75}$

Multiply numerators. Multiply denominators.

2. $\dfrac{3t}{5} \div \dfrac{8t^2}{15} = \dfrac{3t}{5} \cdot \dfrac{15}{8t^2} = \dfrac{(3t)(15)}{(5)(8t^2)} = \dfrac{9}{8t}$

Multiply by the reciprocal of the divisor.

3. $\dfrac{3t}{5} + \dfrac{8t^2}{15} = \dfrac{9t}{15} + \dfrac{8t^2}{15} = \dfrac{9t + 8t^2}{15}$

Change each fraction into an equivalent fraction with the lowest common denominator as the denominator. Add the numerators and put over the lowest common denominator.

4. $\dfrac{t - 5}{5 - t} = \dfrac{t - 5}{-1(t - 5)} = -1$

5. $\dfrac{x^2}{2x + 3} \div \dfrac{5x}{2x + 3} = \dfrac{x^2}{2x + 3} \cdot \dfrac{2x + 3}{5x}$

$\qquad = \dfrac{x^2(2x + 3)}{(2x + 3)5x}$

$\qquad = \dfrac{x}{5}$

6. $\dfrac{x - y}{5x} \div \dfrac{x^2 - y^2}{x^2} = \dfrac{x - y}{5x} \cdot \dfrac{x^2}{(x - y)(x + y)}$

$\qquad = \dfrac{(x - y)x^2}{5x(x - y)(x + y)}$

$\qquad = \dfrac{x}{5(x + y)}$

7. $\dfrac{\dfrac{9}{x}}{\left(\dfrac{6}{x} + 2\right)} \cdot \dfrac{x}{x} = \dfrac{9}{6 + 2x}$

8. $\dfrac{\left(1 + \dfrac{2}{x}\right)}{\left(x - \dfrac{4}{x}\right)} \cdot \dfrac{x}{x} = \dfrac{x + 2}{x^2 - 4} = \dfrac{x + 2}{(x - 2)(x + 2)} = \dfrac{1}{x - 2}$

9. $\dfrac{\dfrac{4}{x^2 - 9} + \dfrac{2}{x - 2}}{\dfrac{1}{x + 3} + \dfrac{1}{x - 3}} \cdot \dfrac{(x - 3)(x + 3)(x - 2)}{(x - 3)(x + 3)(x - 2)} = \dfrac{4(x - 2) + 2(x^2 - 9)}{(x - 3)(x - 2) + (x + 3)(x - 2)}$

$\qquad = \dfrac{4x - 8 + 2x^2 - 18}{x^2 - 5x + 6 + x^2 + x - 6}$

$\qquad = \dfrac{2x^2 + 4x - 26}{2x^2 - 4x}$

$\qquad = \dfrac{2(x^2 + 2x - 13)}{2x(x - 2)}$

$\qquad = \dfrac{x^2 + 2x - 13}{x(x - 2)}$

10. $\dfrac{\left(\dfrac{1}{x + 1} + \dfrac{1}{2}\right)}{\left(\dfrac{3}{2x^2 + 4x + 2}\right)} = \dfrac{\left(\dfrac{1}{x + 1} + \dfrac{1}{2}\right)}{\left(\dfrac{3}{2(x^2 + 2x + 1)}\right)} \cdot \dfrac{2(x + 1)^2}{2(x + 2)^2}$

$\qquad = \dfrac{2(x + 1) + (x + 1)^2}{3}$

$\qquad = \dfrac{2x + 2 + x^2 + 2x + 1}{3}$

$\qquad = \dfrac{x^2 + 4x + 3}{3}$

$\qquad = \dfrac{(x + 1)(x + 3)}{3}$

11. $\dfrac{\dfrac{4x}{3} - \dfrac{x}{2}}{3} \cdot \dfrac{6}{6} = \dfrac{8x - 3x}{18} = \dfrac{5x}{18}$

1st number: $\dfrac{x}{2} + \dfrac{5x}{18} = \dfrac{9x}{18} + \dfrac{5x}{18} = \dfrac{14x}{18} = \dfrac{7x}{9}$

2nd number: $\dfrac{14x}{18} + \dfrac{5x}{18} = \dfrac{19x}{18}$

12. $\dfrac{1}{\left(\dfrac{1}{c_1} + \dfrac{1}{c_2}\right)} = \dfrac{1}{\dfrac{1}{c_1} + \dfrac{1}{c_2}} \cdot \dfrac{c_1 c_2}{c_1 c_2} = \dfrac{c_1 c_2}{c_2 + c_1}$

CHAPTER 10   Quadratic Equations and Inequalities

SECTION 10.1   Factoring and Extracting Square Roots

1. The leading coefficient is -3 because $-3t^3$ is the term of highest degree.

2. $(y^2 - 2)(y^3 + 7) = y^5 + 7y^2 - 2y^3 - 14$

Degree: 5 (the highest power)

3.

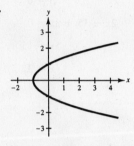

For some values of x there correspond two values of y.

4.

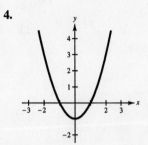

For each value of x there corresponds exactly one value of y.

5. $(x^3 \cdot x^{-2})^{-3} = (x^{3+(-2)})^{-3} = (x^1)^{-3} = x^{-3} = \dfrac{1}{x^3}$

6. $(5x^{-4}y^5)(-3x^2y^{-1}) = -15x^{-4+2}y^{5+(-1)}$

$$= -15x^{-2}y^4 = \dfrac{-15y^4}{x^2}$$

7. $\left(\dfrac{2x}{3y}\right)^{-2} = \left(\dfrac{3y}{2x}\right)^2 = \dfrac{9y^2}{4x^2}$

8. $\left(\dfrac{7u^{-4}}{3v^{-2}}\right)\left(\dfrac{14u}{6v^2}\right)^{-1} = \left(\dfrac{7u^{-4}}{3v^{-2}}\right)\left(\dfrac{6v^2}{14u}\right)$

$$= \dfrac{42u^{-4-1}v^{2-(-2)}}{42} = u^{-5}v^4 = \dfrac{v^4}{u^5}$$

9. $\dfrac{6u^2v^{-3}}{27uv^3} = \dfrac{2u^{2-1}v^{-3-3}}{9} = \dfrac{2u^1v^{-6}}{9} = \dfrac{2u}{9v^6}$

10. $\dfrac{-14r^4s^2}{-98rs^2} = \dfrac{1r^{4-1}s^{2-2}}{7} = \dfrac{r^3s^0}{7} = \dfrac{r^3}{7}$

11. $N = \dfrac{k}{\sqrt{t+1}}$

$300 = \dfrac{k}{\sqrt{0+1}}$

$300 = k$

$N = \dfrac{300}{\sqrt{8+1}}$

$N = 100$ prey

12. $t = \dfrac{k}{r}$

$2 = \dfrac{k}{58}$

$116 = k$

$t = \dfrac{116}{72}$

$t = \dfrac{29}{18} \approx 1.6$ hours

k measures the distance traveled in t hours at r miles per hour.

SECTION 10.2   Completing the Square

1. $(ab)^4 = a^4 b^4$

2. $(a^r)^8 = a^{r \cdot 8} = a^{8r}$

3. $\left(\dfrac{a}{b}\right)^{-r} = \left(\dfrac{b}{a}\right)^r = \dfrac{b^r}{a^r}, \ a \neq 0, b \neq 0$

4. $a^{-r} = \dfrac{1}{a^r}, \quad a \neq 0$

5.
$$\frac{4}{x} - \frac{2}{3} = 0$$
$$(3x)\left(\frac{4}{x} - \frac{2}{3}\right) = (0)(3x)$$
$$12 - 2x = 0$$
$$-2x = -12$$
$$x = 6$$

6. $2x - 3[1 + (4 - x)] = 0$
$$2x - 3[1 + 4 - x] = 0$$
$$2x - 3[5 - x] = 0$$
$$2x - 15 + 3x = 0$$
$$5x = 15$$
$$x = 3$$

7. $3x^2 - 13x - 10 = 0$
$$(3x + 2)(x - 5) = 0$$
$$3x + 2 = 0 \qquad x - 5 = 0$$
$$x = -\tfrac{2}{3} \qquad\quad x = 5$$

8. $x(x - 3) = 40$
$$x^2 - 3x - 40 = 0$$
$$(x - 8)(x + 5) = 0$$
$$x - 8 = 0 \qquad x + 5 = 0$$
$$x = 8 \qquad\quad x = -5$$

9. $g(x) = \tfrac{2}{3}x - 5$

y-intercept:
$$g(0) = \tfrac{2}{3}(0) - 5 = -5$$
x-intercept:
$$0 = \tfrac{2}{3}x - 5$$
$$0 = 2x - 15$$
$$15 = 2x$$
$$\tfrac{15}{2} = x$$
$$7.5 = x$$

10. $h(x) = 5 - \sqrt{x}$

y-intercept:
$$h(0) = 5 - \sqrt{0} = 5$$
x-intercept:
$$0 = 5 - \sqrt{x}$$
$$\sqrt{x} = 5$$
$$x = 5^2 = 25$$

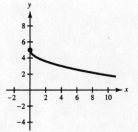

11. $f(x) = \dfrac{4}{x + 2}$

y-intercept:
$$f(0) = \frac{4}{0 + 2} = 2$$
x-intercept:
$$0 = \frac{4}{x + 2}$$
$$0 = 4, \text{ none}$$

Vertical asymptote: $x + 2 = 0$
$$x = -2$$

Horizontal asymptote: $y = 0$ since the degree of the numerator is less than the degree of the denominator.

12. $f(x) = 2x + |x - 1|$

y-intercept:
$$f(0) = 2(0) + |0 - 1| = 1$$
x-intercept:
$$0 = 2x + |x - 1|$$
$$-2x = |x - 1|$$
$$-2x = x - 1 \qquad -2x = -x + 1$$
$$-3x = -1 \qquad\quad -x = 1$$
$$x = \tfrac{1}{3} \qquad\qquad x = -1$$

Check:
$$-2\left(\tfrac{1}{3}\right) \overset{?}{=} \left|\tfrac{1}{3} - 1\right|$$
$$-\tfrac{2}{3} \neq \tfrac{2}{3}$$

Check:
$$-2(-1) = |-1 - 1|$$
$$2 = |-2|$$
$$2 = 2$$

SECTION 10.3   The Quadratic Formula

1. Multiplication Property: $\sqrt{ab} = \sqrt{a}\sqrt{b}$

2. Division Property: $\sqrt{\dfrac{a}{b}} = \dfrac{\sqrt{a}}{\sqrt{b}}, \quad b \neq 0$

3. $\sqrt{72}$ is not in simplest form. A factor (36) of 72 is a perfect square.

$$\sqrt{72} = \sqrt{36 \cdot 2} = 6\sqrt{2}$$

4. $10/\sqrt{5}$ is not in simplest form. There is a radical in the denominator which needs to be rationalized.

$$\frac{10}{\sqrt{5}} = \frac{10}{\sqrt{5}} \cdot \frac{\sqrt{5}}{\sqrt{5}} = \frac{10\sqrt{5}}{5} = 2\sqrt{5}$$

5.
$$\begin{aligned}
\sqrt{128} + 3\sqrt{50} &= \sqrt{64 \cdot 2} + 3\sqrt{25 \cdot 2} \\
&= 8\sqrt{2} + 15\sqrt{2} \\
&= 23\sqrt{2}
\end{aligned}$$

6. $3\sqrt{5}\sqrt{500} = 3\sqrt{5 \cdot 500} = 3\sqrt{2500} = 3 \cdot 50 = 150$

7. $\left(3 + \sqrt{2}\right)\left(3 - \sqrt{2}\right) = 3^2 - \left(\sqrt{2}\right)^2 = 9 - 2 = 7$

8.
$$\begin{aligned}
\left(3 + \sqrt{2}\right)^2 &= 3^2 + 2(3)\sqrt{2} + \left(\sqrt{2}\right)^2 \\
&= 9 + 6\sqrt{2} + 2 \\
&= 11 + 6\sqrt{2}
\end{aligned}$$

9. $\dfrac{8}{\sqrt{10}} = \dfrac{8}{\sqrt{10}} \cdot \dfrac{\sqrt{10}}{\sqrt{10}} = \dfrac{8\sqrt{10}}{10} = \dfrac{4\sqrt{10}}{5}$

10.
$$\begin{aligned}
\frac{5}{\sqrt{12} - 2} &= \frac{5}{\sqrt{12} - 2} \cdot \frac{\sqrt{12} + 2}{\sqrt{12} + 2} \\
&= \frac{5\left(\sqrt{12} + 2\right)}{\left(\sqrt{12}\right)^2 - 2^2} \\
&= \frac{5\left(\sqrt{12} + 2\right)}{12 - 4} \\
&= \frac{5\left(2\sqrt{3} + 2\right)}{8} \\
&= \frac{10\left(\sqrt{3} + 1\right)}{8} \\
&= \frac{5\left(\sqrt{3} + 1\right)}{4}
\end{aligned}$$

11. *Verbal Model:* $\boxed{\text{Perimeter}} = 2 \cdot \boxed{\text{Length}} + 2 \cdot \boxed{\text{Width}}$

$$50 = 2l + 2w$$
$$25 = l + w$$
$$25 - w = l$$

Common Formula: $\quad a^2 + b^2 = c^2$

Equation: $\quad (25 - w)^2 + w^2 = \left(5\sqrt{13}\right)^2$

$$625 - 50w + w^2 + w^2 = (25)(13)$$
$$2w^2 - 50w + 300 = 0$$
$$w^2 - 25w + 150 = 0$$
$$(w - 10)(w - 15) = 0$$
$$w = 10 \qquad\qquad w = 15$$
$$25 - w = 15 \qquad 25 - w = 10$$

10 inches $\times$ 15 inches

12.
$$\begin{aligned}
p &= 75 - \sqrt{1.2(x - 10)} \\
59.90 &= 75 - \sqrt{1.2(x - 10)} \\
-15.10 &= -\sqrt{1.2(x - 10)} \\
(-15.10)^2 &= \left(-\sqrt{1.2(x - 10)}\right)^2 \\
228.01 &= 1.2(x - 10) \\
190.00 &\approx x - 10 \\
200 \text{ units} &\approx x
\end{aligned}$$

SECTION 10.4 Graphs of Quadratic Functions

1. $(x + b)^2 = x^2 + 2bx + b^2$

(Recall $(x + b)^2 = (x + b)(x + b)$ then multiply by FOIL.)

2. $x^2 + 5x + \frac{25}{4}$

To complete the square, take one-half of b and square it. $\left(\frac{1}{2}b\right)^2$

3. $(4x + 3y) - 3(5x + y) = 4x + 3y - 15x - 3y$
$$= -11x$$

4. $(-15u + 4v) + 5(3u - 9v) = -15u + 4v + 15u - 45v$
$$= -41v$$

5. $2x^2 + (2x - 3)^2 + 12x = 2x^2 + 4x^2 - 12x + 9 + 12x$
$$= 6x^2 + 9$$

6. $y^2 - (y + 2)^2 + 4y = y^2 - (y^2 + 4y + 4) + 4y$
$$= y^2 - y^2 - 4y - 4 + 4y$$
$$= -4$$

7. $\sqrt{24x^2y^3} = \sqrt{4 \cdot 6 \cdot x^2 \cdot y^2 \cdot y}$
$$= 2|x|y\sqrt{6y}$$

8. $\sqrt[3]{9} \cdot \sqrt[3]{15} = \sqrt[3]{9 \cdot 15}$
$$= \sqrt[3]{3^3 \cdot 5}$$
$$= 3\sqrt[3]{5}$$

9. $(12a^{-4}b^6)^{1/2} = \sqrt{\dfrac{12b^6}{a^4}} = \sqrt{\dfrac{4 \cdot 3 \cdot b^6}{a^4}} = \dfrac{2b^3}{a^2}\sqrt{3}$

10. $(16^{1/3})^{3/4} = 16^{1/3 \cdot 3/4} = 16^{1/4} = \sqrt[4]{16} = 2$

11. $h = -16t^2 + s_0$

$s_0 = 80 \qquad h = 0$

$0 = -16t^2 + 80$

$16t^2 = 80$

$t^2 = 5$

$t = \pm\sqrt{5}$

Reject $-\sqrt{5}$

$t = \sqrt{5} \approx 2.24$ seconds

12. $h = -16t^2 + s_0$

$s_0 = 150 \qquad h = 0$

$0 = -16t^2 + 150$

$16t^2 = 150$

$t^2 = \dfrac{150}{16}$

$t = \pm\sqrt{\dfrac{150}{16}}$

Reject $-\sqrt{\dfrac{150}{16}}$

$t = \dfrac{\sqrt{25 \cdot 6}}{4} = \dfrac{5\sqrt{6}}{4} \approx 3.06$ seconds

SECTION 10.5 Applications of Quadratic Equations

1. $m = \dfrac{y_2 - y_1}{x_2 - x_1}$

2. (a) Slope-intercept form: $y = mx + b$

(b) Point-slope form: $y - y_1 = m(x - x_1)$

(c) General form: $Ax + By + C = 0$

(d) Horizontal line: $y - b = 0$

3. $(0, 0), (4, -2)$

$$m = \frac{-2 - 0}{4 - 0} = \frac{-2}{4} = \frac{-1}{2}$$

$$y = -\frac{1}{2}x$$

$$2y = -x$$

$$x + 2y = 0$$

4. $(0, 0), (100, 75)$

$$m = \frac{75 - 0}{100 - 0} = \frac{75}{100} = \frac{3}{4}$$

$$y - 0 = \frac{3}{4}(x - 0)$$

$$y = \frac{3}{4}x$$

$$4y = 3x$$

$$0 = 3x - 4y$$

5. $(-1, -2), (3, 6)$

$$m = \frac{6 - (-2)}{3 - (-1)} = \frac{6 + 2}{3 + 1} = \frac{8}{4} = 2$$

$$y - 6 = 2(x - 3)$$

$$y - 6 = 2x - 6$$

$$0 = 2x - y$$

6. $(1, 5), (6, 0)$

$$m = \frac{0 - 5}{6 - 1} = \frac{-5}{5} = -1$$

$$y - 0 = -1(x - 6)$$

$$y = -x + 6$$

$$x + y - 6 = 0$$

7. $\left(\frac{3}{2}, 8\right), \left(\frac{11}{2}, \frac{5}{2}\right)$

$$m = \frac{\frac{5}{2} - 8}{\frac{11}{2} - \frac{3}{2}} \cdot \frac{2}{2} = \frac{5 - 16}{11 - 3} = \frac{-11}{8}$$

$$y - 8 = \frac{-11}{8}\left(x - \frac{3}{2}\right)$$

$$y - 8 = \frac{-11}{8}x + \frac{33}{16}$$

$$16y - 128 = -22x + 33$$

$$22x + 16y - 161 = 0$$

8. $(0, 2), (7.3, 15.4)$

$$m = \frac{15.4 - 2}{7.3 - 0} = \frac{13.4}{7.3} = \frac{134}{73}$$

$$y - 2 = \frac{134}{73}(x - 0)$$

$$y - 2 = \frac{134}{73}x$$

$$73y - 146 = 134x$$

$$0 = 134x - 73y + 146$$

9. $(0, 8), (5, 8)$

$$m = \frac{8 - 8}{5 - 0} = \frac{0}{5} = 0$$

$$y - 8 = 0(x - 0)$$

$$y - 8 = 0$$

10. $(-3, 2), (-3, 5)$

$$m = \frac{5 - 2}{(-3) - (-3)} = \frac{3}{0} = \text{undefined}$$

$$x = -3$$

$$x + 3 = 0$$

11. *Verbal Model:* $\boxed{\begin{array}{c}\text{Cost per person}\\\text{current group}\end{array}} - \boxed{\begin{array}{c}\text{Cost per person}\\\text{new group}\end{array}} = 6250$

Labels: Number current group $= x$

Number new group $= x + 2$

Equation:

$$\frac{250,000}{x} - \frac{250,000}{x + 2} = 6250$$

$$x(x + 2)\left(\frac{250,000}{x} - \frac{250,000}{x + 2}\right) = (6250)x(x + 2)$$

$$250,000(x + 2) - 250,000x = 6250(x^2 + 2x)$$

$$250,000x + 500,000 - 250,000x = 6250x^2 + 12,500x$$

$$0 = 6250x^2 + 12,500x - 500,000$$

$$0 = x^2 + 2x - 80$$

$$0 = (x + 10)(x - 8)$$

$$x = -10 \qquad x = 8 \text{ people}$$

Reject

12. *Verbal Model:* $\boxed{\begin{array}{c}\text{Time}\\\text{upstream}\end{array}} + \boxed{\begin{array}{c}\text{Time}\\\text{downstream}\end{array}} = \boxed{\begin{array}{c}\text{Total}\\\text{time}\end{array}}$

Labels: Speed of the current $= x$

Equation:
$$\frac{35}{18-x} + \frac{35}{18+x} = 4$$

$$(18-x)(18+x)\left(\frac{35}{18-x} + \frac{35}{18+x}\right) = (4)(18-x)(18+x)$$

$$35(18+x) + 35(18-x) = 4(324 - x^2)$$

$$630 + 35x + 630 - 35x = 1296 - 4x^2$$

$$4x^2 - 36 = 0$$

$$4(x^2 - 9) = 0$$

$$x^2 = 9$$

$$x = \pm 3$$

Reject -3

$x = 3$ miles per hour

SECTION 10.6 Quadratic and Rational Inequalities

1. 36.82×10^8 is not written in scientific notation. The number must be between 1 and 10 such as 3.682×10^9.

2. $(n_1 \times 10^2)(n_2 \times 10^4) = n_1 \cdot n_2 \cdot 10^{2+4}$

$$= n_1 \cdot n_2 \cdot 10^6$$

$$1 \le n_1 < 10 \quad \text{and} \quad 1 \le n_2 < 10$$

$$[1 \cdot 1 \le n_1 \cdot n_2 < 10 \cdot 10]10^6$$

$$10^6 \le (n_1 \times 10^2)(n_2 \times 10^4) < 10^8$$

3. $6u^2v - 192v^2 = 6v(u^2 - 32v)$

4. $5x^{2/3} - 10x^{1/3} = 5x^{1/3}(x^{1/3} - 2)$

5. $x(x-10) - 4(x-10) = (x-4)(x-10)$

6. $x^3 + 3x^2 - 4x - 12 = (x^3 + 3x^2) + (-4x - 12)$

$$= x^2(x+3) - 4(x+3)$$

$$= (x+3)(x^2 - 4)$$

$$= (x+3)(x-2)(x+2)$$

7. $16x^2 - 121 = (4x - 11)(4x + 11)$

8. $4x^3 - 12x^2 + 16x = 4x(x^2 - 3x + 4)$

9. Area = Length $\cdot$ Width

$$A = \tfrac{3}{2}h \cdot h$$

$$= \tfrac{3}{2}h^2$$

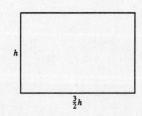

10. Area $= \tfrac{1}{2} \cdot$ Base $\cdot$ Height

$$= \tfrac{1}{2} \cdot b \cdot \tfrac{2}{3}b$$

$$= \tfrac{1}{3}b^2$$

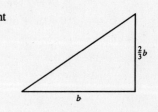

11. Divide figure into 5 congruent squares, each with side length x.

Area = 5 · Area of square

$= 5 \cdot x^2$

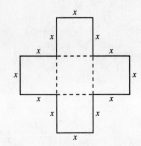

12. Area = Area of rectangle + Area of triangle

$= x \cdot (x + 6) + \frac{1}{2} \cdot x \cdot 4$

$= x^2 + 6x + 2x$

$= x^2 + 8x$

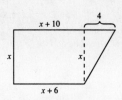

CHAPTER 11   Exponential and Logarithmic Functions

SECTION 11.1   Exponential Functions

1. Graph the line $x + y = 5$. Test one point in each of the half-planes formed by this line. If the point satisfies the inequality, shade the entire half-plane to denote that every point in the region satisfies the inequality.

2. $3x - 5y \le 15$ and $3x - 5y < 15$

The difference between the two graphs is that the first contains the boundary (because of the equal sign) and the second does not.

3. $y > x - 2$

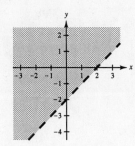

Test point: $(0, 0)$

$0 > 0 - 2$

True

4. $y \le 5 - \frac{3}{2}x$

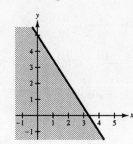

Test point: $(0, 0)$

$0 \le 5 - 0$

True

5. $y < \frac{2}{3}x - 1$

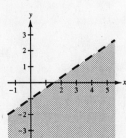

Test point: $(3, -1)$

$-1 < \frac{2}{3}(3) - 1$

$-1 < 1$

True

Test point: $(0, 0)$

$0 < 0 - 1$

False

6. $x > 6 - y$

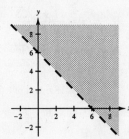

Test point: $(0, 0)$

$0 > 6 - 0$

False

7. $y \le -2$

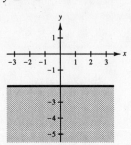

Test point: $(0, -4)$

$-4 \le -2$

True

8. $x > 7$

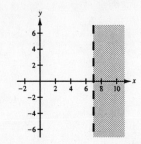

Test point: $(8, 0)$

$8 > 7$

True

9. $2x + 3y \geq 6$

$$3y \geq -2x + 6$$

$$y \geq -\frac{2}{3}x + 2$$

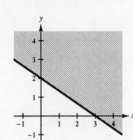

Test point: $(3, 2)$

$2 \geq -\frac{2}{3}(3) + 2$

$2 \geq 0$

True

10. $5x - 2y < 5$

$$-2y < 5 - 5x$$

$$y > \frac{5}{2}x - \frac{5}{2}$$

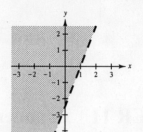

Test point: $(0, 0)$

$0 > 0 - \frac{5}{2}$

True

11. *Verbal Model:*

$$\boxed{\begin{array}{c}\text{Rate for}\\\text{person 1}\end{array}} + \boxed{\begin{array}{c}\text{Rate for}\\\text{person 2}\end{array}} = \boxed{\begin{array}{c}\text{Rate}\\\text{together}\end{array}}$$

Labels: Rate for person 1 $= \dfrac{1}{x}$

Rate for person 2 $= \dfrac{1}{x + 3}$

Rate together $= \dfrac{1}{10}$

Equation:

$$\frac{1}{x} + \frac{1}{x + 3} = \frac{1}{10}$$

$$10x(x + 3)\left(\frac{1}{x} + \frac{1}{x + 3}\right) = \left(\frac{1}{10}\right)10x(x + 3)$$

$$10(x + 3) + 10x = x(x + 3)$$

$$10x + 30 + 10x = x^2 + 3x$$

$$0 = x^2 - 17x - 30$$

$$x = \frac{-(-17) \pm \sqrt{(-17)^2 - 4(1)(-30)}}{2(1)}$$

$$x = \frac{17 \pm \sqrt{289 + 120}}{2}$$

$$x = \frac{17 \pm \sqrt{409}}{2}$$

$$x \approx 18.6 \text{ and } -1.61 \text{ (reject)}$$

$$x + 3 \approx 21.6$$

12. *Formula:* $c^2 = a^2 + b^2$

Labels: $c = $ hypotenuse

$a = 60$ feet

$b = 30$ feet

Equation: $c^2 = 60^2 + 30^2$

$c^2 = 3600 + 900$

$c^2 = 4500$

$c = \sqrt{4500} = \sqrt{900 \cdot 5} = 30\sqrt{5}$

$c \approx 67.1$ feet

SECTION 11.2   Inverse Functions

1. $x - y^2 = 0$

$$x = y^2$$

$$\pm\sqrt{x} = y$$

y is not a function of x because for some values of x there correspond two values of y. For example, $(4, 2)$ and $(4, -2)$ are solution points.

2. $|x| - 2y = 4$

$$-2y = -|x| + 4$$

$$y = \tfrac{1}{2}|x| - 2$$

y is a function of x because for each value of x there corresponds exactly one value of y.

3. $f(x) = \sqrt{4 - x^2}$, $g(x) = \dfrac{6}{\sqrt{4 - x^2}}$

The domain of f is $[-2, 2]$. The domain of g is $(-2, 2)$. g is undefined at $x = \pm 2$.

4. $h(x) = 8 - \sqrt{x}$ over $\{0, 4, 9, 16\}$

Range: $\{4, 5, 6, 8\}$

$$h(0) = 8 - \sqrt{0} = 8$$

$$h(4) = 8 - \sqrt{4} = 6$$

$$h(9) = 8 - \sqrt{9} = 5$$

$$h(16) = 8 - \sqrt{16} = 4$$

5. $-(5x^2 - 1) + (3x^2 - 5) = -5x^2 + 1 + 3x^2 - 5$

$$= -2x^2 - 4$$

6. $(-2x)(-5x)(3x + 4) = 10x^2(3x + 4)$

$$= 30x^3 + 40x^2$$

7. $(u - 4v)(u + 4v) = u^2 - 16v^2$ (multiply by FOIL)

8. $(3a - 2b)^2 = (3a - 2b)(3a - 2b)$

$$= 9a^2 - 6ab - 6ab + 4b^2$$

$$= 9a^2 - 12ab + 4b^2$$

9. $(t - 2)^3 = (t - 2)^2(t - 2)$

$$= (t^2 - 4t + 4)(t - 2)$$

$$= t^3 - 4t^2 + 4t - 2t^2 + 8t - 8$$

$$= t^3 - 6t^2 + 12t - 8$$

10. $\dfrac{6x^3 - 3x^2}{12x} = \dfrac{6x^3}{12x} - \dfrac{3x^2}{12x}$

$$= \dfrac{x^2}{2} - \dfrac{x}{4}$$

11. $v = \sqrt{2gh}$

$$80 = \sqrt{2(32)h}$$

$$80 = \sqrt{64h}$$

$$80^2 = 64h$$

$$\dfrac{80^2}{64} = h = 100 \text{ feet}$$

12. *Verbal Model:* $\boxed{\text{Total cost}} = \boxed{\text{First minute cost}} + \boxed{\text{Additional minute cost}}$

Labels: Total cost $= 5.15$

First minute cost $= 0.95$

Additional minute cost $= 0.35x$

Equation: $5.15 = 0.95 + 0.35x$

$$4.20 = 0.35x$$

$$12 = x$$

13 minutes

SECTION 11.3   Logarithmic Functions

1. $g(x) = (x - 4)^2$

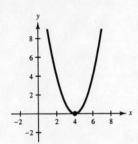

Horizontal shift 4 units right

2. $h(x) = -x^2$

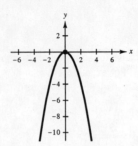

Reflection in the x-axis

3. $j(x) = x^2 + 1$

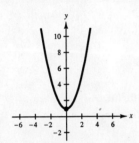

Vertical shift 1 unit up

4. $k(x) = (x + 3)^2 - 5$

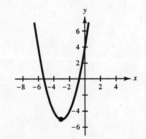

Horizontal shift 3 units left

Vertical shift 5 units down

5. $2x^3 - 6x = 2x(x^2 - 3)$

6. $16 - (y + 2)^2 = [4 - (y + 2)][4 + (y + 2)]$
$$= (4 - y - 2)(4 + y + 2)$$
$$= (2 - y)(6 + y)$$

7. $t^2 + 10t + 25 = (t + 5)(t + 5)$
$$= (t + 5)^2$$

8. $5 - u + 5u^2 - u^3 = 1(5 - u) + u^2(5 - u)$
$$= (5 - u)(1 + u^2)$$

9. $y = 3 - \frac{1}{2}x$

Intercepts:

$y = 3 - \frac{1}{2}(0) = 3, \ (0, 3)$

$0 = 3 - \frac{1}{2}x$

$\frac{1}{2}x = 3$

$x = 6, \ (6, 0)$

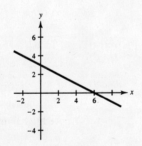

10. $3x - 4y = 6$

Intercepts:

$3(0) - 4y = 6$

$y = -\frac{3}{2}$

$3x - 4(0) = 6$

$3x = 6$

$x = 2$

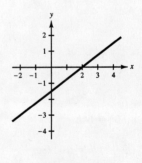

11. $y = x^2 - 6x + 5$

Intercepts:

$y = 5, \ (0, 5)$

$0 = x^2 - 6x + 5$

$0 = (x - 1)(x - 5)$

$x = 1 \quad x = 5$

$(1, 0), (5, 0)$

Vertex:

$y = (x^2 - 6x + 9) + 5 - 9$

$= (x - 3)^2 - 4$

$(3, -4)$

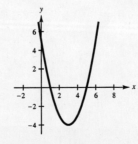

12. $y = -(x - 2)^2 + 1$

Intercepts: Vertex:

$$y = -(0 - 2)^2 + 1 = -3, \ (0, -3) \qquad (2, 1)$$

$$0 = -(x - 2)^2 + 1$$

$$(x - 2)^2 = 1$$

$$x - 2 = \pm 1$$

$$x = 2 \pm 1$$

$$x = 3, 1$$

$$(3, 0), (1, 0)$$

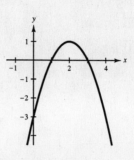

SECTION 11.4   Properties of Logarithms

1. Multiplication Property: $\sqrt[n]{u}\, \sqrt[n]{v} = \sqrt[n]{uv}$

2. Division Property: $\dfrac{\sqrt[n]{u}}{\sqrt[n]{v}} = \sqrt[n]{\dfrac{u}{v}}$

3. $\sqrt{2x}$ and $\sqrt[3]{2x}$ cannot be added because the indices are different.

4. $1/\sqrt{2x}$ is not in simplest form. The radical in the denominator must be rationalized.

$$\frac{1}{\sqrt{2x}} = \frac{1}{\sqrt{2x}} \cdot \frac{\sqrt{2x}}{\sqrt{2x}} = \frac{\sqrt{2x}}{2x}$$

5.
$$\begin{aligned}
25\sqrt{3x} - 3\sqrt{12x} &= 25\sqrt{3x} - 3 \cdot 2\sqrt{3x} \\
&= 25\sqrt{3x} - 6\sqrt{3x} \\
&= (25 - 6)\sqrt{3x} \\
&= 19\sqrt{3x}
\end{aligned}$$

6. $\left(\sqrt{x} + 3\right)\left(\sqrt{x} - 3\right) = x - 9$ (multiply by FOIL)

7.
$$\begin{aligned}
\sqrt{u}\left(\sqrt{20} - \sqrt{5}\right) &= \sqrt{20u} - \sqrt{5u} \\
&= 2\sqrt{5u} - \sqrt{5u} \\
&= (2 - 1)\sqrt{5u} \\
&= \sqrt{5u}
\end{aligned}$$

8.
$$\begin{aligned}
\left(2\sqrt{t} + 3\right)^2 &= \left(2\sqrt{t} + 3\right)\left(2\sqrt{t} + 3\right) \\
&= 4t + 6\sqrt{t} + 6\sqrt{t} + 9 \\
&= 4t + 12\sqrt{t} + 9
\end{aligned}$$

9.
$$\begin{aligned}
\frac{50x}{\sqrt{2}} &= \frac{50x}{\sqrt{2}} \cdot \frac{\sqrt{2}}{\sqrt{2}} \\
&= \frac{50x\sqrt{2}}{2} \\
&= 25x\sqrt{2}
\end{aligned}$$

10.
$$\begin{aligned}
\frac{12}{\sqrt{t + 2} + \sqrt{t}} &= \frac{12}{\sqrt{t + 2} + \sqrt{t}} \cdot \frac{\sqrt{t + 2} - \sqrt{t}}{\sqrt{t + 2} - \sqrt{t}} \\
&= \frac{12\left(\sqrt{t + 2} - \sqrt{t}\right)}{t + 2 - t} \\
&= \frac{12\left(\sqrt{t + 2} - \sqrt{t}\right)}{2} \\
&= 6\left(\sqrt{t + 2} - \sqrt{t}\right)
\end{aligned}$$

11.
$$p = 30 - \sqrt{0.5(x - 1)}$$
$$26.76 = 30 - \sqrt{0.5(x - 1)}$$
$$-3.24 = -\sqrt{0.5(x - 1)}$$
$$3.24 = \sqrt{0.5(x - 1)}$$
$$3.24^2 = 0.5(x - 1)$$
$$10.4976 = 0.5x - 0.5$$
$$10.9976 = 0.5x$$
$$21.9952 = x$$
$$22 \text{ units} \approx x$$

12. *Verbal Model:*

$$\boxed{\frac{\text{Sale}}{\text{price}}} = \boxed{\frac{\text{List}}{\text{price}}} \cdot \boxed{100\% - \text{Discount rate}}$$

Labels: Sale price = 1955

List price = x

$100\% - \text{Discount rate} = 100\% - 15\%$
$$= 85\%$$

Equation: $1955 = x \cdot 0.85$

$$\$2300 = x$$

SECTION 11.5   Solving Exponential and Logarithmic Equations

1. $7x - 2y = 8$

$x + y = 4$

It is not possible for this system to have exactly two solutions. A system of linear equations has no solutions, one solution, or an infinite number of solutions.

2. $8x - 4y = 5$

$-2x + y = 1$

This system has no solution because the equations represent parallel lines and have no point of intersection.

3. $\frac{2}{3}x + \frac{2}{3} = 4x - 6$

$3\left(\frac{2}{3}x + \frac{2}{3}\right) = (4x - 6)3$

$2x + 2 = 12x - 18$

$20 = 10x$

$2 = x$

4. $x^2 - 10x + 17 = 0$

$x^2 - 10x = -17$

$x^2 - 10x + 25 = -17 + 25$

$(x - 5)^2 = 8$

$x - 5 = \pm\sqrt{8}$

$x = 5 \pm 2\sqrt{2}$

(can use quadratic formula also)

5. $\dfrac{5}{2x} - \dfrac{4}{x} = 3$

$2x\left(\dfrac{5}{2x} - \dfrac{4}{x}\right) = (3)2x$

$5 - 8 = 6x$

$-3 = 6x$

$-\dfrac{3}{6} = x$

$-\dfrac{1}{2} = x$

6. $\dfrac{1}{x} + \dfrac{2}{x - 5} = 0$

$x(x - 5)\left(\dfrac{1}{x} + \dfrac{2}{x - 5}\right) = (0)x(x - 5)$

$(x - 5) + 2x = 0$

$3x = 5$

$x = \dfrac{5}{3}$

7. $|x - 4| = 3$

$x - 4 = 3$ or $x - 4 = -3$

$x = 7$ $x = 1$

8. $\sqrt{x + 2} = 7$ **Check:**

$\left(\sqrt{x + 2}\right)^2 = 7^2$ $\sqrt{47 + 2} \overset{?}{=} 7$

$x + 2 = 49$ $\sqrt{49} \overset{?}{=} 7$

$x = 47$ $7 = 7$

9. *Verbal Model:* $\boxed{\text{Distance}} = \boxed{\text{Rate}} \cdot \boxed{\text{Time}}$

Function: $d = 73 \cdot t$

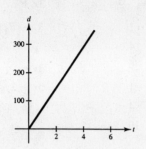

10. $V = \pi r^2 h$

$V = \pi (5)^2 h$

$V = 25\pi h$

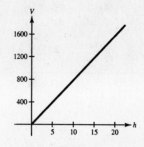

11. $V = \pi r^2 h$

$V = \pi r^2 (10)$

$V = 10\pi r^2$

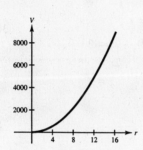

12. $F = kx$

$100 = k(4)$

$25 = k$

$F = 25x$

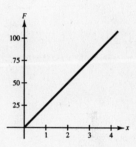

SECTION 11.6 Applications

1. $y = kx^2$

Direct variation as *n*th power

2. $y = \dfrac{k}{x}$

Inverse variation

3. $z = kxy$

Joint variation

4. $z = \dfrac{kx}{y}$

Joint variation

5. $x - y = 0$

$x + 2y = 9$

$\begin{aligned} x - y &= 0 \\ -x - 2y &= -9 \\ \hline -3y &= -9 \\ y &= 3 \end{aligned}$

$x - 3 = 0$

$x = 3$

$(3, 3)$

6. $2x + 5y = 15$

$3x + 6y = 20$

$3(2x + 5y = 15)3$

$-2(3x + 6y = 20) - 2$

$\begin{aligned} 6x + 15y &= 45 \\ -6x - 12y &= -40 \\ \hline 3y &= 5 \\ y &= \tfrac{5}{3} \end{aligned}$

$2x + 5\left(\tfrac{5}{3}\right) = 15$

$2x = 15 - \tfrac{25}{3}$

$2x = \tfrac{45}{3} - \tfrac{25}{3}$

$2x = \tfrac{20}{3}$

$x = \tfrac{20}{3} \cdot \tfrac{1}{2}$

$x = \tfrac{10}{3}$

$\left(\tfrac{10}{3}, \tfrac{5}{3}\right)$

7. $y = x^2$

$-3x + 2y = 2$

$-3x + 2x^2 = 2$

$2x^2 - 3x - 2 = 0$

$(2x + 1)(x - 2) = 0$

$x = -\tfrac{1}{2} \qquad x = 2$

$y = \left(-\tfrac{1}{2}\right)^2 \qquad y = 2^2$

$y = \tfrac{1}{4} \qquad y = 4$

$\left(-\tfrac{1}{2}, \tfrac{1}{4}\right) \qquad (2, 4)$

8.
$$x - \quad y^3 = 0$$
$$x - \quad 2y^2 = 0$$
$$x - \quad y^3 = 0$$
$$\underline{-x + \quad 2y^2 = 0}$$
$$2y^2 - y^3 = 0$$
$$y^2(2 - y) = 0$$

$y = 0$	$y = 2$
$x - 0 = 0$	$x - 2^3 = 0$
$x = 0$	$x = 8$
$(0, 0)$	$(8, 2)$

9.
$$x - y \quad = -1$$
$$x + 2y - 2z = 3$$
$$3x - y + 2z = 3$$

$$x - y \quad = -1$$
$$3y - 2z = 4$$
$$2y + 2z = 6$$

$$x - y \quad = -1$$
$$3y - 2z = 4$$
$$5y \quad = 10$$

$$y = 2$$
$$x - 2 = -1$$
$$x = 1$$

$$1 + 2(2) - 2z = 3$$
$$-2z = -2$$
$$z = 1$$

$$(1, 2, 1)$$

10.
$$2x + y - 2z = 1$$
$$x \quad - z = 1$$
$$3x + 3y + z = 12$$

$$R_1 \leftrightarrow R_2$$

$$x \quad - z = 1$$
$$2x + y - 2z = 1$$
$$3x + 3y + z = 12$$

$$-2R_1 + R_2 \qquad -3R_1 + R_3$$

$$x \quad - z = 1$$
$$y \quad = -1$$
$$3y + 4z = 9$$

$$3(-1) + 4z = 9$$
$$4z = 12$$
$$z = 3$$

$$x - 3 = 1$$
$$x = 4$$

$$(4, -1, 3)$$

11. (a) Graph opens down because $a < 0$.

(b) $0 = -x^2 + 4x$

$0 = -x(x - 4)$

$-x = 0 \qquad x - 4 = 0$

$x = 0 \qquad\quad x = 4$

$(0, 0) \qquad\quad (4, 0)$

(c) $x = \dfrac{-b}{2a}$

$x = \dfrac{-4}{2(-1)} = 2$

$y = -2^2 + 4(2) = -4 + 8 = 4$

$(2, 4)$

12. *Keystrokes:*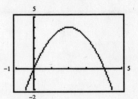

CHAPTER 12   Sequences, Series, and Probability

SECTION 12.1   Sequences and Series

1. Multiplicative Property of Equality

$$-7x = 35$$
$$-\tfrac{1}{7} \cdot -7x = 35 \cdot -\tfrac{1}{7}$$
$$x = -5$$

(Multiply both sides of the equation by the reciprocal of the coefficient of the variable.)

2. Additive Property of Equality

$$7x + 63 = 35$$
$$7x + 63 - 63 = 35 - 63$$
$$7x = -28$$

(Add the opposite of 63 on both sides of the equation.)

3. $t = -3$ is a solution of the equation $t^2 + 4t + 3 = 0$ if the equation is true when -3 is substituted for t.

4. $\dfrac{3}{x} - \dfrac{1}{x + 1} = 10$

The first step in solving this equation is to multiply both sides of the equation by the lowest common denominator $x(x + 1)$.

5. $(x + 10)^{-2} = \dfrac{1}{(x + 10)^2}$

6. $\dfrac{18(x - 3)^5}{(x - 3)^2} = 18(x - 3)^{5-2}$

$$= 18(x - 3)^3$$

7. $(a^2)^{-4} = a^{-8} = \dfrac{1}{a^8}$

8. $(8x^3)^{1/3} = 8^{1/3}x^{3 \cdot 1/3} = 2x$

9. $\sqrt{128x^3} = \sqrt{64 \cdot 2 \cdot x^2 \cdot x}$

$$= 8x\sqrt{2x}$$

10. $\dfrac{5}{\sqrt{x} - 2} = \dfrac{5}{\sqrt{x} - 2} \cdot \dfrac{\sqrt{x} + 2}{\sqrt{x} + 2}$

$$= \dfrac{5(\sqrt{x} + 2)}{(\sqrt{x})^2 - 2^2}$$

$$= \dfrac{5(\sqrt{x} + 2)}{x - 4}$$

11. (a) *Verbal Model:* $\boxed{\text{Area}} = \boxed{\text{Length}} \cdot \boxed{\text{Width}}$

    Equation: $\quad f(x) = (2x - 3) \cdot x$

(b) *Keystrokes:*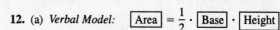

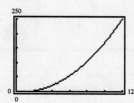

(c) Let $y_2 = 200$ and find the intersection of the two graphs. $x \approx 10.8$

12. (a) *Verbal Model:* $\boxed{\text{Area}} = \dfrac{1}{2} \cdot \boxed{\text{Base}} \cdot \boxed{\text{Height}}$

    Equation: $\quad f(x) = \dfrac{1}{2} \cdot x \cdot (x - 4)$

(b) *Keystrokes:*

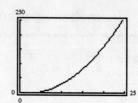

(c) Let $y_2 = 200$ and find the intersection of the two graphs. $x \approx 22.1$

SECTION 12.2   Arithmetic Sequences

1. An algebraic expression is a collection of letters (called variables) and real numbers (called constants) combined with the operations of addition, subtraction, multiplication, and division.

2. The terms of an algebraic expression are those parts separated by addition or subtraction.

3. A trinomial of degree 3 is any polynomial with 3 terms and whose highest exponent on a variable is 3, such as $2x^3 - 3x^2 + 2$.

4. A monomial of degree 4 is any polynomial with only one term and the highest exponent on the variable is 4, such as $7x^4$.

5. $f(x) = x^3 - 2x$

    Domain: $(-\infty, \infty)$

6. $g(x) = \sqrt[3]{x}$

    Domain: $(-\infty, \infty)$

7. $h(x) = \sqrt{16 - x^2}$

Domain: $[-4, 4]$

$16 - x^2 \geq 0$

$(4 - x)(4 + x) \geq 0$

Test intervals: Negative: $(-\infty, -4]$

Positive: $[-4, 4]$

Negative: $[4, \infty)$

Positive: $[-4, 4]$

8. $A(x) = \dfrac{3}{36 - x^2}$

Domain: $(-\infty, -6) \cup (-6, 6) \cup (6, \infty)$

$36 - x^2 \neq 0$

$(6 - x)(6 + x) \neq 0$

$6 - x \neq 0 \qquad 6 + x \neq 0$

$6 \neq x \qquad\quad x \neq -6$

9. $g(t) = \ln(t - 2)$

Domain: $(2, \infty)$

$t - 2 > 0$

$t > 2$

10. $f(s) = 630e^{-0.2s}$

Domain: $(-\infty, \infty)$

11. Formula: $A = P\left(1 + \dfrac{r}{n}\right)^{nt}$

$A = 10{,}000\left(1 + \dfrac{0.075}{365}\right)^{365(15)}$

$A \approx \$30{,}798.61$

12. Formula: $A = P\left(1 + \dfrac{r}{n}\right)^{nt}$

$A = 4000\left(1 + \dfrac{0.06}{12}\right)^{12(5)}$

$A = \$5395.40$

SECTION 12.3   Geometric Sequences and Series

1. The point is 6 units to the left of the y-axis and 4 units above the x-axis.

2.

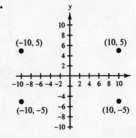

3. The graph of f is the set of ordered pairs $(x, f(x))$, where x is in the domain of f.

4. $f(x) = 2\sqrt{x + 4}$

x-intercept: Let $y = 0$ and solve for x.

$0 = 2\sqrt{x + 4}$

$0 = \sqrt{x + 4}$

$0 = \left(\sqrt{x + 4}\right)^2$

$0 = x + 4$

$-4 = x$

$(-4, 0)$

y-intercept: Let $x = 0$ and solve for y.

$y = 2\sqrt{0 + 4}$

$\quad = 2\sqrt{4}$

$\quad = 2 \cdot 2$

$y = 4$

$(0, 4)$

5. $3x - 5 > 0$

$3x > 5$

$x > \dfrac{5}{3}$

6. $\dfrac{3}{2}y + 11 < 20$

$\dfrac{3}{2}y < 9$

$y < 9 \cdot \dfrac{2}{3}$

$y < 6$

7. $100 < 2x + 30 < 150$

$\quad 70 < 2x < 120$

$\quad 35 < x < 60$

8. $-5 < -\dfrac{x}{6} < 2$

$\quad -30 < -x < 12$

$\quad\quad 30 > x > -12$

$\quad\quad -12 < x < 30$

9. $2x^2 - 7x + 5 > 0$

$\quad (2x - 5)(x - 1) > 0$

Critical numbers: $x = 1, \dfrac{5}{2}$

Positive: $(-\infty, 1)$

Negative: $\left(1, \dfrac{5}{2}\right)$

Positive: $\left(\dfrac{5}{2}, \infty\right)$

Solution: $(-\infty, 1) \cup \left(\dfrac{5}{2}, \infty\right)$

10. $\quad\quad 2x - \dfrac{5}{x} > 3$

$\quad\quad 2x - \dfrac{5}{x} - 3 > 0$

$\quad\quad \dfrac{2x^2 - 5 - 3x}{x} > 0$

$\quad\quad \dfrac{(2x - 5)(x + 1)}{x} > 0$

Critical numbers: $x = -1, 0, \dfrac{5}{2}$

Test intervals:

Negative: $(-\infty, -1) \cup \left(0, \dfrac{5}{2}\right)$

Positive: $(-1, 0) \cup \left(\dfrac{5}{2}, \infty\right)$

Solution: $(-1, 0) \cup \left(\dfrac{5}{2}, \infty\right)$

11. *Formula:* $a^2 + b^2 = c^2$

$\quad$ *Equation:* $a^2 + a^2 = 19^2$

$\quad\quad\quad\quad 2a^2 = 361$

$\quad\quad\quad\quad a^2 = 180.5$

$\quad\quad\quad\quad a = \sqrt{180.5}$

$\quad\quad\quad\quad a \approx 13.4 \text{ inches}$

12. *Formula:* $a^2 + b^2 = c^2$

$\quad$ *Equation:* $\quad 25^2 + 40^2 = c^2$

$\quad\quad\quad\quad 625 + 1600 = c^2$

$\quad\quad\quad\quad\quad\quad 2225 = c^2$

$\quad\quad\quad\quad\quad\quad 47.2 \approx c$

SECTION 12.4   The Binomial Theorem

1. It is not possible to find the determinant of this matrix because it is not square.

2. The three elementary row operations are:

(1) interchange two rows.

(2) multiply a row by a nonzero constant.

(3) add a multiple of one row to another row.

3. This matrix is in row-echelon form.

4. $\det A = \begin{vmatrix} 10 & 25 \\ 6 & -5 \end{vmatrix} = 10(-5) - 6(25)$

$\quad\quad\quad\quad = -50 - 150$

$\quad\quad\quad\quad = -200$

5. $\det A = \begin{vmatrix} 3 & 7 \\ -2 & 6 \end{vmatrix} = 3(6) - (-2)(7)$

$= 18 + 14$

$= 32$

6. $\det A = \begin{vmatrix} 3 & -2 & 1 \\ 0 & 5 & 3 \\ 6 & 1 & 1 \end{vmatrix}$

$= 0 + 5\begin{vmatrix} 3 & 1 \\ 6 & 1 \end{vmatrix} - 3\begin{vmatrix} 3 & -2 \\ 6 & 1 \end{vmatrix}$ (using second row)

$= 5(-3) - 3(15)$

$= -15 - 45$

$= -60$

7. $\det A = \begin{vmatrix} 4 & 3 & 5 \\ 3 & 2 & -2 \\ 5 & -2 & 0 \end{vmatrix}$

$= 5\begin{vmatrix} 3 & 5 \\ 2 & -2 \end{vmatrix} - (-2)\begin{vmatrix} 4 & 5 \\ 3 & -2 \end{vmatrix} + 0$ (using third row)

$= 5(-16) + 2(-23)$

$= -80 - 46$

$= -126$

8. $(x_1, y_1) = (-5, 8), (x_2, y_2) = (10, 0), (x_3, y_3) = (3, -4)$

$\begin{vmatrix} x_1 & y_1 & 1 \\ x_2 & y_2 & 1 \\ x_3 & y_3 & 1 \end{vmatrix} = \begin{vmatrix} -5 & 8 & 1 \\ 10 & 0 & 1 \\ 3 & -4 & 1 \end{vmatrix}$

$= -10\begin{vmatrix} 8 & 1 \\ -4 & 1 \end{vmatrix} + 0 - 1\begin{vmatrix} -5 & 8 \\ 3 & -4 \end{vmatrix}$ (using second row)

$= -10(12) - 1(-4)$

$= -120 + 4$

$= -116$

Area $= -\frac{1}{2}(-116) = 58$

9. $2 = a(0)^2 + b(0) + c \Rightarrow 2 \qquad\qquad + c$

$8 = a(10)^2 + b(10) + c \Rightarrow 8 = 100a + 10b + c$

$0 = a(20)^2 + b(20) + c \Rightarrow 0 = 400a + 20b + c$

$\begin{bmatrix} 0 & 0 & 1 & \vdots & 2 \\ 100 & 10 & 1 & \vdots & 8 \\ 400 & 20 & 1 & \vdots & 0 \end{bmatrix}$

$D = \begin{vmatrix} 0 & 0 & 1 \\ 100 & 10 & 1 \\ 400 & 20 & 1 \end{vmatrix} = 1\begin{vmatrix} 100 & 10 \\ 400 & 20 \end{vmatrix} = (1)(-2000) = -2000$

$a = \dfrac{\begin{vmatrix} 2 & 0 & 1 \\ 8 & 10 & 1 \\ 0 & 20 & 1 \end{vmatrix}}{-2000} = \dfrac{2\begin{vmatrix} 10 & 1 \\ 20 & 1 \end{vmatrix} - 0 + 1\begin{vmatrix} 8 & 10 \\ 0 & 20 \end{vmatrix}}{-2000} = \dfrac{2(-10) + 160}{-2000} = \dfrac{140}{-2000} = -0.07$

—CONTINUED—

9. —CONTINUED—

$$b = \frac{\begin{vmatrix} 0 & 2 & 1 \\ 100 & 8 & 1 \\ 400 & 0 & 1 \end{vmatrix}}{-2000} = \frac{0 - 2\begin{vmatrix} 100 & 1 \\ 400 & 1 \end{vmatrix} + 1\begin{vmatrix} 100 & 8 \\ 400 & 0 \end{vmatrix}}{-2000} = \frac{(-2)(-300) - 3200}{-2000} = \frac{-2600}{-2000} = 1.3$$

$$c = \frac{\begin{vmatrix} 0 & 0 & 2 \\ 100 & 10 & 8 \\ 400 & 20 & 0 \end{vmatrix}}{-2000} = \frac{0 + 0 + 2\begin{vmatrix} 100 & 10 \\ 400 & 20 \end{vmatrix}}{-2000} = \frac{2(-2000)}{-2000} = 2$$

$$y = -0.07x^2 + 1.3x + 2$$

10. $(x_1, y_1) = (2, -1), (x_2, y_2) = (4, 7)$

$$\begin{vmatrix} x & y & 1 \\ 2 & -1 & 1 \\ 4 & 7 & 1 \end{vmatrix} = 0$$

$$x\begin{vmatrix} -1 & 1 \\ 7 & 1 \end{vmatrix} - y\begin{vmatrix} 2 & 1 \\ 4 & 1 \end{vmatrix} + 1\begin{vmatrix} 2 & -1 \\ 4 & 7 \end{vmatrix} = 0 \qquad \text{(using first row)}$$

$$x(-8) - y(-2) + 1(18) = 0$$

$$-8x + 2y + 18 = 0 \qquad \text{(divide by } -2\text{)}$$

$$4x - y - 9 = 0 \text{ or}$$

$$y = 4x - 9$$

SECTION 12.5 Counting Principles

1. $g(x) = 2(5^x)$ is exponential since it has a constant base and a variable exponent.

2. $e^2 \cdot e^{-x^2} = e^{2+(-x^2)} = e^{2-x^2}$ using the law of exponents

$$a^m \cdot a^n = a^{m+n}$$

3. $\log_4 64 = 3$ in exponential form is $4^3 = 64$.

4. $\log_3 \frac{1}{81} = -4$ in exponential form is $3^{-4} = \frac{1}{81}$.

5. $\ln 1 = 0$ in exponential form is $e^0 = 1$.

6. $\ln 5 \approx 1.6094 \ldots$ in exponential form is $e^{1.6094\cdots} \approx 5$.

7. $\quad 3^x = 50$

$$\log_3 3^x = \log_3 50$$

$$x = \frac{\log 50}{\log 3}$$

$$x \approx 3.56$$

8. $\quad e^{x/2} = 8$

$$\ln e^{x/2} = \ln 8$$

$$\frac{x}{2} = \ln 8$$

$$x = 2\ln 8$$

$$x \approx 4.16$$

9. $\log_2(x - 5) = 6$

$$x - 5 = 2^6$$

$$x = 69$$

10. $\ln(x + 3) = 10$

$$x + 3 = e^{10}$$

$$x = e^{10} - 3$$

$$x \approx 22{,}023.47$$

11. (a) *Keystrokes:* [Y=] 22,000 [(] 0.8 [)] [^] [X,T,θ] [GRAPH]   (b) Let $y_2 = 15,000$ and find the intersection. $t = 1.7$

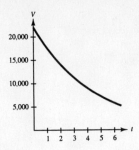

12. $y = Ce^{kt}$ $y = 10e^{kt}$ $y = 10e^{-0.00012097t}$

$10 = Ce^{k(0)}$ $5 = 10e^{k(5730)}$ $y = 10e^{-0.00012097(3000)}$

$10 = C$ $\dfrac{1}{2} = e^{5730k}$ $y \approx 6.96$ grams

$\ln\dfrac{1}{2} = \ln e^{5730k}$

$\ln\dfrac{1}{2} = 5730k$

$\dfrac{\ln\frac{1}{2}}{5730} = k$

$-0.00012097 \approx k$

SECTION 12.6 Probability

1. $\log_a 1 = 0$

2. $\log_a a = 1$

3. $\log_a a^x = x$

4. $\log_a(uv) = \log_a u + \log_a v$

5. $\log_a \dfrac{u}{v} = \log_a u - \log_a v$

6. $\log_a u^n = n \log_a u$

7. $\log_2(x^2 y) = \log_2 x^2 + \log_2 y = 2 \log_2 x + \log_2 y$

8. $\log_2 \sqrt{x^2 + 1} = \log_2(x^2 + 1)^{1/2} = \frac{1}{2} \log_2(x^2 + 1)$

9. $\ln \dfrac{7}{x - 3} = \ln 7 - \ln(x - 3)$

10. $\ln\left(\dfrac{u + 2}{u - 2}\right)^2 = 2 \ln\left(\dfrac{u + 2}{u - 2}\right) = 2[\ln(u + 2) - \ln(u - 2)]$

11. (a) *Keystrokes:*

y_1 [Y=] 10,000 [÷] [(] 1 [+] 4 [e^x] [(] [(-)] [X,T,θ] [÷]

3 [)] [)] [GRAPH]

12. $A = Pe^{rt}$

$A = 1000e^{0.055(1)}$

$A = \$1056.54$

Effective yield $= \dfrac{56.54}{1000} = 0.0565 = 5.65\%$

(b) Let $y_2 = 5000$ and find the intersection of the two graphs. $x \approx 4$ years.

(c) Trace along the graph. The maximum level of annual sales is 10,000.

Part II
Solutions to Odd-Numbered Exercises

CHAPTER 1
The Real Number System

CHAPTER 1
The Real Number System

Section 1.1   Real Numbers: Order and Absolute Value

Solutions to Odd-Numbered Exercises

1. (a) natural numbers: $\left\{2, \frac{9}{3}\right\}$

(b) integers: $\left\{-3, 2, \frac{9}{3}\right\}$

(c) rational numbers: $\left\{-3, 2, -\frac{3}{2}, \frac{9}{3}, 4.5\right\}$

3. (a) natural numbers: $\left\{\frac{8}{4}\right\}$

(b) integers: $\left\{\frac{8}{4}\right\}$

(c) rational numbers: $\left\{-\frac{5}{2}, 6.5, -4.5, \frac{8}{4}, \frac{3}{4}\right\}$

5. $2 < 5$ or $5 > 2$

7. $-4 < -1$ or $-1 > -4$

9. $-2 < \frac{3}{2}$ or $\frac{3}{2} > -2$

11. $-\frac{9}{2} < -3$ or $-3 < -\frac{9}{2}$

13. $3 > -4$ because 3 lies to the *right* of 4.

15. $4 > -\frac{7}{2}$ because 4 lies to the *right* of $-\frac{7}{2}$.

17. $0 > -\frac{7}{16}$ because 0 lies to the *right* of $-\frac{7}{16}$.

19. $-4.6 < 1.5$ because -4.6 lies to the *left* of 1.5.

21. $\frac{7}{16} < \frac{5}{8}$ because $\frac{7}{16}$ lies to the *left* of $\frac{5}{8}$.

Note: $\frac{5}{8} = \frac{10}{16}$

23. $-2\pi > -10$ because -2π lies to the *right* of -10.

Note: $-2\pi \approx -6.28$.

25. 2: The distance between 2 and zero is 2.

27. 4: The distance between -4 and 0 is 4.

29. The opposite of 5 is -5.

The distance from 5 to 0 is five, and the distance from -5 to 0 is also five.

31. The opposite of -3.8 is 3.8

The distance from -3.8 to 0 is 3.8, and the distance from 3.8 to 0 is also 3.8.

33. The opposite of $-\frac{5}{2}$ is $\frac{5}{2}$.

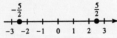

The distance from $-\frac{5}{2}$ to 0 is $\frac{5}{2}$, and the distance from $\frac{5}{2}$ to 0 is also $\frac{5}{2}$.

35. $\left|\frac{5}{2}\right| = \frac{5}{2}$. The distance of $\frac{5}{2}$ from 0 is $\frac{5}{2}$.

37. $\left|-\frac{4}{3}\right| = \frac{4}{3}$. The distance of $-\frac{4}{3}$ from 0 is $\frac{4}{3}$.

39. $|7| = 7$ because the distance between 7 and 0 is 7.

41. $|-3.4| = 3.4$ because the distance between -3.4 and 0 is 3.4.

43. $\left|-\frac{7}{2}\right| = \frac{7}{2}$ because the distance between $-\frac{7}{2}$ and 0 is $\frac{7}{2}$.

45. $-|4.09| = -4.09$

Note: $|4.09| = 4.09$

47. $-|-23.6| = -23.6$

Note: $|-23.6| = 23.6$

49. $|-3.2| = 3.2$ because the distance between -3.2 and 0 is 3.2.

51. $|-15| = |15|$ because $|-15| = 15$, $|15| = 15$, and $15 = 15$.

53. $|-4| > |3|$ because $|-4| = 4$, $|3| = 3$ and 4 is greater than 3.

55. $|32| < |-50|$ because $|32| = 32$, $|-50| = 50$, and $32 < 50$.

57. $|\frac{3}{16}| < |\frac{3}{2}|$ because $|\frac{3}{16}| = \frac{3}{16}$, $|\frac{3}{2}| = |\frac{3}{2}|$, and $\frac{3}{16} < \frac{3}{2}$.

Note: $\frac{3}{2} = \frac{24}{16}$, and $\frac{3}{16} < \frac{24}{16}$

59. $-|-48.5| < |-48.5|$ because $-|-48.5| = -48.5$, $|48.5| = 48.5$, and -48.5 is less than 48.5.

61. $|-\pi| > -|-2\pi|$ because $|-\pi| = \pi$, $-|-2\pi| = -2\pi$, and $\pi > -2\pi$.

63.

65.

67. The number 12.5 units to the right of 8 is 20.5 because $8 + 12.5 = 20.5$.

The number 12.5 units to the left of 8 is -4.5 because $8 - 12.5 = -4.5$.

69. The number 3.5 units to the right of -2 is 1.5 because $-2 + 3.5 = 1.5$.

The number 3.5 units to the left of -2 is -5.5 because $-2 - 3.5 = -5.5$.

71. (a) $A = \{1°, -4°, -15°, 0°, 5°, 8°, 2°, 3°\}$

(b) $B = \{-3°, -5°, -12°, -20°, -6°, -\frac{4}{3}°, 0°, 2°, -1°, -2°, -9°, -10°, -8°\}$

(c) $C = \{0°, 2°\}$

(d) $-15°, -4\frac{1}{2}°, -4°, 0°, 1°, 2°, 2.5°, 3°, 5°, 7.2°, 8°$

(e) $2°, 0°, -1°, -\frac{4}{3}°, -2°, -3°, -5°, -6°, -8°, -9°, -10°, -12°, -20°$

(f) December 11th

73. 2

There are two number on real number line that are 3 units from 0. These two numbers are 3 and -3.

75. 3

The number 3 lies 10 units from -7 on the real number line, but -10 lies only 3 units from -7 on the real number line.

77. 0.35

$\frac{3}{8} = 0.375$, so 0.35 is the smaller number.

79. True. Their absolute values are equal because their distances from zero on the real number line are equal.

81. True. The real number line is a picture used to represent the real numbers. In this picture each point on the line corresponds to exactly one real number and each real number corresponds to exactly one point on the real number line.

83. False. A rational number is a ratio of two integers. Some rational numbers, such as $\frac{6}{2}$ and $-\frac{12}{3}$, are integers, but others are not. For example, the rational numbers $\frac{7}{4}$ and $-\frac{3}{8}$ are not integers.

Section 1.2   Operations with Integers

1. $2 + 7 = 9$

3. $-6 + 4 = -2$

5. $-1 + 0 = -1$

7. $14 + (-14) = 0$

9. $(-14) + 13 = -1$

11. $-23 + 4 = -19$

13. $-18 + (-12) = -30$

15. $-32 + 16 = -16$

17. $5 + |-3| = 5 + 3 = 8$

19. $-|-12| + |-16| = -12 + 16 = 4$

21. $-10 + 6 + 34 = 30$

23. $-82 + (-36) + 82 = -36$

25. $32 + (-32) + (-16) = -16$

27. $1200 + 1300 + (-275) = 2225$

29. $1875 + (-3143) + 5826 = 4558$

31. $|-890| + (-|-82|) + 90 = 890 - 82 + 90 = 898$

33. $12 - 9 = 3$

35. $-4 - (-4) = -4 + 4 = 0$

37. $55 - 20 = 35$

39. $43 - 35 = 10$

41. $-71 - 32 = -103$

43. $-10 - (-4) = -10 + 4 = -6$

45. $-210 - 400 = -610$

47. $-942 - (-942) = 0$

49. $|15| - |-7| = 15 - 7 = 8$

51. $23 - |15| = 23 - 15 = 8$

53. $-32 - (-18) = -32 + 18 = -14$

55. $250 + (-300) = -50$

57. $380 - (-120) = 380 + 120 = 500$

59. $-5 - 10 = -15$

Thus, -15 must be added to 10 to obtain -5.

61. $3 \cdot 2 = 2 + 2 + 2 = 6$

63. $5 \times (-3) = (-3) + (-3) + (-3) + (-3) + (-3)$
$= -15$

65. $7 \times 3 = 21$

67. $4(-8) = -32$

69. $(-6)(-12) = 72$

71. $(310)(-3) = -930$

73. $(5)(-3)(-6) = 90$

75. $(-2)(-3)(-5) = -30$

77. $|3(-5)(6)| = |-90| = 90$

79. $|(-3)4| = |-12| = 12$

81.
$$\begin{array}{r} 26 \\ \times 13 \\ \hline 78 \\ 260 \\ \hline 338 \end{array}$$

83.
$$\begin{array}{r} 63 \\ \times 75 \\ \hline 315 \\ 4410 \\ \hline 4725 \end{array}$$
Thus, $75(-63) = -4725$.

85.
$$\begin{array}{r} 866 \\ \times 72 \\ \hline 1732 \\ 60620 \\ \hline 62352 \end{array}$$
Thus, $(-72)(866) = -62,352$.

87. $27 \div 9 = 3$

89. $72 \div (-12) = -6$

91. $\frac{8}{0}$ is undefined.

Division by zero is undefined.

93. $\dfrac{-81}{-3} = 27$ 　　**95.** $\dfrac{6}{-1} = -6$ 　　**97.** $\dfrac{0}{81} = 0$ 　　**99.** $-180 \div (-45) = 4$

101.
$$\begin{array}{r} 32 \\ 45\,\overline{)\,1440} \\ \underline{135} \\ 90 \\ \underline{90} \\ \end{array}$$
Thus, $1440 \div 45 = 32$.

103.
$$\begin{array}{r} 32 \\ 45\,\overline{)\,1440} \\ \underline{135} \\ 90 \\ \underline{90} \\ \end{array}$$
Thus, $1440 \div (-45) = -32$.

105.
$$\begin{array}{r} 110 \\ 25\,\overline{)\,2750} \\ \underline{25} \\ 25 \\ \underline{25} \\ 0 \\ \underline{0} \\ \end{array}$$
Thus, $2750 \div 25 = 110$.

107. $5(1650) - 3710 = 8250 - 3710 = 4540$ 　　**109.** $\dfrac{44,290}{515} = 86$

111. $\dfrac{169,290}{162} = 1045$

113. $(-2)(532)(500) = -532,000$

You could multipy $-2(500)$ to obtain -1000; multiplying -1000 by 532 yields the result of $-532,000$.

115. 240 is composite; its prime factorization is $2 \cdot 2 \cdot 2 \cdot 2 \cdot 3 \cdot 5$.

117. 643 is prime; the divisibility tests yield no factors of 643. By testing the remaining primes less than or equal to $\sqrt{643} \approx 25$, you can conclude that 643 is a prime number.

119. 3911 is prime; the divisibility tests yield no factors of 3911. By testing the remaining primes less than or equal to $\sqrt{3911} \approx 63$, you can conclude that 3911 is a prime number.

121. 8324 is composite; its prime factorization is $2 \cdot 2 \cdot 2081$.

123. 1321 is prime; the divisibility tests yield no factors of 1321. By testing the remaining primes less than or equal to $\sqrt{1321} \approx 36$, you can conclude that 1321 is a prime number.

125. $12 = 2 \cdot 2 \cdot 3$

127. $210 = 2 \cdot 3 \cdot 5 \cdot 7$ 　　**129.** $192 = 2 \cdot 2 \cdot 2 \cdot 2 \cdot 2 \cdot 2 \cdot 3$ 　　**131.** $525 = 3 \cdot 5 \cdot 5 \cdot 7$

133. $2535 = 3 \cdot 5 \cdot 13 \cdot 13$

135. $-10 + 22 = 12$

The temperature at noon was $12°$ F.

137. $362,000 - (-650,000) = 362,000 + 650,000 = 1,012,000$

Your profit during the second six months was $1,012,000.

139. (a) The increase was approximately $180 million.

    (b) The increase from 1996 to 1997 was approximately $30 million more than the increase from 1995 to 1996.

141. $\left(\frac{8000}{1000}\right)(-3) = 8(-3) = -24°$

The temperature would decrease by approximately $24°$.

143. $(50)(12)(10) = 6000$

You will have deposited a total of $6000.

145. $(160)(360) = 57,600$

The area of the football field is 57,600 square feet.

147. $195 \div 3 = 65$

The average speed of the train is 65 miles per hour.

149. $(9)(6)(11) = 594$

The volume of the rectangular solid is 594 cubic inches.

151. (a) $3 + 2 = 5$

(b) To add two integers with like signs, add their absolute values and attach the common sign to the result.

(c) On these two plays, the team gained 3 yards and then 2 yards for a total of a gain of five yards.

153. The only even prime number is 2. There are no other even prime numbers because every other even number is divisible by itself, by 1, and by 2; all other even numbers are composites because they have more than two factors.

155. To find prime factors of 1997, you need to search among prime numbers less than or equal to $\sqrt{1997} \approx 44.6$. Since 1997 is not divisible by any prime number less than 45, it follows that 1997 is prime.

157. To add two negative numbers, add their absolute values and attach the negative sign.

159. If the factors of a product include an odd number of negative factors, the result will be negative.

161. $3(-5)$ means the sum of three terms of -5.

$$3(-5) = (-5) + (-5) + (-5)$$

163. An even integer has a factor of 2 so the product of this integer and any other integer will also have a factor of 2. Therefore, the product is even.

The product of two odd integers is odd.

165. If an integer n is divided by 2 and the quotient is an even integer, then n must have a factor of 4.

167. The only perfect number less than 25 is 6.

The abundant numbers less than 25 are 12, 18, 20, and 24.

The first perfect number greater than 25 is 28.

Mid-Chapter Quiz for Chapter 1

1. $-2.5 > -4$

2. $\frac{3}{16} < \frac{3}{8}$

3. $-3.1 < 2.7$

4. $2\pi > 6$

Note: $2\pi \approx 6.28$

5. $-|-0.75| = -0.75$

Note: $|-0.75| = 0.75$

6. $|25.2| = 25.2$

7. $\left|\frac{7}{2}\right| = |-3.5|$

Note: $\left|\frac{7}{2}\right| = \frac{7}{2}$ or 3.5, and $|-3.5| = 3.5$.

8. $\left|\frac{3}{4}\right| > -|0.75|$

Note: $\left|\frac{3}{4}\right| = \frac{3}{4}$ or 0.75, and $-|0.75| = -0.75$, $0.75 > -0.75$.

9. The opposite of $-\frac{3}{2}$ is $\frac{3}{2}$; $-\left(-\frac{3}{2}\right) = \frac{3}{2}$.

The opposite of $\frac{5}{2}$ is $-\frac{5}{2}$.

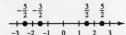

10. The opposite of $-\frac{3}{4}$ is $\frac{3}{4}$; $-\left(-\frac{3}{4}\right) = \frac{3}{4}$.

The opposite of $\frac{9}{4}$ is $-\frac{9}{4}$.

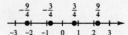

11. $-15 - 12 = -27$

12. $-15 - (-12) = -15 + 12 = -3$

13. $25 + |-75| = 25 + 75 = 100$

14. $-6(10) = -60$

15. $\dfrac{-45}{-3} = 15$

16. $\dfrac{-24}{6} = -4$

17. $513,200 + 136,500 + (-97,750) + (-101,500) = 450,450$

The total profit for the year $450,450.

18. $(8)(4)(4) = 128$

There are 128 cubic feet in a cord of wood.

19. $90 \div 6 = 15$

Each piece of rope is 15 feet long.

20. The statement is false.

Possible changes:

The sum of two integers is *negative*.

The sum of two *positive* integers is positive.

The *product* of two negative integers is positive.

The *quotient* of two negative integers is positive.

Section 1.3   Operations with Rational Numbers

1. By prime factorization, $20 = 2 \cdot 2 \cdot 5$ and $45 = 3 \cdot 3 \cdot 5$. Thus, the greastest common factor is 5.

3. By prime factorization, $28 = 2 \cdot 2 \cdot 7$ and $52 = 2 \cdot 2 \cdot 13$. Thus, the greatest common factor is $2 \cdot 2$, or 4.

5. By prime factorizing, $18 = 2 \cdot 3 \cdot 3$, $84 = 2 \cdot 2 \cdot 3 \cdot 7$, and $90 = 2 \cdot 3 \cdot 3 \cdot 5$. Thus, the greatest common factor is $2 \cdot 3$, or 6.

7. By prime factorization, $240 = 2 \cdot 2 \cdot 2 \cdot 2 \cdot 3 \cdot 5$, $300 = 2 \cdot 2 \cdot 3 \cdot 5 \cdot 5$, and $360 = 2 \cdot 2 \cdot 2 \cdot 3 \cdot 3 \cdot 5$. Thus, the greatest common factor is $2 \cdot 2 \cdot 3 \cdot 5$ or 60.

9. By prime factorizing, $134 = 2 \cdot 67$, $225 = 3 \cdot 3 \cdot 5 \cdot 5$, $315 = 3 \cdot 3 \cdot 5 \cdot 7$, and $945 = 3 \cdot 3 \cdot 3 \cdot 5 \cdot 7$. Because there are no common prime factors, the greatest common factor is 1.

11. $\dfrac{2}{8} = \dfrac{(1)\cancel{(2)}}{(4)\cancel{(2)}} = \dfrac{1}{4}$

13. $\dfrac{12}{18} = \dfrac{2\cancel{(6)}}{3\cancel{(6)}} = \dfrac{2}{3}$

15. $\dfrac{60}{192} = \dfrac{(5)\cancel{(12)}}{(16)\cancel{(12)}} = \dfrac{5}{16}$

Note: This reducing could be done using several steps, such as the following.

$$\dfrac{60}{192} = \dfrac{(30)\cancel{(2)}}{(96)\cancel{(2)}} = \dfrac{(10)\cancel{(3)}}{(32)\cancel{(3)}} = \dfrac{(5)\cancel{(2)}}{(16)\cancel{(2)}} = \dfrac{5}{16}$$

17. $\dfrac{28}{350} = \dfrac{2\cancel{(14)}}{25\cancel{(14)}} = \dfrac{2}{25}$

19. $\dfrac{1}{5} + \dfrac{2}{5} = \dfrac{1 + 2}{5} = \dfrac{3}{5}$

21. $\dfrac{2}{10} + \dfrac{4}{10} = \dfrac{2 + 4}{10} = \dfrac{6}{10} = \dfrac{(3)\cancel{(2)}}{(5)\cancel{(2)}} = \dfrac{3}{5}$

23. $\dfrac{7}{15} + \dfrac{2}{15} = \dfrac{7 + 2}{15} = \dfrac{9}{15} = \dfrac{(3)\cancel{(3)}}{(5)\cancel{(3)}} = \dfrac{3}{5}$

25. $\dfrac{9}{11} + \dfrac{5}{11} = \dfrac{9+5}{11} = \dfrac{14}{11}$

27. $\dfrac{9}{16} - \dfrac{3}{16} = \dfrac{9}{16} + \dfrac{-3}{16}$

$= \dfrac{9 + (-3)}{16} = \dfrac{6}{16} = \dfrac{(3)(2)}{(8)(2)} = \dfrac{3}{8}$

Note: This problem can also be written as follows.

$\dfrac{9}{16} - \dfrac{3}{16} = \dfrac{9-3}{16} = \dfrac{6}{16} = \dfrac{(3)(2)}{(8)(2)} = \dfrac{3}{8}$

29. $\dfrac{-23}{11} + \dfrac{12}{11} = \dfrac{-11}{11}$

$= -1$

31. $\dfrac{3}{4} - \dfrac{5}{4} = \dfrac{3}{4} + \dfrac{-5}{4} = \dfrac{3 + (-5)}{4} = \dfrac{-2}{4} = -\dfrac{(1)(2)}{(2)(2)} = -\dfrac{1}{2}$

33. $\dfrac{13}{15} + \left| -\dfrac{11}{15} \right| - \dfrac{4}{15} = \dfrac{13}{15} + \dfrac{11}{15} - \dfrac{4}{15}$

$= \dfrac{20}{15} = \dfrac{4(5)}{3(5)} = \dfrac{4}{3}$

35. $\dfrac{3}{8} = \dfrac{3(2)}{8(2)} = \dfrac{6}{16}$.

37. Write the original fraction in simplest form, and then find an equivalnt fraction with the indicated denominator.

$\dfrac{6}{15} = \dfrac{2(3)}{5(3)} = \dfrac{2}{5} = \dfrac{2(5)}{5(5)} = \dfrac{10}{25}$

You could do both steps at once by multiplying numerator and denominator by $\frac{5}{3}$.

$\dfrac{6}{15} = \dfrac{6\left(\frac{5}{3}\right)}{15\left(\frac{5}{3}\right)} = \dfrac{10}{25}$

39. $\dfrac{1}{2} + \dfrac{1}{3} = \dfrac{1(3)}{2(3)} + \dfrac{1(2)}{3(2)}$

$= \dfrac{3}{6} + \dfrac{2}{6} = \dfrac{3+2}{6} = \dfrac{5}{6}$

41. $\dfrac{1}{4} - \dfrac{1}{3} = \dfrac{1(3)}{4(3)} - \dfrac{1(4)}{3(4)}$

$= \dfrac{3}{12} - \dfrac{4}{12} = -\dfrac{1}{12}$

43. $\dfrac{3}{16} + \dfrac{3}{8} = \dfrac{3}{16} + \dfrac{3(2)}{8(2)}$

$= \dfrac{3}{16} + \dfrac{6}{16} = \dfrac{3+6}{16} = \dfrac{9}{16}$

45. $-\dfrac{1}{8} - \dfrac{1}{6} = \dfrac{-1}{8} - \dfrac{1}{6}$

$= \dfrac{-1(3)}{8(3)} - \dfrac{1(4)}{6(4)}$

$= \dfrac{-3}{24} - \dfrac{4}{24}$

$= \dfrac{-3-4}{24}$

$= \dfrac{-7}{24} = -\dfrac{7}{24}$

47. $4 - \dfrac{8}{3} = \dfrac{4}{1} + \dfrac{-8}{3}$

$= \dfrac{4(3)}{1(3)} + \dfrac{-8}{3}$

$= \dfrac{12}{3} + \dfrac{-8}{3}$

$= \dfrac{12 + (-8)}{3}$

$= \dfrac{4}{3}$

49. $-\dfrac{7}{8} - \dfrac{5}{6} = -\dfrac{7(3)}{8(3)} - \dfrac{5(4)}{6(4)}$

$= -\dfrac{21}{24} - \dfrac{20}{24}$

$= -\dfrac{41}{24}$

51. $-\dfrac{5}{6} - \left(-\dfrac{3}{4} \right) = -\dfrac{5}{6} + \dfrac{3}{4}$

$= -\dfrac{5(2)}{6(2)} + \dfrac{3(3)}{4(3)}$

$= -\dfrac{10}{12} + \dfrac{9}{12} = -\dfrac{1}{12}$

53. $\dfrac{5}{12} - \dfrac{3}{8} + \dfrac{5}{16} = \dfrac{5(4)}{12(4)} - \dfrac{3(6)}{8(6)} + \dfrac{5(3)}{16(3)}$

$= \dfrac{20}{48} - \dfrac{18}{48} + \dfrac{15}{48}$

$= \dfrac{17}{48}$

55. $2 - \dfrac{25}{6} + \dfrac{3}{4} = \dfrac{2(12)}{1(12)} - \dfrac{25(2)}{6(2)} + \dfrac{3(3)}{4(3)}$

$\qquad\qquad = \dfrac{24}{12} - \dfrac{50}{12} + \dfrac{9}{12}$

$\qquad\qquad = \dfrac{24 - 50 + 9}{12}$

$\qquad\qquad = -\dfrac{17}{12}$

57. $1 + \dfrac{2}{3} - \dfrac{5}{6} = \dfrac{1}{1} + \dfrac{2}{3} + \dfrac{-5}{6}$

$\qquad\qquad = \dfrac{1(6)}{1(6)} + \dfrac{2(2)}{3(2)} + \dfrac{-5}{6}$

$\qquad\qquad = \dfrac{6}{6} + \dfrac{4}{6} + \dfrac{-5}{6}$

$\qquad\qquad = \dfrac{6 + 4 + (-5)}{6}$

$\qquad\qquad = \dfrac{5}{6}$

59. $4\dfrac{3}{5} = \dfrac{4(5) + 3}{5} = \dfrac{23}{5}$

61. $3\dfrac{7}{10} = \dfrac{3(10) + 7}{10} = \dfrac{37}{10}$

63. $8\dfrac{2}{3} = \dfrac{8(3) + 2}{3}$

$\qquad = \dfrac{26}{3}$

65. $-10\dfrac{5}{11} = -\dfrac{10(11) + 5}{11}$

$\qquad\qquad = -\dfrac{115}{11}$

67. $3\dfrac{1}{2} + 5\dfrac{2}{3} = \dfrac{3(2) + 1}{2} + \dfrac{5(3) + 2}{3}$

$\qquad\qquad = \dfrac{7}{2} + \dfrac{17}{3}$

$\qquad\qquad = \dfrac{7(3)}{2(3)} + \dfrac{17(2)}{3(2)}$

$\qquad\qquad = \dfrac{21}{6} + \dfrac{34}{6}$

$\qquad\qquad = \dfrac{21 + 34}{6}$

$\qquad\qquad = \dfrac{55}{6}$

69. $1\dfrac{3}{16} - 2\dfrac{1}{4} = \dfrac{9}{16} - \dfrac{9}{4}$

$\qquad\qquad = \dfrac{19}{16} - \dfrac{9(4)}{4(4)}$

$\qquad\qquad = \dfrac{19}{16} - \dfrac{36}{16}$

$\qquad\qquad = \dfrac{19 - 36}{16}$

$\qquad\qquad = \dfrac{-17}{16}$

$\qquad\qquad = -\dfrac{17}{16}$

71. $15\dfrac{5}{6} - 20\dfrac{1}{4} = \dfrac{15(6) + 5}{6} - \dfrac{20(4) + 1}{4}$

$\qquad\qquad = \dfrac{95}{6} - \dfrac{81}{4}$

$\qquad\qquad = \dfrac{95(2)}{6(2)} - \dfrac{81(3)}{4(3)}$

$\qquad\qquad = \dfrac{190}{12} - \dfrac{243}{12}$

$\qquad\qquad = -\dfrac{53}{12}$

73. $-5\dfrac{2}{3} - 4\dfrac{5}{12} = -\dfrac{17}{3} - \dfrac{53}{12}$

$\qquad\qquad = -\dfrac{17(4)}{3(4)} - \dfrac{53}{12}$

$\qquad\qquad = -\dfrac{68}{12} - \dfrac{53}{12}$

$\qquad\qquad = \dfrac{-68 - 53}{12}$

$\qquad\qquad = \dfrac{-121}{12}$

$\qquad\qquad = -\dfrac{121}{12}$

75. $1 - \dfrac{3}{10} - \dfrac{2}{5} = \dfrac{1}{1} - \dfrac{3}{10} - \dfrac{2}{5}$

$$= \dfrac{1(10)}{1(10)} - \dfrac{3}{10} - \dfrac{2(2)}{5(2)} = \dfrac{10}{10} - \dfrac{3}{10} - \dfrac{4}{10} = \dfrac{10 - 3 - 4}{10} = \dfrac{3}{10}$$

Note: This problem could also be worked in two steps. First, add the two known fractions.

$$\dfrac{3}{10} + \dfrac{2}{5} = \dfrac{3}{10} + \dfrac{2(2)}{5(2)} = \dfrac{3}{10} + \dfrac{4}{10} = \dfrac{3 + 4}{10} = \dfrac{7}{10}$$

Then subtract the sum from 1.

$$1 - \dfrac{7}{10} = \dfrac{1}{1} - \dfrac{7}{10} = \dfrac{1(10)}{1(10)} - \dfrac{7}{10} = \dfrac{10}{10} - \dfrac{7}{10} = \dfrac{10 - 7}{10} = \dfrac{3}{10}$$

77. $\dfrac{1}{2} \times \dfrac{3}{4} = \dfrac{1 \cdot 3}{2 \cdot 4} = \dfrac{3}{8}$

79. $\dfrac{2}{3}\left(-\dfrac{9}{16}\right) = -\dfrac{2 \cdot 9}{3 \cdot 16} = -\dfrac{(2)(3)(3)}{(3)(8)(2)} = -\dfrac{3}{8}$

Note: The reducing could also be written this way.

$$\dfrac{2}{3}\left(-\dfrac{9}{16}\right) = -\dfrac{\overset{1}{\cancel{2}} \cdot \overset{3}{\cancel{9}}}{\underset{1}{\cancel{3}} \cdot \underset{8}{\cancel{16}}} = -\dfrac{3}{8}$$

81. $\left(-\dfrac{7}{16}\right)\left(-\dfrac{12}{5}\right) = \dfrac{7}{16} \cdot \dfrac{12}{5}$

$$= \dfrac{7 \cdot 12}{16 \cdot 5}$$

$$= \dfrac{(7)(3)(4)}{(4)(4)(5)} = \dfrac{21}{20}$$

83. $\left(-\dfrac{3}{2}\right)\left(-\dfrac{15}{16}\right)\left(\dfrac{12}{25}\right) = \dfrac{3}{2} \cdot \dfrac{15}{16} \cdot \dfrac{12}{25} = \dfrac{3 \cdot 15 \cdot 12}{2 \cdot 16 \cdot 25}$

$$= \dfrac{(3)(5)(3)(4)(3)}{(2)(4)(4)(5)(5)} = \dfrac{27}{40}$$

Note: The reducing could also be written this way.

$$\left(-\dfrac{3}{2}\right)\left(-\dfrac{5}{16}\right)\left(\dfrac{12}{25}\right) = \dfrac{3}{2} \cdot \dfrac{15}{16} \cdot \dfrac{12}{25} = \dfrac{3 \cdot \overset{3}{\cancel{15}} \cdot \overset{3}{\cancel{12}}}{2 \cdot \underset{4}{\cancel{16}} \cdot \underset{5}{\cancel{25}}} = \dfrac{27}{40}$$

85. $\left(\dfrac{11}{12}\right)\left(-\dfrac{9}{44}\right) = -\dfrac{11(9)}{12(44)}$

$$= -\dfrac{\cancel{11}(\cancel{3})(3)}{(4)(\cancel{3})(4)(\cancel{11})}$$

$$= -\dfrac{3}{16}$$

87. $9\left(\dfrac{4}{15}\right) = \dfrac{9}{1}\left(\dfrac{4}{15}\right)$

$$= \dfrac{(\cancel{3})(3)(4)}{(\cancel{3})(5)}$$

$$= \dfrac{12}{5}$$

89. $\left(-\dfrac{3}{11}\right)\left(-\dfrac{11}{3}\right) = \dfrac{(-3)(-11)}{(11)(3)}$

$$= \dfrac{33}{33}$$

$$= 1$$

91. $2\dfrac{3}{4} \times 3\dfrac{2}{3} = \left(\dfrac{11}{4}\right)\left(\dfrac{11}{3}\right) = \dfrac{121}{12}$

93. $2\dfrac{4}{5} \times 6\dfrac{2}{3} = \dfrac{14}{5} \cdot \dfrac{20}{3}$

$$= \dfrac{14(20)}{5(3)}$$

$$= \dfrac{14(\cancel{5})(4)}{\cancel{5}(3)}$$

$$= \dfrac{56}{3}$$

95. The reciprocal of 7 is $\dfrac{1}{7}$.

$$7\left(\dfrac{1}{7}\right) = \dfrac{7}{1}\left(\dfrac{1}{7}\right) = \dfrac{7}{7} = 1$$

97. The reciprocal of $\frac{4}{7}$ is $\frac{7}{4}$.

$$\frac{4}{7}\left(\frac{7}{4}\right) = \frac{28}{28} = 1$$

99. $\dfrac{3}{8} \div \dfrac{3}{4} = \dfrac{3}{8} \cdot \dfrac{4}{3} = \dfrac{3 \cdot 4}{8 \cdot 3}$

$$= \frac{\cancel{(3)}(1)\cancel{(4)}}{(2)\cancel{(4)}\cancel{(3)}} = \frac{1}{2}$$

101. $-\dfrac{5}{12} \div \dfrac{45}{32} = -\dfrac{5}{12} \cdot \dfrac{32}{45}$

$$= -\frac{5 \cdot 32}{12 \cdot 45}$$

$$= -\frac{\cancel{(5)}(8)\cancel{(4)}}{(3)\cancel{(4)}(9)\cancel{(5)}}$$

$$= -\frac{8}{27}$$

103. $\dfrac{3}{5} \div 0$ is undefined.

Division by zero is undefined.

105. $-10 \div \dfrac{1}{9} = -\dfrac{10}{1} \cdot \dfrac{9}{1}$

$$= -\frac{90}{1}$$

$$= -90$$

107. $\dfrac{-7/15}{-14/25} = -\dfrac{7}{15} \div \left(-\dfrac{14}{25}\right)$

$$= -\frac{7}{15} \cdot \left(-\frac{25}{14}\right)$$

$$= \frac{7(25)}{15(14)}$$

$$= \frac{7\cancel{(5)}(5)}{\cancel{(5)}(3)\cancel{(7)}(2)}$$

$$= \frac{5}{6}$$

109. $\dfrac{-5}{15/16} = -\dfrac{5}{1} \div \dfrac{15}{16}$

$$= -\frac{5}{1} \cdot \frac{16}{15}$$

$$= -\frac{5 \cdot 16}{1 \cdot 15}$$

$$= -\frac{\cancel{(5)}(16)}{(1)(3)\cancel{(5)}}$$

$$= -\frac{16}{3}$$

111. $3\dfrac{3}{4} \div 1\dfrac{1}{2} = \dfrac{15}{4} \div \dfrac{3}{2}$

$$= \frac{15}{4} \cdot \frac{2}{3}$$

$$= \frac{15(2)}{4(3)}$$

$$= \frac{5\cancel{(3)}\cancel{(2)}}{(2)\cancel{(2)}\cancel{(3)}}$$

$$= \frac{5}{2}$$

113. $3\dfrac{3}{4} \div 2\dfrac{5}{8} = \dfrac{15}{4} \div \dfrac{21}{8}$

$$= \frac{15}{4} \cdot \frac{8}{21}$$

$$= \frac{5\cancel{(3)}\cancel{(4)}(2)}{\cancel{(4)}\cancel{(3)}(7)}$$

$$= \frac{10}{7}$$

115. $\frac{3}{4} = 0.75$

$$
\begin{array}{r}
.75 \\
4\overline{)3.0} \\
\underline{2\,8} \\
20 \\
\underline{20} \\
0
\end{array}
$$

117. $\frac{9}{16} = 0.5625$

$$
\begin{array}{r}
.5625 \\
16\overline{)9.0000} \\
\underline{8\,0} \\
1\,00 \\
\underline{96} \\
40 \\
\underline{32} \\
80 \\
\underline{80} \\
0
\end{array}
$$

119. $\frac{2}{3} = 0.\overline{6}$

$$
\begin{array}{r}
.666\ldots = 0.\overline{6} \\
3\overline{)2.000} \\
\underline{1\,8} \\
20 \\
\underline{18} \\
20 \\
\underline{18} \\
2
\end{array}
$$

121. $\frac{7}{12} = 0.58\overline{3}$

$$
\begin{array}{r}
.58333\ldots = 0.58\overline{3} \\
12\overline{)7.00000} \\
\underline{6\ 0} \\
1\ 00 \\
\underline{96} \\
40 \\
\underline{36} \\
40 \\
\underline{36} \\
40 \\
\underline{36} \\
4
\end{array}
$$

123. $\frac{5}{11} = 0.\overline{45}$

$$
\begin{array}{r}
.4545\ldots = 0.\overline{45} \\
11\overline{)5.0000} \\
\underline{4\ 4} \\
60 \\
\underline{55} \\
50 \\
\underline{44} \\
60 \\
\underline{55} \\
4
\end{array}
$$

125. $1.21 + 4.06 - 3.00 = 2.27$

127. $-0.0005 - 2.01 + 0.111 = -1.8995 \approx -1.90$

129. $(-6.3)(9.05) \approx -57.02$

131. $(-0.09)(-0.45) = 0.0405$

$\qquad\qquad \approx 0.04$

$\qquad$ (rounded to two decimal places)

$$
\begin{array}{r}
0.45 \\
\times\ 0.09 \\
\hline
.0405
\end{array}
$$

133. $4.69 \div 0.12 \approx 39.08$ (rounded to two decimal places)

$$
39.0833\ldots \approx 39.08 \quad \text{(Rounded)}
$$

$$
0.12\overline{)4.69} = 12\overline{)469}
$$

$$
\begin{array}{r}
\underline{36} \\
109 \\
\underline{108} \\
100 \\
\underline{96} \\
40 \\
\underline{36} \\
40 \\
\underline{36} \\
4
\end{array}
$$

135. The sum is approximately 1.

137. $54\frac{1}{4} - 52\frac{5}{8} = \dfrac{54(4) + 1}{4} - \dfrac{52(8) + 5}{8}$

$$
= \frac{217}{4} - \frac{421}{8}
$$

$$
= \frac{434}{8} - \frac{421}{8}
$$

$$
= \frac{13}{8} \text{ or } 1\frac{5}{8}
$$

Thus, the increase in the stock price was $\$1\frac{5}{8}$ per share, or $\$1.625$ per share.

139. $8\frac{3}{4} + 7\frac{1}{5} + 9\frac{3}{8} = \dfrac{35}{4} + \dfrac{36}{5} + \dfrac{75}{8}$

$$
= \frac{35(10)}{4(10)} + \frac{36(8)}{5(8)} + \frac{75(5)}{8(5)}
$$

$$
= \frac{350}{40} + \frac{288}{40} + \frac{375}{40}
$$

$$
= \frac{1013}{40}
$$

$$
= 25\frac{13}{40}
$$

Thus, $25\frac{13}{40}$ tons, or 25.325 tons, of feed were purchased during the first quater of the year.

141. $1 - \dfrac{3}{8} = \dfrac{8}{8} - \dfrac{3}{8} = \dfrac{5}{8}$

Thus, $\frac{5}{8}$ of the gasoline tank is empty.

143. The cost of the milk is $2(2.23) = \$4.46$, and the cost of the bread is $3(1.23) = \$3.69$. Thus, the total cost is $4.46 + 3.69 = \$8.15$. Your change is the difference: $20 - 8.15 = \$11.85$.

145. $60 \div \dfrac{5}{4} = \dfrac{60}{1} \cdot \dfrac{4}{5}$

$\qquad = \dfrac{60(4)}{5}$

$\qquad = \dfrac{12\cancel{(5)}(4)}{\cancel{5}}$

$\qquad = 48$

Thus, you can make 48 breadsticks.

147. The number of gallons needed to drive 12,000 miles in a car which gets 22.3 miles per gallon is

$\dfrac{12,000}{22.3} \approx 538.117$ gallons.

At \$1.259 per gallon, the annual fuel cost is

$(538.117)(1.259) \approx \$677.49.$ (Rounded)

(*Note:* More accurate answers are obtained if you round your answer *only* after all calculations are done.)

149. (a) Two hundred times $23\frac{5}{8}$ is approximately 5000.

Three hundred times $86\frac{1}{4}$ is approximately 26,000.

Thus, the total cost of the stock is approximately \$31,000.

(b) $(200)\left(23\frac{5}{8}\right) + (300)\left(86\frac{1}{4}\right) = (200)\left(\frac{189}{8}\right) + (300)\left(\frac{345}{4}\right)$

$\qquad\qquad = 30,600$

Thus, the actual total cost of the stocks is \$30,600.

Note: You could also use decimals to find the total cost.

$200(23.625) + 300(86.25) = 30,600$

151. (a) Seven times $\frac{2}{3}$ hour is $\frac{14}{3}$ hours; this is approximately equal to 5, or $\frac{15}{3}$, hours. (*Note:* Estimated answers could vary.)

(b) $7\left(\frac{2}{3}\right) = \frac{7}{1}\left(\frac{2}{3}\right)$

$\qquad = \frac{14}{3}$

$\qquad = 4.666\ldots$

$\qquad \approx 4.7$ (rounded to one decimal place)

Thus, the actual number of hours you practiced during the week was approximately 4.7 hours.

153. (g) December 4th

There was a departure of 20 degrees from the monthly average high temperature on this day.

(h) December 4th and 5th

There was a difference of 14 degrees in the high temperature between these two successive days.

(i) December 4th and 5th

There was a difference of 15 degrees in the low temperature between these two successive days.

(j) Average $= \dfrac{2.5 + 1 + (-4) + (-15) + 0 + 5 + 8 + 2 + 7.2 + 3 + 2 + (-4) + (-4.5) + (-4)}{14}$

$\qquad\quad = \dfrac{-0.8}{14}$

$\qquad\quad \approx -0.06$

Thus, the average high temperature of the 14 days was approximately $-0.06°$, just below 0 degrees.

(k) Average $= \dfrac{-3 + (-5) + (-12) + (-20) + (-6) + (-4/3) + 0 + 0 + 2 + (-1) + (-2) + (-9) + (-19) + (-8)}{14}$

$\qquad\quad = \dfrac{-75\frac{1}{3}}{14}$

$\qquad\quad \approx -5.38$

Thus, the average low temperature of 14 days was approximately $-5.38°$.

(l) (g), (h), and (i)

155. No, $\frac{2}{3} + \frac{3}{2} \neq \frac{2+3}{3+2}$.

To add two fractions, determine a common denominator and rewrite each denominator with that denominator. Then add the numerators and write the sum over the common denominator.

$\frac{2}{3} + \frac{3}{2} = \frac{4}{6} + \frac{9}{6} = \frac{13}{6}$

157. (a) $1 - \frac{2}{3} = \frac{1}{3}$

One-third of the pizza was left.

(b) $\frac{1}{2} \cdot \frac{1}{3} = \frac{1}{6}$

You ate one-sixth of the pizza for a midnight snack.

(c)

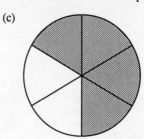

159. $3 = \frac{3}{1} = \frac{3(4)}{1(4)} = \frac{12}{4}$

Using the diagram, you can count that the number of one-fourths in 3 is 12.

You can also divide 3 by 1/4 to find this same result.

$\frac{3}{1/4} = \frac{3}{1} \div \frac{1}{4}$

$= \frac{3}{1} \cdot \frac{4}{1}$

$= 12$

161. False

The reciprocal of the integer 5 is 1/5, but 1/5 is not an integer. In fact, 1 is the only iteger with a reciprocal that is also an integer.

163. True

165. True

167. $\frac{4}{5} + \frac{3}{6} = \frac{13}{10}$

Note:

$\frac{4}{5} + \frac{3}{6} = \frac{4}{5} + \frac{1}{2} = \frac{8}{10} + \frac{5}{10} = \frac{13}{10}$

Section 1.4   Exponents, Order of Operations, and Properties of Real Numbers

1. $2 \cdot 2 \cdot 2 \cdot 2 \cdot 2 = 2^5$

3. $\left(-\frac{1}{4}\right)\left(-\frac{1}{4}\right)\left(-\frac{1}{4}\right) = \left(-\frac{1}{4}\right)^3$

5. $(-3)^6 = (-3)(-3)(-3)(-3)(-3)(-3)$

7. $(9.8)^3 = (9.8)(9.8)(9.8)$

9. $\left(-\frac{1}{2}\right)^5 = \left(-\frac{1}{2}\right)\left(-\frac{1}{2}\right)\left(-\frac{1}{2}\right)\left(-\frac{1}{2}\right)\left(-\frac{1}{2}\right)$

11. $-2^2 = -2 \cdot 2 = -4$

The value is negative.

13. $-5^3 = -5 \cdot 5 \cdot 5 = -125$

The value is negative.

15. $3^2 = 3 \cdot 3 = 9$

17. $2^6 = (2)(2)(2)(2)(2)(2) = 64$

19. $(-5)^3 = (-5)(-5)(-5) = -125$

21. $\left(\frac{1}{4}\right)^3 = \left(\frac{1}{4}\right)\left(\frac{1}{4}\right)\left(\frac{1}{4}\right) = \frac{1}{64}$

23. $(-1.2)^3 = (-1.2)(-1.2)(-1.2)$

$= -1.728$

25. $4 - 6 + 10 = -2 + 10 = 8$

27. $-|2 - (6 + 5)| = -|2 - 11|$
$$= -|-9| = -9$$

29. $15 + 3 \cdot 4 = 15 + 12 = 27$

31. $(16 - 5) \div (3 - 5) = 11 \div (-2)$
$$= -\frac{11}{2}$$

33. $(45 \div 10) \cdot 2 = (4.5)(2) = 9$

35. $5 + (2^2 \cdot 3) = 5 + (4 \cdot 3)$
$$= 5 + 12$$
$$= 17$$

37. $(-6)^2 - (5^2 \cdot 4) = 36 - (25 \cdot 4)$
$$= 36 - 100$$
$$= -64$$

39. $\left(3 \cdot \frac{5}{9}\right) + 1 - \frac{1}{3} = \left(\frac{3}{1} \cdot \frac{5}{9}\right) + \frac{1}{1} - \frac{1}{3}$
$$= \frac{(\cancel{3})(5)}{(\cancel{3})(3)} + \frac{1}{1} - \frac{1}{3}$$
$$= \frac{5}{3} + \frac{3}{3} - \frac{1}{3}$$
$$= \frac{7}{3}$$

41. $4\left(-\frac{2}{3} + \frac{4}{3}\right) = 4\left(\frac{-2 + 4}{3}\right)$
$$= 4\left(\frac{2}{3}\right)$$
$$= \frac{4}{1} \cdot \frac{2}{3} = \frac{4 \cdot 2}{1 \cdot 3}$$
$$= \frac{8}{3}$$

43. $\frac{3}{2}\left(\frac{2}{3} + \frac{1}{6}\right) = \frac{3}{2}\left(\frac{4}{6} + \frac{1}{6}\right)$
$$= \frac{3}{2}\left(\frac{5}{6}\right)$$
$$= \frac{15}{12}$$
$$= \frac{\cancel{3}(5)}{4(\cancel{3})}$$
$$= \frac{5}{4}$$

45. $\frac{3 \cdot 6 - 4 \cdot 6}{5 + 1} = \frac{18 - 24}{6}$
$$= \frac{-6}{6}$$
$$= -1$$

47. $\frac{7}{3}\left(\frac{2}{3}\right) \div \frac{28}{15} = \frac{14}{9} \div \frac{28}{15}$
$$= \frac{14}{9} \cdot \frac{15}{28}$$
$$= \frac{14(\cancel{3})(5)}{3(\cancel{3})(2)(\cancel{14})}$$
$$= \frac{5}{6}$$

49. $\frac{1 - 3^2}{-2} = \frac{1 - 9}{-2} = \frac{-8}{-2} = 4$

51. $\frac{3^2 - 4^2}{0}$ is undefined.

Division by zero is undefined.

53. $\frac{5^2 + 12^2}{13} = \frac{25 + 144}{13}$
$$= \frac{169}{13}$$
$$= 13$$

55. $2.1 \times 10^2 = 2.1 \times 100 = 210$

57. $5.84 \times 10^3 = 5.84(10)(10)(10)$
$$= 5840$$

59. $\frac{8.4}{10^3} = \frac{8.4}{1000} = 0.0084$

61. $\frac{732}{10^2} = \frac{732}{100} = 7.32$

63. $\frac{0}{5^2 + 1} = \frac{0}{5 \cdot 5 + 1} = \frac{0}{25 + 1}$
$$= \frac{0}{26} = 0$$

65. $(3.4)^2 - 6(1.2)^3 = 11.56 - 6(1.728)$

$= 11.56 - 10.368$

$= 1.192$

≈ 1.19 (Rounded to two decimal places)

67. $1000 \div \left(1 + \dfrac{0.09}{4}\right)^8 = 1000 \div (1 + 0.0225)^8$

$= 1000 \div 1.0225^8$

$\approx 1000 \div 1.194831142$

≈ 836.94

(Rounded to 2 decimal places)

69. $4 \cdot 6^2 = 4 \cdot 6 \cdot 6 = 144$

$24^2 = 24 \cdot 24 = 576$

Therefore, $4 \cdot 6^2 \neq 24^2$.

Note: $24^2 = (4 \cdot 6)^2 = 4^2 \cdot 6^2$, *not* $4 \cdot 6^2$.

71. $-3^2 = -(3 \cdot 3)$, or -9

$(-3)(-3) = 9$

Thus, it is true that $-3^2 \neq (-3)(-3)$.

73. Commutative Property of Multiplication

75. Commutative Property of Addition

77. Additive Identity Property

79. Additive Inverse Property

81. Associative Property of Addition

83. Associative Property of Multiplication

85. Multiplicative Inverse Property

87. Distributive Property

89. Distributive Property

91. Associative Property of Addition

93. $5(u + v) = 5(v + u)$ (using the Commutative Property of Addition)

or

$5(u + v) = (u + v)5$ (using the Commutative Property of Multiplication)

95. $3 + x = x + 3$

97. $6(x + 2) = 6x + 6 \cdot 2 = 6x + 12$

99. $(4 + y)25 = 4 \cdot 25 + 25y$

$= 100 + 25y$

101. $3x + (2y + 5) = (3x + 2y) + 5$

103. $(6x)y = 6(xy)$

105. (a) Additive inverse: -50

(b) Multiplicative inverse: $\frac{1}{50}$

107. (a) Additive inverse: 1

(b) Multiplicative inverse: -1

109. (a) Additive Inverse: $-2x$

(b) Multiplicative Inverse: $\dfrac{1}{2x}$

111. (a) Additive Inverse: $-ab$

(b) Multiplicative Inverse: $\dfrac{1}{ab}$

113. $3(6 + 10) = 3(6) + 3(10)$ or $3(6 + 10) = 3(16)$

$= 18 + 30$ $= 48$

$= 48$

115. $\frac{2}{3}(9 + 24) = \frac{2}{3} \cdot 9 + \frac{2}{3} \cdot 24$ or $\frac{2}{3}(9 + 24) = \frac{2}{3}(33)$

$= 6 + 16$ $= 22$

$= 22$

117. $5(x + 3) = 5x + 5 \cdot 3 = 5x + 15$

Therefore, $5(x + 3) \neq 5x + 3$. The 5 should be multiplied by *both* terms in the parenthesis.

119. $\frac{8}{0}$ is undefined because division by 0 is undefined.

Therefore, $\frac{8}{0} \neq 0$.

121. $4(2 + x) = 4(x + 2)$ Commutative Property of Addition

$= 4x + 8$ Distributive Property

123. $7x + 9 + 2x = 7x + 2x + 9$ Commutative Property of Addition

$\qquad\qquad = (7x + 2x) + 9$ Associative Property of Addition

$\qquad\qquad = (7 + 2)x + 9$ Distributive Property

$\qquad\qquad = 9x + 9$ Addition of Real Numbers

$\qquad\qquad = 9(x + 1)$ Distributive Property

125. The total area of the figure is the *sum* of the area of the upper rectangle and the area of the lower rectangle. The upper rectangle has length 3 and width 3. Its area is $3 \cdot 3 = 9$. The lower rectangle has length 9 and width 3. Its area is $9 \cdot 3 = 27$. Therefore, the total area is $9 + 27 = 36$ square units.

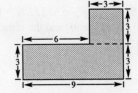

$\qquad$ Area $= 3 \cdot 3 + 9 \cdot 3$

$\qquad\qquad = 9 + 27 = 36$ square units

127. The portion of the expenses for all other categories (except utilities) is the sum

$\qquad 0.06 + 0.09 + 0.35 + 0.15 + 0.13 + 0.08 + 0.07 = 0.93.$

Therefore, the portion of expenses that is spent on utilities is the difference

$\qquad 1 - 0.93 = 0.07$

If the total expenses are \$450,000, the amount spent on utilities is the product

$\qquad 0.07(450,000) = \$31,500.$

The amount spent on utilities is \$31,500.

129. $750 + 48(215) = 750 + 10,320 = 11,070$

Thus, the total amount paid for the car is \$11,070.

131. (a) $x + 0.06x = (1 + 0.06)x$, or $1.06x$

$\qquad$ (b) $1.06(25.95) = 27.507$

$\qquad$ You must pay \$27.51 for the item.

133. $(a - 2) + (b + 11) + (2c + 3) = a + b + 2c - 2 + 11 + 3$

$\qquad\qquad\qquad\qquad\qquad\quad = a + b + 2c + 12$

135. No, because the *order* in which these two actions are performed *does* affect the results.

137. (a) The 3 is called the base.

$\qquad$ (b) The 5 is called the exponent.

139. No

$\qquad 2 \cdot 5^2 = 2 \cdot 25 = 50$

$\qquad 10^2 = 100$

$\qquad$ Thus, $2 \cdot 5^2 \neq 10^2$.

141. $12 + 48 \div 6 - 5 = 12 + (48 \div 6) - 5$

143. (a) Commutative Property of Addition: $a + b = b + a$

$\qquad$ Two real numbers can be added in either order.

$\qquad$ Example: $9 + 1 = 1 + 9$

$\qquad$ (b) Commutative Property of Multiplication: $ab = ba$

$\qquad$ Two real numbers can be multiplied in either order.

$\qquad$ Example: $5(4) = 4(5)$

145. (a) Additive Identity Property: $a + 0 = a$

$\qquad$ The sum of zero and a number equals the number itself.

$\qquad$ Example: $7 + 0 = 7$

$\qquad$ (b) Additive Inverse Property: $a + (-a) = 0$

$\qquad$ The sum of a real number and its opposite is zero.

$\qquad$ Example: $4 + (-4) = 0$

Review Exercises for Chapter 1

1. $-\frac{1}{10} < 4$ because $-\frac{1}{10}$ lies to the *left* of 4.

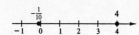

3. $-3 > -7$ because -3 lies to the *right* of -7.

5. The opposite of 152 is -152. The distance from 152 to 0 is 152 units, and the distance from -152 to 0 is also 152 units.

7. The opposite of $-\frac{7}{3}$ is $\frac{7}{3}$. The distance from $-\frac{7}{3}$ to 0 is $\frac{7}{3}$ units, and the distance from $\frac{7}{3}$ to 0 is also $\frac{7}{3}$ units.

9. $|-8.5| = 8.5$

11. $-|-8.5| = 8.5$

Note: $|-8.5| = 8.5$

13. $|-84| = |84|$ because $|-84| = 84$ and $|84| = 84$.

15. $\left|\frac{3}{10}\right| > -\left|\frac{4}{5}\right|$ because $\left|\frac{3}{10}\right| = \frac{3}{10}$, $-\left|\frac{4}{5}\right| = -\frac{4}{5}$, and $\frac{3}{10} > -\frac{4}{5}$.

17. $32 + 68 = 100$

19. $16 + (-5) = 11$

21. $350 - 125 + 15 = 240$

23. $-114 + 76 - 230 = -268$

25. $|-86| - |124| = 86 - 124 = -38$

27. $15 \times 3 = 45$

29. $-300(-5) = 1500$

31. $31(-6)(3) = -558$

33. $\frac{-162}{9} = -18$

35. $815 \div 0$ is undefined. Division by 0 is undefined.

37. $613 - (-549) = 1162$

39. $-27 - 75 = -102$

You must add -102 to 75 to obtain -27.

41.
$$\begin{array}{r} 469 \\ 72\overline{)33768} \\ \underline{288} \\ 496 \\ \underline{432} \\ 648 \\ \underline{648} \end{array}$$

Thus, $33{,}768 \div (-72) = -469$.

43. $7(5207) - 52{,}318 = -15{,}869$

45. $\frac{345{,}582}{438} = 789$

47. 839 is prime; the divisibility tests yield no factors of 839. By testing the remaining primes less than or equal to $\sqrt{839} \approx 29$, you can conclude that 839 is a prime number.

49. 1764 is composite; its prime factorization is $1764 = 2 \cdot 2 \cdot 3 \cdot 3 \cdot 7 \cdot 7$.

51. $378 = 2 \cdot 3 \cdot 3 \cdot 3 \cdot 7$

53. $1612 = 2 \cdot 2 \cdot 13 \cdot 31$

55. By prime factorization, $54 = 2 \cdot 3 \cdot 3 \cdot 3$ and $90 = 2 \cdot 3 \cdot 3 \cdot 5$. Thus, the greatest common factor is $2 \cdot 3 \cdot 3$, or 18.

57. By prime factorization, $63 = 3 \cdot 3 \cdot 7$, $84 = 2 \cdot 2 \cdot 3 \cdot 7$, and $441 = 3 \cdot 3 \cdot 7 \cdot 7$. Thus, the greatest common factor is $3 \cdot 7$, or 21.

59. $\frac{2}{3} = \frac{2(5)}{3(5)} = \frac{10}{15}$

61. $\frac{6}{10} = \frac{3(2)}{5(2)} = \frac{3}{5} = \frac{3(5)}{5(5)} = \frac{15}{25}$

63. $\frac{3}{25} + \frac{7}{25} = \frac{3+7}{25} = \frac{10}{25}$

$$= \frac{(2)(5)}{(5)(5)} = \frac{2}{5}$$

65. $\dfrac{27}{16} - \dfrac{15}{16} = \dfrac{27-15}{16} = \dfrac{12}{16}$

$= \dfrac{(3)(4)}{(4)(4)} = \dfrac{3}{4}$

67. $-\dfrac{5}{9} + \dfrac{2}{3} = \dfrac{-5}{9} + \dfrac{2(3)}{3(3)} = \dfrac{-5}{9} + \dfrac{6}{9}$

$= \dfrac{-5+6}{9} = \dfrac{1}{9}$

69. $\dfrac{25}{32} + \dfrac{7}{24} = \dfrac{25(3)}{32(3)} + \dfrac{7(4)}{24(4)}$

$= \dfrac{75}{96} + \dfrac{28}{96}$

$= \dfrac{75+28}{96} = \dfrac{103}{96}$

71. $5 - \dfrac{15}{4} = \dfrac{5(4)}{1(4)} - \dfrac{15}{4} = \dfrac{20}{4} - \dfrac{15}{4}$

$= \dfrac{20-15}{4} = \dfrac{5}{4}$

73. $5\dfrac{3}{4} - 3\dfrac{5}{8} = \dfrac{23}{4} - \dfrac{29}{8}$

$= \dfrac{23(2)}{4(2)} - \dfrac{29}{8}$

$= \dfrac{46}{8} - \dfrac{29}{8}$

$= \dfrac{46-29}{8} = \dfrac{17}{8}$

75. $\dfrac{5}{8} \cdot \dfrac{-2}{15} = \dfrac{5(-2)}{8 \cdot 15}$

$= -\dfrac{(1)(5)(2)}{(4)(2)(3)(5)} = -\dfrac{1}{12}$

77. $35\left(\dfrac{1}{35}\right) = \dfrac{35}{1} \cdot \dfrac{1}{35} = \dfrac{35 \cdot 1}{1 \cdot 35}$

$= \dfrac{(35)(1)}{(1)(35)} = 1$

79. $\dfrac{5}{14} \div \dfrac{15}{28} = \dfrac{5}{14} \cdot \dfrac{28}{15} = \dfrac{5 \cdot 28}{14 \cdot 15}$

$= \dfrac{(5)(14)(2)}{(14)(5)(3)} = \dfrac{2}{3}$

81. $\dfrac{-\frac{3}{4}}{-\frac{7}{8}} = \dfrac{\frac{3}{4}}{\frac{7}{8}} = \dfrac{3}{4} \div \dfrac{7}{8} = \dfrac{3}{4} \cdot \dfrac{8}{7}$

$= \dfrac{3 \cdot 8}{4 \cdot 7} = \dfrac{(3)(2)(4)}{(4)(7)} = \dfrac{6}{7}$

83. $\dfrac{\frac{5}{2}}{0}$ is undefined. Division by zero is undefined.

85. $\dfrac{5.25}{0.25} = 21$

$$0.25\overline{)5.25} = 25\overline{)525}$$
$$\underline{50}$$
$$25$$
$$\underline{25}$$

87. $(5.8)^4 - (3.2)^5 = 11331.6496 - 335.54432$

$= 796.10528$

≈ 796.11

89. $\dfrac{3000}{(1.05)^{10}} \approx 1841.739761$

≈ 1841.74

91. $7^3 = 7 \cdot 7 \cdot 7 = 343$

93. $(-7)^3 = (-7)(-7)(-7)$

$= -343$

95. $2^2 = 4$

$2^4 = 16$

$4 < 16$

Thus, $2^2 < 2^4$.

97. $\dfrac{3}{4} = \dfrac{12}{16}$

$\left(\dfrac{3}{4}\right)^2 = \dfrac{9}{16}$

$\dfrac{12}{16} > \dfrac{9}{16}$

Thus, $\dfrac{3}{4} > \left(\dfrac{3}{4}\right)^2$.

99. $\left(\dfrac{3}{5}\right)^4 = \left(\dfrac{3}{5}\right)\left(\dfrac{3}{5}\right)\left(\dfrac{3}{5}\right)\left(\dfrac{3}{5}\right)$

$= \dfrac{3 \cdot 3 \cdot 3 \cdot 3}{5 \cdot 5 \cdot 5 \cdot 5} = \dfrac{81}{625}$

101. $240 - (4^2 \cdot 5) = 240 - (16 \cdot 5)$

$= 240 - 80 = 160$

103. $3^2(10 - 2^2) = 9(10 - 4)$

$= 9(6) = 54$

105. $\left(\dfrac{3}{4}\right)\left(\dfrac{5}{6}\right) + 4 = \dfrac{3 \cdot 5}{4 \cdot 6} + 4$

$= \dfrac{(3)(5)}{(4)(2)(3)} + 4 = \dfrac{5}{8} + \dfrac{4}{1} = \dfrac{5}{8} + \dfrac{4(8)}{1(8)}$

$= \dfrac{5}{8} + \dfrac{32}{8} = \dfrac{5+32}{8} = \dfrac{37}{8}$

107. $122 - [45 - (32 + 8) - 23] = 122 - [45 - 40 - 23]$

$= 122 - [-18]$

$= 122 + 18$

$= 140$

109. $\dfrac{6 \cdot 4 - 36}{4} = \dfrac{24 - 36}{4}$

$= \dfrac{-12}{4}$

$= -3$

111. $\dfrac{54 - 4 \cdot 3}{6} = \dfrac{54 - 12}{6}$

$= \dfrac{42}{6}$

$= 7$

113. $\dfrac{78 - |-78|}{5} = \dfrac{78 - 78}{5}$

$= \dfrac{0}{5} = 0$

115. Additive Inverse Property

117. Commutative Property of Multiplication

119. Multiplicative Identity Property

121. Distributive Property

123. The smaller number is 0.6.

Note: $\frac{2}{3} = 0.666 \ldots$ and $0.666 \ldots > 0.6$

125. The statement is false because the sum of a positive and negative integer can be positive, negative or zero.

Note:

(a) If the absolute value of the negative integer is smaller than the positive integer, the sum is positive.

Example: $12 + (-8) = 4$

(b) If the absolute value of the negative integer is larger than the positive integer, the sum is negative.

Example: $12 + (-15) = -3$

(c) If the absolute value of the negative integer is equal to the positive integer, the sum is zero.

Example: $12 + (-12) = 0$

127. $\dfrac{40,000(4)}{5} = \dfrac{160,000}{5} = 32,000$

Thus, each tire has been driven 32,000 miles.

129. (a) Day 1: $25(162) + 10(98) = \$5030$

Day 2: $25(98) + 10(64) = \$3090$

Day 3: $25(148) + 10(81) = \$4510$

Day 4: $25(186) + 10(105) = \$5700$

(b) Adult Tickets: $25(162 + 98 + 148 + 186) = 25(594) = \$14,850$

Student Tickets: $10(98 + 64 + 81 + 105) = 10(348) = \$3480$

(c) Total from (a): $5030 + 3090 + 4510 + 5700 = \$18,330$

Total from (b): $14,850 + 3480 = \$18,330$

The total revenue from tickets sales can be determined by adding the daily tickets sales (from part a) or by adding the sales of the two types of tickets (from part b). These two totals should be the same.

131. $35\frac{1}{4} - \frac{3}{8} - \frac{1}{2} - \frac{1}{8} + 1\frac{1}{4} + \frac{1}{2} = \frac{141}{4} - \frac{3}{8} - \frac{1}{2} - \frac{1}{8} - \frac{5}{4} + \frac{1}{2}$

$= \frac{282}{8} - \frac{3}{8} - \frac{4}{8} - \frac{1}{8} + \frac{10}{8} + \frac{4}{8}$

$= \frac{288}{8}$

$= 36$

Thus, the closing price on Friday was \$36.

Note: These numbers could be written in decimal form.

$35.25 - 0.375 - 0.5 - 0.125 + 1.25 + 0.5 = 36.$

133. The cost of a five-minute call would be the sum of the \$0.64 cost for the first minute plus the \$0.72 cost of each of the *four* additional minutes.

$0.64 + 4(0.72) = 0.64 + 2.88 = \$3.52$

The cost of the call is \$3.52.

135. (a) $16,000\left(\frac{3}{4}\right)^3 = 6750$

(b) $16,000 - 6750 = 9250$

Thus, the car is worth \$6750 after three years. It has depreciated \$9250.

Chapter Test for Chapter 1

1. (a) Natural numbers: $\left\{8, \frac{12}{4}\right\}$

    (b) Integers: $\left\{-10, 8, \frac{12}{4}\right\}$

    (c) Rational numbers:
       $\left\{-10, 8, \frac{3}{4}, \frac{12}{4}, 6.5\right\}$

2. $-\frac{3}{5} > -|-2|$

3. $16 + (-20) = -4$

4. $-50 - (-60) = -50 + 60 = 10$

5. $7 + |-3| = 7 + 3 = 10$

6. $64 - (25 - 8) = 64 - 17 = 47$

7. $-5(32) = -160$

8. $\dfrac{-72}{-9} = 8$

9. $\dfrac{12 + 9}{7} = \dfrac{21}{7} = 3$

10. $-\dfrac{(-2)(5)}{10} = -\dfrac{-10}{10}$
                       $= -(-1)$
                       $= 1$

11. $\dfrac{5}{6} - \dfrac{1}{8} = \dfrac{5(4)}{6(4)} - \dfrac{1(3)}{8(3)}$

       $= \dfrac{20}{24} - \dfrac{3}{24}$

       $= \dfrac{20 - 3}{24}$

       $= \dfrac{17}{24}$

12. $-27\left(\dfrac{5}{6}\right) = -\dfrac{27}{1}\left(\dfrac{5}{6}\right)$

             $= -\dfrac{27(5)}{1(6)}$

             $= -\dfrac{9(3)(5)}{1(3)(2)}$

             $= -\dfrac{45}{2}$

13. $\dfrac{7}{16} \div \dfrac{21}{28} = \dfrac{7}{16} \cdot \dfrac{28}{21}$

             $= \dfrac{7 \cdot 28}{16 \cdot 21}$

             $= \dfrac{(7)(7)(4)}{(4)(4)(3)(7)}$

             $= \dfrac{7}{12}$

14. $\dfrac{-8.1}{0.3} = -\dfrac{8.1}{0.3} = -27$

     $0.3\,\overline{)8.1} = 3\,\overline{)81}$

                  $\underline{6}$
                  21
                  $\underline{21}$

15. $-\left(\dfrac{2}{3}\right)^2 = -\left(\dfrac{2}{3} \cdot \dfrac{2}{3}\right) = -\dfrac{4}{9}$

16. $35 - (50 \div 5^2) = 35 - (50 \div 25)$
                      $= 35 - 2$
                      $= 33$

17. Distributive Property

18. Multiplicative Inverse Property

19. Associative Property of Addition

20. Commutative Property of Multiplication

21. $\dfrac{30}{72} = \dfrac{5(6)}{12(6)} = \dfrac{5}{12}$

22. $-3^4 = -(3 \cdot 3 \cdot 3 \cdot 3) = -81$

     $(-3)^4 = (-3)(-3)(-3)(-3)$

            $= 81$

These expressions are not equal; the bases of the exponents are not the same.

23. $32 - 3 \cdot 2^3 = 32 - 3 \cdot 8$ exponentiation
                 $= 32 - 24$ multiplication
                 $= 8$ subtraction

24.

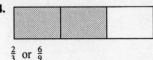

$\frac{2}{3}$ or $\frac{6}{9}$

Other possible answers include $\frac{8}{12}, \frac{10}{15}, \frac{12}{18}$, etc. The answer in reduced form is $\frac{2}{3}$.

C H A P T E R 2
Fundamentals of Algebra

CHAPTER 2
Fundamentals of Algebra

Section 2.1 Writing and Evaluating Algebraic Expressions

Solutions to Odd-Numbered Exercises

1. $60t$

The variable quantity is the number of hours traveled, and this quantity is represented by the letter t. If the average speed is 60 miles per hour, the distance traveled is $60t$.

3. $2.19m$

The variable quantity is the number of pounds of meat, and this quantity is represented by the letter m. If the cost per pound is \$2.19, the cost for the meat is $2.19m$.

5. Variable: x

Constant: 3

7. Variable: x, z

Constant: None

9. $4x, 3$

11. $3x^2, 5$

13. $\dfrac{5}{3}, -3y^3$

15. $2x, -3y, 1$

17. $3(x + 5), 10$

19. $\dfrac{x}{4}, \dfrac{5}{x}$

21. $\dfrac{3}{x + 2}, -3x, 4$

23. -6

25. $-\dfrac{1}{3}$

27. $-\dfrac{3}{2}$

29. 2π

31. 4.7

33. $y^5 = y \cdot y \cdot y \cdot y \cdot y$

35. $2^2 x^4 = 2 \cdot 2 \cdot x \cdot x \cdot x \cdot x$

37. $4y^2 z^3 = 4 \cdot y \cdot y \cdot z \cdot z \cdot z$

39. $(a^2)^3 = a^2 \cdot a^2 \cdot a^2$
$$= a \cdot a \cdot a \cdot a \cdot a \cdot a$$

41. $4x^3 \cdot x^4 = 4 \cdot x \cdot x \cdot x \cdot x \cdot x \cdot x \cdot x$

43. $(ab)^3 = (ab)(ab)(ab) = a \cdot a \cdot a \cdot b \cdot b \cdot b$

45. $(x + y)^2 = (x + y)(x + y)$

47. $\left(\dfrac{a}{3s}\right)^4 = \left(\dfrac{a}{3s}\right)\left(\dfrac{a}{3s}\right)\left(\dfrac{a}{3s}\right)\left(\dfrac{a}{3s}\right)$

49. $[3(r + s)^2][3(r + s)]^2 = 3(r + s)(r + s)[3(r + s)][3(r + s)]$
$$= 3 \cdot 3 \cdot 3(r + s)(r + s)(r + s)(r + s)$$

51. $2 \cdot u \cdot u \cdot u \cdot u = 2u^4$

(2 is *not* a factor of the base.)

53. $(2u) \cdot (2u) \cdot (2u) \cdot (2u) = (2u)^4$

(2 *is* a factor of the base.)

55. $a \cdot a \cdot a \cdot b \cdot b = a^3 b^2$

57. $3 \cdot (x - y) \cdot (x - y) \cdot 3 \cdot 3 = 3 \cdot 3 \cdot 3 \cdot (x - y)(x - y)$
$$= 3^3(x - y)^2$$

59. $\left(\dfrac{x^2}{2}\right)\left(\dfrac{x^2}{2}\right)\left(\dfrac{x^2}{2}\right) = \left(\dfrac{x^2}{2}\right)^3$

61. (a) When $x = \frac{1}{2}$, the value of $2x - 1$ is $2\left(\frac{1}{2}\right) - 1 = 1 - 1 = 0$.

(b) When $x = 4$, the value of $2x - 1$ is $2(4) - 1 = 8 - 1 = 7$.

63. (a) When $x = -2$, the value of $2x^2 - 5$ is
$2(-2)^2 - 5 = 2(4) - 5 = 8 - 5 = 3$.

(b) When $x = 3$, the value of $2x^2 - 5$ is
$2(3)^2 - 5 = 2(9) - 5 = 18 - 5 = 13$.

65. (a) When $x = 4$ and $y = 3$, the value of $3x - 2y$ is
$3(4) - 2(3) = 12 - 6 = 6$.

(b) When $x = \frac{2}{3}$ and $y = 1$, the value of $3x - 2y$ is
$3\left(\frac{2}{3}\right) - 2(1) = 2 - 2 = 0$.

67. (a) When $x = 3$ and $y = 3$, the value of $x - 3(x - y)$ is
$3 - 3(3 - 3) = 3 - 3(0) = 3 - 0 = 3$.

(b) When $x = 4$ and $y = -4$, the value of $x - 3(x - y)$ is
$4 - 3(4 - (-4)) = 4 - 3(4 + 4) = 4 - 3(8)$

$$= 4 - 24 = -20.$$

69. (a) When $a = 2$, $b = -3$, and $c = -1$, the value of
$b^2 - 4ac$ is $(-3)^2 - 4(2)(-1) = 9 + 8 = 17$.

(b) When $a = -4$, $b = 6$, and $c = -2$, the value of
$b^2 - 4ac$ is $6^2 - 4(-4)(-2) = 36 - 32 = 4$.

71. (a) When $x = 4$ and $y = 2$, the value of
$(x - 2y)/(x + 2y)$ is

$$\frac{4 - 2 \cdot 2}{4 + 2 \cdot 2} = \frac{4 - 4}{4 + 4}$$

$$= \frac{0}{8}$$

$$= 0.$$

(b) When $x = 4$ and $y = -2$, the value of
$(x - 2y)/(x + 2y)$ is undefined because

$$\frac{4 - 2(-2)}{4 + 2(-2)} = \frac{4 + 4}{4 - 4} = \frac{8}{0}$$

and division by 0 is undefined.

73. (a) When $x = 2$ and $y = 4$, the value of $\dfrac{5x}{y - 3}$ is
$$\frac{5(2)}{4 - 3} = \frac{10}{1} = 10.$$

(b) When $x = 2$ and $y = 3$, the value of $\dfrac{5x}{y - 3}$ is

undefined because $\dfrac{5(2)}{3 - 3} = \dfrac{10}{0}$, and division by zero

is undefined.

75. (a) When $b = 3$ and $h = 5$, the value of $\frac{1}{2}bh$ is
$\frac{1}{2} \cdot 3 \cdot 5 = \frac{15}{2}$.

(b) When $b = 2$ and $h = 10$, the value of $\frac{1}{2}bh$ is
$\frac{1}{2} \cdot 2 \cdot 10 = 10$.

77. (a) When $r = 50$ and $t = 3.5$, the value of rt is
$(50)(3.5) = 175$.

(b) When $r = 35$ and $t = 4$, the value of rt is 140.

79. (a)

x	-1	0	1	2	3	4
$3x - 2$	-5	-2	1	4	7	10

When $x = -1$, $3x - 2 = 3(-1) - 2 = -3 - 2 = -5$.

When $x = 0$, $3x - 2 = 3 \cdot 0 - 2 = 0 - 2 = -2$.

When $x = 1$, $3x - 2 = 3 \cdot 1 - 2 = 3 - 2 = 1$.

When $x = 2$, $3x - 2 = 3 \cdot 2 - 2 = 6 - 2 = 4$.

When $x = 3$, $3x - 2 = 3 \cdot 3 - 2 = 9 - 2 = 7$.

When $x = 4$, $3x - 2 = 3 \cdot 4 - 2 = 12 - 2 = 10$.

(b) For each one-unit increase in x, the value of the expression $3x - 2$ increases by 3.

(c) You might notice that 3 is the coefficient of x in the expression $3x - 2$. In the expression $\frac{2}{3}x + 4$, the coefficient of x is $\frac{2}{3}$.
You might predict that the value of this expression would increase by $\frac{2}{3}$ for each one-unit increase in the value of x.

x	-1	0	1	2	3	4
$\frac{2}{3}x + 4$	$\frac{10}{3}$	4	$\frac{14}{3}$	$\frac{16}{3}$	6	$\frac{20}{3}$

81. Area $= n^2$

If $n = 8$, the value of $n^2 = 8^2 = 64$.

Thus, the area is 64 square units.

83. Area $= a(a + b)$

If $a = 5$ and $b = 4$, the value of $a(a + b) = 5(5 + 4) = 5(9) = 45$.

Thus, the area is 45 square units.

85. (a) A square has 4 sides.

If $n = 4$, the value of $\dfrac{n(n - 3)}{2}$ is $\dfrac{4(4 - 3)}{2} = \dfrac{4(1)}{2} = \dfrac{4}{2} = 2$.

Thus, a square has 2 diagonals.

(b) A pentagon has 5 sides.

If $n = 5$, the value of $\dfrac{n(n - 3)}{2}$ is $\dfrac{5(5 - 3)}{2} = \dfrac{5(2)}{2} = \dfrac{10}{2} = 5$.

Thus, a pentagon has 5 diagonals.

(c) A hexagon has 6 sides.

If $n = 6$, the value of $\dfrac{n(n - 3)}{2}$ is $\dfrac{6(6 - 3)}{2} = \dfrac{6(3)}{2} = \dfrac{18}{2} = 9$.

Thus, a hexagon has 9 diagonals.

87. (a) 4, 5, 5.5, 5.75, 5.875, 5.9375, 5.96875

The value appears to be approaching 6.

(b) 9, 7.5, 6.75, 6.375, 6.1875, 6.09375, 6.046875

The value appears to be approaching 6.

89. (a) $(15 \cdot 12)c$ or $180c$

Plastic chairs: If $c = \$1.95$, the value of $180c$ is $180(1.95) = \$351$.

Wood chairs: If $c = \$2.95$, the value of $180c$ is $180(2.95) = \$531$.

(b) Canopy 1: $t = (20)(20) = 400$

If $t = 400$, the value of $115 + 0.25t$ is $115 + 0.25(400) = \$215$.

Canopy 2: $t = (20)(30) = 600$

If $t = 600$, the value of $115 + 0.25t$ is $115 + 0.25(600) = \$265$.

Canopy 3: $t = (30)(40) = 1200$

If $t = 1200$, the value of $115 + 0.25t$ is $115 + 0.25(1200) = \$415$.

Canopy 4: $t = (30)(60) = 1800$

If $t = 1800$, the value of $115 + 0.25t$ is $115 + 0.25(1800) = \$565$.

Canopy 5: $t = (40)(60) = 2400$

If $t = 2400$, the value of $115 + 0.25t$ is $115 + 0.25(2400) = \$715$.

91. No, $3x$ is not a term of $4 - 3x$. The terms of this expression are 4 and $-3x$.

93. No, it is not possible to evaluate the expression $\dfrac{x + 2}{y - 3}$ when $x = 5$ and $y = 3$. If $y = 3$, the value of the denominator of $y - 3$ would be 0, and division by 0 is undefined.

Section 2.2 Simplifying Algebraic Expressions

1. $u^2 \cdot u^4 = u^{2+4} = u^6$

3. $3x^3 \cdot x^4 = 3x^{3+4} = 3x^7$

5. $5x(x^6) = 5x^{1+6} = 5x^7$

7. $(-5z^3)(3z^2) = (-5 \cdot 3)(z^3 \cdot z^2)$
$= -15z^{3+2} = -15z^5$

9. $(-xz)(-2y^2z) = 2xy^2z^{1+1}$
$= 2xy^2z^2$

11. $2b^4(-ab)(3b^2) = (2)(-1)(3)(a)(b^4 \cdot b \cdot b^2)$
$= -6ab^{4+1+2} = -6ab^7$

13. $(t^2)^4 = t^{2(4)} = t^8$

15. $5(uv)^5 = 5u^5v^5$

17. $(-2s)^3 = (-2)^3s^3 = -8s^3$

19. $(a^2b)^3(ab^2)^4 = (a^2)^3 \cdot b^3 \cdot a^4(b^2)^4$
$= a^{2\cdot3} \cdot a^4 \cdot b^3 \cdot b^{2\cdot4}$
$= a^6 \cdot a^4 \cdot b^3 \cdot b^8$
$= a^{6+4}b^{3+8}$
$= a^{10}b^{11}$

21. $(u^2v^3)(-2uv^2)^4 = u^2v^3(-2)^4u^4v^{2\cdot4}$
$= u^2v^3 \cdot 16u^4v^8$
$= 16u^{2+4}v^{3+8}$
$= 16u^6v^{11}$

23. $[(x-3)^4]^2 = (x-3)^{4\cdot2} = (x-3)^8$

25. $(x-2y)^3(x-2y)^3 = (x-2y)^{3+3}$
$= (x-2y)^6$

27. $x^5 \cdot x^3 = x^{5+3} = x^8$
$x^5 \cdot x^3 \neq x^{15}$
No, they are *not* equal.

29. $-3x^3 = -3 \cdot x \cdot x \cdot x$
$-27x^3 = -27 \cdot x \cdot x \cdot x$
$-3x^3 \neq -27x^3$
No, they are *not* equal.

31. Commutative Property of Addition

33. Associative Property of Multiplication

35. Multiplicative Identity Property

37. Commutative Property of Multiplication

39. Additive Inverse Property

41. Multiplicative Inverse Property

43. Additive Inverse Property and Additive Identity Property

45. $(x+10) - (x+10) = 0$
Additive Inverse Property

47. $v(2) = 2v$
Commutative Property of Multiplication

49. $5(t-2) = 5(t) - 5(2)$
Distributive Property

51. $5x\left(\dfrac{1}{5x}\right) = 1$
Multiplicative Inverse Property

53. $12 + (8 - x) = (12 + 8) - x$
Associative Property of Addition

55. $-5(2x - y) = -5(2x) - (-5)(y)$
$= -10x + 5y$

57. $(x+2)(3) = x(3) + 2(3)$
$= 3x + 6$

59. $4(x + xy + y^2) = 4(x) + 4(xy) + 4(y^2)$
$= 4x + 4xy + 4y^2$

61. $3(x^2 + x) = 3(x^2) + 3(x)$
$= 3x^2 + 3x$

63. $-4y(3y - 4) = -4y(3y) - (-4y)(4)$
$= -12y^2 + 16y$

65. $-(u - v) = (-1)(u - v)$

$= (-1)(u) - (-1)(v)$

$= -u + v$

67. $x(3x - 4y) = x(3x) - x(4y)$

$= 3x^2 - 4xy$

69. The area of the rectangle on the left is ab. The area of the rectangle on the right is ac. The area of the entire rectangle can be written as $a(b + c)$ and as the sum of the two smaller rectangles $ab + ac$.

Using the Distributive Property, you can see that these expressions are equal. $a(b + c) = ab + ac$

71. The area of the rectangle on the left is $2a$. The area of the rectangle on the right is $2(b - a)$. The area of the entire rectangle can be written as $2b$ and as the sum of the two smaller rectangles $2a + 2(b - a)$.

Using the Distributive Property, you can see that these expressions are equal.
$2a + 2(b - a) = 2a + 2b - 2a = 2b$

73.

Term	Coefficient
$6x^2$	6
$-3xy$	-3
y^2	1

75. In this expression, $16t^3$ and $3t^3$ are like terms, and the constants 4 and -5 are like terms.

77. In this expression, $6x^2y$ and $-4x^2y$ are like terms.

79. $\frac{1}{2}x^2y$ and $-\frac{5}{2}xy^2$ are not like terms because their variable factors are not alike.

Note: $x^2y = x \cdot x \cdot y$ and $xy^2 = x \cdot y \cdot y$.

81. $3y - 5y = (3 - 5)y = -2y$

83. $x + 5 - 3x = x - 3x + 5$

$= (1 - 3)x + 5$

$= -2x + 5$

85. $2x + 9x + 4 = (2 + 9)x + 4$

$= 11x + 4$

87. $5r + 6 - 2r + 1 = 5r - 2r + 6 + 1$

$= (5 - 2)r + (6 + 1)$

$= 3r + 7$

89. $x^2 - 2xy + 4 + xy = x^2 - 2xy + xy + 4$

$= x^2 + (-2 + 1)xy + 4$

$= x^2 + (-1)xy + 4$

$= x^2 - xy + 4$

91. $5z - 5 + 10z + 2z + 16 = 5z + 10z + 2z - 5 + 16$

$= (5 + 10 + 2)z + (-5 + 16)$

$= 17z + 11$

93. $z^3 + 2z^2 + z + z^2 + 2z + 1 = z^3 + 2z^2 + z^2 + z + 2z + 1$

$= z^3 + (2 + 1)z^2 + (1 + 2)z + 1$

$= z^3 + 3z^2 + 3z + 1$

95. $2x^2y + 5xy^2 - 3x^2y + 4xy + 7xy^2 = (2x^2y - 3x^2y) + (5xy^2 + 7xy^2) + 4xy$

$= (2 - 3)x^2y + (5 + 7)xy^2 + 4xy$

$= -x^2y + 12xy^2 + 4xy$

97. $3\left(\dfrac{1}{x}\right) - \dfrac{1}{x} + 8 = (3 - 1)\left(\dfrac{1}{x}\right) + 8$

$= 2\left(\dfrac{1}{x}\right) + 8$ or $\dfrac{2}{x} + 8$

99. $5\left(\dfrac{1}{t}\right) + 6\left(\dfrac{1}{t}\right) - 2t = (5 + 6)\left(\dfrac{1}{t}\right) - 2t$

$= 11\left(\dfrac{1}{t}\right) - 2t$ or $\dfrac{11}{t} - 2t$

101. False

$3(x - 4) = 3 \cdot x - 3 \cdot 4$

$= 3x - 12$

Therefore, $3(x - 4) \neq 3x - 4$.

103. True

$6x - 4x = (6 - 4)x = 2x$

Therefore, $6x - 4x = 2x$.

105. $8(52) = 8(50 + 2)$

$= 8(50) + 8(2)$

$= 400 + 16$

$= 416$

107. $5(7.98) = 5(8 - 0.02)$

$= 5(8) - 5(0.02)$

$= 40 - 0.10$

$= 39.9$

109. $2(6x) = (2 \cdot 6)x = 12x$

111. $-(-4x) = [-1(-4)]x = 4x$

113. $(-2x)(-3x) = (-2)(-3)(x \cdot x)$

$= 6x^2$

115. $(-5z)(2z^2) = (-5)(2)(z \cdot z^2)$

$= -10z^3$

117. $\dfrac{18a}{5} \cdot \dfrac{15}{6} = \dfrac{18a(15)}{5(6)}$

$= \dfrac{9\cancel{(2)}(a)\cancel{(5)}\cancel{(3)}}{\cancel{5}\cancel{(3)}\cancel{(2)}}$

$= 9a$

119. $\left(-\dfrac{3x^2}{2}\right)(4x^3) = -\dfrac{3}{2} \cdot 4(x^2 \cdot x^3)$

$= -6x^5$

121. $(12xy^2)(-2x^3y^2) = 12(-2)(x \cdot x^3)(y^2 \cdot y^2)$

$= -24x^4y^4$

123. $2(x - 2) + 4 = 2x - 4 + 4 = 2x$

125. $6(2s - 1) + s + 4 = 12s - 6 + s + 4$

$= 12s + s - 6 + 4$

$= 13s - 2$

127. $m - 3(m - 5) = m - 3m + 15$

$= -2m + 15$

129. $-6(1 - 2x) + 10(5 - x) = -6 + 12x + 50 - 10x$

$= -6 + 50 + 12x - 10x$

$= 44 + 2x$

131. $\frac{2}{3}(12x + 15) + 16 = 8x + 10 + 16$

$= 8x + 26$

133. $3 - 2[6 + (4 - x)] = 3 - 2[6 + 4 - x]$

$= 3 - 2[10 - x]$

$= 3 - 20 + 2x$

$= 2x - 17 \quad \text{or} \quad 17 + 2x$

Note: This expression may also be simplified as follows:

$3 - 2[6 + (4 - x)] = 3 - 2[6 + 4 - x]$

$= 3 - 12 - 8 + 2x$

$= 2x - 17$

135. $7x(2 - x) - 4x = 14x - 7x^2 - 4x$

$= -7x^2 + 10x$

137. $4x^2 + x(5 - x) = 4x^2 + 5x - x^2$

$= 4x^2 - x^2 + 5x$

$= 3x^2 + 5x$

139. $-3t(4 - t) + t(t + 1) = -12t + 3t^2 + t^2 + t$

$= 4t^2 - 11t$

141. $3t[4 - (t - 3)] + t(t + 5) = 3t[4 - t + 3] + t^2 + 5t$

$= 12t - 3t^2 + 9t + t^2 + 5t$

$= -2t^2 + 26t$

143. $\dfrac{2x}{3} - \dfrac{x}{3} = \dfrac{2}{3}x - \dfrac{1}{3}x$

$\qquad = \left(\dfrac{2}{3} - \dfrac{1}{3}\right)x$

$\qquad = \dfrac{1}{3}x \quad \text{or} \quad \dfrac{x}{3}$

145. $\dfrac{4z}{5} + \dfrac{3z}{5} = \dfrac{4}{5}z + \dfrac{3}{5}z$

$\qquad = \left(\dfrac{4}{5} + \dfrac{3}{5}\right)z$

$\qquad = \dfrac{7}{5}z \quad \text{or} \quad \dfrac{7z}{5}$

147. $\dfrac{x}{3} - \dfrac{5x}{4} = \dfrac{1}{3}x - \dfrac{5}{4}x$

$\qquad = \left(\dfrac{1}{3} - \dfrac{5}{4}\right)x$

$\qquad = \left(\dfrac{1(4)}{3(4)} - \dfrac{5(3)}{4(3)}\right)x$

$\qquad = \left(\dfrac{4}{12} - \dfrac{15}{12}\right)x$

$\qquad = -\dfrac{11}{12}x \quad \text{or} \quad -\dfrac{11x}{12}$

149. $\dfrac{3x}{10} - \dfrac{x}{10} + \dfrac{4x}{5} = \dfrac{3}{10}x - \dfrac{1}{10}x + \dfrac{4}{5}x$

$\qquad = \left(\dfrac{3}{10} - \dfrac{1}{10} + \dfrac{4}{5}\right)x$

$\qquad = \left(\dfrac{3}{10} - \dfrac{1}{10} + \dfrac{4(2)}{5(2)}\right)x$

$\qquad = \left(\dfrac{3}{10} - \dfrac{1}{10} + \dfrac{8}{10}\right)x$

$\qquad = 1 \cdot x$

$\qquad = x$

151. If $P = 10{,}000$, $r = 0.08$, and $t = 10$, the value of $P(1 + r)^t$ is $10{,}000(1 + 0.08)^{10} \approx 21{,}589.25$.

153. Area = (length)(width)

$\qquad = x \cdot x$

$\qquad = x^2$

Volume = (length)(width)(height)

$\qquad = x \cdot x \cdot x$

$\qquad = x^3$

155. $(x - 2) + (x + 11) + (2x + 3) = x - 2 + x + 11 + 2x + 3$

$\qquad\qquad\qquad\qquad\qquad\qquad = (1 + 1 + 2)x + (-2 + 11 + 3)$

$\qquad\qquad\qquad\qquad\qquad\qquad = 4x + 12$

157.

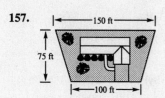

When $b_1 = 150$, $b_2 = 100$, and $h = 75$, the value of $\frac{1}{2}h(b_1 + b_2)$ is

$\frac{1}{2}(75)(150 + 100) = \frac{1}{2}(75)(250) = 9375$.

The area of the lot is 9375 square feet.

159. $(6x)^4 = 6^4 x^4 = 1296x^4$

Thus, $(6x)^4 \neq 6x^4$.

161. (a) $12x^8$ is already in simplified form.

(b) $12(x^3)^5 = 12x^{15}$

(c) $12x^3x^5 = 12x^8$

(d) $3 \cdot 2(x^2)^4 = 3 \cdot 4x^8 = 12x^8$

(e) $3 \cdot 5x^8 = 15x^8$

Thus, (a), (c), and (d) are equivalent.

163. To combine like terms, add their coefficients and attach the common variable factor.

Examples:

$5a + 7b - 2a + b = 3a + 8b$

$2x^2 - 7x + 10x = 2x^2 + 3x$

165. To remove nested symbols of grouping, remove the innermost grouping symbols first and combine like terms.

167. $[x - (3 \cdot 4)] \div 5 = \dfrac{x - 12}{5}$

$[x - 3 \cdot 4] \div 5 = \dfrac{x - 12}{5}$

If the parentheses are removed, the expression is unchanged because multiplication is a higher-order operation than subtraction.

$x - (3 \cdot 4) \div 5 = x - \dfrac{12}{5}$

If the brackets are removed, the expression is changed because division is a higher-order operation than subtraction.

Mid-Chapter Quiz for Chapter 2

1. (a) If $x = 3$, the value of $x^2 - 3x$ is $3^2 - 3(3) = 9 - 9 = 0$.

(b) If $x = -2$, the value of $x^2 - 3x$ is $(-2)^2 - 3(-2) = 4 + 6 = 10$.

(c) If $x = 0$, the value of $x^2 - 3x$ is $0^2 - 3(0) = 0 - 0 = 0$.

2. (a) If $x = 2$ and $y = 4$, the value of $\dfrac{x}{y - 3}$ is $\dfrac{2}{4 - 3} = \dfrac{2}{1} = 2$.

(b) If $x = 0$ and $y = -1$, the value of $\dfrac{x}{y - 3}$ is $\dfrac{0}{-1 - 3} = \dfrac{0}{-4} = 0$.

(c) If $x = 5$ and $y = 3$, the value of $\dfrac{x}{y - 3}$ is undefined because $\dfrac{5}{3 - 3} = \dfrac{5}{0}$ and division by zero is undefined.

3. (a) The coefficient is -5.

(b) The coefficient is $\frac{5}{16}$.

4. (a) $3y \cdot 3y \cdot 3y \cdot 3y = (3y)^4$

(b) $2 \cdot (x - 3)(x - 3)2 \cdot 2 = 2^3(x - 3)^2$

5. $x^4 \cdot x^3 = x^{4+3} = x^7$

6. $(v^2)^5 = v^{2(5)}v^{10}$

7. $(-3y)^2y^3 = (-3)^2y^2y^3$

$= 9y^{2+3}$

$= 9y^5$

8. $8(x - 4)^2(x - 4)^4 = 8(x - 4)^{2+4}$

$= 8(x - 4)^6$

9. $\dfrac{2z^2}{3y} \cdot \dfrac{5z}{7y^3} = \dfrac{2z^2(5z)}{3y(7y^3)} = \dfrac{10z^3}{21y^4}$

10. $\left(\dfrac{x}{y}\right)^2\left(\dfrac{x}{y}\right)^5 = \left(\dfrac{x}{y}\right)^{2+5} = \left(\dfrac{x}{y}\right)^7$

11. Associative Property Multiplication

12. Distributive Property

13. Multiplicative Inverse Property

14. Commutative Property of Addition

15. $2(3x - 1) = 2(3x) - 2(1)$

$\qquad = 6x - 2$

16. $-4(2y - 3) = -4(2y) - (-4)(3)$

$\qquad = -8y + 12$

17. $y^2 - 3xy + y + 7xy = y^2 + (-3 + 7)xy + y$

$\qquad\qquad\qquad\quad = y^2 + 4xy + y$

18. $10\left(\dfrac{1}{u}\right) - 7\left(\dfrac{1}{u}\right) + 3u = (10 - 7)\left(\dfrac{1}{u}\right) + 3u$

$\qquad\qquad\qquad\qquad\quad = 3\left(\dfrac{1}{u}\right) + 3u$

19. $5(a - 2b) + 3(a + b) = 5a - 10b + 3a + 3b$

$\qquad\qquad\qquad\qquad = (5 + 3)a + (-10 + 3)b$

$\qquad\qquad\qquad\qquad = 8a - 7b$

20. $4x + 3[2 - 4(x + 6)] = 4x + 3[2 - 4x - 24]$

$\qquad\qquad\qquad\qquad = 4x + 3[-22 - 4x]$

$\qquad\qquad\qquad\qquad = 4x - 66 - 12x$

$\qquad\qquad\qquad\qquad = (4 - 12)x - 66$

$\qquad\qquad\qquad\qquad = -8x - 66$

21. $\left(\dfrac{1}{3}\pi r^2 h\right)\left(\dfrac{3}{10}r^2\right) = \dfrac{1(\cancel{3})}{\cancel{3}(10)}\pi r^{2+2}h = \dfrac{1}{10}\pi r^4 h$

22. $4 \cdot 10^4 + 5 \cdot 10^3 + 7 \cdot 10^2 = 4 \cdot 10{,}000 + 5 \cdot 1000 + 7 \cdot 100$

$\qquad\qquad\qquad\qquad\qquad\quad = 40{,}000 + 5000 + 700$

$\qquad\qquad\qquad\qquad\qquad\quad = 45{,}700$

Section 2.3   Algebra and Problem Solving

1. (d)

3. (e)

5. (b)

7. $x + 5$

9. $x - 25$

11. $x - 6$

13. $2x$

15. $\dfrac{x}{3}$

17. $\dfrac{x}{50}$

19. $\dfrac{3}{10}x$ or $0.3x$

21. $3x + 5$

23. $8 + 5x$

25. $10(x + 4)$

27. $|x + 4|$

29. $x^2 + 1$

31. A number decreased by ten.

33. A number is tripled and the product is increased by two.

35. A number is multiplied by seven and the product is increased by four.

37. A number is subtracted from 2 and the difference is multiplied by 3
or
Three times the difference of 2 and a number.

39. A number is increased by 1 and the sum is divided by 2.

41. The square of a number is increased by 5.

43. $(x + 3)x = x^2 + 3x$

45. $(25 + x) + x = 25 + x + x$

$\qquad\qquad\qquad = 25 + 2x$

47. $(x - 9)3 = 3x - 27$

49. $\dfrac{8(x + 24)}{2} = \dfrac{8x + 192}{2}$

$$= 4x + 96$$

51. The amount of money is a product.

Verbal model: | Value of dime | $\cdot$ | Number of dimes |

Labels: Value of dime $= 0.10$ (dollars)

Number of dimes $= d$

Algebraic expression: $0.10d$ (dollars)

53. The amount of sales tax is a product.

Verbal model: | Percent of sales tax | $\cdot$ | Amount of purchase |

Labels: Percent of sales tax $= 0.06$ (in decimal form)

Amount of purchase $= L$ (dollars)

Algebraic expression: $0.06L$ (dollars)

55. The travel time is a quotient.

Verbal model: $\dfrac{\boxed{\text{Distance traveled}}}{\boxed{\text{Average speed}}}$

Labels: Distance traveled $= 100$ (miles)

Average speed $= r$ (miles per hour)

Algebraic expression: $\dfrac{100}{r}$ (hours)

57. The camping fee is a sum of products.

Verbal model: | Fee per parent | $\cdot$ | Number of parents | $+$ | Fee per child | $\cdot$ | Number of children |

Labels: Fee per parent $= 15$ (dollars)

Number of parents $= m$

Fee per child $= 2$ (dollars)

Number of children $= n$

Algebraic expression: $15m + 2n$

59. Guesses will vary.

$t = 10.2$ years

61. Guesses will vary.

$t = 11.9$ years

63.

n	0	1	2	3	4	5
$2n - 1$	-1	1	3	5	7	9
Differences		2	2	2	2	2

All entries in the third row are 2's. The value of the expression $2n - 1$ increases by 2 for each increase of 1 in the value of n. Note that 2 is the coefficient of n in the expression $2n - 1$.

65. In Exercise 63, all entries in the third row are 2's, and 2 is the coefficient of n in the expression $2n - 1$. If the algebraic expression were $3n + 5$, all the entries in the third row would be 3's.

67. The difference in the last row are all 5's. Thus, the coefficient of n must be 5. However, $5(0) - 0, 5(1) = 5, 5(2) = 10, 5(3) = 15$, etc. In each instance, the value of $an + b$ is 4 more than 5 times n. Therefore, the expression $an + b$ must be $5n + 4$. This indicates that $a = 5$ and $b = 4$.

69. The area of a rectangle is the product of the length and width of the rectangle.

$3x(6x - 1) = 18x^2 - 3x$

71. The area of a triangle is one-half the product of the base and height of the triangle.

$$\frac{1}{2}(12)(5x^2 + 2) = 6(5x^2 + 2)$$
$$= 30x^2 + 12$$

73.

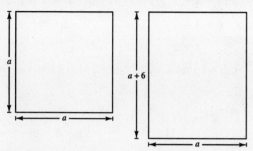

Perimeter of the square: $4a$ centimeters

Area of the square: a^2 square centimeters

Perimeter of the rectangle: $2(a) + 2(a + 6) = 2a + 2a + 12 = 4a + 12$ centimeters

Area of the rectangle: $a(a + 6) = a^2 + 6a$ square centimeters

75. The area of the screen is a product.

Verbal model: [Length of screen] · [Width of screen]

Labels: Length of screen $= s$ (inches)

     Width of screen $= s$ (inches)

Algebraic expression: $s \cdot s$ or s^2 (square inches)

77. The perimeter is a sum of products.

Verbal model: [2] · [Length of frame] + [2] · [Width of frame]

Labels: Length of frame $= 1.5w$

     Width of frame $= w$

Algebraic expression: $2(1.5w) + 2 \cdot w = 3w + 2w = 5w$

79. (a) $m + n$

(b) $1 + 2 = 3$

     $2 + 3 = 5$

     $3 + 5 = 8$

     $5 + 8 = 13$

     $8 + 13 = 21$

So, the next five Fibonacci numbers are 3, 5, 8, 13, and 21.

81. Subtraction

83. (a), (b), and (e)

Note: (c) n less than 4 would be equivalent to $4 - n$

(d) the ratio of n to 4 would be equivalent to $\dfrac{n}{4}$.

85. $3\left(\dfrac{5}{n}\right)$ or $\dfrac{5}{3n}$

Section 2.4 Introduction to Equations

1. (a) $x = 3$

$2(3) - 6 \overset{?}{=} 0$

$6 - 6 \overset{?}{=} 0$

$0 = 0$

3 *is* a solution.

(b) $x = 1$

$2(1) - 6 \overset{?}{=} 0$

$2 - 6 \overset{?}{=} 0$

$-4 \neq 0$

1 *is not* a solution.

3. (a) $x = 0$

$2(0) + 4 \overset{?}{=} 2$

$0 + 4 \overset{?}{=} 2$

$4 \neq 2$

0 *is not* a solution.

(b) $x = -1$

$2(-1) + 4 \overset{?}{=} 2$

$-2 + 4 \overset{?}{=} 2$

$2 = 2$

-1 *is* a solution.

5. (a) $x = -1$

$-1 + 5 \overset{?}{=} 2(-1)$

$4 \neq -2$

-1 *is not* a solution

(b) $x = 5$

$5 + 5 \overset{?}{=} 2(5)$

$10 = 10$

5 *is* a solution.

7. (a) $x = 11$

$11 + 3 \overset{?}{=} 2(11 - 4)$

$14 \overset{?}{=} 2(7)$

$14 = 14$

11 *is* a solution.

(b) $x = -5$

$-5 + 3 \overset{?}{=} 2(-5 - 4)$

$-2 \overset{?}{=} 2(-9)$

$-2 \neq -18$

-5 *is not* a solution.

9. (a) $x = \frac{3}{5}$

$2\left(\frac{3}{5}\right) + 10 \overset{?}{=} 7\left(\frac{3}{5} + 1\right)$

$\frac{6}{5} + \frac{10}{1} \overset{?}{=} 7\left(\frac{3}{5} + \frac{5}{5}\right)$

$\frac{6}{5} + \frac{50}{5} \overset{?}{=} \frac{7}{1}\left(\frac{8}{5}\right)$

$\frac{56}{5} = \frac{56}{5}$

$\frac{3}{5}$ *is* a solution.

(b) $x = \frac{2}{3}$

$2\left(\frac{2}{3}\right) + 10 \overset{?}{=} 7\left(\frac{2}{3} + 1\right)$

$\frac{4}{3} + \frac{10}{1} \overset{?}{=} 7\left(\frac{2}{3} + \frac{3}{3}\right)$

$\frac{4}{3} + \frac{30}{3} \overset{?}{=} \frac{7}{1}\left(\frac{5}{3}\right)$

$\frac{34}{3} \neq \frac{35}{3}$

$\frac{2}{3}$ *is not* a solution.

11. (a) $x = 3$

$3^2 - 4 \overset{?}{=} 3 + 2$

$9 - 4 \overset{?}{=} 3 + 2$

$5 = 5$

3 *is* a solution.

(b) $x = -2$

$(-2)^2 - 4 \overset{?}{=} -2 + 2$

$4 - 4 \overset{?}{=} 0$

$0 = 0$

-2 *is* a solution.

13. (a) $x = 3$

$\frac{2}{3} - \frac{1}{3} \overset{?}{=} 1$

$\frac{2 - 1}{3} \overset{?}{=} 1$

$\frac{1}{3} \neq 1$

3 *is not* a solution

(b) $x = \frac{1}{3}$

$\frac{2}{1/3} - \frac{1}{1/3} \overset{?}{=} 1$

$\left(2 \div \frac{1}{3}\right) - \left(1 \div \frac{1}{3}\right) \overset{?}{=} 1$

$\left(\frac{2}{1} \cdot \frac{3}{1}\right) - \left(\frac{1}{1} \cdot \frac{3}{1}\right) \overset{?}{=} 1$

$6 - 3 \overset{?}{=} 1$

$3 \neq 1$

$\frac{1}{3}$ *is not* a solution.

15. (a) $x = 3$

$\frac{5}{3 - 1} + \frac{1}{3} \overset{?}{=} 5$

$\frac{5}{2} + \frac{1}{3} \overset{?}{=} 5$

$\frac{15}{6} + \frac{2}{6} \overset{?}{=} 5$

$\frac{17}{6} \neq 5$

3 *is not* a solution.

(b) $x = \frac{1}{6}$

$\frac{5}{(1/6) - 1} + \frac{1}{1/6} \overset{?}{=} 5$

$\left[5 \div \left(\frac{1}{6} - 1\right)\right] + \left(1 \div \frac{1}{6}\right) \overset{?}{=} 5$

$\frac{5}{1} \div \left(-\frac{5}{6}\right) + \frac{1}{1} \div \frac{1}{6} \overset{?}{=} 5$

$\frac{5}{1} \cdot \frac{-6}{5} + \frac{1}{1} \cdot \frac{6}{1} \overset{?}{=} 5$

$\frac{-30}{5} + 6 \overset{?}{=} 5$

$-6 + 6 \overset{?}{=} 5$

$0 \neq 5$

$\frac{1}{6}$ *is not* a solution.

17. (a) $x = 1.2$

$1.2 + 3 \overset{?}{=} 3.5$

$4.2 \neq 3.5$

1.2 is *not* a solution.

(b) $x = 4.8$

$4.8 + 3 \overset{?}{=} 3.5$

$7.8 \neq 3.5$

4.8 is *not* a solution.

19. (a) $x = 12.25$

$40(12.25) - 490 \overset{?}{=} 0$

$0 = 0$

12.25 *is* a solution.

(b) $x = -12.25$

$40(-12.25) - 490 \overset{?}{=} 0$

$-980 \neq 0$

-12.25 is *not* a solution.

21. (a) $x = \frac{5}{2}$

$2\left(\frac{5}{2}\right)^2 - \frac{5}{2} - 10 \overset{?}{=} 0$

$0 = 0$

$\frac{5}{2}$ *is* a solution.

(b) $x = -1.09$

$2(-1.09)^2 - (-1.09) - 10 \overset{?}{=} 0$

$-6.5338 \neq 0$

-1.09 is *not* a solution.

23. (a) $x = 0$

$\frac{1}{0}$ is undefined

0 is *not* a solution

(b) $x = -2$

$\frac{1}{-2} - \frac{9}{-2-4} \overset{?}{=} 1$

$1 = 1$

-2 *is* a solution.

25. (a) $x = \frac{6}{5}$

$\left(\frac{6}{5}\right)^3 - 1.728 \overset{?}{=} 0$

$0 = 0$

$\frac{6}{5}$ *is* a solution.

(b) $x = -\frac{6}{5}$

$\left(-\frac{6}{5}\right)^3 - 1.728 \overset{?}{=} 0$

$-3.456 \neq 0$

$-\frac{6}{5}$ is *not* a solution.

27.

$5x + 12 = 22$	Given equation.
$5x + 12 - 12 = 22 - 12$	Subtract 12 from each side.
$5x = 10$	Combine like terms.
$\dfrac{5x}{5} = \dfrac{10}{5}$	Divide both sides by 5.
$x = 2$	Solution.

29.

$\dfrac{2}{3}x = 12$	Given equation
$\dfrac{3}{2}\left(\dfrac{2}{3}x\right) = \dfrac{3}{2}(12)$	Multiply both sides by $\frac{3}{2}$.
$x = 18$	Simplify. (solution)

31.

$2(x - 1) = x + 3$	Given equation.
$2x - 2 = x + 3$	Remove grouping symbols.
$-x + 2x - 2 = -x + x + 3$	Subtract x from each side.
$x - 2 = 3$	Combine like terms.
$x - 2 + 2 = 3 + 2$	Add 2 to both sides.
$x = 5$	Combine like terms. (solution)

33.

$x = -2(x + 3)$	Given equation.
$x = -2x - 6$	Distributive Property.
$2x + x = 2x - 2x - 6$	Add $2x$ to each side.
$3x = 0 - 6$	Combine like terms.
$3x = -6$	Additive identity property.
$\dfrac{3x}{3} = \dfrac{-6}{3}$	Divide each side by 3.
$x = -2$	Simplify. (solution)

35.

$x + 4 = 6$

$x + 4 - 4 = 6 - 4$

$x = 2$

Check: $2 + 4 = 6$

37. $3x = 30$

$\dfrac{3x}{3} = \dfrac{30}{3}$

$x = 10$

Check: $3(10) = 30$

39. The sum of a number and 8 is 25.

or

A number is increased by 8, and the result is 25.

41. The product of 10 and the difference of a number and 3 is 8 times the number.

43. The sum of a number and 1 is divided by 3, and the quotient is equal to 8.

45. *Verbal model:* $\boxed{\begin{array}{c}\text{Original}\\\text{score}\end{array}} + \boxed{\begin{array}{c}\text{Additional}\\\text{points}\end{array}} = \boxed{\begin{array}{c}\text{Final}\\\text{score}\end{array}}$

Labels: Original score = x (points)

Additional points = 6 (points)

Final score = 94 (points)

Equation: $x + 6 = 94$

Note: There are other equivalent ways to write this equation. Here are some other possibilities:

$$94 - x = 6$$
$$94 = x + 6$$
$$x = 94 - 6$$

There are also equivalent ways of writing equations for the *other* exercises in this section.

47. *Verbal model:* $\boxed{\begin{array}{c}\text{Amount}\\\text{saved}\end{array}} + \boxed{\begin{array}{c}\text{Additional}\\\text{savings needed}\end{array}} = \boxed{\begin{array}{c}\text{Computer}\\\text{cost}\end{array}}$

Labels: Amount saved = \$3650

Additional savings needed = x (dollars)

Computer cost = \$4532

Equation: $3650 + x = 4532$

Note: Remember that there are other equivalent ways of writing this equation.

49. *Verbal model:* $\boxed{\text{Unknown number}} + \boxed{12} = \boxed{45}$

Label: Unknown number = x

Equation: $x + 12 = 45$

51. *Verbal model:* $\boxed{4} \cdot \boxed{\text{Sum of number and 6}} = \boxed{100}$

Label: Number = x

Equation: $4(x + 6) = 100$

Note: The parentheses are necessary to indicate that 4 is multiplied by the *sum* of x and 6. (If the equation were written as $4x + 6 = 100$, the 4 would be multiplied by the x only.)

53. *Verbal model:* $2 \cdot \boxed{\text{Number}} - 14 = \dfrac{\boxed{\text{Number}}}{3}$

Labels: Number = x

Equation: $2x - 14 = \dfrac{x}{3}$

55. *Verbal model:* $\boxed{\text{Cost per mile}} \cdot \boxed{\text{Number of miles}} = \boxed{\text{Cost for driving}}$

Label: Cost per mile = 0.32 (dollars per mile)

Number of miles = x

Cost for driving = 135.36 (dollars)

Equation: $0.32x = 135.36$

57. *Verbal model:* $\boxed{2} \cdot \boxed{\text{Length of mirror}} + \boxed{2} \cdot \boxed{\text{Width of mirror}} = \boxed{\text{Perimeter of mirror}}$

Label: Length of mirror = l (inches)

Width of mirror = $\frac{1}{3}l$ (inches)

Perimeter of mirror = 96 (inches)

Equation: $2l + 2\left(\frac{1}{3}l\right) = 96$

59. *Verbal model:*

Distance traveled	+	Remaining distance	=	Total distance

Travel rate	·	Travel time	+	Remaining distance	=	Total distance

Labels: Travel rate $= x$ (miles per hour)

Travel time $= 3$ (hours)

Remaining distance $= 25$ (miles)

Total distance $= 160$ (miles)

Equation: $x \cdot 3 + 25 = 160$ or $3x + 25 = 160$

Note: Remember the formula $d = rt$; distance equals rate times time.

61. *Verbal model:*

Distance for first car	=	Distance for second car

Rate for first car	·	Time for first car	=	Rate for second car	·	Time for second car

Labels: Rate for first car $= 45$ (miles per hour)

Time for first car $= 3$ (hours)

Rate for second car $= x$ (miles per hour)

Time for second car $= 2.5$ (hours)

Equation: $45(3) = x(2.5)$ or $135 = 2.5x$

63. *Verbal model:*

Original price	+	Increase in price	=	Current price

Labels: Original price $= x$ (dollars)

Increase in price $= 45$ (dollars)

Current price $= 375$ (dollars)

Equation: $x + 45 = 375$

65. *Verbal model:*

Total depreciation	=	3	·	Depreciation per year	or

Initial value	−	Final value	=	3	·	Depreciation per year

Labels: Initial value $= \$750,000$

Final value $= \$75,000$

Depreciation per year $= x$ (dollars)

Equation: $750,000 - 75,000 = 3x$

Note: This equation could also be written as $\dfrac{750,000 - 75,000}{3} = x$ or as $750,000 - 3x = 75,000$.

67. *Verbal model:*

Profit per box	·	Number of boxes	=	Amount earned

Labels: Profit per box $= 1.75$ (dollars)

Number of boxes $= x$

Amount earned $= 2000$ (dollars)

Equation: $1.75x = 2000$

69. $\dfrac{3 \text{ dollars}}{\cancel{\text{unit}}} \cdot (5 \text{ } \cancel{\text{units}}) = 15$ dollars

71. $\dfrac{3 \text{ dollars}}{\cancel{\text{pound}}} \cdot (5 \text{ } \cancel{\text{pounds}}) = 15$ dollars

73. $\dfrac{5 \text{ feet}}{\cancel{\text{second}}} \cdot \dfrac{60 \text{ } \cancel{\text{seconds}}}{\cancel{\text{minute}}} \cdot (20 \text{ } \cancel{\text{minutes}}) = 6000$ feet

75. A conditional equation has a solution set that is not the entire set of real numbers, but the solution set for any identity is all real numbers.

77. An expression doesn't contain an equal sign. Simplifying an expression is writing the expression in an equivalent form. Simplifying can involve combining like terms or removing symbols of grouping.

This expression can be simplified: $5x + 2(x + 6)$

$$5x + 2(x + 6) = 5x + 2x + 12 = 7x + 12$$

An equation contains an equal sign, and it can be solved by finding all the values of the variable for which the equation is true.

This equation can be solved: $5x = 2(x + 6)$

$$5x = 2x + 12$$
$$3x = 12$$
$$x = 4$$

79. To transform an equation into an equivalent equation,

(a) Simplify either side by removing symbols of grouping, combining like terms, or reducing fractions on one or both sides.

(b) Add (or subtract) the same quantity to (from) both sides of the equation.

(c) Multiply (or divide) both sides of the equation by the same nonzero quantity.

(d) Interchange the sides of the equation.

Review Exercises for Chapter 2

1. Terms: $4, -\frac{1}{2}x^3$

Coefficient: $-\frac{1}{2}$

3. Terms: $y^2, -10yz, \frac{2}{3}z^2$

Coefficients: $1, -10, \frac{2}{3}$

5. $5z \cdot 5z \cdot 5z = (5z)^3$ or 5^3z^3

7. $a(b - c) \cdot a(b - c) = [a(b - c)]^2$ or $a^2(b - c)^2$

9. (a) When $x = 0$, the value of $x^2 - 2x + 5$ is
$0^2 - 2 \cdot 0 + 5 = 0 - 0 + 5 = 5.$

(b) When $x = 2$, the value of $x^2 - 2x + 5$ is
$2^2 - 2 \cdot 2 + 5 = 4 - 4 + 5 = 5.$

11. When $x = 2$ and $y = -1$, the value of $x^2 - x(y + 1) = 2^2 - 2(-1 + 1) = 4 - 2(0) = 4 - 0 = 4$

When $x = 1$ and $y = 2$, the value of $x^2 - x(y + 1) = 1^2 - 1[2 + 1] = 1 - 1(3) = 1 - 3 = -2$

13. $x^2 \cdot x \cdot x = x^{2+1+4} = x^7$

15. $(x^3)^2 = x^{3 \cdot 2} = x^6$

17. $t^4(-2t^2) = -2t^{4+2} = -2t^6$

19. $(xy)(-5x^2y^3) = -5x^{1+2}y^{1+3}$
$= -5x^3y^4$

21. $(-2y^2)^3(8y) = (-2)^3(y^2)^3(8y)$
$= -8y^{2 \cdot 3}(8y)$
$= -8 \cdot 8 \cdot y^6 \cdot y$
$= -64y^{6+1}$
$= -64y^7$

23. Multiplicative Inverse Property

25. Commutative Property of Multiplication

27. Associative Property of Addition

29. $4(x + 3y) = 4x + 4 \cdot 3y$

$= 4x + 12y$

31. $-5(2u - 3v) = (-5)(2u) - (-5)(3v) = -10u + 15v$

33. $x(8x + 5y) = x(8x) + x(5y) = 8x^2 + 5xy$

35. $-(-a + 3b) = a - 3b$

Note: The sign of *each* term is changed.

The expression $-(-a + 3b)$ can be written as $(-1)(-a + 3b) = (-1)(-a) + (-1)(3b) = a - 3b$.

37. $3a - 5a = (3 - 5)a = -2a$

39. $3p - 4q + q + 8p = 3p + 8p - 4q + q$

$= (3 + 8)p + (-4 + 1)q$

$= 11p + (-3)q$

$= 11p - 3q$

41. $\frac{1}{4}s - 6t + \frac{7}{2}s + t = \frac{1}{4}s + \frac{7}{2}s - 6t + t$

$= \left(\frac{1}{4} + \frac{14}{4}\right)s + (-6 + 1)t$

$= \frac{15}{4}s - 5t$

43. $x^2 + 3xy - xy + 4 = x^2 + (3 - 1)xy + 4$

$= x^2 + 2xy + 4$

45. $5x - 5y + 3xy - 2x + 2y = 5x - 2x + 3xy - 5y + 2y$

$= (5 - 2)x + 3xy + (-5 + 2)y$

$= 3x + 3xy - 3y$

47. $5\left(1 + \frac{r}{n}\right)^2 - 2\left(1 + \frac{r}{n}\right)^2 = (5 - 2)\left(1 + \frac{r}{n}\right)^2$

$= 3\left(1 + \frac{r}{n}\right)^2$

49. $5(u - 4) + 10 = 5u - 5 \cdot 4 + 10$

$= 5u - 20 + 10$

51. $3s - (r - 2s) = 3s - r + 2s$

$= 3s + 2s - r$

$= 5s - r$

53. $-3(1 - 10z) + 2(1 - 10z) = (-3 + 2)(1 - 10z)$

$= -1(1 - 10z)$

$= -1 + 10z$

Note: The parentheses could be removed first.

$-3(1 - 10z) + 2(1 - 10z) = -3 \cdot 1 - (-3)(10z) + 2 \cdot 1 - 2(10z)$

$= -3 + 30z + 2 - 20z = -3 + 2 + 30z - 20z$

$= -1 + (30 - 20)z = -1 + 10z$

55. $\frac{1}{3}(42 - 18z) - 2(8 - 4z) = 14 - 6z - 16 + 8z$

$= -6z + 8z + 14 - 16$

$= (-6 + 8)z + 14 - 16$

$= 2z - 2$

57. $10 - [8(5 - x) + 2] = 10 - [40 - 8x + 2]$

$= 10 - 40 + 8x - 2$

$= 8x + 10 - 40 - 2$

$= 8x - 32$

59. $2[x + 2(y - x)] = 2[x + 2y - 2x]$

$= 2x + 4y - 4x$

$= (2 - 4)x + 4y$

$= -2x + 4y$

61. $\frac{2}{3}x + 5$

63. $2x - 10$

65. $50 + 7x$

67. $\dfrac{x + 10}{8}$

69. $x^2 + 64$

71. The sum of a number and three *or* a number increased by three.

73. A number is decreased by two and the result is divided by three *or* two is subtracted from a number and the result is divided by three.

75. (a) $x = 3$

$5(3) + 6 \overset{?}{=} 36$

$15 + 6 \overset{?}{=} 16$

$21 \neq 36$

3 is *not* a solution.

(b) $x = 6$

$5(6) + 6 \overset{?}{=} 36$

$30 + 6 \overset{?}{=} 36$

$36 = 36$

6 *is* a solution.

77. (a) $x = -1$

$3(-1) - 12 \overset{?}{=} -1$

$-3 - 12 \overset{?}{=} -1$

$-15 \neq -1$

-1 is *not* a solution.

(b) $x = 6$

$3(6) - 12 \overset{?}{=} 6$

$18 - 12 \overset{?}{=} 6$

$6 = 6$

6 *is* a solution.

79. (a) $x = \frac{2}{7}$

$4\left(2 - \frac{2}{7}\right) \overset{?}{=} 3\left(2 + \frac{2}{7}\right)$

$4\left(\frac{14}{7} - \frac{2}{7}\right) \overset{?}{=} 3\left(\frac{14}{7} + \frac{2}{7}\right)$

$4\left(\frac{12}{7}\right) \overset{?}{=} 3\left(\frac{16}{7}\right)$

$\frac{48}{7} = \frac{48}{7}$

$\frac{2}{7}$ *is* a solution.

(b) $x = -\frac{2}{3}$

$4\left(2 - \left(-\frac{2}{3}\right)\right) \overset{?}{=} 3\left(2 + \left(-\frac{2}{3}\right)\right)$

$4\left(\frac{6}{3} + \frac{2}{3}\right) \overset{?}{=} 3\left(\frac{6}{3} - \frac{2}{3}\right)$

$4\left(\frac{8}{3}\right) \overset{?}{=} 3\left(\frac{4}{3}\right)$

$\frac{32}{3} \neq 4$

$-\frac{2}{3}$ is *not* a solution.

81. (a) $x = -1$

$\dfrac{4}{-1} - \dfrac{2}{-1} \overset{?}{=} 5$

$-4 - (-2) \overset{?}{=} 5$

$-4 + 2 \overset{?}{=} 5$

$-2 \neq 5$

-1 is *not* a solution.

(b) $x = \dfrac{2}{5}$

$\dfrac{4}{2/5} - \dfrac{2}{2/5} \overset{?}{=} 5$

$4\left(\dfrac{5}{2}\right) - 2\left(\dfrac{5}{2}\right) \overset{?}{=} 5$

$10 - 5 \overset{?}{=} 5$

$5 = 5$

$\dfrac{2}{5}$ *is* a solution.

83. (a) $x = 3$

$3(3 - 7) \overset{?}{=} -12$

$3(-4) \overset{?}{=} -12$

$-12 = -12$

3 *is* a solution.

(b) $x = 4$

$4(4 - 7) \overset{?}{=} -12$

$4(-3) \overset{?}{=} -12$

$-12 = -12$

4 *is* a solution.

85. $P\left(\frac{9}{10}\right)\left(\frac{9}{10}\right)\left(\frac{9}{10}\right)\left(\frac{9}{10}\right)\left(\frac{9}{10}\right) = P\left(\frac{9}{10}\right)^5$

87. The amount of tax is a product.

Verbal model: | Percent of income tax | · | Taxable income |

Labels: Percent of income tax $= 0.28$ (in decimal form)

Taxable income $= I$ (dollars)

Algebraic expression: $0.28I$ (dollars)

89. The area is a difference of products.

Verbal model: | Length of larger rectangle | · | Width of larger rectangle | $-$ | Length of smaller rectangle | · | Width of smaller rectangle |

Labels: Length of larger rectangle $= 6x$

Width of larger rectangle $= 2x$

Length of smaller rectangle $= 3x$

Width of smaller rectangle $= x$

Algebraic expression: $6x \cdot 2x - 3x \cdot x$ or $12x^2 - 3x^2$ or $9x^2$ (square units)

91. *Verbal model:* | Time | · | Average speed |

Labels: Time $= 10$

Average speed $= s$

Algebraic expression: $10s$

93. *Verbal model:* | Monthly rent | · | Number of months |

Labels: Monthly rent $= \$625$

Number of months $= n$

Algebraic expression: $625n$

95. (a)

n	0	1	2	3	4	5
$n^2 + 3n + 2$	2	6	12	20	30	42
Differences:		4	6	8	10	12
Differences:			2	2	2	2

When $n = 0$, the value of $n^2 + 3n + 2$ is $0^2 + 3 \cdot 0 + 2 = 0 + 0 + 2 = 2$.

When $n = 1$, the value of $n^2 + 3n + 2$ is $1^2 + 3 \cdot 1 + 2 = 1 + 3 + 2 = 6$.

When $n = 2$, the value of $n^2 + 3n + 2$ is $2^2 + 3 \cdot 2 + 2 = 4 + 6 + 2 = 12$.

When $n = 3$, the value of $n^2 + 3n + 2$ is $3^2 + 3 \cdot 3 + 2 = 9 + 9 + 2 = 20$.

When $n = 4$, the value of $n^2 + 3n + 2$ is $4^2 + 3 \cdot 4 + 2 = 16 + 12 + 2 = 30$.

When $n = 5$, the value of $n^2 + 3n + 2$ is $5^2 + 3 \cdot 5 + 2 = 25 + 15 + 2 = 42$.

(b) In third row, the differences are consecutive even integers. In the fourth row, each difference is two.

97. *Verbal model:* | Unknown number | $+$ | Reciprocal of number | $=$ | $\dfrac{37}{6}$ |

Labels: Unknown number $= x$

Reciprocal $= \dfrac{1}{x}$

Equation: $x + \dfrac{1}{x} = \dfrac{37}{6}$

99. *Verbal model:* | Area of triangle | $-$ | Area of triangle | $=$ | Area of shaded region |

Labels: Area of rectangle $= x(6)$ or $6x$ (square inches)

Area of triangle $\frac{1}{2}(x)(6)$ or $3x$ (square inches)

Shaded area $= 24$ (square inches)

Equation: $6x - \frac{1}{2}(x)(6) = 24$ or $6x - 3x = 24$

This equation could also be simplified to $3x = 24$.

Note: The area of a rectangle is the product of its length and width $(A = lw)$. The area of a triangle is one-half the product of its base and height $\left(A = \frac{1}{2}bh\right)$.

Chapter Test for Chapter 2

1. Terms $2x^2, -7xy, 3y^3$

Coefficients: $2, -7, 3$

2. $x \cdot (x + y) \cdot x \cdot (x + y) \cdot x = x^3(x + y)^2$

3. Associative Property of Multiplication

4. Commutative Property of Addition

5. Additive Inverse Property

6. Multiplicative Identity Property

7. $3(x + 8) = 3x + 24$

8. $-y(3 - 2y) = -3y + 2y^2$

9. $(c^2)^4 = c^{2 \cdot 4}$

$= c^8$

10. $-5uv(2u^3) = -5 \cdot 2 \cdot u \cdot u^3 \cdot v$

$= -10u^{1+3}v$

$= -10u^4v$

11. $3b - 2a + a - 10b = (-2 + 1)a + (3 - 10)b$

$= -a - 7b$

12. $15(u - v) - 7(u - v) = (15 - 7)(u - v)$

$= 8(u - v)$

$= 8u - 8v$

Note: This problem can also be worked as follows.

$15(u - v) - 7(u - v) = 15u - 15v - 7u + 7v$

$= (15 - 7)u + (-15 + 7)u$

$= 8u - 8v$

13. $3z - (4 - z) = 3z - 4 + z$

$= 3z + z - 4$

$= 4z - 4$

14. $2[10 - (t + 1)] = 2[10 - t - 1]$

$\qquad\qquad\qquad\;\; = 2[10 - 1 - t]$

$\qquad\qquad\qquad\;\; = 2[9 - t]$

$\qquad\qquad\qquad\;\; = 18 - 2t$

15. (a) When $x = 3$, the value of $x^3 - 2$ is
$3^3 - 2 = 27 - 2 = 25$.

(b) When $x = 3$ and $y = -12$, the value of $x^2 + 4(y + 2)$
is $3^2 + 4(-12 + 2) = 9 + 4(-10) = 9 - 40 = -31$.

16. When $a = 2$ and $b = 6$, the value of $(a + 2b)/(3a - b)$
is undefined because

$$\frac{2 + 2(6)}{3(2) - 6} = \frac{2 + 12}{6 - 6} = \frac{14}{0}$$

and division by zero is undefined.

17. $\dfrac{1}{5}n + 2$ or $\dfrac{n}{5} + 2$

18. (a) The perimeter of a rectangle is twice the length plus twice the width.

Perimeter: $2(2w - 4) + 2(w)$

The area of a rectangle is the product of the length and the width.

Area: $(2w - 4)w$

(b) Perimeter: $2(2w - 4) + 2w = 4w - 8 - 2w = 6w - 8$

Area: $(2w - 4)w = 2w^2 - 4w$

(c) The perimeter is measured in units of length, such as feet, meters, inches, etc. The area is measured in square units, such as square centimeters, square inches, square yards, etc.

(d) When $w = 12$ feet, the perimeter $6w - 8 = 6(12) - 8$

$$= 72 - 8$$

$$= 64 \text{ feet.}$$

When $w = 12$ feet, the area $2w^2 - 4w = 2(12)^2 - 4(12)$

$$= 2(144) - 48$$

$$= 288 - 48$$

$$= 240 \text{ square feet.}$$

19. The concert income is a sum of products.

Verbal description: | Price of adult's ticket | $\cdot$ | Number of adults | $+$ | Price of child's ticket | $\cdot$ | Number of children |

Labels: Price of adults ticket $= 3$ (dollars)

Number of adults $= n$

Price of child's ticket $= 2$ (dollars)

Number of children $= m$

Algebraic expression: $3n + 2m$ (dollars)

20. (a) $6(3 - (-2)) - 5(2(-2) - 1) \overset{?}{=} 7$

$6(3 + 2) - 5(-4 - 1) \overset{?}{=} 7$

$6(5) - 5(-5) \overset{?}{=} 7$

$30 + 25 \overset{?}{=} 7$

$55 \neq 7$

-2 is *not* a solution.

(b) $6(3 - 1) - 5(2 \cdot 1 - 1) \overset{?}{=} 7$

$6(2) - 5(2 - 1) \overset{?}{=} 7$

$12 - 5(1) \overset{?}{=} 7$

$12 - 5 \overset{?}{=} 7$

$7 = 7$

1 *is* a solution.

CHAPTER 3
Linear Equations and Problem Solving

CHAPTER 3
Linear Equations and Problem Solving

Section 3.1   Solving Linear Equations

Solutions to Odd-Numbered Exercises

1. $x = 8$

3. $x = 13$

5. $y = 4$

7. $s = 3$

9.

$5x + 15 = 0$	Original equation.
$5x + 15 - 15 = 0 - 15$	Subtract 15 from both sides.
$5x = -15$	Combine like terms.
$\dfrac{5x}{5} = \dfrac{-15}{5}$	Divide both sides by 5.
$x = -3$	Simplify.

11.

$-2x + 5 = 13$	Original equation.
$-2x + 5 - 5 = 13 - 5$	Subtract 5 from both sides.
$-2x = 8$	Combine like terms.
$\dfrac{-2x}{-2} = \dfrac{8}{-2}$	Divide both sides by -2.
$x = -4$	Simplify.

13. $5x = 30$

$\dfrac{5x}{5} = \dfrac{30}{5}$

$x = 6$

15. $9x = -21$

$\dfrac{9x}{9} = \dfrac{-21}{9}$

$x = -\dfrac{21}{9}$

$x = -\dfrac{7}{3}$

17. $8x - 4 = 20$

$8x - 4 + 4 = 20 + 4$

$8x = 24$

$x = 3$

19. $25x - 4 = 46$

$25x - 4 + 4 = 46 + 4$

$25x = 50$

$\dfrac{25x}{25} = \dfrac{50}{25}$

$x = 2$

21. $10 - 4x = -6$

$10 - 10 - 4x = -6 - 10$

$-4x = -16$

$\dfrac{-4x}{-4} = \dfrac{-16}{-4}$

$x = 4$

23. $6x - 4 = 0$

$6x - 4 + 4 = 0 + 4$

$6x = 4$

$\dfrac{6x}{6} = \dfrac{4}{6}$

$x = \dfrac{4}{6}$

$x = \dfrac{2}{3}$

25. $3y - 2 = 2y$

$3y - 3y - 2 = 2y - 3y$

$-2 = -y$

$(-1)(-2) = (-1)(-y)$

$2 = y$

27. $4 - 7x = 5x$

$4 - 7x + 7x = 5x + 7x$

$4 = 12x$

$\dfrac{4}{12} = \dfrac{12x}{12}$

$\dfrac{1}{3} = x$

29. $4 - 5t = 16 + t$

$4 - 5t - t = 16 + t - t$

$4 - 6t = 16$

$4 - 4 - 6t = 16 - 4$

$-6t = 12$

$\dfrac{-6t}{-6} = \dfrac{12}{-6}$

$t = -2$

31. $-3t + 5 = -3t$

$-3t + 3t + 5 = 3t + 3t$

$5 = 0$ (False)

The original equation has no solution.

33.
$$15x - 3 = 15 - 3x$$
$$15x + 3x - 3 = 15 + 3x - 3x$$
$$18x - 3 = 15$$
$$18x - 3 + 3 = 15 + 3$$
$$18x = 18$$
$$\frac{18x}{18} = \frac{18}{18}$$
$$x = 1$$

35. $4z = 10$
$$\frac{4z}{4} = \frac{10}{4}$$
$$z = \frac{10}{4}$$
$$z = \frac{5}{2}$$

37. $8t - 4 = -6$
$$8t - 4 + 4 = -6 + 4$$
$$8t = -2$$
$$\frac{8t}{8} = \frac{-2}{8}$$
$$t = -\frac{1}{4}$$

39.
$$4x - 6 = 4x - 6$$
$$4x - 4x - 6 = 4x - 4x - 6$$
$$-6 = -6 \qquad \text{Identity}$$

The original equation has infinitely many solutions.

41.
$$2x + 4 = -3x + 6$$
$$2x + 3x + 4 = -3x + 3x + 6$$
$$5x + 4 = 6$$
$$5x + 4 - 4 = 6 - 4$$
$$5x = 2$$
$$\frac{5x}{5} = \frac{2}{5}$$
$$x = \frac{2}{5}$$

43.
$$2x = -3x$$
$$2x + 3x = -3x + 3x$$
$$5x = 0$$
$$\frac{5x}{5} = \frac{0}{5}$$
$$x = 0$$

45. $2x - 5 + 10x = 3$
$$12x - 5 = 3$$
$$12x - 5 + 5 = 3 + 5$$
$$12x = 8$$
$$\frac{12x}{12} = \frac{8}{12}$$
$$x = \frac{2}{3}$$

47. $\dfrac{x}{3} = 10$
$$3\left(\frac{x}{3}\right) = 3 \cdot 10$$
$$x = 30$$
Note: $3\left(\dfrac{x}{3}\right) = \dfrac{3}{1} \cdot \dfrac{x}{3} = \dfrac{3x}{3} = x$

49. $x - \dfrac{1}{3} = \dfrac{4}{3}$
$$x - \frac{1}{3} + \frac{1}{3} = \frac{4}{3} + \frac{1}{3}$$
$$x = \frac{5}{3}$$

51. $t - \dfrac{1}{3} = \dfrac{1}{2}$
$$t - \frac{1}{3} + \frac{1}{3} = \frac{1}{2} + \frac{1}{3}$$
$$t = \frac{3}{6} + \frac{2}{6}$$
$$t = \frac{5}{6}$$

53. $3t + 1 - 2t = t + 1$
$$t + 1 = t + 1 \qquad \text{Identity}$$
This equation has infinitely many solutions.

55.
$$2y - 18 = -5y - 4$$
$$2y + 5y - 18 = -5y + 5y - 4$$
$$7y - 18 = -4$$
$$7y - 18 + 18 = -4 + 18$$
$$7y = 14$$
$$\frac{7y}{7} = \frac{14}{7}$$
$$y = 2$$

57. *Verbal model:*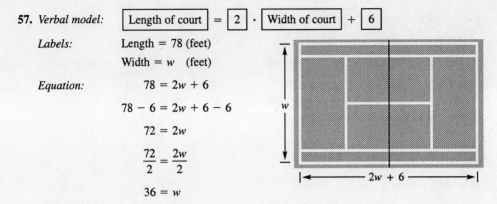

Labels: Length = 78 (feet)

Width = w (feet)

Equation: $78 = 2w + 6$

$78 - 6 = 2w + 6 - 6$

$72 = 2w$

$\dfrac{72}{2} = \dfrac{2w}{2}$

$36 = w$

The width of the court is 36 feet.

59. *Verbal model:*   First board length $+$ Second board length $+$ Third board length $=$ Original board length

Labels: First board length $= x$ (feet)

Second board length $= x$ (feet)

Third board length $= 2x$ (feet)

Original board length $= 12$ (feet)

Equation: $x + x + 2x = 12$

$4x = 12$

$x = 3$ and $2x = 6$

Thus, the first two pieces are 3 feet long and the third piece is 6 feet long.

61. *Verbal model:*   Cost of parts $+$ Hourly cost of labor $\cdot$ Hours of labor $=$ Total cost

Labels: Cost of parts $= 285$ (dollars)

Hourly cost of labor $= 32$ (dollars per hour)

Hours of labor $= t$ (hours)

Total cost $= 357$ (dollars)

Equation: $285 + 32t = 357$

$285 + 32t - 285 = 357 - 285$

$32t = 72$

$\dfrac{32t}{32} = \dfrac{72}{32}$

$t = 2.25$

The time spent on labor was 2.25 hours (or 2 hours and 15 minutes).

63. *Verbal model:* | Total revenue | = | Revenue from main floor seats | + | Revenue from balcony seats |

Labels: Total revenue = 5200 (dollars)

Price per main floor seat = 10 (dollars per seat)

Number of main floor seats = 400 (seats)

Price per balcony seat = 8 (dollars per seat)

Number of balcony seats = x (seats)

Equation:
$$5200 = 400(10) + 8x$$
$$5200 = 4000 + 8x$$
$$5200 - 4000 = 4000 - 4000 + 8x$$
$$1200 = 8x$$
$$\frac{1200}{8} = \frac{8x}{8}$$
$$150 = x$$

There were 150 balcony seats sold.

65. *Verbal model:* | Total wages | = | Rate for job 1 | · | Time at job 1 | + | Rate for job 2 | · | Time at job 2 |

Labels: Total wages = 425 (dollars)

Rate for job 1 = 9.25 (dollars per hour)

Time at job 1 = 40 (hours)

Rate for job 2 = 7.50 (dollars per hour)

Time at job 2 = x (hours)

Equation:
$$425 = 9.25(40) + 7.50x$$
$$425 = 370 + 7.50x$$
$$55 = 7.50x$$
$$\frac{55}{7.50} = \frac{7.50x}{7.50}$$
$$7\frac{1}{3} = x$$

You must work $7\frac{1}{3}$ hours, or 7 hours and 20 minutes, at the second job.

67. *Verbal model:* | Number | + | 45 | = | 75 |

Labels: Number = x

Equation:
$$x + 45 = 75$$
$$x + 45 - 45 = 75 - 45$$
$$x = 30$$

The number is 30.

69. *Verbal model:*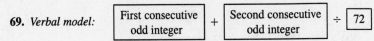

$\boxed{\text{First consecutive odd integer}} + \boxed{\text{Second consecutive odd integer}} = \boxed{72}$

Labels: First consecutive odd integer $= 2n + 1$

Second consecutive odd integer $= 2n + 3$

Equation:
$$(2n + 1) + (2n + 3) = 72$$
$$2n + 1 + 2n + 3 = 72$$
$$4n + 4 = 72$$
$$4n + 4 - 4 = 72 - 4$$
$$4n = 68$$
$$\frac{4n}{4} = \frac{68}{4}$$
$$n = 17$$

$2n + 1 = 2(17) + 1 = 34 + 1 = 35$

$2n + 3 = 2(17) + 3 = 34 + 3 = 37$

The integers are 35 and 37.

71. (a)

t	1	1.5	2	3	4	5
Width	300	240	200	150	120	100
Length	300	360	400	450	480	500
Area	90,000	86,400	80,000	67,500	57,600	50,000

$t = 1$:
$$P = 2w + 2(tw)$$
$$1200 = 2w + 2(1 \cdot w)$$
$$1200 = 2w + 2w$$
$$1200 = 4w$$
$$\frac{1200}{4} = \frac{4w}{4}$$
$$300 = w$$

Length $= tw = 1(300) = 300$

Area $= (300)(300) = 90,000$

$t = 1.5$:
$$1200 = 2w + 2(1.5w)$$
$$1200 = 2w + 3w$$
$$1200 = 5w$$
$$\frac{1200}{5} = \frac{5w}{5}$$
$$240 = w$$

Length $= tw = 1.5(240) = 360$

Area $= (240)(360) = 86,400$

$t = 2$:
$$1200 = 2w + 2(2w)$$
$$1200 = 2w + 4w$$
$$1200 = 6w$$
$$\frac{1200}{6} = \frac{6w}{6}$$
$$200 = w$$

Length $= tw = 2(200) = 400$

Area $= (400)(200) = 80,000$

$t = 3$:
$$1200 = 2w + 2(3w)$$
$$1200 = 2w + 6w$$
$$1200 = 8w$$
$$\frac{1200}{8} = \frac{8w}{8}$$
$$150 = w$$

Length $= tw = 3(150) = 450$

Area $= (450)(150) = 67,500$

—CONTINUED—

71. —CONTINUED—

$t = 4$: $1200 = 2w + 2(4w)$ $t = 5$: $1200 = 2w + 2(5w)$

$1200 = 2w + 8w$ $1200 = 2w + 10w$

$1200 = 10w$ $1200 = 12w$

$\dfrac{1200}{10} = \dfrac{10w}{10}$ $\dfrac{1200}{12} = \dfrac{12w}{12}$

$120 = w$ $100 = w$

Length $= tw = 4(120) = 480$ Length $= tw = 5(100) = 500$

Area $= 480(120) = 57{,}600$ Area $= 500(100) = 50{,}000$

(b) The area of a rectangle of a given perimeter *decreases* as its length increases relative to its width.

73. *Verbal model:* $\boxed{\begin{array}{c}\text{Weight of}\\\text{red box}\end{array}} + 3 \cdot \boxed{\begin{array}{c}\text{Weight of}\\\text{blue box}\end{array}} = 9 \cdot \boxed{\begin{array}{c}\text{Weight of}\\\text{blue box}\end{array}}$

Labels: Weight of red box $= R$ (ounces)

Weight of blue box $= 1$ (ounce)

Equation: $R + 3(1) = 9(1)$

$R + 3 = 9$

$R + 3 - 3 = 9 - 3$

$R = 6$

The red box weighs 6 ounces.

If you remove three blue boxes from each side, the scale would still balance. The red box would balance the remaining six blue boxes, showing that the red box weighs 6 ounces.

This illustrates the addition (or subtraction) property of equality.

75. Subtract 5 from each side of the equation.

This is using the addition (or subtraction) property of equality.

$x + 5 = 32$

$x + 5 - 5 = 32 - 5$

$x = 27$

77. $134.5 \text{ km/hr} = 134.5 \dfrac{\text{km}}{\text{hr}} \cdot \dfrac{0.621}{1 \text{ km}} \approx 83.5 \text{ mi/hr}$

No, the answer of 134.5 kilometers per hour cannot be correct; 83.5 miles per hour is too fast an average speed for a moving van.

79. True.

Subtracting 0 from each side of an equation produces an equivalent equation.

Section 3.2   Equations That Reduce to Linear Form

1. $-5(t + 3) = 0$ **3.** $2(y - 4) = 12$ **5.** $2(x - 3) = 4$

$-5t - 15 = 0$ $2y - 8 = 12$ $2x - 6 = 4$

$-5t - 15 + 15 = 0 + 15$ $2y - 8 + 8 = 12 + 8$ $2x - 6 + 6 = 4 + 6$

$-5t = 15$ $2y = 20$ $2x = 10$

$\dfrac{-5t}{-5} = \dfrac{15}{-5}$ $\dfrac{2y}{2} = \dfrac{20}{2}$ $\dfrac{2x}{2} = \dfrac{10}{2}$

$t = -3$ $y = 10$ $x = 5$

7.
$$7(x + 5) = 49$$
$$7x + 35 = 49$$
$$7x + 35 - 35 = 49 - 35$$
$$7x = 14$$
$$\frac{7x}{7} = \frac{14}{7}$$
$$x = 2$$

9. $4 - (z + 6) = 8$
$$4 - z - 6 = 8$$
$$-z - 2 = 8$$
$$-z - 2 + 2 = 8 + 2$$
$$-z = 10$$
$$(-1)(-z) = (-1)(10)$$
$$z = -10$$

11. $3 - (2x - 4) = 3$
$$3 - 2x + 4 = 3$$
$$-2x + 7 = 3$$
$$-2x + 7 - 7 = 3 - 7$$
$$-2x = -4$$
$$\frac{-2x}{-2} = \frac{-4}{-2}$$
$$x = 2$$

13.
$$-3(t + 5) = 0$$
$$-3t - 15 = 0$$
$$-3t - 15 + 15 = 0 + 15$$
$$-3t = 15$$
$$\frac{-3t}{-3} = \frac{15}{-3}$$
$$t = -5$$

15. $-4(t + 5) = -2(2t + 10)$
$$-4t - 20 = -4t - 20 \quad \text{Identity}$$
This equation has infinitely many solutions.

17.
$$3(x + 4) = 10(x + 4)$$
$$3x + 12 = 10x + 40$$
$$3x - 10x + 12 = 10x - 10x + 40$$
$$-7x + 12 = 40$$
$$-7x + 12 - 12 = 40 - 12$$
$$-7x = 28$$
$$\frac{-7x}{-7} = \frac{28}{-7}$$
$$x = -4$$

19. $7 = 3(x + 2) - 3(x - 5)$
$$7 = 3x + 6 - 3x + 15$$
$$7 = 21 \qquad \text{(False)}$$
Thus, this equation has no solution.

21. $7x - 2(x - 2) = 12$
$$7x - 2x + 4 = 12$$
$$5x + 4 = 12$$
$$5x + 4 - 4 = 12 - 4$$
$$5x = 8$$
$$\frac{5x}{5} = \frac{8}{5}$$
$$x = \frac{8}{5}$$

23.
$$6 = 3(y + 1) - 4(1 - y)$$
$$6 = 3y + 3 - 4 + 4y$$
$$6 = 7y - 1$$
$$6 + 1 = 7y - 1 + 1$$
$$7 = 7y$$
$$\frac{7}{7} = \frac{7y}{7}$$
$$1 = y$$

25.
$$7(2x - 1) = 4(1 - 5x) + 6$$
$$14x - 7 = 4 - 20x + 6$$
$$14x - 7 = 10 - 20x$$
$$14x + 20x - 7 = 10 - 20x + 20x$$
$$34x - 7 = 10$$
$$34x - 7 + 7 = 10 + 7$$
$$34x = 17$$
$$\frac{34x}{34} = \frac{17}{34}$$
$$x = \frac{1}{2}$$

27. $2[(3x + 5) - 7] = 3(5x - 2)$
$$2[3x - 2] = 3(5x - 2)$$
$$6x - 4 = 15x - 6$$
$$6x - 15x - 4 = 15x - 15x - 6$$
$$-9x - 4 = -6$$
$$-9x - 4 + 4 = -6 + 4$$
$$-9x = -2$$
$$\frac{-9x}{-9} = \frac{-2}{-9}$$
$$x = \frac{2}{9}$$

29. $4x + 3[x - 2(2x - 1)] = 4 - 3x$

$$4x + 3[x - 4x + 2] = 4 - 3x$$
$$4x + 3[-3x + 2] = 4 - 3x$$
$$4x - 9x + 6 = 4 - 3x$$
$$-5x + 6 = 4 - 3x$$
$$-5x + 3x + 6 = 4 - 3x + 3x$$
$$-2x + 6 = 4$$
$$-2x + 6 - 6 = 4 - 6$$
$$-2x = -2$$
$$\frac{-2x}{-2} = \frac{-2}{-2}$$
$$x = 1$$

31. $\dfrac{x}{2} = \dfrac{3}{2}$

$$2\left(\frac{x}{2}\right) = 2\left(\frac{3}{2}\right)$$
$$x = 3$$

33. $\dfrac{y}{5} = \dfrac{3}{5}$

$$5\left(\frac{y}{5}\right) = 5\left(\frac{3}{5}\right)$$
$$y = 3$$

35. $\dfrac{y}{5} = -\dfrac{3}{10}$

$$5\left(\frac{y}{5}\right) = 5\left(-\frac{3}{10}\right)$$
$$y = -\frac{3}{2}$$

37. $\dfrac{6x}{25} = \dfrac{3}{5}$

$$30x = 75 \text{ (Cross-}$$
$$\text{multiply)}$$
$$\frac{30x}{30} = \frac{75}{30}$$
$$x = \frac{75}{30}$$
$$x = \frac{5}{2}$$

39. $\dfrac{5x}{4} + \dfrac{1}{2} = 0$

$$4\left(\frac{5x}{4} + \frac{1}{2}\right) = 4(0)$$
$$4\left(\frac{5x}{4}\right) + 4\left(\frac{1}{2}\right) = 0$$
$$5x + 2 = 0$$
$$5x + 2 - 2 = 0 - 2$$
$$5x = -2$$
$$\frac{5x}{5} = \frac{-2}{5}$$
$$x = -\frac{2}{5}$$

41. $\dfrac{x}{5} - \dfrac{x}{2} = 1$

$$10\left(\frac{x}{5} - \frac{x}{2}\right) = 10(1)$$
$$10\left(\frac{x}{5}\right) - 10\left(\frac{x}{2}\right) = 10$$
$$2x - 5x = 10$$
$$-3x = 10$$
$$\frac{-3x}{-3} = \frac{10}{-3}$$
$$x = -\frac{10}{3}$$

43. $2s + \dfrac{3}{2} = 2s + 2$

$$2s - 2s + \frac{3}{2} = 2s - 2s + 2$$
$$\frac{3}{2} = 2 \qquad \text{False}$$

This equation has no solution.

45. $3x + \dfrac{1}{4} = \dfrac{3}{4}$

$$4\left(3x + \frac{1}{4}\right) = 4\left(\frac{3}{4}\right)$$
$$12x + 1 = 3$$
$$12x + 1 - 1 = 3 - 1$$
$$12x = 2$$
$$\frac{12x}{12} = \frac{2}{12}$$
$$x = \frac{1}{6}$$

47.
$$\frac{1}{5}x + 1 = \frac{3}{10}x - 4$$

$$10\left(\frac{1}{5}x + 1\right) = 10\left(\frac{3}{10}x - 4\right)$$

$$10\left(\frac{1}{5}x\right) + 10(1) = 10\left(\frac{3}{10}x\right) - 10(4)$$

$$2x + 10 = 3x - 40$$

$$2x - 3x + 10 = 3x - 3x - 40$$

$$-x + 10 = -40$$

$$-x + 10 - 10 = -40 - 10$$

$$-x = -50$$

$$-1(-x) = -1(-50)$$

$$x = 50$$

49.
$$\frac{2}{3}(z + 5) - \frac{1}{4}(z + 24) = 0$$

$$12\left[\frac{2}{3}(z + 5) - \frac{1}{4}(z + 24)\right] = 12(0)$$

$$12 \cdot \frac{2}{3}(z + 5) - 12 \cdot \frac{1}{4}(z + 24) = 0$$

$$8(z + 5) - 3(z + 24) = 0$$

$$8x + 40 - 3z - 72 = 0$$

$$5z - 32 = 0$$

$$5z - 32 + 32 = 0 + 32$$

$$5z = 32$$

$$\frac{5z}{5} = \frac{32}{5}$$

$$z = \frac{32}{5}$$

51.
$$\frac{100 - 4u}{3} = \frac{5u + 6}{4} + 6$$

$$12\left(\frac{100 - 4u}{3}\right) = 12\left(\frac{5u + 6}{4} + 6\right)$$

$$\frac{12}{1}\left(\frac{100 - 4u}{3}\right) = \frac{12}{1}\left(\frac{5u + 6}{4}\right) + 12(6)$$

$$4(100 - 4u) = 3(5u + 6) + 72$$

$$400 - 16u = 15u + 18 + 72$$

$$400 - 16u = 15u + 90$$

$$400 - 16u - 15u = 15u - 15u + 90$$

$$400 - 31u = 90$$

$$400 - 400 - 31u = 90 - 400$$

$$-31u = -310$$

$$\frac{-31u}{-31} = \frac{-310}{-31}$$

$$u = 10$$

53.
$$\frac{t + 4}{6} = \frac{2}{3}$$

$$3(t + 4) = 12 \qquad \text{(Cross-multiply)}$$

$$3t + 12 = 12$$

$$3t + 12 - 12 = 12 - 12$$

$$3t = 0$$

$$\frac{3t}{3} = \frac{0}{3}$$

$$t = 0$$

55.
$$\frac{x - 2}{5} = \frac{2}{3}$$

$$3(x - 2) = 5(2) \quad \text{(Cross-multiply)}$$

$$3x - 6 = 10$$

$$3x - 6 + 6 = 10 + 6$$

$$3x = 16$$

$$\frac{3x}{3} = \frac{16}{3}$$

$$x = \frac{16}{3}$$

57.
$$\frac{5x - 4}{4} = \frac{2}{3}$$

$$3(5x - 4) = 4(2)$$

$$15x - 12 = 8$$

$$15x - 12 + 12 = 8 + 12$$

$$15x = 20$$

$$\frac{15x}{15} = \frac{20}{15}$$

$$x = \frac{4}{3}$$

59.
$$\frac{x}{4} = \frac{1 - 2x}{3}$$

$$3(x) = 4(1 - 2x)$$

$$3x = 4 - 8x$$

$$3x + 8x = 4 - 8x + 8x$$

$$11x = 4$$

$$\frac{11x}{11} = \frac{4}{11}$$

$$x = \frac{4}{11}$$

61. $\dfrac{10 - x}{2} = \dfrac{x + 4}{5}$

$5(10 - x) = 2(x + 4)$

$50 - 5x = 2x + 8$

$50 - 5x - 2x = 2x - 2x + 8$

$50 - 7x = 8$

$50 - 50 - 7x = 8 - 50$

$-7x = -42$

$\dfrac{-7x}{-7} = \dfrac{-42}{-7}$

$x = 6$

63. $0.2x + 5 = 6$

$0.2x + 5 - 5 = 6 - 5$

$0.2x = 1$

$10(0.2x) = 10(1)$

$2x = 10$

$\dfrac{2x}{2} = \dfrac{10}{2}$

$x = 5$

65. $0.234x + 1 = 2.805$

$0.234x + 1 - 1 = 2.805 - 1$

$0.234x = 1.085$

$\dfrac{0.234x}{0.234} = \dfrac{1.805}{0.234}$

$x \approx 7.71$ (Rounded)

67. $0.02x - 0.96 = 1.50$

$0.02x - 0.96 + 0.96 = 1.50 + 0.96$

$0.02x = 2.46$

$\dfrac{0.02x}{0.02} = \dfrac{2.46}{0.02}$

$x = 123.00$

69. $\dfrac{x}{3.25} + 1 = 2.08$

$\dfrac{x}{3.25} + 1 - 1 = 2.08 - 1$

$\dfrac{x}{3.25} = 1.08$

$\dfrac{x}{3.25}(3.25) = 1.08(3.25)$

$x = 3.51$

71. $\dfrac{x}{3.155} = 2.850$

$3.155\left(\dfrac{x}{3.155}\right) = 3.155(2.850)$

$x \approx 8.99$ (Rounded)

73. $\dfrac{t}{10} + \dfrac{t}{15} = 0.8$

$30\left(\dfrac{t}{10} + \dfrac{t}{15}\right) = 30(0.8)$

$30\left(\dfrac{t}{10}\right) + 30\left(\dfrac{t}{15}\right) = 24$

$3t + 2t = 24$

$5t = 24$

$\dfrac{5t}{5} = \dfrac{24}{5}$

$t = \dfrac{24}{5}$ or 4.8 hours

75. (a) $\dfrac{87 + 92 + 84 + x}{4} = 90$

$\dfrac{263 + x}{4} = 90$

$263 + x = 4(90)$

$263 + x = 360$

$263 + x - 263 = 360 - 263$

$x = 97$

(b) $\dfrac{87 + 69 + 89 + x}{4} = 90$

$\dfrac{245 + x}{4} = 90$

$245 + x = 4(90)$

$245 + x = 360$

$245 + x - 245 = 360 - 245$

$x = 115$

No, it would not be possible to get an A with these three scores. You would need more than the possible 100 points on the fourth test.

If you earned the maximum 100 points on test four, your average would be 86.25%.

$\dfrac{87 + 69 + 89 + 100}{4} = \dfrac{345}{4} = 86.25$

77. $p_1 x + p_2(a - x) = p_3 a$

$0.1x + 0.3(100 - x) = 0.25(100)$

$0.1x + 30 - 0.3x = 25$

$-0.2x + 30 = 25$

$-0.2x + 30 - 30 = 25 - 30$

$-0.2x = -5$

$\dfrac{-0.2x}{-0.2} = \dfrac{-5}{-0.2}$

$x = 25$ quarts

79. $p_1 x + p_2(a - x) = p_3 a$

$1(x) + 0.4(8 - x) = 0.5(8)$

$x + 3.2 - 0.4x = 4$

$0.6x + 3.2 = 4$

$0.6x + 3.2 - 3.2 = 4 - 3.2$

$0.6x = 0.8$

$\dfrac{0.6x}{0.6} = \dfrac{0.8}{0.6}$

$x = \dfrac{4}{3}$ or $1.\overline{3}$ quarts

81. Use the equation $W_1 w = W_2(a - x)$.

$W_1 x = W_2(a - x)$

$90x = 60(10 - x)$

$90x = 600 - 60x$

$90x + 60x = 600 - 60x + 60x$

$150x = 600$

$\dfrac{150x}{150} = \dfrac{600}{150}$

$x = 4$ feet

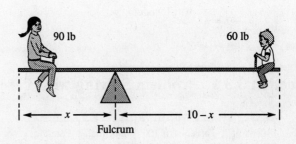

83. (a) Each brick is 8 inches long, so the *n* bricks in each row have a combined length of 8*n* inches. There is $\frac{1}{2}$ inch of mortar between adjoining bricks and there are $n - 1$ mortar joints between the *n* bricks; therefore, the mortar joints in each row have a combined length of $\frac{1}{2}(n - 1)$ inches. Thus, the 8*n* inches of bricks plus the $\frac{1}{2}(n - 1)$ inches of mortar joints equal the 93 inches of the width of the fireplace.

(b) $8n + \frac{1}{2}(n - 1) = 93$

$\qquad 8n + \frac{1}{2}n - \frac{1}{2} = 93$

$\quad 2\left(8n + \frac{1}{2}n - \frac{1}{2}\right) = 2(93)$

$\qquad\quad 16n + n - 1 = 186$

$\qquad\qquad 17n - 1 = 186$

$\qquad 17n - 1 + 1 = 186 + 1$

$\qquad\qquad\quad 17n = 187$

$\qquad\qquad\qquad n = 11$

Thus, there are 11 bricks per row in the fireplace.

85. Use the Distributive Property to remove symbols of grouping. Remove the innermost symbols first and combine like terms. Symbols of grouping preceded by a minus sign can be removed by changing the sign of each term within the symbols.

Examples: $4(x + 7) = 4x - 28$ $\qquad$ $2x - [3 + (x - 1)] = 2x - [3 + x - 1]$

$\qquad\qquad\qquad\qquad\qquad\qquad\qquad\qquad\qquad = 2x - [2 + x]$

$\qquad\qquad\qquad\qquad\qquad\qquad\qquad\qquad\qquad = 2x - 2 - x$

$\qquad\qquad\qquad\qquad\qquad\qquad\qquad\qquad\qquad = x - 2$

87. You could begin by dividing both sides of the equation by 3.

$3(x - 7) = 15$

$\dfrac{3(x - 7)}{3} = \dfrac{15}{3}$

$x - 7 = 5$

$x = 12$

89. Multiplying both sides of the equation by the least common multiple of the denominators clears the equation of fractions.

Section 3.3   Problem Solving with Percents

1.

Percent	Parts out of 100	Decimal	Fraction
40%	40	0.4	$\frac{2}{5}$

(a) 40% means 40 parts out of 100.

(b) *Verbal model:* $\boxed{\text{Decimal}} \cdot \boxed{100\%} = \boxed{\text{Percent}}$

Label: $\qquad$ Decimal $= x$

Equation: $\qquad x(100\%) = 40\%$

$\qquad\qquad\qquad x = \dfrac{40\%}{100\%}$

$\qquad\qquad\qquad x = 0.4$

(c) *Verbal model:* $\boxed{\text{Fraction}} \cdot \boxed{100\%} = \boxed{\text{Percent}}$

Label: $\qquad$ Fraction $= x$

Equation: $\qquad x(100\%) = 40\%$

$\qquad\qquad\qquad x = \dfrac{40\%}{100\%}$

$\qquad\qquad\qquad x = \dfrac{2}{5}$

3.

Percent	Parts out of 100	Decimal	Fraction
7.5%	7.5	0.075	$\frac{3}{40}$

(a) 7.5% means 7.5 parts out of 100.

(b) *Verbal model:* | Decimal | $\cdot$ | 100% | = | Percent |

    Label: Decimal $= x$

    Equation: $x(100\%) = 7.5\%$

$$x = \frac{7.5\%}{100\%}$$

$$x = 0.075$$

(c) *Verbal model:* | Fraction | $\cdot$ | 100% | = | Percent |

    Label: Fraction $= x$

    Equation: $x(100\%) = 7.5\%$

$$x = \frac{7.5\%}{100\%}$$

$$x = \frac{75}{1000}$$

$$x = \frac{3}{40}$$

5.

Percent	Parts out of 100	Decimal	Fraction
63%	63	0.63	$\frac{63}{100}$

(a) 63 parts out of 100 means 63%.

(b) *Verbal model:* | Decimal | $\cdot$ | 100% | = | Percent |

    Label: Decimal $= x$

    Equation: $x(100\%) = 63\%$

$$x = \frac{63\%}{100\%}$$

$$x = 0.63$$

(c) *Verbal model:* | Fraction | $\cdot$ | 100% | = | Percent |

    Label: Fraction $= x$

    Equation: $x(100\%) = 63\%$

$$x = \frac{63\%}{100\%}$$

$$x = \frac{63}{100}$$

7.

Percent	Parts out of 100	Decimal	Fraction
15.5%	15.5	0.155	$\frac{31}{200}$

(a) *Verbal model:* | Decimal | $\cdot$ | 100% | = | Percent |

    Label: Percent $= x$

    Equation: $(0.155)(100\%) = x$

$$15.5\% = x$$

(b) 15.5% means 15.5 parts out of 100.

(c) *Verbal model:* | Fraction | $\cdot$ | 100% | = | Percent |

    Label: Fraction $= x$

    Equation: $x(100\%) = 15.5\%$

$$x = \frac{15.5\%}{100\%}$$

$$x = \frac{155\%}{1000\%} = \frac{31}{200}$$

9.

Percent	Parts out of 100	Decimal	Fraction
60%	60	0.6	$\frac{3}{5}$

(a) *Verbal model:* | Fraction | $\cdot$ | 100% | = | Percent |

    Label: Percent $= x$

    Equation: $\frac{3}{5}(100\%) = x$

$$60\% = x$$

(b) 60% means 60 parts out of 100.

(c) *Verbal model:* | Decimal | $\cdot$ | 100% | = | Percent |

    Label: Decimal $= x$

    Equation: $x(100\%) = 60\%$

$$x = \frac{60\%}{100\%}$$

$$x = 0.6$$

11.

Percent	Parts out of 100	Decimal	Fraction
150%	150	1.5	$\frac{3}{2}$

(a) 150% means 150 parts out of 100.

(b) *Verbal model:* $\boxed{\text{Decimal}} \cdot \boxed{100\%} = \boxed{\text{Percent}}$

   Label: Decimal $= x$

   Equation: $x(100\%) = 150\%$

$$x = \frac{150\%}{100\%}$$

$$x = 0.15$$

(c) *Verbal model:* $\boxed{\text{Fraction}} \cdot \boxed{100\%} = \boxed{\text{Percent}}$

   Label: Fraction $= x$

   Equation: $x(100\%) = 150\%$

$$x = \frac{150\%}{100\%}$$

$$x = \frac{3}{2}$$

13. *Verbal model:* $\boxed{\text{Decimal}} \cdot \boxed{100\%} = \boxed{\text{Percent}}$

   Label: Percent $= x$

   Equation: $0.62(100\%) = x$

$$62\% = x$$

15. *Verbal model:* $\boxed{\text{Decimal}} \cdot \boxed{100\%} = \boxed{\text{Percent}}$

   Label: Decimal $= x$

   Equation: $0.20(100\%) = x$

$$20\% = x$$

17. *Verbal model:* $\boxed{\text{Decimal}} \cdot \boxed{100\%} = \boxed{\text{Percent}}$

   Label: Percent $= x$

   Equation: $0.075(100\%) = x$

$$7.5\% = x$$

19. *Verbal model:* $\boxed{\text{Decimal}} \cdot \boxed{100\%} = \boxed{\text{Percent}}$

   Label: Decimal $= x$

   Equation: $2.5(100\%) = x$

$$250\% = x$$

21. *Verbal model:* $\boxed{\text{Decimal}} \cdot \boxed{100\%} = \boxed{\text{Percent}}$

   Label: Decimal $= x$

   Equation: $x(100\%) = 12.5\%$

$$x = \frac{12.5\%}{100\%}$$

$$x = 0.125$$

23. *Verbal model:* $\boxed{\text{Decimal}} \cdot \boxed{100\%} = \boxed{\text{Percent}}$

   Label: Decimal $= x$

   Equation: $x(100\%) = 125\%$

$$x = \frac{125\%}{100\%}$$

$$x = 1.25$$

25. *Verbal model:* $\boxed{\text{Decimal}} \cdot \boxed{100\%} = \boxed{\text{Percent}}$

   Label: Decimal $= x$

   Equation: $x(100\%) = 250\%$

$$x = \frac{250\%}{100\%}$$

$$x = 2.5$$

27. *Verbal model:* $\boxed{\text{Decimal}} \cdot \boxed{100\%} = \boxed{\text{Percent}}$

Label: Decimal = x

Equation: $x(100\%) = \frac{3}{4}\%$

$$x = \frac{\frac{3}{4}\%}{100\%}$$

$$x = \frac{0.75\%}{100\%}$$

$$x = 0.0075$$

29. *Verbal model:* $\boxed{\text{Fraction}} \cdot \boxed{100\%} = \boxed{\text{Percent}}$

Label: Percent = x

Equation: $\frac{4}{5}(100\%) = x$

$$80\% = x$$

31. *Verbal model:* $\boxed{\text{Fraction}} \cdot \boxed{100\%} = \boxed{\text{Percent}}$

Label: Percent = x

Equation: $\left(\frac{5}{4}\right)(100\%) = x$

$$125\% = x$$

33. *Verbal model:* $\boxed{\text{Fraction}} \cdot \boxed{100\%} = \boxed{\text{Percent}}$

Label: Decimal = x

Equation: $\frac{5}{6}(100\%) = x$

$$83\frac{1}{3}\% = x$$

35. *Verbal model:* $\boxed{\text{Fraction}} \cdot \boxed{100\%} = \boxed{\text{Percent}}$

Label: Percent = x

Equation: $\frac{7}{20}(100\%) = x$

$$35\% = x$$

37. $\frac{3}{8}$ of the figure is shaded.

$\frac{3}{8}(100\%) = 37\frac{1}{2}\%$ or 37.5%

39. $\frac{150}{360}$ of the figure is shaded.

$$\frac{150}{360} = \frac{5}{12}$$

$\frac{5}{12}(100\%) = 41\frac{2}{3}\% \approx 41.67\%$

41. *Verbal model:* $\boxed{\text{What number}} = \boxed{\begin{array}{c}30\% \\ \text{of }150\end{array}}$ $(a = pb)$

Label: a = unknown number

Percent equation: $a = 0.30(150)$

$$a = 45$$

Therefore, 45 is 30% of 150.

43. *Verbal model:* $\boxed{\text{What number}} = \boxed{\begin{array}{c}9.5\% \\ \text{of }816\end{array}}$ $(a = pb)$

Label: a = unknown number

Percent equation: $a = 0.095(816)$

$$a = 77.52$$

Therefore, 77.52 is 9.5% of 816.

45. *Verbal model:* $\boxed{\text{What number}} = \boxed{\begin{array}{c}\frac{3}{4}\% \text{ of} \\ 56\end{array}}$ $(a = pb)$

Label: a = unknown number

Percent equation: $a = 0.0075(56)$ *Note:* $\frac{3}{4}\% = 0.75\% = 0.0075$

$$a = 0.42$$

Therefore, 0.42 is $\frac{3}{4}\%$ of 56.

47. *Verbal model:* $\boxed{\text{What number}} = \boxed{\begin{array}{c}200\% \\ \text{of } 88\end{array}} \quad (a = pb)$

Label: a = unknown number

Percent equation: $a = (2.00)(88)$

$\qquad\qquad\qquad a = 176$

Therefore, 176 is 200% of 88.

49. *Verbal model:* $\boxed{903} = \boxed{\begin{array}{c}43\% \text{ of} \\ \text{what number}\end{array}} \quad (a = pb)$

Label: b = unknown number

Percent equation: $903 = 0.43b$

$\qquad\qquad\qquad \dfrac{903}{0.43} = b$

$\qquad\qquad\qquad 2100 = b$

Therefore, 903 is 43% of 2100.

51. *Verbal model:* $\boxed{275} = \boxed{\begin{array}{c}12\frac{1}{2}\% \text{ of} \\ \text{what number}\end{array}} \quad (a = pb)$

Label: b = unknown number

Percent equation: $275 = 0.125b$

$\qquad\qquad\qquad \dfrac{275}{0.125} = b$

$\qquad\qquad\qquad 2200 = b$

Therefore, 275 is $12\frac{1}{2}$% of 2200.

53. *Verbal model:* $\boxed{594} = \boxed{\begin{array}{c}450\% \text{ of} \\ \text{what number}\end{array}} \quad (a = pb)$

Label: b = unknown number

Percent equation: $594 = 4.50b$

$\qquad\qquad\qquad \dfrac{594}{4.50} = b$

$\qquad\qquad\qquad 132 = b$

Therefore, 594 is 450% of 132.

55. *Verbal model:* $\boxed{2.16} = \boxed{\begin{array}{c}0.6\% \text{ of} \\ \text{what number}\end{array}} \quad (a = pb)$

Label: b = unknown number

Percent equation: $2.16 = 0.006b$

$\qquad\qquad\qquad \dfrac{2.16}{0.006} = b$

$\qquad\qquad\qquad 360 = b$

Therefore, 2.16 is 0.6% of 360.

57. *Verbal model:* $\boxed{576} = \boxed{\begin{array}{c}\text{What percent} \\ \text{of } 800\end{array}} \quad (a = pb)$

Label: p = unknown percent (in decimal form)

Percent equation: $576 = p(800)$

$\qquad\qquad\qquad \dfrac{576}{800} = p$

$\qquad\qquad\qquad 0.72 = p$

Therefore, 576 is 72% of 800.

59. *Verbal model:* $\boxed{45} = \boxed{\begin{array}{c}\text{What percent} \\ \text{of } 360\end{array}} \quad (a = pb)$

Label: p = unknown percent (in decimal form)

Percent equation: $45 = p(360)$

$\qquad\qquad\qquad \dfrac{45}{360} = p$

$\qquad\qquad\qquad 0.125 = p$

Therefore, 45 is 12.5% of 360.

61. *Verbal model:* $\boxed{22} = \boxed{\begin{array}{c}\text{What percent} \\ \text{of } 800\end{array}} \quad (a = pb)$

Label: p = unknown percent (in decimal form)

Percent equation: $22 = p(800)$

$\qquad\qquad\qquad \dfrac{22}{800} = p$

$\qquad\qquad\qquad 0.0275 = p$

Therefore, 22 is 2.75% of 800.

63. *Verbal model:* $\boxed{1000} = \boxed{\begin{array}{c}\text{What percent} \\ \text{of } 200\end{array}} \quad (a = pb)$

Label: p = unknown percent (in decimal form)

Percent equation: $1000 = p(200)$

$\qquad\qquad\qquad \dfrac{1000}{200} = p$

$\qquad\qquad\qquad 5 = p$

Therefore, 1000 is 500% of 200.

65. *Verbal model:* $\boxed{\text{Selling price}} = \boxed{\text{Cost}} + \boxed{\text{Markup}}$

Labels: Selling price = \$49.95

Cost = \$26.97

Markup = x (dollars)

Equation: $49.95 = 26.97 + x$

$49.95 - 26.97 = x$

$22.98 = x$

The markup is \$22.98.

Verbal model: $\boxed{\text{Markup}} = \boxed{\text{Markup rate}} \cdot \boxed{\text{Cost}}$

Labels: Markup = \$22.98

Markup rate = p (percent in decimal form)

Cost = \$26.97

Equation: $22.98 = p(26.97)$

$\dfrac{22.98}{26.97} = p$

$0.852 \approx p$

The markup rate is approximately 85.2%.

69. *Verbal model:* $\boxed{\text{Selling price}} = \boxed{\text{Cost}} + \boxed{\text{Markup}}$

Labels: Selling price = \$125.98

Cost = c

Markup = \$56.69

Markup rate = p (percent in decimal form)

Equation: $125.98 = c + 56.69$

$125.98 - 56.69 = c$

$69.29 = c$

Verbal model: $\boxed{\text{Markup}} = \boxed{\text{Markup rate}} \cdot \boxed{\text{Cost}}$

Equation: $56.69 = p(69.29)$

$\dfrac{56.69}{69.29} = p$

$0.818 \approx p$

The cost is \$69.29 and the markup rate is approximately 81.8%.

67. *Verbal model:* $\boxed{\text{Selling price}} = \boxed{\text{Cost}} + \boxed{\text{Markup}}$

Labels: Selling price = \$74.38

Cost = c (dollars)

Markup rate = 0.815

Markup = $0.815c$

Equation: $74.38 = c + 0.815c$

$74.38 = 1.815c$

$\dfrac{74.38}{1.815} = c$

$40.98 \approx c$

$0.815 \approx 0.815(40.98) \approx 33.40$

The cost is approximately \$40.98 and the markup is \$33.40.

71. *Verbal model:* $\boxed{\text{Selling price}} = \boxed{\text{Cost}} + \boxed{\text{Markup}}$

Labels: Selling price = \$15,900

Cost = c (dollars)

Markup rate = p (percent in decimal form)

Markup = \$2650

Equation: $15,900 = c + 2650$

$15,900 - 2650 = c$

$13,250 = c$

Verbal model: $\boxed{\text{Markup}} = \boxed{\text{Markup rate}} \cdot \boxed{\text{Cost}}$

Equation: $2650 = p(13,250)$

$\dfrac{2650}{13,250} = p$

$0.2 = p$

The cost is \$13,250 and the markup rate is 20%.

73. *Verbal model:* | Selling price | = | Cost | + | Markup |

Labels: Selling price = x (dollars)

Cost = $107.97

Markup rate = 0.852 (percent in decimal form)

Markup = $0.852(107.97) \approx \$91.99$

Equation: $x = 107.97 + 91.99 = 199.96$

The markup is $91.99 and the selling price is $199.96.

75. *Verbal model:* | Sale price | = | List price | − | Discount |

Labels: Sale price = $29.95

List price = $39.95

Discount rate = p (percent in decimal form)

Discount = $p(39.95)$ (dollars)

Equation: $29.95 = 39.95 - p(39.95)$

$$29.95 - 39.95 = -p(39.95)$$

$$-10.00 = -p(39.95)$$

$$\frac{-10.00}{-39.95} = p$$

$$0.2503 \approx p$$

The discount is $39.95 − $29.95 = $10.00; the discount rate is approximately 25.03% (approximately 25%).

77. *Verbal model:* | Sale price | = | List price | − | Discount |

Labels: Sale price = $18.95

List price = x (dollars)

Discount rate = 0.2 (percent in decimal form)

Discount = $0.2x$ (dollars)

Equation: $18.95 = x - 0.2x$ *Note:* $x - 0.2x = (1 - 0.2)x = 0.8x$

$$18.95 = 0.8x$$

$$\frac{18.95}{0.8} = x$$

$$23.69 \approx x$$

The list price is $23.69 and the discount is $23.69 − $18.95 = $4.74.

79. *Verbal model:* | Sale price | = | List price | − | Discount |

Labels: Sale price = x (dollars)

List price = $189.99

Discount rate = p (percent in decimal form)

Discount = $30.00

Equation: $x = 189.99 - 30.00 = 159.99$

Verbal model: | Discount | = | Discount rate | · | List price |

Equation: $30.00 = p(189.99)$

$$\frac{30.00}{189.99} = p$$

$$0.158 \approx p$$

The sale price is $159.99 and the discount rate is approximately 15.8%.

81. *Verbal model:* | Sale price | = | List price | − | Discount |

Labels: Sale price = x (dollars)

List price = $119.96

Discount rate = 0.50 (percent in decimal form)

Discount = $0.50(119.96) \approx 59.98$ (dollars)

Equation: $x = 119.96 - 59.98 = 59.98$

The sale price is $59.98 and the discount is $59.98.

83. *Verbal model:* $\boxed{\text{Sale price}} = \boxed{\text{List price}} - \boxed{\text{Discount}}$

Labels: Sale price = \$695.00

List price = x (dollars)

Discount rate = p (percent in decimal form)

Discount = \$300

Equation: $695.00 = x - 300$

$695.00 + 300 = x$

$995 = x$

Verbal model: $\boxed{\text{Discount}} = \boxed{\text{Discount rate}} \cdot \boxed{\text{List price}}$

Equation: $300 = p(995)$

$\dfrac{300}{995} = p$

$0.302 \approx p$

The list price is \$995.00 and the discount rate is approximately 30.2%.

87. *Verbal model:* $\boxed{\begin{array}{c}\text{Annual retirement} \\ \text{fund contribution}\end{array}} = \boxed{\begin{array}{c}\text{Percent of annual} \\ \text{gross income}\end{array}}$

Labels: Annual retirement fund contribution = x (dollars)

Percent = 0.075   (in decimal form)

Gross income = 45,800   (dollars)

Equation: $x = 0.075(45,800)$

$x = 3435$

Thus, each year you put in \$3435.

89. *Verbal model:* $\boxed{\begin{array}{c}\text{December} \\ \text{snowfall}\end{array}} = \boxed{\begin{array}{c}\text{Percent of} \\ \text{winter snowfall}\end{array}}$ $(a = pb)$

Labels: a = December snowfall = 86   (inches)

p = unknown percent   (in decimal form)

b = winter snowfall = 120   (inches)

Percent equation: $86 = p(120)$

$\dfrac{86}{120} = p$

$0.71\overline{6} = p$

Therefore, $71.\overline{6}\%$ or $71\frac{2}{3}\%$ of the snow fell in December.

85. *Verbal model:* $\boxed{\text{Rent}} = \boxed{\begin{array}{c}\text{Percent of} \\ \text{income}\end{array}}$ $(a = pb)$

Labels: a = rent (dollars)

p = 0.17 (percent in decimal form)

b = income = \$3200

Percent equation: $a = 0.17(3200)$

$a = 544$

Therefore, the monthly rent payment is \$544.

91. *Verbal model:* $\boxed{\begin{array}{c}\text{Unemployed}\\\text{workers}\end{array}} = \boxed{\begin{array}{c}\text{Percent of}\\\text{workers}\end{array}}$ $(a = pb)$

Labels: a = unemployed workers = 72

p = unknown percent   (in decimal form)

b = workers = 1000

Equation: $72 = p(1000)$

$$\frac{72}{1000} = p$$

$$0.072 = p$$

Therefore, 7.2% of the workers were unemployed.

95. *Verbal model:* $\boxed{\begin{array}{c}\text{New}\\\text{Price}\end{array}} = \boxed{\begin{array}{c}\text{Percent of price}\\\text{3 years ago}\end{array}}$ $(a = pb)$

Labels: a = price of new van = 26,850   (dollars)

$p = 1.10$ (percent in decimal form)

b = price of van 3 years ago   (dollars)

Percent equation: $26,850 = 1.10b$

$$\frac{26,850}{1.10} = b$$

$$24,409 \approx b$$

Thus, the price of the van 3 years ago was approximately $24,409.

99. *Verbal model:* $\boxed{\begin{array}{c}\text{Points needed}\\\text{for a } B\end{array}} = \boxed{\begin{array}{c}\text{Percent of}\\\text{total points}\end{array}}$ $(a = pb)$

Labels: a = Points needed for a B = 394 + 6 = 400

$p = 0.80$ (percent in decimal form)

b = total points in course

Equation: $400 = 0.80(b)$

$$\frac{400}{0.80} = b$$

$$500 = b$$

Therefore, there were 500 possible points.

93. *Verbal model:* $\boxed{\begin{array}{c}\text{Original}\\\text{price}\end{array}} - \boxed{\begin{array}{c}\text{Percent of}\\\text{original price}\end{array}} = \boxed{\begin{array}{c}\text{Sale}\\\text{price}\end{array}}$

Labels: Original price = x (dollars)

Percent = 0.20   (in decimal form)

Sale price = 250   (dollars)

Equation: $x - 0.20 = 250$

$$0.80x = 250$$

$$\frac{0.80x}{0.80} = \frac{250}{0.80}$$

$$x = 312.50$$

Thus, the original price of the coat was $312.50.

97. *Verbal model:* $\boxed{\begin{array}{c}\text{Votes}\\\text{cast}\end{array}} = \boxed{\begin{array}{c}\text{Percent of}\\\text{eligible voters}\end{array}}$ $(a = pb)$

Labels: a = votes cast = 6432

$p = 0.63$ (percent in decimal form)

b = eligible voters

Percent equation: $6432 = 0.63b$

$$\frac{6432}{0.63} = b$$

$$10,210 \approx b$$

There are approximately 10,210 voters in the district.

101. *Verbal model:* $\boxed{\begin{array}{c}\text{People in age-group}\\\text{visiting physicians}\end{array}} = \boxed{\begin{array}{c}\text{Percent of all people}\\\text{visiting physicians}\end{array}}$ $(a = pb)$

Labels: a = people in age-group visiting physicians

p = percent in decimal form

b = people visiting physicians = 697,100,000

Equations: 75 years old and older:

$a = 0.111(697,100,000)$

$a = 77,378,100$ or approximately 77.4 million

65-74 years old:

$a = 0.13(697,100,000)$

$a = 90,623,000$ or approximately 90.6 million

46-64 years old:

$a = 0.229(697,100,000)$

$a = 159,635,900$ or approximately 159.6 million

25-44 years old:

$a = 0.261(697,100,000)$

$a = 181,943,100$ or approximately 181.9 million

15-24 years old:

$a = 0.081(697,100,000)$

$a = 56,465,100$ or approximately 56.5 million

Under 15 years old:

$a = 0.189(697,100,000)$

$a = 131,751,900$ or approximately 131.8 million

103. (a) *Verbal model:* $\boxed{\begin{array}{c}\text{Number of female}\\\text{math/computer scientists}\end{array}} = \boxed{\begin{array}{c}\text{Percent of all}\\\text{math/computer scientists}\end{array}}$ $(a = pb)$

Labels: a = female math/computer scientists = 411,600

$p = 0.306$ (percent in decimal form)

b = unknown number of math/computer scientists

Equations: $411,600 = 0.306(b)$

$\dfrac{411,600}{0.306} = b$

$1,345,098 \approx b$

Thus, the total number of mathematical and computer scientists in the U.S. in 1996 was approximately 1,345,098.

(b) *Verbal model:* $\boxed{\begin{array}{c}\text{Number of female}\\\text{chemists in 1983}\end{array}} = \boxed{\begin{array}{c}\text{Percent of all}\\\text{chemists in 1983}\end{array}}$ $(a = pb)$

Labels: a = female chemists = 22,800

$p = 0.233$ (percent in decimal form)

b = unknown number of chemists

Equations: $22,800 = 0.233(b)$

$\dfrac{22,800}{0.233} = b$

$97,854 \approx b$

Thus, the total number of chemists in the U.S. in 1983 was approximately 97,854.

(c) The number of women in biology increased from 1983 to 1996, but the percentage of women in biology decreased during the same period. This is because the number of men in biology increased *faster* than the number of women during this period.

105. (a) *Verbal model:* $\boxed{\begin{array}{c}\text{Cost of}\\\text{phone}\end{array}} + 3 \cdot \boxed{\begin{array}{c}\text{Monthly}\\\text{service cost}\end{array}} = \boxed{\begin{array}{c}\text{Total}\\\text{cost}\end{array}}$

Labels: Cost of phone $= x$ (dollars)

Monthly service cost $= 19.50$ (dollars)

Total cost $= 99$ (dollars)

Equation: $x + 3(19.50) = 99$

$x + 58.50 = 99$

$x = 40.50$

The cost of the phone in the first package is \$40.50.

(b) *Verbal model:* $\boxed{\begin{array}{c}\text{Cost of}\\\text{phone}\end{array}} + \boxed{\begin{array}{c}\text{Monthly}\\\text{service cost}\end{array}} = \boxed{\begin{array}{c}\text{Total}\\\text{cost}\end{array}}$

Labels: Cost of phone $= x$ (dollars)

Monthly service cost $= 24$ (dollars)

Total cost $= 80$

Equation: $x + 24 = 80$

$x = 56$

The cost of the phone in the second package is \$56. Thus, the phone in the second package costs more.

(c) *Verbal model:* $\boxed{\begin{array}{c}\text{Cost of}\\\text{phone}\end{array}} = \boxed{\begin{array}{c}\text{Percent of}\\\text{cost of package}\end{array}}$ $(a = pb)$

Labels: $a =$ Cost of phone $= 40.50$ (dollars)

$p =$ Unknown percent   (in decimal form)

$b =$ Cost of package $= 99$ (dollars)

Equation: $40.50 = p(99)$

$\dfrac{40.50}{99} = p$

$0.409 \approx p$

Approximately, 40.9% of the purchase price of the first package goes toward the price of the phone.

(d) *Verbal model:* $\boxed{\begin{array}{c}\text{Cost of}\\\text{phone}\end{array}} = \boxed{\begin{array}{c}\text{Percent of}\\\text{cost of package}\end{array}}$ $(a = pb)$

Labels: $a =$ Cost of phone $= 56$ (dollars)

$p =$ Unknown percent   (in decimal form)

$b =$ Cost of package $= 80$ (dollars)

Equation: $56 = p(80)$

$\dfrac{56}{80} = p$

$0.70 = p$

Thus, 70% of the purchase price of the second package goes toward the price of the phone.

—CONTINUED—

105. —CONTINUED—

(e) *Verbal model:* | Total cost | = | Price of package | + | Tax on package price |

Labels: Total cost $= x$ (dollars)

Price of package $= 99$ (dollars)

Tax on package price $= 5\%(99)$ (dollars)

Equation: $x = 99 + 0.05(99)$

$x = 99 + 4.95$

$x = 103.95$

The total cost of purchasing the first phone package is \$103.95.

(f) *Verbal model:* | Monthly bill | = | Service charge | + | Number of minutes | · | Per minute usage rate |

Labels: Monthly bill $= 92.46$ (dollars)

Service charge $= 19.50$ (dollars)

Number of minutes $= 3.2(60) = 192$ (minutes)

Per minute usage rate $= x$ (dollars per minute)

Equation: $92.46 = 19.50 + 192x$

$72.96 = 192x$

$\dfrac{72.96}{196} = x$

$0.38 = x$

The usage rate is \$0.38 per minute.

107. A rate is a fixed ratio.

109. To change a decimal to a percent, multiply the decimal by 100 and affix the percent sign.

111. Yes, any positive decimal can be written as a percent. Multiply the decimal by 100 and affix the percent sign.

Section 3.4   Ratios and Proportions

1. 36 to 9 $= \frac{36}{9} = \frac{4}{1}$

3. 27 to 54 $= \frac{27}{54} = \frac{1}{2}$

5. 14 : 21 $= \frac{14}{21} = \frac{2}{3}$

7. 144 : 16 $= \frac{144}{16} = \frac{9}{1}$

9. $\dfrac{36 \text{ inches}}{24 \text{ inches}} = \dfrac{36}{24} = \dfrac{3}{2}$

11. $\dfrac{40 \text{ dollars}}{60 \text{ dollars}} = \dfrac{40}{60} = \dfrac{2}{3}$

13. $\dfrac{1 \text{ quart}}{1 \text{ gallon}} = \dfrac{1 \text{ quart}}{4 \text{ quarts}} = \dfrac{1}{4}$

15. $\dfrac{7 \text{ nickels}}{3 \text{ quarters}} = \dfrac{7 \text{ nickels}}{15 \text{ nickels}} = \dfrac{7}{15}$

17. $\dfrac{3 \text{ hours}}{90 \text{ minutes}} = \dfrac{180 \text{ minutes}}{90 \text{ minutes}} = \dfrac{180}{90} = \dfrac{2}{1}$

Note: This problem could also be done using cents as the common ratio.

$\dfrac{7 \text{ nickels}}{3 \text{ quarters}} = \dfrac{35 \text{ cents}}{75 \text{ cents}} = \dfrac{35}{75} = \dfrac{7}{15}$

19. $\dfrac{75 \text{ centimeters}}{2 \text{ meters}} = \dfrac{75 \text{ centimeters}}{200 \text{ centimeters}}$

$= \dfrac{75}{200}$

$= \dfrac{3}{8}$

21. $\dfrac{60 \text{ milliliters}}{1 \text{ liters}} = \dfrac{60 \text{ milliliters}}{1000 \text{ milliliters}}$

$= \dfrac{60}{1000}$

$= \dfrac{3}{50}$

23. $\dfrac{90 \text{ minutes}}{2 \text{ hours}} = \dfrac{90 \text{ minutes}}{120 \text{ minutes}}$

$= \dfrac{90}{120}$

$= \dfrac{3}{4}$

25. $\dfrac{3000 \text{ pounds}}{5 \text{ tons}} = \dfrac{3000 \text{ pounds}}{10,000 \text{ pounds}} = \dfrac{3}{10}$

27. *Verbal model:* $\boxed{\begin{array}{c}\text{Unit}\\\text{price}\end{array}} = \boxed{\dfrac{\text{Total price}}{\text{Total units}}}$

Unit price: $\dfrac{79 \text{ cents}}{20 \text{ oz}} = \dfrac{\$0.79}{20}$

$= \$0.0395 \text{ per ounce}$

29. *Verbal model:* $\boxed{\begin{array}{c}\text{Unit}\\\text{price}\end{array}} = \boxed{\dfrac{\text{Total price}}{\text{Total units}}}$

Total units: $\quad 1 \text{ pound} + 4 \text{ ounces} = 1(16 \text{ oz}) + 4\text{oz}$

$= 16 \text{ oz} + 4 \text{ oz}$

$= 20 \text{ oz}$

Unit price: $\dfrac{\$1.29}{20 \text{ oz}} = \$0.0645 \text{ per ounce}$

31. The unit price for the can is

$\text{Unit price} = \dfrac{\text{Total price}}{\text{Total units}} = \dfrac{\$1.19}{27\frac{3}{4} \text{ oz}} = \dfrac{\$1.19}{27.75 \text{ oz}}$

$\approx \$0.0429 \text{ per ounce.}$

The unit price for the jar is

$\text{Unit price} = \dfrac{\text{Total price}}{\text{Total units}} = \dfrac{\$1.45}{32 \text{ oz}} \approx \$0.0453 \text{ per ounce.}$

The can has the smaller unit price.

33. The unit price for the smaller package is

$\text{Unit price} = \dfrac{\text{Total price}}{\text{Total units}} = \dfrac{\$0.59}{10 \text{ oz}} = \$0.059 \text{ per ounce.}$

The unit price for the larger package is

$\text{Unit price} = \dfrac{\text{Total price}}{\text{Total units}} = \dfrac{\$0.89}{16 \text{ oz}} \approx \$0.056 \text{ per ounce.}$

The 16-ounce package has the smaller unit price.

35. The unit price for the 2-liter bottle is

$\text{Unit price} = \dfrac{\text{Total price}}{\text{Total units}} = \dfrac{\$1.09}{67.6 \text{ oz}}$

$\approx \$0.016 \text{ per ounce.}$

The unit price for the cans is

$\text{Unit price} = \dfrac{\text{Total price}}{\text{Total units}} = \dfrac{\$1.69}{6(12) \text{ oz}} = \dfrac{\$1.69}{72 \text{ oz}}$

$\approx \$0.023 \text{ per ounce.}$

The 2-liter bottle has the smaller unit price.

37. $\dfrac{5}{3} = \dfrac{20}{y}$

$5y = 60 \quad \text{(Cross-multiply)}$

$y = \dfrac{60}{5}$

$y = 12$

39. $\dfrac{4}{t} = \dfrac{2}{25}$

$100 = 2t \quad \text{(Cross-multiply)}$

$\dfrac{100}{2} = t$

$50 = t$

41. $\dfrac{5}{x} = \dfrac{3}{2}$

$10 = 3x \quad \text{(Cross-multiply)}$

$\dfrac{10}{3} = x$

43. $\dfrac{8}{3} = \dfrac{t}{6}$

$6\left(\dfrac{8}{3}\right) = 6\left(\dfrac{t}{6}\right)$

$16 = t$

45. $\dfrac{0.5}{0.8} = \dfrac{n}{0.3}$

$0.15 = 0.8n$ (Cross-multiply)

$\dfrac{0.15}{0.8} = n$

$\dfrac{15}{80} = n$

$\dfrac{3}{16} = n$

The answer could also be written as $n = 0.1875$.

47. $\dfrac{x+1}{5} = \dfrac{3}{10}$

$10(x+1) = 15$ (Cross-multiply)

$10x + 10 = 15$

$10x + 10 - 10 = 15 - 10$

$10x = 5$

$\dfrac{10x}{10} = \dfrac{5}{10}$

$x = \dfrac{1}{2}$

49. $\dfrac{x+6}{3} = \dfrac{x-5}{2}$

$2(x+6) = 3(x-5)$ (Cross-multiply)

$2x + 12 = 3x - 15$

$2x - 3x + 12 = 3x - 3x - 15$

$-x + 12 = -15$

$-x + 12 - 12 = -15 - 12$

$-x = -27$

$(-1)(-x) = (-1)(-27)$

$x = 27$

51. $\dfrac{x+2}{8} = \dfrac{x-1}{3}$

$3(x+2) = 8(x-1)$

$3x + 6 = 8x - 8$

$3x - 8x + 6 = 8x - 8x - 8$

$-5x + 6 = -8$

$-5x + 6 - 6 = -8 - 6$

$-5x = -14$

$\dfrac{-5x}{-5} = \dfrac{-14}{-5}$

$x = \dfrac{14}{5}$

53. $\dfrac{6 \text{ hours}}{3 \text{ hours}} = \dfrac{6}{3} = \dfrac{2}{1}$

55. $\dfrac{78 \text{ dollars}}{6.50 \text{ dollars}} = \dfrac{78}{6.50} = \dfrac{12}{1}$

57. $\dfrac{345 \text{ cubic centimeters}}{17.25 \text{ cubic centimeters}} = \dfrac{345}{17.25}$

$= \dfrac{20}{1}$

59. $\dfrac{45 \text{ teeth}}{30 \text{ teeth}} = \dfrac{45}{30} = \dfrac{3}{2}$

61. $\dfrac{\pi(10)^2 \text{ square inches}}{\pi(7)^2 \text{ square inches}} = \dfrac{100\pi}{49\pi}$

$= \dfrac{100}{49}$

63. *Verbal model:* $\boxed{\dfrac{\text{Gallons for shorter trip}}{\text{Miles for shorter trip}}} = \boxed{\dfrac{\text{Gallons for longer trip}}{\text{Miles for longer trip}}}$

Labels: Gallons for shorter trip $= x$ (gallons)

Miles for shorter trip $= 400$ (miles)

Gallons for longer trip $= 20$ (gallons)

Miles for longer trip $= 500$ (miles)

Proportion: $\dfrac{x}{400} = \dfrac{20}{500}$

$$400\left(\dfrac{x}{400}\right) = 400\left(\dfrac{20}{500}\right)$$

$$x = \dfrac{8000}{500}$$

$$x = 16$$

On a trip of 400 miles, 16 gallons of gas would be used.

65. *Verbal model:* $\boxed{\dfrac{\text{Blocks for smaller wall}}{\text{Length of smaller wall}}} = \boxed{\dfrac{\text{Blocks for larger wall}}{\text{Length of larger wall}}}$

Labels: Blocks for smaller wall $= 100$

Length of smaller wall $= 16$ (feet)

Blocks for larger wall $= x$

Length of larger wall $= 40$ (feet)

Proportion: $\dfrac{100}{16} = \dfrac{x}{40}$

$$40\left(\dfrac{100}{16}\right) = 40\left(\dfrac{x}{40}\right)$$

$$250 = x$$

Thus, 250 blocks are needed to build a 40-foot wall.

67. *Verbal model:* $\boxed{\dfrac{\text{Tax on first property}}{\text{Value of first property}}} = \boxed{\dfrac{\text{Tax on second property}}{\text{Value of second property}}}$

Labels: Tax on first property $= \$825$

Value of first property $= \$65{,}000$

Tax on second property $= x$ (dollars)

Value of second property $= \$90{,}000$

Proportion: $\dfrac{825}{65{,}000} = \dfrac{x}{90{,}000}$

$$90{,}000\left(\dfrac{825}{65{,}000}\right) = 90{,}000\left(\dfrac{x}{90{,}000}\right)$$

$$\$1142 \approx x$$

The taxes on the property with an assessed value of $90,00 are approximately $1142.

69. *Verbal model:*
$$\boxed{\frac{\text{Poll voters for candidate}}{\text{Voters in poll}}} = \boxed{\frac{\text{Election voters for candidate}}{\text{Voters in election}}}$$

Labels:
Poll voters for candidate = 624

Voters in poll = 1100

Election voters for candidate = x

Voters in election = 40,000

Proportion:
$$\frac{624}{1100} = \frac{x}{40,000}$$

$$40,000\left(\frac{624}{1100}\right) = 40,00\left(\frac{x}{40,000}\right)$$

$$22,691 \approx x$$

The candidate can expect 22,691 votes.

71. *Verbal model:*
$$\boxed{\frac{\text{Gallons in first tank}}{\text{Gallons in second tank}}} = \boxed{\frac{\text{Time to fill first tank}}{\text{Time to fill second tank}}}$$

Labels:
Gallons in first tank = 750

Gallons in second tank = 1000

Time to fill first tank = 35   (minutes)

Time to fill second tank = x (minutes)

Proportion:
$$\frac{750}{1000} = \frac{35}{x}$$

$$750x = 35,000$$

$$x = \frac{35,000}{750}$$

$$x = 46\tfrac{2}{3}$$

It would take $46\tfrac{2}{3}$ minutes to pump fill the second tank.

73. *Verbal model:*
$$\boxed{\text{Gas to oil ratio}} = \boxed{\frac{\text{Pints of gas in mixture}}{\text{Pints of oil in mixture}}}$$

Labels:
Gas to oil ratio = $\dfrac{40}{1}$

Pints of gas in mixture = x

Pints of oil in mixture = $\dfrac{1}{2}$

Proportion:
$$\frac{40}{1} = \frac{x}{\frac{1}{2}}$$

$$40 = x\left(\frac{2}{1}\right) \qquad Note: \; x \div \frac{1}{2} = x\left(\frac{2}{1}\right).$$

$$40 = 2x$$

$$\frac{40}{2} = \frac{2x}{2}$$

$$20 = x$$

Thus, 20 pints of gasoline $\left(\text{or } 2\tfrac{1}{2} \text{ gallons}\right)$ are required for the mixture.

75. *Verbal model:* $\boxed{\dfrac{\text{Inches on map scale}}{\text{Miles represented on scale}}} = \boxed{\dfrac{\text{Inches between cities on map}}{\text{Miles between cities}}}$

Labels: Inches on map scale $\approx \dfrac{11}{16}$

Miles represented on scale $= 100$

Inches between cities on map $\approx 1\dfrac{11}{16}$

Miles between cities $= x$

Proportion: $\dfrac{\frac{11}{16}}{100} = \dfrac{1\frac{11}{16}}{x}$

$\dfrac{0.6875}{100} = \dfrac{1.6875}{x}$

$0.6875x = 1.6875$ (Cross-multiply)

$x = \dfrac{1.6875}{0.6875}$

$x \approx 245$

The approximate distance between the cities is 245 miles.

77. Corresponding sides of similar triangles are proportional.

$\dfrac{1}{2} = \dfrac{x}{5}$

$5 = 2x$ (Cross-multiply)

$\dfrac{5}{2} = \dfrac{2x}{2}$

$\dfrac{5}{2} = x$

Note: There are several ways to set up this proportion. It could also be written as $\dfrac{2}{1} = \dfrac{5}{x}$, $\dfrac{2}{5} = \dfrac{1}{x}$, or $\dfrac{5}{2} = \dfrac{x}{1}$.

79. *Verbal model:* $\boxed{\dfrac{\text{Height of smaller triangle}}{\text{Base of smaller triangle}}} = \boxed{\dfrac{\text{Height of larger triangle}}{\text{Base of larger triangle}}}$

Labels: Height of smaller triangle (person's height) $= 6$ (feet)

Base of smaller triangle (length of shadow) $= x$ (feet)

Height of larger triangle (height of streetlight) $= 15$ (feet)

Base of larger triangle $= 10 + x$ (feet)

Proportion: $\dfrac{6}{x} = \dfrac{15}{10 + x}$

$6(10 + x) = 15x$ (Cross-multiply)

$60 + 6x = 15x$

$60 + 6x - 6x = 15x - 6x$

$60 = 9x$

$\dfrac{60}{9} = \dfrac{9x}{9}$

$\dfrac{20}{3} = x$

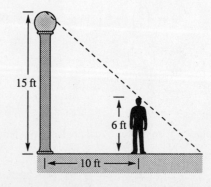

The length of the shadow is $6\frac{2}{3}$ feet (or 6 feet, 8 inches).

81. *Verbal model:* $\boxed{\dfrac{\text{New length}}{\text{Old length}}} = \boxed{\dfrac{\text{New percent}}{\text{Old percent}}}$

   Labels: New length = 2 (inches)
          Old length = 10 (inches)
          New percent = x
          Old percent = 100

   Proportion: $\dfrac{2}{10} = \dfrac{x}{100}$

$$100\left(\dfrac{2}{10}\right) = 100\left(\dfrac{x}{100}\right)$$

$$20 = x$$

The new photo needs to be 20% of the size of the original. (In other words, the original photo needs to be reduced by 80%.)

83. *Verbal model:* $\boxed{\dfrac{\text{Price in 1988}}{\text{Price in 1978}}} = \boxed{\dfrac{\text{Index in 1988}}{\text{Index in 1978}}}$

   Labels: Price in 1988 = x (dollars)
          Price in 1978 = 2875 (dollars)
          Index in 1988 = 118.3
          Index in 1978 = 65.2

   Proportion: $\dfrac{x}{2875} = \dfrac{118.3}{65.2}$

$$2875\left(\dfrac{x}{2875}\right) = 2875\left(\dfrac{118.3}{65.2}\right)$$

$$x \approx 5216$$

The 1988 price of the lawn tractor was approximately $5216.

85. *Verbal model:* $\boxed{\dfrac{\text{Price in 1960}}{\text{Price in 1996}}} = \boxed{\dfrac{\text{Index in 1960}}{\text{Index in 1996}}}$

   Labels: Price in 1960 = x (dollars)
          Price in 1996 = 2.75 (dollars)
          Index in 1960 = 29.6
          Index in 1996 = 156.9

   Proportion: $\dfrac{x}{2.75} = \dfrac{29.6}{156.9}$

$$2.75\left(\dfrac{x}{2.75}\right) = 2.75\left(\dfrac{29.6}{156.9}\right)$$

$$x \approx 0.52$$

The 1960 price of the gallon of milk was approximately $0.52.

87. *Verbal model:* $\boxed{\dfrac{\text{Cost of calls in 5th month}}{\text{Cost of calls in 6th month}}} = \boxed{\dfrac{\text{Time of calls in 5th month}}{\text{Time of calls in 6th month}}}$

   Labels: Cost of calls in 5th month = 47.50 (dollars)
          Cost of calls in 6th month = x (dollars)
          Time of calls in 5th month = 125 (minutes)
          Time of calls in 6th month = 150 (minutes)

   Proportion: $\dfrac{47.50}{x} = \dfrac{125}{150}$

$$47.50(150) = 125x$$

$$7125 = 125x$$

$$\dfrac{7125}{125} = x$$

$$57 = x$$

The charge for 150 minutes of use is $57.

89. No, the ratio of men to women is not enough information to tell you the total number of people in the class. To determine the total number of people in the class, you also need to know the number of men or the number of women in the class.

91. A proportion is a statement that two ratios are equal.

Mid-Chapter Quiz for Chapter 3

1.
$$120 - 3y = 0$$
$$120 - 120 - 3y = 0 - 120$$
$$-3y = -120$$
$$\frac{-3y}{-3} = \frac{-120}{-3}$$
$$y = 40$$

2.
$$10(y - 8) = 0$$
$$10y - 80 = 0$$
$$10y - 80 + 80 = 0 + 80$$
$$10y = 80$$
$$\frac{10y}{10} = \frac{80}{10}$$
$$y = 8$$

3.
$$3x + 1 = x + 20$$
$$3x - x + 1 = x - x + 20$$
$$2x + 1 = 20$$
$$2x + 1 - 1 = 20 - 1$$
$$2x = 19$$
$$\frac{2x}{2} = \frac{19}{2}$$
$$x = \frac{19}{2}$$

4.
$$6x + 8 = 8 - 2x$$
$$6x + 2x + 8 = 8 - 2x + 2x$$
$$8x + 8 = 8$$
$$8x + 8 - 8 = 8 - 8$$
$$8x = 0$$
$$\frac{8x}{8} = \frac{0}{8}$$
$$x = 0$$

5.
$$-10x + \frac{2}{3} = \frac{7}{3} - 5x$$
$$3\left(-10x + \frac{2}{3}\right) = 3\left(\frac{7}{3} - 5x\right)$$
$$-30x + 2 = 7 - 15x$$
$$-30x + 15x + 2 = 7 - 15x + 15x$$
$$-15x + 2 = 7$$
$$-15x + 2 - 2 = 7 - 2$$
$$-15x = 5$$
$$\frac{-15x}{15} = \frac{5}{-15}$$
$$x = -\frac{1}{3}$$

6.
$$\frac{x}{5} + \frac{x}{8} = 1$$
$$40\left(\frac{x}{5} + \frac{x}{8}\right) = 40(1)$$
$$40\left(\frac{x}{5}\right) + 40\left(\frac{x}{8}\right) = 40$$
$$8x + 5x = 40$$
$$13x = 40$$
$$\frac{13x}{13} = \frac{40}{13}$$
$$x = \frac{40}{13}$$

7.
$$\frac{9 + x}{3} = 15$$
$$3\left(\frac{9 + x}{3}\right) = 3(15)$$
$$9 + x = 45$$
$$9 - 9 + x = 45 - 9$$
$$x = 36$$

8.
$$4 - 0.3(1 - x) = 7$$
$$4 - 0.3 + 0.3x = 7$$
$$3.7 + 0.3x = 7$$
$$10(3.7 + 0.3x) = 10(7)$$
$$37 + 3x = 70$$
$$37 - 37 + 3x = 70 - 37$$
$$3x = 33$$
$$\frac{3x}{3} = \frac{33}{3}$$
$$x = 11$$

9.
$$\frac{x + 3}{6} = \frac{4}{3}$$
$$3(x + 3) = 24$$
$$3x + 9 = 24$$
$$3x + 9 - 9 = 24 - 9$$
$$3x = 15$$
$$x = \frac{15}{3}$$
$$x = 5$$

10.
$$\frac{x + 7}{5} = \frac{x + 9}{7}$$
$$7(x + 7) = 5(x + 9)$$
$$7x + 49 = 5x + 45$$
$$7x - 5x + 49 = 5x - 5x + 45$$
$$2x + 49 = 45$$
$$2x + 49 - 49 = 45 - 49$$
$$2x = -4$$
$$x = -2$$

11.
$$32.86 - 10.5 = 11.25$$
$$32.86 - 32.86 - 10.5x = 11.25 - 32.86$$
$$-10.5x = -21.61$$
$$\frac{-10.5x}{-10.5} = \frac{-21.61}{-10.5}$$
$$x \approx 2.06$$

Comment: Replace the variable x by the number 2.06. Simplify the left-hand side of the equation, and check whether it is *approximately* equal to 11.25, the number on the right-hand side. (The two numbers will not be *equal* because of the rounding.)

12.
$$\frac{x}{5.45} + 3.2 = 12.6$$
$$\frac{x}{5.45} + 3.2 - 3.2 = 12.6 - 3.2$$
$$\frac{x}{5.45} = 9.4$$
$$5.45\left(\frac{x}{5.45}\right) = 5.45(9.4)$$
$$x = 51.23$$

Comment: Replace the variable x by the number 51.23. Simplify the left-hand side of the equation, and check whether it is equal to 12.6, the number on the right-hand side.

13. *Verbal model:* | What number | $=$ | 62% of 25 | $(a = pb)$

Label: Unknown number $= a$

Equation: $a = 0.62(25)$

$$a = 15.5$$

Therefore, 62% of 25 is 15.5.

14. *Verbal model:* | What number | $=$ | $\frac{1}{2}$% of 8400 | $(a = pb)$

Label: $a =$ unknown number

Equation: $a = 0.005(8400)$

$$a = 42$$

Therefore, 42 is $\frac{1}{2}$%, or 0.5%, of 8400.

15. *Verbal model:* | 300 | $=$ | What percent of 150 | $(a = pb)$

Label: $p =$ unknown percent

Equation: $300 = p(150)$

$$\frac{300}{150} = p$$
$$2 = p$$
$$200\% = p$$

Therefore, 300 is 200% of 150.

16. *Verbal model:* | 145.6 | $=$ | 32% of what number | $(a = pb)$

Label: $b =$ unknown number

Equation: $145.6 = 0.32(b)$

$$\frac{145.6}{0.32} = b$$
$$455 = b$$

Therefore, 145.6 is 32% of 455.

17. *Verbal model:* 2| Length | $+$ 2| Width | $=$ | Perimeter

Labels: Width $= w$ (meters)

Length $= \frac{3}{2}w$ (meters)

Perimeter $= 60$ (meters)

Equation: $2\left(\frac{3}{2}\right)w + 2w = 60$

$$3w + 2w = 60$$
$$5w = 60$$
$$w = 12 \text{ and } \frac{3}{2}w = 18$$

Therefore, the length of the rectangle is 18 meters and the width is 12 meters.

18. *Verbal model:*

Hourly pay at job 1	$\cdot$	Hours at job 1	$+$	Hourly pay at job 2	$\cdot$	Hours at job 2	$=$	Weekly pay

Labels: Hourly pay at job 1 = 7.50   (dollars per hour)

Hours at job 1 = 40   (hours)

Hourly pay at job 2 = 6   (dollars per hour)

Hours at job 2 = x (hours)

Weekly pay = 360   (dollars)

Equation: $7.50(40) + 6(x) = 360$

$$300 + 6x = 360$$

$$6x = 60$$

$$x = 10$$

Therefore, you must work 10 hours per week at the second job.

19. *Verbal model:*

First area	$+$	Second area	$+$	Third area	$=$	Total area

Labels: First area = x (square meters)

Second area = $2x$ (square meters)

Third area = $2(2x) = 4x$ (square meters)

Total area = 42   (square meters)

Equation: $x + 2x + 4x = 42$

$$7x = 42$$

$$x = 6 \quad \text{and } 2x = 12, 4x = 24$$

Therefore, the area of the first subregion is 6 square meters, the area of the second subregion is 12 square meters, and the area of the third subregion is 24 square meters.

20. *Verbal model:* $\frac{1}{3}\Bigg($

First score	$+$	Second score	$+$	Third score

$\Bigg) = 0.90(100)$

Labels: First score = 84   (points)

Second score = 93   (points)

Third score = x (points)

Equation: $\frac{1}{3}(84 + 93 + x) = 0.90(100)$

$$\frac{1}{3}(84) + \frac{1}{3}(93) + \frac{1}{3}x = 90$$

$$28 + 31 + \frac{1}{3}x = 90$$

$$59 + \frac{1}{3}x = 90$$

$$\frac{1}{3}x = 31$$

$$3\left(\frac{1}{3}x\right) = 3(31)$$

$$x = 93$$

Therefore, you must score 93 on the third test to earn a 90% average.

21. *Verbal model:* | Current price | = | 108% of price two years ago | $(a = pb)$

Labels: Current price = 535   (dollars)

Price two years ago = b (dollars)

Equation: $535 = 1.08(b)$

$$\frac{535}{1.08} = b$$

$$495.37 \approx b$$

Therefore, two years ago the price was approximately $495.37.

22. *Verbal model:* | Religious giving | = | What percent of charitable giving | $(a = pb)$

Labels: Religious giving = 63.5   (billions of dollars)

Unknown percent = p (in decimal form)

Charitable giving = 63.5 + 17.9 + 20.6 + 11.7 + 12.6 + 7.6 + 10

$\qquad\qquad\qquad = 143.9$ (billions of dollars)

Equation: $63.5 = p(143.9)$

$$\frac{63.5}{143.9} = p$$

$$0.441 \approx p$$

Approximately 44.1% of the charitable giving in 1996 went to religious organizations.

23. $\dfrac{\text{Area of large pizza}}{\text{Area of small pizza}} = \dfrac{\pi(15)^2 \text{ square inches}}{\pi(8)^2 \text{ square inches}}$

$$= \frac{225 \,\pi}{64 \,\pi}$$

$$= \frac{225}{64}$$

Section 3.5   Geometric and Scientific Applications

1. $A = \dfrac{1}{2}bh$

$2A = 2 \cdot \dfrac{1}{2}bh$

$2A = bh$

$\dfrac{2A}{b} = \dfrac{bh}{b}$

$\dfrac{2A}{b} = h$

3. $E = IR$

$\dfrac{E}{I} = \dfrac{IR}{I}$

$\dfrac{E}{I} = R$

5. $V = lwh$

$\dfrac{V}{wh} = \dfrac{lwh}{wh}$

$\dfrac{V}{wh} = l$

7. $A = P + Prt$

$A - P = P - P + Prt$

$A - P = Prt$

$\dfrac{A - P}{Pt} = \dfrac{Prt}{Pt}$

$\dfrac{A - P}{Pt} = r$

9. $S = C + RC$

$S = C(1 + R)$

$\dfrac{S}{1 + R} = \dfrac{C(1 + R)}{1 + R}$

$\dfrac{S}{1 + R} = C$

11. $A = \dfrac{1}{2}(a + b)h$

$2A = 2 \cdot \dfrac{1}{2}(a + b)h$

$2A = (a + b)h$

$2A = ah + bh$

$2A - ah = bh$

$\dfrac{2A - ah}{h} = \dfrac{bh}{h}$

$\dfrac{2A - ah}{h} = b$

13. $V = \dfrac{1}{3}\pi h^2(3r - h)$

$3V = \pi h^2(3r - h)$

$3V = 3\pi h^2 r - \pi h^3$

$3V + \pi h^3 = 3\pi h^2 r$

$\dfrac{3V + \pi h^3}{3\pi h^2} = \dfrac{3\pi h^2 r}{3\pi h^2}$

$\dfrac{3V + \pi h^3}{3\pi h^2} = r$

Note: This answer could also be written as

$r = \dfrac{V}{\pi h^2} + \dfrac{h}{3}.$

15. $h = v_0 t + \dfrac{1}{2}at^2$

$h - v_0 t = \dfrac{1}{2}at^2$

$2(h - v_0 t) = 2 \cdot \dfrac{1}{2}at^2$

$2(h - v_0 t) = at^2$

$\dfrac{2(h - v_0 t)}{t^2} = a$ or $a = \dfrac{2h - 2v_0 t}{t^2}$

17. $V = \pi r^2 h$

$V = \pi (5)^2 4$

$V = \pi (25) 4$

$V = 100\pi$ cubic meters

or approximately 314.2 cubic meters

19. *Verbal model:* $\boxed{\text{Distance}} = \boxed{\text{Rate}} \cdot \boxed{\text{Time}}$

Labels: Distance $= d$ (miles)

Rate $= 55$ (miles per hour)

Time $= 3$ (hours)

Equation: $d = 55(3)$

$d = 165$

The distance is 165 miles.

21. *Verbal model:* $\boxed{\text{Distance}} = \boxed{\text{Rate}} \cdot \boxed{\text{Time}}$

Labels: Distance $= 500$ (kilometers)

Rate $= 90$ (kilometers per hour)

Time $= t$ (hours)

Equation: $500 = 90t$

$\dfrac{500}{90} = t$

$5.6 \approx t$

The time is approximately 5.6 hours (or five hours and approximately 36 minutes).

23. *Verbal model:* $\boxed{\text{Distance}} = \boxed{\text{Rate}} \cdot \boxed{\text{Time}}$

Labels: Distance $= 5280$ (feet)

Rate $= r$ (feet per second)

Time $= \dfrac{5}{2}$ (seconds)

Equation: $5280 = r\left(\dfrac{5}{2}\right)$

$5280(2) = r\left(\dfrac{5}{2}\right)(2)$

$10{,}560 = 5r$

$\dfrac{10{,}560}{5} = r$

$2112 = r$

The rate is 2112 feet per second.

25. The living room is a square. The area of a square is given by the formula $A = s^2$, where s is the side of the square. To find the area of the living room, you need to find the side of the living room.

Verbal model: | Side of living room | $=$ | Side of bathroom | $+$ | Side of kitchen |

Labels: Side of living room $= s$ (feet)

                   Side of bathroom $= b$ (feet)

                   Side of kitchen $= k$ (feet)

The perimeter of the bathroom is 32 feet.

Common formula: $P = 4b$

Label: $P = 32$ (feet)

Equation: $32 = 4b$

                    $\frac{32}{4} = b$

                    $8 = b$

The side of the bathroom is 8 feet.

The perimeter of the kitchen is 80 feet.

Common formula: $P = 4k$

Label: $P = 80$ (feet)

Equation: $80 = 4k$

                    $\frac{80}{4} = k$

                    $20 = k$

The side of the kitchen is 20 feet.

Equation: $s = b + k$

                    $s = 8 + 20$

                    $s = 28$

The side of the living room is 28 feet. Therefore, the area of the living room $A = s^2 = 28^2$.

   $A = (28)(28) = 784$

The living room has an area of 784 square feet.

27. *Common formula:* $A = \frac{1}{2}bh$

   Labels: $A = 48$ (square meters)

              $b =$ base   (meters)

              $h = 12$ (meters)

   Equation: $A = \frac{1}{2}bh$

                 $48 = \frac{1}{2}b(12)$

                 $48 = 6b$

                 $8 = b$

   The base is 8 meters.

29. *Common formula:* $C = 2\pi r$

   Labels: $C = 30\pi$ (inches)

              $2r =$ diameter $= d$ (inches)

   Equation: $C = 2\pi r$

                 $C = 2r\pi$

                 $C = d\pi$

                 $30\pi = d\pi$

                 $\frac{30\pi}{\pi} = d$

                 $30 = d$

   The diameter is 30 inches.

31. *Common formula:* $C = 2\pi r$

    Labels: $C = 25$ (meters)

                 $r = $ radius   (meters)

    Equation: $C = 2\pi r$

                  $25 = 2\pi r$

                  $\dfrac{25}{2\pi} = r$

The radius is $\dfrac{25}{2\pi} \approx 3.98$ meters.

Common formula: $A = \pi r^2$

    Labels: $A = $ area   (square meters)

                  $r = \dfrac{25}{2\pi}$ (meters)

    Equation: $A = \pi r^2$

                  $A = \pi\left(\dfrac{25}{2\pi}\right)^2$

                  $A = \dfrac{625\pi}{4\pi^2}$

                  $A = \dfrac{625}{4\pi}$

The area is $\dfrac{625}{4\pi} \approx 49.74$ square meters.

33. *Common formula:* $A = lw$

    Labels: $A = $ area in rectangular base   (square inches)

                  $l = 8$ (inches)

                  $w = 3$ (inches)

    Equation: $A = lw$

                  $A = (8)(3)$

                  $A = 24$

Thus, the area is 24 square inches.

35. *Common formula:* $V = lwh$

    Labels: $V = $ volume of box   (cubic inches)

                  $l = 8$ (inches)

                  $w = 3$ (inches)

                  $h = 4$ (inches)

    Equation: $V = lwh$

                  $V = (8)(3)(4)$

                  $V = 96$

Thus, the volume of the box is 96 cubic inches.

37. *Common formula:* $I = Prt$

    Labels: $I = $ interest   (dollars)

                  $P = \$1000$

                  $r = 0.09$ (annual interest rate)

                  $t = 6$ (years)

    Equation: $I = Prt$

                  $I = (1000)(0.09)(6)$

                  $I = 540$

The interest is $540.

39. *Common formula:* $A = P + Prt$

    Labels: $A = $ amount of payment   (dollars)

                  $P = \$15,000$

                  $r = 0.13$

                  $t = \frac{1}{2}$ (year)

    Equation: $A = 15,000 + 15,000(0.13)\left(\frac{1}{2}\right)$

                  $A = 15,000 + 975$

                  $A = 15,975$

The amount of the payment is $15,975.

41. *Common formula:* $I = Prt$

    Labels: $I = \$110$

                  $P = \$1000$

                  $r = $ (annual interest rate)

                  $t = 1$ (year)

    Equation: $I = Prt$

                  $110 = (1000)(r)(1)$

                  $\frac{110}{1000} = r$

                  $0.11 = r$

The annual interest rate is 11%.

43. *Common formula:* $I = Prt$

Labels: $I = \$408$

$P = \text{principal}$ (dollars)

$r = 0.085$

$t = 4$ (years)

Equation: $I = Prt$

$408 = P(0.85)(4)$

$\dfrac{408}{(0.085)(4)} = P$

$\dfrac{408}{0.34} = P$

$1200 = P$

The required principal is $1200.

45. *Verbal model:* $\boxed{\begin{array}{c}\text{Total}\\\text{interest}\end{array}} = \boxed{\begin{array}{c}\text{Interest from}\\\text{investment } A\end{array}} + \boxed{\begin{array}{c}\text{Interest from}\\\text{investment } B\end{array}}$

Labels: Total interest $= \$500$

Principal for investment $A = P$ (dollars)

Annual interest rate for investment $A = 0.07$

Time in investment $A = 1$ (year)

Interest from investment $A = 0.07P$ (dollars)

Principal for investment $B = 6000 - P$ (dollars)

Annual interest rate for investment $B = 0.09$

Time in investment $B = 1$ (year)

Interest from investment $B = 0.09(6000 - P)$ (dollars)

Equation: $500 = 0.07P + 0.09(6000 - P)$

$500 = 0.07P + 540 - 0.09P$

$-40 = -0.02P$

$\dfrac{-40}{-0.02} = P$

$2000 = P$ and $6000 - P = 4000$

The smallest amount you can invest at 9% to meet your objective is $4,000.

47. *Verbal model:* $\boxed{\text{Distance}} = \boxed{\text{Rate}} \cdot \boxed{\text{Time}}$

Labels: Distance $= 3000$ (miles)

Rate $= 17,000$ (miles per hour)

Time $= t$ (hours)

Equation: $3000 = 17,000t$

$\dfrac{3000}{17,000} = t$

$0.176 \approx t$

The time is approximately 0.176 hours (or approximately 10.6 minutes).

49. *Verbal model:* | Faster car's distance | $-$ | Slower car's distance | $=$ | Distance between cars |

Labels: Faster car's rate $= 52$ (miles per hour)

Slower car's rate $= 45$ (miles per hour)

Time $= 4$ (hour)

Distance between cars $= x$ (miles)

Equation: $52(4) - 45(4) = x$

$208 - 180 = x$

$28 = x$

The cars will be 28 miles apart.

51. *Verbal model:* | Distance | $=$ | Rate | $\cdot$ | Time |

Labels: Distance $= 3000$ (miles)

Rate $= r$ (miles per hour)

Time $= 2.6$ (hours)

Equation: $3000 = r(2.6)$

$\dfrac{3000}{2.6} = r$

$1154 \approx r$

The rate is approximately 1154 miles per hour.

53. *Verbal model:* | Time traveled by car 1 | $=$ | Time traveled by car 2 |

Labels: Distance traveled by car 1 $= d$ (miles)

Rate of car 1 $= 40$ mph

Distance traveled by car 2 $= d + 5$ (miles)

Rate of car 2 $= 55$ mph

Time traveled by car 1 $= \dfrac{d}{40}$

Time traveled by car 2 $= \dfrac{d + 5}{55}$

Equation: $\dfrac{d}{40} = \dfrac{d + 5}{55}$

$55d = 40(d + 5)$

$55d = 40d + 200$

$15d = 200$

$d = \dfrac{200}{15}$

$d = 13\frac{1}{3}$ miles

—CONTINUED—

53. —CONTINUED—

Verbal model: | Distance | = | Rate | · | Time |

Labels: Distance = $13\frac{1}{3}$ (miles)

   Rate = 40   (miles per hour)

   Time = t (hours)

Equation: $13\frac{1}{3} = 40t$

   $\frac{\frac{40}{3}}{40} = t$

   $\frac{1}{3} = t$

The cars are 5 miles apart after $\frac{1}{3}$ hr.

55. (a) Answers will vary. A common guess is 50 miles per hour.

   (b) *Common formula:* | Distance | = | Rate | · | Time |

   Labels: Distance = Distance of first trip + Distance of return

   = 200 + 200 = 400   (miles)

   Rate = r (miles per hour)

   Time = Time for first trip + Time for return

   $= \frac{200}{60} + \frac{200}{40} = \frac{10}{3} + 5 = \frac{25}{3}$ (hours)

   Equation: $400 = r\left(\frac{25}{3}\right)$

   $400\left(\frac{3}{25}\right) = r\left(\frac{25}{3}\right)\left(\frac{3}{25}\right)$

   $48 = r$

The average speed for the round trip was 48 miles per hour.

57. *Verbal model:* | Amount of alcohol in solution 1 | + | Amount of alcohol in solution 2 | = | Amount of alcohol in final solution |

   Labels: Solution 1: percent alcohol = 0.10, amount = x (gallons)

   Solution 2: percent alcohol = 0.30, amount = 100 − x (gallons)

   Final solution: percent alcohol = 0.25, amount = 100   (gallons)

   Equation: $0.10x + 0.30(100 - x) = 0.25(100)$

   $0.10x + 30 - 0.30x = 25$

   $-0.20x + 30 = 25$

   $-0.20x = -5$

   $x = \frac{-5}{-0.20} = 25$

   $100 - x = 100 - 25 = 75$

The final solution will contain 25 gallons of solution 1 and 75 gallons of solution 2.

59. *Verbal model:* | Amount of alcohol in solution 1 | $+$ | Amount of alcohol in solution 2 | $=$ | Amount of alcohol in final solution |

Labels: Solution 1: percent alcohol $= 0.15$, amount $= x$ (quarts)

Solution 2: percent alcohol $= 0.45$, amount $= 10 - x$ (quarts)

Final solution: percent alcohol $= 0.30$, amount $= 10$ (quarts)

Equation: $0.15x + 0.45(10 - x) = 0.30(10)$

$$0.15x + 4.5 - 0.45x = 3$$

$$-0.3x + 4.5 = 3$$

$$-0.3x = -1.5$$

$$x = \frac{-1.5}{-0.3} = 5$$

$$10 - x = 5$$

The final solution will contain 5 quarts of solution 1 and 5 quarts of solution 2.

61. *Verbal model:* | Value of 20¢ stamps | $+$ | Value of 33¢ stamps | $=$ | Total value |

Labels: Mixed stamps: total worth $= \$27.80$, number of stamps $= 100$

20¢ stamps: value per stamp $= 0.20$, number of stamps $= x$

33¢ stamps: value per stamp $= 0.33$, number of stamps $= 100 - x$

Equation: $0.20x + 0.33(100 - x) = 27.80$

$$0.20x + 33 - 0.33x = 27.80$$

$$-0.13x + 33 = 27.80$$

$$-0.13x = -5.20$$

$$x = \frac{-5.20}{-0.13}$$

$$x = 40 \text{ and } 100 - x = 60$$

There are 40 stamps worth 20¢ each and 60 stamps worth 33¢ each.

63. *Verbal model:* | Value of nickels | $+$ | Value of dimes | $=$ | Total value |

Labels: Mixed coins: total value $= \$1.60$, number of coins $= 20$

Nickels: value per coin $= \$0.05$, number of coins $= x$

Dimes: value per coin $= \$0.10$, number of coins $= 20 - x$

Equation: $0.05x + 0.10(20 - x) = 1.60$

$$0.05x + 2 - 0.10x = 1.60$$

$$-0.05x + 2 = 1.60$$

$$-0.05x = -0.40$$

$$x = \frac{-0.40}{-0.05}$$

$$x = 8$$

$$20 - x = 20 - 8 = 12$$

There are 8 nickels and 12 dimes.

65. *Verbal model:* | Total cost of first nuts | + | Total cost of second nuts | = | Total cost of mixture |

Labels: First nuts: price per pound = \$2.49, number of pounds = x

Second nuts: price per pound = \$3.89, number of pounds = $100 - x$

Mixture: price per pound = \$3.47, number of pounds = 100

Equation: $2.49(x) + 3.89(100 - x) = 3.47(100)$

$$2.49 + 389 - 3.89x = 347$$

$$-1.40x + 389 = 347$$

$$-1.40x = -42$$

$$x = \frac{-42}{-1.40}$$

$$x = 30$$

$$100 - x = 100 - 30 = 70$$

The mixture contained 30 pounds of the nuts costing \$2.49 per pound and 70 pounds of the nuts costing \$3.89 per pound.

67. *Verbal model:* | Amount of antifreeze in first solution | + | Amount of antifreeze in pure solution | = | Amount of antifreeze in final solution |

Labels: Original solution: percent antifreeze = 0.30, amount = $4 - x$ (gallons)

Pure antifreeze: percent antifreeze = 1.00, amount = x (gallons)

Final solution: percent antifreeze = 0.50, amount = 4   (gallons)

Equation: $0.30(4 - x) + 1.00(x) = 0.50(4)$

$$1.2 - 0.3x + x = 2$$

$$1.2 + 0.7x = 2$$

$$0.7x = 0.8$$

$$x = \frac{0.8}{0.7} \approx 1.14$$

Approximately 1.14 gallons $\left(\text{or } 1\frac{1}{7} \text{ gallons}\right)$ must be withdrawn and replaced.

69. *Verbal model:* | Value of corn | + | Value of soybeans | = | Value of mixture |

Labels: Corn: price per ton = \$125, number of tons = x, value of corn = \$125x$

Soybeans: price per ton = \$200, number of tons = $100 - x$, value of soybeans = $\$200(100 - x)$

Mixture: price per ton = m (dollars), number of tons = 100, value of mixture = \$100m$

Equation: $125x + 200(100 - x) = 100m$

$$\frac{125x + 200(100 - x)}{100} = m$$

—CONTINUED—

69. —CONTINUED—

Corn weight, x	Soybeans weight, $100 - x$	Price/ton of the mixture, m
0	100	$\dfrac{125(0) + 200(100)}{100} = \dfrac{20,000}{100} = \$200$
20	80	$\dfrac{125(20) + 200(80)}{100} = \dfrac{18,500}{100} = \$185$
40	60	$\dfrac{125(40) + 200(60)}{100} = \dfrac{17,000}{100} = \$170$
60	40	$\dfrac{125(60) + 200(40)}{100} = \dfrac{15,500}{100} = \$155$
80	20	$\dfrac{125(80) + 200(20)}{100} = \dfrac{14,000}{100} = \$140$
100	0	$\dfrac{125(100) + 200(0)}{100} = \dfrac{12,500}{100} = \$125$

(a) As the number of tons of corn increases, the number of tons of soybeans *decreases* by the same amount.

(b) As the number of tons of corn increases, the price per ton of the mixture *decreases* ad gets closer to the price per ton of the corn.

(c) If there were equal numbers of tons of corn and soybeans in the mixture, the price per ton of the mixture would be halfway between the prices of the two components. In other words, the price per ton of the mixture would be the *average of the two prices,* 125 and 200, which is

$$\frac{125 + 200}{2} = \frac{325}{2} = 162.50.$$

Note: This result can be verified with the equation above, using $x = 50$ and $100 - x = 50$.

$$m = \frac{125(50) + 200(50)}{100} = \$162.50.$$

71. *Verbal model:* $\boxed{\begin{array}{c}\text{Work}\\\text{done}\end{array}} = \boxed{\begin{array}{c}\text{Portion done}\\\text{by you}\end{array}} + \boxed{\begin{array}{c}\text{Portion done}\\\text{by friend}\end{array}}$

Labels: Both persons: work done = 1 job, time = t (hours)

Your work: rate = $\frac{1}{2}$ job per hour, time = t (hours)

Friend's work: rate = $\frac{1}{3}$ job per hour, time = t (hours)

Equation: $1 = \frac{1}{2}(t) + \frac{1}{3}(t)$

$1 = \left(\frac{1}{2} + \frac{1}{3}\right)t$

$1 = \left(\frac{3}{6} + \frac{2}{6}\right)t$

$1 = \frac{5}{6}t$

$\frac{6}{5}(1) = \frac{6}{5}\left(\frac{5}{6}t\right)$

$\frac{6}{5} = t$

It would take $\frac{6}{5}$ hours (or 1 hour, 12 minutes) to mow the lawn.

73. *Verbal model:* $\boxed{\text{Work done}} = \boxed{\text{Portion done by first worker}} + \boxed{\text{Portion done by second worker}}$

Labels: Both persons: work done = 1 task, time = t (hours)

First worker: rate = $\dfrac{1}{h}$ tasks per hour, time = t (hours)

Second worker: rate = $\dfrac{1}{3h}$ tasks per hour, time = t (hours)

Equation:

$$1 = \frac{1}{h}(t) + \frac{1}{3h}(t)$$

$$1 = \left(\frac{1}{h} + \frac{1}{3h}\right)t$$

$$1 = \left(\frac{3}{3h} + \frac{1}{3h}\right)t$$

$$1 = \frac{4}{3h}t$$

$$\frac{3h}{4}(1) = \frac{3h}{4}\left(\frac{4}{3h}t\right)$$

$$\frac{3h}{4} = t$$

Working together, the two people can complete the task in $3h/4$ hours.

75. *Verbal model:* $\boxed{\text{Son's age}} = \frac{1}{3} \cdot \boxed{\text{Mother's age}}$

Labels: Son's age = x (years)

Mother's age = $x + 30$ (years)

Equation:

$$x = \tfrac{1}{3}(x + 30)$$

$$3 \cdot x = 3 \cdot \tfrac{1}{3}(x + 30)$$

$$3x = x + 30$$

$$2x = 30$$

$$x = \tfrac{30}{2}$$

$$x = 15$$

The son will be 15 years old. (His mother will be $15 + 30$ or 45 years old.)

77. *Verbal model:* $\boxed{\text{Votes for candidate } A} + \boxed{\text{Votes for candidate } B} + \boxed{\text{Votes for candidate } C} = \boxed{\text{Total votes}}$

Labels: Votes for candidate $A = x$

Votes for candidate $B = x$

Votes for candidate $C = 2x$

Total votes = 1000

Equation:

$$x + x + 2x = 1000$$

$$4x = 1000$$

$$x = \tfrac{1000}{4}$$

$$x = 250$$

$$2x = 2(250) = 500$$

Candidates A and B received 250 votes each, and candidate C received 500 votes.

79. The units of measure for perimeter are linear measures, such as inches, feet, yards, miles, meters, etc.

The units of measure for area are square units, such as square inches, square yards, square miles, etc.

The units of measure for volume are cubic units, such as cubic inches, cubic feet, cubic meters, etc.

81. Yes, if the radius of a circle is doubled, the circumference is doubled.

$c = 2\pi r$

$2\pi(2r) = 4\pi r = 2(2\pi r) = 2c$

No, if the radius is doubled, the area is not doubled; the area is multiplied by 4.

$A = \pi r^2$

$\pi(2r)^2 = \pi(4r^2) = 4\pi r^2 = 4A$

83. If it takes you 5 hours to complete a job, then you complete $\frac{1}{5}$ of the job each hour.

Section 3.6   Linear Inequalities

1. x is greater than or equal to 3.

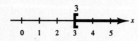

3. x is less than or equal to 10.

5. y is greater than $-\frac{3}{2}$ and less than or equal to 5.

7. (a) $x = 3$

$5(3) - 12 \overset{?}{>} 0$

$15 - 12 \overset{?}{>} 0$

$3 > 0$

3 *is* a solution.

(b) $x = -3$

$5(-3) - 12 \overset{?}{>} 0$

$-15 - 12 \overset{?}{>} 0$

$-27 \not> 0$

-3 *is not* a solution.

(c) $x = \frac{5}{2}$

$5\left(\frac{5}{2}\right) - 12 \overset{?}{>} 0$

$\frac{25}{2} - \frac{24}{2} \overset{?}{>} 0$

$\frac{1}{2} > 0$

$\frac{5}{2}$ *is* a solution.

(d) $x = \frac{3}{2}$

$5\left(\frac{3}{2}\right) - 12 \overset{?}{>} 0$

$\frac{15}{2} - \frac{24}{2} \overset{?}{>} 0$

$-\frac{9}{2} \not> 0$

$\frac{3}{2}$ *is not* a solution.

9. (a) $x = 10$

$3 - \frac{1}{2}(10) \overset{?}{>} 0$

$3 - 5 \overset{?}{>} 0$

$-2 \not> 0$

10 *is not* a solution.

(b) $x = 6$

$3 - \frac{1}{2}(6) \overset{?}{>} 0$

$3 - 3 \overset{?}{>} 0$

$0 > 0$

6 *is not* solution.

(c) $x = -\frac{3}{4}$

$3 - \frac{1}{2}\left(-\frac{3}{4}\right) \overset{?}{>} 0$

$3 + \frac{3}{8} \overset{?}{>} 0$

$3\frac{3}{8} > 0$

$-\frac{3}{4}$ *is* solution.

(d) $x = 0$

$3 - \frac{1}{2}(0) \overset{?}{>} 0$

$3 - 0 \overset{?}{>} 0$

$3 > 0$

0 *is* a solution.

11. (a) $x = 4$

$0 \overset{?}{<} \frac{4 - 2}{4} \overset{?}{<} 2$

$0 \overset{?}{<} \frac{2}{4} \overset{?}{<} 2$

$0 < \frac{1}{2} < 2$

4 *is* a solution.

(b) $x = 10$

$0 \overset{?}{<} \frac{10 - 2}{4} \overset{?}{<} 2$

$0 \overset{?}{<} \frac{8}{4} \overset{?}{<} 2$

$0 < 2 \not< 2$

10 *is not* a solution.

(c) $x = 0$

$0 \overset{?}{<} \frac{0 - 2}{4} \overset{?}{<} 2$

$0 \overset{?}{<} \frac{-2}{4} \overset{?}{<} 2$

$0 \not< -\frac{1}{2} < 2$

0 *is not* a solution.

(d) $x = \frac{7}{2}$

$0 \overset{?}{<} \frac{\frac{7}{2} - 2}{4} \overset{?}{<} 2$

$0 \overset{?}{<} \frac{\frac{3}{2}}{4} \overset{?}{<} 2$

$0 < \frac{3}{8} < 2$

$\frac{7}{2}$ *is* a solution.

13. (a) $x = -5$

$$-12 \stackrel{?}{\leq} 3(-5 + 4) \stackrel{?}{\leq} 6$$
$$-12 \stackrel{?}{\leq} 3(-1) \stackrel{?}{\leq} 6$$
$$-12 \leq -3 \leq 6$$

-5 *is* a solution.

(b) $x = -1$

$$-12 \stackrel{?}{\leq} 3(-1 + 4) \stackrel{?}{\leq} 6$$
$$-12 \stackrel{?}{\leq} 3(3) \stackrel{?}{\leq} 6$$
$$-12 \leq 9 \nleq 6$$

-1 *is not* a solution.

(c) $x = -7$

$$-12 \stackrel{?}{\leq} 3(-7 + 4) \stackrel{?}{\leq} 6$$
$$-12 \stackrel{?}{\leq} 3(-3) \stackrel{?}{\leq} 6$$
$$-12 \leq -9 \leq 6$$

-7 *is* a solution.

(d) $x = \frac{2}{3}$

$$-12 \stackrel{?}{\leq} 3\left(\frac{2}{3} + 4\right) \stackrel{?}{\leq} 6$$
$$-12 \stackrel{?}{\leq} 3\left(\frac{14}{3}\right) \stackrel{?}{\leq} 6$$
$$-12 \leq 14 \nleq 6$$

$\frac{2}{3}$ *is not* a solution.

15. (b)

17. (c)

19. (d)

21. $t - 3 \geq 2$

$$t - 3 + 3 \geq 2 + 3$$
$$t \geq 5$$

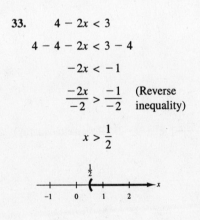

23. $x + 4 \leq 6$

$$x + 4 - 4 \leq 6 - 4$$
$$x \leq 2$$

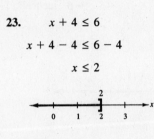

25. $4x < 12$

$$\frac{4x}{4} < \frac{12}{4}$$
$$x < 3$$

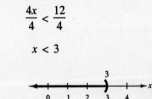

27. $-10x < 40$

$$\frac{-10x}{-10} > \frac{40}{-10} \quad \text{(Reverse inequality)}$$
$$x > -4$$

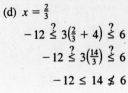

29. $\frac{2}{3}x \leq 12$

$$3\left(\frac{2}{3}\right)x \leq 3(12)$$
$$2x < 36$$
$$\frac{2x}{2} \leq \frac{36}{2}$$
$$x \leq 18$$

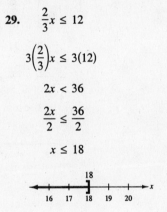

31. $2x - 5 > 7$

$$2x - 5 + 5 > 7 + 5$$
$$2x > 12$$
$$\frac{2x}{2} > \frac{12}{2}$$
$$x > 6$$

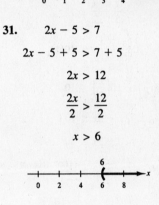

33. $4 - 2x < 3$

$$4 - 4 - 2x < 3 - 4$$
$$-2x < -1$$
$$\frac{-2x}{-2} > \frac{-1}{-2} \quad \text{(Reverse inequality)}$$
$$x > \frac{1}{2}$$

35. $2x - 5 > -x + 6$

$$2x + x - 5 > -x + x + 6$$
$$3x - 5 > 6$$
$$3x - 5 + 5 > 6 + 5$$
$$3x > 11$$
$$\frac{3x}{3} > \frac{11}{3}$$
$$x > \frac{11}{3}$$

37. $6 < 3(y + 1) - 4(1 - y)$

$$6 < 3y + 3 - 4 + 4y$$
$$6 < 7y - 1$$
$$6 + 1 < 7y - 1 + 1$$
$$7 < 7y$$
$$\frac{7}{7} < \frac{7y}{7}$$
$$1 < y \text{ or } y > 1$$

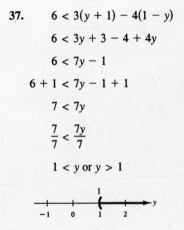

39. $-2(z + 1) \geq 3(z + 1)$

$-2z - 2 \geq 3z + 3$

$-2z - 2 + 2z \geq 3z + 3 + 2z$

$-2 \leq 5z + 3$

$-2 - 3 \leq 5z + 3 - 3$

$-5 \geq 5z$

$\dfrac{-5}{5} \geq \dfrac{5z}{5}$

$-1 \geq z$ or $z \leq -1$

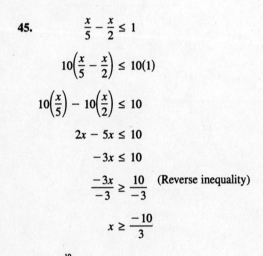

41. $10(1 - y) < -4(y - 2)$

$10 - 10y < -4y + 8$

$10 - 10y + 4y < -4y + 4y + 8$

$10 - 6y < 8$

$10 - 6y - 10 < 8 - 10$

$-6y < -2$

$\dfrac{-6y}{-6} > \dfrac{-2}{-6}$ (Reverse inequality)

$y > \dfrac{1}{3}$

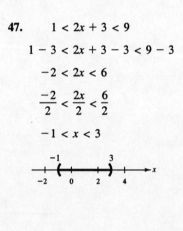

43. $\dfrac{x}{4} + \dfrac{1}{2} > 0$

$4\left(\dfrac{x}{4} + \dfrac{1}{2}\right) > 4(0)$

$4\left(\dfrac{x}{4}\right) + 4\left(\dfrac{1}{2}\right) > 0$

$x + 2 > 0$

$x + 2 - 2 > 0 - 2$

$x > -2$

45. $\dfrac{x}{5} - \dfrac{x}{2} \leq 1$

$10\left(\dfrac{x}{5} - \dfrac{x}{2}\right) \leq 10(1)$

$10\left(\dfrac{x}{5}\right) - 10\left(\dfrac{x}{2}\right) \leq 10$

$2x - 5x \leq 10$

$-3x \leq 10$

$\dfrac{-3x}{-3} \geq \dfrac{10}{-3}$ (Reverse inequality)

$x \geq \dfrac{-10}{3}$

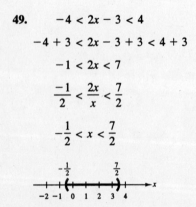

47. $1 < 2x + 3 < 9$

$1 - 3 < 2x + 3 - 3 < 9 - 3$

$-2 < 2x < 6$

$\dfrac{-2}{2} < \dfrac{2x}{2} < \dfrac{6}{2}$

$-1 < x < 3$

49. $-4 < 2x - 3 < 4$

$-4 + 3 < 2x - 3 + 3 < 4 + 3$

$-1 < 2x < 7$

$\dfrac{-1}{2} < \dfrac{2x}{x} < \dfrac{7}{2}$

$-\dfrac{1}{2} < x < \dfrac{7}{2}$

51. $6 > \dfrac{x - 2}{-3} > -2$

$-3(6) < -3\left(\dfrac{x - 2}{-3}\right) < -3(-2)$ (Reverse inequality)

$-18 < x - 2 < 6$

$-18 + 2 < x - 2 + 2 < 6 + 2$

$-16 < x < 8$

53.
$$\frac{3}{4} > x + 1 > \frac{1}{4}$$

$$4\left(\frac{3}{4}\right) > 4(x + 1) > 4\left(\frac{1}{4}\right)$$

$$3 > 4x + 4 > 1$$

$$3 - 4 > 4x + 4 - 4 > 1 - 4$$

$$-1 > 4x > -3$$

$$\frac{-1}{4} > \frac{4x}{4} > \frac{-3}{4}$$

$$-\frac{1}{4} > x > -\frac{3}{4} \text{ or } -\frac{3}{4} < x < -\frac{1}{4}$$

Alternate Solution:

$$\frac{3}{4} > x + 1 > \frac{1}{4}$$

$$\frac{3}{4} - 1 > x + 1 - 1 > \frac{1}{4} - 1$$

$$\frac{3}{4} - \frac{4}{4} > x > \frac{1}{4} - \frac{1}{4}$$

$$-\frac{1}{4} > x > -\frac{3}{4} \text{ or } -\frac{3}{4} < x < -\frac{1}{4}$$

55. $-5 < 2x + 3$ and $2x + 3 \le 9$

$$-5 < 2x + 3 \le 9$$

$$-5 - 3 < 2x + 3 - 3 \le 9 - 3$$

$$-8 < 2x \le 6$$

$$-4 < x \le 3$$

57. $4 - 3x < -8$ or $4 - 3x > 7$

$4 - 4 - 3x < -8 - 4$ $4 - 4 - 3x > 7 - 4$

$-3x < -12$ $-3x > 3$

$x > 4$ or $x < -1$

59. $A = \{x | x < -5\}$

$B = \{x | x > 3\}$

The solution set of the compound inequality is $A \cup B$.

61. $A = \{x | x < 6\}$

$B = \{x | x > 0\}$

The solution set of compound inequality is $A \cap B$.

63. $A = \{x | x \ge -3\}$

$B = \{x | x \le 7\}$

The solution set of the compound inequality is $A \cap B$.

65. $A = \{x | x < 7\}$

$B = \{x | x > 8\}$

The solution set of the compound inequality is $A \cup B$.

67. $x \ge 0$

69. $y > -6$

71. $x \ge 4$

73. $y \le 25$

75. $0 < x \le 6$

77. Mars is farther from the sun than Mercury.
(This illustrates the transitive property of inequalities.)

79. *Verbal model:* $\boxed{\begin{array}{c}\text{Cost for}\\\text{first minute}\end{array}} + \boxed{\begin{array}{c}\text{Cost for}\\\text{additional minutes}\end{array}} \le \boxed{\$4}$

   Label: Cost for first minute = \$0.46

   Number of *additional* minutes = x

   Cost per additional minute = \$0.31

   Cost for additional minutes = $0.31x$ (dollars)

—CONTINUED—

79. **—CONTINUED—**

Inequality:

$$0.46 + 0.31x \leq 4$$

$$0.46 - 0.46 + 0.31x \leq 4 - 0.46$$

$$0.31x \leq 3.54$$

$$\frac{0.31x}{0.31} \leq \frac{3.54}{0.31}$$

$$x \leq 11.4 \quad \text{(Rounded)}$$

The variable x represents the number of *additional* minutes *after* the first minute. Thus, $x + 1$ represents the *total* length of the call. If $x \leq 11.4$, then $x + 1 \leq 12.4$.

Therefore, the call will cost no more than \$4 if the length of the call, t, lies this interval:

$0 \leq t \leq 12.4$ minutes

Note: If any portion of a minute is billed at the full minute rate, the call must be no more than 12 minutes long to keep the cost from exceeding \$4.

81. Given equation:

$$C = 0.45m + 3200$$

$$c < 12{,}000$$

$$0.45m + 3200 < 12{,}000$$

$$0.45m < 8800$$

$$\frac{0.45m}{0.45} < \frac{8800}{0.45}$$

$$m < 19{,}555.55$$

If $m \leq 19{,}555$ miles, the annual operating cost for the van will be less than \$12,000.

83. $a + b > c$

85. *Verbal model:* | Minimum distance | $\leq$ | Distance | $\leq$ | Maximum distance |

Labels:

Minimum rate $= 45$	(miles per hour)
Maximum rate $= 65$	(miles per hour)
Time $= 4$	(hours)
Minimum distance $= 45(4) = 180$	(miles)
Distance $= d$	(miles)
Maximum distance $= 65(4) = 260$	(miles)

Inequality: $180 \leq d \leq 260$ miles

87. *Verbal model:* | Service charge | $+$ | 0.35 | $\cdot$ | Usage time | $<$ | 75 |

Label:

Service charge $= 19.50$ (dollars)

Usage time $= x$ (minutes)

Inequality:

$$19.50 + 0.35x \leq 75$$

$$0.35x \leq 56.50$$

$$\frac{0.35x}{0.35} \leq \frac{56.50}{0.35}$$

$$x \leq 158.57$$

You can spend up to 158.57 minutes talking on the cellular phone each month. Therefore, $0 \leq x \leq 158.57$ minutes.

89. Yes, adding -5 to both sides of an inequality is the same as subtracting 5 from both sides of an inequality. The addition and subtraction properties of inequalities indicate that either of these steps will produce an inequality that is equivalent to the original inequality.

91. Often there are infinitely many numbers in the solution set of linear inequality. For example, the solution set of $2x + 6 < 16$ is the set of all real numbers $x < 5$. There are infinitely many such numbers.

93. The process of solving linear equations is similar to the process for solving linear inequalities except that the direction of the inequality must be reversed if both sides of the inequality are multiplied or divided by a negative number.

95. $<$

The inequality symbol $<$ is equivalent to the symbol $\not\geq$.

97. False.

$$-\tfrac{1}{2}x + 6 > 0$$
$$-\tfrac{1}{2}x > -6$$
$$x < 12$$

99. True.

These two statements are equivalent.

Section 3.7   Absolute Value Equations and Inequalities

1. $|x + 3| = 10$
$$|-13 + 3| \stackrel{?}{=} 10$$
$$|-10| \stackrel{?}{=} 10$$
$$10 = 10$$
Yes, -13 *is* a solution.

3. $|3 - 2y| = 2$
$$|3 - 2(1)| \stackrel{?}{=} 2$$
$$|3 - 2| \stackrel{?}{=} 2$$
$$|1| \stackrel{?}{=} 2$$
$$1 \neq 2$$
No, 1 *is not* a solution.

5. $|u - 3| = 7$
$$u - 3 = 7 \text{ or } u - 3 = -7$$

7. $\left|\tfrac{1}{2}x + 7\right| = \tfrac{3}{2}$
$$\tfrac{1}{2}x + 7 = \tfrac{3}{2} \text{ or } \tfrac{1}{2}x + 7 = -\tfrac{3}{2}$$

9. $|x| = 7$
$$x = 7 \text{ or } x = -7$$

11. $|v| = 15$
$$v = 15 \text{ or } v = -15$$

13. $|m| = -3$

This equation has no solution because it is not possible for the absolute value of a real number to be negative.

15. $|3x| = 18$
$$3x = 18 \text{ or } 3x = -18$$
$$x = 6 \qquad x = -6$$

17. $|x - 12| = 4$
$$x - 12 = 4 \quad \text{or } x - 12 = -4$$
$$x = 16 \qquad x = 8$$

19. $|s + 3| = 11$
$$s + 3 = 11 \text{ or } s + 3 = -11$$
$$s = 8 \qquad s = -14$$

21. $|16 - y| = 3$
$$16 - y = 3 \quad \text{or} \quad 16 - y = -3$$
$$-y = -13 \qquad -y = -19$$
$$y = 13 \qquad y = 19$$

23. $|2x + 4| = -8$

This equation has no solution because it is not possible for the absolute value of a real number to be negative.

25. $|2x - 3| = 21$
$$2x - 3 = 21 \text{ or } 2x - 3 = -21$$
$$2x = 24 \qquad 2x = -18$$
$$x = 12 \qquad x = -9$$

27. $|3x + 2| = 5$
$$3x + 2 = 5 \text{ or } 3x + 2 = -5$$
$$3x = 3 \qquad 3x = -7$$
$$x = 1 \qquad x = -\tfrac{7}{3}$$

29. $|5x - 9| - 4 = 0$
$$|5x - 9| = 4$$
$$5x - 9 = 4 \quad \text{or } 5x - 9 = -4$$
$$5x = 13 \qquad 5x = 5$$
$$x = \tfrac{13}{5} \qquad x = 1$$

31. $\left|5 - \frac{2}{3}x\right| = 3$

$5 - \frac{2}{3}x = 3$ or $5 - \frac{2}{3}x = -3$

$-\frac{2}{3}x = -2$ $\quad -\frac{1}{2}x = -8$

$-2x = -6$ $\quad -2x = -24$

$x = 3$ $\quad x = 12$

33. $|0.25x - 2| = 4$

$0.25x - 2 = 4$ or $0.25x - 2 = -4$

$0.25x = 6$ $\quad 0.25x = -2$

$x = \dfrac{6}{0.25}$ $\quad x = \dfrac{-2}{0.25}$

$x = 24$ $\quad x = -8$

35. $|x + 3| = 8$

37. $|x - 4| = 2$

39. $|x| < 2$

(a) $x = 1$

$|1| \overset{?}{<} 2$

$1 < 2$

Yes, 1 *is* a solution.

(b) $x = -3$

$|-3| \overset{?}{<} 2$

$3 \not< 2$

No, -3 *is not* a solution.

(c) $x = 5$

$|5| \overset{?}{<} 2$

$5 \not< 2$

No, 5 *is not* a solution.

(d) $x = -\frac{1}{2}$

$\left|-\frac{1}{2}\right| \overset{?}{<} 2$

$\frac{1}{2} < 2$

Yes, $-\frac{1}{2}$ *is* a solution.

41. $|x| \geq 5$

(a) $x = 2$

$|2| \overset{?}{\geq} 5$

$2 \not\geq 5$

No, 2 *is not* a solution.

(b) $x = -7$

$|-7| \overset{?}{\geq} 5$

$7 \geq 5$

Yes, -7 *is* a solution.

(c) $x = 25$

$|25| \overset{?}{\geq} 5$

$25 \geq 5$

Yes, 25 *is* a solution.

(d) $x = -3$

$|-3| \overset{?}{\geq} 5$

$3 \not\geq 5$

No, -3 *is not* a solution.

43. $|x - 4| < 2$

(a) $x = 2$

$|2 - 4| \overset{?}{<} 2$

$|-2| \overset{?}{<} 2$

$2 \not< 2$

No, 2 *is not* a solution.

(b) $x = 1.5$

$|1.5 - 4| \overset{?}{<} 2$

$|-2.5| \overset{?}{<} 2$

$2.5 \not< 2$

No, 1.5 *is not* solution.

(c) $x = 0$

$|0 - 4| \overset{?}{<} 2$

$|-4| \overset{?}{<} 2$

$4 \not< 2$

No, 0 *is not* a solution.

(d) $x = 5$

$|5 - 4| \overset{?}{<} 2$

$|1| \overset{?}{<} 2$

$1 < 2$

Yes, 5 *is* a solution.

45. $|x + 5| \geq 3$

(a) $x = -8$

$|-8 + 5| \overset{?}{\geq} 3$

$|-3| \overset{?}{\geq} 3$

$3 \geq 3$

Yes, -8 *is* a solution.

(b) $x = -5$

$|-5 + 5| \overset{?}{\geq} 3$

$|0| \overset{?}{\geq} 3$

$0 \not\geq 3$

No, -5 *is not* a solution.

(c) $x = -2$

$|-2 + 5| \overset{?}{\geq} 3$

$|3| \overset{?}{\geq} 3$

$3 \geq 3$

Yes, -2 is a solution.

(d) $x = -4$

$|-4 + 5| \overset{?}{\geq} 3$

$|1| \overset{?}{\geq} 3$

$1 \not\geq 3$

No, -4 *is not* a solution.

47. $|z + 2| < 1$

$-1 < z + 2 < 1$

49. $|5 - h| \geq 2$

$5 - h \geq 2$ or $5 - h \leq -2$

51.

53.

55. $|y| < 3$

$-3 < y < 3$

57. $|y| \geq 2$

$y \leq -2$ or $y \geq 2$

59. $|t| > 5$

$t < -5$ or $t > 5$

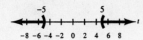

61. $|y - 2| \leq 2$

$-2 \leq y - 2 \leq 2$

$-2 + 2 \leq y - 2 + 2 \leq 2 + 2$

$0 \leq y \leq 4$

63. $|x + 1| < 4$

$-4 < x + 1 < 4$

$-4 - 1 < x + 1 - 1 < 4 - 1$

$-5 < x < 3$

65. $|2x| < 12$

$-12 < 2x < 12$

$-6 < x < 6$

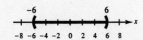

67. $|y - 5| > 2$

$y - 5 < -2$ or $y - 5 > 2$

$y - 5 + 5 < -2 + 5$ $y - 5 + 5 > 2 + 5$

$y < 3$ $y > 7$

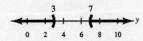

69. $|m + 2| \geq 3$

$m + 2 \leq -3$ or $m + 2 \geq 3$

$m + 2 - 2 \leq -3 - 2$ $m + 2 - 2 \geq 3 - 2$

$m \leq -5$ $m \geq 1$

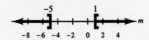

71. Graph (c)

73. Graph (a)   **75.** $|x - 1| < 2$ **77.** $|x + 5| \geq 10$ **79.** $|x| < 2$ **81.** $|x| > 5$

83. $|x - 4| > 2$

85. $|t - 72| < 1.5$

$-1.5 < t - 72 < 1.5$

$-1.5 + 72 < t - 72 + 72 < 1.5 + 72$

$70.5 < t < 73.5$

87. (a) $|s - x| \leq 0.005$

(b) $|3.5 - x| \leq 0.005$

$-0.005 \leq 3.5 - x \leq 0.005$

$-0.005 - 3.5 \leq 3.5 - 3.5 - x \leq 0.005 - 3.5$

$-3.5005 \leq -x \leq -3.495$

$-1(-3.5005) \geq -1(-x) \geq -1(-3.495)$

$3.5005 \geq x \geq 3.495$

The acceptable diameters for this piece satisfy this double inequality: $3.495 \leq x \leq 3.5005$.

89. $|a|$ represents the two real numbers that are a units from 0 on the real number line.

91. To solve an absolute value equation, first write the equation in standard form $|ax + b| = c$. Then form two linear equations $ax + b = c$ and $ax + b = -c$. The solutions of these two linear equations are the solutions to the absolute value equation.

Example: $|2x - 3| + 1 = 8$

$$|2x - 3| = 7$$

$$2x - 3 = 7 \text{ or } 2x - 3 = -7$$

$$2x = 10 \qquad 2x = -4$$

$$x = 5 \qquad x = -2$$

Review Exercises for Chapter 3

1. $y = 35$

3. $x = 28$

5.

$10x - 12 = 18$	Original equation.
$10x - 12 + 12 = 18 + 12$	Add 12 to each side.
$10x = 30$	Combine like terms.
$\dfrac{10x}{10} = \dfrac{30}{10}$	Divide by 10 on both sides.
$x = 3$	

7.

$$x + 10 = 13$$

$$x + 10 - 10 = 13 - 10$$

$$x = 3$$

9.

$$5 - x = 2$$

$$5 - 5 - x = 2 - 5$$

$$-x = -3$$

$$-1(-x) = -1(-3)$$

$$x = 3$$

11.

$$10x = 50$$

$$\frac{10x}{10} = \frac{50}{10}$$

$$x = 5$$

13.

$$8x + 7 = 39$$

$$8x + 7 - 7 = 39 - 7$$

$$8x = 32$$

$$\frac{8x}{8} = \frac{32}{8}$$

$$x = 4$$

15.

$$24 - 7x = 3$$

$$24 - 24 - 7x = 3 - 24$$

$$-7x = -21$$

$$\frac{-7x}{-7} = \frac{-21}{-7}$$

$$x = 3$$

17.

$$15x - 4 = 16$$

$$15x - 4 + 4 = 16 + 4$$

$$15x = 20$$

$$\frac{15x}{15} = \frac{20}{15}$$

$$x = \frac{20}{15}$$

$$x = \frac{4}{3}$$

19.

$$\frac{x}{5} = 4$$

$$5\left(\frac{x}{5}\right) = 5(4)$$

$$x = 20$$

21. $3x - 2(x + 5) = 10$

$$3x - 2x - 10 = 10$$

$$x - 10 = 10$$

$$x - 10 + 10 = 10 + 10$$

$$x = 20$$

23.

$$2x + 3 = 5x - 2$$
$$2x - 5x + 3 = 5x - 5x - 2$$
$$-3x + 3 = -2$$
$$-3x + 3 - 3 = -2 - 3$$
$$-3x = -5$$
$$\frac{-3x}{-3} = \frac{-5}{-3}$$
$$x = \frac{5}{3}$$

25.

$$\frac{2}{3}x - \frac{1}{6} = \frac{9}{2}$$
$$6\left(\frac{2}{3}x - \frac{1}{6}\right) = 6\left(\frac{9}{2}\right)$$
$$6\left(\frac{2}{3}x\right) - 6\left(\frac{1}{6}\right) = 27$$
$$4x - 1 = 27$$
$$4x - 1 + 1 = 27 + 1$$
$$4x = 28$$
$$\frac{4x}{4} = \frac{28}{4}$$
$$x = 7$$

27.

$$\frac{x}{3} - \frac{1}{9} = 2$$
$$9\left(\frac{x}{3} - \frac{1}{9}\right) = 9(2)$$
$$9\left(\frac{x}{3}\right) - 9\left(\frac{1}{9}\right) = 18$$
$$3x - 1 = 18$$
$$3x - 1 + 1 = 18 + 1$$
$$3x = 19$$
$$\frac{3x}{3} = \frac{19}{3}$$
$$x = \frac{19}{3}$$

29.

$$\frac{u}{10} + \frac{u}{5} = 6$$
$$10\left(\frac{u}{10} + \frac{u}{5}\right) = 10(6)$$
$$10\left(\frac{u}{10}\right) + 10\left(\frac{u}{5}\right) = 60$$
$$u + 2u = 60$$
$$3u = 60$$
$$\frac{3u}{3} = \frac{60}{3}$$
$$u = 20$$

31.

$$516x - 875 = 3250$$
$$516x - 875 + 875 = 3250 + 875$$
$$516x = 4125$$
$$\frac{516x}{516} = \frac{4125}{516}$$
$$x \approx 7.99$$
(Rounded)

33.

$$\frac{x}{4.625} = 48.5$$
$$4.625\left(\frac{x}{4.625}\right) = 4.625(48.5)$$
$$x \approx 224.31$$
(Rounded)

35.

Percent	Parts out of 100	Decimal	Fraction
35%	35	0.35	$\frac{7}{20}$

Parts out of 100: 35% means 35 parts out of 100.

Decimal: *Verbal model:* | Decimal | · | 100% | = | Percent |

Label: Decimal = x

Equation: $x(100\%) = 35\%$

$$x = \frac{35\%}{100\%}$$
$$x = 0.35$$

Fraction: *Verbal model:* | Fraction | · | 100% | = | Percent |

Label: Fraction = x

Equation: $x(100\%) = 35\%$

$$x = \frac{35\%}{100\%}$$
$$x = \frac{35}{100}$$
$$x = \frac{7}{20}$$

37. *Verbal model:* $\boxed{\text{What number}} = \boxed{125\% \text{ of } 16}$ $(a = pb)$

Label: $a = $ unknown number

Percent equation: $a = (1.25)(16)$

$a = 20$

Therefore, 20 is 125% of 16.

39. *Verbal model:* $\boxed{150} = \boxed{37\frac{1}{2}\% \text{ of what number}}$ $(a = pb)$

Label: $b = $ unknown number

Percent equation: $150 = 0.375b$

$\dfrac{150}{0.375} = b$

$400 = b$

Therefore, 150 is $37\frac{1}{2}\%$ of 400.

41. *Verbal model:* $\boxed{150} = \boxed{\text{What percent of } 250}$ $(a = pb)$

Label: $p = $ unknown percent (in decimal form)

Percent equation: $150 = p(250)$

$\dfrac{150}{250} = p$

$0.6 = p$

Therefore, 150 is 60% of 250.

43. $\dfrac{18 \text{ inches}}{4 \text{ yards}} = \dfrac{18 \cancel{\text{ inches}}}{4(36) \cancel{\text{ inches}}} = \dfrac{18}{144} = \dfrac{1}{8}$

45. $\dfrac{2 \text{ hours}}{90 \text{ minutes}} = \dfrac{2(60) \cancel{\text{ minutes}}}{90 \cancel{\text{ minutes}}} = \dfrac{120}{90} = \dfrac{4}{3}$

47. $\dfrac{7}{16} = \dfrac{z}{8}$

$(7)(8) = 16z$

$56 = 16z$

$\dfrac{15}{16} = z$

$\dfrac{7}{2} = z$

49. $\dfrac{x+2}{4} = -\dfrac{1}{3}$

$12\left(\dfrac{x+2}{4}\right) = 12\left(-\dfrac{1}{3}\right)$

$3(x+2) = 4(-1)$

$3x + 6 = -4$

$3x + 6 - 6 = -4 - 6$

$3x = -10$

$\dfrac{3x}{3} = \dfrac{-10}{3}$

$x = -\dfrac{10}{3}$

51. $\dfrac{x-3}{2} = \dfrac{x+6}{5}$

$5(x-3) = 2(x+6)$

$5x - 15 = 2x + 12$

$5x - 2x - 15 = 2x + 12 - 2x$

$3x - 15 = 12$

$3x - 15 + 15 = 12 + 15$

$3x = 27$

$\dfrac{3x}{3} = \dfrac{27}{3}$

$x = 9$

53. $A = \dfrac{r^2\theta}{2}$

$2A = 2\left(\dfrac{r^2\theta}{2}\right)$

$2A = r^2\theta$

$\dfrac{2A}{r^2} = \dfrac{r^2\theta}{r^2}$

$\dfrac{2A}{r^2} = \theta$

55. *Verbal model:* | Distance | = | Rate | · | Time |

Labels: Distance = d (miles)

Rate = 65   (miles per hour)

Time = 8   (hours)

Equation: $d = 65(8)$

$d = 520$

The distance is 520 miles.

57. *Verbal model:* | Distance | = | Rate | · | Time |

Labels: Distance = 400   (miles)

Rate = 50   (miles per hour)

Time = t (hours)

Equation: $400 = 50t$

$\dfrac{400}{50} = \dfrac{50t}{50}$

$8 = t$

The time is 8 hours.

59. *Verbal model:* | Distance | = | Rate | · | Time |

Labels: Distance = 3000   (miles)

Rate = r (miles per hour)

Time = 50   (hours)

Equation: $3000 = r(50)$

$\dfrac{3000}{50} = \dfrac{r(50)}{50}$

$60 = r$

The rate is 60 miles per hour.

61. $1 \le x < 4$

63. $x < 3$

65. $x + 5 \ge 7$

$x + 5 - 5 \ge 7 - 5$

$x \ge 2$

![Number line showing closed dot at 2 with shading to the right, labeled 2]

67. $3x - 8 < 1$

$3x - 8 + 8 < 1 + 8$

$3x < 9$

$\dfrac{3x}{3} < \dfrac{9}{3}$

$x < 3$

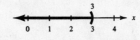

69. $-11x \le -22$

$\dfrac{-11x}{-11} \ge \dfrac{-22}{-11}$ (Reverse inequality)

$x \ge 2$

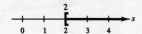

71. $\dfrac{4}{5}x > 8$

$5\left(\dfrac{4}{5}x\right)5(8)$

$4x > 40$

$\dfrac{4x}{4} > \dfrac{40}{4}$

$x > 10$

Note: This could be done in one step.

$\dfrac{4}{5}x > 8$

$\dfrac{5}{4}\left(\dfrac{4}{5}x\right) > \dfrac{5}{4}(8)$

$x > 10$

73. $14 - \dfrac{1}{2}t < 12$

$2\left(14 - \dfrac{1}{2}t\right) < 2(12)$

$28 - t < 24$

$28 - 28 - t < 24 - 28$

$-t < -4$

$(-1)(-t) > (-1)(-4)$ (Reverse inequality)

$t > 4$

75. $3(1 - y) \geq 2(4 + y)$

$3 - 3y \geq 8 + 2y$

$3 - 3y - 2y \geq 8 + 2y - 2y$

$3 - 5y \geq 8$

$3 - 3 - 5y \geq 8 - 3$

$-5y \geq 5$

$\dfrac{-5y}{-5} \leq \dfrac{5}{-5}$ (Reverse inequality)

$y \leq -1$

77. $-2 < 2x + 6 \leq 2$

$-2 - 6 < 2x + 6 - 6 \leq 2 - 6$

$-8 < 2x \leq -4$

$\dfrac{-8}{2} < \dfrac{2x}{2} \leq \dfrac{-4}{2}$

$-4 < x \leq -2$

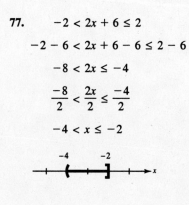

79. $3 > \dfrac{x + 1}{-2} > 0$

$-2(3) < -2\left(\dfrac{x + 1}{-2}\right) < -2(0)$ (Reverse inequality)

$-6 < x + 1 < 0$

$-6 - 1 < x + 1 - 1 < 0 - 1$

$-7 < x < -1$

81. $A = \{x \mid x < -3\}$

$B = \{x \mid x > 7\}$

The solution set of the compound inequality is $A \cup B$.

83. $A = \{x \mid x < 0\}$

$B = \{x \mid x > -6\}$

The solution set of the compound inequality is $A \cap B$.

85. $A = \{x \mid x \geq -8\}$

$B = \{x \mid x \leq -5\}$

The solution set of the compound inequality is $A \cap B$.

87. $A = \{x \mid x < 2\}$

$B = \{x \mid x > 3\}$

The solution set of the compound inequality is $A \cup B$.

89. $z \geq 10$

91. $8 < y < 12$

93. $V < 12$

95. $|x - 25| = 5$

$$x - 25 = 5 \qquad \text{or} \qquad x - 25 = -5$$

$$x - 25 + 25 = 5 + 25 \qquad x - 25 + 25 = -5 + 25$$

$$x = 30 \qquad\qquad x = 20$$

97. $|7t| = 42$

$$7t = 42 \text{ or } 7t = -42$$

$$t = \frac{42}{7} \qquad t = -\frac{42}{7}$$

$$t = 6 \qquad t = -6$$

99. $\quad |3u + 24| = 0$

$$3u + 24 = 0$$

$$3u + 24 - 24 = 0 - 24$$

$$3u = -24$$

$$\frac{3u}{3} = \frac{-24}{3}$$

$$u = -8$$

101. $\left|\frac{1}{2}v\right| \leq 3$

$$-3 \leq \frac{1}{2}v \leq 3$$

$$-6 \leq v \leq 6$$

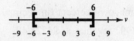

103. $|y - 4| > 3$

$$y - 4 < -3 \qquad \text{or} \qquad y - 4 > 3$$

$$y - 4 + 4 < -3 + 4 \qquad y - 4 + 4 > 3 + 4$$

$$y < 1 \qquad\qquad y > 7$$

105. $|2n - 3| < 5$

$$-5 < 2n - 3 < 5$$

$$-5 + 3 < 2n - 3 + 3 < 5 + 3$$

$$-2 < 2n < 8$$

$$\frac{-2}{2} < \frac{2n}{2} < \frac{8}{2}$$

$$-1 < n < 4$$

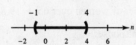

107. *Verbal model:* | Distance that you drive | $+$ | Distance that your friend drives | $=$ | Total distance |

Labels: Distance that you drive $= x$ (miles)

Distance that your friend drives $= \frac{3}{2}x$ (miles)

Total distance $= 1200$ (miles)

Equation: $x + \frac{3}{2}x = 1200$

$$2\left(x + \frac{3}{2}x\right) = 2(1200)$$

$$2x + 3x = 2400$$

$$5x = 2400$$

$$x = 480 \text{ and } \frac{3}{2}x = 720$$

Therefore, you drive 480 miles and your friend drives 720 miles.

109. *Verbal model:* $2 \cdot \boxed{\text{Length}} + 2 \cdot \boxed{\text{Width}} + \boxed{\text{Perimeter}}$

  Labels: Width $= w$ (meters)

           Length $= w + 30$ (meters)

           Perimeter $= 260$ (meters)

  Equation: $2(w + 30) + 2w = 260$

          $2w + 60 + 2w = 260$

             $4w + 60 = 260$

          $4w + 60 - 60 = 260 - 60$

               $4w = 200$

               $\dfrac{4w}{4} = \dfrac{200}{4}$

             $w = 50$ and $w + 30 = 80$

The length of the rectangle is 80 meters and the width is 50 meters.

111. *Verbal model:* $\boxed{\text{Revenue increase}} = \boxed{\text{What percent of 1997 revenue}}$ $(a = pb)$

  Labels: $a =$ revenue increase $= 4679.00 - 4521.40 = \$157.60$

           $p =$ unknown percent (in decimal form)

           $b =$ 1997 revenue $= \$4521.40$

  Percent equation: $157.60 = p(4521.40)$

             $\dfrac{157.60}{4521.40} = p$

             $0.035 \approx p$

Therefore, the percentage increase in revenue was approximately 3.5%.

113. *Verbal model:* $\boxed{\begin{array}{c}\text{Women in} \\ \text{each category}\end{array}} = \boxed{\begin{array}{c}\text{Percent of all} \\ \text{women over 65}\end{array}}$ $(a = pb)$

  Labels: $a =$ women in each category

           $p =$ percent in decimal form

           $b =$ number of women over 65 $= 19{,}844{,}000$

  Equation: living with spouse:

           $a = 0.48(19{,}844{,}000)$

           $a = 9{,}525{,}120$

         living with other relatives:

           $a = 0.08(19{,}844{,}000)$

           $a = 1{,}587{,}520$

         living alone or with nonrelatives:

           $a = 0.44(19{,}844{,}000)$

           $a = 8{,}731{,}360$

115. *Verbal model:* $\boxed{\dfrac{\text{Tax on first property}}{\text{Value of first property}}} = \boxed{\dfrac{\text{Tax on second property}}{\text{Value of second property}}}$

Labels:　Tax on first property = \$1150

　　　　　Value of first property = \$75,000

　　　　　Tax on second property = x (dollars)

　　　　　Value of second property = \$110,000

Equation:

$$\frac{1150}{75,000} = \frac{x}{110,000}$$

$$110,000\left(\frac{1150}{75,000}\right) = \left(\frac{x}{110,000}\right)110,000$$

$$1687 \approx x$$

The taxes would be approximately \$1687.

117. *Verbal model:* $\boxed{\dfrac{\text{Inches on map scale}}{\text{Miles represented on scale}}} = \boxed{\dfrac{\text{Inches between cities on map}}{\text{Miles between cities}}}$

Labels:　Inches on map scale $\approx \dfrac{7}{16}$

　　　　　Miles represented on scale = 100

　　　　　Inches between cities on map $\approx \dfrac{15}{16}$

　　　　　Miles between cities = x

Proportion:

$$\frac{7/16}{100} = \frac{15/16}{x}$$

$$\frac{7}{16}x = 100\left(\frac{15}{16}\right) \quad \text{(Cross-multiply)}$$

$$0.4375x = 93.75$$

$$x = \frac{93.75}{0.4375}$$

$$x \approx 214$$

The cities are approximately 214 miles apart.

119. *Verbal model:* $\boxed{\text{Distance}} = \boxed{\text{Rate}} \cdot \boxed{\text{Time}}$

Labels:　Distance = 562　(miles)

　　　　　Rate = 60　　　(miles per hour)

　　　　　Time = x　　　(hours)

Equation:　$562 = 60x$

$$\frac{562}{60} = x$$

$$9.37 \approx x$$

It would take the train approximately 9.37 hours (or approximately 9 hours and 22 minutes).

121. *Verbal model:* $\boxed{\text{Distance}} = \boxed{\text{Rate}} \cdot \boxed{\text{Time}}$

Labels:　Distance = 20　(kilometers)

　　　　　Rate = r　　　(kilometers per hour)

　　　　　Time = $3\dfrac{47}{60}$　(hours)

Equation:　$20 = r\left(3\dfrac{47}{60}\right)$

$$20 = r\left(\frac{227}{60}\right)$$

$$1200 = r(227)$$

$$\frac{1200}{227} = \frac{r(227)}{227}$$

$$5.3 \approx r$$

The rate is approximately 5.3 kilometers per hour.

123. *Verbal model:* | Value of dimes | + | Value of quarters | = | Total value |

Labels: Mixed coins: total value = $5.55, number of coins = 30

Dimes: value per coin = 0.10, number of coins = x

Quarters: value per coin = 0.25, number of coins = 30 − x

Equation: $0.10x + 0.25(30 - x) = 5.55$

$0.10x + 7.50 - 0.25x = 5.55$

$-0.05x + 7.50 = 5.55$

$-0.15x + 7.50 - 7.50 = 5.55 - 7.50$

$-0.15x = -1.95$

$x = \dfrac{-1.95}{-0.15}$

$x = 13$ and $30 - x = 17$

You have 13 dimes and 17 quarters.

125. *Common formula:* $P = 2l + 2w$

Labels:

$P = 112$	(feet)
$l = $ length	(feet)
$w = $ width $= l - 4$	(feet)

Equation:

$112 = 2l + 2(l - 4)$

$112 = 2l + 2l - 8$

$112 = 4l - 8$

$120 = 4l$

$\dfrac{120}{4} = l$

$30 = l$

$26 = l - 4$

The length of the pool is 30 feet and the width is 26 feet.

127. *Verbal model:* | Interest | = | Principal | · | Rate | · | Time |

Labels:

Interest $= I$	(dollars)
Principal $= 1000$	(dollars)
Rate $= 0.095$	(percent in decimal form)
Time $= 5$	(years)

Equation: $I = 1000(0.095)(5)$

$I = 475$

The total interest for the 5 years is $475.

129. *Common formula:* $I = Prt$

Labels:

$I = \$25,000$	
$P = $ principal	(dollars)
$r = 0.0875$	(percent in decimal form)
$t = 1$	

Equation:

$25,000 = P(0.0875)1$

$25,000 = P(0.0875)$

$\dfrac{25,000}{0.0875} = P$

$285,714 \approx P$

The required principal is approximately $285,714.

131. *Verbal model:*

| Work done | = | Portion done by first person | + | Portion done by second person |

Labels: Both persons: work done = 1 complete task, time = t (hours)

First person: rate = $\frac{1}{5}$ task per hour, time = t (hours)

Second person: rate = $\frac{1}{6}$ task per hour, time = t (hours)

Equation:
$$1 = \left(\tfrac{1}{5}\right)t + \left(\tfrac{1}{6}\right)t$$
$$1 = \left(\tfrac{1}{5} + \tfrac{1}{6}\right)t$$
$$1 = \left(\tfrac{6}{30} + \tfrac{5}{30}\right)t$$
$$1 = \left(\tfrac{11}{30}\right)t$$
$$\tfrac{30}{11}(1) = \tfrac{30}{11}\left(\tfrac{11}{30}\right)t$$
$$\tfrac{30}{11} = t$$

It would take $\frac{30}{11}$ hours (or approximately 2.7 hours) for both people to complete the task.

Chapter Test for Chapter 3

1.
$$4x - 3 = 18$$
$$4x - 3 + 3 = 18 + 3$$
$$4x = 21$$
$$\frac{4x}{4} = \frac{21}{4}$$
$$x = \frac{21}{4}$$

2.
$$10 - (2 - x) = 2x - 1$$
$$10 - 2 + x = 2x + 1$$
$$8 + x = 2x + 1$$
$$8 + x - 2x = 2x - 2x + 1$$
$$8 - x = 1$$
$$8 - 8 - x = 1 - 8$$
$$-x = -7$$
$$(-1)(-x) = (-1)(-7)$$
$$x = 7$$

3.
$$\frac{3x}{4} = \frac{5}{2} + x$$
$$4\left(\frac{3x}{4}\right) = 4\left(\frac{5}{2} + x\right)$$
$$3x = 4\left(\frac{5}{2}\right) + 4x$$
$$3x = 10 + 4x$$
$$3x - 4x = 10 + 4x - 4x$$
$$-x = 10$$
$$(-1)(-x) = (-1)(10)$$
$$x = -10$$

4.
$$\frac{t + 2}{3} = \frac{2t}{5}$$
$$5(t + 2) = 3(2t)$$
$$5t + 10 = 6t$$
$$5t - 5t + 10 = 6t - 5t$$
$$10 = t$$

5. $|x - 5| = 2$

$x - 5 = 2$ or $x - 5 = -2$
$x - 5 + 5 = 2 + 5$ $x - 5 + 5 = -2 + 5$
$x = 7$ $x = 3$

6. $|3x - 4| = 5$

$3x - 4 = 5$ or $3x - 4 = -5$
$3x - 4 + 4 = 5 + 4$ $3x - 4 + 4 = -5 + 4$
$3x = 9$ $3x = -1$
$x = \dfrac{9}{3}$ $x = \dfrac{-1}{3}$
$x = 3$ $x = -\dfrac{1}{3}$

7.
$$4.08(x + 10) = 9.50(x - 2)$$
$$4.08x + 40.8 = 9.50x - 19$$
$$4.08x - 9.50x + 40.8 = 9.50x - 9.50x - 19$$
$$-5.42x + 40.8 = -19$$
$$-5.42x + 40.8 - 40.8 = -19 - 40.8$$
$$-5.42x = -59.8$$
$$\frac{-5.42x}{-5.42} = \frac{-59.8}{-5.42}$$
$$x \approx 11.03 \quad \text{(Rounded)}$$

8. *Verbal model:* $\boxed{\text{Cost for parts}}$ + $\boxed{\text{Labor cost per hour}}$ · $\boxed{\text{Hours of labor}}$ = $\boxed{\text{Total bill}}$

   Labels: Cost for parts $= 62$ (dollars)

   Labor cost per hour $= 32$ (dollars/hour)

   Hours of labor $= x$ (hours)

   Total bill $= 142$ (dollars)

Equation: $62 + 32x = 142$

$62 - 62 + 32x = 142 - 62$

$32x = 80$

$$\frac{32x}{32} = \frac{80}{32}$$

$$x = 2\frac{1}{2}$$

Therefore, $2\frac{1}{2}$ hours were spent repairing the appliance.

9. (a) *Verbal model:* $\boxed{\text{Fraction}}$ · $\boxed{100\%}$ = $\boxed{\text{Percent}}$

   Label: Percent $= x$

   Equation: $\dfrac{3}{8}(100\%) = x$

$37.5\% = x$

   (b) *Verbal model:* $\boxed{\text{Decimal}}$ · $\boxed{100\%}$ = $\boxed{\text{Percent}}$

   Label: Decimal $= x$

   Equation: $x(100\%) = 37.5\%$

$$x = \frac{37.5\%}{100\%} = 0.375$$

10. *Verbal model:* $\boxed{324}$ = $\boxed{27\% \text{ of what number}}$ $(a = pb)$

   Label: $b =$ unknown number

   Equation: $324 = 0.27b$

$$\frac{324}{0.27} = b$$

$1200 = b$

Therefore, 324 is 27% of 1200.

11. *Verbal model:* $\boxed{90}$ = $\boxed{\text{What percent of 250}}$ $(a = pb)$

   Labels: $p =$ unknown percent (in decimal form)

   Percent equation: $90 = p(250)$

$$\frac{90}{250} = p$$

$0.36 = p$

Therefore, 90 is 36% of 250.

12. $\dfrac{40 \text{ inches}}{2 \text{ yards}} = \dfrac{40 \text{ inches}}{2(36) \text{ inches}} = \dfrac{40}{72} = \dfrac{5}{9}$

13. $\dfrac{2x}{3} = \dfrac{x + 4}{5}$

$5(2x) = 3(x + 4)$

$10x = 3x + 12$

$10x - 3x = 3x + 12 - 3x$

$7x = 12$

$$\frac{7x}{7} = \frac{12}{7}$$

$$x = \frac{12}{7}$$

14. The distance between Columbus and Akron measures 2 centimeters.

$$\frac{1 \text{ centimeters}}{55 \text{ miles}} = \frac{2 \text{ centimeters}}{x \text{ miles}}$$

$$\frac{1}{55} = \frac{2}{x}$$

$$1 \cdot x = 55 \cdot 2 \qquad \text{(Cross-multiply)}$$

$$x = 110$$

The distance between Columbus and Akron is approximately 110 miles.

15. *Verbal model:* $\boxed{\text{Distance}} = \boxed{\text{Rate}} \cdot \boxed{\text{Time}}$

Label: Distance = 264   (miles)

         Rate = x (miles per hour)

         Time = $5\frac{1}{2}$ (hours)

Equation: $264 = x\left(5\frac{1}{2}\right)$

$$\frac{264}{5.5} = x$$

$$48 = x$$

The average speed was 48 miles per hour.

16. *Verbal model:* $\boxed{\begin{array}{c}\text{Work}\\\text{done}\end{array}} = \boxed{\begin{array}{c}\text{Portion done}\\\text{by you}\end{array}} + \boxed{\begin{array}{c}\text{Portion done}\\\text{by your friend}\end{array}}$

Labels: Both persons: work done = 1 task, time = t (hours)

        Your work: rate = $\frac{1}{9}$ task per hour, time = t (hours)

        Friend's work: rate = $\frac{1}{12}$ task per hour, time = t (hours)

Equation: $1 = \left(\frac{1}{9}\right)t + \left(\frac{1}{12}\right)t$

$$1 = \left(\frac{1}{9} + \frac{1}{12}\right)t$$

$$1 = \left(\frac{4}{36} + \frac{3}{36}\right)t$$

$$1 = \frac{7}{36}t$$

$$\frac{36}{7}(1) = \frac{36}{7}\left(\frac{7}{36}\right)t$$

$$\frac{36}{7} = t$$

It would take $\frac{36}{7}$ hours (or approximately 5.1 hours) to paint the building.

17. $S = C + RC$

$$S - C = C - C + RC$$

$$S - C = RC$$

$$\frac{S - C}{C} = \frac{RC}{C}$$

$$\frac{S - C}{C} = R$$

Note: This answer could also be written as

$$\frac{S}{C} - 1 = R.$$

18. *Common formula:* $I = Prt$

Labels: $I = \$500$

         P = principal   (dollars)

         $r = 0.08$ (percent in decimal form)

         $t = 1$ (year)

Equation: $500 = P(0.08)(1)$

$$500 = P(0.08)$$

$$\frac{500}{0.08} = P$$

$$6250 = P$$

The required principal is \$6250.

19. $x + 3 \leq 7$

$$x + 3 - 3 \leq 7 - 3$$

$$x \leq 4$$

20. $-\dfrac{2x}{3} > 4$

$3\left(-\dfrac{2x}{3}\right) > 3(4)$

$-2x > 12$

$\dfrac{-2x}{-2} < \dfrac{12}{-2}$ (Reverse inequality)

$x < -6$

21. $-3 < 2x - 1 \le 3$

$-3 + 1 < 2x - 1 + 1 \le 3 + 1$

$-2 < 2x \le 4$

$\dfrac{-2}{2} < \dfrac{2x}{2} \le \dfrac{4}{2}$

$-1 < x \le 2$

22. $2 \ge \dfrac{3 - x}{2} > -1$

$2(2) \ge 2\left(\dfrac{3 - x}{2}\right) > 2(-1)$

$4 \ge 3 - x > -2$

$4 - 3 \ge 3 - 3 - x > -2 - 3$

$1 \ge -x > -5$

$(-1)(1) \le (-1)(-x) < (-1)(-5)$

$-1 \le x < 5$

23. $|x + 4| \le 3$

$-3 \le x + 4 \le 3$

$-3 - 4 \le x + 4 - 4 \le 3 - 4$

$-7 \le x \le -1$

24. $|2x - 1| > 3$

$\begin{array}{lll} 2x - 1 < -3 & \text{or} & 2x - 1 > 3 \\ 2x - 1 + 1 < -3 + 1 & & 2x - 1 + 1 > 3 + 1 \\ 2x < -2 & & 2x > 4 \\ x < \dfrac{-2}{2} & & x > \dfrac{4}{2} \\ x < -1 & & x > 2 \end{array}$

Cumulative Test for Chapters 1–3

1. $-\dfrac{3}{4} < \left|-\dfrac{7}{8}\right|$

Note: $\left|-\dfrac{7}{8}\right| = \dfrac{7}{8}$ and

$-\dfrac{3}{4} < \dfrac{7}{8}$

2. $(-200)(2)(-3) = 1200$

3. $\dfrac{3}{8} - \dfrac{5}{6} = \dfrac{3 \cdot 3}{8 \cdot 3} - \dfrac{5 \cdot 4}{6 \cdot 4}$

$= \dfrac{9}{24} - \dfrac{20}{24}$

$= \dfrac{9 - 20}{24}$

$= -\dfrac{11}{24}$

4. $-\dfrac{2}{9} \div \dfrac{8}{75} = -\dfrac{2}{9} \cdot \dfrac{75}{8}$

$= -\dfrac{2 \cdot 75}{9 \cdot 8}$

$= \dfrac{(2)(3)(25)}{(3)(3)(2)(4)}$

$= -\dfrac{25}{12}$

5. $-(-2)^3 = -(-8) = 8$

6. $3 + 2(6) - 1 = 3 + 12 - 1$

$= 15 - 1$

$= 14$

7. $24 + 12 \div 3 = 24 + 4$

$= 28$

8. When $x = -2$ and $y = 3$, the expression

$$2x + y^2 = 2(-2) + 3^2$$
$$= 2(-2) + 9$$
$$= -4 + 9$$
$$= 5$$

9. When $x = -2$ and $y = 3$, the expression

$$4y - x^3 = 4(3) - (-2)^3$$
$$= 4(3) - (-8)$$
$$= 12 - (-8)$$
$$= 12 + 8$$
$$= 20$$

10. $3 \cdot (x + y) \cdot (x + y) \cdot 3 \cdot 3 = 3^3(x + y)^2$

11. $-2x(x - 3) = -2x^2 + 6x$

12. Associative Property of Addition

13. $(3x^3)(5x^4) = 15x^{3+4} = 15x^7$

14. $(a^3b^2)(ab)^5 = (a^3b^2)(a^5b^5)$
$$= a^{3+5}b^{2+5}$$
$$= a^8b^7$$

15. $2x^2 - 3x + 5x^2 - (2 + 3x)$
$$= 2x^2 - 3x + 5x^2 - 2 - 3x$$
$$= 7x^2 - 6x - 2$$

16. $12x - 3 = 7x + 27$
$$5x - 3 = 27$$
$$5x = 30$$
$$x = \frac{30}{5}$$
$$x = 6$$

17. $2x - \dfrac{5x}{4} = 13$

$$4\left(2x - \frac{5x}{4}\right) = 4(13)$$
$$8x - 5x = 52$$
$$3x = 52$$
$$x = \frac{52}{3}$$

18. $2(x - 3) + 3 = 12 - x$
$$2x - 6 + 3 = 12 - x$$
$$2x - 3 = 12 - x$$
$$3x - 3 = 12$$
$$3x = 15$$
$$x = 5$$

19. $-1 \le \dfrac{x + 3}{2} < 2$

$$2(-1) \le 2\left(\frac{x + 3}{2}\right) < 2(2)$$
$$-2 \le x + 3 < 4$$
$$-2 - 3 \le x < 4 - 3$$
$$-5 \le x < 1$$

20. The annual fuel cost can be estimated by (a) dividing the 15,000 miles by the 28.3 miles per gallon to determine how many gallons of fuel will be needed, and then (b) multiplying the result by the \$1.179 cost per gallon of fuel.

Verbal model: $\boxed{\begin{array}{c}\text{Annual} \\ \text{cost}\end{array}} = \left(\boxed{\begin{array}{c}\text{Miles driven} \\ \text{per year}\end{array}} \div \boxed{\begin{array}{c}\text{Miles per} \\ \text{gallon}\end{array}}\right) \cdot \boxed{\begin{array}{c}\text{Cost per} \\ \text{gallon}\end{array}}$

Label:

Annual cost = x (dollars)

Miles driven per year = 15,000   (miles)

Miles per gallon = 28.3   (miles per gallon)

Cost per gallon = 1.179   (dollars per gallon)

Equation: $x = \dfrac{15,000}{28.3}(1.179)$

$$x \approx 624.91$$

Therefore, the annual fuel cost is approximately \$624.91.   *Note:* $\dfrac{15,000 \text{ miles}}{1 \text{ year}} \cdot \dfrac{1 \text{ gallon}}{28.3 \text{ miles}} \cdot \dfrac{\$1.179}{1 \text{ gallon}} \approx \$624.91 \text{ per year}$

21. $\dfrac{24 \text{ ounces}}{2 \text{ pounds}} = \dfrac{24 \text{ ounces}}{2(16) \text{ ounces}}$

$= \dfrac{24}{32}$

$= \dfrac{3}{4}$

22. *Verbal model:* | First consecutive even integer | $+$ | Second consecutive even integer | $=$ | 494 |

Labels: First consecutive even integer $= 2n$

Second consecutive even integer $= 2n + 2$

Equation: $2n + (2n + 2) = 494$

$4n + 2 = 494$

$4n = 492$

$n = \dfrac{492}{4}$

$n = 123$

$2n = 246$

$2n + 2 = 248$

The two consecutive even integers are 246 and 248.

23. *Verbal model:* | Sale price | $=$ | List price | $-$ | Discount |

Labels: Sale price $= x$ (dollars)

List price $= \$1150$

Discount rate $= 0.20$ (percent in decimal form)

Discount $= 0.20(1150) = \$230$

Equation: $x = 1150 - 230$

$x = 920$

The sale price of the camcorder is $920.

24. *Verbal model:* | $\dfrac{\text{Assessed value of larger piece}}{\text{Area of larger piece}}$ | $=$ | $\dfrac{\text{Assessed value of smaller piece}}{\text{Area of smaller piece}}$ |

Labels: Assessed value of larger piece $= 95{,}000$ (dollars)

Area of larger piece $= (100)(80) = 8000$ (square units)

Assessed value of smaller piece $= x$ (dollars)

Area of smaller piece $= (60)(80) = 4800$ (square units)

Proportion: $\dfrac{95{,}000}{8000} = \dfrac{x}{4800}$

$8000x = 95{,}000(4800)$

$\dfrac{8000x}{8000} = \dfrac{95{,}000(4800)}{8000}$

$x = 57{,}000$

Therefore, the assessed value of the smaller piece is $57,000.

CHAPTER 4
Graphs and Functions

CHAPTER 4
Graphs of Functions

Section 4.1   Ordered Pairs and Graphs

Solutions to Odd-Numbered Exercises

1.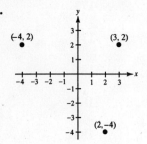

3.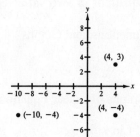

5.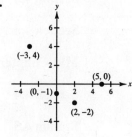

7.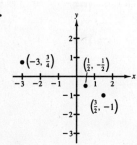

9.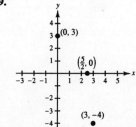

11. (a) $(5, 2)$

(b) $(-3, 4)$

(c) $(2, -5)$

(d) $(-2, -2)$

13. (a) $(-1, 3)$

(b) $(5, -3)$

(c) $(2, 1)$

(d) $(-1, -2)$

15. The point $(-3, 1)$ is located 3 units to the left of the vertical axis and 1 unit above the horizontal axis; it is in Quadrant II.

17. The point $\left(-\frac{1}{8}, -\frac{2}{7}\right)$ is located $\frac{1}{8}$ unit to the left of the vertical axis and $\frac{2}{7}$ unit below the horizontal axis; it is in Quadrant III.

19. The point $(-100, -365.6)$ is located 100 units to the left of the vertical axis and 365.6 units below the horizontal axis; it is in Quadrant III.

21. Quadrants II or III.

This point is located 5 units to the left of the vertical axis. If y is positive, the point would be above the horizontal axis and in the second quadrant. If y is negative, the point would be below the horizontal axis and in the third quadrant. (If y is zero, the point would be on the horizontal axis between the second and third quadrants.)

23. Quadrants III or IV.

This point is located 2 units below the horizontal axis. If x is positive, the point would be to the right of the vertical axis and in the fourth quadrant. If x is negative, the point would be to the left of the vertical axis and in the third quadrant. (If x is zero, the point would be on the vertical axis between the third and fourth quadrants.)

25. Quadrants II or IV.

The two coordinated must have opposite signs if $xy < 0$. If x is positive and y is negative, the point would be located to the right of the vertical axis and below the horizontal axis in the fourth quadrant. If x is negative and y is positive, the point would be located to the left of the vertical axis and above the horizontal axis in the second quadrant.

27. $(0, 3)$

29. $(-5, -10)$

31.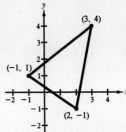

33.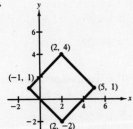

35.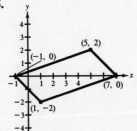

37.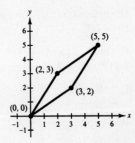

39.

x	-2	0	2	4	6
$y = 3x - 4$	-10	-4	2	8	14

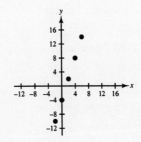

When $x = -2$, $y = 3(-2) - 4 = -6 - 4 = -10$.

When $x = 0$, $y = 3(0) - 4 = 0 - 4 = -4$.

When $x = 2$, $y = 3(2) - 4 = 6 - 4 = 2$.

When $x = 4$, $y = 3(4) - 4 = 12 - 4 = 8$.

When $x = 6$, $y = 3(6) - 4 = 18 - 4 = 14$.

41.

x	-4	-2	4	6	8
$y = -\frac{3}{2}x + 5$	11	8	-1	-4	-7

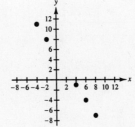

When $x = -4$, $y = -\frac{3}{2}(-4) + 5 = 6 + 5 = 11$.

When $x = -2$, $y = -\frac{3}{2}(-2) + 5 = 3 + 5 = 8$.

When $x = 4$, $y = -\frac{3}{2}(4) + 5 = -6 + 5 = -1$.

When $x = 6$, $y = -\frac{3}{2}(6) + 5 = -9 + 5 = -4$.

When $x = 8$, $y = -\frac{3}{2}(8) + 5 = -12 + 5 = -7$.

43.

x	-2	-1	0	1	2
$y = 2x - 1$	-5	-3	-1	1	3

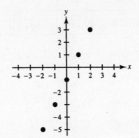

When $x = -2$, $y = 2(-2) - 1 = -4 - 1 = -5$.

When $x = -1$, $y = 2(-1) - 1 = -2 - 1 = -3$.

When $x = 0$, $y = 2(0) - 1 = 0 - 1 = -1$.

When $x = 1$, $y = 2(1) - 1 = 2 - 1 = 1$.

When $x = 2$, $y = 2(2) - 1 = 4 - 1 = 3$.

45.
$$6x - 3y = 3$$
$$6x - 6x - 3y = 3 - 6x$$
$$-3y = -6x + 3$$
$$\frac{-3y}{-3} = \frac{-6x + 3}{-3}$$
$$y = 2x - 1$$

47.
$$x + 4y = 8$$
$$x - x + 4y = 8 - x$$
$$4y = -x + 8$$
$$\frac{4y}{4} = \frac{-x + 8}{4}$$
$$y = -\frac{1}{4}x + 2$$

49.
$$4x - 5y = 3$$
$$4x - 4x - 5y = 3 - 4x$$
$$-5y = -4x + 3$$
$$\frac{-5y}{-5} = \frac{-4x + 3}{-5}$$
$$y = \frac{4}{5}x - \frac{3}{5}$$

51. (a) The ordered pair $(3, 10)$ *is* a solution because
$10 = 2(3) + 4$.

    (b) The ordered pair $(-1, 3)$ *is not* a solution because
$3 \neq 2(-1) + 4$.

    (c) The ordered pair $(0, 0)$ *is not* a solution because
$0 \neq 2(0) + 4$.

    (d) The ordered pair $(-2, 0)$ *is* a solution because
$0 = 2(-2) + 4$.

53. (a) The ordered pair $(1, 1)$ *is* a solution because
$2(1) - 3(1) + 1 = 2 - 3 + 1 = 0$.

    (b) The ordered pair $(5, 7)$ *is* a solution because
$2(7) - 3(5) + 1 = 14 - 15 + 1 = 0$.

    (c) The ordered pair $(-3, -1)$ *is not* a solution because
$2(-1) - 3(-3) + 1 = -2 + 9 + 1 = 6 \neq 0$.

    (d) The ordered pair $(-3, -5)$ *is* a solution because
$2(-5) - 3(-3) + 1 = 0$.

55. (a) The ordered pair $(6, 6)$ *is not* a solution because
$6 \neq \frac{2}{3}(6)$.

    (b) The ordered pair $(-9, -6)$ *is* a solution because
$-6 = \frac{2}{3}(-9)$.

    (c) The ordered pair $(0, 0)$ *is* a solution because $0 = \frac{2}{3}(0)$.

    (d) The ordered pair $\left(-1, \frac{2}{3}\right)$ *is not* a solution because
$\frac{2}{3} \neq \frac{2}{3}(-1)$.

57. (a) The ordered pair $\left(-\frac{1}{2}, 5\right)$ *is* a solution because
$5 = 3 - 4\left(-\frac{1}{2}\right)$.

    (b) The ordered pair $(1, 7)$ *is not* a solution because
$7 \neq 3 - 4(1)$.

    (c) The ordered pair $(0, 0)$ *is not* a solution because
$0 \neq 3 - 4(0)$.

    (d) The ordered pair $\left(-\frac{3}{4}, 0\right)$ *is not* a solution because
$0 \neq 3 - 4\left(-\frac{3}{4}\right)$.

59.

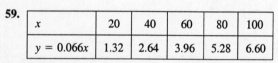

x	20	40	60	80	100
$y = 0.066x$	1.32	2.64	3.96	5.28	6.60

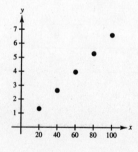

When $x = 20$, $y = 0.066(20) = 1.32$.

When $x = 40$, $y = 0.066(40) = 2.64$.

When $x = 60$, $y = 0.066(60) = 3.96$.

When $x = 80$, $y = 0.066(80) = 5.28$.

When $x = 100$, $y = 0.066(100) = 6.60$.

61.

x	100	150	200	250	300
$y = 35x + 5000$	8500	10,250	12,000	13,750	15,500

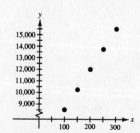

When $x = 100$, $y = 35(100) + 5000 = 8500$.

When $x = 150$, $y = 35(150) + 5000 = 10,250$.

When $x = 200$, $y = 35(200) + 5000 = 12,000$.

When $x = 250$, $y = 35(250) + 5000 = 13,750$.

When $x = 300$, $y = 35(300) + 5000 = 15,500$.

63. (a)

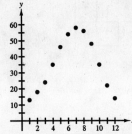

(b) No. On the graph shown above, each mark on the horizontal axis represents 1 unit and each mark on the vertical axis represents 5 units. *Note:* Answers many vary on this question.

(c) Normal temperatures change the least during the months of June, July, and August.

65. (a)

(b) As the number of hours of study increases, the exam score increases.

67. There were approximately 1,800,000 new housing starts in 1986.

69. The increase in housing starts from 1993 to 1994 was approximately 170,000. This was an increase of about 13%.

71. The per capita personal income was approximately $20,750 in 1992.

73. The percent increase in per capita personal income from 1996 to 1997 was approximately 5%.

75. Approximately 7% of the gross domestic product was spent on health care in Sweden in 1995.

77. (a) & (b)

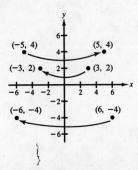

Original Points	New Points
(3, 2)	(−3, 2)
(−5, 4)	(5, 4)
(6, −4)	(−6, −4)

(c) When the sign of the x-coordinate is changed, the location of the point is reflected about the y-axis.

79. A pair of coordinates are referred to as an "ordered" pair because the order of the two numbers is important. The points (3, 5) and (5, 3) are not the same; the points (8, 0) and (0, 8) are not the same. The first number shows how far to the right or left the point is from the vertical axis, and the second number tells how far up or down the point is from the horizontal axis.

81. The x-coordinate of any point on the y-axis is 0.

The y-coordinate of any point on the x-axis is 0.

83. Points that lie in the third quadrant have negative first and second coordinates. Points that lie in the fourth quadrant have positive first coordinates and negative second coordinates.

85. The y-coordinates increase if the coefficient of x is positive, and the y-coordinates decrease if the coefficient of x is negative.

Section 4.2   Graphs of Equations in Two Variables

1. Graph (g)   **3.** Graph (a)   **5.** Graph (h)   **7.** Graph (d)

9.

x	-2	-1	0	1	2
y	11	10	9	8	7

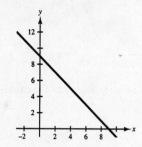

11.

x	-2	-1	0	1	2
y	-10	-6	-2	2	6

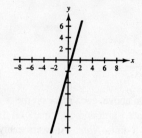

13.

x	-2	0	2	4	6
y	3	2	1	0	-1

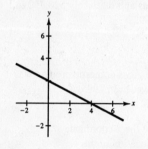

15.

x	-3	-2	-1	0	1
y	2	1	0	1	2

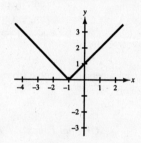

17. *Graphical solution:* It appears that the x-intercept is $(-2, 0)$ and the y-intercept is $(0, 4)$.

> *Algebraic check:* $4x - 2y = -8$ $4x - 2y = -8$
>
> Let $y = 0$.   Let $x = 0$.
>
> $4x - 2(0) = -8$ $4(0) - 2y = -8$
>
> $4x = -8$ $-2y = -8$
>
> $x = -2$ $y = 4$
>
> x-intercept: $(-2, 0)$ y-intercept: $(0, 4)$

19. *Graphical solution:* It appears that the x-intercept is $(6, 0)$ and the y-intercept is $(0, 2)$.

> *Algebraic check:* $x + 3y = 6$ $x + 3y = 6$
>
> Let $y = 0$.   Let $x = 0$.
>
> $x + 3(0) = 6$ $0 + 3y = 6$
>
> $x = 6$ $3y = 6$
>
> x-intercept: $(6, 0)$ $y = 2$
>
> y-intercept: $(0, 2)$

21. *Graphical solution:* It appears that the x-intercepts are $(-3, 0)$ and $(3, 0)$; it appears that the y-intercept is $(0, -3)$.

> *Algebraic check:* $y = |x| - 3$ $y = |x| - 3$
>
> Let $y = 0$.   Let $x = 0$.
>
> $0 = |x| - 3$ $y = |0| - 3$
>
> $3 = |x|$ $y = -3$
>
> $x = -3$ or $x = 3$ y-intercept: $(0, -3)$
>
> x-intercepts: $(-3, 0), (3, 0)$

23. *Graphical solution:* It appears that the x-intercepts are $(4, 0)$ and $(-4, 0)$ and the y-intercept is $(0, 16)$.

Algebraic check: $y = 16 - x^2$ $y = 16 - x^2$

$\qquad\qquad$ Let $y = 0$.   Let $x = 0$.

$\qquad\qquad$ $0 = 16 - x^2$ $y = 16 - 0^2$

$\qquad\qquad$ $0 = (4 + x)(4 - x)$ $y = 16$

$\qquad\qquad$ $4 + x = 0 \implies x = -4$ y-intercept: $(0, 16)$

$\qquad\qquad$ $4 - x = 0 \implies x = 4$

$\qquad\qquad$ x-intercepts: $(-4, 0)$ and $(4, 0)$

25. *x-intercept* $\qquad$ *y-intercept*

Let $y = 0$. $\qquad\qquad$ Let $x = 0$.

$\quad 0 = 6x + 2 \qquad\quad y = 6(0) + 2$

$\ -2 = 6x \qquad\qquad\quad y = 0 + 2$

$\ -\frac{1}{3} = x \qquad\qquad\quad y = 2$

$\left(-\frac{1}{3}, 0\right) \qquad\qquad\ (0, 2)$

The graph has one x-intercept at $\left(-\frac{1}{3}, 0\right)$ and one y-intercept at $(0, 2)$.

27. *x-intercept* $\qquad$ *y-intercept*

Let $y = 0$. $\qquad\qquad$ Let $x = 0$.

$\quad 0 = \frac{1}{2}x - 1 \qquad\quad y = \frac{1}{2}(0) - 1$

$2(0) = 2\left(\frac{1}{2}x - 1\right) \quad\ y = 0 - 1$

$\quad 0 = x - 2 \qquad\qquad y = -1$

$\quad 2 = 0 \qquad\qquad\qquad (0, -1)$

$(2, 0)$

The graph has one x-intercept at $(2, 0)$ and one y-intercept at $(0, -1)$.

29. *x-intercept* $\qquad$ *y-intercept*

Let $y = 0$. $\qquad\qquad$ Let $x = 0$.

$x - 0 = 1 \qquad\qquad 0 - y = 1$

$\quad\ x = 1 \qquad\qquad\quad -y = 1$

$(1, 0) \qquad\qquad\qquad\ y = -1$

$\qquad\qquad\qquad\qquad\ (0, -1)$

The graph has one x-intercept at $(1, 0)$ and one y-intercept at $(0, -1)$.

31. *x-intercept* $\qquad$ *y-intercept*

Let $y = 0$. $\qquad\qquad$ Let $x = 0$.

$2x + 0 = 4 \qquad\qquad 2(0) + y = 4$

$\quad\ 2x = 4 \qquad\qquad\qquad\ y = 4$

$\quad\ \ x = 2 \qquad\qquad\qquad (0, 4)$

$(2, 0)$

The graph has one x-intercept $(2, 0)$ and one y-intercept $(0, 4)$.

33. *x-intercept* $\qquad$ *y-intercept*

Let $y = 0$. $\qquad\qquad$ Let $x = 0$.

$2x + 6(0) - 9 = 0 \qquad 2(0) + 6y - 9 = 0$

$\quad\ 2x - 9 = 0 \qquad\qquad\ 6y - 9 = 0$

$\qquad 2x = 9 \qquad\qquad\qquad 6y = 9$

$\qquad\ \ x = \frac{9}{2} \qquad\qquad\qquad\ y = \frac{9}{6}$

$\left(\frac{9}{2}, 0\right) \qquad\qquad\qquad\qquad\ y = \frac{3}{2}$

$\qquad\qquad\qquad \left(0, \frac{3}{2}\right)$

The graph has one x-intercept at $\left(\frac{9}{2}, 0\right)$ and one y-intercept at $\left(0, \frac{3}{2}\right)$.

35. *x-intercept*

Let $y = 0$.

$\frac{3}{4}x - \frac{1}{2}(0) = 3$

$\frac{3}{4}x = 3$

$4\left(\frac{3}{4}x\right) = 4(3)$

$3x = 12$

$x = 4$

$(4, 0)$

y-intercept

Let $x = 0$.

$\frac{3}{4}(0) - \frac{1}{2}y = 3$

$-\frac{1}{2}y = 3$

$2\left(-\frac{1}{2}y\right) = 2(3)$

$-y = 6$

$y = -6$

$(0, -6)$

The graph has one x-intercept at $(4, 0)$ and one y-intercept at $(0, -6)$.

37.

x	0	2	3
$y = 2 - x$	2	0	-1

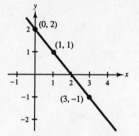

39.

x	0	1	3
$y = x - 1$	-1	0	2

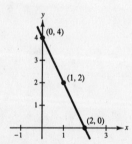

41.

x	-1	0	2
$y = 3x$	-3	0	6

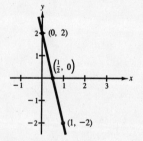

43. $2x - y = 4$

$-y = -2x + 4$

$y = 2x - 4$

x	0	1	2
$y = 2x - 4$	-4	-2	0

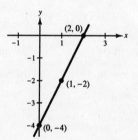

45. $10x + 5y = 20$

$5y = -10x + 20$

$y = -2x + 4$

x	0	1	2
$y = -2x + 4$	4	2	0

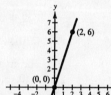

47. $4x + y = 2$

$y = -4x + 2$

x	0	$\frac{1}{2}$	1
$y = -4x + 2$	2	0	-2

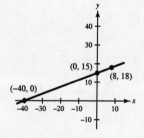

49.

x	-40	0	8
$y = \frac{3}{8}x + 15$	0	15	18

51.

x	0	3	$\frac{15}{2}$
$y = \frac{2}{3}x - 5$	-5	-3	0

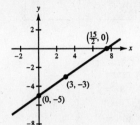

53.

x	-2	-1	0	1	2
$y = x^2$	4	1	0	1	4

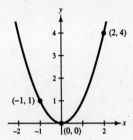

55.

x	-3	-2	0	1	3
$y = -x^2 + 9$	0	5	9	6	0

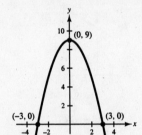

57.

x	1	3	5
$y = (x - 3)^2$	4	0	4

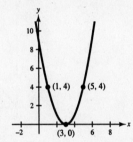

59.

x	0	4	5	6
$y = |x - 5|$	5	1	0	1

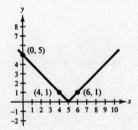

61.

x	-5	0	5
$y = 5 - |x|$	0	5	0

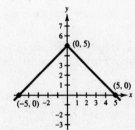

63.

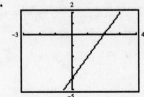

Yes, the graphs are identical. By the Commutative Property of Addition, $\frac{1}{3}x - 1 = -1 + \frac{1}{3}x$.

65.

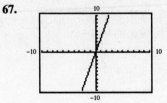

Yes, the graphs are identical. By the Distributive Property, $2(x - 2) = 2x - 4$.

67.

69.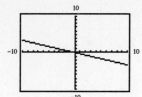

71.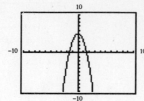

73.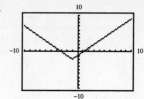

75.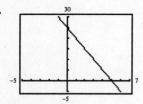

77.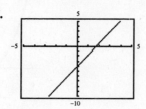

79.
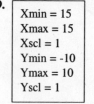
Xmin = 15
Xmax = 15
Xscl = 1
Ymin = -10
Ymax = 10
Yscl = 1

81.
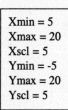
Xmin = 5
Xmax = 20
Xscl = 5
Ymin = -5
Ymax = 20
Yscl = 5

83. $y = 35t$

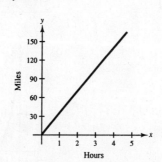

85. (a)

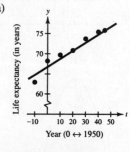

(b) $y = 0.2t + 66.7$

$y = 0.2(60) + 66.7$

$y = 12 + 66.7$

$y = 78.7$

87. The graph on the left is the declining balances depreciation graph. The graph on the right is the straight-line depreciation graph.

89. The straight-line depreciation is easier to compute. The declining balances method yields a more realistic approximation of the faster rate of depreciation early in the useful lifetime equipment.

91. The graph of an equation is the set of all solutions of the equation plotted on a rectangular coordinate system.

93. The point-plotting method of sketching the graph of an equation involves making a table of values showing several solution points, then plotting these points on a rectangular coordinate system, and connecting the points with a smooth curve or line.

95. To find the x-intercept(s), set $y = 0$ and solve the equation for x. To find the y-intercept(s), set $x = 0$ and solve the equation for y.

97. Answers will vary.

Section 4.3   Relations, Functions, and Graphs

1. Domain: $\{-4, 1, 2, 4\}$

Range: $\{-3, 2, 3, 5\}$

3. Domain: $\{-9, \frac{1}{2}, 2\}$

Range: $\{-10, 0, 16\}$

5. Domain: $\{-1, 1, 5, 8\}$

Range: $\{-7, -2, 3, 4\}$

7. Yes. No first component has two different second components, so this relation *is* a function.

9. No. Since one of the first components, -2, is paired with two second components, this relation is *not* a function.

11. Yes. No first component has two different second components, so this relation *is* a function.

13. No. One first component, 0, is paired with two second components, so the relation is *not* a function.

15. No. Since first components, CBS and ABC, are paired with three second components, this relation is *not* a function.

17. Yes. No first component has two different second components, so this relation *is* a function.

19. Yes. No first component has two different second components, so the relation *is* a function.

21. No. Two first components, 1 and 3, are paired with two second components, so the relation is *not* a function.

23. Yes. No first component has two different second components, so the relation *is* a function.

25. Yes. This graph indicates that *y is* a function of *x* because *no* vertical would intersect the graph more than once.

27. Yes, *y is* a function of *x*. No vertical line would intersect the graph at more than one point.

29. No, *y* is *not* a function of *x*. Some vertical lines could intersect the graph at more than one point.

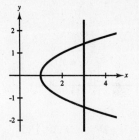

31. Yes. This graph indicates that *y is* a function of *x* because *no* vertical line would intersect the graph more than once.

33. No. This graph indicates that *y* is *not* a function of *x* because it *is* possible for a vertical line to intersect the graph more than once.

35. Yes, *y is* a function of *x*. No vertical line would intersect the graph at more than one point.

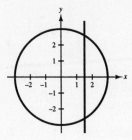

37. (a) $f(2) = \frac{1}{2}(2) = 1$

(b) $f(5) = \frac{1}{2}(5) = \frac{5}{2}$

(c) $f(-4) = \frac{1}{2}(-4) = -2$

(d) $f\left(-\frac{2}{3}\right) = \frac{1}{2}\left(-\frac{2}{3}\right) = -\frac{1}{3}$

39. (a) $f(0) = 2(0) - 1 = -1$

(b) $f(3) = 2(3) - 1 = 5$

(c) $f(-3) = 2(-3) - 1 = -7$

(d) $f\left(-\frac{1}{2}\right) = 2\left(-\frac{1}{2}\right) - 1 = -1 - 1 = -2$

41. (a) $f(1) = 4(1) + 1 = 5$

(b) $f(-1) = 4(-1) + 1 = -3$

(c) $f(-4) = 4(-4) + 1 = -15$

(d) $f\left(-\frac{4}{3}\right) = 4\left(-\frac{4}{3}\right) + 1 = -\frac{13}{3}$

43. (a) $h(200) = \frac{1}{4}(200) - 1 = 49$

(b) $h(-12) = \frac{1}{4}(-12) - 1 = -4$

(c) $h(8) = \frac{1}{4}(8) - 1 = 1$

(d) $h\left(-\frac{5}{2}\right) = \frac{1}{4}\left(-\frac{5}{2}\right) - 1 = -\frac{13}{8}$

45. (a) $f(-4) = \frac{1}{2}(-4)^2 = \frac{1}{2}(16) = 8$

(b) $f(4) = \frac{1}{2}(4)^2 = \frac{1}{2}(16) = 8$

(c) $f(0) = \frac{1}{2}(0)^2 = \frac{1}{2}(0) = 0$

(d) $f(2) = \frac{1}{2}(2)^2 = \frac{1}{2}(4) = 2$

47. (a) $g(0) = 2(0)^2 - 3(0) + 1 = 0 - 0 + 1 = 1$

(b) $g(-2) = 2(-2)^2 - 3(-2) + 1 = 8 + 6 + 1 = 15$

(c) $g(1) = 2(1)^2 - 3(1) + 1 = 2 - 3 + 1 = 0$

(d) $g\left(\frac{1}{2}\right) = 2\left(\frac{1}{2}\right)^2 - 3\left(\frac{1}{2}\right) + 1 = \frac{1}{2} - \frac{3}{2} + 1 = 0$

49. (a) $g(2) = |2 + 2| = |4| = 4$

(b) $g(-2) = |-2 + 2| = |0| = 0$

(c) $g(10) = |10 + 2| = |12| = 12$

(d) $g\left(-\frac{5}{2}\right) = \left|-\frac{5}{2} + 2\right| = \left|-\frac{1}{2}\right| = \frac{1}{2}$

51. (a) $h(0) = 0^3 - 1 = -1$

(b) $h(1) = 1^3 - 1 = 0$

(c) $h(3) = 3^3 - 1 = 26$

(d) $h\left(\frac{1}{2}\right) = \left(\frac{1}{2}\right)^3 - 1 = \frac{1}{8} - 1 = -\frac{7}{8}$

53. $R = \{4 - 0, 4 - 1, 4 - 2, 4 - 3, 4 - 4\}$

$= \{4, 3, 2, 1, 0\}$

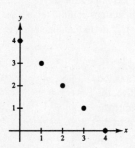

55. $R = \{100\}$

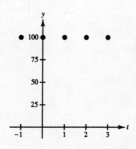

57. $R = \{(-2)^3, (-1)^3, 0^3, 1^3, 2^3\}$

$= \{-8, -1, 0, 1, 8\}$

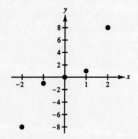

59. $R = \{|-2|, |-1|, |0|, |1|, |2|\}$

$= \{2, 1, 0, 1, 2\}$

$= \{2, 1, 0\}$

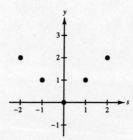

61. (a) $f(10) = 20 - 0.5(10)$ $f(15) = 20 - 0.5(15)$

$= 20 - 5$ $= 20 - 7.5$

$= 15$ $= 12.5$

(b) As the price increases, the demand decreases.

63. (a) $d(2) = 50(2) = 100$

When $t = 2$ hours, the car will travel 100 miles.

(b) $d(4) = 50(4) = 200$

When $t = 4$ hours, the car will travel 200 miles.

(c) $d(10) = 50(10) = 500$

When $t = 10$ hours, the car will travel 500 miles.

65. Yes, the high school enrollment is a function of the year.

67. $f(1992) \approx 12,900,000$

69. $P = 4s$

Yes, *P is* a function of *s*.

71. (a) Yes. No day of the year, *t*, has two different lengths of time between sunrise and sunset, *L*, so the relation *is* a function.

(b) *R:* $9.5 \leq L \leq 15$

73. A relation is any set of ordered pairs. A function is a relation in which no two ordered pairs have the same first component and different second components. Here are two examples of relations that are not functions:

$\{(5, 2), (5, 0), (7, 3)\}$

Domain	Range
1	4
2	5
3	6

75. The domain of a function is the set of inputs of that function; the range of a function is the set of outputs of the function.

77. If no vertical line intersects the graph of an equation at two or more points, then the equation represents *y* as a function of *x*.

79. Yes, the number of elements in the domain can be greater than the number of elements in the range. Here is an example:

$\{(1, 5), (2, 5), (3, 5), (4, 5)\}$

Mid-Chapter Quiz for Chapter 4

1.

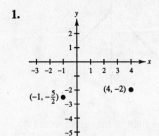

2. Quadrants I or II. This point is located 5 units *above* the horizontal axis. If *x* is positive, the point would be to the *right* of the vertical axis and in the first quadrant. If *x* is negative, the point would be to the *left* of the vertical axis and in the second quadrant. (If *x* is zero, the point would be on the vertical axis *between* the second and third quadrants.)

3. (a) Yes, the ordered pair (2, 7) *is* a solution because $7 = 9 - |2|$.

(b) No, the ordered pair $(-3, 12)$ is *not* a solution because $12 \neq 9 - |-3|$.

(c) Yes, the ordered pair $(-9, 0)$ *is* a solution because $0 = 9 - |-9|$.

(d) No, the ordered pair $(0, -9)$ is *not* a solution because $-9 \neq 9 - |0|$.

4. 1991: Approximately 180 million

1992: Approximately 200 million

1993: Approximately 270 million

1994: Approximately 290 million

1995: Approximately 350 million

1996: Approximately 410 million

1997: Approximately 530 million

5. *x-intercept*

Let $y = 0$.

$x - 3(0) = 12$

$x = 12$

$(12, 0)$

y-intercept

Let $x = 0$.

$0 - 3y = 12$

$-3y = 12$

$\dfrac{-3y}{-3} = \dfrac{12}{-3}$

$y = -4$

$(0, -4)$

The graph has one *x*-intercept at $(12, 0)$ and one *y*-intercept at $(0, -4)$.

6. *x-intercept*

Let $y = 0$.

$0 = 6 - 4x$

$4x = 6$

$x = \dfrac{6}{4}$

$x = \dfrac{3}{2}$

$\left(\dfrac{3}{2}, 0\right)$

y-intercept

Let $x = 0$.

$y = 6 - 4(0)$

$y = 6$

$(0, 6)$

The graph has one *x*-intercept at $\left(\dfrac{3}{2}, 0\right)$ and one *y*-intercept at $(0, 6)$.

7.

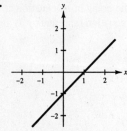

8.

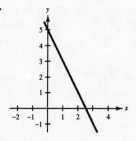

9.

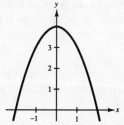

10.

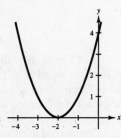

11.

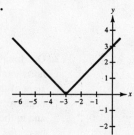

12.

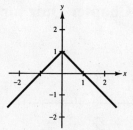

13. Yes. No first componenet has two different second components, so this relation *is* a function.

14. No. Any vertical line to the right of the origin would intersect the graph at more than one point, so the graph does *not* represent *y* as a function of *x*.

15. (a) $f(-2) = 3(-2) - 2 = -6 - 2 = -8$

(b) $f(0) = 3(0) - 2 = 0 - 2 = -2$

(c) $f(5) = 3(5) - 2 = 15 - 2 = 13$

(d) $f\left(-\frac{1}{3}\right) = 3\left(-\frac{1}{3}\right) - 2 = -1 - 2 = -3$

16. (a) $g(-2) = 2(-2)^2 - |-2|$

$= 2(4) - 2$

$= 8 - 2$

$= 6$

(b) $g(2) = 2(2)^2 - |2|$

$= 2(4) - 2$

$= 8 - 2$

$= 6$

(c) $g(0) = 2(0)^2 - |0|$

$= 2(0) - 0$

$= 0 - 0$

$= 0$

(d) $g\left(-\frac{1}{2}\right) = 2\left(-\frac{1}{2}\right)^2 - \left|-\frac{1}{2}\right|$

$= 2\left(\frac{1}{4}\right) - \frac{1}{2}$

$= \frac{1}{2} - \frac{1}{2}$

$= 0$

17. $R = \{(-2)^2 - (-2), (-1)^2 - (-1), 0^2 - 0, 1^2 - 1, 2^2 - 2\}$

$= \{4 + 2, 1 + 1, 0 - 0, 1 - 1, 4 - 2\}$

$= \{6, 2, 0\}$

18. The domain for the side is $s > 0$.

19. The intercepts are $(-1, 0)$, $\left(\frac{7}{3}, 0\right)$, and $(0, -7)$.
To verify the x-intercepts algebraically, set $h(x) = 0$,
or $3x^2 - 4x - 7 = 0$, and solve for x. To verify
the y-intercept algebraically, set $x = 0$, or
$h(x) = 3(0)^2 - 4(0) - 7 = -7$.

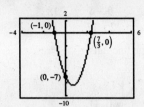

20. $V = 3000 - 500t$
Domain: $0 \le t \le 4$

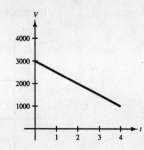

Section 4.4   Slope and Graphs of Linear Equations

1. $m = \dfrac{1}{1} = 1$

3. $m = 0$

5. $m = \dfrac{-1}{3} = -\dfrac{1}{3}$

7. m is undefined.

9. $m = \dfrac{5}{4}$

11. (a) L_2 has slope $m = \dfrac{3}{2}$

(b) L_3 has slope $m = 0$

(c) L_4 has slope $m = -\dfrac{2}{3}$

(d) L_1 has slope $m = -2$

13. $m = \dfrac{5 - 0}{4 - 0} = \dfrac{5}{4}$

The line rises.

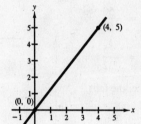

15. $m = \dfrac{-4 - 0}{8 - 0}$

$= \dfrac{-4}{8}$

$= -\dfrac{1}{2}$

The line falls.

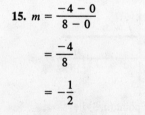

17. $m = \dfrac{0 - 6}{8 - 0} = \dfrac{-6}{8} = -\dfrac{3}{4}$

The line falls.

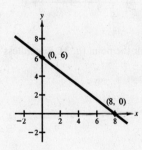

19. $m = \dfrac{6 - (-2)}{1 - (-3)} = \dfrac{8}{4} = 2$

The line rises.

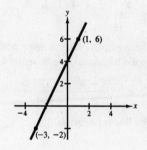

21. $m = \dfrac{4 - (-1)}{-6 - (-6)} = \dfrac{5}{0}$

m is undefined. The line is vertical.

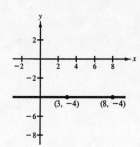

23. $m = \dfrac{-4 - (-4)}{8 - 3} = \dfrac{0}{5} = 0$

The line is horizontal.

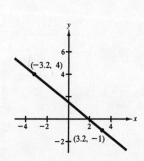

25. $m = \dfrac{-3 - \frac{3}{2}}{\frac{9}{2} - \frac{1}{4}} = \dfrac{-\frac{6}{2} - \frac{3}{2}}{\frac{18}{4} - \frac{1}{4}}$

$= \dfrac{-\frac{9}{2}}{\frac{17}{4}} = -\dfrac{9}{2} \cdot \dfrac{4}{17}$

$= -\dfrac{9 \cdot 4}{2 \cdot 17} = -\dfrac{18}{17}$

The line falls.

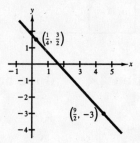

27. $m = \dfrac{4 - (-1)}{-3.2 - 3.2}$

$= \dfrac{5}{-6.4}$

$= -\dfrac{50}{64}$

$= -\dfrac{25}{32}$

The line falls.

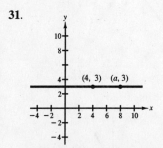

29. $m = \dfrac{4.25 - (-1)}{5.75 - 3.5}$

$= \dfrac{5.25}{2.25}$

$= \dfrac{525}{225} = \dfrac{7}{3}$

The line rises.

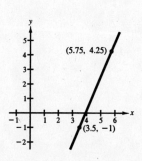

31.

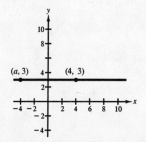

These graphs illustrate some possible locations for the point $(a, 3)$. Regardless of the value of a, the point $(a, 3)$ lies on the horizontal line through $(4, 3)$.

$$m = \dfrac{3 - 3}{a - 4} = \dfrac{0}{a - 4} = 0$$

The line is horizontal.

33.

x	-2	0	2	4
$y = -2x - 2$	2	-2	-6	-10
Solution points	$(-2, 2)$	$(0, -2)$	$(2, -6)$	$(4, -10)$

Using $(-2, 2)$ and $(2, -6)$: $m = \dfrac{-6 - 2}{2 - (-2)} = \dfrac{-8}{4} = -2$

Using $(0, -2)$ and $(4, -10)$: $m = \dfrac{-10 - (-2)}{4 - 0} = \dfrac{-8}{4} = -2$

35. $\dfrac{y - (-2)}{0 - 3} = -8$

$\dfrac{y + 2}{-3} = -8$

$y + 2 = (-3)(-8)$

$y + 2 = 24$

$y = 22$

37. $\dfrac{6 - y}{7 - (-4)} = \dfrac{5}{2}$

$\dfrac{6 - y}{11} = \dfrac{5}{2}$

$2(6 - y) = 55$ (Cross-multiply)

$12 - 2y = 55$

$-2y = 43$

$y = \dfrac{-43}{2}$

39.

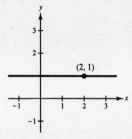

$(3, 1), (10, 1), (-2, 1)$, etc.

41.

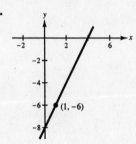

$(2, -4), (3, -2), (4, 0)$, etc.

43.

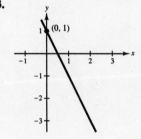

$(1, -1), (2, -3), (3, -5)$, etc.

45.

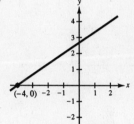

$(-1, 2), (2, 4), (5, 6)$, etc.

47.

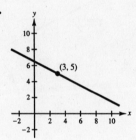

$(5, 4), (7, 3), (9, 2), (11, 1)$, etc.

49.

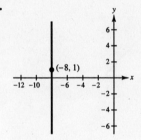

$(-8, 2), (-8, 10), (-8, -3)$, etc.

51.

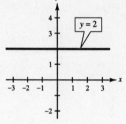

53.

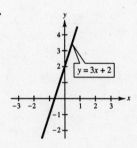

55.

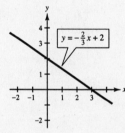

57. *x-intercept*

Let $y = 0$.

$2x - 3(0) + 6 = 0$

$2x + 6 = 0$

$2x = -6$

$x = -3$

$(-3, 0)$

y-intercept

Let $x = 0$.

$2(0) - 3y + 6 = 0$

$-3y + 6 = 0$

$-3y = -6$

$y = 2$

$(0, 2)$

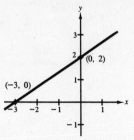

59. *x-intercept*

Let $y = 0$.

$-5x + 2(0) - 10 = 0$

$-5x - 10 = 0$

$-5x = 10$

$x = -2$

$(-2, 0)$

y-intercept

Let $x = 0$.

$-5(0) + 2y - 10 = 0$

$2y - 10 = 0$

$2y = 10$

$y = 5$

$(0, 5)$

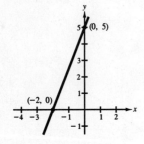

61. *x-intercept*

Let $y = 0$.

$6x - 4(0) + 12 = 0$

$6x + 12 = 0$

$6x = -12$

$x = -2$

$(-2, 0)$

y-intercept

Let $x = 0$.

$6(0) - 4y + 12 = 0$

$-4y + 12 = 0$

$-4y = -12$

$y = 3$

$(0, 3)$

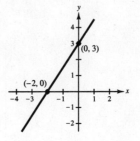

63. $x + y = 0$

$y = -x$

$m = -1$

y-intercept: $(0, b) = (0, 0)$

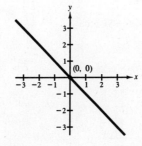

65. $\frac{1}{2}x + y = 0$

$y = -\frac{1}{2}x$

$m = -\frac{1}{2}$

y-intercept: $(0, b) = (0, 0)$

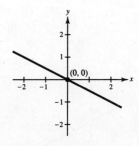

67. $2x - y - 3 = 0$

$-y = -2x + 3$

$y = 2x - 3$

$m = 2$

y-intercept: $(0, b) = (0, -3)$

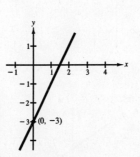

69. $x - 3y + 6 = 0$

$-3y = -x - 6$

$\frac{-3y}{-3} = \frac{-x}{-3} + \frac{-6}{-3}$

$y = \frac{1}{3}x + 2$

$m = \frac{1}{3}$

y-intercept: $(0, b) = (0, 2)$

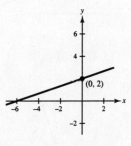

71. $x + 2y - 2 = 0$

$2y = -x + 2$

$y = -\frac{1}{2}x + 1$

$m = -\frac{1}{2}$

y-intercept: $(0, b) = (0, 1)$

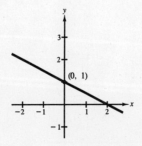

73. $3x - 4y + 2 = 0$

$-4y = -3x - 2$

$y = \frac{3}{4}x + \frac{1}{2}$

$m = \frac{3}{4}$

y-intercept: $(0, b) = \left(0, \frac{1}{2}\right)$

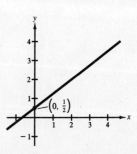

75. $y + 5 = 0$

$y = -5$

$m = 0$

y-intercept: $(0, b) = (0, -5)$

77. L_1: $(0, -1)$, $(5, 9)$

$$m_1 = \frac{9 - (-1)}{5 - 0} = \frac{10}{5} = 2$$

L_2: $(0, 3)$, $(4, 1)$

$$m_2 = \frac{1 - 3}{4 - 0} = \frac{-2}{4} = -\frac{1}{2}$$

Since the two slopes are negative reciprocals of each other, L_1 and L_2 are *perpendicular*.

79. L_1: $(3, 6)$, $(-6, 0)$

$$m_1 = \frac{0 - 6}{-6 - 3} = \frac{-6}{-9} = \frac{2}{3}$$

L_2: $(0, -1)$, $\left(5, \frac{7}{3}\right)$

$$m_2 = \frac{\frac{7}{3} - (-1)}{5 - 0} = \frac{\frac{7}{3} + \frac{3}{3}}{5} = \frac{\frac{10}{3}}{5} = \frac{10}{3} \cdot \frac{1}{5} = \frac{10}{15} = \frac{2}{3}$$

Since the two slopes are equal, L_1 and L_2 are *parallel*.

81. L_1: $y = 2x - 3$

$m_1 = 2$

L_2: $y = 2x + 1$

$m_2 = 2$

Since the two slopes are equal, the two lines are *parallel*.

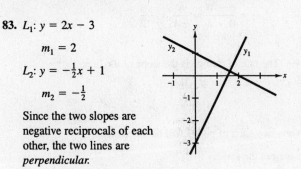

83. L_1: $y = 2x - 3$

$m_1 = 2$

L_2: $y = -\frac{1}{2}x + 1$

$m_2 = -\frac{1}{2}$

Since the two slopes are negative reciprocals of each other, the two lines are *perpendicular*.

85. The vertical change, or rise, in the roof is $26 - 20$ feet. The horizontal change, or run, in the roof is half of 30 feet. The slope of the roof is the ratio of 6 feet to 15 feet.

$$m = \frac{26 - 20}{(1/2)(30)} = \frac{6}{15} = \frac{2}{5}$$

The slope, or pitch, of the roof is $\frac{2}{5}$.

87. (a)

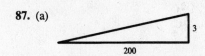

(b) $m = \frac{3}{200}$

(c) Yes. If the track rose 3 feet over a distance of 100 feet, the track would rise a total of 6 feet over the 200-foot distance. This is twice as much rise over that distance, and thus it would be a steeper slope.

89. (a) 1993-1994: $m = \dfrac{67.3 - 55.5}{1994 - 1993} = \dfrac{11.8}{1} = 11.8$

1994-1995: $m = \dfrac{82.5 - 67.3}{1995 - 1994} = \dfrac{15.2}{1} = 15.2$

1995-1996: $m = \dfrac{93.6 - 82.5}{1996 - 1995} = \dfrac{11.1}{1} = 11.1$

1996-1997: $m = \dfrac{104.9 - 93.6}{1997 - 1996} = \dfrac{11.3}{1} = 11.3$

(b) 1993-1997: $m = \dfrac{104.9 - 55.5}{1997 - 1993} = \dfrac{49.4}{4} = 12.35$

The slope of 12.35 indicates the average annual increase in net sales during the four years.

91. (a) No. The line appears to be much steeper when using the second range setting.

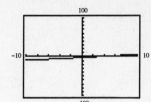 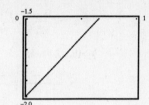

(b) No. Using the first range setting, the slope appears to be much smaller than 0.75. Using the second range setting, the slope appears to be greater than 0.75. To make the line appear to have a slope of 0.75, use the SQUARE feature.

(c) Answers vary. The following are sample answers.

If the function represents the price of a product over a given number of years, using the first setting makes it appear as if the price of your product has changed very little over time. This may be an attractive feature to customers.

If the function represents the number of students enrolled in Biology courses, using the second setting makes it appear as if the number of students taking Biology courses is increasing rapidly. This may be a good way to present the data if you are requesting funds for more lab equipment.

93. Yes, the slope of a line is the ratio of the change in y to the change in x.

95. False. For example, if the x-intercept is $(4, 0)$ and the y-intercept is $(0, 6)$, then the slope

$$m = \frac{6 - 0}{0 - 4} = \frac{6}{-4} = -\frac{3}{2}.$$

97. No. The slopes of nonvertical perpendicular lines have opposite signs. The slopes are the negative reciprocals of each other.

99. The rate of change is the slope of the line.

101. If the points lie on the same line, the slopes of the lines between any two pairs of points will be the same.

These three points do not lie on the same line because the slopes are not the same.

$(-2, -3)$ and $(1, 1)$: $m = \dfrac{1 - (-3)}{1 - (-2)} = \dfrac{4}{3}$

$(1, 1)$ and $(3, 4)$: $m = \dfrac{4 - 1}{3 - 1} = \dfrac{3}{2}$

$(-2, -3)$ and $(3, 4)$: $\dfrac{4 - (-3)}{3 - (-2)} = \dfrac{7}{5}$

Section 4.5   Equations of Lines

1.
$$m = \frac{y_2 - y_1}{x_2 - x_1}$$

$$-2 = \frac{y - 0}{x - 0}$$

$$-2 = \frac{y}{x}$$

$$-2x = y$$

$$-2x - y = 0$$

$$2x + y = 0$$

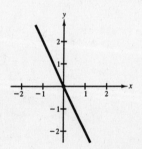

3.
$$m = \frac{y_2 - y_1}{x_2 - x_1}$$

$$\frac{1}{2} = \frac{y - 0}{x - 6}$$

$$\frac{1}{2} = \frac{y}{x - 6}$$

$$x - 6 = 2y$$

$$x - 2y - 6 = 0$$

$$x - 2y = 6$$

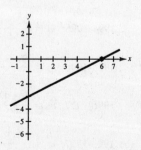

5. $m = \dfrac{y_2 - y_1}{x_2 - x_1}$

$$2 = \frac{y - 1}{x - (-2)}$$

$$2 = \frac{y - 1}{x + 2}$$

$$2(x + 2) = y - 1$$

$$2x + 4 = y - 1$$

$$2x - y + 4 = -1$$

$$2x - y = -5$$

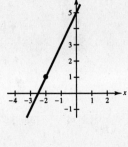

7. $m = \dfrac{y_2 - y_1}{x_2 - x_1}$

$$-\frac{1}{4} = \frac{y - (-1)}{x - (-8)}$$

$$\frac{-1}{4} = \frac{y + 1}{x + 8}$$

$$-1(x + 8) = 4(y + 1)$$

$$-x - 8 = 4y + 4$$

$$-x - 4y - 8 = 4$$

$$-x - 4y = 12$$

$$x + 4y = -12$$

9. $m = \dfrac{y_2 - y_1}{x_2 - x_1}$

$$0 = \frac{y - (-3)}{x - \frac{1}{2}}$$

$$0 = \frac{y + 3}{x - \frac{1}{2}}$$

$$0 = y + 3$$

$$y = -3$$

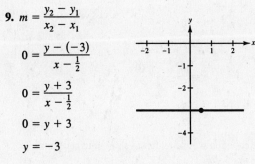

11. $m = \dfrac{y_2 - y_1}{x_2 - x_1}$

$$\frac{2}{3} = \frac{y - \frac{3}{2}}{x - 0}$$

$$\frac{2}{3} = \frac{y - \frac{3}{2}}{x}$$

$$2x = 3\left(y - \frac{3}{2}\right)$$

$$2x = 3y - \frac{9}{2}$$

$$2x - 3y = -\frac{9}{2}$$

$$4x - 6y = -9$$

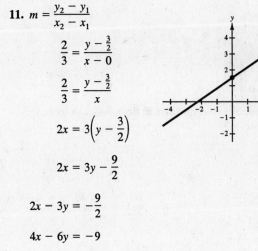

13. $m = \dfrac{y_2 - y_1}{x_2 - x_1}$

$$-0.8 = \frac{y - 4}{x - 2}$$

$$\frac{-4}{5} = \frac{y - 4}{x - 2}$$

$$-4(x - 2) = 5(y - 4)$$

$$-4x + 8 = 5y - 20$$

$$-4x - 5y + 8 = -20$$

$$-4x - 5y = -28$$

$$4x + 5y = 28$$

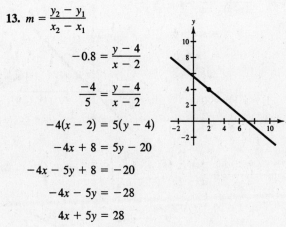

15. $y - y_1 = m(x - x_1)$

$$y - (-4) = 3(x - 0)$$

$$y + 4 = 3x$$

$$y = 3x - 4$$

17. $y - y_1 = m(x - x_1)$

$$y - 6 = -2(x + 3)$$

$$y - 6 = -2x - 6$$

$$y = -2x$$

19. $y - y_1 = m(x - x_1)$

$$y - 0 = -\frac{1}{3}(x - 9)$$

$$y = -\frac{1}{3}x + 3$$

21. The slope $m = 0$ indicates a horizontal line with an equation in the form $y = b$. Thus, the equation is $y = 4$.

$$y = 4$$

23. $y - y_1 = m(x - x_1)$

$y - 1 = -\frac{3}{4}(x - 8)$

$y - 1 = -\frac{3}{4}x + 6$

$y = -\frac{3}{4}x + 7$

25. $y - y_1 = m(x - x_1)$

$y - 1 = \frac{2}{3}[x - (-2)]$

$y - 1 = \frac{2}{3}(x + 2)$

$y - 1 = \frac{2}{3}x + \frac{4}{3}$

$y = \frac{2}{3}x + \frac{7}{3}$

27. $y = \frac{3}{8}x - 4$

$m = \frac{3}{8}$

The equation is in slope-intercept form, $y = mx + b$.

29. $y - 2 = 5(x + 3)$

$m = 5$

The equation is in point-slope form, $y - y_1 = m(x - x_1)$

31. $y + \frac{5}{6} = \frac{2}{3}(x + 4)$

$m = \frac{2}{3}$

The equation is in point-slope form, $y - y_1 = m(x - x_1)$.

33. $3x + y = 0$

$y = -3x$

$m = -3$

The equation was rewritten in slope-intercept form, $y = mx + b$.

35. $2x - y = 0$

$-y = -2x$

$y = 2x$

$m = 2$

The equation was rewritten in slope-intercept form, $y = mx + b$.

37. $3x - 2y + 10 = 0$

$-2y = -3x - 10$

$y = \frac{-3x}{-2} - \frac{10}{-2} = \frac{3}{2}x + 5$

$m = \frac{3}{2}$

The equation was rewritten in slope-intercept form, $y = mx + b$.

39. $y = mx + b$

$y = \frac{1}{2}x + 2$

41. $y = mx + b$

$y = -3x - 1$

43. $y - y_1 = m(x - x_1)$

$y - 2 = -\frac{1}{2}(x + 1)$

45. $m = \frac{1 - (-1)}{4 - (-2)} = \frac{2}{6} = \frac{1}{3}$

$y - y_1 = m(x - x_1)$

$y - 1 = \frac{1}{3}(x - 4) \text{ or } y + 1 = \frac{1}{3}(x + 2)$

47. $m = \frac{4 - 0}{4 - 0} = \frac{4}{4} = 1$

$y = mx + b$

$y = 1x + 0$

$y = x$

$-x + y = 0 \text{ or }$

$x - y = 0$

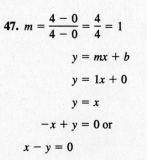

49. $m = \frac{-4 - 0}{2 - 0} = \frac{-4}{2} = -2$

$y = mx + b$

$y = -2x + 0$

$y = -2x$

$2x + y = 0$

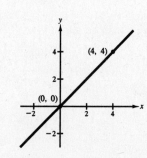

51. $m = \frac{5 - 3}{6 - 2} - \frac{2}{4} = \frac{1}{2}$

$y - y_1 = m(x - x_1)$

$y - 3 = \frac{1}{2}(x - 2)$

$y - 3 = \frac{1}{2}x - 1$

$2(y - 3) = 2\left(\frac{1}{2}x - 1\right)$

$2y - 6 = x - 2$

$-x + 2y - 4 = 0$

$x - 2y + 4 = 0$

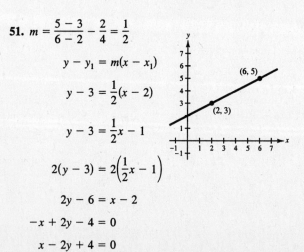

53. $m = \dfrac{5-2}{3+6} = \dfrac{3}{9} = \dfrac{1}{3}$

$$y - y_1 = m(x - x_1)$$

$$y - 2 = \dfrac{1}{3}(x + 6)$$

$$y - 2 = \dfrac{1}{3}x + 2$$

$$3(y - 2) = 3\left(\dfrac{1}{3}x + 2\right)$$

$$3y - 6 = x + 6$$

$$-x + 3y - 12 = 0 \text{ or } x - 3y + 12 = 0$$

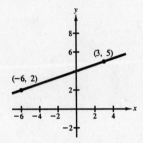

55. $m = \dfrac{2-(-1)}{3-5} = \dfrac{3}{-2} = -\dfrac{3}{2}$

$$y - y_1 = m(x - x_1)$$

$$y - (-1) = -\dfrac{3}{2}(x - 5)$$

$$y + 1 = -\dfrac{3}{2}x + \dfrac{15}{2}$$

$$2(y + 1) = 2\left(-\dfrac{3}{2}x + \dfrac{15}{2}\right)$$

$$2y + 2 = -3x + 15$$

$$3x + 2y - 13 = 0$$

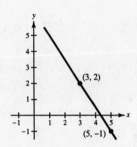

57. $m = \dfrac{7-(-1)}{9/2 - 5/2} = \dfrac{8}{4/2} = \dfrac{8}{2} = 4$

$$y - y_1 = m(x - x_1)$$

$$y - 7 = 4\left(x - \dfrac{9}{2}\right)$$

$$y - 7 = 4x - 18$$

$$-4x + y + 11 = 0 \text{ or } 4x - y - 11 = 0$$

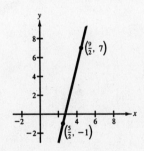

59. $m = \dfrac{0-3}{3-0} = \dfrac{-3}{3} = -1$

$$y - y_1 = m(x - x_1)$$

$$y - 3 = -1(x - 0)$$

$$y - 3 = -x$$

$$x + y - 3 = 0$$

61. $m = \dfrac{5+1}{-5-5} = \dfrac{6}{-10} = -\dfrac{3}{5}$

$$y - y_1 = m(x - x_1)$$

$$y + 1 = -\dfrac{3}{5}(x - 5)$$

$$y + 1 = -\dfrac{3}{5}x + 3$$

$$5(y + 1) = 5\left(-\dfrac{3}{5}x + 3\right)$$

$$5y + 5 = -3x + 15$$

$$3x + 5y - 10 = 0$$

63. $m = \dfrac{-4-4}{1-5} = \dfrac{-8}{-4} = 2$

$$y - y_1 = m(x - x_1)$$

$$y - 4 = 2(x - 5)$$

$$y - 4 = 2x - 10$$

$$-2x + y + 6 = 0$$

$$2x - y - 6 = 0$$

65. $m = \dfrac{-4 - (-1)}{7 - 5} = \dfrac{-3}{2} = -\dfrac{3}{2}$

$$y - y_1 = m(x - x_1)$$

$$y - (-1) = -\frac{3}{2}(x - 5)$$

$$y + 1 = -\frac{3}{2}x + \frac{15}{2}$$

$$2(y + 1) = 2\left(-\frac{3}{2}x + \frac{15}{2}\right)$$

$$2y + 2 = -3x + 15$$

$$3x + 2y - 13 = 0$$

67. $m = \dfrac{5 - 8}{2 - (-3)} = \dfrac{-3}{5} = -\dfrac{3}{5}$

$$y - y_1 = m(x - x_1)$$

$$y - 8 = -\frac{3}{5}(x + 3)$$

$$y - 8 = -\frac{3}{5}x - \frac{9}{5}$$

$$5(y - 8) = 5\left(-\frac{3}{5}x - \frac{9}{5}\right)$$

$$5y - 40 = -3x - 9$$

$$3x + 5y - 31 = 0$$

69. $m = \dfrac{\frac{5}{2} - \frac{1}{2}}{\frac{1}{2} - 2} = \dfrac{2}{-\frac{3}{2}} = -\dfrac{4}{3}$

$$y - y_1 = m(x - x_1)$$

$$y - \frac{1}{2} = -\frac{4}{3}(x - 2)$$

$$y - \frac{1}{2} = -\frac{4}{3}x + \frac{8}{3}$$

$$6\left(y - \frac{1}{2}\right) = 6\left(-\frac{4}{3}x + \frac{8}{3}\right)$$

$$6y - 3 = -8x + 16$$

$$8x + 6y - 19 = 0$$

71. $m = \dfrac{-0.6 - 0.6}{2 - 1} = \dfrac{-1.2}{1} = -1.2$

$$y - y_1 = m(x - x_1)$$

$$y - 0.6 = -1.2(x - 1)$$

$$y - 0.6 = -1.2x + 1.2$$

$$1.2x + y - 1.8 = 0$$

Note: The equation could also be written as

$$12x + 10y - 18 = 0 \text{ or } 6x + 5y - 9 = 0.$$

73. $x - y = 3$

$$-y = -x + 3$$

$$y = x - 3$$

$$m = 1$$

(a) Parallel line: $m = 1$; $(2, 1)$

$$y - y_1 = m(x - x_1)$$

$$y - 1 = 1(x - 2)$$

$$y - 1 = x - 2$$

$$-x + y + 1 = 0$$

$$x - y - 1 = 0$$

(b) Perpendicular line: $m = -1$; $(2, 1)$

$$y - y_1 = m(x - x_1)$$

$$y - 1 = -1(x - 2)$$

$$y - 1 = -x + 2$$

$$x + y - 3 = 0$$

75. $3x + 4y = 7$

$$4y = -3x + 7$$

$$y = -\frac{3}{4}x + \frac{7}{4}$$

(a) Parallel line: $m = -\frac{3}{4}$; $(-12, 4)$

$$y - y_1 = m(x - x_1)$$

$$y - 4 = -\frac{3}{4}(x + 12)$$

$$y - 4 = -\frac{3}{4}x - 9$$

$$4(y - 4) = 4\left(-\frac{3}{4}x - 9\right)$$

$$4y - 16 = -3x - 36$$

$$3x + 4y + 20 = 0$$

(b) Perpendicular line: $m = \frac{4}{3}$; $(-12, 4)$

$$y - y_1 = m(x - x_1)$$

$$y - 4 = \frac{4}{3}(x + 12)$$

$$y - 4 = \frac{4}{3}x + 16$$

$$3(y - 4) = 3\left(\frac{4}{3}x + 16\right)$$

$$3y - 12 = 4x + 48$$

$$-4x + 3y - 60 = 0$$

$$4x - 3y + 60 = 0$$

77. $2x + y = 0$

$$y = -2x$$

$$m = -2$$

(a) Parallel line: $m = -2$; $(1, 3)$

$$y - y_1 = m(x - x_1)$$

$$y - 3 = -2(x - 1)$$

$$y - 3 = -2x + 2$$

$$2x + y - 5 = 0$$

(b) Perpendicular line: $m = \frac{1}{2}$; $(1, 3)$

$$y - y_1 = m(x - x_1)$$

$$y - 3 = \frac{1}{2}(x - 1)$$

$$y - 3 = \frac{1}{2}x - \frac{1}{2}$$

$$2y - 6 = x - 1$$

$$0 = x - 2y + 5$$

79. $y + 3 = 0$

$$y = -3$$

or

$$y = 0x - 3$$

$$m = 0 \qquad \text{(Horizontal line)}$$

(a) Parallel line: $m = 0$; $(-1, 0)$

$$y = 0 \qquad \text{(Horizontal line)}$$

(b) Perpendicular line: m undefined; $(-1, 0)$

$$x = -1 \quad \text{(Vertical line)}$$

$$x + 1 = 0 \qquad \text{(General form)}$$

81. $3y - 2x = 7$

$$3y = 2x + 7$$

$$y = \frac{2}{3}x + \frac{7}{3}$$

$$m = \frac{2}{3}$$

(a) Parallel line: $m = \frac{2}{3}$; $(4, -1)$

$$y - y_1 = m(x - x_1)$$

$$y - (-1) = \frac{2}{3}(x - 4)$$

$$y + 1 = \frac{2}{3}x - \frac{8}{3}$$

$$3(y + 1) = 3\left(\frac{2}{3}x - \frac{8}{3}\right)$$

$$3y + 3 = 2x - 8$$

$$-2x + 3y + 11 = 0$$

$$2x - 3y - 11 = 0$$

(b) Perpendicular line: $m = -\frac{3}{2}$; $(4, -1)$

$$y - y_1 = m(x - x_1)$$

$$y + 1 = -\frac{3}{2}(x - 4)$$

$$y + 1 = -\frac{3}{2}x + 6$$

$$2(y + 1) = 2\left(-\frac{3}{2}x + 6\right)$$

$$2y + 2 = -3x + 12$$

$$3x + 2y - 10 = 0$$

83. Because the line is vertical and passes through the point $(-2, 4)$, every point on the line has an x-coordinate of -2. So, the equation of the line is $x = -2$.

85. Because the line horizontal and passes through the point $\left(\frac{1}{2}, \frac{2}{3}\right)$, every point on the line has a y-coordinate of $\frac{2}{3}$. So, the equation of the line is $y = \frac{2}{3}$.

87. Because both points have the same x-coordinate, the line through $(4, 1)$ and $(4, 8)$ is vertical. So its equation is $x = 4$.

89. Because both points have the same y-coordinate, the line through $(1, -8)$ and $(7, -8)$ is horizontal. So its equation is $y = -8$.

91.

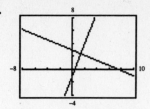

The lines are perpendicular.

93.

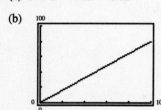

The lines are neither parallel nor perpendicular.

95.

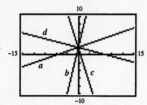

Graphs (a) and (c) are perpendicular. (Note that their slopes are negative reciprocals of each other.) Graphs (b) and (d) are perpendicular. (Note that their slopes are negative reciprocals of each other.)

97. $C = 225 + 0.28x$

99. $W = 2300 + 0.03S$

101. (a) $S = L - 0.2L = 0.8L$

(b)

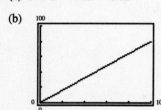

(c) The sale price is approximately $40. You can estimate it more accurately with your graphing calculator, and you can confirm it algebraically.

$S = 0.8(49.98) \approx 39.98$

103. (a) During the year, the machine was depreciated by $200,000 - $170,000 = $30,000.

$V = 200,000 - 30,000t$

(b) $V = 200,000 - 30,000t$

$= 200,000 - 30,000(5)$

$= 50,000$

After 5 years, the value of the machine was $50,000.

105. (a) $N = 50t + 1200$

(b) The variable $t = 14$ for the year 2004.

$N = 50(14) + 1200$

$N = 700 + 1200$

$N = 1900$

(c) The variable $t = 8$ for the year 1998.

$N = 50(8) + 1200$

$= 400 + 1200$

$= 1600$

107. (a) (f); $m = -10$, The amount of the loan decreases $10 each week.

(b) (e); $m = 1.50$; The employee's pay increases $1.50 for every unit produced.

(c) (g); $m = 0.32$; For each mile traveled, the representative receives $0.32.

(d) (h); $m = -100$; Each year the value of the typewriter decreases $100.

109. Yes. When different pairs of points are selected, the change in y and the change in x are the lengths of the sides of similar triangles. Corresponding sides of similar triangles are proportional, so the ratios, or slopes, are equal.

111. In the equation $y = mx + b$, m indicates the slope of the line and b indicates that the y-intercept is $(0, b)$.

113. To find the *x*-intercept, set $y = 0$ and solve for *x*.

$$y = mx + b$$

$$0 = mx + b$$

$$-b = mx$$

$$-\frac{b}{m} = x$$

The *x*-intercept is $\left(-\dfrac{b}{m}, 0\right)$.

115. Answers will vary.

Section 4.6   Graphs of Linear Inequalities

1. (a) $(0, 0)$

$$0 + 0 \overset{?}{>} 5$$

$$0 \not> 5$$

The point $(0, 0)$ is *not* a solution.

(b) $(3, 6)$

$$3 + 6 \overset{?}{>} 5$$

$$9 > 5$$

The point $(3, 6)$ *is* a solution.

(c) $(-6, 20)$

$$-6 + 20 \overset{?}{>} 5$$

$$14 > 5$$

The point $(-6, 20)$ *is* a solution.

(d) $(3, 2)$

$$3 + 2 \overset{?}{>} 5$$

$$5 \not> 5$$

The point $(3, 2)$ is *not* a solution.

3. (a) $(1, 2)$

$$-3(1) + 5(2) \overset{?}{\leq} 12$$

$$-3 + 10 \overset{?}{\leq} 12$$

$$7 \leq 12$$

The point $(1, 2)$ *is* a solution.

(b) $(2, -3)$

$$-3(2) + 5(-3) \overset{?}{\leq} 12$$

$$-6 - 15 \overset{?}{\leq} 12$$

$$-21 \leq 12$$

The point $(2, -3)$ *is* a solution.

(c) $(1, 3)$

$$-3(1) + 5(3) \overset{?}{\leq} 12$$

$$-3 + 15 \overset{?}{\leq} 12$$

$$12 \leq 12$$

The point $(1, 3)$ *is* a solution.

(d) $(2, 8)$

$$-3(2) + 5(8) \overset{?}{\leq} 12$$

$$-6 + 40 \overset{?}{\leq} 12$$

$$34 \not\leq 12$$

The point $(2, 8)$ is *not* a solution.

5. (a) $(1, 3)$

$$3(1) - 2(3) \overset{?}{<} 2$$

$$3 - 6 \overset{?}{<} 2$$

$$-3 < 2$$

The point $(1, 3)$ *is* a solution.

(b) $(2, 0)$

$$3(2) - 2(0) \overset{?}{<} 2$$

$$6 - 0 \overset{?}{<} 2$$

$$6 \not< 2$$

The point $(2, 0)$ is *not* a solution.

(c) $(0, 0)$

$$3(0) - 2(0) \overset{?}{<} 2$$

$$0 - 0 \overset{?}{<} 2$$

$$0 < 2$$

The point $(0, 0)$ *is* a solution.

(d) $(3, -5)$

$$3(3) - 2(-5) \overset{?}{<} 2$$

$$9 + 10 \overset{?}{<} 2$$

$$19 \not< 2$$

The point $(3, -5)$ is *not* a solution.

7. (a) $(-2, 4)$

$$5(-2) + 4(4) \overset{?}{\geq} 6$$

$$-10 + 16 \overset{?}{\geq} 6$$

$$6 \geq 6$$

The point $(-2, 4)$ *is* a solution.

(c) $(7, 0)$

$$5(7) + 4(0) \overset{?}{\geq} 6$$

$$35 + 0 \overset{?}{\geq} 6$$

$$35 \geq 6$$

The point $(7, 0)$ *is* a solution.

(b) $(5, 5)$

$$5(5) + 4(5) \overset{?}{\geq} 6$$

$$25 + 20 \overset{?}{\geq} 6$$

$$45 \geq 6$$

The point $(5, 5)$ *is* a solution.

(d) $(-2, 5)$

$$5(-2) + 4(5) \overset{?}{\geq} 6$$

$$-10 + 20 \overset{?}{\geq} 6$$

$$10 \geq 6$$

The point $(-2, 5)$ *is* a solution.

9. The boundary of the graph of $2x + 3y < 6$ should be *dashed*.

11. The boundary of the graph of $2x + 3y \geq 6$ should be *solid*.

13. Graph (b)   **15.** Graph (d)   **17.** Graph (c)   **19.** Graph (b)

21.

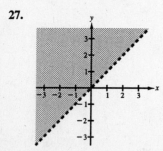

23.

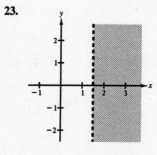

25.

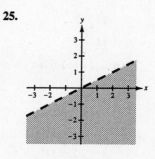

27.

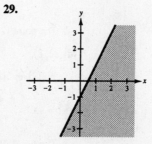

29.

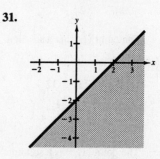

31.

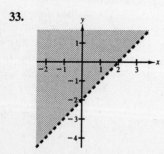

33.

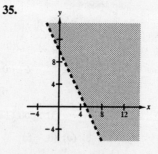

35.

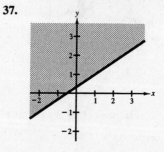

37.

39.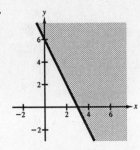

41.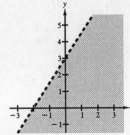

43.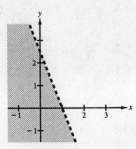

45.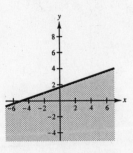

47.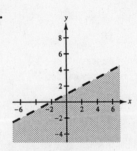

49. $\dfrac{x}{3} + \dfrac{y}{4} < 1$

$12\left(\dfrac{x}{3} + \dfrac{y}{4}\right) < 12(1)$

$4x + 3y < 12$

This could be written in slope-intercept form as $y < -\frac{4}{3}x + 4$.

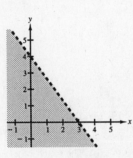

51.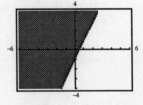

53.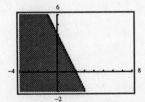

55.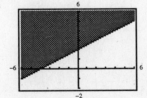

57.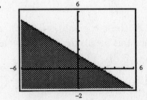

59. The graph shows the points on and above the horizontal line $y = 2$.

$y \geq 2$

61. The inequality can be written in slope-intercept form. The slope is -2 and the y-intercept is 2.

$y \leq -2x + 2$

(This could also be written as $2x + y \leq 2$ or as $2x + y - 2 \leq 0$.)

63. The inequality can be written in slope-intercept form. The slope is 2 and the y-intercept is 0.

$y < 2x$

(This could also be written as $2x - y > 0$.)

65. $7x + 5y \geq 140$

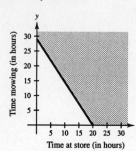

Representative solutions

(20, 0)

(0, 28)

(10, 15)

(5, 25)

(15, 7)

67. $T + \frac{3}{2}C \leq 12$

$2T + 3C \leq 24$ (Multiply both sides 2.)

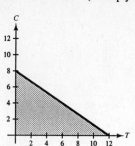

Representative solutions

(12, 0)

(0, 8)

(6, 4)

(9, 2)

(6, 2)

(4, 4)

(2, 6)

69. *Labels:* $x = $ number of ounces of food X

$y = $ number of ounces of food Y

Inequality: $20x + 10y \geq 300$

Representative solutions

(10, 20)

(12, 10)

(20, 0)

(0, 30)

(5, 25)

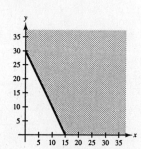

71. $ax + by < c$

$ax + by \leq c$

$ax + by > c$

$ax + by \geq c$

73. Graphs that have dashed lines represent inequalities with either of these inequality symbols: $<$ or $>$.

Graphs that have solid lines represent inequalities with either of these inequality symbols: $\leq$ or $\geq$.

75. (a) The graph of $x \geq 1$ on the real number line is an unbounded interval on the number line.

(b) The graph of $x \geq 1$ on a rectanglular coordinate system is a half-plane.

77. Yes, the two graphs are the same. If $2x < 2y$, then $x < y$, and this means the same thing as $y > x$.

Review Exercises for Chapter 4

1.

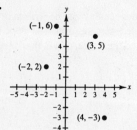

3.

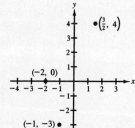

5. Quadrant II

The point $(-5, 3)$ is located 5 units to the *left* of the vertical axis and 3 units *above* the horizontal axis in Quadrant II.

7. The point $(4, 0)$ is located on the x-axis between the first and fourth quadrants.

9. Quadrant II

The point $(x, 5)$, $x < 0$, is located to the left of the vertical axis because x is negative; it is 5 units above the horizontal axis. Therefore, the point is in Quadrant II.

11. Quadrant II or III

The point $(-6, y)$ is located to the left of the vertical axis. If y is positive, the point is above the horizontal axis in the second quadrant, and if y is negative, the point is below the horizontal axis in the third quadrant. If $y = 0$, the point is on the x-axis between the second and third quadrants.

13. $3x + 4y = 12$

$$4y = -3x + 12$$
$$y = -\tfrac{3}{4}x + 3$$

15. $x - 2y = 8$

$$-2y = -x + 8$$
$$y = \tfrac{1}{2}x - 4$$

17. $2x - y = 1$

$$-y = -2x + 1$$
$$y = 2x - 1$$

x	-1	0	1	2
$y = 2x - 1$	-3	-1	1	3

19. (a) $(1, -1)$ is a solution because $1 - 3(-1) = 4$.

(b) $(0, 0)$ is not a solution because $1 - 3(0) \neq 4$.

(c) $(2, 1)$ is not a solution because $2 - 3(1) \neq 4$.

(d) $(5, -2)$ is not a solution because $5 - 3(-2) \neq 4$.

21. (a) $(3, 5)$ is a solution because $5 = \tfrac{2}{3}(3) + 3$.

(b) $(-3, 1)$ is a solution because $1 = \tfrac{2}{3}(-3) + 3$.

(c) $(-6, 0)$ is not a solution because $0 \neq \tfrac{2}{3}(-6) + 3$.

(d) $(0, 3)$ is a solution because $3 = \tfrac{2}{3}(0) + 3$.

23.

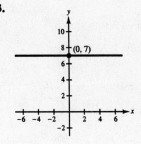

25.

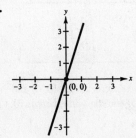

27.

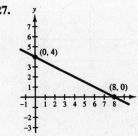

29.

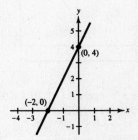

31.

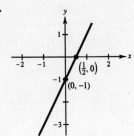

33.

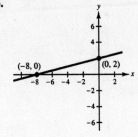

35.

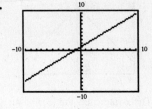

37.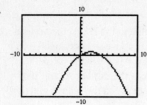

39.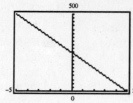

41.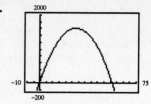

43. Domain: $\{-2, 3, 5, 8\}$

Range: $\{1, 3, 7, 8\}$

45. Domain: $\{-4, -2, 2, 7\}$

Range: $\{-3, -2, 0, 3\}$

47. No, the relation is not a function because the input value 2 has two different output values.

49. Yes. No first componenet has two different second components, so this relation is a function.

51. No. A vertical line could cross the graph more than once so this relation does not represent y as a function of x.

53. Yes. No vertical line could cross this graph more than once, so the relation represents y as a function of x.

55. Yes. No vertical line could cross this graph more than once, so the relation represents y as a function of x.

57. (a) $f(0) = |2(0) + 3| = |3| = 3$

(b) $f(5) = |2(5) + 3| = |13| = 13$

(c) $f(-4) = |2(-4) + 3| = |-5| = 5$

(d) $f\left(-\frac{3}{2}\right) = \left|2\left(-\frac{3}{2}\right) + 3\right| = |0| = 0$

59. (a) $h(0) = 0(0 - 3)^2 = 0(-3)^2 = 0(9) = 0$

(b) $h(3) = 3(3 - 3)^2 = 3(0)^2 = 3(0) = 0$

(c)
$h(-1) = -1(-1 - 3)^2 = -1(-4)^2 = -1(16) = -16$

(d) $h\left(\frac{3}{2}\right) = \frac{3}{2}\left(\frac{3}{2} - 3\right)^2 = \frac{3}{2}\left(-\frac{3}{2}\right)^2 = \frac{3}{2}\left(\frac{9}{4}\right) = \frac{27}{8}$

61. (a) $f(-1) = 2(-1) - 7 = -2 - 7 = -9$

(b) $f(3) = 2(3) - 7 = 6 - 7 = -1$

(c) $f\left(\frac{1}{2}\right) = 2\left(\frac{1}{2}\right) - 7 = 1 - 7 = -6$

(d) $f(-4) = 2(-4) - 7 = -8 - 7 = -15$

63.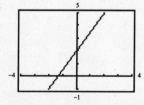

The x-intercept is $\left(-\frac{4}{3}, 0\right)$ and the y-intercept is $(0, 2)$.

65.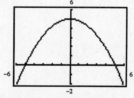

The x-intercepts are $(5, 0)$ and $(-5, 0)$ and the y-intercept is $(0, 5)$.

67.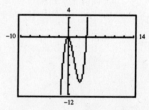

The x-intercepts are $(0, 0)$ and $(4, 0)$ and the y-intercept is $(0, 0)$.

69.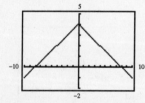

The x-intercepts are $(8, 0)$ and $(-8, 0)$ and the y-intercept is $(0, 4)$.

71. The slope $m = \frac{1}{2}$.

73. Graph (c)

75. Graph (b)

77. $m = \dfrac{6-1}{14-2} = \dfrac{5}{12}$

79. $m = \dfrac{2-0}{6-(-1)} = \dfrac{2}{7}$

81. $m = \dfrac{6-0}{4-4} = \dfrac{6}{0}$ (undefined)

m is undefined.

83. $m = \dfrac{1-5}{1-(-2)} = -\dfrac{4}{3}$

85. $m = \dfrac{10-(-4)}{5-1} = \dfrac{14}{4} = \dfrac{7}{2}$

87. $m = \dfrac{0 - \frac{5}{2}}{\frac{5}{6} - 0} = \dfrac{-\frac{5}{2}}{\frac{5}{6}}$

$= -\dfrac{5}{2} \div \dfrac{5}{6} = -\dfrac{5}{2} \cdot \dfrac{6}{5}$

$= -\dfrac{30}{10} = -3$

89. $3x + 6y = 12$

$6y = -3x + 12$

$y = -\tfrac{1}{2}x + 2$

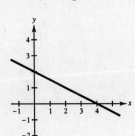

91. $5y - 2x = 5$

$5y = 2x + 5$

$y = \tfrac{2}{5}x + 1$

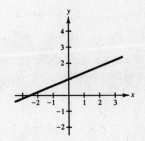

93. $y = 4x - 8 \implies m = 4$

$y = -\tfrac{1}{4}x + 2 \implies m = -\tfrac{1}{4}$

The slopes are negative reciprocals so the lines are perpendicular.

95. $4x - y = 7$

$-y = -4x + 7$

$y = 4x - 7 \implies m = 4$

$8x - 2y = 3$

$-2y = -8x + 3$

$y = 4x - \tfrac{3}{2} \implies m = 4$

The slopes are equal so the lines are parallel.

97. $(4, -3), (5, -5), (6, -7),$ etc.

99. $(6, 6), (10, 9), (14, 12),$ etc.

101. $(0, 4), (10, 4), (-5, 4),$ etc.

103. $y + 1 = 2(x - 4)$

$y + 1 = 2x - 8$

$-2x + y + 9 = 0$ or

$2x - y - 9 = 0$

105. $y - 2 = -4(x - 1)$

$y - 2 = -4x + 4$

$4x + y - 6 = 0$

107. $y + 2 = \tfrac{4}{5}(x + 5)$

$y + 2 = \tfrac{4}{5}x + 4$

$5(y + 2) = 5\left(\tfrac{4}{5}x + 4\right)$

$5y + 10 = 4x + 20$

$-4x + 5y - 10 = 0$

$4x - 5y + 10 = 0$

109. $y - 3 = -\tfrac{8}{3}(x + 1)$

$y - 3 = -\tfrac{8}{3}x - \tfrac{8}{3}$

$3(y - 3) = 3\left(-\tfrac{8}{3}x - \tfrac{8}{3}\right)$

$3y - 9 = -8x - 8$

$8x + 3y - 1 = 0$

111. The undefined slope indicates that this is a vertical line through $(3, 8)$. All the points have the same x-coordinate of 3. The equation of the line is $x = 3$, or, in general form, $x - 3 = 0$.

113. $y = 5x + 5 \implies m = 5$

The equation is in slope-intercept form $y = mx + b$.

115. $y + 10 = \frac{4}{3}(x - 4) \implies m = \frac{4}{3}$

The equation is in point-slope form $y - y_1 = m(x - x_1)$.

117. $m = \dfrac{-2 - 0}{0 - (-4)} = \dfrac{-2}{4} = -\dfrac{1}{2}$

$$y - 0 = -\frac{1}{2}(x + 4)$$

$$y = -\frac{1}{2}x - 2$$

$$2y = 2\left(-\frac{1}{2}x - 2\right)$$

$$2y = -x - 4$$

$$x + 2y + 4 = 0$$

Note: We could also obtain the solution using $m = -\frac{1}{2}$ and the y-intercept $(0, -2)$.

$$y = mx + b$$

$$y = -\frac{1}{2}x - 2$$

$$2y = -x - 4$$

$$x + 2y + 4 = 0$$

119. $m = \dfrac{8 - 8}{6 - 0} = \dfrac{0}{6} = 0$

$$y - 8 = 0(x - 0)$$

$$y - 8 = 0$$

Note: The slope of 0 indicates a horizontal line with an equation of the form $y = b$. The horizontal line through $(0, 8)$ has the equation $y = 8$ or $y - 8 = 0$.

121. $m = \dfrac{7 - 2}{4 - (-1)} = \dfrac{5}{5} = 1$

$$y - 7 = 1(x - 4)$$

$$y - 7 = x - 4$$

$$0 = x - y + 3$$

123. $m = \dfrac{7.8 - 3.3}{6 - 2.4} = \dfrac{4.5}{3.6} = \dfrac{45}{36} = \dfrac{5}{4}$

$$y - 7.8 = \frac{5}{4}(x - 6)$$

$$y - \frac{39}{5} = \frac{5}{4}x - \frac{15}{2}$$

$$20\left(y - \frac{39}{5}\right) = 20\left(\frac{5}{4}x - \frac{15}{2}\right)$$

$$20y - 156 = 25x - 150$$

$$-25x + 20y - 6 = 0$$

$$25x - 20y + 6 = 0$$

Note: You could use decimals instead of fractions.

$$m = \frac{7.8 - 3.3}{6 - 2.4} = \frac{4.5}{3.6} = 1.25$$

$$y - 7.8 = 1.25(x - 6)$$

$$y - 7.8 = 1.25x - 7.5$$

$$-1.25x + y - 0.3 = 0$$

$$-125x + 100y - 30 = 0 \qquad \text{(Multiply both sides by 100.)}$$

$$25x - 20y + 6 = 0 \qquad \text{(Divide both sides by } -5.\text{)}$$

125. $2x + 3y = 1$

$$3y = -2x + 1$$

$$y = -\frac{2}{3}x + \frac{1}{3}$$

$m = -\frac{2}{3}$

(a) Parallel line: $m = -\frac{2}{3}$

$$y - 3 = -\frac{2}{3}(x + 6)$$

$$y - 3 = -\frac{2}{3}x - 4$$

$$3(y - 3) = 3\left(-\frac{2}{3}x - 4\right)$$

$$3y - 9 = -2x - 12$$

$$2x + 3y + 3 = 0$$

(b) Perpendicular line: $m = \frac{3}{2}$

$$y - 3 = \frac{3}{2}(x + 6)$$

$$y - 3 = \frac{3}{2}x + 9$$

$$2(y - 3) = 2\left(\frac{3}{2}x + 9\right)$$

$$2y - 6 = 3x + 18$$

$$-3x + 2y - 24 = 0$$

$$3x - 2y + 24 = 0$$

127. $4x + 3y = 16$

$$3y = -4x + 16$$

$$y = -\frac{4}{3}x + \frac{16}{3}$$

$m = -\frac{4}{3}$

(a) Parallel line: $m = -\frac{4}{3}$

$$y - 4 = -\frac{4}{3}\left(x - \frac{3}{8}\right)$$

$$y - 4 = -\frac{4}{3}x + \frac{1}{2}$$

$$6(y - 4) = 6\left(-\frac{4}{3}x + \frac{1}{2}\right)$$

$$6y - 24 = -8x + 3$$

$$8x + 6y - 27 = 0$$

(b) Perpendicular line: $m = \frac{3}{4}$

$$y - 4 = \frac{3}{4}\left(x - \frac{3}{8}\right)$$

$$y - 4 = \frac{3}{4}x - \frac{9}{32}$$

$$32(y - 4) = 32\left(\frac{3}{4}x - \frac{9}{32}\right)$$

$$32y - 128 = 24x - 9$$

$$-24x + 32y - 119 = 0$$

$$24x - 32y + 119 = 0$$

129.

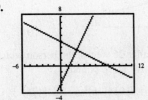

The lines are perpendicular.

131.

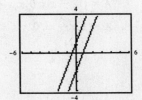

The lines are parallel.

133. (a) $(-1, -5)$ is not a solution because $-1 - (-5) = -1 + 5 \not> 4$.

(b) $(0, 0)$ is not a solution because $0 - 0 \not> 4$.

(c) $(3, -2)$ is a solution because $3 - (-2) = 3 + 2 > 4$.

(d) $(8, 1)$ is a solution because $8 - 1 > 4$.

135.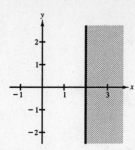

137.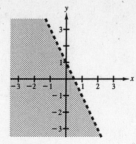

139.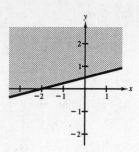

141. The graph shows the points below the horizontal line $y = 2$. The inequality is $y < 2$.

143. The inequality can be written in slope-intercept form. The slope is 1 and the y-intercept is $(0, 1)$. The inequality is $y \leq x + 1$.

145. (a)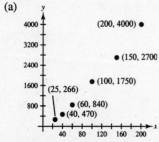

(b) The relationship between x and y is almost linear. As x gets larger so does y.

(c) When $x = 125$, y is approximately 2225 (lumens). This can be seen most easily from the graph.

147. $C = 2.25 + 0.75x$

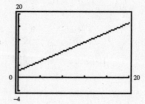

149. The perimeter of the triangle is 24. Thus, $24 = 2l + 2w$ or:

$$24 = 2l + 2w \qquad A = lw$$
$$24 - 2x = 2l \qquad A = (12 - x)x,\ 0 < x < 12$$
$$\frac{24 - 2x}{2} = l \qquad A = 12x - x^2,\ 0 < x < 12$$
$$12 - x = l$$

151. (a)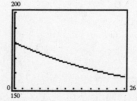

(b) Using the graph, after 26 weeks, the person will weigh approximately 157.8 pounds.

Algebraically, $y = 0.014(26)^2 - 1.218(26) + 180$
$$y = 0.014(676) - 31.668 + 180$$
$$y = 9.464 - 31.668 + 180$$
$$y \approx 157.8 \text{ pounds}$$

(c) Graphically it appears to take about 4.3 weeks to lose 5 pounds.

153. $m = -\dfrac{15,000}{10(5280)} = \dfrac{-15,000}{52,800} = -\dfrac{25(600)}{88(600)} = -\dfrac{25}{88}$

155. (a) $C = 16,000 + 5.35x$

(b) Profit = Revenue − Cost
$$P = 8.20x - (16,000 + 5.35x)$$
$$P = 8.20x - 16,000 - 5.35x$$
$$P = 2.85x - 16,000$$

Chapter Test for Chapter 4

1.

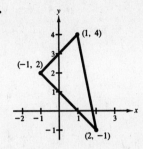

2. (a) No, $(0, -2)$ is not a solution because
$-2 \neq |0| + |0 - 2|$ or $0 + 2$.

(b) Yes, $(0, 2)$ is a solution because $2 = |0| + |0 - 2|$ or
$0 + 2$.

(c) Yes, $(-4, 10)$ is a solution because
$10 = |-4| + |-4 - 2|$ or $4 + 6$.

(d) No, $(-2, -2)$ is not a solution because
$-2 \neq |-2| + |-2 - 2|$ or $2 + 4$.

3. The y-coordinate of any point on the x-axis is 0.

4. *x-intercept*

Let $y = 0$.

$3x - 4(0) + 12 = 0$

$3x + 12 = 0$

$3x = -12$

$x = -4$

$(-4, 0)$

y-intercept

Let $x = 0$.

$3(0) - 4y + 12 = 0$

$-4y + 12 = 0$

$-4y = -12$

$y = 3$

$(0, 3)$

The graph has *one* x-intercept at $(-4, 0)$ and *one* y-intercept at $(0, 3)$.

5. No. Some first components, 0 and 1, have two different second components, so the table does not represent y as a function of x.

6. Yes. No vertical line would cross the graph more than once.

7. (a) $f(0) = 0^3 - 2(0)^2$

$= 0 - 2(0)$

$= 0$

(b) $f(2) = 2^3 - 2(2)^2$

$= 8 - 2(4)$

$= 8 - 8$

$= 0$

(c) $f(-2) = (-2)^3 - 2(-2)^2$

$= -8 - 2(4)$

$= -8 - 8$

$= -16$

(d) $f\left(\frac{1}{2}\right) = \left(\frac{1}{2}\right)^3 - 2\left(\frac{1}{2}\right)^2$

$= \frac{1}{8} - 2\left(\frac{1}{4}\right)$

$= \frac{1}{8} - \frac{1}{2}$

$= \frac{1}{8} - \frac{4}{8}$

$= -\frac{3}{8}$

8. $m = \dfrac{\frac{3}{2} - 0}{2 - (-5)} = \dfrac{\frac{3}{2}}{7} = \dfrac{3}{2} \cdot \dfrac{1}{7} = \dfrac{3}{14}$

9. $(-2, 2), (-1, 0), (0, -2)$, etc.

10. $3x - 5y + 2 = 0$

$-5y = -3x - 2$

$y = \dfrac{-3}{-5}x - \dfrac{2}{-5}$

$y = \dfrac{3}{5}x + \dfrac{2}{5}$

$m = \dfrac{3}{5}$

A perpendicular line would have slope $m = -\frac{5}{3}$
(the negative reciprocal of $\frac{3}{5}$).

11.

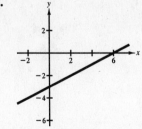

12.

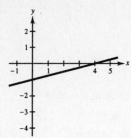

13.

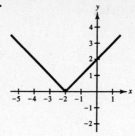

14.

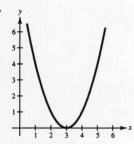

15.
$$y = mx + b$$
$$m = -\tfrac{3}{8} \text{ and } (0, b) = (0, 6)$$
$$y = -\tfrac{3}{8}x + 6 \text{ or}$$
$$3x + 8y - 48 = 0$$

16. (a) $(2, 2)$

$3 \cdot 2 + 5 \cdot 2 \overset{?}{\leq} 16$

$6 + 10 \overset{?}{\leq} 16$

$16 \leq 16$

The point $(2, 2)$ is a solution.

(b) $(6, -1)$

$3 \cdot 6 + 5(-1) \overset{?}{\leq} 16$

$18 + (-5) \overset{?}{\leq} 16$

$18 - 5 \overset{?}{\leq} 16$

$13 \leq 16$

The point $(6, -1)$ *is* a solution.

(c) $(-2, 4)$

$3(-2) + 5 \cdot 4 \overset{?}{\leq} 16$

$-6 + 20 \overset{?}{\leq} 16$

$14 \leq 16$

The point $(-2, 4)$ is a solution.

(d) $(7, -1)$

$3 \cdot 7 + 5(-1) \overset{?}{\leq} 16$

$21 - 5 \overset{?}{\leq} 16$

$16 \leq 16$

The point $(7, -1)$ *is* a solution.

17.

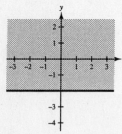

18.

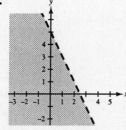

19.

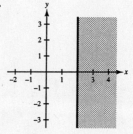

20.

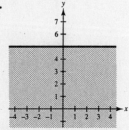

21. The slope of 230 indicates that sales are increasing at the rate of 230 units per year.

CHAPTER 5
Exponents and Polynomials

CHAPTER 5
Exponents and Polynomials

Section 5.1   Adding and Subtracting Polynomials

Solutions to Odd-Numbered Exercises

1. Yes, the expression $9 - z$ is a polynomial.

3. No, the expression $x^{2/3} + 8$ is not a polynomial because the exponent in the first term, $x^{2/3}$, is not an integer.

5. No, the expression $6x^{-1}$ is not a polynomial because the exponent is negative.

7. No, the expression $z^{-1} + z^2 - 2$ is not a polynomial because the exponent in the first term, z^{-1}, is negative.

9. *Polynomial:* $5 - 32x$

   Standard Form: $-32x + 5$

   Degree: 1

   Leading Coefficient: -32

11. *Polynomial:* $x^3 - 4x^2 + 9$

   Standard Form: $x^3 - 4x^2 + 9$

   Degree: 3

   Leading Coefficient: 1

13. *Polynomial:* $8x + 2x^5 - x^2 - 1$

   Standard Form: $2x^5 - x^2 + 8x - 1$

   Degree: 5

   Leading Coefficient: 2

15. *Polynomial:*   10

   Standard Form: 10

   Degree: 0

   Leading Coefficient: 10

17. *Polynomial:* $v_0 t - 16t^2$

   Standard Form: $-16t^2 + v_0 t$

   Degree: 2

   Leading Coefficient: -16

19. The polynomial $x^2 - 2x + 3$ has three terms; it is a trinomial.

21. The polynomial $x^3 - 4$ has two terms; it is a binomial.

23. The polynomial 5 has one term; it is a monomial.

25. $8x^3 + 5x$ or $-x^3 + 2$

27. $10x^2$ or $-2x^2$

29. $x^6 - 4x^3 - 2$ or $10x^6 - x^5 + 3x^4$

31. $(11x - 2) + (3x + 8) = (11x + 3x) + (-2 + 8)$

$= 14x + 6$

33. $(3z^2 - z + 2) + (z^2 - 4) = (3z^2 + z^2) + (-z) + (2 - 4)$

$= 4z^2 - z - 2$

35. $b^2 + (b^3 - 2b^2 + 3) + (b^3 - 3) = (b^3 + b^3) + (b^2 - 2b^2) + (3 - 3)$

$= 2b^3 - b^2$

37. $(12 - 3t - 7t^2) + (1 + 3t - t^2) = (-7t^2 - t^2) + (-3t + 3t) + (12 + 1)$

$= -8t^2 + 13$

39. $(2ab - 3) + (a^2 + 2ab) + (4b^2 - a^2) = (a^2 - a^2) + (2ab - 2ab) + 4b^2 - 3$

$= 4b^2 - 3$

41. $\left(\frac{2}{3}y^2 - \frac{3}{4}\right) + \left(\frac{5}{6}y^2 + 2\right) = \left(\frac{2}{3}y^2 + \frac{5}{6}y^2\right) + \left(-\frac{3}{4} + 2\right)$

$\qquad\qquad\qquad\qquad = \left(\frac{4}{6}y^2 + \frac{5}{6}y^2\right) + \left(-\frac{3}{4} + \frac{8}{4}\right)$

$\qquad\qquad\qquad\qquad = \left(\frac{9}{6}y^2\right) + \left(\frac{5}{4}\right)$

$\qquad\qquad\qquad\qquad = \frac{3}{2}y^2 + \frac{5}{4}$

43. $(0.1t^3 - 3.4t^2) + (1.5t^3 - 7.3) = (0.1t^3 + 1.5t^3) - 3.4t^2 - 7.3$

$\qquad\qquad\qquad\qquad\qquad = 1.6t^3 - 3.4t^2 - 7.3$

45. $\begin{aligned} 2x + \ 5 \\ \underline{3x + \ 8} \\ 5x + 13 \end{aligned}$

47. $\begin{aligned} -2x + 10 \\ \underline{x - 38} \\ -x - 28 \end{aligned}$

49. $\begin{aligned} -x^3 \qquad\ + 3 \\ \underline{3x^3 + 2x^2 + 5} \\ 2x^3 + 2x^2 + 8 \end{aligned}$

51. $\begin{aligned} 3x^4 - 2x^3 - 4x^2 + 2x - 5 \\ \underline{\qquad\qquad\ x^2 - 7x + 5} \\ 3x^4 - 2x^3 - 3x^2 - 5x \end{aligned}$

53. $\begin{aligned} x^2 - 4 \\ \underline{2x^2 + 6} \\ 3x^2 + 2 \end{aligned}$

55. $\begin{aligned} \ -3y + 2 \\ \underline{y^4 + 3y + 2} \\ y^4 \qquad + 4 \end{aligned}$

57. $\begin{aligned} x^2 - 2x + 2 \\ x^2 + 4x \\ \underline{2x^2 \qquad\qquad} \\ 4x^2 + 2x + 2 \end{aligned}$

59. $\begin{aligned} -3y^3 + \ 5 \\ \underline{8y^3 + \ 7} \\ 5y^3 + 12 \end{aligned}$

61. (a) $(6x^2 + 5) + (3 - 2x^2) = (6x^2 - 2x^2) + (5 + 3)$

$\qquad\qquad\qquad\qquad\qquad = 4x^2 + 8$

$\quad$ (b) $\begin{aligned} 6x^2 + 5 \\ \underline{-2x^2 + 3} \\ 4x^2 + 8 \end{aligned}$

$\quad$ Answers regarding format preference will vary.

63. $(11x - 8) - (2x + 3) = 11x - 8 - 2x - 3$

$\qquad\qquad\qquad\qquad = (11x - 2x) + (-8 - 3)$

$\qquad\qquad\qquad\qquad = 9x - 11$

65. $(x^2 - x) - (x - 2) = x^2 - x - x + 2$

$\qquad\qquad\qquad\qquad = (x^2) + (-x - x) + 2$

$\qquad\qquad\qquad\qquad = x^2 - 2x + 2$

67. $(4 - 2x - x^3) - (3 - 2x + 2x^3) = 4 - 2x - x^3 - 3 + 2x - 2x^3$

$\qquad\qquad\qquad\qquad\qquad\qquad = (-x^3 - 2x^3) + (-2x + 2x) + (4 - 3)$

$\qquad\qquad\qquad\qquad\qquad\qquad = -3x^3 + 1$

69. $10 - (u^2 + 5) = 10 - u^2 - 5$

$\qquad\qquad\qquad = (-u^2) + (10 - 5)$

$\qquad\qquad\qquad = -u^2 + 5$

71. $(x^5 - 3x^4 + x^3 - 5x + 1) - (4x^5 - x^3 + x - 5) = x^5 - 3x^4 + x^3 - 5x + 1 - 4x^5 + x^3 - x + 5$

$\qquad\qquad\qquad\qquad\qquad\qquad\qquad\qquad = (x^5 - 4x^5) - 3x^4 + (x^3 + x^3) + (-5x - x) + (1 + 5)$

$\qquad\qquad\qquad\qquad\qquad\qquad\qquad\qquad = -3x^5 - 3x^4 + 2x^3 - 6x + 6$

73.
$$2x - 2 \implies 2x - 2$$
$$\underline{- (x - 1) \implies -x + 1}$$
$$x - 1$$

75.
$$2x^2 - x + 2 \implies 2x^2 - x + 2$$
$$\underline{- (3x^2 + x - 1) \implies -3x^2 - x + 1}$$
$$-x^2 - 2x + 3$$

77.
$$-3x^3 - 4x^2 + 2x - 5 \implies -3x^3 - 4x^2 + 2x - 5$$
$$\underline{- (2x^4 + 2x^3 \qquad - 4x + 5) \implies -2x^4 - 2x^3 \qquad + 4x - 5}$$
$$-2x^4 - 5x^3 - 4x^2 + 6x - 10$$

79.
$$-x^3 + 2 \implies -x^3 + 2$$
$$\underline{- (x^3 + 2) \implies -x^3 - 2}$$
$$-2x^3$$

81.
$$4t^3 - 3t + 5 \implies 4t^3 \qquad - 3t + 5$$
$$\underline{- (3t^2 - 3t - 10) \implies \qquad - 3t^2 + 3t + 10}$$
$$4t^3 - 3t^2 \qquad + 15$$

83.
$$x^3 + 3x^2 \qquad + 3$$
$$\underline{\qquad \qquad x - 3}$$
$$x^3 + 3x^2 + x$$

$$6x^3 - 3x^2 + x \implies 6x^3 - 3x^2 + x$$
$$\underline{- (x^3 + 3x^2 + x) \implies -x^3 - 3x^2 - x}$$
$$5x^3 - 6x^2$$

85.
$$10x^3 \qquad + 15 \implies 10x^3 \qquad + 15$$
$$\underline{- (7x^3 - 4x + 5) \implies -7x^3 + 4x - 5}$$
$$3x^3 + 4x + 10$$

87. $(6x - 5) - (8x + 15) = 6x - 5 - 8x - 15$
$$= (6x - 8x) + (-5 - 15)$$
$$= -2x - 20$$

89. $-(x^3 - 2) + (4x^3 - 2x) = -x^3 + 2 + 4x^3 - 2x$
$$= (-x^3 + 4x^3) + (-2x) + 2$$
$$= 3x^3 - 2x + 2$$

91. $2(x^4 + 2x) + (5x + 2) = 2x^4 + 4x + 5x + 2$
$$= 2x^4 + (4x + 5x) + 2$$
$$= 2x^4 + 9x + 2$$

93. $(15x^2 - 6) - (-8x^3 - 14x^2 - 17) = 15x^2 - 6 + 8x^3 + 14x^2 + 17$
$$= (8x^3) + (15x^2 + 14x^2) + (-6 + 17)$$
$$= 8x^3 + 29x^2 + 11$$

95. $5z - [3z - (10z + 8)] = 5z - [3z - 10z - 8]$
$$= 5z - [-7z - 8]$$
$$= 5z + 7z + 8$$
$$= 12z + 8$$

97. $2(t^2 + 5) - 3(t^2 + 5) + 5(t^2 + 5) = 2t^2 + 10 - 3t^2 - 15 + 5t^2 + 25$
$$= (2t^2 - 3t^2 + 5t^2) + (10 - 15 + 25)$$
$$= 4t^2 + 20$$

99. $8v - 6(3v - v^2) + 10(10v + 3) = 8v - 18v + 6v^2 + 100v + 30$
$$= (6v^2) + (8v - 18v + 100v) + 30$$
$$= 6v^2 + 90v + 30$$

101. Perimeter $= 2z + 4z + 2z + z + 1 + 2 + 1 + z$

$$= 10z + 4$$

103. *Verbal model:* | Area of shaded region | $=$ | Area of larger rectangle | $-$ | Area of smaller rectangle |

     Labels: Area of shaded region $= A$

                    Length of larger rectangle $= 2x$

                    Width of larger rectangle $= x$

                    Area of larger rectangle $= 2x(x) = 2x^2$

                    Length of smaller rectangle $= 4$

                    Width of smaller rectangle $= \dfrac{x}{2}$

                    Area of smaller rectangle $= 4\left(\dfrac{x}{2}\right) = 2x$

     Solution: $A = 2x^2 - 2x$

The area of the shaded region is $2x^2 - 2x$.

Note: To find the area of each rectangle, we use the formula $A = lw$ or Area $=$ (length)(width).

105. *Verbal model:* | Area of shaded region | $=$ | Area of larger rectangle | $-$ | Area of smaller rectangle |

     Labels: Area of shaded region $= A$

                    Length of larger rectangle $= 6x$

                    Width of larger rectangle $= \frac{7}{2}x$

                    Area of larger rectangle $= 6x\left(\frac{7}{2}x\right)$

                    Length of smaller rectangle $= 10$

                    Width of smaller rectangle $= \frac{4}{5}x$

                    Area of smaller rectangle $= 10\left(\frac{4}{5}x\right)$

     Solution: $A = (6x)\left(\frac{7}{2}x\right) - (10)\left(\frac{4}{5}x\right)$

                          $= 21x^2 - 8x$

The area of the shaded region is $21x^2 - 8x$.

Note: To find the area of each rectangle, we use the formula $A = lw$ or Area $=$ (length)(width).

107. *Verbal model:* | Area of shaded region | $=$ | Area of larger triangle | $-$ | Area of smaller triangle |

     Labels: Area of shaded region $= A$

                    Base of larger triangle $= 4x$

                    Height of larger triangle $= 5$

                    Area of larger triangle $= \frac{1}{2}(4x)(5)$

                    Base of smaller triangle $= 4x$

                    Height of smaller triangle $= 2$

                    Area of smaller triangle $= \frac{1}{2}(4x)(2)$

     Solution: $A = \frac{1}{2}(4x)(5) - \frac{1}{2}(4x)(2)$

                        $= 10x - 4x$

                        $= 6x$

The area of the shaded region is $6x$.

Note: To find the area of a triangle, we use the formula Area $= \frac{1}{2}$(Base)(Height).

109. (a) $T = B + C$

$= (0.29t^2 - 1.43t + 64.11) + (-0.32t^2 + 2.98 + 42.17)$

$= -0.03t^2 + 1.55t + 106.28$

(b)

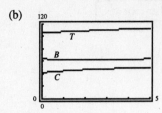

(c) T is increasing.

111. (a) Length: $2x^2$

Width: $3x + 5$

(b) Perimeter = 2(Length) + 2(Width)

$= 2(2x^2) + 2(3x + 5)$

$= 4x^2 + 6x + 10$

(c) Girth = 2(Height + Width)

$= 2[x + (3x + 5)]$

$= 2(4x + 5)$

$= 8x + 10$

Length + Girth = $2x^2 + 8x + 10$

Yes.

If $x = 5$, Length = $2x^2$

$= 2(5)^2$

$= 50$

If $x = 5$, Length + Girth = $2x^2 + 8x + 10$

$= 2(5)^2 + 8(5) + 10$

$= 50 + 40 + 10$

$= 100$

If $x = 5$, the length of 50 inches meets the second restriction of being less than 108 inches. Also, if $x = 5$, the length plus girth of 100 inches meets the third restriction of being less than 130 inches.

113. Statement (b) is true; a trinomial is a polynomial.

Statement (a) is sometimes true; the polynomial $x^3 - 2x + 1$ is a trinomial, but the polynomial $x^3 - 2x^2 + x + 1$ is not a trinomial.

115. Like terms are combined by adding or subtracting their coefficients.

117. To subtract one polynomial from another, change the sign of each term of the polynomial that is being subracted and then combine the resulting like terms.

Example: $(5x^2 + 8x - 2) - (2x^2 - 5x - 4) = 5x^2 + 8x - 2 - 2x^2 + 5x + 4$

$= 3x^2 + 13x + 2$

Section 5.2   Multiplying Polynomials: Special Products

1. $x(-2x) = -2x^2$ **3.** $t^2(4t) = 4t^3$ **5.** $\left(\dfrac{x}{4}\right)(10x) = \dfrac{5}{2}x^2$ or $\dfrac{5x^2}{2}$ **7.** $(-2b^2)(-3b) = 6b^3$

9. $y(3 - y) = (y)(3) - (y)(y)$ **11.** $-x(x^2 - 4) = (-x)(x^2) - (-x)(4)$ **13.** $3t(2t - 5) = (3t)(2t) - (3t)(5)$

$\qquad = 3y - y^2$ or $-y^2 + 3y$ $\qquad = -x^3 + 4x$ $\qquad = 6t^2 - 15t$

15. $-4x(3 + 3x^2 - 6x^3) = (-4x)(3) + (-4x)(3x^2) - (-4x)(6x^3)$

$\qquad = -12x - 12x^3 + 24x^4$ or $24x^4 - 12x^3 - 12x$

17. $3x(x^2 - 2x + 1) = (3x)(x^2) - (3x)(2x) + (3x)(1)$ **19.** $2x(x^2 - 2x + 8) = (2x)(x^2) - (2x)(2x) + (2x)(8)$

$\qquad = 3x^3 - 6x^2 + 3x$ $\qquad = 2x^3 - 4x^2 + 16x$

21. $4t^3(t - 3) = (4t^3)(t) - (4t^3)(3)$ **23.** $x^2(4x^2 - 3x + 1) = (x^2)(4x^2) - (x^2)(3x) + (x^2)(1)$

$\qquad = 4t^4 - 12t^3$ $\qquad = 4x^4 - 3x^3 + x^2$

25. $-3x^3(4x^2 - 6x + 2) = -3x^3(4x^2) - (-3x^3)(6x) + (-3x^3)(2)$

$\qquad = -12x^5 + 18x^4 - 6x^3$

27. $-2x(-3x)(5x + 2) = [(-2x)(-3x)](5x + 2)$ **29.** $(2x)(6x^4) - 3x^2(2x^2) = 12x^5 - 6x^4$

$\qquad = (6x^2)(5x + 2)$

$\qquad = (6x^2)(5x) + (6x^2)(2)$

$\qquad = 30x^3 + 12x^2$

$$\text{F} \quad \text{O} \quad \text{I} \quad \text{L} \qquad\qquad\qquad\qquad \text{F} \quad \text{O} \quad \text{I} \quad \text{L}$$

31. $(x + 3)(x + 4) = x^2 + 4x + 3x + 12$ **33.** $(3x - 5)(2x + 1) = 6x^2 + 3x - 10x - 5$

$\qquad = x^2 + 7x + 12$ $\qquad = 6x^2 - 7x - 5$

$$\text{F} \quad \text{O} \quad \text{I} \quad \text{L} \qquad\qquad\qquad\qquad \text{F} \quad \text{O} \quad \text{I} \quad \text{L}$$

35. $(2x - y)(x - 2y) = 2x^2 - 4xy - xy + 2y^2$ **37.** $(2x + 4)(x + 1) = 2x^2 + 2x + 4x + 4$

$\qquad = 2x^2 - 5xy + 2y^2$ $\qquad = 2x^2 + 6x + 4$

$$\text{F} \quad \text{O} \quad \text{I} \quad \text{L} \qquad\qquad\qquad\qquad \text{F} \quad \text{O} \quad \text{I} \quad \text{L}$$

39. $(6 - 2x)(4x + 3) = 24x + 18 - 8x^2 - 6x$ **41.** $(3x - 2y)(x - y) = 3x^2 - 3xy - 2xy + 2y^2$

$\qquad = -8x^2 + 18x + 18$ $\qquad = 3x^2 - 5xy + 2y^2$

$$\text{F} \quad \text{O} \quad \text{I} \quad \text{L} \qquad\qquad\qquad\qquad \text{F} \quad \text{O} \quad \text{I} \quad \text{L}$$

43. $(3x^2 - 4)(x + 2) = 3x^3 + 6x^2 - 4x - 8$ **45.** $(2x^3 + 4)(x^2 + 6) = 2x^5 + 12x^3 + 4x^2 + 24$

47. $(3s + 1)(3s + 4) - (3s)^2 = 9s^2 + 12s + 3s + 4 - 9s^2$ **49.** $(4x^2 - 1)(2x + 8) + (-x)^3 = 8x^3 + 32x^2 - 2x - 8 - x^3$

$\qquad = 15s + 4$ $\qquad = 7x^3 + 32x^2 - 2x - 8$

51. $(x + 10)(x + 2) = x(x + 2) + 10(x + 2)$
$\qquad\qquad\qquad\quad = x^2 + 2x + 10x + 20$
$\qquad\qquad\qquad\quad = x^2 + 12x + 20$

53. $(2x - 5)(x + 2) = 2x(x + 2) - 5(x + 2)$
$\qquad\qquad\qquad\quad = 2x^2 + 4x - 5x - 10$
$\qquad\qquad\qquad\quad = 2x^2 - x - 10$

55. $(x + 1)(x^2 + 2x - 1) = x(x^2 + 2x - 1) + 1(x^2 + 2x - 1)$
$\qquad\qquad\qquad\qquad\quad = x^3 + 2x^2 - x + x^2 + 2x - 1$
$\qquad\qquad\qquad\qquad\quad = x^3 + 3x^2 + x - 1$

57. $(x^3 - 2x + 1)(x - 5) = x^3(x - 5) - 2x(x - 5) + 1(x - 5)$
$\qquad\qquad\qquad\qquad\quad = x^4 - 5x^3 - 2x^2 + 10x + x - 5$
$\qquad\qquad\qquad\qquad\quad = x^4 - 5x^3 - 2x^2 + 11x - 5$

59. $(x - 2)(x^2 + 2x + 4) = x(x^2 + 2x + 4) - 2(x^2 + 2x + 4)$
$\qquad\qquad\qquad\qquad\quad = x^3 + 2x^2 + 4x - 2x^2 - 4x - 8$
$\qquad\qquad\qquad\qquad\quad = x^3 - 8$

61. $(x^2 + 3)(x^2 - 6x + 2) = x^2(x^2 - 6x + 2) + 3(x^2 - 6x + 2)$
$\qquad\qquad\qquad\qquad\qquad = x^4 - 6x^3 + 2x^2 + 3x^2 - 18x + 6$
$\qquad\qquad\qquad\qquad\qquad = x^4 - 6x^3 + 5x^2 - 18x + 6$

63. $(3x^2 + 1)(x^2 - 4x - 2) = 3x^2(x^2 - 4x - 2) + 1(x^2 - 4x - 2)$
$\qquad\qquad\qquad\qquad\qquad = 3x^4 - 12x^3 - 6x^2 + x^2 - 4x - 2$
$\qquad\qquad\qquad\qquad\qquad = 3x^4 - 12x^3 - 5x^2 - 4x - 2$

65.
$$
\begin{array}{r}
x + 3 \\
\times \quad x - 2 \\
\hline
-2x - 6 \\
x^2 + 3x \quad\;\; \\
\hline
x^2 + x - 6
\end{array}
$$

67.
$$
\begin{array}{r}
x^2 - 3x + 9 \\
\times \qquad\quad x + 3 \\
\hline
3x^2 - 9x + 27 \\
x^3 - 3x^2 + 9x \qquad\;\; \\
\hline
x^3 \qquad\qquad + 27
\end{array}
$$

69.
$$
\begin{array}{r}
x^2 + x - 2 \\
\times \quad x^2 - x + 2 \\
\hline
2x^2 + 2x - 4 \\
- x^3 - x^2 + 2x \qquad\;\; \\
x^4 + x^3 - 2x^2 \qquad\qquad \\
\hline
x^4 \qquad - x^2 + 4x - 4
\end{array}
$$

71.
$$
\begin{array}{r}
x^3 + x + 3 \\
\times \qquad\quad x^2 + 5x - 4 \\
\hline
- 4x^3 \qquad - 4x - 12 \\
5x^4 \qquad + 5x^2 + 15x \qquad\;\; \\
x^5 \qquad + x^3 + 3x^2 \qquad\qquad \\
\hline
x^5 + 5x^4 - 3x^3 + 8x^2 + 11x - 12
\end{array}
$$

73.
$$
\begin{array}{r}
x - 2 \\
\times \quad x - 2 \\
\hline
- 2x + 4 \\
x^2 - 2x \qquad\;\; \\
\hline
x^2 - 4x + 4
\end{array}
$$
$(x - 2)^3 = (x - 2)^2(x - 2) = (x^2 - 4x + 4)(x - 2)$
$$
\begin{array}{r}
x^2 - 4x + 4 \\
\times \qquad\quad x - 2 \\
\hline
- 2x^2 + 8x - 8 \\
x^3 - 4x^2 + 4x \qquad\;\; \\
\hline
x^3 - 6x^2 + 12x - 8
\end{array}
$$

75.
$$\begin{array}{r} x - 1 \\ \times \quad x - 1 \\ \hline - x + 1 \\ x^2 - x \quad\;\; \\ \hline x^2 - 2x + 1 \end{array}$$

$(x - 1)^2(x - 1)^2 = (x^2 - 2x + 1)(x^2 - 2x + 1)$

$$\begin{array}{r} x^2 - 2x + 1 \\ \times \quad x^2 - 2x + 1 \\ \hline x^2 - 2x + 1 \\ - 2x^3 + 4x^2 - 2x \quad\;\; \\ x^4 - 2x^3 + \; x^2 \quad\quad\quad\;\; \\ \hline x^4 - 4x^3 + 6x^2 - 4x + 1 \end{array}$$

77.
$$\begin{array}{r} x + 2 \\ \times \quad x + 2 \\ \hline 2x + 4 \\ x^2 + 2x \quad\;\; \\ \hline x^2 + 4x + 4 \end{array}$$

$(x + 2)^2(x - 4) = (x^2 + 4x + 4)(x - 4)$

$$\begin{array}{r} x^2 + \; 4x + \; 4 \\ \times \quad x - \; 4 \\ \hline - 4x^2 - 16x - 16 \\ x^3 + 4x^2 + \; 4x \quad\quad\;\; \\ \hline x^3 \quad\quad\; - 12x - 16 \end{array}$$

79.
$$\begin{array}{r} u - 1 \\ \times \quad 2u + 3 \\ \hline 3u - 3 \\ 2u^2 - 2u \quad\;\; \\ \hline 2u^2 + \; u - 3 \end{array}$$

$(u - 1)(2u + 3)(2u + 1) = (2u^2 + u - 3)(2u + 1)$

$$\begin{array}{r} 2u^2 + \; u - 3 \\ \times \quad 2u + 1 \\ \hline 2u^2 + \; u - 3 \\ 4u^3 + 2u^2 - 6u \quad\quad\;\; \\ \hline 4u^3 + 4u^2 - 5u - 3 \end{array}$$

81. $(x + 3)(x - 3) = x^2 - 3^2$
$\qquad\qquad\qquad\; = x^2 - 9$

83. $(x + 4)(x - 4) = x^2 - 4^2$
$\qquad\qquad\qquad\; = x^2 - 16$

85. $(2u + 3)(2u - 3) = (2u)^2 - (3)^2$
$\qquad\qquad\qquad\qquad = 4u^2 - 9$

87. $(4t - 6)(4t + 6) = (4t)^2 - 6^2$
$\qquad\qquad\qquad\quad = 16t^2 - 36$

89. $(2x + 3y)(2x - 3y) = (2x)^2 - (3y)^2$
$\qquad\qquad\qquad\qquad = 4x^2 - 9y^2$

91. $(4u - 3v)(4u + 3v) = (4u)^2 - (3v)^2$
$\qquad\qquad\qquad\qquad = 16u^2 - 9v^2$

93. $(2x^2 + 5)(2x^2 - 5) = (2x^2)^2 - 5^2$
$\qquad\qquad\qquad\qquad = 4x^4 - 25$

95. $(x + 6)^2 = (x)^2 + 2(x)(6) + (6)^2$
$\qquad\qquad\;\; = x^2 + 12x + 36$

97. $(t - 3)^2 = t^2 - 2(t)(3) + (3)^2$
$\qquad\qquad\; = t^2 - 6t + 9$

99. $(3x + 2)^2 = (3x)^2 + 2(3x)(2) + 2^2$
$\qquad\qquad\quad = 9x^2 + 12x + 4$

101. $(8 - 3z)^2 = (8)^2 - 2(8)(3z) + (3z)^2$
$\qquad\qquad\quad = 64 - 48z + 9z^2 \quad$ or $\quad 9z^2 - 48z + 64$

103. $(2x - 5y)^2 = (2x)^2 - 2(2x)(5y) + (5y)^2$
$\qquad\qquad\qquad = 4x^2 - 20xy + 25y^2$

105. $(6t + 5s)^2 = (6t)^2 + 2(6t)(5s) + (5s)^2$
$\qquad\qquad\qquad = 36t^2 + 60st + 25s^2$

107. $[(x + 1) + y]^2 = (x + 1)^2 + 2(x + 1)(y) + (y)^2$
$\qquad\qquad\qquad\;\; = (x + 1)^2 + 2y(x + 1) + y^2$
$\qquad\qquad\qquad\;\; = x^2 + 2x + 1 + 2y(x + 1) + y^2$
$\qquad\qquad\qquad\;\; = x^2 + 2x + 1 + 2xy + 2y + y^2$
$\qquad\qquad\;$ or $x^2 + y^2 + 2xy + 2x + 2y + 1$

109. $[u - (v - 3)]^2 = (u)^2 - 2(u)(v - 3) + (v - 3)^2$
$\qquad\qquad\qquad\;\; = u^2 - 2u(v - 3) + (v - 3)^2$
$\qquad\qquad\qquad\;\; = u^2 - 2u(v - 3) + v^2 - 6v + 9$
$\qquad\qquad\qquad\;\; = u^2 - 2uv + 6u + v^2 - 6v + 9$
$\qquad\qquad\;$ or $u^2 + v^2 - 2uv + 6u - 6v + 9$

111. $(x + 2)^2 - (x - 2)^2 = [(x)^2 + 2(x)(2) + (2)^2] - [(x)^2 - 2(x)(2) + (2)^2]$

$$= [x^2 + 4x + 4] - [x^2 - 4x + 4]$$

$$= x^2 + 4x + 4 - x^2 + 4x - 4$$

$$= 8x$$

113. Yes, this is an identity.

$(x + y)^3 = (x + y)(x + y)(x + y)$

$$= (x^2 + 2xy + y^2)(x + y)$$

$$= x^3 + x^2y + 2x^2y + 2xy^2 + xy^2 + y^3$$

$$= x^3 + 3x^2y + 3xy^2 + y^3$$

115. $(x + 2)^3 = (x)^3 + 3(x)^2(2) + 3(x)(2)^2 + (2)^3$

$$= x^3 + 6x^2 + 12x + 8$$

Pattern: $(a + b)^3 = a^3 + 3a^2b + 3ab^2 + b^3$

117. (a) $(x - 1)(x + 1) = x^2 - 1$

(b) $(x - 1)(x^2 + x + 1) = x(x^2 + x + 1) - 1(x^2 + x + 1)$

$$= x^3 + x^2 + x - x^2 - x - 1$$

$$= x^3 - 1$$

(c) $(x - 1)(x^3 + x^2 + x + 1) = x(x^3 + x^2 + x + 1) - 1(x^3 + x^2 + x + 1)$

$$= x^4 + x^3 + x^2 + x - x^3 - x^2 - x - 1$$

$$= x^4 - 1$$

(d) $(x - 1)(x^4 + x^3 + x^2 + x + 1) = x^5 - 1$

$(x - 1)(x^4 + x^3 + x^2 + x + 1) = x(x^4 + x^3 + x^2 + x + 1) - 1(x^4 + x^3 + x^2 + x + 1)$

$$= x^5 + x^4 + x^3 + x^2 + x - x^4 - x^3 - x^2 - x - 1$$

$$= x^5 - 1$$

119. (a) *Common formula:* $P = 2l + 2w$

      Labels: Width $= w$

                     Length (height) $= 2w$

      Perimeter: $P = 2(2w) + 2(w)$

$$= 4w + 2w$$

$$= 6w$$

The perimeter of the sign is $6w$.

(b) *Common formula:* $A = l \cdot w$

      Labels: Width $= w$

                     Length (height) $= 2w$

      Area: $A = (2w)(w)$

$$= 2w^2$$

The area of the sign is $2w^2$.

121. $x^2 + 7x + 12 = (x + 3)(x + 4)$

123. $2x^2 + 4x = 2x(x + 2)$

125. Area $=$ Area 1 $+$ Area 2 $+$ Area 3 $+$ Area 4

$(x + a)(x + b) = x^2$ $+$ bx $+$ ax $+$ ab

127. Area $=$ (Width)(Length)

The width is $x + x$, or $2x$, and the length is $(x + 1) + (x + 1)$ or $2(x + 1)$.

$$2x[2(x + 1)] = 2x(2x + 2) = 4x^2 + 4x$$

129. Area = (Length)(Width)

$$= (z + 4)(z)$$

$$= z^2 + 4z$$

Area = Area of large rectangle − Area of 2 smaller rectangles

$$= (z + 5)(z + 4) - 5z - 4(5)$$

Therefore, $z^2 + 4z = (z + 5)(z + 4) - 5z - 20$

Note: You can verify algebraically that the two expressions are the same.

131. (a)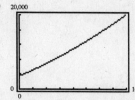

(b) $(P)(S) = (2.37t + 226.94)(16.99t^2 + 767.47t + 3525.39)$

$$= 2.37t(16.99t^2 + 767.47t + 3525.39) + 226.94(16.99t^2 + 767.47t + 3525.39)$$

$$= 40.2663t^3 + 1818.9039t^2 + 8355.1743t + 3855.7106t^2 + 174169.6418t + 800052.0066$$

$$= 40.2663t^3 + 5674.6145t^2 + 182,524.8161t + 800,052.0066$$

$$\approx 40.27t^3 + 5674.61t^2 + 182,524.82t + 800,052.01$$

(c) In this model, $t = 0$ represents 1980. Therefore, for 1990, $t = 10$. Substituting $t = 10$ in the expression for the product PS yields

$$40.27(10)^3 + 5674.61(10)^2 + 182,524.82(10) + 800,052.01$$

$$= 40,270 + 567,461 + 1,825,248.2 + 800,052.01$$

$$\approx 3,233,031$$

The federal debt for 1990 was approximately $3,233,031 million.

133. $500(1 + r)^2 = 500[(1)^2 + 2(1)(r) + (r)^2]$

$$= 500[1 + 2r + r^2]$$

$$= 500 + 1000r + 500r^2$$

135. (d) Surface area = 2(Length)(Width) + 2(Length)(Height) + 2(Width)(Height)

$$= 2(2x^2)(3x + 5) + 2(2x^2)(x) + 2(3x + 5)(x)$$

$$= 4x^2(3x + 5) + 4x^2(x) + (6x + 10)(x)$$

$$= 12x^3 + 20x^2 + 4x^3 + 6x^2 + 10x$$

$$= 16x^3 + 26x^2 + 10x$$

The surface area is $16x^3 + 26x^2 + 10x$ square inches.

(e) Area of base = (Length)(Width)

$$= (3x + 5)^2$$

$$= (3x)^2 + 2(3x)(5) + 5^2$$

$$= 9x^2 + 30x + 25$$

The area of the base is $9x^2 + 30x + 25$ square inches.

(f) Volume = (Area of new base)(Height)

$$= (3x + 5)^2(x)$$

$$= (9x^2 + 30x + 25)(x)$$

$$= 9x^3 + 25x^2 + 25x$$

The volume is $9x^3 + 25x^2 + 25x$ cubic inches.

137. $a^m \cdot a^n = a^{m+n}$ Example: $x^8(x^7 + x^3) = x^{15} + x^{11}$

     $(ab)^n = a^n b^n$ Example: $(3x + 5y)(3x - 5y) = (3x)^2 - (5y)^2 = 9x^2 - 25y^2$

     $(a^m)^n = a^{mn}$ Example: $(x^5 - 3)^2 = (x^5)^2 - 2(x^5)(3) + 3^2 = x^{10} - 6x^5 + 9$

139. F: **First**
     O: **Outer**
     I: **Inner**
     L: **Last**

141. Each of the m terms of the first factor must be multiplied by each of the n terms of the second factor, so there will be mn products.

143. False. $(x + 2)^2 = x^2 + 4x + 4$

Mid-Chapter Quiz for Chapter 5

1. The expression $x^2 + 2x - 3x^{-1}$ is not a polynomial because the exponent of the third term, $-3x^{-1}$, is negative.

2. *Degree:* 4
     Leading Coefficient: -3

3. $6x^5 + x^2 - 1$

Note: There are many correct answers.

4. False. For example, $(x - 2)(x + 7) = x^2 + 5x - 14$.

5. $(y^2 + 3y - 1) + (4 + 3y) = y^2 + (3y + 3y) + (-1 + 4)$
                            $= y^2 + 6y + 3$

6. $(3v^2 - 5) - (v^3 + 2v^2 - 6v) = 3v^2 - 5 - v^3 - 2v^2 + 6v$
                                     $= -v^3 + v^2 + 6v - 5$

7. $9s - [6 - (s - 5) + 7s] = 9s - [6 - s + 5 + 7s]$
                          $= 9s - 6 + s - 5 - 7s$
                          $= (9s + s - 7s) + (-6 - 5)$
                          $= 3s - 11$

8. $-3(4 - x) + 4(x^2 + 2) - (x^2 - 2x) = -12 + 3x + 4x^2 + 8 - x^2 + 2x$
                                       $= (4x^2 - x^2) + (3x + 2x) + (-12 + 8)$
                                       $= 3x^2 + 5x - 4$

9. $2r^2(5r) = 10r^3$

10. $m^3(-2m) = -2m^4$

11. $(2y - 3)(y + 5) = 2y^2 + 10y - 3y - 15$
                        $= 2y^2 + 7y - 15$

12. $(x + 4)(2x^2 - 3x - 2) = x(2x^2 - 3x - 2) + 4(2x^2 - 3x - 2)$
                               $= 2x^3 - 3x^2 - 2x + 8x^2 - 12x - 8$
                               $= 2x^3 + (-3x^2 + 8x^2) + (-2x - 12x) - 8$
                               $= 2x^3 + 5x^2 - 14x - 8$

13. $(4 - 3x)^2 = 4^2 - 2(4)(3x) + (3x)^2$
                  $= 16 - 24x + 9x^2$

14. $(2u - 3)(2u + 3) = (2u)^2 - 3^2$
                       $= 4u^2 - 9$

15.
$$
\begin{array}{r}
5x^4 \quad\ + 2x^2 +\ x - 3 \\
+\quad\ \ 3x^3 - 2x^2 - 3x + 5 \\
\hline
5x^4 + 3x^3 \qquad\ - 2x + 2
\end{array}
$$

16.
$$
\begin{array}{r}
2x^3 + x^2 \qquad\ - 8 \\
-\quad (5x^2 - 3x - 9)
\end{array}
\Rightarrow
\begin{array}{r}
2x^3 + \ x^2 \qquad\ - 8 \\
- 5x^2 + 3x + 9 \\
\hline
2x^3 - 4x^2 + 3x + 1
\end{array}
$$

17.
$$
\begin{array}{r}
3x^2 +\ 7x + 1 \\
\times \qquad\quad 2x - 5 \\
\hline
- 15x^2 - 35x - 5 \\
6x^3 + 14x^2 +\ 2x \\
\hline
6x^3 -\quad x^2 - 33x - 5
\end{array}
$$

18.
$$
\begin{array}{r}
5x^3 - 6x^2 +\ 3 \\
\times \qquad\quad x^2 - 3x \\
\hline
- 15x^4 + 18x^3 \qquad\ - 9x \\
5x^5 -\ 6x^4 \qquad\ + 3x^2 \\
\hline
5x^5 - 21x^4 + 18x^3 + 3x^2 - 9x
\end{array}
$$

19. Perimeter $= 5x + 18 + 2x + 2x + 3x + (18 - 2x)$

$\qquad = (5x + 2x + 2x + 3x - 2x) + (18 + 18)$

$\qquad = 10x + 36$

20. Area = Area of rectangle on left + Area of rectangle on right

$\qquad = x(x + 1) + 3(x)$

$\qquad = x^2 + x + 3x$

$\qquad = x^2 + 4x$

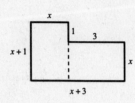

or

Area = Area of upper rectangle + Area of lower rectangle

$\qquad = x(1) + (x + 3)x$

$\qquad = x + x^2 + 3x$

$\qquad = x^2 + 4x$

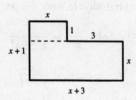

Section 5.3   Negative Exponents and Scientific Notation

1. $3^{-3} = \dfrac{1}{3^3}$

3. $y^{-5} = \dfrac{1}{y^5}$

5. $8x^{-7} = \dfrac{8}{x^7}$

7. $7x^{-4}y^{-1} = \dfrac{7}{x^4 y}$

9. $\dfrac{1}{2z^{-4}} = \dfrac{1z^4}{2} = \dfrac{z^4}{2}$

11. $\dfrac{2x}{3y^{-2}} = \dfrac{2xy^2}{3}$

13. $\dfrac{1}{4} = 4^{-1}$

15. $\dfrac{1}{x^2} = x^{-2}$

17. $\dfrac{10}{t^5} = 10t^{-5}$

19. $\dfrac{5}{x^n} = 5x^{-n}$

21. $\dfrac{2x^2}{y^4} = 2x^2 y^{-4}$

23. $3^{-2} = \dfrac{1}{3^2} = \dfrac{1}{9}$

25. $(-4)^{-3} = \dfrac{1}{(-4)^3} = \dfrac{1}{-64}$

$\qquad\qquad = -\dfrac{1}{64}$

27. $\dfrac{1}{4^{-2}} = 4^2 = 16$

29. $\dfrac{2}{3^{-4}} = 2(3^4) = 2(81)$

$\qquad\qquad = 162$

31. $\dfrac{2^{-4}}{3^{-2}} = \dfrac{3^2}{2^4} = \dfrac{9}{16}$

33. $\dfrac{4^{-2}}{3^{-4}} = \dfrac{3^4}{4^2} = \dfrac{81}{16}$

35. $\left(\dfrac{2}{3}\right)^{-2} = \dfrac{2^{-2}}{3^{-2}}$

$\qquad\qquad = \dfrac{3^2}{2^2}$

$\qquad\qquad = \dfrac{9}{4}$

37. $3.8^{-4} \approx 0.0048$

39. $100(1.06)^{-15} \approx 41.7265$

41. $4^{-2} \cdot 4^3 = 4^{-2+3} = 4^1 = 4$

43. $x^{-4} \cdot x^6 = x^{-4+6} = x^2$

45. $u^{-6} \cdot u^3 = u^{-6+3} = u^{-3} = \dfrac{1}{u^3}$

47. $xy^{-3} \cdot y^2 = xy^{-3+2} = xy^{-1} = \dfrac{x}{y}$

49. $\dfrac{x^2}{x^{-3}} = x^2 \cdot x^3 = x^{2+3} = x^5$

Note: We could also use the rule
$$\frac{a^m}{a^n} = a^{m-n}.$$

$$\frac{x^2}{x^{-3}} = x^{2-(-3)} = x^{2+3} = x^5$$

51. $\dfrac{y^{-5}}{y} = \dfrac{1}{y \cdot y^5}$

$$= \frac{1}{y^{1+5}}$$

$$= \frac{1}{y^6}$$

Note: We could also use the rule
$$\frac{a^m}{a^n} = a^{m-n}.$$

$$\frac{y^{-5}}{y} = y^{-5-1}$$

$$= y^{-6}$$

$$= \frac{1}{y^6}$$

53. $\dfrac{x^{-4}}{x^{-2}} = \dfrac{x^2}{x^4} = \dfrac{1}{x^2}$

Note: There are several ways to work this problem.

$$\frac{x^{-4}}{x^{-2}} = x^{-4-(-2)} \quad \text{or} \quad \frac{x^{-4}}{x^{-2}} = \frac{1}{x^{-2} \cdot x^4} \quad \text{or} \quad \frac{x^{-4}}{x^{-2}} = x^{-4}x^2$$

$$= x^{-4+2} \qquad\qquad = \frac{1}{x^{-2+4}} \qquad\qquad = x^{-4+2}$$

$$= x^{-2} \qquad\qquad\qquad = \frac{1}{x^2} \qquad\qquad\qquad = x^{-2}$$

$$= \frac{1}{x^2} \qquad\qquad\qquad\qquad\qquad\qquad\qquad\qquad = \frac{1}{x^2}$$

55. $(y^{-3})^2 = y^{-6} = \dfrac{1}{y^6}$

57. $(s^2)^{-1} = s^{-2} = \dfrac{1}{s^2}$

59. $(2x^{-2})^0 = 1$

61. $\dfrac{b^2 b^{-3}}{b^4} = \dfrac{b^{2+(-3)}}{b^4}$

$$= \frac{b^{-1}}{b^4}$$

$$= b^{-1-4}$$

$$= b^{-5}$$

$$= \frac{1}{b^5}$$

Note: Here is another way to work this problem.

$$\frac{b^2 b^{-3}}{b^4} = \frac{b^2}{b^3 b^4}$$

$$= \frac{b^2}{b^{3+4}}$$

$$= \frac{b^2}{b^7}$$

$$= b^{2-7}$$

$$= b^{-5}$$

$$= \frac{1}{b^5}$$

63. $(3x^2y)^{-2} = 3^{-2}(x^2)^{-2}y^{-2}$

$= 3^{-2}x^{-4}y^{-2}$

$= \dfrac{1}{3^2x^4y^2}$

$= \dfrac{1}{9x^4y^2}$

65. $(4a^{-2}b^3)^{-3} = 4^{-3}(a^{-2})^{-3}(b^3)^{-3}$

$= 4^{-3}a^6b^{-9}$

$= \dfrac{a^6}{4^3b^9}$

$= \dfrac{a^6}{64b^9}$

67. $(-2x^2)(4x^{-3}) = -8x^{2-3}$

$= -8x^{-1}$

$= -\dfrac{8}{x}$

69. $\left(\dfrac{x}{10}\right)^{-1} = \dfrac{x^{-1}}{10^{-1}} = \dfrac{10}{x}$

71. $\left(\dfrac{3z^2}{x}\right)^{-2} = \dfrac{3^{-2}(z^2)^{-2}}{x^{-2}}$

$= \dfrac{3^{-2}z^{-4}}{x^{-2}}$

$= \dfrac{x^2}{3^2z^4}$

$= \dfrac{x^2}{9z^4}$

73. $\dfrac{(2y)^{-4}}{(2y)^{-4}} = (2y)^{-4-(-4)} = (2y)^0 = 1$

75. $\dfrac{3}{2} \cdot \left(\dfrac{-2}{3}\right)^{-3} = \dfrac{3}{2} \cdot \dfrac{(-2)^{-3}}{3^{-3}} = \dfrac{3}{2} \cdot \dfrac{3^3}{(-2)^3}$

$= \dfrac{3}{2} \cdot \dfrac{27}{-8} = \dfrac{3(27)}{2(-8)}$

$= \dfrac{81}{-16} = -\dfrac{81}{16}$

77. $\dfrac{(-2x)^{-3}}{-4x^{-2}} = \dfrac{(-2)^{-3}x^{-3}}{-4x^{-2}} = \dfrac{x^{-3-(-2)}}{(-4)(-2)^3}$

$= \dfrac{x^{-1}}{-4(-8)} = \dfrac{1}{32x}$

79. $(5x^2y^4z^6)^3(5x^2y^4z^6)^{-3} = (5x^2y^4z^6)^{3+(-3)}$

$= (5x^2y^4z^6)^0$

$= 1$

81. $(x + y)^{-8}(x + y)^8 = (x + y)^{-8+8}$

$= (x + y)^0 = 1$

83. $93{,}000{,}000 = 9.3 \times 10^7$

85. $1{,}637{,}000{,}000 = 1.637 \times 10^9$

87. $0.000435 = 4.35 \times 10^{-4}$

89. $0.004392 = 4.392 \times 10^{-3}$

91. $16{,}000{,}000 = 1.6 \times 10^7$

93. $1.09 \times 10^6 = 1{,}090{,}000$

95. $8.67 \times 10^{-2} = 0.0867$

97. $8.52 \times 10^{-3} = 0.00852$

99. $6.21 \times 10^0 = 6.21$

Note: $10^0 = 1$

101. $(8 \times 10^3) + (3 \times 10^0) + (5 \times 10^{-2})$

$= 8000 + 3 + 0.05$

$= 8003.05$

103. $8{,}000{,}000 \times 623{,}000 = (8 \times 10^6)(6.23 \times 10^5)$

$= 4.984 \times 10^{12}$

105. $0.000345 \times 8{,}980{,}000{,}000 = (3.45 \times 10^{-4})(8.98 \times 10^9)$

$= 3{,}098{,}100 \text{ or } 3.0981 \times 10^6$

107. $3{,}200{,}000^5 = (3.2 \times 10^6)^5 \approx 3.3554 \times 10^{32}$

109. $(3.28 \times 10^{-6})^4 \approx 1.1574 \times 10^{-22}$

111. $\dfrac{848{,}000{,}000}{1{,}620{,}000} = \dfrac{8.48 \times 10^8}{1.62 \times 10^6} = \left(\dfrac{8.48}{1.62}\right) \times 10^2$

$\approx 5.2346 \times 10^2$

113. $(4.85 \times 10^5)(2.04 \times 10^8) = (4.85)(2.04) \times 10^{13}$
$$= 9.894 \times 10^{13}$$

115. $\dfrac{9.3 \times 10^7}{1.1 \times 10^7} = 8.\overline{45}$ minutes

117.

Planet	Mercury	Saturn	Neptune	Pluto
Kilometers	5.83×10^7	1.43×10^9	4.50×10^9	5.90×10^9

Mercury: $0.39 \times 149{,}503{,}000 = 58{,}306{,}170 \approx 5.83 \times 10^7$

Saturn: $9.56 \times 149{,}503{,}000 = 1{,}429{,}248{,}680 \approx 1.43 \times 10^9$

Neptune: $30.13 \times 149{,}503{,}000 = 4{,}504{,}525{,}390 \approx 4.50 \times 10^9$

Pluto: $39.47 \times 149{,}503{,}000 = 5{,}900{,}883{,}410 \approx 5.90 \times 10^9$

119. (a)

x	-1	-2	-3	-4	-5
2^x	$\frac{1}{2}$	$\frac{1}{4}$	$\frac{1}{8}$	$\frac{1}{16}$	$\frac{1}{32}$

(b)

(c) As n gets very large, 2^{-n} approaches 0. 2^{-n} will never be negative. It will continue to be a smaller positive fraction.

121. $k = \dfrac{8.31 \times 10^7}{6.01 \times 10^{23}} = \dfrac{8.31}{6.01} \times 10^{7-23}$
$$\approx 1.38 \times 10^{-16}$$

123. True. $x^{-1}y^{-1} = \dfrac{1}{xy}$

125. False. If $x = 2$, $\dfrac{x^{-4}}{x^{-3}} = \dfrac{2^{-4}}{2^{-3}} = \dfrac{\frac{1}{16}}{\frac{1}{8}} = \dfrac{1}{16} \cdot \dfrac{8}{1} = \dfrac{1}{2} \neq 2$.

127. True. $\dfrac{2x \times 10^{-5}}{x \times 10^{-3}} = 2 \times 10^{-5-(-3)} = 2 \times 10^{-2}$

129. Examples will vary.

$3.4 \times 10^7 = 34{,}000{,}000$

$3.4 \times 10^{-6} = 0.0000034$

131.

$(3 \times 10^5)(4 \times 10^6) = (3 \times 10^5)(10^6 \times 4)$	Commutative Property of Mutiplication
$= 3(10^5 \times 10^6)(4)$	Associative Property of Multiplication
$= 3(10^{5+6})(4)$	Property of exponents
$= (3 \cdot 4)10^{11}$	Commutative Property of Multiplication
$= 12 \times 10^{11}$	Multiplication
$= 1.2 \times 10^{12}$	Scientific notation

Section 5.4   Dividing Polynomials

1. *By Cancellation*

$$\frac{x^5}{x^2} = \frac{x \cdot x \cdot x \cdot \cancel{x} \cdot \cancel{x}}{\cancel{x} \cdot \cancel{x}} = x^3$$

By Subtracting Exponents

$$\frac{x^5}{x^2} = x^{5-2} = x^3$$

3. *By Cancellation*

$$\frac{x^2}{x^5} = \frac{\cancel{x} \cdot \cancel{x}}{\cancel{x} \cdot \cancel{x} \cdot x \cdot x \cdot x} = \frac{1}{x^3}$$

By Subtracting Exponents

$$\frac{x^2}{x^5} = x^{2-5} = x^{-3} = \frac{1}{x^3}$$

5. *By Cancellation*

$$\frac{z^4}{z^7} = \frac{\cancel{z} \cdot \cancel{z} \cdot \cancel{z} \cdot \cancel{z}}{z \cdot z \cdot z \cdot \cancel{z} \cdot \cancel{z} \cdot \cancel{z} \cdot \cancel{z}} = \frac{1}{z^3}$$

By Subtracting Exponents

$$\frac{z^4}{z^7} = z^{4-7} = z^{-3} = \frac{1}{z^3}$$

7. *By Cancellation*

$$\frac{3u^4}{u^3} = \frac{3 \cdot u \cdot \cancel{u} \cdot \cancel{u} \cdot \cancel{u}}{\cancel{u} \cdot \cancel{u} \cdot \cancel{u}} = 3u$$

By Subtracting Exponents

$$\frac{3u^4}{u^3} = 3u^{4-3} = 3u$$

9. *By Cancellation*

$$\frac{2^3y^4}{2^2y^2} = \frac{2 \cdot \cancel{2} \cdot \cancel{2} \cdot y \cdot y \cdot \cancel{y} \cdot \cancel{y}}{\cancel{2} \cdot \cancel{2} \cdot \cancel{y} \cdot \cancel{y}} = 2y^2$$

By Subtracting Exponents

$$\frac{2^3y^4}{2^2y^2} = 2^{3-2}y^{4-2} = 2y^2$$

11. *By Cancellation*

$$\frac{4^5x^3}{4x^5} = \frac{\cancel{4} \cdot 4 \cdot 4 \cdot 4 \cdot 4 \cdot \cancel{x} \cdot \cancel{x} \cdot \cancel{x}}{\cancel{4} \cdot \cancel{x} \cdot \cancel{x} \cdot \cancel{x} \cdot x \cdot x} = \frac{4^4}{x^2} \quad \text{or} \quad \frac{256}{x^2}$$

By Subtracting Exponents

$$\frac{4^5x^3}{4x^5} = 4^{5-1}x^{3-5} = 4^4x^{-2} = \frac{256}{x^2}$$

13. *By Cancellation*

$$\frac{3^4(ab)^2}{3(ab)^3} = \frac{\cancel{3} \cdot 3 \cdot 3 \cdot 3 \cdot \cancel{(ab)}\cancel{(ab)}}{\cancel{3}\cancel{(ab)}\cancel{(ab)}(ab)}$$

$$= \frac{3^3}{ab}$$

$$= \frac{27}{ab}$$

By Subtracting Exponents

$$\frac{3^4(ab)^2}{3(ab)^3} = 3^{4-1}(ab)^{2-3}$$

$$= 3^3(ab)^{-1}$$

$$= \frac{27}{ab}$$

15. $\dfrac{-3x^2}{x} = -3x^{2-1} = -3x$

17. $\dfrac{4}{x^3}$ This expression is already in simplified form.

19. $\dfrac{-12z^3}{-3z} = \left(\dfrac{-12}{-3}\right)(z^{3-1}) = 4z^2$

21. $\dfrac{32b^4}{12b^3} = \left(\dfrac{32}{12}\right)(b^{4-3}) = \dfrac{8}{3}b \quad \text{or} \quad \dfrac{8b}{3}$

23. $\dfrac{-22y^2}{4y} = \left(\dfrac{-22}{4}\right)y^{2-1} = -\dfrac{11}{2}y \quad \text{or} \quad -\dfrac{11y}{2}$

25. $\dfrac{-18s^4}{-12r^2s} = \left(\dfrac{-18}{-12}\right) \cdot \dfrac{s^{4-1}}{r^2}$

$$= \dfrac{3}{2} \cdot \dfrac{s^3}{r^2} = \dfrac{3s^3}{2r^2}$$

27. $\dfrac{(-3z)^2}{18z^3} = \dfrac{9z^2}{18z^3} = \dfrac{9}{18}z^{2-3}$

$$= \dfrac{1}{2}z^{-1} = \dfrac{1}{2z}$$

29. $\dfrac{(2x^2y)^3}{(4y^2)^2x^4} = \dfrac{2^3x^6y^3}{4^2y^4x^4} = \dfrac{8}{16}x^{6-4}y^{3-4}$

$$= \dfrac{1}{2}x^2y^{-1} = \dfrac{x^2}{2y}$$

31. $\dfrac{24u^2v^4}{18u^2v^6} = \dfrac{24}{18}u^{2-2}v^{4-6} = \dfrac{4}{3}u^0v^{-2}$

$$= \dfrac{4}{3}(1)v^{-2} = \dfrac{4}{3v^2}$$

33. $\dfrac{3z+3}{3} = \dfrac{3z}{3} + \dfrac{3}{3} = z + 1$

35. $\dfrac{4z-12}{4} = \dfrac{4z}{4} - \dfrac{12}{4} = z - 3$

37. $\dfrac{9x - 5}{3} = \dfrac{9x}{3} - \dfrac{5}{3} = 3x - \dfrac{5}{3}$

39. $\dfrac{b^2 - 2b}{b} = \dfrac{b^2}{b} - \dfrac{2b}{b} = b - 2$

41. $(5x^2 - 2x) \div x = \dfrac{5x^2}{x} - \dfrac{2x}{x}$
$$= 5x - 2$$

43. $\dfrac{25z^3 + 10z^2}{-5z} = \dfrac{25z^3}{-5z} + \dfrac{10z^2}{-5z}$
$$= -5z^2 - 2z$$

45. $\dfrac{8z^3 + 3z^2 - 2z}{2z} = \dfrac{8z^3}{2z} + \dfrac{3z^2}{2z} - \dfrac{2z}{2z}$
$$= 4z^2 + \dfrac{3z}{2} - 1$$

47. $\dfrac{m^3 + 3m - 4}{m} = \dfrac{m^3}{m} + \dfrac{3m}{m} - \dfrac{4}{m}$
$$= m^2 + 3 - \dfrac{4}{m}$$

49. $\dfrac{4x^2 - 12x}{4x^2} = \dfrac{4x^2}{4x^2} - \dfrac{12x}{4x^2} = 1 - \dfrac{3}{x}$

51. $\dfrac{6x^4 - 2x^3 + 3x^2 - x + 4}{2x^3} = \dfrac{6x^4}{2x^3} - \dfrac{2x^3}{2x^3} + \dfrac{3x^2}{2x^3} - \dfrac{x}{2x^3} + \dfrac{4}{2x^3}$
$$= 3x - 1 + \dfrac{3}{2x} - \dfrac{1}{2x^2} + \dfrac{2}{x^3}$$

53.
$$
\begin{array}{r}
x - 2 \\
x + 1\overline{)\,x^2 - x - 2} \\
\underline{x^2 + x} \\
-2x - 2 \\
\underline{-2x - 2} \\
0
\end{array}
$$

Thus, $\dfrac{x^2 - x - 2}{x + 1} = x - 2.$

55.
$$
\begin{array}{r}
x + 5 \\
x + 4\overline{)\,x^2 + 9x + 20} \\
\underline{x^2 + 4x} \\
5x + 20 \\
\underline{5x + 20} \\
0
\end{array}
$$

Thus, $(x^2 + 9x + 20) \div (x + 4) = x + 5.$

57.
$$
\begin{array}{r}
y + 2 \\
3y - 2\overline{)\,3y^2 + 4y - 4} \\
\underline{3y^2 - 2y} \\
6y - 4 \\
\underline{6y - 4} \\
0
\end{array}
$$

Thus, $\dfrac{3y^2 + 4y - 4}{3y - 2} = y + 2.$

59.
$$
\begin{array}{r}
6t + 1 \\
3t - 4\overline{)\,18t^2 - 21t - 4} \\
\underline{18t^2 - 24t} \\
3t - 4 \\
\underline{3t - 4} \\
0
\end{array}
$$

Thus, $\dfrac{18t^2 - 21t - 4}{3t - 4} = 6t + 1.$

61.
$$
\begin{array}{r}
x^2 - 2x + 5 + \dfrac{3}{x - 2} \\
x - 2\overline{)\,x^3 - 4x^2 + 9x - 7} \\
\underline{x^3 - 2x^2} \\
-2x^2 + 9x \\
\underline{-2x^2 + 4x} \\
5x - 7 \\
\underline{5x - 10} \\
3
\end{array}
$$

Thus, $\dfrac{x^3 - 4x^2 + 9x - 7}{x - 2} = x^2 - 2x + 5 + \dfrac{3}{x - 2}.$

63.
$$
\begin{array}{r}
7 - \dfrac{11}{x + 2} \\
x + 2\overline{)\,7x + 3} \\
\underline{7x + 14} \\
-11
\end{array}
$$

Thus, $\dfrac{7x + 3}{x + 2} = 7 - \dfrac{11}{x + 2}.$

65.

$$
\begin{array}{r}
x^2 + 2x + 4 \\
x - 2 \overline{)\, x^3 + 0x^2 + 0x - 8} \\
\underline{x^3 - 2x^2} \\
2x^2 \\
\underline{2x^2 - 4x} \\
4x - 8 \\
\underline{4x - 8} \\
0
\end{array}
$$

Thus, $\dfrac{x^3 - 8}{x - 2} = x^2 + 2x + 4.$

67.

$$
\begin{array}{r}
x - 3 + \dfrac{18}{x + 3} \\
x + 3 \overline{)\, x^2 + 0x + 9} \\
\underline{x^2 + 3x} \\
-3x + 9 \\
\underline{-3x - 9} \\
18
\end{array}
$$

Thus, $\dfrac{x^2 + 9}{x + 3} = x - 3 + \dfrac{18}{x + 3}.$

69.

$$
\begin{array}{r}
3x - 1 \\
3x + 1 \overline{)\, 9x^2 + 0x - 1} \\
\underline{9x^2 + 3x} \\
-3x - 1 \\
\underline{-3x - 1} \\
0
\end{array}
$$

Thus, $\dfrac{9x^2 - 1}{3x + 1} = 3x - 1.$

71.

$$
\begin{array}{r}
x^3 + x^2 + x + 1 \\
x - 1 \overline{)\, x^4 + 0x^3 + 0x^2 + 0x - 1} \\
\underline{x^4 - x^3} \\
x^3 \\
\underline{x^3 - x^2} \\
x^2 \\
\underline{x^2 - x} \\
x - 1 \\
\underline{x - 1} \\
0
\end{array}
$$

Thus, $\dfrac{x^4 - 1}{x - 1} = x^3 + x^2 + x + 1.$

73.
$$
\begin{array}{r|rrr}
-1 & 4 & 3 & 1 \\
 & & -4 & 1 \\
\hline
 & 4 & -1 & 2
\end{array}
$$

$\dfrac{4x^2 + 3x + 1}{x + 1} = 4x - 1 + \dfrac{2}{x + 1}$

75.
$$
\begin{array}{r|rrrr}
2 & 1 & 0 & -7 & 6 \\
 & & 2 & 4 & -6 \\
\hline
 & 1 & 2 & -3 & 0
\end{array}
$$

$\dfrac{x^3 - 7x + 6}{x - 2} = x^2 + 2x - 3$

77.
$$
\begin{array}{r|rrrr}
-2 & 3 & 7 & 3 & -2 \\
 & & -6 & -2 & -2 \\
\hline
 & 3 & 1 & 1 & -4
\end{array}
$$

$\dfrac{3t^3 + 7t^2 + 3t - 2}{t + 2} = 3t^2 + t + 1 - \dfrac{4}{t + 2}$

79. $\dfrac{4x^3}{x^2} - \dfrac{8x}{4} = 4x - 2x = 2x$

81. $\dfrac{8u^2v}{2u} + \dfrac{(uv)^2}{uv} = \dfrac{8u^2v}{2u} + \dfrac{u^2v^2}{uv} = 4uv + uv = 5uv$

83.

$$
\begin{array}{r}
x + 1 \\
x + 1 \overline{)\, x^2 + 2x + 1} \\
\underline{x^2 + x} \\
x + 1 \\
\underline{x + 1} \\
0
\end{array}
$$

$\dfrac{x^2 + 2x + 1}{x + 1} - (3x - 4) = x + 1 - (3x - 4)$

$$= x + 1 - 3x + 4 = -2x + 5$$

85. No, the cancellation is *not* valid because a term (rather than a factor) was cancelled from the numerator. Only common *factors* can be cancelled from the numerator and denominator.

87. Yes, the cancellation *is* valid.

89. (a) Yes, the graphs are the same.

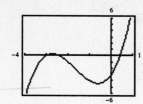

(b) $(x + 3)(x^2 + 2x - 1) = x(x^2 + 2x - 1) + 3(x^2 + 2x - 1)$

$$= x^3 + 2x^2 - x + 3x^2 + 6x - 3$$

$$= x^3 + 5x^2 + 5x - 3$$

(c)
$$
\begin{array}{r}
x^2 + 2x - 1 \\
x + 3 \overline{)\,x^3 + 5x^2 + 5x - 3} \\
\underline{x^3 + 3x^2} \\
2x^2 + 5x \\
\underline{2x^2 + 6x} \\
-x - 3 \\
\underline{-x - 3}
\end{array}
$$

$$\frac{x^3 + 5x^2 + 5x - 3}{x + 3} = x^2 + 2x - 1$$

91. (a)
$$
\begin{array}{r}
1 \\
t + 8 \overline{)\,t + 18} \\
\underline{t + 8} \\
10
\end{array}
$$
    Thus, $\dfrac{t + 18}{t + 8} = 1 + \dfrac{10}{t + 8}$.

(b)

t	0	10	20	30	40	50	60
$\dfrac{t + 18}{t + 8}$	$\dfrac{18}{8} = 2.25$	$\dfrac{28}{18} \approx 1.56$	$\dfrac{38}{28} \approx 1.36$	$\dfrac{48}{38} \approx 1.26$	$\dfrac{58}{48} \approx 1.21$	$\dfrac{68}{58} \approx 1.17$	$\dfrac{78}{68} \approx 1.15$

(c) As the value of t increases, the value of the ratio approaches 1. The radio $\dfrac{t + 18}{t + 8} = 1 + \dfrac{10}{t + 8}$ approaches 1 as t increases because the fraction $\dfrac{10}{t + 8}$ approaches 0.

93. (Length)(Width) = Area

$$\text{Length} = \frac{\text{Area}}{\text{Width}}$$

$$
\begin{array}{r}
x - 3 \\
x + 5 \overline{)\,x^2 + 2x - 15} \\
\underline{x^2 + 5x} \\
-3x - 15 \\
\underline{-3x - 15}
\end{array}
$$

$$\text{Length} = \frac{x^2 + 2x - 15}{x + 5} = x - 3$$

95. (a) Dividend: $x^2 + 2$

(b) Divisor: $x - 3$

(c) Quotient: $x + 3$

(d) Remainder: 11

97. $\dfrac{3x^8}{2x^3} = \dfrac{3}{2}x^{8-3} = \dfrac{3}{2}x^5$

99. When the divisor divides evenly into the dividend, the remainder is 0 and the divisor is a factor of the dividend.

Review Exercises for Chapter 5

1. *Polynomial:* $10x - 4 - 5x^3$

Standard form: $-5x^3 + 10x - 4$

Degree: 3

Leading coefficient: -5

3. *Polynomial:* $4x^3 - 2x + 5x^4 - 7x^2$

Standard form: $5x^4 + 4x^3 - 7x^2 - 2x$

Degree: 4

Leading coefficient: 5

5. *Polynomial:* $7x^4 - 1$

Standard form: $7x^4 - 1$

Degree: 4

Leading coefficient: 7

7. *Polynomial:* -2

Standard form: -2

Degree: 0

Leading coefficient: -2

9. $8x^4 + 3x - 2$

11. $-2x + 3$

13. $(2x + 3) + (x - 4) = (2x + x) + (3 - 4)$
$$= 3x - 1$$

15. $(t - 5) - (3t - 1) = t - 5 - 3t + 1$
$$= -2t - 4$$

17. $(2x^3 - 4x^2 + 3) + (x^3 + 4x^2 - 2x) = 3x^3 - 2x + 3$

19. $3(2x^2 - 4) - (2x^2 - 5) = 6x^2 - 12 - 2x^2 + 5$
$$= 4x^2 - 7$$

21. $(5x^4 - 7x^3 + x) - (4x^3 + 2x^2 - 4) + (4x + 8x^3 - 2x^4)$
$$= 5x^4 - 7x^3 + x - 4x^3 - 2x^2 + 4 + 4x + 8x^3 - 2x^4$$
$$= 3x^4 - 3x^3 - 2x^2 + 5x + 4$$

23. $(4 - x^2) + 2(x - 2) = 4 - x^2 + 2x - 4$
$$= -x^2 + 2x$$

25. $(-x^3 - 3x) - 2(2x^3 + x + 1) = -x^3 - 3x - 4x^3 - 2x - 2$
$$= -5x^3 - 5x - 2$$

27. $4y^2 - [y - 3(y^2 + 2)] = 4y^2 - [y - 3y^2 - 6]$
$$= 4y^2 - y + 3y^2 + 6$$
$$= 7y^2 - y + 6$$

29.
$$\begin{array}{r} -x^4 - 2x^2 + 3 \\ + \ 3x^4 - 5x^2 \\ \hline 2x^4 - 7x^2 + 3 \end{array}$$

31.
$$\begin{array}{l} 5x^2 + 2x - 27 \implies 5x^2 + 2x - 27 \\ -\ (2x^2 - 2x - 13) \implies -2x^2 + 2x + 13 \\ \hline 3x^2 + 4x - 14 \end{array}$$

33. $2x(x + 4) = 2x^2 + 8x$

35.
$$\text{F}\text{O}\text{I}\text{L}$$
$$(x - 4)(x + 6) = x^2 + 6x - 4x - 24 = x^2 + 2x - 24$$

37.
$$\text{F}\text{O}\text{I}\text{L}$$
$$(x + 3)(2x - 4) = 2x^2 - 4x + 6x - 12$$
$$= 2x^2 + 2x - 12$$

39.
$$\text{F}\text{O}\text{I}\text{L}$$
$$(4x - 3)(3x + 4) = 12x^2 + 16x - 9x - 12$$
$$= 12x^2 + 7x - 12$$

41. $(x^2 + 5x + 2)(2x + 3) = x^2(2x + 3) + 5x(2x + 3) + 2(2x + 3)$

$$= 2x^3 + 3x^2 + 10x^2 + 15x + 4x + 6$$

$$= 2x^3 + 13x^2 + 19x + 6$$

43. $(2t - 1)(t^2 - 3t + 3) = 2t(t^2 - 3t + 3) - 1(t^2 - 3t + 3)$

$$= 2t^3 - 6t^2 + 6t - t^2 + 3t - 3$$

$$= 2t^3 - 7t^2 + 9t - 3$$

45. $2u(u - 5) - (u + 1)(u - 5) = 2u^2 - 10u - (u^2 - 4u - 5)$

$$= 2u^2 - 10u - u^2 + 4u + 5$$

$$= u^2 - 6u + 5$$

47. $(x + 3)^2 = (x)^2 + 2(x)(3) + (3)^2$ **49.** $(4x - 7)^2 = (4x)^2 - 2(4x)(7) + (7)^2$ **51.** $\left(\frac{1}{2}x - 4\right)^2 = \left(\frac{1}{2}x\right)^2 - 2\left(\frac{1}{2}x\right)(4) + 4^2$

$\qquad = x^2 + 6x + 9$ $\qquad\qquad\qquad\qquad = 16x^2 - 56x + 49$ $\qquad\qquad\qquad\qquad = \frac{1}{4}x^2 - 4x + 16$

53. $(u - 6)(u + 6) = (u)^2 - (6)^2$ **55.** $(3t - 1)(3t + 1) = (3t)^2 - (1)^2$ **57.** $(2x - y)^2 = (2x)^2 - 2(2x)(y) + y^2$

$\qquad = u^2 - 36$ $\qquad\qquad\qquad\qquad = 9t^2 - 1$ $\qquad\qquad\qquad\qquad = 4x^2 - 4xy + y^2$

59. $(2x - 4y)(2x + 4y) = (2x)^2 - (4y)^2 = 4x^2 - 16y^2$

61. $4^{-2} = \frac{1}{4^2} = \frac{1}{16}$ **63.** $6^{-4}6^2 = 6^{-4+2} = 6^{-2}$ **65.** $\frac{1}{3^{-2}} = 3^2 = 9$ **67.** $\frac{4}{4^{-2}} = 4(4^2) = 4^3 = 64$

$\qquad\qquad\qquad\qquad\qquad = \frac{1}{6^2} = \frac{1}{36}$

69. $\left(\frac{3}{5}\right)^{-3} = \frac{3^{-3}}{5^{-3}} = \frac{5^3}{3^3} = \frac{125}{27}$ **71.** $\left(-\frac{2}{5}\right)^3\left(\frac{5}{2}\right)^2 = -\frac{2^3}{5^3} \cdot \frac{5^2}{2^2} = -\frac{2^{3-2}}{5^{3-2}} = -\frac{2}{5}$

73. $(3 \times 10^3)^2 = 3^2 \times 10^6 = 9 \times 10^6 = 9{,}000{,}000$ **75.** $\frac{1.85 \times 10^9}{5 \times 10^4} = 3.7 \times 10^4 = 37{,}000$

77. $y^{-4} = \frac{1}{y^4}$ **79.** $6t^{-2} = \frac{6}{t^2}$ **81.** $\frac{1}{7x^{-6}} = \frac{x^6}{7}$ **83.** $2x^{-1}y^{-3} = \frac{2}{x^1y^3} = \frac{2}{xy^3}$

85. $t^{-4} \cdot t^2 = t^{-4+2} = t^{-2} = \frac{1}{t^2}$ **87.** $4x^{-6}y^2 \cdot x^6 = 4x^{-6+6}y^2 = 4x^0y^2$ **89.** $(-3a^2)^{-2} = (-3)^{-2}(a^2)^{-2}$

$\qquad\qquad\qquad\qquad\qquad\qquad\qquad = 4(1)y^2 = 4y^2$ $\qquad\qquad\qquad\qquad\qquad = (-3)^{-2}a^{-4}$

$\qquad\qquad\qquad\qquad\qquad\qquad\qquad\qquad\qquad\qquad\qquad\qquad\qquad\qquad\qquad = \frac{1}{(-3)^2a^4} = \frac{1}{9a^4}$

91. $(x^2y^{-3})^2 = (x^2)^2(y^{-3})^2 = x^4y^{-6} = \frac{x^4}{y^6}$ **93.** $\frac{t^{-4}}{t^{-1}} = \frac{t}{t^4} = \frac{1}{t^3}$

95. $\frac{u^5 \cdot u^{-8}}{u^{-3}} = \frac{u^{5-8}}{u^{-3}} = \frac{u^{-3}}{u^{-3}} = u^{-3-(-3)} = u^0 = 1$ **97.** $\left(\frac{y}{5}\right)^{-2} = \frac{y^{-2}}{5^{-2}} = \frac{5^2}{y^2} = \frac{25}{y^2}$

99. $(2u^{-2}v)^3(4u^{-5}v^4)^{-1} = 2^3u^{-6}v^3(4^{-1})u^5v^{-4}$

$\qquad\qquad\qquad\quad = \dfrac{8}{4}u^{-6+5}v^{3-4}$

$\qquad\qquad\qquad\quad = 2u^{-1}v^{-1}$

$\qquad\qquad\qquad\quad = \dfrac{2}{uv}$

101. $\dfrac{8x^3 - 12x}{4x^2} = \dfrac{8x^3}{4x^2} - \dfrac{12x}{4x^2} = 2x - \dfrac{3}{x}$

103. $(5x^2 + 15x) \div (5x) = \dfrac{5x^2}{5x} + \dfrac{15x}{5x}$

$\qquad\qquad\qquad\qquad\quad = x + 3$

105.

$$\begin{array}{r} x + 2 \\ x - 3 \overline{)\, x^2 - x - 6} \\ \underline{x^2 - 3x} \\ 2x - 6 \\ \underline{2x - 6} \\ 0 \end{array}$$

Thus, $\dfrac{x^2 - x - 6}{x - 3} = x + 2.$

107.

$$\begin{array}{r} 8x + 5 + \dfrac{2}{3x-2} \\ 3x - 2 \overline{)\, 24x^2 - x - 8} \\ \underline{24x^2 - 16x} \\ 15x - 8 \\ \underline{15x - 10} \\ 2 \end{array}$$

Thus, $\dfrac{24x^2 - x - 8}{3x - 2} = 8x + 5 + \dfrac{2}{3x - 2}.$

109.

$$\begin{array}{r} 2x^2 + 4x + 3 + \dfrac{5}{x-1} \\ x - 1 \overline{)\, 2x^3 + 2x^2 - x + 2} \\ \underline{2x^3 - 2x^2} \\ 4x^2 - x \\ \underline{4x^2 - 4x} \\ 3x + 2 \\ \underline{3x - 3} \\ 5 \end{array}$$

Thus, $\dfrac{2x^3 + 2x^2 - x + 2}{x - 1} = 2x^2 + 4x + 3 + \dfrac{5}{x - 1}.$

111.

$$\begin{array}{r} x^2 - 2 \\ x^2 - 1 \overline{)\, x^4 + 0x^3 - 3x^2 + 0x + 2} \\ \underline{x^4 - x^2} \\ -2x^2 \\ \underline{-2x^2 + 2} \\ 0 \end{array}$$

Thus, $\dfrac{x^4 - 3x^2 + 2}{x^2 - 1} = x^2 - 2.$

113.

Area of shaded region	=	Area of entire rectangle	−	Area of two small squares

Area of entire rectangle $= 10(8) = 80$

Area of two small squares $= 2(x \cdot x) = 2x^2$

Area of shaded region $= 80 - 2x^2$

115.

Area of shaded region	$= \frac{1}{2}$	Area of square

Area of square $= (2x + 4)^2$

Area of shaded region $= \frac{1}{2}(2x + 4)^2$

$\qquad\qquad\qquad\quad = \frac{1}{2}(4x^2 + 16x + 16)$

$\qquad\qquad\qquad\quad = 2x^2 + 8x + 8$

117. (a) *Common formula:* $P = 2l + 2w$

     Labels: $P = $ perimeter of wall

                   $l = x$

                   $w = x - 3$

     Equation: $P = 2x + 2(x - 3)$

$$= 2x + 2x - 6$$

$$= 4x - 6$$

The perimeter of the wall is $4x - 6$ units.

(b) *Common formula:* $A = lw$

     Labels: $A = $ area of wall

                       $l = x$

                       $w = x - 3$

     Equation: $A = x(x - 3)$

$$A = x^2 - 3x$$

The area of the wall is $x^2 - 3x$ square units.

119. (Length)(Width) = Area

$$\text{Length} = \frac{\text{Area}}{\text{Width}}$$

$$\begin{array}{r} 2x + 3 \\ x - 4 \overline{\smash{\big)}\ 2x^2 - 5x - 12} \\ \underline{2x^2 - 8x} \\ 3x - 12 \\ \underline{3x - 12} \end{array}$$

$$\text{Length} = \frac{2x^2 - 5x - 12}{x - 4} = 2x + 3$$

121. Increase $= 100(150)(10 \times 10^{-6})$

$$= (100 \cdot 150 \cdot 10) \times 10^{-6}$$

$$= 150{,}000 \times 10^{-6}$$

$$= 1.5 \times 10^{-1}$$

$$= 0.15 \text{ feet}$$

123. (a) The area of the larger square is x^2, and the area of the smaller square is y^2. If the smaller square is removed, the area of the remaining figure is $x^2 - y^2$.

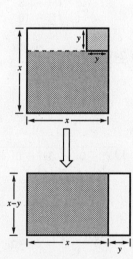

(b) If the remaining figure is rearranged as shown, the resulting rectangle has the length $x + y$ and width $x - y$. The area of this resulting rectangle is the product of its length and width or $(x + y)(x - y)$.

The area of the remaining figure in part (a) and the area of the resulting rectangle in part (b) are equal. Thus, $x^2 - y^2 = (x + y)(x - y)$. This demonstrates geometrically the special product formula $(x + y)(x - y) = x^2 - y^2$.

Chapter Test for Chapter 5

1. Degree: 4

Leading Coefficient: -3

The degree of a polynomial is the degree of the term with the highest power, and the coefficient of this term is the leading coefficient of the polynomial.

2. $-9x^4 + 8x^3 + 5$ or $y^4 - y^2 - y$

Note: There are many correct answers.

3. $(3z^2 - 3z + 7) + (8 - z^2) = 3z^2 - 3z + 7 + 8 - z^2$

$$= 2z^2 - 3z + 15$$

4. $(8u^3 + 3u^2 - 2u - 1) - (u^3 + 3u^2 - 2u) = 8u^3 + 3u^2 - 2u - 1 - u^3 - 3u^2 + 2u$
$$= 7u^3 - 1$$

5. $6y - [2y - (3 + 4y - y^2)] = 6y - [2y - 3 - 4y + y^2]$
$$= 6y - [y^2 - 2y - 3]$$
$$= 6y - y^2 + 2y + 3$$
$$= -y^2 + 8y + 3$$

6. $-5(x^2 - 1) + 3(4x + 7) - (x^2 + 26) = -5x^2 + 5 + 12x + 21 - x^2 - 26$
$$= -6x^2 + 12x$$

7.   **F   O   I   L**
$$(5b + 3)(2b - 1) = 10b^2 - 5b + 6b - 3 = 10b^2 + b - 3$$

8. $4x\left(\dfrac{3x}{2}\right)^2 = 4x\left(\dfrac{3^2 x^2}{2^2}\right) = \dfrac{4x}{1} \cdot \dfrac{9x^2}{4} = \dfrac{\cancel{4} \cdot 9x^3}{\cancel{4}} = 9x^3$

9. $(z + 2)(2z^2 - 3z + 5) = 2z^3 - 3z^2 + 5z + 4z^2 - 6z + 10$
$$= 2z^3 + z^2 - z + 10$$

10. $(x - 5)^2 = (x)^2 - 2(x)(5) + (5)^2 = x^2 - 10x + 25$

11. $(2x - 3)(2x + 3) = (2x)^2 - (3)^2 = 4x^2 - 9$

12. $\dfrac{15x + 25}{5} = \dfrac{15x}{5} + \dfrac{25}{5} = 3x + 5$

13.
$$
\begin{array}{r}
x^2 + 2x + 3 \\
x - 2 \overline{\smash{)}\, x^3 + 0x^2 - x - 6} \\
\underline{x^3 - 2x^2} \\
2x^2 - x \\
\underline{2x^2 - 4x} \\
3x - 6 \\
\underline{3x - 6} \\
0
\end{array}
$$

Thus, $\dfrac{x^2 - x - 6}{x - 2} = x^2 + 2x + 3.$

14.
$$
\begin{array}{r}
2x^2 + 4x - 3 - \dfrac{2}{2x + 1} \\
2x + 1 \overline{\smash{)}\, 4x^3 + 10x^2 - 2x - 5} \\
\underline{4x^3 + 2x^2} \\
8x^2 - 2x \\
\underline{8x^2 + 4x} \\
-6x - 5 \\
\underline{-6x - 3} \\
-2
\end{array}
$$

Thus, $\dfrac{4x^3 + 10x^2 - 2x - 5}{2x + 1} = 2x^2 + 4x - 3 - \dfrac{2}{2x + 1}.$

15. $\dfrac{-6a^2 b}{-9ab} = \dfrac{2a}{3}$

16. $(3x^{-2} y^3)^{-2} = 3^{-2}(x^{-2})^{-2}(y^3)^{-2}$
$$= 3^{-2} x^4 y^{-6}$$
$$= \dfrac{x^4}{3^2 y^6}$$
$$= \dfrac{x^4}{9y^6}$$

17. (a) $4^{-3} = \dfrac{1}{4^3} = \dfrac{1}{64}$

(b) $\dfrac{2^{-3}}{3^{-1}} = \dfrac{3^1}{2^3} = \dfrac{3}{8}$

(c) $(1.5 \times 10^5)^2 = (1.5)^2 \times (10^5)^2$
$$= 2.25 \times 10^{10}$$
$$= 22{,}500{,}000{,}000$$

18.

$$\boxed{\begin{array}{c}\text{Area of}\\\text{shaded region}\end{array}} = \boxed{\begin{array}{c}\text{Area of}\\\text{large rectangle}\end{array}} - \boxed{\begin{array}{c}\text{Area of small}\\\text{rectangle}\end{array}}$$

$$\begin{aligned}\text{Area of shaded region} &= (3x + 2)(2x) - (2x + 5)(x)\\ &= 6x^2 + 4x - 2x^2 - 5x\\ &= 4x^2 - x\end{aligned}$$

19.
$$\begin{aligned}\text{Area of triangle} &= \tfrac{1}{2}(\text{Base})(\text{Height})\\ &= \tfrac{1}{2}(4x - 2)(x + 6)\\ &= (2x - 1)(x + 6)\\ &= 2x^2 + 12x - x - 6\\ &= 2x^2 + 11x - 6\end{aligned}$$

20. $3.84 \times 10^8 = 384{,}000{,}000$

21. $101{,}300 = 1.013 \times 10^5$

22. $(\text{Length})(\text{Width}) = \text{Area}$

$$\text{Width} = \frac{\text{Area}}{\text{Length}}$$

$$\begin{array}{r}x - 3 \\ x + 1 \overline{)\, x^2 - 2x - 3}\\ \underline{x^2 + x}\\ -3x - 3\\ \underline{-3x - 3}\end{array}$$

$$\text{Width} = \frac{x^2 - 2x - 3}{x + 1} = x - 3$$

CHAPTER 6
Factoring and Solving Equations

CHAPTER 6
Factoring and Solving Equations

Section 6.1　Factoring Polynomials with Common Factors

Solutions to Odd-Numbered Exercises

1. $24 = 2 \cdot 2 \cdot 2 \cdot 3 = 6(4)$

$90 = 2 \cdot 3 \cdot 3 \cdot 5 = 6(15)$

The greatest common factor is 6.

3. $18 = 2 \cdot 3 \cdot 3 = 2(9)$

$150 = 2 \cdot 3 \cdot 5 \cdot 5 = 2(75)$

$100 = 2 \cdot 2 \cdot 5 \cdot 5 = 2(50)$

The greatest common factor is 2.

5. $z^2 = z \cdot z = z^2(1)$

$-z^6 = -(z \cdot z \cdot z \cdot z \cdot z \cdot z) = z^2(-z^4)$

The greatest common factor is z^2.

7. $2x^2 = 2 \cdot x \cdot x = 2x(x)$

$12x = 2 \cdot 2 \cdot 3 \cdot x = 2x(6)$

The greatest common factor is $2x$.

9. $u^2v = u \cdot u \cdot v = u^2v(1)$

$u^3v^2 = u \cdot u \cdot u \cdot v \cdot v = u^2v(uv)$

The greatest common factor is u^2v.

11. $9yz^2 = 3 \cdot 3 \cdot y \cdot z \cdot z = 3yz^2(3)$

$-12y^2z^3 = -(2 \cdot 2 \cdot 3 \cdot y \cdot y \cdot z \cdot z \cdot z)$

$= 3yz^2(-4yz)$

The greatest common factor is $3yz^2$.

13. $14x^2 = 2 \cdot 7 \cdot x \cdot x = 1(14x^2)$

$1 = 1(1)$

$7x^4 = 7 \cdot x \cdot x \cdot x \cdot x = 1(7x^4)$

The greatest common factor is 1.

15. $28a^4b^2 = 2 \cdot 2 \cdot 7 \cdot a \cdot a \cdot a \cdot a \cdot b \cdot b$

$= 14a^2b^2(2a^2)$

$14a^3b^3 = 2 \cdot 7 \cdot a \cdot a \cdot a \cdot b \cdot b \cdot b$

$= 14a^2b^2(ab)$

$42a^2b^5 = 2 \cdot 3 \cdot 7 \cdot a \cdot a \cdot b \cdot b \cdot b \cdot b \cdot b$

$= 14a^2b^2(3b^3)$

The greatest common factor is $14a^2b^2$.

17. $3x + 3 = 3(x + 1)$

19. $6z - 6 = 6(z - 1)$

21. $8t - 16 = 8(t - 2)$

23. $-25x - 10 = -5(5x + 2)$

or $5(-5x - 2)$

25. $24y^2 - 18 = 6(4y^2 - 3)$

27. $x^2 + x = x(x + 1)$

29. $25u^2 - 14u = u(25u - 14)$

31. $2x^4 + 6x^3 = 2x^3(x + 3)$

33. $7s^2 + 9t^2$　(No common factor)

35. $12x^2 - 2x = 2x(6x - 1)$

37. $-10r^3 - 35r = -5r(2r^2 + 7)$

39. $16a^3b^3 + 24a^4b^3 = 8a^3b^3(2 + 3a)$

41. $10ab + 10a^2b = 10ab(1 + a)$

43. $12x^2 + 16x - 8 = 4(3x^2 + 4x - 2)$

45. $100 + 75z - 50z^2$

$= 25(4 + 3z - 2z^2)$

47. $9x^4 + 6x^3 + 18x^2$

$\quad = 3x^2(3x^2 + 2x + 6)$

49. $5u^2 + 5u^2 + 5u = 10u^2 + 5u$

$\quad = 5u(2u + 1)$

51. $x(x - 3) + 5(x - 3)$

$\quad = (x - 3)(x + 5)$

53. $t(s + 10) - 8(s + 10) = (s + 10)(t - 8)$

55. $a^2(b + 2) - b(b + 2) = (b + 2)(a^2 - b)$

57. $z^3(z + 5) + z^2(z + 5) = [z^2(z + 5)](z + 1)$

$\quad = z^2(z + 5)(z + 1)$

59. $(a + b)(a - b) + a(a + b) = (a + b)[(a - b) + a]$

$\quad = (a + b)(a - b + a)$

$\quad = (a + b)(2a - b)$

61. $5 - 10x = -5(-1 + 2x)$

$\quad = -5(2x - 1)$

63. $3000 - 3x = -3(-1000 + x)$

$\quad = -3(x - 1000)$

65. $4 + 2x - x^2 = -1(-4 - 2x + x^2)$

$\quad = -1(x^2 - 2x - 4) \text{ or } -(x^2 - 2x - 4)$

67. $4 + 12x - 2x^2 = -2(-2 - 6x + x^2)$

$\quad = -2(x^2 - 6x - 2)$

69. $x^2 + 10x + x + 10 = x(x + 10) + 1(x + 10)$

$\quad = (x + 10)(x + 1)$

71. $a^2 - 4a + a - 4 = a(a - 4) + 1(a - 4)$

$\quad = (a - 4)(a + 1)$

73. $ky^2 - 4ky + 2y - 8 = ky(y - 4) + 2(y - 4)$

$\quad = (y - 4)(ky + 2)$

75. $t^3 - 3t^2 + 2t - 6 = t^2(t - 3) + 2(t - 3)$

$\quad = (t - 3)(t^2 + 2)$

77. $x^3 + 2x^2 + x + 2 = x^2(x + 2) + 1(x + 2)$

$\quad = (x + 2)(x^2 + 1)$

79. $6z^3 + 3z^2 - 2z - 1 = 3z^2(2z + 1) - 1(2z + 1)$

$\quad = (2z + 1)(3z^2 - 1)$

81. $x^3 - 3x - x^2 + 3 = x(x^2 - 3) - 1(x^2 - 3)$

$\quad = (x^2 - 3)(x - 1)$

83. $4x^2 - x^3 - 8 + 2x = x^2(4 - x) - 2(4 - x)$

$\quad = (4 - x)(x^2 - 2)$

85. $\frac{1}{4}x + \frac{3}{4} = \frac{1}{4}(x + 3)$

The missing factor is $x + 3$.

87. $2y - \frac{1}{5} = \frac{10}{5}y - \frac{1}{5}$

$\quad = \frac{1}{5}(10y - 1)$

The missing factor is $10y - 1$.

89. $\frac{7}{8}x + \frac{5}{16}y = \frac{14}{16}x + \frac{5}{16}y$

$\quad = \frac{1}{16}(14x + 5y)$

The missing factor is $14x + 5y$.

91. $y_1 = y_2$

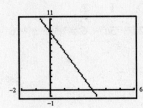

93. $y_1 = y_2$

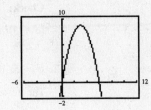

95. Area = (Length)(Width)

Area = $2x^2 + 2x$ and Width = $2x$

$2x^2 + 2x = 2x(x + 1)$

Therefore, the length of the rectangle must be $x + 1$.

97. Area of large rectangle $= 4x(2x) = 8x^2$

Area of small rectangle $= 2x(x) = 2x^2$

Shaded area = Area of large rectangle − Area of small rectangle

$$= 8x^2 - 2x^2$$

$$= 6x^2$$

99. Area of entire rectangle $= (9x)(6x) = 54x^2$

Area of semicircle $= \frac{1}{2}\pi(3x)^2 = \frac{9}{2}\pi x^2$

Shaded area = Area of entire rectangle − Area of semicircle

$$= 54x^2 - \frac{9}{2}\pi x^2$$

$$= 9x^2\left(6 - \frac{\pi}{2}\right)$$

101. $S = 2\pi r^2 - 2\pi rh$

$\quad = 2\pi r(r + h)$

103. $kQx - kx^2 = kx(Q - x)$

105. $5x^2 + 15x = 5x(x + 3)$

There are many correct answers.

107. Determine the prime factorization of each term. The greatest common factor contains each common prime factor, repeated the minimum number of times it occurs in any one of the factorizations.

109. "Factor" as a noun means any one of the expressions that, when multiplied together, yield the product. For example, you could say that $5x$ is a "factor" of the polynomial $5x^2 + 15x$.

"Factor" as a verb means to find the expressions that, when multiplied together yield the given product. For example, you could say that you can "factor" the polynomial $5x^2 + 15x$.

111. $x^3 - 3x^2 - 5x + 15 = x^2(x - 3) - 5(x - 3)$

$\qquad\qquad\qquad\quad = (x - 3)(x^2 - 5)$

There are many correct answers.

You can find such a polynomial by using the Distributive Property to multiply two binomials together. The result could be factored by grouping.

Section 6.2   Factoring Trinomials

1. $x^2 + 4x + 3 = (x + 3)(x + 1)$

The missing factor is $x + 1$.

Check: F   O   I   L

$(x + 3)(x + 1) = x^2 + x + 3x + 3 = x^2 + 4x + 3$

3. $a^2 + a - 6 = (a + 3)(a - 2)$

The missing factor is $a - 2$.

Check: F   O   I   L

$(a + 3)(a - 2) = a^2 - 2a + 3a - 6 = a^2 + a - 6$

5. $y^2 - 2y - 15 = (y + 3)(y - 5)$

The missing factor is $y - 5$.

Check: F   O   I   L

$(y + 3)(y - 5) = y^2 - 5y + 3y - 15 = y^2 - 2y - 15$

7. $z^2 - 5z + 6 = (z - 3)(z - 2)$

The missing factor is $z - 2$.

Check: F   O   I   L

$(z - 3)(z - 2) = z^2 - 2z - 3z + 6 = z^2 - 5z + 6$

9. $(x + 11)(x + 1)$

$(x - 11)(x - 1)$

11. $(x + 12)(x + 1)$ $(x - 12)(x - 1)$

$(x + 6)(x + 2)$ $(x - 6)(x - 2)$

$(x + 4)(x + 3)$ $(x - 4)(x - 3)$

13. $x^2 + 6x + 8 = (x + 4)(x + 2)$

15. $x^2 - 13x + 40 = (x - 5)(x - 8)$ **17.** $z^2 - 7z + 12 = (z - 3)(z - 4)$ **19.** $y^2 + 5y + 11$

The trinomial is prime.

21. $x^2 - x - 6 = (x - 3)(x + 2)$ **23.** $x^2 + 2x - 15 = (x + 5)(x - 3)$ **25.** $y^2 - 6y + 10$

This trinomial is prime.

27. $u^2 - 22u - 48 = (u - 24)(u + 2)$ **29.** $x^2 + 19x + 60 = (x + 15)(x + 4)$ **31.** $x^2 - 17x + 72 = (x - 8)(x - 9)$

33. $x^2 - 8x - 240 = (x - 20)(x + 12)$ **35.** $x^2 + xy - 2y^2 = (x + 2y)(x - y)$

37. $x^2 + 8xy + 15y^2 = (x + 3y)(x + 5y)$ **39.** $x^2 - 7xz - 18z^2 = (x - 9z)(x + 2z)$

41. $a^2 + 2ab - 15b^2 = (a + 5b)(a - 3b)$ **43.** $3x^2 + 21x + 30 = 3(x^2 + 7x + 10) = 3(x + 5)(x + 2)$

45. $4y^2 - 8y - 12 = 4(y^2 - 2y - 3)$ **47.** $3z^2 + 5z + 6$ **49.** $9x^2 + 18x - 18 = 9(x^2 + 2x - 2)$

$= 4(y - 3)(y + 1)$ This trinomial is prime.

51. $x^3 - 13x^2 + 30x = x(x^2 - 13x + 30)$ **53.** $x^4 - 5x^3 + 6x^2 = x^2(x^2 - 5x + 6)$

$= x(x - 3)(x - 10)$ $= x^2(x - 3)(x - 2)$

55. $-3y^2x - 9xy + 54x = -3x(y^2 + 3y - 18)$ **57.** $x^3 + 5x^2y + 6xy^2 = x(x^2 + 5xy + 6y^2)$

$= -3x(y + 6)(y - 3)$ $= x(x + 3y)(x + 2y)$

59. $2x^3y + 4x^2y^2 - 6xy^3 = 2xy(x^2 + 2xy - 3y^2)$ **61.** $b = 8$ $(x + 3)(x + 5)$

$= 2xy(x + 3y)(x - y)$ $b = -8$ $(x - 3)(x - 5)$

$b = 16$ $(x + 15)(x + 1)$

$b = -16$ $(x - 15)(x - 1)$

63. $b = 4$ $(x + 7)(x - 3)$ **65.** $b = 12$ $(x + 6)(x + 6)$

$b = -4$ $(x - 7)(x + 3)$ $b = -12$ $(x - 6)(x - 6)$

$b = 20$ $(x + 21)(x - 1)$ $b = 13$ $(x + 9)(x + 4)$

$b = -20$ $(x - 21)(x + 1)$ $b = -13$ $(x - 9)(x - 4)$

$b = 15$ $(x + 12)(x + 3)$

$b = -15$ $(x - 12)(x - 3)$

$b = 20$ $(x + 18)(x + 2)$

$b = -20$ $(x - 18)(x - 2)$

$b = 37$ $(x + 36)(x + 1)$

$b = -37$ $(x - 36)(x - 1)$

67. $c = 2$ $(x + 2)(x + 1)$

$c = -4$ $(x + 4)(x - 1)$

$c = -10$ $(x + 5)(x - 2)$

$c = -18$ $(x + 6)(x - 3)$

$c = -28$ $(x + 7)(x - 4)$

There are many correct answers.

69. $c = 5$ $(x - 5)(x - 1)$

$c = 8$ $(x - 4)(x - 2)$

$c = 9$ $(x - 3)(x - 3)$

$c = -7$ $(x - 7)(x + 1)$

$c = -16$ $(x - 8)(x + 2)$

There are many correct answers.

71. $c = 20$ $(x - 4)(x - 5)$

$c = 18$ $(x - 6)(x - 3)$

$c = 14$ $(x - 7)(x - 2)$

$c = 8$ $(x - 8)(x - 1)$

$c = -10$ $(x - 10)(x + 1)$

$c = -22$ $(x - 11)(x + 2)$

$c = -36$ $(x - 12)(x + 3)$

There are many correct answers.

73. $y_1 = y_2$

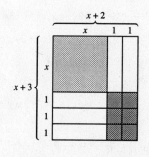

The graphs are identical, and the two expressions are equivalent.

75. $y_1 = y_2$

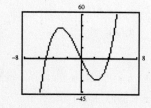

The graphs are identical, and the two expressions are equivalent.

77. $x^2 + 4x + 3 = (x + 3)(x + 1)$

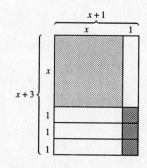

79. $x^2 + 5x + 6 = (x + 3)(x + 2)$

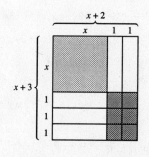

81. (a) $V = 4x^3 - 20x^2 + 24x,$ $0 < x < 2$

$= 4x(x^2 - 5x + 6)$ or $x(2)(x - 3)(2)(x - 2)$ Note: $4 = 2(2)$

$= 4x(x - 3)(x - 2)$ or $x(2x - 6)(2x - 4)$

                   or $x(2x - 6)(-1)(-1)(2x - 4)$ Note: $(-1)(-1) = 1$

                   or $x[(2x - 6)(-1)][(-1)(2x - 4)]$

                   or $x(-2x + 6)(-2x + 4)$

                   or $x(6 - 2x)(4 - 2x)$

(b)

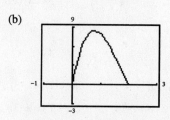

The volume is greatest when x is approximately 0.785 feet.

The volume is the product of the length, width, and height. The length is $6 - 2x$ feet, the width is $4 - 2x$ feet, and the height is x feet.

Note: Here is an alternate factoring approach which also points out the dimensions of the box.

$V = 4x^3 - 20x^2 + 24x$

$= 4x(x^2 - 5x + 6)$

$= 4x(6 - 5x + x^2)$

$= 4x(3 - x)(2 - x)$

$= x(2)(3 - x)(2)(2 - x)$

$= x[2(3 - x)][2(2 - x)]$

$= x(6 - 2x)(4 - 2x)$

83. $x^2 + 17x + 70 = (x + 7)(x + 10)$

Width of rectangle: $x + 7$

Length of rectangle: $x + 10$

Width of shaded region: 7

Length of shaded region: $x + 10 - x = 10$

Area of shaded region: $7(10) = 70$ square units

85. The constant 3 in the trinomial $x^2 - 4x + 3$ is positive, so the signs of the constants in the binomial factors must be the same. It is unnecessary to test $(x - 1)(x + 3)$ or $(x + 1)(x - 3)$ because both of these factorizations would yield a negative constant term in the product. The correct factorization is $(x - 1)(x - 3)$.

87. A prime trinomial is a polynomial with three terms that cannot be factored with integer coefficients.

89. The process of factoring $x^2 + bx + c$ is easier if c is a prime number because there are fewer factorizations to examine.

Section 6.3 More About Factoring Trinomials

1. $5x^2 + 18x + 9 = (x + 3)(5x + 3)$

The missing factor is $5x + 3$.

3. $5a^2 + 12a - 9 = (a + 3)(5a - 3)$

The missing factor is $5a - 3$.

5. $2y^2 - 3y - 27 = (y + 3)(2y - 9)$

The missing factor is $2y - 9$.

7. $4z^2 - 13z + 3 = (z - 3)(4z - 1)$

The missing factor is $4z - 1$.

9. $(5x + 3)(x + 1)$

$(5x - 3)(x - 1)$

$(5x + 1)(x + 3)$

$(5x - 1)(x - 3)$

11. $(5x + 12)(x + 1)$ $(5x - 12)(x - 1)$

$(5x + 1)(x + 12)$ $(5x - 1)(x - 12)$

$(5x + 6)(x + 2)$ $(5x - 6)(x - 2)$

$(5x + 2)(x + 6)$ $(5x - 2)(x - 6)$

$(5x + 4)(x + 3)$ $(5x - 4)(x - 3)$

$(5x + 3)(x + 4)$ $(5x - 3)(x - 4)$

13. $2x^2 + 5x + 3 = (2x + 3)(x + 1)$

15. $4y^2 + 5y + 1 = (4y + 1)(y + 1)$

17. $2y^2 - 3y + 1 = (2y - 1)(y - 1)$

19. $2x^2 - x - 3 = (2x - 3)(x + 1)$

21. $5x^2 - 2x + 1$

This trinomial is prime.

23. $2x^2 + x + 3$

This trinomial is prime.

25. $5s^2 - 10s + 6$

This trinomial is prime.

27. $4x^2 + 13x - 12 = (4x - 3)(x + 4)$

29. $9x^2 - 18x + 8 = (3x - 2)(3x - 4)$

31. $18u^2 - 9u - 2 = (6u + 1)(3u - 2)$

33. $15a^2 + 14a - 8 = (5a - 2)(3a + 4)$

35. $10t^2 - 3t - 18 = (5t + 6)(2t - 3)$

37. $15m^2 + 16m - 15 = (5m - 3)(3m + 5)$

39. $16z^2 - 34z + 15 = (8z - 5)(2z - 3)$

41. $-2x^2 + x + 3 = (-1)(2x^2 - x - 3)$

$= -(2x - 3)(x + 1)$

or $(-2x + 3)(x + 1)$

or $(2x - 3)(-x - 1)$

43. $4 - 4x - 3x^2 = (2 - 3x)(2 + x)$ or

$\quad 4 - 4x - 3x^2 = -1(-4 + 4x + 3x^2)$

$\qquad\qquad\qquad = -1(3x^2 + 4x - 4)$

$\qquad\qquad\qquad = -(3x - 2)(x + 2)$

$\qquad\qquad\qquad$ or $(-3x + 2)(x + 2)$

$\qquad\qquad\qquad$ or $(3x - 2)(-x - 2)$

45. $-6x^2 + 7x + 10 = (-1)(6x^2 - 7x - 10)$

$\qquad\qquad\qquad = -(6x + 5)(x - 2)$

$\qquad\qquad$ or $(-6x - 5)(x - 2)$

$\qquad\qquad$ or $(6x + 5)(-x + 2)$

47. $1 - 4x - 60x^2 = (1 - 10x)(1 + 6x)$ or

$\quad 1 - 4x - 60x^2 = (-1)(60x^2 + 4x - 1)$

$\qquad\qquad\qquad = -(10x - 1)(6x + 1)$

$\qquad\qquad\qquad$ or $(-10x + 1)(6x + 1)$

$\qquad\qquad\qquad$ or $(10x - 1)(-6x - 1)$

49. $16 - 8x - 15x^2 = (4 + 3x)(4 - 5x)$ or

$\quad 16 - 8x - 15x^2 = -15x^2 - 8x + 16$

$\qquad\qquad\qquad = -1(15x^2 + 8x - 16)$

$\qquad\qquad\qquad = -(5x - 4)(3x + 4)$

$\qquad\qquad\qquad$ or $(-5x + 4)(3x + 4)$

$\qquad\qquad\qquad$ or $(5x - 4)(-3x - 4)$

51. $6x^2 - 3x = 3x(2x - 1)$

53. $15y^2 + 18y = 3y(5y + 6)$

55. $u(u - 3) + 9(u - 3) = (u - 3)(u + 9)$

57. $2v^2 + 8v - 42 = 2(v^2 + 4v - 21)$

$\qquad\qquad\qquad = 2(v + 7)(v - 3)$

59. $-3x^2 - 3x - 60 = -3(x^2 + x + 20)$

61. $9z^2 - 24z + 15 = 3(3z^2 - 8z + 5)$

$\qquad\qquad\qquad = 3(3z - 5)(z - 1)$

63. $4x^2 + 4x + 2 = 2(2x^2 + 2x + 1)$

65. $-15x^4 - 2x^3 + 8x^2 = -x^2(15x^2 + 2x - 8)$

$\qquad\qquad\qquad = -x^2(3x - 2)(5x + 4)$

67. $3x^3 + 4x^2 + 2x = x(3x^2 + 4x + 2)$

69. $6x^3 + 24x^2 - 192x = 6x(x^2 + 4x - 32)$

$\qquad\qquad\qquad = 6x(x + 8)(x - 4)$

71. $18u^4 + 18u^3 - 27u^2 = 9u^2(2u^2 + 2u - 3)$

73.

$b = 13$	$(3x + 10)(x + 1)$
$b = -13$	$(3x - 10)(x - 1)$
$b = 31$	$(3x + 1)(x + 10)$
$b = -31$	$(3x - 1)(x - 10)$
$b = 11$	$(3x + 5)(x + 2)$
$b = -11$	$(3x - 5)(x - 2)$
$b = 17$	$(3x + 2)(x + 5)$
$b = -17$	$(3x - 2)(x - 5)$

75.

$b = 4$	$2(x + 3)(x - 1)$ or $(2x + 6)(x - 1)$ or $(2x - 2)(x + 3)$
$b = -4$	$2(x - 3)(x + 1)$ or $(2x - 6)(x + 1)$ or $(2x + 2)(x - 3)$
$b = -11$	$(2x + 1)(x - 6)$
$b = 11$	$(2x - 1)(x + 6)$
$b = -1$	$(2x + 3)(x - 2)$
$b = 1$	$(2x - 3)(x + 2)$

77. $b = 26$ $(6x + 20)(x + 1)$ or $(3x + 10)(2x + 2)$

$b = -26$ $(6x - 20)(x - 1)$ or $(3x - 10)(2x - 2)$

$b = 22$ $(6x + 10)(x + 2)$ or $(3x + 5)(2x + 4)$

$b = -22$ $(6x - 10)(x - 2)$ or $(3x - 5)(2x - 4)$

$b = 34$ $(6x + 4)(x + 5)$ or $(3x + 2)(2x + 10)$

$b = -34$ $(6x - 4)(x - 5)$ or $(3x - 2)(2x - 10)$

$b = 62$ $(6x + 2)(x + 10)$ or $(3x + 1)(2x + 20)$

$b = -62$ $(6x - 2)(x - 10)$ or $(3x - 1)(2x - 20)$

$b = 121$ $(6x + 1)(x + 20)$

$b = -121$ $(6x - 1)(x - 20)$

$b = 29$ $(6x + 5)(x + 4)$

$b = -29$ $(6x - 5)(x - 4)$

$b = 43$ $(3x + 20)(2x + 1)$

$b = -43$ $(3x - 20)(2x - 1)$

$b = 23$ $(3x + 4)(2x + 5)$

$b = -23$ $(3x - 4)(2x - 5)$

79. $c = -1$ $(4x - 1)(x + 1)$

$c = -10$ $(4x - 5)(x + 2)$

$c = -27$ $(4x - 9)(x + 3)$

$c = -7$ $(4x + 7)(x - 1)$

$c = -22$ $(4x + 11)(x - 2)$

$c = -45$ $(4x + 15)(x - 3)$

There are many correct answers.

81. $c = 8$ $(3x - 4)(x - 2)$

$c = 7$ $(3x - 7)(x - 1)$

$c = 3$ $(3x - 1)(x - 3)$

$c = -8$ $(3x + 2)(x - 4)$

$c = -25$ $(3x + 5)(x - 5)$

$c = -13$ $(3x - 13)(x + 1)$

There are many correct answers.

83. $c = -1$ $(6x + 1)(x - 1)$

$c = -14$ $(6x + 7)(x - 2)$

$c = -11$ $(6x - 11)(x + 1)$

$c = 1$ $(3x - 1)(2x - 1)$

$c = -6$ $(3x + 2)(2x - 3)$

$c = -4$ $(3x - 4)(2x + 1)$

There are many correct answers.

85. $ac = 3 \cdot 2 = 6$

$b = 7$

The two numbers with a product of 6 and a sum of 7 are 6 and 1.

$$3x^2 + 7x + 2 = 3x^2 + 6x + x + 2$$
$$= (3x^2 + 6x) + (x + 2)$$
$$= 3x(x + 2) + (x + 2)$$
$$= (x + 2)(3x + 1)$$

87. $ac = 2(-3) = -6$

$b = 1$

The two numbers with a product of -6 and a sum of 1 are 3 and -2.

$$2x^2 + x - 3 = 2x^2 + 3x - 2x - 3$$
$$= (2x^2 + 3x) - (2x + 3)$$
$$= x(2x + 3) - (2x + 3)$$
$$= (2x + 3)(x - 1)$$

89. $ac = 6(-4) = -24$

$b = 5$

The two numbers with a product of -24 and a sum of 5 are -3 and 8.

$$6x^2 + 5x - 4 = 6x^2 - 3x + 8x - 4$$
$$= (6x^2 - 3x) + (8x - 4)$$
$$= 3x(2x - 1) + 4(2x - 1)$$
$$= (2x - 1)(3x + 4)$$

91. $ac = 15(2) = 30$

$b = -11$

The two numbers with a product of 30 and a sum of -11 are -6 and -5.

$$15x^2 - 11x + 2 = 15x^2 - 6x - 5x + 2$$
$$= (15x^2 - 6x) + (-5x + 2)$$
$$= 3x(5x - 2) - 1(5x - 2)$$
$$= (5x - 2)(3x - 1)$$

93. $ac = 3(10) = 30$

$b = 11$

The two numbers with a product of 30 and a sum of 11 are 6 and 5.

$$3a^2 + 11a + 10 = 3a^2 + 6a + 5a + 10$$
$$= 3a(a + 2) + 5(a + 2)$$
$$= (a + 2)(3a + 5)$$

95. $ac = 16(-3) = -48$

$b = 2$

The two numbers with a product of -48 and a sum of 2 are 8 and -6.

$$16x^2 + 2x - 3 = 16x^2 + 8x - 6x - 3$$
$$= (16x^2 + 8x) - (6x + 3)$$
$$= 8x(2x + 1) - 3(2x + 1)$$
$$= (2x + 1)(8x - 3)$$

97. $ac = 12(6) = 72$

$b = -17$

The two numbers with a product of 72 and a sum of -17 are -9 and -8.

$$12x^2 - 17x + 6 = 12x^2 - 9x - 8x + 6$$
$$= (12x^2 - 9x) + (-8x + 6)$$
$$= 3x(4x - 3) - 2(4x - 3)$$
$$= (4x - 3)(3x - 2)$$

99. $ac = 6(-14) = -84$

$b = -5$

The two numbers with a product of -84 and a sum of -5 are -12 and 7.

$$6u^2 - 5u - 14 = 6u^2 - 12u + 7u - 14$$
$$= (6u^2 - 12u) + (7u - 14)$$
$$= 6u(u - 2) + 7(u - 2)$$
$$= (u - 2)(6u + 7)$$

101. $2x^2 + 5x + 2 = (2x + 1)(x + 2)$

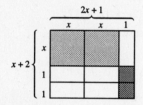

103. Volume = (Length)(Width)(Height)

$$2x^3 + 7x^2 + 6x = x(2x^2 + 7x + 6)$$
$$= x(2x + 3)(x + 2)$$
$$= (2x + 3)(x + 2)x$$

Thus, $2x + 3$ is the length of the box.

105. $2x^2 + 9x + 10 = (2x + 5)(x + 2)$

Length of largest rectangle:	$2x + 5$
Width of largest rectangle:	$x + 2$
Length of shaded region:	$(2x + 5) - x = x + 5$
Width of shaded region:	2
Area of shaded region:	$(x + 5)2 = 2x + 10$

107. (a) $y_1 = 2x^3 + 3x^2 - 5x$

$$= x(2x^2 + 3x - 5)$$
$$= x(2x + 5)(x - 1)$$

$y_1 = y_2$

(b)

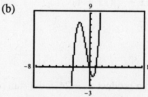

(c) The x-intercepts are $(0, 0)$, $(1, 0)$, and $\left(-\frac{5}{2}, 0\right)$. The y-intercept is $(0, 0)$.

109. (a) Volume $= 12x^3 + 64x^2 - 48x$

$\qquad\qquad = 4x(3x^2 + 16x - 12)$

$\qquad\qquad = $ (Height)(Area of Base)

The area of the base of the bin is $(3x^2 + 16x - 12)$.

(b) Volume $= 12x^3 + 64x^2 - 48x$

$\qquad\qquad = 4x(3x^2 + 16x - 12)$

$\qquad\qquad = 4x(3x - 2)(x + 6)$

The dimensions of the base of the bin must be $3x - 2$ and $x + 6$.

111. The constant of the trinomial is -15, but the product of the last terms of the binomials is 15.

113. Four

These are the four factorizations that need to be tested:

$(ax + 1)(x + c)$ $\qquad$ $(ax + c)(x + 1)$

$(ax - 1)(x - c)$ $\qquad$ $(ax - c)(x - 1)$

115. Examples of third-degree trinomials that have a common factor of $2x$:

$\qquad 2x(x^2 + x + 1) = 2x^3 + 2x^2 + 2x$

$\qquad 2x(x^2 - 7x + 3) = 2x^3 - 14x^2 + 6x$

$\qquad 2x(x + 5)(x - 1) = 2x(x^2 + 4x - 5) = 2x^3 + 8x^2 - 10x$

$\qquad 2x(x + 3)(x + 3) = 2x(x^2 + 6x + 9) = 2x^3 + 12x^2 + 18x$

$\qquad 2x(3x - 7)(x - 2) = 2x(3x^2 - 13x + 14) = 6x^3 - 26x^2 + 28x$

There are many correct answers.

Mid-Chapter Quiz for Chapter 6

1. $\frac{2}{3}x - 1 = \frac{2}{3}x - \frac{3}{3}$

$\qquad = \frac{1}{3}(2x - 3)$

The missing factor is $2x - 3$.

2. $x^2y - xy^2 = xy(x - y)$

The missing factor is $x - y$.

3. $y^2 + y - 42 = (y + 7)(y - 6)$

The missing factor is $y - 6$.

4. $2x^2 - x - 1 = (x - 1)(2x + 1)$

The missing factor is $2x + 1$.

5. $10x^2 + 70 = 10(x^2 + 7)$

6. $2a^3b - 4a^2b^2 = 2a^2b(a - 2b)$

7. $x(x + 2) - 3(x + 2) = (x + 2)(x - 3)$

8. $t^3 - 3t^2 + t - 3 = t^2(t - 3) + (t - 3)$

$\qquad\qquad = t^2(t - 3) + 1(t - 3)$

$\qquad\qquad = (t - 3)(t^2 + 1)$

9. $y^2 + 11y + 30 = (y + 6)(y + 5)$

10. $u^2 + u - 30 = (u + 6)(u - 5)$

11. $x^3 - x^2 - 30x = x(x^2 - x - 30)$

$\qquad\qquad = x(x - 6)(x + 5)$

12. $2x^2y + 8xy - 64y = 2y(x^2 + 4x - 32)$

$\qquad\qquad = 2y(x + 8)(x - 4)$

13. $3v^2 - 4v - 2$

This trinomial is prime.

14. $6 - 13z - 5z^2 = (3 + z)(2 - 5z)$ or

$6 - 13z - 5z^2 = -1(5z^2 + 13z - 6)$

$= -(5z - 2)(z + 3)$

or $(-5z + 2)(z + 3)$

or $(5z - 2)(-z - 3)$

15. $6x^2 - x - 2 = (3x - 2)(2x + 1)$

16. $10s^4 - 14s^3 + 2s^2 = 2s^2(5s^2 - 7s + 1)$

17. $(x + 3)(x + 4)$ $b = 7$

$(x - 3)(x - 4)$ $b = -7$

$(x + 6)(x + 2)$ $b = 8$

$(x - 6)(x - 2)$ $b = -8$

$(x + 12)(x + 1)$ $b = 13$

$(x - 12)(x - 1)$ $b = -13$

18. $(x - 7)(x - 3)$ $c = 21$

$(x - 6)(x - 4)$ $c = 24$

$(x - 2)(x - 8)$ $c = 16$

$(x + 2)(x - 12)$ $c = -24$

$(x - 11)(x + 1)$ $c = -11$

$(x - 5)(x - 5)$ $c = 25$

These are some of the possible values of c. There are many correct answers.

19. $(3x + 2)(x + 3)$ $(3x - 2)(x - 3)$

$(3x + 3)(x + 2)$ $(3x - 3)(x - 2)$

$(3x + 6)(x + 1)$ $(3x - 6)(x - 1)$

$(3x + 1)(x + 6)$ $(3x - 1)(x - 6)$

20. $3x^2 + 38x + 80 = (3x + 8)(x + 10)$

Width of the large rectangle: $x + 10$

Length of the large rectangle: $3x + 8$

Width of shaded region: 10

Length of shaded region: $3x + 8 - x = 2x + 8$

Area of shaded region: $10(2x + 8)$

21. $y_1 = y_2$

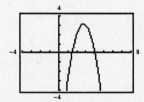

The graphs are identical, and the two expressions are equivalent.

Section 6.4   Factoring Polynomials with Special Forms

1. $x^2 - 36 = x^2 - 6^2$

$= (x + 6)(x - 6)$

3. $u^2 - 64 = u^2 - 8^2$

$= (u + 8)(u - 8)$

5. $49 - x^2 = 7^2 - x^2$

$= (7 + x)(7 - x)$

7. $u^2 - \frac{1}{4} = u^2 - \left(\frac{1}{2}\right)^2$

$= \left(u + \frac{1}{2}\right)\left(u - \frac{1}{2}\right)$

9. $t^2 - \frac{1}{16} = t^2 - \left(\frac{1}{4}\right)^2$

$= \left(t + \frac{1}{4}\right)\left(t - \frac{1}{4}\right)$

11. $16y^2 - 9 = (4y)^2 - 3^2$

$= (4y + 3)(4y - 3)$

13. $100 - 49x^2 = 10^2 - (7x)^2$

$= (10 - 7x)(10 + 7x)$

15. $(x - 1)^2 - 4 = (x - 1)^2 - 2^2$

$\qquad = [(x - 1) - 2][(x - 1) + 2]$

$\qquad = (x - 3)(x + 1)$

17. $25 - (z + 5)^2 = 5^2 - (z + 5)^2$

$\qquad = [5 + (z + 5)][5 - (z + 5)]$

$\qquad = [5 + z + 5][5 - z - 5]$

$\qquad = (z + 10)(-z)$

$\qquad = -z(z + 10)$

19. $2x^2 - 72 = 2(x^2 - 36)$

$\qquad = 2(x^2 - 6^2)$

$\qquad = 2(x + 6)(x - 6)$

21. $8 - 50x^2 = 2(4 - 25x^2)$

$\qquad = 2(2^2 - (5x)^2)$

$\qquad = 2(2 + 5x)(2 - 5x)$

23. $y^4 - 81 = (y^2)^2 - 9^2$

$\qquad = (y^2 + 9)(y^2 - 9)$

$\qquad = (y^2 + 9)(y^2 - 3^2)$

$\qquad = (y^2 + 9)(y + 3)(y - 3)$

25. $1 - x^4 = 1^2 - (x^2)^2$

$\qquad = (1 + x^2)(1 - x^2)$

$\qquad = (1 + x^2)(1^2 - x^2)$

$\qquad = (1 + x^2)(1 + x)(1 - x)$

27. $3x^4 - 48 = 3(x^4 - 16)$

$\qquad = 3[(x^2)^2 - 4^2]$

$\qquad = 3(x^2 + 4)(x^2 - 4)$

$\qquad = 3(x^2 + 4)(x^2 - 2^2)$

$\qquad = 3(x^2 + 4)(x + 2)(x - 2)$

29. $81x^4 - 16 = (9x^2)^2 - 4^2$

$\qquad = (9x^2 + 4)(9x^2 - 4)$

$\qquad = (9x^2 + 4)[(3x)^2 - 2^2]$

$\qquad = (9x^2 + 4)(3x + 2)(3x - 2)$

31. $x^2 - 4x + 4 = x^2 - 2(2x) + 2^2$

$\qquad = (x - 2)^2$

33. $z^2 + 6z + 9 = z^2 + 2(3z) + 3^2$

$\qquad = (z + 3)^2$

35. $4t^2 + 4t + 1 = (2t)^2 + 2(2t) + 1^2$

$\qquad = (2t + 1)^2$

37. $25y^2 - 10y + 1 = (5y)^2 - 2(5y) + 1^2$

$\qquad = (5y - 1)^2$

39. $b^2 + b + \frac{1}{4} = b^2 + 2\left(\frac{1}{2}b\right) + \left(\frac{1}{2}\right)^2$

$\qquad = \left(b + \frac{1}{2}\right)^2$

41. $4x^2 - x + \frac{1}{16} = (2x)^2 - (2)(2x)\left(\frac{1}{4}\right) + \left(\frac{1}{4}\right)^2$

$\qquad = \left(2x - \frac{1}{4}\right)^2$

43. $x^2 - 6xy + 9y^2 = x^2 - 2(3xy) + (3y)^2$

$\qquad = (x - 3y)^2$

45. $4y^2 + 20yz + 25z^2 = (2y)^2 + 2(10yz) + (5z)^2$

$\qquad = (2y)^2 + 2(2y)(5z) + (5z)^2$

$\qquad = (2y + 5z)^2$

47. $9a^2 - 12ab + 4b^2 = (3a)^2 - 2(3a)(2b) + (2b)^2$

$\qquad = (3a - 2b)^2$

49. $x^2 + bx + 1 = x^2 + bx + 1^2$

$\quad (x + 1)^2 = x^2 + 2(x)(1) + 1^2$

$\qquad = x^2 + 2x + 1$

$\quad (x - 1)^2 = x^2 - 2(x)(1) + 1^2$

$\qquad = x^2 - 2x + 1$

Thus, $b = 2$ or $b = -2$.

51. $x^2 + bx + \frac{16}{25} = x^2 + bx + \left(\frac{4}{5}\right)^2$

$\quad \left(x + \frac{4}{5}\right)^2 = x^2 + 2(x)\left(\frac{4}{5}\right) + \left(\frac{4}{5}\right)^2$

$\qquad = x^2 + \frac{8}{5}x + \frac{16}{25}$

$\quad \left(x - \frac{4}{5}\right)^2 = x^2 - 2(x)\left(\frac{4}{5}\right) + \left(\frac{4}{5}\right)^2$

$\qquad = x^2 - \frac{8}{5}x + \frac{16}{25}$

Thus, $b = \frac{8}{5}$ or $b = -\frac{8}{5}$.

53. $4x^2 + bx + 81 = (2x)^2 + bx + 9^2$

$(2x + 9)^2 = (2x)^2 + 2(2x)(9) + 9^2$

$\qquad = 4x^2 + 36x + 81$

$(2x - 9)^2 = (2x)^2 - 2(2x)(9) + 9^2$

$\qquad = 4x^2 - 36x + 81$

Thus, $b = 36$ or $b = -36$.

55. $x^2 + 6x + c = x^2 + 2(x)(3) + c$

$(x + 3)^2 = x^2 + 2(x)(3) + 3^2$

$\qquad = x^2 + 6x + 9$

Thus, $c = 9$.

57. $y^2 - 4y + c = (y)^2 - 2(y)(2) + c$

$(y - 2)^2 = y^2 - 2(y)(2) + 2^2$

$\qquad = y^2 - 4y + 4$

Thus, $c = 4$.

59. $x^3 - 8 = x^3 - 2^3$

$\qquad = (x - 2)(x^2 + (x)(2) + 2^2)$

$\qquad = (x - 2)(x^2 + 2x + 4)$

61. $y^3 + 64 = y^3 + 4^3$

$\qquad = (y + 4)(y^2 - (y)(4) + 4^2)$

$\qquad = (y + 4)(y^2 - 4y + 16)$

63. $1 + 8t^3 = 1^3 + (2t)^3$

$\qquad = (1 + 2t)[1^2 - (1)(2t) + (2t)^2]$

$\qquad = (1 + 2t)(1 - 2t + 4t^2)$

65. $27u^3 + 8 = (3u)^2 + 2^3$

$\qquad = (3u + 2)[(3u)^2 - (3u)(2) + 2^2]$

$\qquad = (3u + 2)(9u^2 - 6u + 4)$

67. $6x - 36 = 6(x - 6)$

69. $u^2 + 3u = u(u + 3)$

71. $5y^2 - 25y = 5y(y - 5)$

73. $5y^2 - 125 = 5(y^2 - 25)$

$\qquad = 5(y^2 - 5^2)$

$\qquad = 5(y + 5)(y - 5)$

75. $y^4 - 25y^2 = y^2(y^2 - 25)$

$\qquad = y^2(y^2 - 5^2)$

$\qquad = y^2(y + 5)(y - 5)$

77. $1 - 4x + 4x^2 = 1^2 - 2(2x) + (2x)^2$

$\qquad = (1 - 2x)^2$

Alternate Method:

$1 - 4x + 4x^2 = 4x^2 - 4x + 1$

$\qquad = (2x)^2 - 2(2x) + 1^2$

$\qquad = (2x - 1)^2$

79. $x^2 - 2x + 1 = x^2 - 2(x) + 1^2$

$\qquad = (x - 1)^2$

81. $9x^2 + 10x + 1 = (9x + 1)(x + 1)$

83. $2x^2 + 4x - 2x^3 = 2x(x + 2 - x^2)$

$\qquad = 2x(-x^2 + x + 2)$

$\qquad = -2x(x^2 - x - 2)$

$\qquad = -2x(x - 2)(x + 1)$

Note: This answer could also be written in other equivalent forms, such as $2x(-x + 2)(x + 1)$, $2x(x - 2)(-x - 1)$, $2x(2 - x)(x + 1)$, or $2x(2 - x)(1 + x)$.

85. $9t^2 - 16 = (3t)^2 - 4^2$

$\qquad = (3t + 4)(3t - 4)$

87. $36 - (z + 6)^2 = 6^2 - (z + 6)^2$

$\qquad = [6 - (z + 6)][6 + (z + 6)]$

$\qquad = -z(z + 12)$

89. $(t - 1)^2 - 121 = (t - 1)^2 - 11^2$

$\qquad = [(t - 1) + 11][(t - 1) - 11]$

$\qquad = (t + 10)(t - 12)$

91. $u^3 + 2u^2 + 3u = u(u^2 + 2u + 3)$

93. $x^2 + 81$

This polynomial is prime.

95. $2t^3 - 16 = 2(t^3 - 8)$

$\qquad = 2(t^3 - 2^3)$

$\qquad = 2(t - 2)(t^2 + (t)(2) + 2^2)$

$\qquad = 2(t - 2)(t^2 + 2t + 4)$

97. $2 - 16x^3 = 2(1 - 8x^3)$

$\qquad = 2[1^3 - (2x)^3]$

$\qquad = 2(1 - 2x)[1^2 + (1)(2x) + (2x)^2]$

$\qquad = 2(1 - 2x)(1 + 2x + 4x^2)$

99. $x^4 - 81 = (x^2)^2 - 9^2$

$\qquad = (x^2 + 9)(x^2 - 9)$

$\qquad = (x^2 + 9)(x^2 - 3^2)$

$\qquad = (x^2 + 9)(x + 3)(x - 3)$

101. $1 - x^4 = 1^2 - (x^2)^2$

$\qquad = (1 + x^2)(1 - x^2)$

$\qquad = (1 + x^2)(1^2 - x^2)$

$\qquad = (1 + x^2)(1 + x)(1 - x)$

Note: This answer could be written in other forms, such as $-(x^2 + 1)(x + 1)(x - 1)$.

103. $x^3 - 4x^2 - x + 4 = x^2(x - 4) - 1(x - 4)$

$\qquad = (x - 4)(x^2 - 1)$

$\qquad = (x - 4)(x^2 - 1^2)$

$\qquad = (x - 4)(x + 1)(x - 1)$

105. $x^4 + 3x^3 - 16x^2 - 48x = x(x^3 + 3x^2 - 16x - 48)$

$\qquad = x[x^2(x + 3) - 16(x + 3)]$

$\qquad = x[(x + 3)(x^2 - 16)]$

$\qquad = x(x + 3)(x^2 - 4^2)$

$\qquad = x(x + 3)(x + 4)(x - 4)$

107. $64 - y^6 = 8^2 - (y^3)^2$

$\qquad = (8 - y^3)(8 + y^3)$

$\qquad = (2^3 - y^3)(2^3 + y^3)$

$\qquad = (2 - y)(2^2 + 2y + y^2)(2 + y)(2^2 - 2y + y^2)$

$\qquad = (2 - y)(4 + 2y + y^2)(2 + y)(4 - 2y + y^2)$

or $(2 - y)(2 + y)(y^2 + 2y + 4)(y^2 - 2y + 4)$

109. The two graphs are the same because

$\qquad 1x^2 - 36 = (x + 6)(x - 6)$

$\qquad y_1 = y_2.$

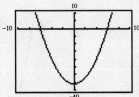

111. The two graphs are the same because

$\qquad x^3 - 6x^2 + 9x = x(x^2 - 6x + 9)$

$\qquad = x(x - 3)^2$

$\qquad y_1 = y_2$

113. $21^2 = (20 + 1)^2$

$= 20^2 + 2(20)(1) + 1^2$

$= 400 + 40 + 1$

$= 441$

115. $59 \cdot 61 = (60 - 1)(60 + 1)$

$= 60^2 - 1^2$

$= 3600 - 1$

$= 3599$

117. $\pi R^2 - \pi r^2 = \pi(R^2 - r^2)$

$= \pi(R + r)(R - r)$

119. $x^2 + 8x + 12 = (x^2 + 8x + 16) - 4$

$= (x + 4)^2 - 2^2$

$= [(x + 4) + 2][(x + 4) - 2]$

$= (x + 6)(x + 2)$

121. $2x^2 + 4x + 2 = (2x + 2)(x + 1)$

123. Side of square garden $= x$

Length of rectangular garden $= x + 1$

Width of rectangular garden $= x - 1$

Area of rectangular garden $= (x + 1)(x - 1) = 224$ square feet

In Exercise 122, we saw that one less than the square of a number n is equal to the product of $n + 1$ and $n - 1$; $n^2 - 1 = (n + 1)(n - 1)$. If 224 is one less than the square of x, then the square of x is 225, or 15^2. Each side of the square garden in the original design was 15 feet long.

125. A difference of two squares involves two terms, one being subtracted from the other; each term has a coefficient that is a perfect square and an even power of the variable. A difference of two squares according to the following pattern: $a^2 - b^2 = (a + b)(a - b)$.

127. No, $x(x + 2) - 2(x + 2)$ is not in factored form.

$x(x + 2) - 2(x + 2) = (x + 2)(x - 2)$

129. False.

The sum of two squares cannot be factored; however, the sum of two cubes can be factored using this pattern: $a^3 + b^3 = (a + b)(a^2 - ab + b^2)$.

Section 6.5   Polynomial Equations and Applications

1. $x(x - 5) = 0$

$x = 0$

$x - 5 = 0 \Rightarrow x = 5$

Note: There are *two* solutions, 0 and 5.

3. $(y - 2)(y - 3) = 0$

$y - 2 = 0 \Rightarrow y = 2$

$y - 3 = 0 \Rightarrow y = 3$

5. $(a + 1)(a - 2) = 0$

$a + 1 = 0 \Rightarrow a = -1$

$a - 2 = 0 \Rightarrow a = 2$

7. $(2t - 5)(3t + 1) = 0$

$2t - 5 = 0 \Rightarrow 2t = 5 \Rightarrow t = \frac{5}{2}$

$3t + 1 = 0 \Rightarrow 3t = -1 \Rightarrow t = -\frac{1}{3}$

9. $\left(\frac{2}{3}x - 4\right)(x + 2) = 0$

$\frac{2}{3}x - 4 = 0 \Rightarrow \frac{2}{3}x = 4 \Rightarrow x = 6$

$x + 2 = 0 \Rightarrow x = -2$

11. $(0.2y - 12)(0.7y + 10) = 0$

$0.2y - 12 = 0 \Rightarrow 0.2y = 12 \Rightarrow y = \frac{12}{0.2} \Rightarrow y = 60$

$0.7y + 10 = 0 \Rightarrow 0.7y = -10 \Rightarrow y = \frac{-10}{-0.7} \Rightarrow y = \frac{100}{7}$

13. $3x(x + 8)(4x - 5) = 0$

$$3x = 0 \Rightarrow x = 0$$

$$x + 8 = 0 \Rightarrow x = -8$$

$$4x - 5 = 0 \Rightarrow 4x = 5 \Rightarrow x = \frac{5}{4}$$

15. $(y - 1)(2y + 3)(y + 12) = 0$

$$y - 1 = 0 \Rightarrow y = 1$$

$$2y + 3 = 0 \Rightarrow 2y = -3 \Rightarrow y = -\frac{3}{2}$$

$$y + 12 = 0 \Rightarrow y = -12$$

17. $x^2 - 16 = 0$

$$(x + 4)(x - 4) = 0$$

$$x + 4 = 0 \Rightarrow x = -4$$

$$x - 4 = 0 \Rightarrow x = 4$$

19. $100 - v^2 = 0$

$$10^2 - v^2 = 0$$

$$(10 + v)(10 - v) = 0$$

$$10 + v = 0 \Rightarrow v = -10$$

$$10 - v = 0 \Rightarrow -v = -10 \Rightarrow v = 10$$

21. $3y^2 - 27 = 0$

$$3(y^2 - 9) = 0$$

$$3(y + 3)(y - 3) = 0$$

$$3 \neq 0$$

$$y + 3 = 0 \Rightarrow y = -3$$

$$y - 3 = 0 \Rightarrow y = 3$$

23. $(t - 3)^2 - 25 = 0$

$$[(t - 3) + 5][(t - 3) - 5] = 0$$

$$[t + 2][t - 8] = 0$$

$$t + 2 = 0 \Rightarrow t = -2$$

$$t - 8 = 0 \Rightarrow t = 8$$

Note: Here is an alternative solution.

$$(t - 3)^2 - 25 = 0$$

$$t^2 - 6t + 9 - 25 = 0$$

$$t^2 - 6t - 16 = 0$$

$$(t + 2)(t - 8) = 0$$

$$t + 2 = 0 \Rightarrow t = -2$$

$$t - 8 = 0 \Rightarrow t = 8$$

25. $81 - (u + 4)^2 = 0$

$$9^2 - (u + 4)^2 = 0$$

$$[9 + (u + 4)][9 - (u + 4)] = 0$$

$$(u + 13)(-u + 5) = 0$$

$$u + 13 = 0 \Rightarrow u = -13$$

$$-u + 5 = 0 \Rightarrow -u = -5 \Rightarrow u = 5$$

27. $2x^2 + 4x = 0$

$$2x(x + 2) = 0$$

$$2x = 0 \Rightarrow x = 0$$

$$x + 2 = 0 \Rightarrow x = -2$$

29. $4x^2 - x = 0$

$$x(4x - 1) = 0$$

$$x = 0$$

$$4x - 1 = 0 \Rightarrow 4x = 1 \Rightarrow x = \frac{1}{4}$$

31. $y(y - 4) + 3(y - 4) = 0$

$\qquad (y - 4)(y + 3) = 0$

$\qquad\qquad y - 4 = 0 \Rightarrow y = 4$

$\qquad\qquad y + 3 = 0 \Rightarrow y = -3$

33. $x(x - 8) + 2(x - 8) = 0$

$\qquad (x - 8)(x + 2) = 0$

$\qquad\qquad x - 8 = 0 \Rightarrow x = 8$

$\qquad\qquad x + 2 = 0 \Rightarrow x = -2$

35. $m^2 - 2m + 1 = 0$

$\qquad (m - 1)^2 = 0$

$\qquad m - 1 = 0 \Rightarrow m = 1$

37. $x^2 + 14x + 49 = 0$

$\qquad (x + 7)^2 = 0$

$\qquad x + 7 = 0 \Rightarrow x = -7$

39. $4t^2 - 12t + 9 = 0$

$\qquad (2t - 3)^2 = 0$

$\qquad 2t - 3 = 0 \Rightarrow 2t = 3 \Rightarrow t = \frac{3}{2}$

41. $x^2 - 2x - 8 = 0$

$\qquad (x - 4)(x + 2) = 0$

$\qquad\qquad x - 4 = 0 \Rightarrow x = 4$

$\qquad\qquad x + 2 = 0 \Rightarrow x = -2$

43. $3 + 5x - 2x^2 = 0$

$\qquad (3 - x)(1 + 2x) = 0$

$\qquad\qquad 3 - x = 0 \Rightarrow -x = -3 \Rightarrow x = 3$

$\qquad\qquad 1 + 2x = 0 \Rightarrow 2x = -1 \Rightarrow x = -\frac{1}{2}$

45. $6x^2 + 4x - 10 = 0$

$\qquad 2(3x^2 + 2x - 5) = 0$

$\qquad 2(3x + 5)(x - 1) = 0$

$\qquad\qquad 2 \neq 0$

$\qquad\qquad 3x + 5 = 0 \Rightarrow 3x = -5 \Rightarrow x = -\frac{5}{3}$

$\qquad\qquad x - 1 = 0 \Rightarrow x = 1$

47. $z(z + 2) = 15$

$\qquad z^2 + 2z = 15$

$\qquad z^2 + 2z - 15 = 0$

$\qquad (z + 5)(z - 3) = 0$

$\qquad\qquad z + 5 = 0 \Rightarrow z = -5$

$\qquad\qquad z - 3 = 0 \Rightarrow z = 3$

49. $x(x - 5) = 14$

$\qquad x^2 - 5x = 14$

$\qquad x^2 - 5x - 14 = 0$

$\qquad (x - 7)(x + 2) = 0$

$\qquad\qquad x - 7 = 0 \Rightarrow x = 7$

$\qquad\qquad x + 2 = 0 \Rightarrow x = -2$

51. $y(2y + 1) = 3$

$\qquad 2y^2 + y = 3$

$\qquad 2y^2 + y - 3 = 0$

$\qquad (2y + 3)(y - 1) = 0$

$\qquad\qquad 2y + 3 = 0 \Rightarrow 2y = -3 \Rightarrow y = -\frac{3}{2}$

$\qquad\qquad y - 1 = 0 \Rightarrow y = 1$

53. $(x - 3)(x - 6) = 4$

$\qquad x^2 - 9x + 18 = 4$

$\qquad x^2 - 9x + 14 = 0$

$\qquad (x - 7)(x - 2) = 0$

$\qquad\qquad x - 7 = 0 \Rightarrow x = 7$

$\qquad\qquad x - 2 = 0 \Rightarrow x = 2$

55. $(x + 1)(x + 4) = 4$

$\qquad x^2 + 5x + 4 = 4$

$\qquad x^2 + 5x = 0$

$\qquad x(x + 5) = 0$

$\qquad\qquad x = 0$

$\qquad\qquad x + 5 = 0 \Rightarrow x = -5$

57. $x^3 + 5x^2 + 6x = 0$

$\qquad x(x^2 + 5x + 6) = 0$

$\qquad x(x + 3)(x + 2) = 0$

$\qquad\qquad x = 0$

$\qquad\qquad x + 3 = 0 \Rightarrow x = -3$

$\qquad\qquad x + 2 = 0 \Rightarrow x = -2$

59. $2t^3 + 5t^2 - 12t = 0$

$t(2t^2 + 5t - 12) = 0$

$t(2t - 3)(t + 4) = 0$

$t = 0$

$2t - 3 = 0 \implies 2t = 3 \implies t = \frac{3}{2}$

$t + 4 = 0 \implies t = -4$

61. $x^2(x - 2) - 9(x - 2) = 0$

$(x - 2)(x^2 - 9) = 0$

$(x - 2)(x + 3)(x - 3) = 0$

$x - 2 = 0 \implies x = 2$

$x + 3 = 0 \implies x = -3$

$x - 3 = 0 \implies x = 3$

63. $x^3 - x^2 - 16x + 16 = 0$

$x^2(x - 1) - 16(x - 1) = 0$

$(x - 1)(x^2 - 16) = 0$

$(x - 1)(x - 4)(x + 4) = 0$

$x - 1 = 0 \implies x = 1$

$x - 4 = 0 \implies x = 4$

$x + 4 = 0 \implies x = -4$

65. $u^4 + 2u^3 - u^2 - 2u = 0$

$u(u^3 + 2u^2 - u - 2) = 0$

$u[u^2(u + 2) - 1(u + 2)] = 0$

$u(u + 2)(u^2 - 1) = 0$

$u(u + 2)(u + 1)(u - 1) = 0$

$u = 0$

$u + 2 = 0 \implies u = -2$

$u + 1 = 0 \implies u = -1$

$u - 1 = 0 \implies u = 1$

67. The x-intercepts are $(1, 0)$ and $(-3, 0)$.

$y = x^2 + 2x - 3$

$0 = x^2 + 2x - 3$

$0 = (x + 3)(x - 1)$

$x + 3 = 0 \implies x = -3$

$x - 1 = 0 \implies x = 1$

x-intercepts: $(-3, 0)$, $(1, 0)$

The number of solutions is equal to the number of x-intercepts.

69. The x-intercepts are $(-3, 0)$ and $(4, 0)$.

$y = 12 + x - x^2$

$0 = 12 + x - x^2$

$0 = (4 - x)(3 + x)$

$4 - x = 0 \implies -x = -4 \implies x = 4$

$3 + x = 0 \implies x = -3$

x-intercepts: $(4, 0)$ and $(-3, 0)$

The number of solutions is equal to the number of x-intercepts.

71. The x-intercepts are $(0, 0)$ and $(3, 0)$.

$y = x^3 - 6x^2 + 9x$

$0 = x^3 - 6x^2 + 9x$

$0 = x(x^2 - 6x + 9)$

$0 = x(x - 3)^2$

$x = 0$

$x - 3 = 0 \implies x = 3$

x-intercepts: $(0, 0)$, $(3, 0)$

The number of solutions is equal to the number of x-intercepts.

73. The x-intercepts are $(0, 0)$, $(1, 0)$ and $\left(-\frac{5}{2}, 0\right)$.

$y = 5x - 3x^2 - 2x^3$

$0 = 5x - 3x^2 - 2x^3$

$0 = x(5 - 3x - 2x^2)$

$0 = x(5 + 2x)(1 - x)$

$x = 0$

$5 + 2x = 0 \implies 2x = -5 \implies x = -\frac{5}{2}$

$1 - x = 0 \implies -x = -1 \implies x = 1$

x-intercepts: $(0, 0)$, $(1, 0)$ and $\left(-\frac{5}{2}, 0\right)$

The number of solutions is equal to the number of x-intercepts.

75.

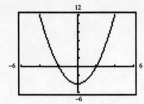

The x-intercepts are $(2, 0)$ and $(-2, 0)$.

Check: $y = x^2 - 4$ $y = x^2 - 4$

$y = 2^2 - 4$ $y = (-2)^2 - 4$

$y = 4 - 4$ $y = 4 - 4$

$y = 0$ $y = 0$

77.

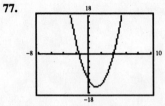

The x-intercepts are $\left(-\frac{3}{2}, 0\right)$ and $(4, 0)$.

Check: $y = 2x^2 - 5x - 12$

$y = 2\left(-\frac{3}{2}\right)^2 - 5\left(-\frac{3}{2}\right) - 12$ $y = 2x^2 - 5x - 12$

$y = 2\left(\frac{9}{4}\right) + \frac{15}{2} - 12$ $y = 2(4)^2 - 5(4) - 12$

$y = \frac{9}{2} + \frac{15}{2} - 12$ $y = 2(16) - 20 - 12$

$y = \frac{24}{2} - 12$ $y = 32 - 20 - 12$

$y = 12 - 12$ $y = 0$

$y = 0$

79.

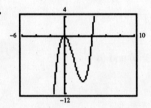

The x-intercepts are $(0, 0)$ and $(4, 0)$.

Check: $y = x^3 - 4x^2$ $y = x^3 - 4x^2$

$y = 0^3 - 4(0)^2$ $y = 4^3 - 4(4)^2$

$y = 0 - 4(0)$ $y = 64 - 4(16)$

$y = 0 - 0$ $y = 64 - 64$

$y = 0$ $y = 0$

81.

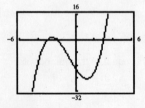

The x-intercepts are $(-2, 0)$, $(-3, 0)$, and $(3, 0)$.

Check:

$y = x^2(x + 2) - 9(x + 2)$ $y = x^2(x + 2) - 9(x + 2)$ $y = x^2(x + 2) - 9(x + 2)$

$y = (-2)^2(-2 + 2) - 9(-2 + 2)$ $y = (-3)^2(-3 + 2) - 9(-3 + 2)$ $y = 3^2(3 + 2) - 9(3 + 2)$

$y = 4(0) - 9(0)$ $y = 9(-1) - 9(-1)$ $y = 9(5) - 9(5)$

$y = 0 - 0$ $y = -9 + 9$ $y = 45 - 45$

$y = 0$ $y = 0$ $y = 0$

83. *Verbal model:*

$$\boxed{\begin{array}{c}\text{First consecutive}\\\text{positive integer}\end{array}} \cdot \boxed{\begin{array}{c}\text{Second consecutive}\\\text{positive integer}\end{array}} = \boxed{72}$$

Labels: First integer $= n$

Second integer $= n + 1$

Equation: $n(n + 1) = 72$

$n^2 + n = 72$

$n^2 + n - 72 = 0$

$(n + 9)(n - 8) = 0$

$n + 9 = 0 \implies n = -9$

$n - 8 = 0 \implies n = 8$ and $n + 1 = 9$

The original problem required that the integers be positive, so we discard -9 as a solution. Thus, $n = 8$ and $n + 1 = 9$; the two consecutive integers are 8 and 9.

85. *Verbal model:*

$$\boxed{\begin{array}{c}\text{First consecutive}\\\text{even integer}\end{array}} \cdot \boxed{\begin{array}{c}\text{Second consecutive}\\\text{even integer}\end{array}} = \boxed{440}$$

Labels: First consecutive even integer $= 2n$

Second consecutive even integer $= 2n + 2$

Equation: $2n(2n + 2) = 440$

$4n^2 + 4n = 440$

$4n^2 + 4n - 440 = 0$

$4(n^2 + n - 110) = 0$

$4(n + 11)(n - 10) = 0$

$4 \neq 0$

$n + 11 = 0 \implies n = -11$

$n - 10 = 0 \implies n = 10$ so $2(n) = 20$ and $2n + 2 = 22$

We discard the negative solution. The two consecutive positive even integers are $2(10) = 20$ and $2(10) + 2 = 22$.

87. *Verbal model:*

$$\boxed{\begin{array}{c}\text{Length of}\\\text{frame}\end{array}} \cdot \boxed{\begin{array}{c}\text{Width of}\\\text{frame}\end{array}} = \boxed{\begin{array}{c}\text{Area of}\\\text{frame}\end{array}}$$

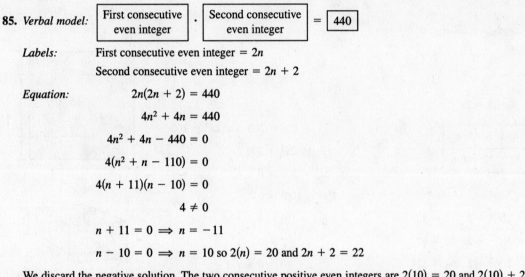

Labels: Width $= x$ (inches)

Length $= x + 3$ (inches)

Area $= 108$ (square inches)

Equation: $(x + 3)x = 108$

$x^2 + 3x = 108$

$x^2 + 3x - 108 = 0$

$(x + 12)(x - 9) = 0$

$x + 12 = 0 \implies x = -12$

$x - 9 = 0 \implies x = 9$ and $x + 3 = 12$

The width of the frame could not be -12, so we discard this answer. Thus, $x = 9$ and $x + 3 = 12$; the width of the frame is 9 inches and the length is 12 inches.

89. *Verbal model:*   ┌─────┐ ═ ┌────────┐ · ┌───────┐
                           │Area │   │ Length │   │ Width │
                           └─────┘   └────────┘   └───────┘

   Labels: Area = 450 square inches

                Length = $2x$

                Width = x

   Equation: $450 = (2x)(x)$

                  $450 = 2x^2$

                  $0 = 2x^2 - 450$

                  $0 = 2(x^2 - 225)$

                  $0 = 2(x - 15)(x + 15)$

                  $x - 15 = 0 \Rightarrow x = 15$

                  $x + 15 = 0 \Rightarrow x = -15$

The width cannot be -15 inches, so we discard the negative answer. Thus, the rectangle is $(2)(15) = 30$ inches by 15 inches.

91. (a) Volume = (length)(width)(height)

          Length = x

          Width = x

          Height = 2

          Volume = $(x)(x)(2) = 2x^2$

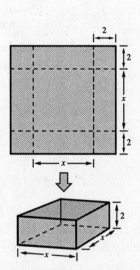

   (b)

x	2	4	6	8
V	8	32	72	128

   $x = 2,\quad V = 2(2)^2 = 2 \cdot 4 = 8$

   $x = 4,\quad V = 2(4)^2 = 2 \cdot 16 = 32$

   $x = 6,\quad V = 2(6)^2 = 2 \cdot 36 = 72$

   $x = 8,\quad V = 2(8)^2 = 2 \cdot 64 = 128$

   (c) $V = 200 \Rightarrow 2x^2 = 200$

          $2x^2 = 200$

          $2x^2 - 200 = 0$

          $2(x^2 - 100) = 0$

          $2(x + 10)(x - 10) = 0$

          $2 \neq 0$

          $x + 10 = 0 \Rightarrow x = -10$

          $x - 10 = 0 \Rightarrow x = 10$

The length of the box could not be -10, so we discard this negative solution. Thus, $x = 10$. The original price of cardboard is a square with each side $2 + x + 2$ or $x + 4$ inches long. We know $x = 10$, so $x + 4 = 14$. Therefore, the original piece of cardboard is a 14 inch × 14 inch square.

93.   Height = $-16t^2 + 1600$

                  $0 = -16t^2 + 1600$ (Set height equal to 0)

          $16t^2 - 1600 = 0$

          $16(t^2 - 100) = 0$

   $16(t + 10)(t - 10) = 0$

                  $16 \neq 0$

          $t + 10 = 0 \Rightarrow t = -10$

          $t - 10 = 0 \Rightarrow t = 10$

The time could not be -10 seconds, so we discard this negative answer. Thus, $t = 10$; the object reaches the ground in 10 seconds.

95. (a)

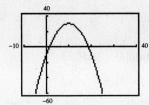

(b) The profit is $20 for two x-values, $x = 5$ and $x = 15$.

(c)
$$P = -0.4x^2 + 8x - 10$$
$$20 = -0.4x^2 + 8x - 10$$
$$0 = -0.4x^2 + 8x - 30$$
$$10(0) = 10(-0.4x^2 + 8x - 30)$$
$$0 = -4x^2 + 80x - 300$$
$$0 = -4(x^2 + 20x + 75)$$
$$0 = -4(x - 5)(x - 15)$$
$$-4 \neq 0$$
$$x - 5 = 0 \Rightarrow x = 5$$
$$x - 15 = 0 \Rightarrow x = 15$$

97.
$$\text{Height} = -16t^2 + 16t + 32$$
$$0 = -16t^2 + 16t + 32$$
$$16t^2 - 16t - 32 = 0$$
$$16(t^2 - t - 2) = 0$$
$$16(t - 2)(t + 1) = 0$$
$$16 \neq 0$$
$$t - 2 = 0 \Rightarrow t = 2$$
$$t + 1 = 0 \Rightarrow t = -1$$

Time cannot be $t = -1$ seconds, so discard the negative answer. The diver reaches the water after $t = 2$ seconds.

99. (a)

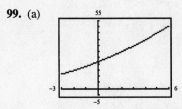

(b) From the graph, it appears that the number of passengers would reach 100 million when $t = 15$, or in 2005.

101. $ax^2 - ax = 0$
$$ax(x - 1) = 0$$
$$ax = 0 \Rightarrow x = 0$$
$$x - 1 = 0 \Rightarrow x = 1$$

The two solutions are 0 and 1.

103. If $ab = 0$, then $a = 0$ or $b = 0$.

105. The degree of a linear equation is 1, and the degree of a quadratic equation is 2.

107. The maximum number of solutions of an nth-degree polynomial equation is n.

109. False.

The zero-factor property can be used only when the equation is set equal to 0. There are an unlimited number of factors whose product is 21. Here is the correct solution:
$$(5x - 1)(x + 3) = 21$$
$$5x^2 + 14x - 3 = 21$$
$$5x^2 + 14x - 24 = 0$$
$$(5x - 6)(x + 4) = 0$$
$$5x - 6 = 0 \Rightarrow 5x = 6 \Rightarrow x = \tfrac{6}{5}$$
$$x + 4 = 0 \Rightarrow x = -4$$

Review Exercises for Chapter 6

1. $20 = (2)(2)(5) = 10(2)$

$60 = (2)(2)(3)(5) = 10(6)$

$150 = (2)(3)(5)(5) = 10(15)$

The greatest common factor is $2(5)$, or 10.

3. $18ab^2 = (2)(3)(3)(a)(b)(b) = 9ab(2b)$

$27a^2b = (3)(3)(3)(a)(a)(b) = 9ab(3a)$

The greatest common factor is $(3)(3)(a)(b)$, or $9ab$.

5. $3x - 6 = 3(x - 2)$

7. $3t - t^2 = t(3 - t)$

9. $5x^2 + 10x^3 = 5x^2(1 + 2x)$

11. $8a - 12a^3 = 4a(2 - 3a^2)$

13. $x(x + 1) - 3(x + 1) = (x + 1)(x - 3)$

15. $y^3 + 3y^2 + 2y + 6 = y^2(y + 3) + 2(y + 3)$

$\qquad = (y + 3)(y^2 + 2)$

17. $x^3 + 2x^2 + x + 2 = (x^3 + 2x^2) + (x + 2)$

$\qquad = x^2(x + 2) + 1(x + 2)$

$\qquad = (x + 2)(x^2 + 1)$

19. $x^2 - 4x + 3x - 12 = x(x - 4) + 3(x - 4)$

$\qquad = (x - 4)(x + 3)$

21. $x^2 - 3x - 28 = (x - 7)(x + 4)$

23. $u^2 + 5u - 36 = (u + 9)(u - 4)$

25. $x^2 + 9xy - 10y^2 = (x + 10y)(x - y)$

27. $y^2 - 6xy - 27x^2 = (y - 9x)(y + 3x)$

29. $4x^2 - 24x + 32 = 4(x^2 - 6x + 8)$

$\qquad = 4(x - 4)(x - 2)$

31. $b = 6 \qquad (x + 3)(x + 3)$

$b = -6 \qquad (x - 3)(x - 3)$

$b = 10 \qquad (x + 9)(x + 1)$

$b = -10 \qquad (x - 9)(x - 1)$

33. $b = 12 \qquad (z + 11)(z + 1)$

$b = -12 \qquad (z - 11)(z - 1)$

35. $5 - 2x - 3x^2 = (5 + 3x)(1 - x)$

37. $50 - 5x - x^2 = (10 + x)(5 - x)$ or

$50 - 5x - x^2 = -1(-50 + 5x + x^2)$

$\qquad = -1(x^2 + 5x - 50)$

$\qquad = -1(x + 10)(x - 5)$

39. $6x^2 + 7x + 2 = (3x + 2)(2x + 1)$

41. $6u^3 + 3u^2 - 30u = 3u(2u^2 + u - 10)$

$\qquad = 3u(2u + 5)(u - 2)$

43. $2x^2 - 3x + 1 = (2x - 1)(x - 1)$

45. $b = 23$ $(b + 24)(b - 1)$

$b = -23$ $(b - 24)(b + 1)$

$b = 10$ $(b + 12)(b - 2)$

$b = -10$ $(b - 12)(b + 2)$

$b = 5$ $(b + 8)(b - 3)$

$b = -5$ $(b - 8)(b + 3)$

$b = 2$ $(b + 6)(b - 4)$

$b = -2$ $(b - 6)(b + 4)$

47. $b = 59$ $(3x - 1)(x + 20)$

$b = -59$ $(3x + 1)(x - 20)$

$b = 17$ $(3x + 20)(x - 1)$

$b = -17$ $(3x - 20)(x + 1)$

$b = 28$ $(3x - 2)(x + 10)$

$b = -28$ $(3x + 2)(x - 10)$

$b = 4$ $(3x + 10)(x - 2)$

$b = -4$ $(3x - 10)(x + 2)$

$b = 11$ $(3x - 4)(x + 5)$

$b = -11$ $(3x + 4)(x - 5)$

$b = 7$ $(3x - 5)(x + 4)$

$b = -7$ $(3x + 5)(x - 4)$

49. $c = 2$ $2(x - 1)^2$

$c = -6$ $2(x + 1)(x - 3)$

$c = -16$ $2(x + 2)(x - 4)$

$c = -30$ $2(x + 3)(x - 5)$

$c = -48$ $2(x + 4)(x - 6)$

(There are many other correct answers.)

51. $x^3 - x = x(x^2 - 1)$

$\qquad = x(x^2 - 1^2)$

$\qquad = x(x - 1)(x + 1)$

The two missing factors are $(x - 1)(x + 1)$.

53. $a^2 - 100 = a^2 - 10^2$

$\qquad = (a + 10)(a - 10)$

55. $25 - 4y^2 = 5^2 - (2y)^2$

$\qquad = (5 + 2y)(5 - 2y)$

57. $(u + 1)^2 - 4 = (u + 1)^2 - 2^2$

$\qquad = [(u + 1) + 2][(u + 1) - 2]$

$\qquad = (u + 3)(u - 1)$

59. $x^2 - 8x + 16 = x^2 - 2(x)(4) + 4^2$

$\qquad = (x - 4)^2$

61. $x^2 + 6x + 9 = x^2 + 2(x)(3) + 3^2$

$\qquad = (x + 3)^2$

63. $9s^2 + 12s + 4 = (3s)^2 + 2(3s)(2) + 2^2$

$\qquad = (3s + 2)^2$

65. $s^3t - st^3 = st(s^2 - t^2)$

$\qquad = st(s + t)(s - t)$

67. $a^3 + 1 = a^3 + 1^3$

$\qquad = (a + 1)[a^2 - a(1) + 1^2]$

$\qquad = (a + 1)(a^2 - a + 1)$

69. $27 - 8t^3 = 3^3 - (2t)^3$

$\qquad = (3 - 2t)(3^2 + (3)(2t) + (2t)^2)$

$\qquad = (3 - 2t)(9 + 6t + 4t^2)$

71. $-16a^3 - 16a^2 - 4a = -4a(4a^2 + 4a + 1)$

$\qquad = -4a[(2a)^2 + 2(2a)(1) + 1^2]$

$\qquad = -4a(2a + 1)^2$

73. $\qquad x^2 - 81 = 0$

$(x + 9)(x - 9) = 0$

$x + 9 = 0 \Rightarrow x = -9$

$x - 9 = 0 \Rightarrow x = 9$

75. $x^2 - 12x + 36 = 0$

$(x - 6)^2 = 0$

$x - 6 = 0 \Rightarrow x = 6$

77. $4s^2 + s - 3 = 0$

$(4s - 3)(s + 1) = 0$

$4s - 3 = 0 \implies 4s = 3 \implies s = \frac{3}{4}$

$s + 1 = 0 \implies s = -1$

79. $x(2x - 3) = 0$

$x = 0$

$2x - 3 = 0 \implies 2x = 3 \implies x = \frac{3}{2}$

81.
$$(z - 2)^2 - 4 = 0$$
$$(z - 2)^2 - 2^2 = 0$$
$$[(z - 2) + 2][(z - 2) - 2] = 0$$
$$z(z - 4) = 0$$
$$z = 0$$
$$z - 4 = 0 \implies z = 4$$

83.
$$x(7 - x) = 12$$
$$7x - x^2 = 12$$
$$-x^2 + 7x - 12 = 0$$
$$x^2 - 7x + 12 = 0 \quad \text{(Multiply both sides of equation by } -1\text{)}$$
$$(x - 3)(x - 4) = 0$$
$$x - 3 = 0 \implies x = 3$$
$$x - 4 = 0 \implies x = 4$$

85.
$$u^3 + 5u^2 - u = 5$$
$$u^3 + 5u^2 - u - 5 = 0$$
$$u^2(u + 5) - 1(u + 5) = 0$$
$$(u + 5)(u^2 - 1) = 0$$
$$(u + 5)(u^2 - 1^2) = 0$$
$$(u + 5)(u + 1)(u - 1) = 0$$
$$u + 5 = 0 \implies u = -5$$
$$u + 1 = 0 \implies u = -1$$
$$u - 1 = 0 \implies u = 1$$

87. $3x^3 + 4x^2 + x = x(3x^2 + 4x + 1)$

$= x(3x + 1)(x + 1)$

Volume = (Length)(Width)(Height)

The height of the cake box is x and the width is $x + 1$. The remaining factor, $3x + 1$, is the length of the box.

89. Area of entire rectangle: $4x(5x) = 20x^2$

Area of semicircle: $\frac{1}{2}\pi(2x)^2 = \frac{1}{2}\pi(4x^2) = 2\pi x^2$

Area of shaded region $= 20x^2 - 2\pi x^2$

$= 2x^2(10 - \pi)$

91. (a)

(b) From the graph, it appears that $R = \$120$ when $x = 20$ units.

(c)
$$R = 12x - 0.3x^2$$
$$120 = 12x - 0.3x^2$$
$$0.3x^2 - 12x + 120 = 0$$
$$\frac{0.3x^2 - 12x + 120}{0.3} = \frac{0}{0.3}$$
$$x^2 - 40x + 400 = 0$$
$$(x - 20)(x - 20) = 0$$
$$x - 20 = 0 \implies x = 20$$

The revenue is $\$120$ when $x = 20$ units.

93. *Common formula:* $lw = A$

Labels: $A = 2400$ (square inches)

$w = $ width (inches)

$l = $ height $= \frac{3}{2}w$

Equation: $\left(\frac{3}{2}w\right)(w) = 2400$

$\frac{3}{2}w^2 = 2400$

$2\left(\frac{3}{2}w^2\right) = 2(2400)$

$3w^2 = 4800$

$3w^2 - 4800 = 0$

$3(w^2 - 1600) = 0$

$3(w + 40)(w - 40) = 0$

$3 \neq 0$

$w + 40 = 0 \implies w = -40$

$w - 40 = 0 \implies w = 40$ and $\frac{3}{2}w = 60$

The width of the window could not be a negative number, so we discard the negative answer. Thus, $w = 40$ and $\frac{3}{2}w = 60$. Thus, the window is 60 inches by 40 inches.

95. *Verbal model:* $\boxed{\begin{array}{c} \text{First positive consecutive} \\ \text{even integer} \end{array}} \cdot \boxed{\begin{array}{c} \text{Second positive consecutive} \\ \text{even integer} \end{array}} = 168$

Labels: First positive consecutive even integer $= 2n$

Second positive consecutive even integer $= 2n + 2$

Equation: $2n(2n + 2) = 168$

$4n^2 + 4n = 168$

$4n^2 + 4n - 168 = 0$

$4(n^2 + n - 42) = 0$

$4(n + 7)(n - 6) = 0$

$4 \neq 0$

$n + 7 = 0 \implies n = -7$

$n - 6 = 0 \implies n = 6$ and $2n + 2 = 2(6) + 2 = 14$

If $n = -7$, the integers are not positive, so we discard the negative answer. Thus, the two positive even integers are $2(6) = 12$ and $2(6) + 2 = 12 + 2 = 14$.

Chapter Test for Chapter 6

1. $7x^2 - 14x^3 = 7x^2(1 - 2x)$

2. $z(z + 7) - 3(z + 7) = (z + 7)(z - 3)$

3. $t^2 - 4t - 5 = (t - 5)(t + 1)$

4. $6x^2 - 11x + 4 = (3x - 4)(2x - 1)$

5. $6y^3 + 45y^2 + 75y = 3y(2y^2 + 15y + 25)$

$= 3y(2y + 5)(y + 5)$

6. $4 - 25v^2 = 2^2 - (5v)^2$

$= (2 + 5v)(2 - 5v)$

7. $4x^2 - 20x + 25 = (2x)^2 - 2(2x)(5) + 5^2$
$= (2x - 5)^2$

8. $16 - (z + 9)^2 = 4^2 - (z + 9)^2$
$= [4 - (z + 9)][(4 + (z + 9)]$
$= [4 - z - 9][4 + z + 9]$
$= (-z - 5)(z + 13)$
or $-(z + 5)(z + 13)$

9. $x^3 + 2x^2 - 9x - 18 = (x^3 + 2x^2) - (9x + 18)$
$= x^2(x + 2) - 9(x + 2)$
$= (x + 2)(x^2 - 9)$
$= (x + 2)(x + 3)(x - 3)$

10. $16 - z^4 = [4^2 - (z^2)^2]$
$= (4 + z^2)(4 - z^2)$
$= (4 + z^2)(2^2 - z^2)$
$= (4 + z^2)(2 + z)(2 - z)$

or

$16 - z^4 = -1(z^2 + 4)(z + 2)(z - 2)$

11. $\frac{2}{5}x - \frac{3}{5} = \frac{1}{5}(2x - 3)$

The missing factor is $2x - 3$.

12. $b = 6$ $(x + 5)(x + 1)$
$b = -6$ $(x - 5)(x - 1)$

13. $x^2 + 12x + c = x^2 + 2(6x) + c$
$= x^2 + 2(x)(6) + c$
$(x + 6)^2 = x^2 + 2(x)(6) + 6^2$
$= x^2 + 12x + 36$

Thus, if $c = 36$, $x^2 + 12x + c$ is a perfect square trinomial.

14. This factorization is not complete because the second factor, $3x - 6$, has a common factor of 3. The complete factoriztion of the polynomial is $3(x + 1)(x - 2)$.

15. $(x + 4)(2x - 3) = 0$
$x + 4 = 0 \Rightarrow x = -4$
$2x - 3 = 0 \Rightarrow 2x = 3 \Rightarrow x = \frac{3}{2}$

16. $7x^2 - 14x = 0$
$7x(x - 2) = 0$
$7x = 0 \Rightarrow x = 0$
$x - 2 = 0 \Rightarrow x = 2$

17. $3x^2 + 7x - 6 = 0$
$(3x - 2)(x + 3) = 0$
$3x - 2 = 0 \Rightarrow 3x = 2 \Rightarrow x = \frac{2}{3}$
$x + 3 = 0 \Rightarrow x = -3$

18. $y(2y - 1) = 6$
$2y^2 - y - 6 = 0$
$2y + 3 = 0 \Rightarrow 2y = -3 \Rightarrow y = -\frac{3}{2}$
$y - 2 = 0 \Rightarrow y = 2$

19. *Common formula:* $\qquad lw = A$

    Labels: $\qquad\qquad l = $ length (inches)

$$w = \text{width} = l - 5 \text{ (inches)}$$

$$A = \text{area} = 84 \text{ (square inches)}$$

    Equation: $\qquad\qquad l(l - 5) = 84$

$$l^2 - 5l = 84$$

$$l^2 - 5l - 84 = 0$$

$$(l - 12)(l + 7) = 0$$

$$l - 12 = 0 \implies l = 12 \text{ and } l - 5 = 7$$

$$l + 7 = 0 \implies l = -7$$

We discard the negative answer for the length of the rectangle. Thus, $l = 12$ and $l - 5 = 7$. The dimensions of the rectangle are 12 inches by 7 inches.

20. $\qquad\qquad$ Height $= -16t^2 + 64$

$$0 = -16t^2 + 64 \qquad \text{Set height equal to 0.}$$

$$16t^2 - 64 = 0$$

$$16(t^2 - 4) = 0$$

$$16(t^2 - 2^2) = 0$$

$$16(t + 2)(t - 2) = 0$$

$$16 \neq 0$$

$$t + 2 = 0 \implies t = -2$$

$$t - 2 = 0 \implies t = 2$$

We discard the negative answer for the time. Thus, it will take the object 2 seconds to hit the ground.

$$\text{Height} = -16t^2 + 64$$

$$28 = -16t^2 + 64 \qquad \text{Set height equal to 28.}$$

$$16t^2 - 36 = 0$$

$$4(4t^2 - 9) = 0$$

$$4[(2t)^2 - 3^2] = 0$$

$$4(2t + 3)(2t - 3) = 0$$

$$4 \neq 0$$

$$2t + 3 = 0 \implies 2t = -3 \implies t = -\tfrac{3}{2}$$

$$2t - 3 = 0 \implies 2t = 3 \implies t = \tfrac{3}{2}$$

We discard the negative answer for the time. Thus, it will take the object $\frac{3}{2}$ seconds to fall to the height of 28 feet.

21. *Verbal model:*

First positive consecutive even integer	$\cdot$	Second positive consecutive even integer	$= 624$

    Labels: First positive consecutive even integer $= 2n$

            Second positive consecutive even integer $= 2n + 2$

—CONTINUED—

21. —CONTINUED—

Equation: $2n(2n + 2) = 624$

$4n^2 + 4n = 624$

$4n^2 + 4n - 624 = 0$

$4(n^2 + n - 156) = 0$

$4(n + 13)(n - 12) = 0$

$4 \neq 0$

$n + 13 = 0 \Rightarrow n = -13$

$n - 12 = 0 \Rightarrow n = 12$ so $2n = 24$ and $2n + 2 = 26$

If $n = -13$, the integers are not positive, so we discard the negative answer. Thus, the two positive even integers are $2(12) = 24$ and $2(12) + 2 = 24 + 2 = 26$.

Cumulative Test for Chapters 4–6

1. $(-2, y)$, y is a real number.

Quadrant II or III.

This point is located 2 units to the *left* of the vertical axis. If y is positive, the point would be *above* the horizontal axis and in the second quadrant. If y is negative, the point would be *below* the horizontal axis and in the third quadrant. (If y is zero, the point would be on the horizontal axis *between* the second and third quadrants.

2. (a) The ordered pair $(-1, -1)$ is *not* a solution because $9(-1) - 4(-1) + 36 = -9 + 4 + 36 \neq 0$.

(b) The ordered pair $(8, 27)$ *is* a solution because $9(8) - 4(27) + 36 = 72 - 108 + 36 = 0$.

(c) The ordered pair $(-4, 0)$ *is* a solution because $9(-4) - 4(0) + 36 = -36 - 0 + 36 = 0$.

(d) The ordered pair $(3, -2)$ is *not* a solution because $9(3) - 4(-2) + 36 = 27 + 8 + 36 \neq 0$.

3.

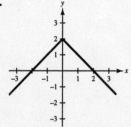

The x-intercepts are $(2, 0)$ and $(-2, 0)$, and the y-intercept is $(0, 2)$.

4.

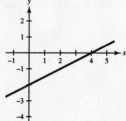

The x-intercept is $(4, 0)$, and the y-intercept is $(0, -2)$.

5. Slope $m = -\dfrac{1}{4}$

$\dfrac{\text{rise}}{\text{run}} = \dfrac{-1}{4}$

Here are some other points on the line with $(2, 1)$:

$(6, 0)$ $(10, -1)$ $(14, -2)$

$(-2, 2)$ $(-6, 3)$ $(-10, 4)$

There are many correct answers because there are an unlimited number of points on the line.

6. Point: $\left(0, -\dfrac{3}{2}\right)$ *Note:* This is the y-intercept.

Slope: $m = \dfrac{5}{6}$

$y = mx + b$

$y = \dfrac{5}{6}x - \dfrac{3}{2}$

7. $y = \frac{2}{3}x - 3 \Rightarrow m_1 = \frac{2}{3}$

$y = -\frac{3}{2}x + 1 \Rightarrow m_2 = -\frac{3}{2}$

The slopes are negative reciprocals, so the lines are perpendicular.

8. $y = 2 - 0.4x$ or $y = -0.4x + 2 \Rightarrow m_1 = -0.4$

$y = -\frac{2}{5}x \Rightarrow m_2 = -\frac{2}{5}$

The slopes, -0.4 and $-\frac{2}{5}$, are equal, so the lines are parallel.

9. $(x^3 - 3x^2) - (x^3 + 2x^2 - 5)$

$= x^3 - 3x^2 - x^3 - 2x^2 + 5$

$= (x^3 - x^3) + (-3x^2 - 2x^2) + 5$

$= -5x^2 + 5$

10. $(6z)(-7z)(z^2) = (6)(-7)z \cdot z \cdot z^2$

$= -42z^{1+1+2}$

$= -42z^4$

11.   F   O   I   L

$(3x + 5)(x - 4) = 3x^2 - 12x + 5x - 20 = 3x^2 - 7x - 20$

12. $(5x - 3)(5x + 3) = (5x)^2 - 3^2$ Pattern: $(a - b)(a + b) = a^2 - b^2$

$= 25x^2 - 9$

13. $(5x + 6)^2 = (5x)^2 + 2(5x)(6) + 6^2$ Pattern: $(a + b)^2 = a^2 + 2ab + b^2$

$= 25x^2 + 60x + 36$

14. $(6x^2 + 72x) \div 6x = \dfrac{6x^2}{6x} + \dfrac{72x}{6x} = x + 12$

15.

$$
\begin{array}{r}
x + 1 + \dfrac{2}{x-4} \\
x - 4 \overline{)\, x^2 - 3x - 2\,} \\
\underline{x^2 - 4x} \\
x - 2 \\
\underline{x - 4} \\
2
\end{array}
$$

Thus, $\dfrac{x^2 - 3x - 2}{x - 4} = x + 1 + \dfrac{2}{x - 4}$.

16. $(3^2 \cdot 4^{-1})^2 = 3^{2 \cdot 2} \cdot 4^{-1 \cdot 2}$

$= 3^4 \cdot 4^{-2}$

$= 3^4 \cdot \dfrac{1}{4^2}$

$= \dfrac{3^4}{4^2}$

$= \dfrac{81}{16}$

17. $2u^2 - 6u = 2u(u - 3)$

18. $(x - 2)^2 - 16 = (x - 2)^2 - 4^2$

$= [(x - 2) + 4][(x - 2) - 4]$

$= (x - 2 + 4)(x - 2 - 4)$

$= (x + 2)(x - 6)$

19. $x^3 + 8x^2 + 16x = x(x^2 + 8x + 16)$

$= x(x^2 + 2(x)(4) + 4^2)$

$= x(x + 4)^2$

20. $x^3 + 2x^2 - 4x - 8 = (x^3 + 2x^2) - (4x + 8)$

$= x^2(x + 2) - 4(x + 2)$

$= (x + 2)(x^2 - 4)$

$= (x + 2)(x^2 - 2^2)$

$= (x + 2)(x + 2)(x - 2)$

$= (x + 2)^2(x - 2)$

21. $u(u - 12) = 0$

$\qquad u = 0$

$\qquad u - 12 = 0 \implies u = 12$

22. $5x^2 - 12x - 9 = 0$

$\qquad (5x + 3)(x - 3) = 0$

$\qquad\qquad 5x + 3 = 0 \implies 5x = -3 \implies x = -\frac{3}{5}$

$\qquad\qquad x - 3 = 0 \implies x = 3$

The solutions are $-\frac{3}{5}$ and 3.

23. $\left(\dfrac{x}{2}\right)^{-2} = \dfrac{x^{-2}}{2^{-2}} = \dfrac{2^2}{x^2} = \dfrac{4}{x^2}$

24. $\quad g(t) = 2t^2 - |t|$

(a) $g(-2) = 2(-2)^2 - |-2|$

$\qquad\quad = 2(4) - 2$

$\qquad\quad = 8 - 2$

$\qquad\quad = 6$

(b) $g(2) = 2(2)^2 - |2|$

$\qquad\quad = 2(4) - 2$

$\qquad\quad = 8 - 2$

$\qquad\quad = 6$

(c) $g(0) = 2(0)^2 - |0|$

$\qquad\quad = 2(0) - 0$

$\qquad\quad = 0 - 0$

$\qquad\quad = 0$

(d) $g\left(-\frac{1}{2}\right) = 2\left(-\frac{1}{2}\right)^2 - \left|-\frac{1}{2}\right|$

$\qquad\quad = 2\left(\frac{1}{4}\right) - \frac{1}{2}$

$\qquad\quad = \frac{1}{2} - \frac{1}{2}$

$\qquad\quad = 0$

25. $\qquad\qquad C = 125 + 0.35x$

$\quad x = 70 \implies C = 125 + 0.35(70)$

$\qquad\qquad\quad C = 125 + 24.5$

$\qquad\qquad\quad C = \$149.50$

CHAPTER 7
Systems of Equations

C H A P T E R 7
Systems of Equations

Section 7.1   Solving Systems of Equations by Graphing and Substitution

Solutions to Odd-Numbered Exercises

1. (a) $(1, 4)$

$1 + 2(4) \stackrel{?}{=} 9$

$9 = 9$

$-2(1) + 3(4) \stackrel{?}{=} 10$

$-2 + 12 \stackrel{?}{=} 10$

$10 = 10$

Solution

(b) $(3, -1)$

$3 + 2(-1) \stackrel{?}{=} 9$

$3 - 2 \stackrel{?}{=} 9$

$1 \neq 9$

Not a solution

3. (a) $(-3, 2)$

$-2(-3) + 7(2) \stackrel{?}{=} 46$

$6 + 14 \stackrel{?}{=} 46$

$20 \neq 46$

Not a solution

(b) $(-2, 6)$

$-2(-2) + 7(6) \stackrel{?}{=} 46$

$4 + 42 \stackrel{?}{=} 46$

$46 = 46$

$3(-2) + 6 \stackrel{?}{=} 0$

$-6 + 6 \stackrel{?}{=} 0$

$0 = 0$

Solution

5. (a) $(8, 4)$

$4(8) - 5(4) \stackrel{?}{=} 12$

$12 = 12$

$3(8) + 2(4) \stackrel{?}{=} -2.5$

$32 \neq -2.5$

Not a solution

(b) $\left(\frac{1}{2}, -2\right)$

$4\left(\frac{1}{2}\right) - 5(-2) \stackrel{?}{=} 12$

$12 = 12$

$3\left(\frac{1}{2}\right) + 2(-2) \stackrel{?}{=} -2.5$

$-2.5 = -2.5$

Solution

7. (a) $(5, -12)$

$5^2 + (-12)^2 \stackrel{?}{=} 169$

$169 = 169$

$17(5) - 7(-12) \stackrel{?}{=} 169$

$169 = 169$

Solution

(b) $(-7, 10)$

$(-7)^2 + (10)^2 \stackrel{?}{=} 169$

$149 \neq 169$

Not a solution

9. Solve each equation for y.

$x + 2y = 6$

$2y = -x + 6$

$y = -\frac{1}{2}x + 3$

$x + 2y = 3$

$2y = -x + 3$

$y = -\frac{1}{2}x + \frac{3}{2}$

Slopes are equal; therefore the system is inconsistent.

11. Solve each equation for y.

$2x - 3y = -12$

$-3y = -2x - 12$

$y = \frac{2}{3}x + 4$

$-8x + 12y = -12$

$12y = 8x - 12$

$y = \frac{2}{3}x - 1$

Slopes are equal; therefore the system is inconsistent.

13. Solve each equation for y.

$-x + 4y = 7$

$4y = x + 7$

$y = \frac{1}{4}x + \frac{7}{4}$

$3x - 12y = -21$

$-12y = -3x - 21$

$y = \frac{1}{4}x + \frac{7}{4}$

Lines are the same; therefore the system is consistent and dependent.

15. Solve each equation for y.

$5x - 3y = 1$

$-3y = -5x + 1$

$y = \frac{5}{3}x - \frac{1}{3}$

$6x - 4y = -3$

$-4y = -6x - 3$

$y = \frac{3}{2}x + \frac{3}{4}$

Slopes are not equal; therefore the system is consistent.

17. Solve each equation for *y*.

$$\tfrac{1}{3}x - \tfrac{1}{2}y = 1 \qquad\qquad -2x + 3y = 6$$

$$-\tfrac{1}{2}y = -\tfrac{1}{3}x + 1 \qquad\qquad 3y = 2x + 6$$

$$y = \tfrac{2}{3}x - 2 \qquad\qquad y = \tfrac{2}{3}x + 2$$

Keystrokes:

y_1 [Y=] [(] 2 [÷] 3 [)] [X,T,θ] [−] 2 [ENTER]

y_2 [(] 2 [÷] 3 [)] [X,T,θ] [+] 2 [GRAPH]

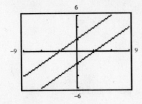

Inconsistent

19. Solve each equation for *y*.

$$-2x + 3y = 6 \qquad\qquad x - y = -1$$

$$3y = 2x + 6 \qquad\qquad -y = -x - 1$$

$$y = \tfrac{2}{3}x + 2 \qquad\qquad y = x + 1$$

Keystrokes:

y_1 [Y=] [(] 2 [÷] 3 [)] [X,T,θ] [+] 2 [ENTER]

y_2 [X,T,θ] [+] 1 [GRAPH]

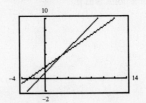

One solution

21. No solution

Solve first equation for *y*.

$$x + y = 4 \qquad\qquad x + y = -1$$

$$y = -x + 4 \qquad\qquad y = -x - 1$$

Slopes are the same.

23. One solution

Solve first equation for *y*.

$$3y = 3 - 2x$$

$$y = \frac{3 - 2x}{3}$$

Substitute into second equation.

$$5x - 3\left(\frac{3 - 2x}{3}\right) = 4$$

$$5x - 3 + 2x = 4$$

$$7x - 3 = 4$$

$$7x = 7$$

$$x = 1$$

$$y = \frac{3 - 2(1)}{3}$$

$$y = \frac{1}{3}$$

$$\left(1, \frac{1}{3}\right)$$

25. Infinite number of solutions

Solve each equation for *y*.

$$x - 2y = -4 \qquad\qquad -0.5x + y = 2$$

$$-2y = -x - 4 \qquad\qquad y = 0.5x + 2$$

$$y = \tfrac{1}{2}x + 2$$

Slopes are the same; lines are the same.

27. No solution

Solve second equation for *y*.

$$x^2 - y = 0$$

$$x^2 = y$$

Substitute into first equation.

$$x - 2x^2 = 4$$

$$0 = 2x^2 - x + 4$$

no real solution

29.

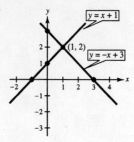

The two lines intersect in a point and the coordinates are (1, 2).

31. Solve each equation for *y*.

$$x - y = 2 \qquad\qquad x + y = 2$$

$$-y = -x + 2 \qquad\qquad y = -x + 2$$

$$y = x - 2$$

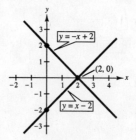

The two lines intersect in a point and the coordinates are (2, 0).

33. Solve first equation for *y*.

$$3x - 4y = 5$$

$$-4y = -3x + 5$$

$$y = \tfrac{3}{4}x - \tfrac{5}{4}$$

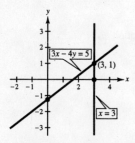

The two lines intersect in a point and the coordinates are (3, 1).

35. Solve each equation for *y*.

$$4x + 5y = 20 \qquad\qquad \tfrac{4}{5}x + y = 4$$

$$5y = -4x + 20 \qquad\qquad y = -\tfrac{4}{5}x + 4$$

$$y = -\tfrac{4}{5}x + 4$$

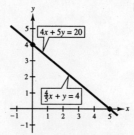

The lines representing the two equations are the same. System is dependent and has infinitely many solutions.

37. Solve each equation for *y*.

$$2x - 5y = 20 \qquad\qquad 4x - 5y = 40$$

$$-5y = -2x + 20 \qquad\qquad -5y = -4x + 40$$

$$y = \tfrac{2}{5}x - 4 \qquad\qquad y = \tfrac{4}{5}x - 8$$

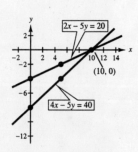

The two lines intersect in a point and the coordinates are (10, 0).

39. Solve each equation for *y*.

$$x + y = 2 \qquad\qquad 3x + 3y = 6$$

$$y = -x + 2 \qquad\qquad 3y = -3x + 6$$

$$y = -x + 2$$

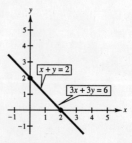

The lines representing the two equations are the same. System is dependent and has infinitely many solutions.

41. Solve each equation for y.

$$4x + 5y = 7 \qquad\qquad 2x - 3y = 9$$

$$5y = -4x + 7 \qquad\qquad -3y = -2x + 9$$

$$y = -\frac{4}{5}x + \frac{7}{5} \qquad\qquad y = \frac{2}{3}x - 3$$

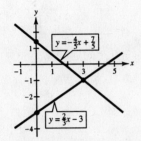

The two lines intersect in a point and the coordinates are $(3, -1)$.

43. *Keystrokes:*

y_1 [Y=] [X,T,θ] [x^2] [ENTER]

y_2 4 [X,T,θ] [−] [X,T,θ] [x^2] [GRAPH]

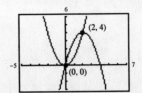

Points of intersection are $(0, 0)$ and $(2, 4)$.

45. *Keystrokes:*

y_1 [Y=] [X,T,θ] [∧] 3 [ENTER]

y_2 [X,T,θ] [∧] 3 [−] 3 [X,T,θ] [x^2] [+] 3 [X,T,θ] [GRAPH]

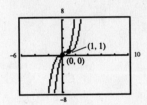

Points of intersection are $(0, 0)$ and $(1, 1)$.

47. Solve for x in first equation.

$$x = 2y$$

Substitute into second equation.

$$3(2y) + 2y = 8$$

$$6y + 2y = 8$$

$$8y = 8$$

$$y = 1$$

$$x = 2(1)$$

$$= 2$$

$$(2, 1)$$

49. $x = 4$

Substitute into second equation.

$$4 - 2y = -2$$

$$-2y = -6$$

$$y = 3$$

$$(4, 3)$$

51. Solve for y.

$$y = 3 - x$$

Substitute into second equation.

$$2x - (3 - x) = 0$$

$$2x - 3 + x = 0$$

$$3x = 3$$

$$x = 1$$

$$y = 3 - 1$$

$$y = 2$$

$$(1, 2)$$

53. Solve for x.

$x = 2 - y$

Substitute into second equation.

$2 - y - 4y = 12$

$-5y = 10$

$y = -2$

$x = 2 - (-2)$

$x = 4$

$(4, -2)$

55. Solve for x.

$x = -7 + 7y$

Substitute into first equation.

$-7 + 7y + 6y = 19$

$13y = 26$

$y = 2$

$x = -7 + 7(2) = 7$

$(7, 2)$

57. Solve for y.

$5y = -8x + 100$

$y = -\frac{8}{5}x + 20$

Substitute into second equation.

$9x - 10\left(-\frac{8}{5}x + 20\right) = 50$

$9x + 16x - 200 = 50$

$25x = 250$

$x = 10$

$y = -\frac{8}{5}(10) + 20$

$y = 4$

$(10, 4)$

59. Solve for y.

$16y = 13x + 10$

$y = \frac{13}{16}x + \frac{10}{16}$

Substitute into second equation.

$5x + 16\left(\frac{13}{16}x + \frac{10}{16}\right) = -26$

$5x + 13x + 10 = -26$

$18x = -36$

$x = -2$

$y = \frac{13}{16}(-2) + \frac{10}{16}$

$y = -\frac{13}{8} + \frac{5}{8}$

$y = -1$

$(-2, -1)$

61. Solve for x.

$4x = -15 + 14y$

$x = \frac{-15 + 14y}{4}$

Substitute into second equation.

$18\left(\frac{-15 + 14y}{4}\right) - 12y = 9$

$18(-15 + 14y) - 48y = 36$

$-270 + 252y - 48y = 36$

$204y = 306$

$y = \frac{3}{2}$

$x = \frac{-15 + 14\left(\frac{3}{2}\right)}{4}$

$= \frac{-15 + 21}{4} = \frac{3}{2}$

$\left(\frac{3}{2}, \frac{3}{2}\right)$

63. Solve for y.

$y = -x + 20$

Substitute into first equation.

$\frac{1}{5}x + \frac{1}{2}(-x + 20) = 8$

$\frac{1}{5}x - \frac{1}{2}x + 10 = 8$

$\frac{1}{5}x - \frac{1}{2}x = -2$

$2x - 5x = -20$

$-3x = -20$

$x = \frac{20}{3}$

$y = -\frac{20}{3} + 20$

$y = -\frac{20}{3} + \frac{60}{3}$

$y = \frac{40}{3}$

$\left(\frac{20}{3}, \frac{40}{3}\right)$

65. Substitute into second equation.

$$y = -2x + 12$$
$$2x^2 = -2x + 12$$
$$2x^2 + 2x - 12 = 0$$
$$x^2 + x - 6 = 0$$
$$(x + 3)(x - 2) = 0$$

$x = -3$	$x = 2$
$y = 2(-3)^2$	$y = 2(2)^2$
$= 2(9)$	$= 2(4)$
$= 18$	$= 8$

$(-3, 18)$ and $(2, 8)$

67. $y = 3x^2$

Substitute into first equation.

$$3x + 2(3x^2) = 30$$
$$3x + 6x^2 = 30$$
$$6x^2 + 3x - 30 = 0$$
$$2x^2 + x - 10 = 0$$
$$(2x + 5)(x - 2) = 0$$

$x = -\frac{5}{2}$	$x = 2$
$y = 3\left(-\frac{5}{2}\right)^2$	$y = 3(2)^2$
$y = 3\left(\frac{25}{4}\right)$	$y = 3(4)$
$= \frac{75}{4}$	$y = 12$

$\left(-\frac{5}{2}, \frac{75}{4}\right)$ and $(2, 12)$

69. Solve for x.

$$x = -3 + y$$

Substitute into first equation.

$$4(-3 + y)^2 + y = 9$$
$$9 - 6y + y^2 + y - 9 = 0$$
$$y^2 - 5y = 0$$
$$y(y - 5) = 0$$

$y = 0$	$y = 5$
$x = -3 + 0$	$x = -3 + 5$
$x = -3$	$x = 2$

$(-3, 0)$ and $(2, 5)$

71. Solve for y.

$$y = -x + 2$$

Substitute into first equation.

$$x^2 + (-x + 2)^2 = 100$$
$$x^2 + x^2 - 4x + 4 = 100$$
$$2x^2 - 4x - 96 = 0$$
$$x^2 - 2x - 48 = 0$$
$$(x - 8)(x + 6) = 0$$

$x = 8$	$x = -6$
$y = -8 + 2$	$y = -(-6) + 2$
$y = -6$	$y = 8$

$(8, -6)$ and $(-6, 8)$

73. Solve for y.

$$y = -3x + 2$$

Substitute into first equation.

$$x^2 - (-3x + 2) = 2$$
$$x^2 + 3x - 2 = 2$$
$$x^2 + 3x - 4 = 0$$
$$(x + 4)(x - 1) = 0$$

$x = -4$	$x = 1$
$y = -3(-4) + 2$	$y = -3(1) + 2$
$y = 14$	$y = -1$

$(-4, 14)$ and $(1, -1)$

75. Solve for y.

$$-y = -5 - 2x$$
$$y = 5 + 2x$$

Substitute into first equation.

$$x^2 + (5 + 2x)^2 = 25$$
$$x^2 + 25 + 20x + 4x^2 - 25 = 0$$
$$5x^2 + 20x = 0$$
$$5x(x + 4) = 0$$

$x = 0$	$x = -4$
$y = 5 + 2(0)$	$y = 5 + 2(-4)$
$y = 5$	$y = -3$
$(0, 5)$	$(-4, -3)$

77. $x = -200 + 100y$

$$3(-200 + 100y) - 275y = 198$$
$$-600 + 300y - 275y = 198$$
$$25y = 798$$
$$y = \frac{798}{25}$$
$$y = 31.92$$

$\left.\rule{0pt}{3.5em}\right\}$ by substitution

$x = -200 + 100(31.92) = -200 + 3192 = 2992$

$(2.992, 31.92)$ or $\left(2992, \frac{798}{25}\right)$

Solve each equation for y.

$x - 100y = -200$ 　　　　　$3x - 275y = 198$

$y = \dfrac{1}{100}x + 2$ 　　　　$-275y = -3x + 198$

$y = 0.01x + 2$ 　　　　　$y = \dfrac{-3}{-275}x + \dfrac{198}{-275}$

　　　　　　　　　　　　$y = \dfrac{3}{275}x - 0.72$

Keystrokes:

y_1 [Y=] .01 [X,T,θ] [+] 2 [ENTER]

y_2 [(] 3 [÷] 275 [)] [X,T,θ] [−] .72 [GRAPH]

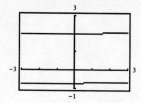

79. Answers will vary. Write equations so that $(4, 5)$ satisfies each equation.

$2x - 3y = -7$ 　　　or 　　　$x - y = -1$

$x + y = 9$ 　　　　　　$2x + 3y = 23$

81. Answers will vary. Write equations so that $(-1, -2)$ satisfies each equation.

$7x + y = -9$ 　　　or 　　　$x + y = -3$

$-x + 3y = -5$ 　　　　　　$x - y = 1$

83. *Verbal Model:*

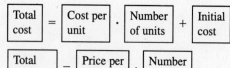

| Total Revenue | = | Price per unit | · | Number of units |

Labels: 　　Total cost $= C$

　　　　　Cost per unit $= 1.20$

　　　　　Number of units $= x$

　　　　　Initial cost $= 8000$

　　　　　Total revenue $= R$

　　　　　Price per unit $= 2.00$

System: 　　$C = 1.20x + 8000$

　　　　　$R = 2.00x$

Break-even point occurs when $R = C$ so

$$1.20x + 8000 = 2.00x$$
$$8000 = 0.80x$$
$$10,000 = x$$

10,000 items

85. *Verbal Model:*

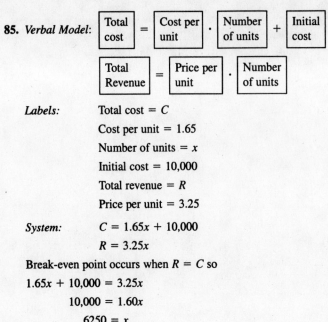

$$\boxed{\text{Total cost}} = \boxed{\text{Cost per unit}} \cdot \boxed{\text{Number of units}} + \boxed{\text{Initial cost}}$$

$$\boxed{\text{Total Revenue}} = \boxed{\text{Price per unit}} \cdot \boxed{\text{Number of units}}$$

Labels: Total cost $= C$

Cost per unit $= 1.65$

Number of units $= x$

Initial cost $= 10,000$

Total revenue $= R$

Price per unit $= 3.25$

System: $C = 1.65x + 10,000$

$R = 3.25x$

Break-even point occurs when $R = C$ so

$1.65x + 10,000 = 3.25x$

$10,000 = 1.60x$

$6250 = x$

6250 units

87. *Verbal Model:*

$$\boxed{\text{Amount at 8\%}} + \boxed{\text{Amount at 9.5\%}} = \boxed{20,000}$$

$$\boxed{8\%} \cdot \boxed{\text{Amount at 8\%}} + \boxed{9.5\%} \cdot \boxed{\text{Amount at 9.5\%}} = \boxed{1675}$$

Labels: Amount at 8% $= x$

Amount at 9.5% $= y$

System: $x + y = 20,000$

$0.08x + 0.095y = 1675$

Solve for x.

$x = 20,000 - y$

Substitute into second equation.

$0.08(20,000 - y) + 0.095y = 1675$

$1600 - 0.08y + 0.095y = 1675$

$0.015y = 75$

$y = \$5000$ at 9.5%

$x = 20,000 - 5000 = \$15,000$ at 8%

89. *Verbal Model:* $\boxed{\text{Amount in 8\% fund}} + \boxed{\text{Amount in 8.5\% fund}} = \boxed{25{,}000}$

$\boxed{8\%} \cdot \boxed{\text{Amount in 8\% fund}} + \boxed{8.5\%} \cdot \boxed{\text{Amount in 8.5\% fund}} = \boxed{2060}$

Labels: Amount in 8% fund = x

Amount in 9.5% fund = y

System: $x + y = 25{,}000$

$0.08x + 0.085y = 2060$

Solve for x.

$x = 25{,}000 - y$

Substitute into second equation.

$0.08(25{,}000 - y) + 0.085y = 2060$

$2000 - 0.08y + 0.085y = 2060$

$0.005y = 60$

$y = \$12{,}000$ at 8.5%

$x = 25{,}000 - 12{,}000 = \$13{,}000$ at 8%

91. *Verbal Model:* $\boxed{\text{Larger number}} + \boxed{2} \cdot \boxed{\text{Smaller number}} = \boxed{61}$

$\boxed{\text{Larger number}} - \boxed{\text{Smaller number}} = \boxed{7}$

Labels: Larger number = x

Smaller number = y

System: $x + 2y = 61$

$x - 7 = 7$

Solve for x.

$x = y + 7$

Substitute into second equation.

$y + 7 + 2y = 61$

$3y + 7 = 61$

$3y = 54$

$y = 18$

$x = 18 + 7$

$x = 25$

$(25, 18)$

93. *Verbal Model:*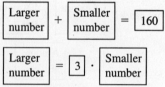

Labels: Larger number = x

Smaller number = y

System: $x + y = 160$

$x = 3y$

Substitute into first equation.

$3y + y = 160$

$4y = 160$

$y = 40$

$x = 3(40)$

$x = 120$

$(120, 40)$

95. *Verbal Model:* $\boxed{2} \cdot \boxed{\text{Length}} + \boxed{2} \cdot \boxed{\text{Width}} = \boxed{320}$

$\boxed{\text{Width}} = \boxed{\text{Length}} - \boxed{20}$

Labels: Length $= x$

Width $= y$

System: $2x + 2y = 320$

$y = x - 20$

Substitute into first equation.

$2x + 2(x - 20) = 320$

$2x + 2x - 40 = 320$

$4x = 360$

$x = 90$

$y = 90 - 20$

$y = 70$

length $= 90$ inches   width $= 70$ inches

97. *Verbal Model:* $\boxed{2} \cdot \boxed{\text{Length}} + \boxed{2} \cdot \boxed{\text{Width}} = \boxed{90}$

$\boxed{\text{Length}} = \boxed{1\frac{1}{2}} \cdot \boxed{\text{Width}}$

Labels: Length $= x$

Width $= y$

System: $2x + 2y = 90$

$x = \left(1\frac{1}{2}\right)y$

$x = \frac{3}{2}y$

Substitute into first equation.

$2\left(\frac{3}{2}y\right) + 2y = 90$

$3y + 2y = 90$

$5y = 90$

$y = 18$

$x = \frac{3}{2}(18)$

$x = 27$

length $= 27$ meters   width $= 18$ meters

99. *Keystrokes:*

y_1 $\boxed{Y=}$ 49,088.2 $\boxed{+}$ 194.6 $\boxed{X,T,\theta}$ $\boxed{-}$ 2.2 $\boxed{X,T,\theta}$ $\boxed{x^2}$ $\boxed{\text{ENTER}}$

y_2 43,132.9 $\boxed{+}$ 977.3 $\boxed{X,T,\theta}$ $\boxed{\text{GRAPH}}$

The point of intersection is $(7.45, 50416.28)$, so $t \approx 7$.

1987

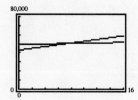

101. (a) Solve one of the equations for one variable in terms of the other.

(b) Substitute the expression found in Step (a) into the other equation to obtain an equation in one variable.

(c) Solve the equation obtained in Step (b).

(d) Back-substitute the solution from Step (c) into the expression obtained in Step (a) to find the value of the other variable.

(e) Check the solution in each of the original equations of the system.

103. After finding a value for one of the variables, substitute this value back into one of the original equations. This is called back-substitution.

105. The graphical method usually yields only approximate solutions.

Section 7.2   Solving Systems of Equations by Elimination

1. $2x + y = 4$

$\underline{x - y = 2}$

$3x = 6$

$x = 2$

$2 - y = 0$

$-y = 0$

$y = 0$

$(2, 0)$

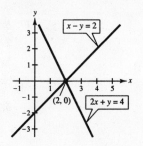

3. $-x + 2y = 1$

$\underline{x - y = 2}$

$y = 3$

$x - 3 = 2$

$x = 5$

$(5, 3)$

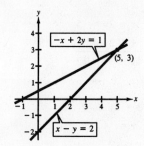

5. $3x + y = 3$
$\underline{2x - y = 7}$
$5x = 10$
$x = 2$

$3(2) + y = 3$
$6 + y = 3$
$y = -3$
$(2, -3)$

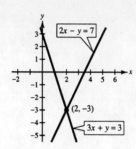

7. $x - y = 1 \Longrightarrow 3x - 3y = 3$
$-3x + 3y = 8 \Longrightarrow \underline{-3x + 3y = 8}$
$0 \neq 11$

No solution

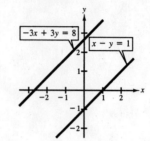

9. $x - 3y = 5 \Longrightarrow 2x - 6y = 10$
$-2x + 6y = -10 \Longrightarrow \underline{-2x + 6y = -10}$
$0 = 0$

All solutions to $x - 3y = 5$

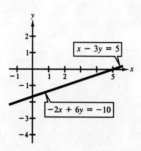

11. $2x - 8y = -11 \Longrightarrow 6x - 24y = -33$
$5x + 3y = 7 \Longrightarrow \underline{40x + 24y = 56}$
$46x = 23$
$x = \dfrac{23}{46}$
$x = \dfrac{1}{2}$

$2\left(\dfrac{1}{2}\right) - 8y = -11$
$-8y = -12$
$y = \dfrac{-12}{-8}$
$y = \dfrac{3}{2}$

$\left(\dfrac{1}{2}, \dfrac{3}{2}\right)$

13. $3x - 2y = 5$
$\underline{x + 2y = 5}$
$4x = 12$
$x = 3$

$3 + 2y = 7$
$2y = 4$
$y = 2$
$(3, 2)$

15. $4x + y = -3$
$\underline{-4x + 3y = 23}$
$4y = 20$
$y = 5$

$4x + 5 = -3$
$4x = -8$
$x = -2$
$(-2, 5)$

17. $3x - 5y = 1$
$\underline{2x + 5y = 9}$
$5x = 10$
$x = 2$

$3(2) - 5y = 1$
$-5y = -5$
$y = 1$
$(2, 1)$

19. $5x + 2y = 7 \Longrightarrow 5x + 2y = 7$

$3x - y = 13 \Longrightarrow \underline{6x - 2y = 26}$

$11x = 33$

$x = 3$

$3(3) - y = 13$

$-y = 4$

$y = -4$

$(3, -4)$

21. $x - 3y = 2 \Longrightarrow -3x + 9y = -6$

$3x - 7y = 4 \Longrightarrow \underline{3x - 7y = 4}$

$2y = -2$

$y = -1$

$x - 3(-1) = 2$

$x = -1$

$(-1, -1)$

23. $2x + y = 9$

$\underline{3x - y = 16}$

$5x = 25$

$x = 5$

$2(5) + y = 9$

$10 + y = 9$

$y = -1$

$(5, -1)$

25. $2u + 3v = 8 \Longrightarrow -6u - 9v = 24$

$3u + 4v = 13 \Longrightarrow \underline{6u + 8v = 26}$

$-v = 2$

$v = -2$

$2u + 3(-2) = 8$

$2u = 14$

$u = 7$

$(7, -2)$

27. $12x - 5y = 2 \Longrightarrow 24x - 10y = 4$

$-24x + 10y = 6 \Longrightarrow \underline{-24x + 10y = 6}$

$0 \neq 10$

Inconsistent

29. $\frac{2}{3}r - s = 0 \Longrightarrow 2r - 3s = 0 \Longrightarrow 8r - 12s = 0$

$10r + 4s = 19 \Longrightarrow 10r + 4s = 19 \Longrightarrow \underline{30r + 12s = 57}$

$38r = 57$

$r = \frac{57}{38}$

$r = \frac{3}{2}$

$\frac{2}{3}\left(\frac{3}{2}\right) - s = 0$

$-s = -1$

$s = 1$

$\left(\frac{3}{2}, 1\right)$

31. $0.05x - 0.03y = 0.21 \Longrightarrow 5x - 3y = 21$

$x + y = 9 \Longrightarrow \underline{3x + 3y = 27}$

$8x = 48$

$x = 6$

$x + y = 9$

$6 + y = 9$

$y = 3$

$(6, 3)$

33. $0.7u - v = -0.4 \Longrightarrow 7u - 10v = -4 \Longrightarrow 21u - 30v = -12$

$0.3u - 0.8v = 0.2 \Longrightarrow 3u - 8v = 2 \Longrightarrow \underline{-21u + 56v = -14}$

$26v = -26$

$v = -1$

$7u - 10(1) = -4$

$7u = -14$

$u = -2$

$(-2, -1)$

35.
$$5x + 7y = 25 \implies 5x + 7y = 25$$
$$x + 1.4y = 5 \implies \underline{-5x - 7y = -25}$$
$$0 = 0$$

All solutions of the form $x + 1.4y = 5$

37.
$$\tfrac{3}{2}x - y = 4 \implies 3x - 2y = 8$$
$$-x + \tfrac{2}{3}y = -1 \implies \underline{-3x + 2y = -3}$$
$$0 \neq 5$$

Inconsistent

39.
$$2x \quad = 25 \implies -4x \quad = -50$$
$$4x - 10y = 0.52 \implies \underline{4x - 10y = 0.52}$$
$$-10y = -49.48$$
$$y = 4.948$$

$$4x - 10(4.948) = 0.52$$
$$4x - 49.48 = 0.52$$
$$4x = 50$$
$$x = 12.5$$

$(12.5, 4.948)$

41.
$$3x + 2y = 5$$
$$y = 2x + 13$$
$$3x + 2(2x + 13) = 5$$
$$3x + 4x + 26 = 5$$
$$7x = -21$$
$$x = -3$$
$$y = 2(-3) + 13$$
$$= -6 + 13$$
$$= 7$$

$(-3, 7)$

43.
$$y = 5x - 3 \implies y = 5x - 3$$
$$y = -2x + 11 \implies \underline{-y = 2x - 11}$$
$$0 = 7x - 14$$
$$14 = 7x$$
$$2 = x$$

$$y = 5(2) - 3$$
$$y = 10 - 3$$
$$y = 7$$

$(2, 7)$

45.
$$2x - y = 20$$
$$\underline{-x + y = -5}$$
$$x = 15$$
$$-15 + y = -5$$
$$y = 10$$

$(15, 10)$

47.
$$y = 4 - \tfrac{1}{4}x$$
$$\tfrac{3}{2}x + 2\left(4 - \tfrac{1}{4}x\right) = 12$$
$$\tfrac{3}{2}x + 8 - \tfrac{1}{2}x = 12$$
$$x = 4$$
$$\tfrac{1}{4}(4) + y = 4$$
$$y = 3$$

$(4, 3)$

49.
$$4x - 5y = 3 \implies -5y = -4x + 3 \implies y = \tfrac{4}{5}x - \tfrac{3}{5}$$
$$-8x + 10y = -6 \implies 10y = 8x - 6 \implies y = \tfrac{4}{5}x - \tfrac{3}{5}$$

Many solutions $\implies$ consistent

51.
$$-2x + 5y = 3 \implies 5y = 2x + 3 \implies y = \tfrac{2}{5}x + \tfrac{3}{5}$$
$$5x + 2y = 8 \implies 2y = -5x + 8 \implies y = -\tfrac{5}{2}x + 4$$

One solution $\implies$ consistent

53.
$$-10x + 5y = 25 \implies 15y = 10x + 25 \implies y = \tfrac{2}{3}x + \tfrac{5}{3}$$
$$2x - 3y = -24 \implies -3y = -2x - 24 \implies y = \tfrac{2}{3}x + 8$$

No solution $\implies$ inconsistent

55. $5x - 10y = 40 \implies y = \dfrac{1}{2}x - 4$

$-2x + ky = 30 \implies y = \dfrac{2}{k}x + \dfrac{30}{16}$

so $\dfrac{2}{k} = \dfrac{1}{2} \implies k = 4$

$5x - 10y = 40 \implies 10x - 20y = 80$

$-2x + 4y = 30 \implies -10x + 20y = 150$

$ 0 \neq 230$

Inconsistent; no solution

57. Answers will vary. Write equations so that $\left(3, -\dfrac{3}{2}\right)$ satisfies each equation.

$x + 2y = 0$

$x - 4y = 9$

$3 + 2\left(-\dfrac{3}{2}\right) \overset{?}{=} 0 \qquad\qquad 3 - 4\left(-\dfrac{3}{2}\right) \overset{?}{=} 9$

$0 = 0 \qquad\qquad\qquad\qquad 9 = 9$

59. *Verbal Model:*

Total cost	=	Cost per unit	·	Number of units	+	Initial cost

Total revenue	=	Price per unit	·	Number of units

Labels: Total cost $= C$

Cost per unit $= 7400$

Number of weeks $= x$

Initial cost $= 85,000$

Total revenue $= R$

Price per unit $= 8100$

System: $C = 7400x + 85,000$

$R = 8100x$

Break-even point occurs when $R = C$

$7400x + 85,000 = 8100x$

$85,000 = 700x$

$121.4285 \approx x$

122 weeks

61. *Verbal Model:*

Amount in 8% bond	+	Amount in 9.5% bond	=	Total investment

Interest in 8% bond	+	Interest in 9.5% bond	=	Total interest

Labels: Amount in 8% bond $= x$

Amount in 9.5% bond $= y$

System: $x + y = 20,000 \implies -0.08x - 0.08y = -1600$

$0.08x + 0.095y = 1675 \implies \underline{0.08x + 0.095y = 1675}$

$ 0.015y = 75$

$ y = 5000$

$x + 5000 = 20,000 \qquad \$15,000 \text{ at } 8\%$

$ x = 15,000 \qquad \$5,000 \text{ at } 9.5\%$

63. *Verbal Model:* $\boxed{\text{Distance}} = \boxed{\text{Rate}} \cdot \boxed{\text{Time}}$

Labels:

Time at 55 mph $= x$

$D_1 =$ distance at 40 mph for 2 hours $+$ at 55 mph for x hours

$D_2 =$ distance at 50 mph for $2 + x$ hours

$D_1 = 40(2) + 55(x)$

$D_2 = 50(2 + x)$

System: Since $D_1 = D_2$

$$40(2) + 55(x) = 50(2 + x)$$

$$80 + 55x = 100 + 50x$$

$$5x = 20$$

$$x = 4 \text{ hours}$$

65. *Verbal Model:*

$\boxed{\begin{array}{c}\text{Plane speed} \\ \text{(still air)}\end{array}} - \boxed{\begin{array}{c}\text{Speed} \\ \text{of air}\end{array}} = \boxed{\begin{array}{c}\text{Speed into} \\ \text{head wind}\end{array}}$

$\boxed{\begin{array}{c}\text{Plane speed} \\ \text{(still air)}\end{array}} + \boxed{\begin{array}{c}\text{Speed} \\ \text{of air}\end{array}} = \boxed{\begin{array}{c}\text{Speed into} \\ \text{head wind}\end{array}}$

Labels:

Plane speed $= x$

Speed of air $= y$

System:

$$x - y = \tfrac{1800}{3.6} \implies x - y = 500$$

$$\underline{x + y = \tfrac{1800}{3} \implies x + y = 600}$$

$$2x \qquad = 1100$$

$$x \qquad = 550 \text{ mph}$$

$$550 - y = 500$$

$$-y = -50$$

$$y = 50 \text{ mph}$$

67. *Verbal Model:*

$\boxed{\begin{array}{c}\text{Number of} \\ \text{adult tickets}\end{array}} + \boxed{\begin{array}{c}\text{Number of} \\ \text{children tickets}\end{array}} = \boxed{500}$

$\boxed{\begin{array}{c}\text{Value of} \\ \text{adult tickets}\end{array}} + \boxed{\begin{array}{c}\text{Value of} \\ \text{children tickets}\end{array}} = \boxed{3312.50}$

Labels:

Number of adult tickets $= x$

Number of children tickets $= y$

System:

$$x + y = 500$$

$$7.50x + 4.00y = 3312.50$$

$$y = 500 - x$$

$$7.50x + 4.00(500 - x) = 3312.50$$

$$7.50x + 2000 - 4.00x = 3312.50$$

$$3.5x = 1312.50$$

$$x = 375 \text{ adult tickets}$$

$$y = 500 - 375 = 125 \text{ children tickets}$$

69. *Verbal Model*: $12 \left(\boxed{\begin{array}{l} \text{Cost of regular} \\ \text{gasoline} \end{array}} \right) + 8 \left(\boxed{\begin{array}{l} \text{Cost of premium} \\ \text{gasoline} \end{array}} \right) = \boxed{\$23.08}$

$\boxed{\begin{array}{l} \text{Cost of premium} \\ \text{gasoline} \end{array}} = \boxed{\$0.11} + \boxed{\begin{array}{l} \text{Cost of regular} \\ \text{gasoline} \end{array}}$

Labels: Cost of regular gasoline $= x$

Cost of premium gasoline $= y$

System: $12x + 8y = 23.08$

$y = 0.11 + x$

$12x + 8(0.11 + x) = 23.08$

$12x + 0.88 + 8x = 23.08$

$20x = 22.20$

$x = \$1.11$ regular

$y = 0.11 + 1.11 = \$1.22$ premium

71. *Verbal Model*: $\boxed{\begin{array}{l} \text{Number of liters} \\ \text{Solution 1} \end{array}} + \boxed{\begin{array}{l} \text{Number of liters} \\ \text{Solution 2} \end{array}} = \boxed{20}$

$\boxed{\begin{array}{l} \text{Value of} \\ \text{Solution 1} \end{array}} + \boxed{\begin{array}{l} \text{Value of} \\ \text{Solution 2} \end{array}} = \boxed{20(0.50)}$

Labels: Number liters Solution 1 $= x$

Number liters Solution 2 $= y$

System: $x + \quad y = 20$

$0.40x + 0.65y = 20(0.50)$

$x = 20 - y$

$40(20 - y) + 65y = 20(50)$

$800 - 40y + 65y = 100$

$25y = 200$

$y = 8$ liters at 65% alcohol solution

$x = 20 - 8 = 12$ liters at 40% alcohol solution

73. *Verbal Model:*

$$\boxed{\begin{array}{c}\text{Amount of}\\ \$5.65 \text{ variety}\end{array}} + \boxed{\begin{array}{c}\text{Amount of}\\ \$8.95 \text{ variety}\end{array}} = \boxed{10}$$

$$\boxed{\begin{array}{c}\text{Cost for}\\ \$5.65 \text{ variety}\end{array}} + \boxed{\begin{array}{c}\text{Cost for}\\ \$8.95 \text{ variety}\end{array}} = \boxed{\text{Total cost}}$$

Labels: Amount of \$5.65 variety $= x$

Amount of \$8.95 variety $= y$

System: $x + y = 10$

$5.65x + 8.95y = 6.95(10)$

$y = 10 - x$

$5.65x + 8.95(10 - x) = 69.5$

$5.65x + 89.5 - 8.95x = 69.5$

$-3.3x = -20$

$x \approx 6.1$ lbs of \$5.65 variety

$y = 10 - x = 10 - 6.1 = 3.9$ lbs of \$8.95 variety

75. (a) $3b + 3m = 7$ (b)

$\underline{3b + 5m = 4}$

$-2m = 3$

$\boxed{m = -\frac{3}{2}}$

$3b + 5\left(-\frac{3}{2}\right) = 4$

$3b = 4 + \frac{15}{2}$

$3b = \frac{23}{2}$

$\boxed{b = \frac{23}{6}}$

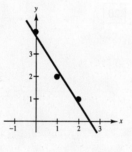

Therefore, $y = -1.5x + 3\frac{5}{6}$

77. (a) $y = \frac{2}{25}x - 10$ Solve by substitution.   (b) $0 = \frac{2}{25}x - 10$ $0 = \frac{5}{61}x - 10$

$y = -\frac{5}{61}x - 10$ $10 = \frac{2}{25}x$ $10 = \frac{5}{61}x$

$\frac{2}{25}x - 10 = -\frac{5}{61}x - 10$ $y = \frac{2}{25}(0) - 10$ $125 = x$ $122 = x$

$\frac{2}{25}x = -\frac{5}{61}x$ $y = -10$ 122 feet and 125 feet

$x = 0$ $(0, -10)$

The memorial is 10 feet deep.

79. When solving a system by elimination, you can recognize that it has infinitely many solutions when adding a nonzero multiple of one equation to another equation to eliminate a variable, you get $0 = 0$ for the second equation.

81. (a) Obtain coefficients for x or y that differ only in sign by multiplying all terms of one or both equations by suitable chosen constants.

   (b) Add the equations to eliminate one variable, and solve the resulting equation.

   (c) Back-substitute the value obtained in Step (b) into either of the original equations and solve for the other variable.

   (d) Check your solution in both of the original equations.

83. Substitution may be better than elimination when it is easy to solve for one of the variables in one of the equations of the system.

Section 7.3   Linear Systems in Three Variables

1. (a) $(0, 3, -2)$

$$0 + 3(3) + 2(-2) \overset{?}{=} 1$$
$$9 - 4 \neq 1$$

not a solution

(b) $(12, 5, -13)$

$$12 + 3(5) + 2(-13) \overset{?}{=} 1$$
$$12 + 15 - 26 = 1$$
$$1 = 1$$

solution

(c) $(1, -2, 3)$

$$1 + 3(-2) + 2(3) \overset{?}{=} 1$$
$$1 - 6 + 6 = 1$$
$$1 = 1$$

solution

(d) $(-2, 5, -3)$

$$-2 + 3(5) + 2(-3) \overset{?}{=} 1$$
$$-2 + 15 - 6 = 1$$
$$7 \neq 1$$

not a solution

3. $3y - (-5) = 2$

$$3y = -3$$
$$y = -1$$
$$x - 2(-1) + 4(-5) = 4$$
$$x + 2 - 20 = 4$$
$$x - 18 = 4$$
$$x = 22$$

$(22, -1, -5)$

5. $3 + z = 2$

$$z = -1$$
$$x - 2(3) + 4(-1) = 4$$
$$x - 6 - 4 = 4$$
$$x - 10 = 4$$
$$x = 14$$

$(14, 3, -1)$

7. The two systems are not equivalent because when the first equation was multiplied by -2 and added to the second equation the constant term should have been -11.

9.
$$\begin{array}{r} x - 2y = 8 \\ \underline{-x + 3y = 6} \\ y = 14 \end{array}$$

This operation eliminated the x-term from the second equation.

11.
$$\begin{array}{r} x - 2y + 3z = 5 \\ \underline{-x + y + 5z = 4} \\ -y + 8z = 9 \end{array}$$

This operation eliminated the x-term in Equation 2.

13. $x \quad + z = \quad 4$

$\qquad y \qquad = \quad 2$

$4x \quad + z = \quad 7$

$x \quad + z = \quad 4$

$\qquad y \qquad = \quad 2$

$\qquad -3z = -9$

$x \quad + z = \quad 4$

$\qquad y \qquad = \quad 2$

$\qquad z = \quad 3$

$x \qquad = \quad 1$

$\qquad y \qquad = \quad 1$

$\qquad z = \quad 3$

$(1, 2, 3)$

15. $x + y + \quad z = \quad 6$

$2x - y + \quad z = \quad 3$

$3x \qquad - z = \quad 0$

$x + y + \quad z = \quad 6$

$\qquad -3y - \quad z = -9$

$\qquad -3y - 4z = -18$

$x + y + \quad z = \quad 6$

$\qquad y + \frac{1}{3}z = \quad 3$

$\qquad -3y - 4z = -18$

$x + y + \quad z = \quad 6$

$\qquad y + \frac{1}{3}z = \quad 3$

$-3x \qquad = -9$

$x + y + \quad z = \quad 6$

$\qquad y + \frac{1}{3}z = \quad 3$

$\qquad z = \quad 3$

$y + \frac{1}{3}(3) = \quad 3$

$\qquad y \qquad = \quad 2$

$x + 2 + \quad 3 = \quad 6$

$x \qquad = \quad 1$

$(1, 2, 3)$

17. $x + y + z = -3$

$\qquad -3y - 7z = 23$

$\qquad -5y = 15$

$x + y + z = -3$

$\qquad y + \frac{7}{3}z = -\frac{23}{3}$

$\qquad y = -3$

$x + y + z = -3$

$\qquad y + \frac{7}{3}z = -\frac{23}{3}$

$\qquad -\frac{7}{3}z = \frac{14}{3}$

$x + y + z = -3$

$\qquad y + \frac{7}{3}z = -\frac{23}{3}$

$\qquad z = -2$

$y + \frac{7}{3}(-2) = \frac{-23}{3}$

$y - \frac{14}{3} = \frac{-23}{3}$

$y = -\frac{9}{3}$

$y = -3$

$x + (-3) + (-2) = -3$

$x - 5 = -3$

$x = 2$

$(2, -3, -2)$

19. $x + 2y + 6z = 5$
$-x + y - 2z = 3$
$x - 4y - 2z = 1$

$x + 2y + 6z = 5$
$3y + 4z = 8$
$x - 4y - 2z = 1$

$x + 2y + 6z = 5$
$3y + 4z = 8$
$-6y - 8z = -4$

$x + 2y + 6z = 5$
$3y + 4z = 8$
$0 = 12$

No solution

Inconsistent

21. $2x \quad + 2z = 2$
$5x + 3y \quad = 4$
$3y - 4z = 4$

$x \quad + z = 1$
$5x + 3y \quad = 4$
$3y - 4z = 4$

$x \quad + z = 1$
$3y - 5z = -1$
$3y - 4z = 4$

$x \quad + z = 1$
$y - \frac{5}{3}z = -\frac{1}{3}$
$3y - 4z = 4$

$x \quad + z = 1$
$y - \frac{5}{3}z = -\frac{1}{3}$
$z = 5$

$y - \frac{5}{3}(5) = -\frac{1}{3}$
$y \quad = 8$

$x \quad + 5 = 1$
$x \quad = -4$

$(-4, 8, 5)$

23. $x + y + 8z = 3$
$2x + y + 11z = 4$
$x \quad + 3z = 0$

$x + y + 8z = 3$
$-y - 5z = -2$
$-y - 5z = -3$

$x + y + 8z = 3$
$y + 5z = 2$
$y + 5z = \frac{3}{5}$

No solution

Inconsistent

25. $2x + y + 3z = 1$
$2x + 6y + 8z = 3$
$6x + 8y + 18z = 5$

$2x + y + 3z = 1$
$5y + 5z = 2$
$5y + 9z = 2$

$2x + y + 3z = 1$
$5y + 5z = 2$
$4z = 0$

$x + \frac{1}{2}y + \frac{3}{2}z = \frac{1}{2}$
$y + z = \frac{2}{5}$
$z = 0$

$y + 0 = \frac{2}{5}$
$y = \frac{2}{5}$

$x + \frac{1}{2}\left(\frac{2}{5}\right) + \frac{3}{2}(0) = \frac{1}{2}$
$x + \frac{1}{5} = \frac{1}{2}$
$x = \frac{5}{10} - \frac{2}{10} = \frac{3}{10}$

$\left(\frac{3}{10}, \frac{2}{5}, 0\right)$

27.
$$y + z = 5$$
$$2x \qquad + 4z = 4$$
$$2x - 3y \qquad = -14$$

$$y + z = 5$$
$$2x \qquad + 4z = 4$$
$$-3y - 4z = -18$$

$$y + z = 5$$
$$2x \qquad + 4z = 4$$
$$-z = -3$$

$$y + z = 5$$
$$2x \qquad + 4z = 4$$
$$z = 3$$

$$y + 3 = 5$$
$$y \qquad = 2$$

$$2x + 4(3) \qquad = 4$$
$$2x + 12 \qquad = 4$$
$$2x \qquad = -8$$
$$x \qquad = -4$$

$$(-4, 2, 3)$$

29.
$$2x + 6y - 4z = 8$$
$$3x + 10y - 7z = 12$$
$$-2x - 6y + 5z = -3$$

$$x + 3y - 2z = 4$$
$$3x + 10y - 7z = 12$$
$$-2x - 6y + 5z = -3$$

$$x + 3y - 2z = 4$$
$$y - z = 0$$
$$z = 5$$

$$y - 5 = 0$$
$$y = 5$$

$$x + 3(5) - 2(5) = 4$$
$$x + 15 - 10 = 4$$
$$x + 5 = 4$$
$$x = -1$$

$$(-1, 5, 5)$$

31.
$$2x \qquad + z = 3$$
$$5y - 3z = 2$$
$$6x + 20y - 9z = 11$$

$$x \qquad + \tfrac{1}{2}z = \tfrac{1}{2}$$
$$5y - 3z = 2$$
$$6x + 20y - 9z = 11$$

$$x \qquad + \tfrac{1}{2}z = \tfrac{1}{2}$$
$$5y - 3z = 2$$
$$20y - 12z = 8$$

$$x \qquad + \tfrac{1}{2}z = \tfrac{1}{2}$$
$$y - \tfrac{3}{5}z = \tfrac{2}{5}$$
$$20y - 12z = 8$$

$$x \qquad + \tfrac{1}{2}z = \tfrac{1}{2}$$
$$y - \tfrac{3}{5}z = \tfrac{2}{5}$$
$$0 = 0$$

$$y = \tfrac{3}{5}z + \tfrac{2}{5}$$
$$x + \tfrac{1}{2}z = \tfrac{1}{2}$$
$$x = \tfrac{1}{2} - \tfrac{1}{2}z$$

$$\text{let } a = z \quad \left(\tfrac{1}{2} - \tfrac{1}{2}a, \tfrac{3}{5}a + \tfrac{2}{5}, a\right)$$

33.
$$3x + y + z = 2$$
$$4x \qquad + 2z = 1$$
$$5x - y + 3z = 0$$

$$x + \tfrac{1}{3}y + \tfrac{1}{3}z = \tfrac{2}{3}$$
$$4x \qquad + 2z = 1$$
$$5x - y + 3z = 0$$

$$x + \tfrac{1}{3}y + \tfrac{1}{3}z = \tfrac{2}{3}$$
$$-\tfrac{4}{3}y + \tfrac{2}{3}z = -\tfrac{5}{3}$$
$$-\tfrac{8}{3}y + \tfrac{4}{3}z = -\tfrac{10}{3}$$

$$x + \tfrac{1}{3}y + \tfrac{1}{3}z = \tfrac{2}{3}$$
$$y - \tfrac{1}{2}z = \tfrac{5}{4}$$
$$-8y + 4z = -10$$

$$x + \tfrac{1}{3}y + \tfrac{1}{3}z = \tfrac{2}{3}$$
$$y - \tfrac{1}{2}z = \tfrac{5}{4}$$
$$0 = 0$$

$$y = \tfrac{1}{2}z + \tfrac{5}{4}$$
$$x + \tfrac{1}{3}\left(\tfrac{1}{2}z + \tfrac{5}{4}\right) + \tfrac{1}{3}z = \tfrac{2}{3}$$
$$x + \tfrac{1}{6}z + \tfrac{5}{12} + \tfrac{1}{3}z = \tfrac{2}{3}$$
$$x + \tfrac{1}{2}z = \tfrac{1}{4}$$
$$= \tfrac{1}{4} - \tfrac{1}{2}z$$

$$\text{let } a = z \quad \left(\tfrac{1}{4} - \tfrac{1}{2}a, \tfrac{1}{2}a + \tfrac{5}{4}, a\right)$$

35. $0.2x + 1.3y + 0.6y = 0.1$

   $0.1x \qquad + 0.3z = 0.7$

   $2x + 10y + 8z = 8$

   $2x + 13y + 6z = 1$

   $1x \qquad + 3z = 7$

   $2x + 10y + 8z = 8$

   $1x \qquad + 3z = 7$

   $\qquad 13y \qquad = -13$

   $\qquad 10y + 2z = -6$

   $1x \qquad + 3z = 7$

   $\qquad y \qquad = -1$

   $\qquad 10y + 2z = -6$

   $x \qquad + 3z = 7$

   $\qquad y \qquad = -1$

   $\qquad\qquad 2z = 4$

   $x + \qquad 3z = 7 \qquad x + 3(2) = 7$

   $\qquad y \qquad = -1 \qquad\qquad x = 1$

   $\qquad\qquad z = 2 \qquad (1, -1, 2)$

37. $x + 4y - 2z = 2$

   $-3x + y + z = -2$

   $5x + 7y - 5z = 6$

   $x + 4y - 2z = 2$

   $\qquad 13y - 5z = 4$

   $\qquad -13y + 5z = -4$

   $x + 4y - 2z = 2$

   $\qquad y - \frac{5}{13}z = \frac{4}{13}$

   $\qquad -13y + 5z = -4$

   $x + 4y - 2z = 2$

   $\qquad y - \frac{5}{13}z = \frac{4}{13}$

   $\qquad\qquad 0 = 0$

   $y = \frac{5}{13}z + \frac{4}{13}$

   $x + 4\left(\frac{5}{13}x + \frac{4}{13}\right) - 2z = 2$

   $x + \frac{20}{13}x + \frac{16}{13} - \frac{26}{13}z = \frac{26}{13}$

   $x - \frac{6}{13}z = \frac{10}{13} \qquad$ Let $z = a$

   $x = \frac{6}{13}z + \frac{10}{13} \quad \left(\frac{6}{13}a + \frac{10}{13}, \frac{5}{13}a + \frac{4}{13}, a\right)$

39. $\qquad -4x + y + 0.2z = 6$

   $\qquad 6x - 3y + 0.5z = -4$

   $\qquad -8x + 2y + 0.6z = 14$

   $\qquad -4x + y + 0.2z = 6$

   $\qquad -6x \qquad + 1.1z = 14$

   $\qquad\qquad 0.2z = 2$

   $\qquad\qquad z = 10$

   $-6x \qquad + 1.1(10) = 14$

   $\qquad -6x = 3$

   $\qquad x = -\frac{1}{2}$

   $-4\left(-\frac{1}{2}\right) + y + 0.2(10) = 6$

   $2 + y + 2 = 6$

   $y = 2$

   $\left(-\frac{1}{2}, 2, 10\right)$

41. $x + y + z = 3 \qquad x + 2y - z = -4$

   $2x + y + 2z = 9 \;$ or $\;\qquad y + 2z = 1$

   $x \quad - 2z = 0 \qquad 3x + y + 3z = 15$

Many correct answers. Write equations so that $(4, -3, 2)$ satisfies each equation.

43.

$$128 = \tfrac{1}{2}a(1)^2 + v_0(1) + s_0$$
$$80 = \tfrac{1}{2}a(2)^2 + v_0(2) + s_0$$
$$0 = \tfrac{1}{2}a(3)^2 + v_0(3) + s_0$$

$$128 = \tfrac{1}{2}a + v_0 + s_0$$
$$80 = 2a + 2v_0 + s_0$$
$$0 = \tfrac{9}{2}a + 3v_0 + s_0$$

$$256 = a + 2v_0 + 2s_0$$
$$80 = 2a + 2v_0 + s_0$$
$$0 = \tfrac{9}{2}a + 3v_0 + s_0$$

$$256 = a + 2v_0 + 2s_0$$
$$-432 = - 2v_0 - 3s_0$$
$$-1152 = - 6v_0 - 8s_0$$

$$256 = a + 2v_0 + 2s_0$$
$$216 = v_0 + \tfrac{3}{2}s_0$$
$$1152 = - 6v_0 - 8s_0$$

$$256 = a + 2v_0 + 2s_0$$
$$216 = v_0 + \tfrac{3}{2}s_0$$
$$144 = + s_0$$

$$216 = v_0 + \tfrac{3}{2}(144)$$
$$0 = v_0$$
$$256 = a + 0 + 288$$
$$-32 = a$$
$$s = -16t^2 + 144$$

45.

$$32 = \tfrac{1}{2}a(1)^2 + v_0(1) + s_0$$
$$32 = \tfrac{1}{2}a(2)^2 + v_0(2) + s_0$$
$$0 = \tfrac{1}{2}a(3)^2 + v_0(3) + s_0$$

$$64 = a + 2v_0 + 2s_0$$
$$32 = 2a + 2v_0 + s_0$$
$$0 = 9a + 6v_0 + s_0$$

$$64 = a + 2v_0 + 2s_0$$
$$-96 = - 2v_0 - 3s_0$$
$$-576 = - 12v_0 - 16s_0$$

$$64 = a + 2v_0 + 2s_0$$
$$48 = v_0 + \tfrac{3}{2}s_0$$
$$-576 = - 12v_0 - 16s_0$$

$$64 = a + 2v_0 + 2s_0$$
$$48 = v_0 + \tfrac{3}{2}s_0$$
$$0 = + 2s_0$$

$$0 = s_0$$
$$48 = v_0 + 0$$
$$64 = a + 2(48) + 0$$
$$-32 = a$$
$$s = -16t^2 + 48t$$

47.

$$-4 = a(0)^2 + b(0) + c$$
$$1 = a(1)^2 + b(1) + c$$
$$10 = a(2)^2 + b(2) + c$$

$$-4 = c$$
$$1 = a + b + c$$
$$10 = 4a + 2b + c$$

$$-4 = c$$
$$1 = a + b + c$$
$$6 = -2b - 3c$$

$$c = -4$$
$$6 = -2b - 3(-4)$$
$$6 = -2b + 12$$
$$-6 = -2b$$
$$3 = b$$
$$1 = a + 3 + (-4)$$
$$1 = a - 1$$
$$2 = a$$
$$y = 2x^2 + 3x - 4$$

49.

$$0 = a(1)^2 + b(1) + c \implies 0 = a + b + c$$
$$-1 = a(2)^2 + b(2) + c \implies -1 = 4a + 2b + c$$
$$0 = a(3)^2 + b(3) + c \implies 0 = 9a + 3b + c$$

$$a + b + c = 0$$
$$-2b - 3c = -1$$
$$-6b - 8c = 0$$

$$a + b + c = 0$$
$$b + \tfrac{3}{2}c = \tfrac{1}{2}$$
$$3b + 4c = 0$$

$$a + \quad -\tfrac{1}{2}c = -\tfrac{1}{2}$$
$$b + \tfrac{3}{2}c = \tfrac{1}{2}$$
$$-\tfrac{1}{2}c = -\tfrac{3}{2}$$

$$a \quad -\tfrac{1}{2}c = -\tfrac{1}{2}$$
$$b + \tfrac{3}{2}c = \tfrac{1}{2}$$
$$c = 3$$

$$a \quad = 1$$
$$b \quad = -4$$
$$c = 3$$

$$y = x^2 - 4x + 3$$

51.

$$-3 = a(-1)^2 + b(-1) + c$$
$$1 = a(1)^2 + b(1) + c$$
$$0 = a(2)^2 + b(2) + c$$

$$-3 = a - b + c$$
$$1 = a + b + c$$
$$0 = 4 + 2b + c$$

$$-3 = a - b + c$$
$$4 = +2b$$
$$12 = +6b - 3c$$

$$2 = b$$
$$12 = 6(2) - 3c$$
$$0 = -3c$$

$$0 = c$$
$$-3 = a - 2 + 0$$
$$-1 = a$$

$$y = -1x^2 + 2x + 0$$

53.

$$3 = a(3)^2 + b(3) + c$$
$$6 = a(4)^2 + b(4) + c$$
$$10 = a(5)^2 + b(5) + c$$

$$3 = 9a + 3b + c \qquad\qquad 3 = 9a + 3b + c \qquad\quad 3 = 9a + 3b + c$$
$$6 = 16a + 4b + c \implies \tfrac{2}{3} = -\tfrac{4}{3}b - \tfrac{7}{9}c \implies \tfrac{2}{3} = -\tfrac{4}{3}b - \tfrac{7}{9}c$$
$$10 = 25a + 5b + c \qquad \tfrac{5}{3} = -\tfrac{10}{3}b - \tfrac{16}{9}c \qquad 0 = \tfrac{3}{18}c$$

$$\tfrac{2}{3} = -\tfrac{4}{3}b - \tfrac{7}{9}(0) \qquad 3 = 9a + 3\left(-\tfrac{1}{2}\right) + 0$$
$$0 = \tfrac{3}{18}c \qquad\qquad \tfrac{2}{3} = -\tfrac{4}{3}b \qquad\qquad 3 = 9a - \tfrac{3}{2}$$
$$0 = c \qquad\qquad -\tfrac{1}{2} = b \qquad\qquad \tfrac{9}{2} = 9a$$
$$\tfrac{1}{2} = a$$

$$y = \tfrac{1}{2}x^2 - \tfrac{1}{2}x = \tfrac{1}{2}x(x - 1)$$
$$y = \tfrac{1}{2}(6)^2 - \tfrac{1}{2}(6) = \tfrac{1}{2}(36) - \tfrac{1}{2}(6) = 18 - 3 = 15 \text{ yes}$$

55. $0^2 + \quad 0^2 + D(0) + \quad E(0) + F = 0$

$2^2 + (-2)^2 + D(2) + E(-2) + F = 0$

$4^2 + \quad 0^2 + D(4) + \quad E(0) + F = 0$

$$F = 0$$

$$2D - 2E + F = -8$$

$$4D + F = -16$$

$$4D + 0 = -16$$

$$4D = -16$$

$$D = -4$$

$$2(-4) - 2E + 0 = -8$$

$$-2E = 0$$

$$E = 0$$

$$x^2 + y^2 - 4x = 0$$

57. $3^2 + (-1)^2 + D(3) + E(-1) + F = 0$

$(-2)^2 + 4^2 + D(-2) + E(4) + F = 0$

$6^2 + 8^2 + D(6) + E(8) + F = 0$

$$3D - E + F = -10$$

$$-2D + 4E + F = -20$$

$$6D + 8E + F = -100$$

$$F + 3D - E = -10$$

$$F - 2D + 4E = -20$$

$$F + 6D + 8E = -100$$

$$F + 3D - E = -10$$

$$-5D + 5E = -10$$

$$3D + 9E = -90$$

$$F + 3D - E = -10$$

$$D + 3E = -30$$

$$-5D + 5E = -10$$

$$F + 3D - E = -10$$

$$D + 3E = -30$$

$$+20E = -160$$

$$F + 3D - E = -10$$

$$D + 3 = -30$$

$$E = -8$$

$$D + 3(-8) = -30$$

$$D = -6$$

$$F + 3(-6) - (-8) = -10$$

$$F = 0$$

$$x^2 + y^2 - 6x - 8y = 0$$

59. $(-3)^2 + 5^2 + D(-3) + E(5) + F = 0$

$4^2 + 6^2 + \quad D(4) + E(6) + F = 0$

$5^2 + 5^2 + \quad D(5) + E(5) + F = 0$

$$-3D + 5E + \quad F = -34$$

$$4D + 6E + \quad F = -52$$

$$5D + 5E + \quad F = -50$$

$$D + \quad E + \tfrac{1}{5}F = -10$$

$$4D + 6E + \quad F = -52$$

$$-3D + 5E + \quad F = -34$$

$$D + \quad E + \tfrac{1}{5}F = -10$$

$$2E + \tfrac{1}{5}F = -12$$

$$8E + \tfrac{8}{5}F = -64$$

—CONTINUED—

59. —CONTINUED—

$$D + \quad E + \tfrac{1}{5}F = -10$$

$$E + \tfrac{1}{10}F = \ -6$$

$$8E + \tfrac{8}{5}F = -64$$

$$D \quad\quad + \tfrac{1}{10}F = \ -4$$

$$E + \ \tfrac{1}{10}F = \ -6$$

$$\tfrac{4}{5}F = -16$$

$$D \quad\quad + \tfrac{1}{10}F = \ -4$$

$$E + \ \tfrac{1}{10}F = \ -6$$

$$F = -20$$

$$D \quad\quad\quad\quad = -2$$

$$E \quad\quad\quad = -4$$

$$F = -20$$

$$x^2 + y^2 - 2x - 4y - 20 = 0$$

61.
$$.20x \quad\quad + .50z = 12$$
$$.40x \quad\quad + .50z = 16$$
$$.40x + 1y \quad\quad = 26$$

$$x \quad + 2.5z = 60 \quad\quad x \ \ + 2.5z = 60 \quad\quad x \ + 2.5z = 60 \quad\quad x \ = 20$$
$$.4x \quad + .5z = 16 \implies \quad\quad -.5z = -8 \implies y - \quad z = 2 \implies y = 18$$
$$.4x + 1y \quad = 26 \quad\quad 1y - \quad 1z = \ 2 \quad\quad z = 16 \quad\quad z = 16$$

Spray X: 20 gal

Spray Y: 18 gal

Spray Z: 16 gal

63.
$$.40x + .30y + .50z = 30$$
$$.20x + .25y + .25z = 17$$
$$.10x + .15y + .25z = 10$$

$$x + .75y + 1.25z = 75 \quad\quad x + \ .75y + 1.25z = 75 \quad\quad x + \ .75y + 1.25z = 75$$
$$.20x + .25y + \ .25z = 17 \implies \quad\quad .1y \quad\quad = \ 2 \implies \quad\quad y \quad\quad = 20 \implies$$
$$.10x + .15y + \ .25z = 10 \quad\quad 0.75y + .125z = 2.5 \quad\quad .075y + .125z = 2.5$$

$$x \ + 1.25z = 60 \quad\quad x \ + 1.25z = 60 \quad\quad x \ = 50$$
$$y \quad\quad = 20 \implies y \quad\quad = 20 \implies y = 20$$
$$+ .125z = \ 1 \quad\quad z = \ 8 \quad\quad z = \ 8$$

String: 50

Wind: 20

Percussion: 8

65. (d) $\begin{aligned} x + \quad y + \quad z &= 200 \\ 8x + 15y + 100z &= 4995 \\ x \quad\quad - \quad 4z &= 0 \end{aligned}$

$\begin{aligned} x + \quad y + \quad z &= 200 \\ 7y + \quad 92z &= 3395 \\ -y - \quad 5z &= -200 \end{aligned}$

$\begin{aligned} x + \quad y + \quad z &= 200 \\ y + \quad 5z &= 200 \\ 57z &= 1995 \end{aligned}$

$\begin{aligned} z &= 35 \\ y + 5(35) &= 200 \\ y &= 25 \end{aligned}$

$\begin{aligned} x + 25 + \quad 35 &= 200 \\ x &= 140 \end{aligned}$

(e) Students: 140; Nonstudents: 25; Major contributors: 35

(f) $\begin{aligned} x + \quad y + \quad z &= 200 \\ 8x + 15y + 100z &= 4995 \\ z &= 18 \end{aligned}$

$x + \quad y = 182$

$8x + 15y = 3195$

$y = 182 - x$

$8x + 15(182 - x) = 3195$

$8x + 2730 - 15x = 3195$

$-7x = 465$

$x = -\dfrac{465}{7}$ (not possible)

67. Substitute $y = 3$ into the first equation to obtain $x + 2(3) = 2$ or $x = 2 - 6 = -4$.

69. Answers will vary.

Mid-Chapter Quiz for Chapter 7

1. $(1, -2)$ $\qquad 5(1) - 12(-2) \overset{?}{=} 2$

$\qquad\qquad\qquad 5 + 24 \neq 2$

This is not the solution.

$(10, 4) \qquad 5(10) - 12(4) \overset{?}{=} 2$

$\qquad\qquad\qquad 50 - 48 = 2$

$\qquad\qquad\qquad\qquad 2 = 2$

$\qquad\qquad 2(10) + 1.5(4) \overset{?}{=} 26$

$\qquad\qquad\qquad 20 + 6 = 26$

$\qquad\qquad\qquad\qquad 26 = 26$

This is a solution.

2.

No solution

3.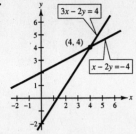

$3x - 2y = 4$

$(4, 4)$

$x - 2y = -4$

One solution

4.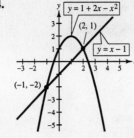

$y = 1 + 2x - x^2$

$(2, 1)$

$y = x - 1$

$(-1, -2)$

Two solutions

5.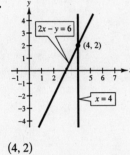

$2x - y = 6$

$(4, 2)$

$x = 4$

$(4, 2)$

6.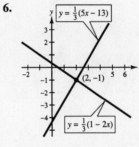

$y = \frac{1}{3}(5x - 13)$

$(2, -1)$

$y = \frac{1}{3}(1 - 2x)$

$(2, -1)$

7.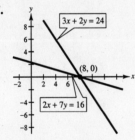

$3x + 2y = 24$

$(8, 0)$

$2x + 7y = 16$

$(8, 0)$

8.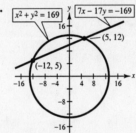

$x^2 + y^2 = 169$

$7x - 17y = -169$

$(5, 12)$

$(-12, 5)$

$(5, 12), (-12, 5)$

9. $2x - 3y = 4$

$y = 2$

$2x - 3(2) = 4$

$2x - 6 = 4$

$2x = 10$

$x = 5$

$(5, 2)$

10. $y = 5 - x^2$

$y = 2(x + 1)$

$5 - x^2 = 2(x + 1)$

$5 - x^2 = 2x + 2$

$0 = x^2 + 2x - 3$

$0 = (x + 3)(x - 1)$

$x = -3 \qquad x = 1$

$y = -4 \qquad y = 4$

$(-3, -4), (1, 4)$

11. $5x - y = 32 \implies -y = -5x + 32$

$6x - 9y = 18 \qquad\qquad y = 5x - 32$

$6x - 9(5x - 32) = 18$

$6x - 45x + 288 = 18$

$\qquad\qquad -39x = -270$

$\qquad\qquad x = \dfrac{-270}{-39} = \dfrac{90}{13}$

$y = 5\left(\dfrac{90}{13}\right) - 32$

$\quad = \dfrac{450}{13} - \dfrac{416}{13}$

$\quad = \dfrac{34}{13}$

$\left(\dfrac{90}{13}, \dfrac{34}{13}\right)$

12. $0.2x + 0.7y = 8$

$\quad -x + 2y = 15 \implies -x = -2y + 15$

$0.2(2y - 15) + 0.7y = 8 \qquad x = 2y - 15$

$0.4y - 3 + 0.7y = 8$

$\qquad\qquad 1.1y = 11$

$\qquad\qquad y = 10$

$x = 2(10) - 15$

$\quad = 20 - 15$

$\quad = 5$

$(5, 10)$

13. $x + 10y = 18$

$5x + 2y = 42$

$x + 10y = 18$

$\quad -48y = -48$

$x + 10y = 18$

$\qquad\quad y = 1$

$x \qquad\quad = 8$

$\qquad\quad y = 1$

$(8, 1)$

14. $3x + 11y = 38$

$7x - 5y = -34$

$x + \dfrac{11}{3}y = \dfrac{38}{3}$

$7x - 5y = -34$

$x + \dfrac{11}{3}y = \dfrac{38}{3}$

$\quad -\dfrac{92}{3}y = -\dfrac{368}{3}$

$x + \dfrac{11}{3}y = \dfrac{38}{3}$

$\qquad\quad y = 4$

$x \qquad\quad = -2$

$\qquad\quad y = 4$

$(-2, 4)$

15.
$$a + b + c = 1$$
$$4a + 2b + c = 2$$
$$9a + 3b + c = 4$$

$$a + b + c = 1$$
$$-2b - 3c = -2$$
$$-6b - 8c = -5$$

$$a + b + c = 1$$
$$b + \tfrac{3}{2}c = 1$$
$$-6b - 8c = -5$$

$$a \quad -\tfrac{1}{2}c = 0$$
$$b + \tfrac{3}{2}c = 1$$
$$c = 1$$

$$a \quad = \tfrac{1}{2}$$
$$b \quad = -\tfrac{1}{2}$$
$$c = 1$$

$$\left(\tfrac{1}{2}, -\tfrac{1}{2}, 1\right)$$

16.
$$x \quad + 4z = 17$$
$$-3x + 2y - z = -20$$
$$x - 5y + 3z = 19$$

$$x \quad + 4z = 17$$
$$2y + 11z = 31$$
$$-5y - z = 2$$

$$x \quad + 4z = 17$$
$$y + \tfrac{11}{2}z = \tfrac{31}{2}$$
$$\tfrac{53}{2}z = \tfrac{159}{2}$$

$$x \quad + 4z = 17$$
$$y + \tfrac{11}{2}z = \tfrac{31}{2}$$
$$z = 3$$

$$x \quad = 5$$
$$y \quad = -1$$
$$z = 3$$

$$(5, -1, 3)$$

17.
$$x + y = -2$$
$$2x - y = 32$$

18.
$$x + y - z = 11$$
$$x + 2y - z = 14$$
$$-2x + y + z = -6$$

19. *Verbal model:*

$$\boxed{\begin{array}{c}\text{Amount}\\\text{Solution 1}\end{array}} + \boxed{\begin{array}{c}\text{Amount}\\\text{Solution 2}\end{array}} = \boxed{\begin{array}{c}\text{Amount}\\\text{Mixture}\end{array}}$$

$$0.20\;\boxed{\begin{array}{c}\text{Amount}\\\text{Solution 1}\end{array}} + 0.50\;\boxed{\begin{array}{c}\text{Amount}\\\text{Solution 2}\end{array}} = 0.30 \cdot 20$$

Labels:
Amount Solution 1 = x
Amount Solution 2 = y

System of equations:
$$x + y = 20$$
$$0.20x + 0.50y = 0.30(20)$$
$$x + y = 20$$
$$20x + 50y = 600$$

By substitution $y = 20 - x$
$$20x + 50(20 - x) = 600$$
$$20x + 1000 - 50x = 600$$
$$-30x = -400$$
$$x = 13\tfrac{1}{3} \text{ gallons at 20\% solution}$$
$$20 - x = 6\tfrac{2}{3} \text{ galllons at 50\% solution}$$

20.
$$2 = a(1)^2 + b(1) + c \implies a + b + c = 2$$
$$-4 = a(-1)^2 + b(-1) + c \implies a - b + c = -4$$
$$8 = a(2)^2 + b(2) + c \implies 4a + 2b + c = 8$$

$$a + b + c = 2$$
$$-2b = -6$$
$$-2b - 3c = 0$$

$$a + b + c = 2$$
$$b = 3$$
$$-2b - 3c = 0$$

$$a + c = -1$$
$$b = 3$$
$$-3c = 6$$

$$a + c = -1$$
$$b = 3$$
$$c = -2$$

$$a = 1$$
$$b = 3$$
$$c = -2$$

$$y = x^2 + 3x - 2$$

Section 7.4   Matrices and Linear Systems

1. 4×2

3. 2×3

5. 4×1

7. $\begin{bmatrix} 4 & -5 & \vdots & -2 \\ -1 & 8 & \vdots & 10 \end{bmatrix}$

9. $\begin{bmatrix} 1 & 10 & -3 & \vdots & 2 \\ 5 & -3 & 4 & \vdots & 0 \\ 2 & 4 & 0 & \vdots & 6 \end{bmatrix}$

11. $\begin{bmatrix} 5 & 1 & -3 & \vdots & 7 \\ 0 & 2 & 4 & \vdots & 12 \end{bmatrix}$

13.
$$4x + 3y = 8$$
$$x - 2y = 3$$

15.
$$x \quad\quad + 2z = -10$$
$$3y - z = 5$$
$$4x + 2y \quad\quad = 3$$

17.
$$5x + 8y + 2z \quad\quad = -1$$
$$-2x + 15y + 5z + w = 9$$
$$x + 6y - 7z \quad\quad = -3$$

19.
$$\begin{bmatrix} 1 & 4 & 3 \\ 2 & 10 & 5 \end{bmatrix}$$
$$-2R_1 + R_2 \begin{bmatrix} 1 & 4 & 3 \\ 0 & 2 & -1 \end{bmatrix}$$

21.
$$\begin{bmatrix} 9 & -18 & 6 \\ 2 & 8 & 15 \end{bmatrix}$$
$$\tfrac{1}{9}R_1 \begin{bmatrix} 1 & -2 & \tfrac{2}{3} \\ 2 & 8 & 15 \end{bmatrix}$$

23.
$$\begin{bmatrix} 1 & 1 & 4 & -1 \\ 3 & 8 & 10 & 3 \\ -2 & 1 & 12 & 6 \end{bmatrix}$$
$$\begin{matrix} -3R_1 + R_2 \\ \\ 2R_1 + R_3 \end{matrix} \begin{bmatrix} 1 & 1 & 4 & -1 \\ 0 & 5 & -2 & 6 \\ 0 & 3 & 20 & 4 \end{bmatrix}$$
$$\tfrac{1}{5}R_2 \begin{bmatrix} 1 & 1 & 4 & -1 \\ 0 & 1 & -\tfrac{2}{5} & \tfrac{6}{5} \\ 0 & 3 & 20 & 4 \end{bmatrix}$$

25. $\begin{bmatrix} 1 & 2 & 3 \\ 2 & -1 & -4 \end{bmatrix}$

$-2R_1 + R_2 \begin{bmatrix} 1 & 2 & 3 \\ 0 & -5 & -10 \end{bmatrix}$

$-\frac{1}{5}R_2 \begin{bmatrix} 1 & 2 & 3 \\ 0 & 1 & 2 \end{bmatrix}$

27. $\begin{bmatrix} 4 & 6 & 1 \\ -2 & 2 & 5 \end{bmatrix}$

$\frac{1}{4}R_1 \begin{bmatrix} 1 & \frac{3}{2} & \frac{1}{4} \\ -2 & 2 & 5 \end{bmatrix}$

$2R_1 + R_2 \begin{bmatrix} 1 & \frac{3}{2} & \frac{1}{4} \\ 0 & 5 & \frac{11}{2} \end{bmatrix}$

$\frac{1}{5}R_2 \begin{bmatrix} 1 & \frac{3}{2} & \frac{1}{4} \\ 0 & 1 & \frac{11}{10} \end{bmatrix}$

$-\frac{3}{2}R_3 + R_2 \begin{bmatrix} 1 & 0 & -\frac{7}{5} \\ 0 & 1 & \frac{11}{10} \end{bmatrix}$

29. $\begin{bmatrix} 1 & 1 & 0 & 5 \\ -2 & -1 & 2 & -10 \\ 3 & 6 & 7 & 14 \end{bmatrix}$

$\begin{matrix} 2R_1 + R_2 \\ -3R_1 + R_3 \end{matrix} \begin{bmatrix} 1 & 1 & 0 & 5 \\ 0 & 1 & 2 & 0 \\ 0 & 3 & 7 & -1 \end{bmatrix}$

$-3R_2 + R_3 \begin{bmatrix} 1 & 1 & 0 & 5 \\ 0 & 1 & 2 & 0 \\ 0 & 0 & 1 & -1 \end{bmatrix}$

31. $\begin{bmatrix} 1 & -1 & -1 & 1 \\ 4 & -4 & 1 & 8 \\ -6 & 8 & 18 & 0 \end{bmatrix}$

$\begin{matrix} -4R_1 + R_2 \\ 6R_1 + R_3 \end{matrix} \begin{bmatrix} 1 & -1 & -1 & 1 \\ 0 & 0 & 5 & 4 \\ 0 & 2 & 12 & 6 \end{bmatrix}$

$\begin{matrix} R_3 \\ R_2 \end{matrix} \begin{bmatrix} 1 & -1 & -1 & 1 \\ 0 & 2 & 12 & 6 \\ 0 & 0 & 5 & 4 \end{bmatrix}$

$\begin{matrix} \frac{1}{2}R_2 \\ \frac{1}{5}R_3 \end{matrix} \begin{bmatrix} 1 & -1 & -1 & 1 \\ 0 & 1 & 6 & 3 \\ 0 & 0 & 1 & \frac{4}{5} \end{bmatrix}$

33. $\begin{bmatrix} 1 & 1 & -1 & 3 \\ 2 & 1 & 2 & 5 \\ 3 & 2 & 1 & 8 \end{bmatrix}$

$\begin{matrix} -2R_1 + R_2 \\ -3R_1 + R_3 \end{matrix} \begin{bmatrix} 1 & 1 & -1 & 3 \\ 0 & -1 & 4 & -1 \\ 0 & 1 & 4 & -1 \end{bmatrix}$

$R_2 + R_3 \begin{bmatrix} 1 & 1 & -1 & 3 \\ 0 & -1 & 4 & -1 \\ 0 & 0 & 8 & -2 \end{bmatrix}$

$-R_2 \begin{bmatrix} 1 & 1 & -1 & 3 \\ 0 & 1 & -4 & 1 \\ 0 & 0 & 8 & -2 \end{bmatrix}$

35.
$$x - 2y = 4 \qquad x - 2(-3) = 4$$
$$y = -3 \qquad x + 6 = 4$$
$$x = -2$$

$(-2, -3)$

37.
$$x + 5y = 3$$
$$y = -2$$
$$x + 5(-2) = 3$$
$$x - 10 = 3$$
$$x = 13$$

$(13, -2)$

39.
$$x - y + 2z = 4 \qquad y - (-2) = 2$$
$$y - z = 2 \qquad y + 2 = 2$$
$$z = -2 \qquad y = 0$$
$$x - 0 + 2(-2) = 4$$
$$x - 4 = 4$$
$$x = 8$$

$(8, 0, -2)$

41. $\begin{bmatrix} 1 & 2 & \vdots & 7 \\ 3 & 1 & \vdots & 8 \end{bmatrix}$

$-3R_1 + R_2 \begin{bmatrix} 1 & 2 & \vdots & 7 \\ 0 & -5 & \vdots & -13 \end{bmatrix}$

$-\frac{1}{5}R_2 \begin{bmatrix} 1 & 2 & \vdots & 7 \\ 0 & 1 & \vdots & \frac{13}{5} \end{bmatrix}$

$-2R_2 + R_1 \begin{bmatrix} 1 & 0 & \vdots & \frac{9}{5} \\ 0 & 1 & \vdots & \frac{13}{5} \end{bmatrix}$

$\left(\frac{9}{5}, \frac{13}{5}\right)$

43. $\begin{bmatrix} 6 & -4 & \vdots & 2 \\ 5 & 2 & \vdots & 7 \end{bmatrix}$

$\frac{1}{6}R_1 \begin{bmatrix} 1 & -\frac{2}{3} & \vdots & \frac{1}{3} \\ 5 & 2 & \vdots & 7 \end{bmatrix}$

$-5R_1 + R_2 \begin{bmatrix} 1 & -\frac{2}{3} & \vdots & \frac{1}{3} \\ 0 & \frac{16}{3} & \vdots & \frac{16}{3} \end{bmatrix}$

$\frac{3}{16}R_2 \begin{bmatrix} 1 & -\frac{2}{3} & \vdots & \frac{1}{3} \\ 0 & 1 & \vdots & 1 \end{bmatrix}$

$\frac{2}{3}R_2 + R_1 \begin{bmatrix} 1 & 0 & \vdots & 1 \\ 0 & 1 & \vdots & 1 \end{bmatrix}$

$(1, 1)$

45. $\begin{bmatrix} -1 & 2 & \vdots & 1.5 \\ 2 & -4 & \vdots & 3 \end{bmatrix}$

$-R_1 \begin{bmatrix} 1 & -2 & \vdots & -1.5 \\ 2 & -4 & \vdots & 3 \end{bmatrix}$

$-2R_1 + R_2 \begin{bmatrix} 1 & -2 & \vdots & -1.5 \\ 0 & 0 & \vdots & 6 \end{bmatrix}$

Inconsistent; no solution

47.
$$\begin{bmatrix} 1 & -2 & -1 & \vdots & 6 \\ 0 & 1 & 4 & \vdots & 5 \\ 4 & 2 & 3 & \vdots & 8 \end{bmatrix}$$

$$-4R_1 + R_3 \begin{bmatrix} 1 & -2 & -1 & \vdots & 6 \\ 0 & 1 & 4 & \vdots & 5 \\ 0 & 10 & 7 & \vdots & -16 \end{bmatrix}$$

$$-10R_2 + R_3 \begin{bmatrix} 1 & -2 & -1 & \vdots & 6 \\ 0 & 1 & 4 & \vdots & 5 \\ 0 & 0 & -33 & \vdots & -66 \end{bmatrix}$$

$$\frac{1}{-33}R_3 \begin{bmatrix} 1 & -2 & -1 & \vdots & 6 \\ 0 & 1 & 4 & \vdots & 5 \\ 0 & 0 & 1 & \vdots & 2 \end{bmatrix}$$

$z = 2 \qquad y + 4(2) = 5 \qquad x - 2(-3) - (2) = 6$

$\qquad\qquad\qquad\quad y = -3 \qquad\qquad x + 6 - 2 = 6$

$\qquad\qquad\qquad\qquad\qquad\qquad\qquad\quad x = 2$

$(2, -3, 2)$

49.
$$\begin{bmatrix} 1 & 1 & -5 & \vdots & 3 \\ 1 & 0 & -2 & \vdots & 1 \\ 2 & -1 & -1 & \vdots & 0 \end{bmatrix}$$

$$\begin{matrix} -R_1 + R_2 \\ -2R_1 + R_3 \end{matrix} \begin{bmatrix} 1 & 1 & -5 & \vdots & 3 \\ 0 & -1 & 3 & \vdots & -2 \\ 0 & -3 & 9 & \vdots & -6 \end{bmatrix}$$

$$-R_2 \begin{bmatrix} 1 & 1 & -5 & \vdots & 3 \\ 0 & 1 & -3 & \vdots & 2 \\ 0 & -3 & 9 & \vdots & -6 \end{bmatrix}$$

$$3R_2 + R_3 \begin{bmatrix} 1 & 1 & -5 & \vdots & 3 \\ 0 & 1 & -3 & \vdots & 2 \\ 0 & 0 & 0 & \vdots & 0 \end{bmatrix}$$

$y - 3z = 2 \qquad\qquad x + (2 + 3z) - 5z = 3$

$\quad y = 2 + 3z \qquad\qquad\qquad\quad x = 1 + 2z$

let $a = z$ (a is any real number)

$(1 + 2a, 2 + 3a, a)$

51.
$$\begin{bmatrix} 2 & 4 & 0 & \vdots & 10 \\ 2 & 2 & 3 & \vdots & 3 \\ -3 & 1 & 2 & \vdots & -3 \end{bmatrix}$$

$$-R_1 + R_2 \begin{bmatrix} 2 & 4 & 0 & \vdots & 10 \\ 0 & -2 & 3 & \vdots & -7 \\ -3 & 1 & 2 & \vdots & -3 \end{bmatrix}$$

$$\frac{1}{2}R_1 \begin{bmatrix} 1 & 2 & 0 & \vdots & 5 \\ 0 & -2 & 3 & \vdots & -7 \\ -3 & 1 & 2 & \vdots & -3 \end{bmatrix}$$

$$3R_1 + R_3 \begin{bmatrix} 1 & 2 & 0 & \vdots & 5 \\ 0 & -2 & 3 & \vdots & -7 \\ 0 & 7 & 2 & \vdots & 12 \end{bmatrix}$$

$$-\frac{1}{2}R_2 \begin{bmatrix} 1 & 2 & 0 & \vdots & 5 \\ 0 & 1 & -\frac{3}{2} & \vdots & \frac{7}{2} \\ 0 & 7 & 2 & \vdots & 12 \end{bmatrix}$$

$$-7R_2 + R_3 \begin{bmatrix} 1 & 2 & 0 & \vdots & 5 \\ 0 & 1 & -\frac{3}{2} & \vdots & \frac{7}{2} \\ 0 & 0 & \frac{25}{2} & \vdots & -\frac{25}{2} \end{bmatrix}$$

$$\frac{2}{25}R_3 \begin{bmatrix} 1 & 2 & 0 & \vdots & 5 \\ 0 & 1 & -\frac{3}{2} & \vdots & \frac{7}{2} \\ 0 & 0 & 1 & \vdots & -1 \end{bmatrix}$$

$z = -1 \qquad y - \frac{3}{2}(-1) = \frac{7}{2} \qquad x + 2(2) = 5$

$\qquad\qquad\qquad\qquad y = \frac{4}{2} \qquad\qquad x + 4 = 5$

$\qquad\qquad\qquad\qquad y = 2 \qquad\qquad\quad x = 1$

$(1, 2, -1)$

53.
$$\begin{bmatrix} 1 & -3 & 2 & \vdots & 8 \\ 0 & 2 & -1 & \vdots & -4 \\ 1 & 0 & 1 & \vdots & 3 \end{bmatrix}$$

$$\begin{matrix} \frac{1}{2}R_2 \\ -R_1 + R_3 \end{matrix} \begin{bmatrix} 1 & -3 & 2 & \vdots & 8 \\ 0 & 1 & -\frac{1}{2} & \vdots & -2 \\ 0 & 3 & -1 & \vdots & -5 \end{bmatrix}$$

$$\begin{matrix} 3R_2 + R_1 \\ \\ -3R_2 + R_3 \end{matrix} \begin{bmatrix} 1 & 0 & \frac{1}{2} & \vdots & 2 \\ 0 & 1 & -\frac{1}{2} & \vdots & -2 \\ 0 & 0 & \frac{1}{2} & \vdots & 1 \end{bmatrix}$$

$$2R_3 \begin{bmatrix} 1 & 0 & \frac{1}{2} & \vdots & 2 \\ 0 & 1 & -\frac{1}{2} & \vdots & -2 \\ 0 & 0 & 1 & \vdots & 2 \end{bmatrix}$$

$$\begin{matrix} -\frac{1}{2}R_3 + R_1 \\ \frac{1}{2}R_3 + R_2 \end{matrix} \begin{bmatrix} 1 & 0 & 0 & \vdots & 1 \\ 0 & 1 & 0 & \vdots & -1 \\ 0 & 0 & 1 & \vdots & 2 \end{bmatrix}$$

$(1, -1, 2)$

55.
$$\begin{bmatrix} -2 & -2 & -15 & \vdots & 0 \\ 1 & 2 & 2 & \vdots & 18 \\ 3 & 3 & 22 & \vdots & 2 \end{bmatrix}$$

$$\begin{matrix} R_2 \\ R_1 \end{matrix} \begin{bmatrix} 1 & 2 & 2 & \vdots & 18 \\ -2 & -2 & -15 & \vdots & 0 \\ 3 & 3 & 22 & \vdots & 2 \end{bmatrix}$$

$$\begin{matrix} \\ 2R_1 + R_2 \\ -3R_1 + R_3 \end{matrix} \begin{bmatrix} 1 & 2 & 2 & \vdots & 18 \\ 0 & 2 & -11 & \vdots & 36 \\ 0 & -3 & 16 & \vdots & -52 \end{bmatrix}$$

$$\begin{matrix} \\ \\ \frac{3}{2}R_2 + R_3 \end{matrix} \begin{bmatrix} 1 & 2 & 2 & \vdots & 18 \\ 0 & 2 & -11 & \vdots & 36 \\ 0 & 0 & -\frac{1}{2} & \vdots & 2 \end{bmatrix}$$

$$\begin{matrix} \frac{1}{2}R_2 \\ \\ -2R_3 \end{matrix} \begin{bmatrix} 1 & 2 & 2 & \vdots & 18 \\ 0 & 1 & -\frac{11}{2} & \vdots & 18 \\ 0 & 0 & 1 & \vdots & -4 \end{bmatrix}$$

$z = -4$ $y - \frac{11}{2}(-4) = 18$ $x + 2(-4) + 2(-4) = 18$

$y + 22 = 18$ $x - 8 - 8 = 18$

$y = -4$ $x - 16 = 18$

$x = 34$

$(34, -4, -4)$

57.
$$\begin{bmatrix} 2 & 0 & 4 & \vdots & 1 \\ 1 & 1 & 3 & \vdots & 0 \\ 1 & 3 & 5 & \vdots & 0 \end{bmatrix}$$

$$\begin{matrix} R_2 \\ R_1 \end{matrix} \begin{bmatrix} 1 & 1 & 3 & \vdots & 0 \\ 2 & 0 & 4 & \vdots & 1 \\ 1 & 3 & 5 & \vdots & 0 \end{bmatrix}$$

$$\begin{matrix} \\ -2R_1 + R_2 \\ -R_1 + R_3 \end{matrix} \begin{bmatrix} 1 & 1 & 3 & \vdots & 0 \\ 0 & -2 & -2 & \vdots & 1 \\ 0 & 2 & 2 & \vdots & 0 \end{bmatrix}$$

$$\begin{matrix} \\ -\frac{1}{2}R_2 \\ \\ \end{matrix} \begin{bmatrix} 1 & 1 & 3 & \vdots & 0 \\ 0 & 1 & 1 & \vdots & -\frac{1}{2} \\ 0 & 2 & 2 & \vdots & 0 \end{bmatrix}$$

$$\begin{matrix} \\ \\ -2R_2 + R_3 \end{matrix} \begin{bmatrix} 1 & 1 & 3 & \vdots & 0 \\ 0 & 1 & 1 & \vdots & -\frac{1}{2} \\ 0 & 0 & 0 & \vdots & 1 \end{bmatrix}$$

Inconsistent; no solution

59.
$$\begin{bmatrix} 1 & 3 & 0 & \vdots & 2 \\ 2 & 6 & 0 & \vdots & 4 \\ 2 & 5 & 4 & \vdots & 3 \end{bmatrix}$$

$$\begin{matrix} \\ -2R_1 + R_2 \\ -2R_1 + R_3 \end{matrix} \begin{bmatrix} 1 & 3 & 0 & \vdots & 2 \\ 0 & 0 & 0 & \vdots & 0 \\ 0 & -1 & 4 & \vdots & -1 \end{bmatrix}$$

$$\begin{matrix} \\ -R_3 \\ R_2 \end{matrix} \begin{bmatrix} 1 & 3 & 0 & \vdots & 2 \\ 0 & 1 & -4 & \vdots & 1 \\ 0 & 0 & 0 & \vdots & 0 \end{bmatrix}$$

let $z = a$

then $y = 1 + 4a$

$x = 2 - 3(1 + 4a)$

$= 2 - 3 - 12a$

$= -1 - 12a$

$(-12a - 1, 1 + 4a, a)$

61.
$$\begin{bmatrix} 2 & 1 & -2 & \vdots & 4 \\ 3 & -2 & 4 & \vdots & 6 \\ -4 & 1 & 6 & \vdots & 12 \end{bmatrix}$$

$$\begin{matrix} \\ -\frac{3}{2}R_1 + R_2 \\ 2R_1 + R_3 \end{matrix} \begin{bmatrix} 2 & 1 & -2 & \vdots & 4 \\ 0 & -\frac{7}{2} & 7 & \vdots & 0 \\ 0 & 3 & 2 & \vdots & 20 \end{bmatrix}$$

$$\begin{matrix} \frac{1}{2}R_1 \\ -\frac{2}{7}R_2 \\ \\ \end{matrix} \begin{bmatrix} 1 & \frac{1}{2} & -1 & \vdots & 2 \\ 0 & 1 & -2 & \vdots & 0 \\ 0 & 3 & 2 & \vdots & 20 \end{bmatrix}$$

$$\begin{matrix} \\ \\ -3R_2 + R_3 \end{matrix} \begin{bmatrix} 1 & \frac{1}{2} & -1 & \vdots & 2 \\ 0 & 1 & -2 & \vdots & 0 \\ 0 & 0 & 8 & \vdots & 20 \end{bmatrix}$$

$$\begin{matrix} \\ \\ \frac{1}{8}R_3 \end{matrix} \begin{bmatrix} 1 & \frac{1}{2} & -1 & \vdots & 2 \\ 0 & 1 & -2 & \vdots & 0 \\ 0 & 0 & 1 & \vdots & \frac{5}{2} \end{bmatrix}$$

$z = \frac{5}{2}$ $y - 2\left(\frac{5}{2}\right) = 0$ $x + \frac{1}{2}(5) - \left(\frac{5}{2}\right) = 2$

$y - 5 = 0$ $x + \frac{5}{2} - \frac{5}{2} = 2$

$y = 5$ $x = 2$

$\left(2, 5, \frac{5}{2}\right)$

63. *Verbal model:*

Money 1 + Money 2 + Money 3 = 1,500,000

0.08 · Money 1 + 0.09 · Money 2 + 0.12 · Money 3 = 113,000

Money 1 = 4 · Money 3

Labels: $x = $ Money 1

$y = $ Money 2

$z = $ Money 3

System of equations:

$$x + y + z = 1,500,000$$
$$0.08x + 0.09y + 0.12y = 133,000$$
$$x = 4z$$

$$\begin{bmatrix} 1 & 1 & 1 & : & 1,500,000 \\ 8 & 9 & 12 & : & 13,300,000 \\ 1 & 0 & -4 & : & 0 \end{bmatrix}$$

$$\begin{array}{c} \\ -8R_1 + R_2 \\ -R_1 + R_3 \end{array} \begin{bmatrix} 1 & 1 & 1 & : & 1,500,000 \\ 0 & 1 & 4 & : & 1,300,000 \\ 0 & -1 & -5 & : & -1,500,000 \end{bmatrix}$$

$$\begin{array}{c} \\ \\ R_2 + R_3 \end{array} \begin{bmatrix} 1 & 1 & 1 & : & 1,500,000 \\ 0 & 1 & 4 & : & 1,300,000 \\ 0 & 0 & -1 & : & -200,000 \end{bmatrix}$$

$$\begin{array}{c} \\ \\ -R_3 \end{array} \begin{bmatrix} 1 & 1 & 1 & : & 1,500,000 \\ 0 & 1 & 4 & : & 1,300,000 \\ 0 & 0 & 1 & : & 200,000 \end{bmatrix}$$

$z = 200,000 \qquad y + 4(200,000) = 1,300,000 \qquad x + 500,000 + 200,000 = 1,500,000$

$$y = 500,000 \qquad\qquad x = 800,000$$

$800,000 at 8%, $500,000 at 90%, $200,000 at 12%

65. *Verbal model:* $0.10 \cdot$ CDs $+ 0.08$ Bonds $+ 0.12$ BC stocks $+ 0.13$ G stocks $= 50,000$

BC stocks + G stocks $= 125,000$

CDs + Bonds $= 375,00$

Labels: $x = $ certificates of deposit

$y = $ municipal bonds

$z = $ blue-chip stocks

$w = $ growth stocks

System of equations:

$$0.10x + 0.08y + 0.12z + 0.13w = 50,000$$
$$z + w = 125,000$$
$$x + y = 375,000$$

$$\begin{bmatrix} 10 & 8 & 12 & 13 & : & 5,000,000 \\ 0 & 0 & 1 & 1 & : & 125,000 \\ 1 & 1 & 0 & 0 & : & 375,000 \end{bmatrix}$$

$$\begin{array}{c} R_1 \\ R_2 \\ \\ \end{array} \begin{bmatrix} 1 & 1 & 0 & 0 & : & 375,000 \\ 0 & 0 & 1 & 1 & : & 125,000 \\ 10 & 8 & 12 & 13 & : & 5,000,000 \end{bmatrix}$$

$$\begin{array}{c} \\ \\ -10R_1 + R_3 \end{array} \begin{bmatrix} 1 & 1 & 0 & 0 & : & 375,000 \\ 0 & 0 & 1 & 1 & : & 125,000 \\ 0 & -2 & 12 & 13 & : & 1,250,000 \end{bmatrix}$$

$$\begin{array}{c} \\ \\ -\frac{1}{2}R_3 \end{array} \begin{bmatrix} 1 & 1 & 0 & 0 & : & 375,000 \\ 0 & 0 & 1 & 1 & : & 125,000 \\ 0 & 1 & -6 & -\frac{13}{2} & : & -625,000 \end{bmatrix}$$

—CONTINUED—

65. —CONTINUED—

$$-R_3 + R_1 \begin{bmatrix} 1 & 0 & 6 & \frac{13}{2} & \vdots & 1,000,000 \\ 0 & 0 & 1 & 1 & \vdots & 125,000 \\ 0 & 1 & -6 & -\frac{13}{2} & \vdots & -625,000 \end{bmatrix}$$

$$\begin{matrix} -6R_2 + R_1 \\ 6R_2 + R_3 \end{matrix} \begin{bmatrix} 1 & 0 & 0 & .5 & \vdots & 250,000 \\ 0 & 0 & 1 & 1 & \vdots & 125,000 \\ 0 & 1 & 0 & -.5 & \vdots & 125,000 \end{bmatrix}$$

so let $w = s$

then $x + .5w = 250,000$ Certificates of deposit: $250,000 - .5s$

$\qquad\qquad x = -.5s + 250,000$ Municipal bonds: $125,000 + .5s$

Blue-chip stocks: $125,000 - s$

$\qquad z + w = 125,000$ Growth stocks: s

$\qquad\qquad z = -s + 125,000$

$\qquad y - .5w = 125,000$

$\qquad\qquad y = .5s + 125,000$

If $s = \$100,000$

CD $= \$200,000$

M Bonds $= \$175,000$

BC Stocks $= \$25,000$

G Stocks $= \$100,000$

67. *Verbal model:*

$$\boxed{\begin{array}{c}\text{Pounds}\\\text{Nut 1}\end{array}} + \boxed{\begin{array}{c}\text{Pounds}\\\text{Nut 2}\end{array}} + \boxed{\begin{array}{c}\text{Pounds}\\\text{Nut 3}\end{array}} = \boxed{\text{50 pounds}}$$

$$\boxed{3.50\ (\text{Nut 1})} + \boxed{4.50\ (\text{Nut 2})} + \boxed{6.00\ (\text{Nut 3})} = \boxed{50(4.95)}$$

$$\boxed{\begin{array}{c}\text{Pounds}\\\text{Nut 1}\end{array}} + \boxed{\begin{array}{c}\text{Pounds}\\\text{Nut 2}\end{array}} = \boxed{\text{25 pounds}}$$

Labels: Pounds Nut 1 $= x$

Pounds Nut 2 $= y$

Pounds Nut 3 $= z$

System of equations:

$$x + y + z = 50$$
$$3.50x + 4.50y + 6.00z = 50(4.95)$$
$$x + y = 25$$

$$\begin{bmatrix} 1 & 1 & 1 & \vdots & 50 \\ 350 & 450 & 600 & \vdots & 24,750 \\ 1 & 1 & 0 & \vdots & 25 \end{bmatrix}$$

$$\begin{matrix} -350R_1 + R_2 \\ -R_1 + R_3 \end{matrix} \begin{bmatrix} 1 & 1 & 1 & \vdots & 50 \\ 0 & 100 & 250 & \vdots & 7250 \\ 0 & 0 & -1 & \vdots & -25 \end{bmatrix}$$

$$\begin{matrix} \frac{1}{100}R_2 \\ -R_3 \end{matrix} \begin{bmatrix} 1 & 1 & 1 & \vdots & 50 \\ 0 & 1 & 2.5 & \vdots & 72.5 \\ 0 & 0 & 1 & \vdots & 25 \end{bmatrix}$$

$z = 25 \qquad y + 2.5(25) = 72.5 \qquad x + 10 + 25 = 50$

$\qquad\qquad\qquad y = 10 \qquad\qquad\qquad x = 15$

15 pounds at \$3.50, 10 pounds at \$4.50, 25 pounds at \$6.00

69. *Verbal model:* $\boxed{\text{Number 1}}$ + $\boxed{\text{Number 2}}$ + $\boxed{\text{Number 3}}$ = $\boxed{33}$

$\boxed{\text{Number 2}}$ = 3 + $\boxed{\text{Number 1}}$

$\boxed{\text{Number 3}}$ = 4 · $\boxed{\text{Number 1}}$

Labels: Number 1 = x

Number 2 = y

Number 3 = z

System of equations: $x + y + z = 33$

$y = 3 + x$

$z = 4x$

$$\begin{bmatrix} 1 & 1 & 1 & \vdots & 33 \\ -1 & 1 & 0 & \vdots & 3 \\ -4 & 0 & 1 & \vdots & 0 \end{bmatrix}$$

$$\begin{matrix} \\ R_1 + R_2 \\ 4R_1 + R_3 \end{matrix} \begin{bmatrix} 1 & 1 & 1 & \vdots & 33 \\ 0 & 2 & 1 & \vdots & 36 \\ 0 & 4 & 5 & \vdots & 132 \end{bmatrix}$$

$$\begin{matrix} \\ \frac{1}{2}R_2 \\ \\ \end{matrix} \begin{bmatrix} 1 & 1 & 1 & \vdots & 33 \\ 0 & 1 & \frac{1}{2} & \vdots & 18 \\ 0 & 4 & 5 & \vdots & 132 \end{bmatrix}$$

$$\begin{matrix} \\ \\ -4R_2 + R_3 \end{matrix} \begin{bmatrix} 1 & 1 & 1 & \vdots & 33 \\ 0 & 1 & \frac{1}{2} & \vdots & 18 \\ 0 & 0 & 3 & \vdots & 60 \end{bmatrix}$$

$$\begin{matrix} \\ \\ \frac{1}{3}R_3 \end{matrix} \begin{bmatrix} 1 & 1 & 1 & \vdots & 33 \\ 0 & 1 & \frac{1}{2} & \vdots & 18 \\ 0 & 0 & 1 & \vdots & 20 \end{bmatrix}$$

$z = 20$ $y + \frac{1}{2}(20) = 18$ $x + 8 + 20 = 33$

$y = 8$ $x = 5$

$(5, 8, 20)$

71. $7 = a(1)^2 + b(1) + c \implies 7 = a + b + c$

$12 = a(2)^2 + b(2) + c \implies 12 = 4a + 2b + c$

$19 = a(3)^2 + b(3) + c \implies 19 = 9a + 3b + c$

$$\begin{bmatrix} 1 & 1 & 1 & \vdots & 7 \\ 4 & 2 & 1 & \vdots & 12 \\ 9 & 3 & 1 & \vdots & 19 \end{bmatrix}$$

$$\begin{matrix} \\ -4R_1 + R_2 \\ -9R_1 + R_3 \end{matrix} \begin{bmatrix} 1 & 1 & 1 & \vdots & 7 \\ 0 & -2 & -3 & \vdots & -16 \\ 0 & -6 & -8 & \vdots & -44 \end{bmatrix}$$

$$\begin{matrix} \\ -\frac{1}{2}R_2 \\ -\frac{1}{2}R_3 \end{matrix} \begin{bmatrix} 1 & 1 & 1 & \vdots & 7 \\ 0 & 1 & \frac{3}{2} & \vdots & 8 \\ 0 & 3 & 4 & \vdots & 22 \end{bmatrix}$$

$$\begin{matrix} -R_2 + R_1 \\ \\ -3R_2 + R_3 \end{matrix} \begin{bmatrix} 1 & 0 & -\frac{1}{2} & \vdots & -1 \\ 0 & 1 & \frac{3}{2} & \vdots & 8 \\ 0 & 0 & -\frac{1}{2} & \vdots & -2 \end{bmatrix}$$

—CONTINUED—

71. —CONTINUED —

$$-2R_3 \begin{bmatrix} 1 & 0 & -\frac{1}{2} & \vdots & -1 \\ 0 & 1 & \frac{3}{2} & \vdots & 8 \\ 0 & 0 & 1 & \vdots & 4 \end{bmatrix}$$

$$\begin{array}{c} \frac{1}{2}R_3 + R_1 \\ -\frac{3}{2}R_3 + R_2 \\ \\ \end{array} \begin{bmatrix} 1 & 0 & 0 & \vdots & 1 \\ 0 & 1 & 0 & \vdots & 2 \\ 0 & 0 & 1 & \vdots & 4 \end{bmatrix}$$

$a = 1, b = 2, c = 4$

$y = x^2 + 2x + 4$

73. $8 = a(1)^2 + b(1) + c \implies 8 = a + b + c$

$2 = a(2)^2 + b(2) + c \implies 2 = 4a + 2b + c$

$-25 = a(3)^2 + b(3) + c \implies -25 = 9a + 3b + c$

$$\begin{bmatrix} 1 & 1 & 1 & \vdots & 8 \\ 4 & 2 & 1 & \vdots & 2 \\ 9 & 3 & 1 & \vdots & -25 \end{bmatrix}$$

$$\begin{array}{c} \\ -4R_1 + R_2 \\ -9R_1 + R_3 \end{array} \begin{bmatrix} 1 & 1 & 1 & \vdots & 8 \\ 0 & -2 & -3 & \vdots & -30 \\ 0 & -6 & -81 & \vdots & -97 \end{bmatrix}$$

$$\begin{array}{c} \\ -\frac{1}{2}R_2 \\ -\frac{1}{2}R_3 \end{array} \begin{bmatrix} 1 & 1 & 1 & \vdots & 8 \\ 0 & 1 & 1.5 & \vdots & 15 \\ 0 & 3 & 4 & \vdots & 48.5 \end{bmatrix}$$

$$\begin{array}{c} -R_2 + R_1 \\ \\ -3R_2 + R_3 \end{array} \begin{bmatrix} 1 & 0 & -0.5 & \vdots & -7 \\ 0 & 1 & 1.5 & \vdots & 15 \\ 0 & 0 & -0.5 & \vdots & 3.5 \end{bmatrix}$$

$$\begin{array}{c} \\ \\ -2R_3 \end{array} \begin{bmatrix} 1 & 0 & -0.5 & \vdots & -7 \\ 0 & 1 & 1.5 & \vdots & 15 \\ 0 & 0 & 1 & \vdots & -7 \end{bmatrix}$$

$z = -7 \qquad y + 1.5(-7) = 15 \qquad x + -0.5(-7) = -7$

$\qquad\qquad\qquad\qquad y = 25.5 \qquad\qquad\qquad x = -10.5$

$y = -10.5x^2 + 25.5x - 7$

75. $1^2 + 1^2 + D(1) + E(1) + F = 0 \implies D + E + F = -2$

$3^2 + 3^2 + D(3) + E(3) + F = 0 \implies 3D + 3E + F = -18$

$4^2 + 2^2 + D(4) + E(2) + F = 0 \implies 4D + 2E + F = -20$

$$\begin{bmatrix} 1 & 1 & 1 & \vdots & -2 \\ 3 & 3 & 1 & \vdots & -18 \\ 4 & 2 & 1 & \vdots & -20 \end{bmatrix}$$

$$\begin{array}{c} \\ -3R_1 + R_2 \\ -4R_1 + R_3 \end{array} \begin{bmatrix} 1 & 1 & 1 & \vdots & -2 \\ 0 & 0 & -2 & \vdots & -12 \\ 0 & -2 & -3 & \vdots & -12 \end{bmatrix}$$

—CONTINUED —

75. —CONTINUED —

$$\begin{matrix} R_2 \\ R_3 \end{matrix} \begin{bmatrix} 1 & 1 & 1 & \vdots & -2 \\ 0 & -2 & -3 & \vdots & -12 \\ 0 & 0 & -2 & \vdots & -12 \end{bmatrix}$$

$$\begin{matrix} \\ -\frac{1}{2}R_2 \\ -\frac{1}{2}R_3 \end{matrix} \begin{bmatrix} 1 & 1 & 1 & \vdots & -2 \\ 0 & 1 & \frac{3}{2} & \vdots & 6 \\ 0 & 0 & 1 & \vdots & 6 \end{bmatrix}$$

$F = 6 \qquad E + \frac{3}{2}(6) = 6 \qquad D + (-3) + 6 = -2$

$\qquad\qquad\qquad E + 9 = 6 \qquad\qquad D + 3 = -2$

$\qquad\qquad\qquad E = -3 \qquad\qquad\qquad D = -5$

$x^2 + y^2 - 5x - 3y + 6 = 0$

77. (a) $\quad 6 = a(0)^2 + b(0) + c \implies \quad 6 = \qquad\qquad c$

$\qquad 18.5 = a(25)^2 + b(25) + c \implies 18.5 = 625a + 25b + c$

$\qquad 26 = a(50)^2 + b(50) + c \implies \quad 26 = 2500a + 50b + c$

$$\begin{bmatrix} 0 & 0 & 1 & \vdots & 6 \\ 625 & 25 & 1 & \vdots & 18.5 \\ 2500 & 50 & 1 & \vdots & 26 \end{bmatrix}$$

$$\begin{matrix} R_1 \\ R_2 \\ R_3 \end{matrix} \begin{bmatrix} 625 & 25 & 1 & \vdots & 18.5 \\ 2500 & 50 & 1 & \vdots & 26 \\ 0 & 0 & 1 & \vdots & 6 \end{bmatrix}$$

$$\frac{1}{625}R_1 \begin{bmatrix} 1 & 0.04 & 0.0016 & \vdots & 0.0296 \\ 2500 & 50 & 1 & \vdots & 26 \\ 0 & 0 & 1 & \vdots & 6 \end{bmatrix}$$

$$-2500R_1 + R_2 \begin{bmatrix} 1 & 0.04 & 0.0016 & \vdots & 0.0296 \\ 0 & -50 & -3 & \vdots & -48 \\ 0 & 0 & 1 & \vdots & 6 \end{bmatrix}$$

$$-\frac{1}{50}R_2 \begin{bmatrix} 1 & 0.04 & 0.0016 & \vdots & 0.0296 \\ 0 & 1 & 0.06 & \vdots & 0.96 \\ 0 & 0 & 1 & \vdots & 6 \end{bmatrix}$$

$$-.04R_2 + R_1 \begin{bmatrix} 1 & 0 & -0.0008 & \vdots & -0.0088 \\ 0 & 1 & 0.06 & \vdots & 0.96 \\ 0 & 0 & 1 & \vdots & 6 \end{bmatrix}$$

$$\begin{matrix} .0008R_3 + R_1 \\ -.06R_3 + R_2 \end{matrix} \begin{bmatrix} 1 & 0 & 0 & \vdots & -0.004 \\ 0 & 1 & 0 & \vdots & 0.6 \\ 0 & 0 & 1 & \vdots & 6 \end{bmatrix}$$

$\quad$ so $a = -0.004$

$\qquad b = 0.6$

$\qquad c = 6$

$\quad y = -0.004x^2 + 0.6x + 6$

(b) *Keystrokes:*

$\boxed{Y=}\ \boxed{(-)}\ .004\ \boxed{X,T,\theta}\ \boxed{x^2}\ \boxed{+}\ .6\ \boxed{X,T,\theta}\ \boxed{+}\ 6\ \boxed{GRAPH}$

(c) Maximum height $= 28.5$ feet

$\qquad$ Point at which the ball struck the ground $= 159.4$ feet

79.

$$\begin{bmatrix} 4 & -2 & 1 & \vdots & 0 \\ -4 & 1 & 0 & \vdots & -9 \\ 1 & 0 & 0 & \vdots & 2 \end{bmatrix}$$

$$\begin{matrix} R_1 \\ \\ R_3 \end{matrix} \begin{bmatrix} 1 & 0 & 0 & 2 \\ -4 & 1 & 0 & -9 \\ 4 & -2 & 1 & 0 \end{bmatrix}$$

$$\begin{matrix} \\ 4R_1 + R_2 \\ -4R_1 + R_2 \end{matrix} \begin{bmatrix} 1 & 0 & 0 & \vdots & 2 \\ 0 & 1 & 0 & \vdots & -1 \\ 0 & -2 & 1 & \vdots & -8 \end{bmatrix}$$

$$\begin{matrix} \\ \\ 2R_2 + R_3 \end{matrix} \begin{bmatrix} 1 & 0 & 0 & \vdots & 2 \\ 0 & 1 & 0 & \vdots & -1 \\ 0 & 0 & 1 & \vdots & -10 \end{bmatrix}$$

$$\frac{2x^2 - 9x}{(x-2)^3} = \frac{2}{x-2} - \frac{1}{(x-2)^2} - \frac{10}{(x-2)^3}$$

81. (a) Interchange two rows.

(b) Multiply a row by a nonzero constant.

(c) Add a multiple of a row to another row.

83. The one matrix can be obtained from the other by using the elementary row operations.

85. There will be a row in the matrix with all zero entries except in the last column.

Section 7.5   Determinants and Linear Systems

1. $\det(A) = \begin{vmatrix} 2 & 1 \\ 3 & 4 \end{vmatrix} = 2(4) - 3(1) = 8 - 3 = 5$

3. $\det(A) = \begin{vmatrix} 5 & 2 \\ -5 & 3 \end{vmatrix} = 5(3) - (-6)(2) = 15 + 12 = 27$

5. $\det(A) = \begin{vmatrix} 5 & -4 \\ -10 & 8 \end{vmatrix} = 5(8) - (-10)(-4)$

$= 40 - 40 = 0$

7. $\det(A) = \begin{vmatrix} 2 & 6 \\ 0 & 3 \end{vmatrix} = 2(3) - 0(6) = 6 - 0 = 6$

9. $\det(A) = \begin{vmatrix} -7 & 3 \\ \frac{1}{2} & 6 \end{vmatrix} = (-7)(3) - (\frac{1}{2})(6)$

$= -21 - 3 = -24$

11. $\det(A) = \begin{vmatrix} 0.3 & 0.5 \\ 0.5 & 0.3 \end{vmatrix} = (0.3)(0.3) - (0.5)(0.5)$

$= .09 - .25 = -0.16$

13. $\det(A) = \begin{vmatrix} 2 & 3 & -1 \\ 6 & 0 & 0 \\ 4 & 1 & 1 \end{vmatrix}$

$= -(6) \begin{vmatrix} 3 & -1 \\ 1 & 1 \end{vmatrix} + 0 + 0 \text{ (second row)}$

$= (-6)(4)$

$= -24$

15. $\det(A) = \begin{vmatrix} 1 & 1 & 2 \\ 3 & 1 & 0 \\ -2 & 0 & 3 \end{vmatrix}$

$= (2)\begin{vmatrix} 3 & 1 \\ -2 & 0 \end{vmatrix} - (0)\begin{vmatrix} 1 & 1 \\ -2 & 0 \end{vmatrix} + (3)\begin{vmatrix} 1 & 1 \\ 3 & 1 \end{vmatrix}$ (third column)

$= (2)(2) - 0 + (3)(-2)$

$= 4 - 6 = -2$

17. $\det(A) = \begin{vmatrix} 2 & 4 & 6 \\ 0 & 3 & 1 \\ 0 & 0 & -5 \end{vmatrix}$

$= (2)\begin{vmatrix} 3 & 1 \\ 0 & -5 \end{vmatrix} - 0 + 0$ (first column)

$= (2)(-15) = -30$

19. $\det(A) = \begin{vmatrix} -2 & 2 & 3 \\ 1 & -1 & 0 \\ 0 & 1 & 4 \end{vmatrix}$

$= -(1)\begin{vmatrix} 2 & 3 \\ 1 & 4 \end{vmatrix} + (-1)\begin{vmatrix} -2 & 3 \\ 0 & 4 \end{vmatrix} - 0$ (second row)

$= (-1)(5) + (-1)(-8)$

$= -5 + 8 = 3$

21. $\det(A) = \begin{vmatrix} 1 & 4 & -2 \\ 3 & 6 & -6 \\ -2 & 1 & 4 \end{vmatrix}$

$= (1)\begin{vmatrix} 6 & -6 \\ 1 & 4 \end{vmatrix} - (4)\begin{vmatrix} 3 & -6 \\ -2 & 4 \end{vmatrix} + (-2)\begin{vmatrix} 3 & 6 \\ -2 & 1 \end{vmatrix}$ (first row)

$= (1)(30) - (4)(0) + (-2)(15)$

$= 30 - 0 - 30 = 0$

23. $\det(A) = \begin{vmatrix} -3 & 2 & 1 \\ 4 & 5 & 6 \\ 2 & 3 & 1 \end{vmatrix}$

$= (1)\begin{vmatrix} 4 & 5 \\ 2 & -3 \end{vmatrix} - (6)\begin{vmatrix} -3 & 2 \\ 2 & -3 \end{vmatrix} + (1)\begin{vmatrix} -3 & 2 \\ 4 & 5 \end{vmatrix}$ (third column)

$= (1)(-22) - (6)(5) + (1)(-3)$

$= -22 - 30 - 23 = -75$

25. $\det(A) = \begin{vmatrix} 1 & 4 & -2 \\ 3 & 2 & 0 \\ -1 & 4 & 3 \end{vmatrix}$

$= -(3)\begin{vmatrix} 4 & -2 \\ 4 & 3 \end{vmatrix} + (2)\begin{vmatrix} 1 & -2 \\ -1 & 3 \end{vmatrix} - 0$ (second row)

$= (-3)(20) + (2)(1)$

$= -60 + 2$

$= -58$

27. $\det(A) = \begin{vmatrix} 2 & -5 & 0 \\ 4 & 7 & 0 \\ -7 & 25 & 3 \end{vmatrix}$

$= 0 - 0 + 3\begin{vmatrix} 2 & -5 \\ 4 & 7 \end{vmatrix}$ (third column)

$= (3)(34)$

$= 102$

29. $\det(A) = \begin{vmatrix} 0.1 & 0.2 & 0.3 \\ -0.3 & 0.2 & 0.2 \\ 5 & 4 & 4 \end{vmatrix}$

$= (5)\begin{vmatrix} 0.2 & 0.3 \\ 0.2 & 0.2 \end{vmatrix} - (4)\begin{vmatrix} 0.1 & 0.3 \\ -0.3 & 0.2 \end{vmatrix} + (4)\begin{vmatrix} 0.1 & 0.2 \\ -0.3 & 0.2 \end{vmatrix}$ (third row)

$= (5)(-0.02) - (4)(0.11) + (4)(0.08)$

$= -0.1 - 0.44 + 0.32$

$= -0.22$

31. $\det(A) = \begin{vmatrix} x & y & 1 \\ 3 & 1 & 1 \\ -2 & 0 & 1 \end{vmatrix}$

$= (-2)\begin{vmatrix} y & 1 \\ 1 & 1 \end{vmatrix} - 0 + (1)\begin{vmatrix} x & y \\ 3 & 1 \end{vmatrix}$ (third row)

$= (-2)(y - 1) + (1)(x - 3y)$

$= -2y + 2 + x - 3y$

$= x - 5y + 2$

33. *Keystrokes:*

[MATRX] [EDIT 1] 3 [ENTER] 3 [ENTER] 5 [ENTER] [(−)] 3 [ENTER] 2 [ENTER] 7 [ENTER] 5 [ENTER] [(−)] 7 [ENTER]

0 [ENTER] 6 [ENTER] [(−)] 1 [ENTER] [QUIT]

[MATRX] [MATH 1] [MATRX 1] [ENTER]

Solution is 248.

35. *Keystrokes:*

[MATRX] [EDIT 1] 3 [ENTER] 3 [ENTER] 3 [ENTER] [(−)] 1 [ENTER]

2 [ENTER] 1 [ENTER] [(−)] 1 [ENTER] 2 [ENTER] [(−)] 2 [ENTER] 3 [ENTER] 10 [ENTER] [QUIT]

[MATRX] [MATH 1] [MATRX 1] [ENTER]

Solution is −32.

37. *Keystrokes:*

[MATRX] [EDIT 1] 3 [ENTER] 3 [ENTER] .2 [ENTER] .8 [ENTER] [(−)] [ENTER] .3

.1 [ENTER] .8 [ENTER] .6 [ENTER] [(−)] 10 [ENTER] [(−)] 5 [ENTER]

1 [ENTER] [QUIT]

[MATRX] [MATH 1] [MATRX 1] [ENTER]

Solution is −6.37

39. $\begin{bmatrix} 1 & 2 & \vdots & 5 \\ -1 & 1 & \vdots & 1 \end{bmatrix}$

$$D = \begin{vmatrix} 1 & 2 \\ -1 & 1 \end{vmatrix} = 1 - (-2) = 3$$

$$x = \frac{D_x}{D} = \frac{\begin{vmatrix} 5 & 2 \\ 1 & 1 \end{vmatrix}}{3} = \frac{5 - 2}{3} = \frac{3}{3} = 1$$

$$y = \frac{D_y}{D} = \frac{\begin{vmatrix} 1 & 5 \\ -1 & 1 \end{vmatrix}}{3} = \frac{1 - (-5)}{3} = \frac{6}{3} = 2$$

$(1, 2)$

41. $\begin{bmatrix} 3 & 4 & \vdots & -2 \\ 5 & 3 & \vdots & 4 \end{bmatrix}$

$$D = \begin{vmatrix} 3 & 4 \\ 5 & 3 \end{vmatrix} = 9 - 20 = -11$$

$$x = \frac{D_x}{D} = \frac{\begin{vmatrix} -2 & 4 \\ 4 & 3 \end{vmatrix}}{-11} = \frac{-6 - 16}{-11} = \frac{-22}{-11} = 2$$

$$y = \frac{D_y}{D} = \frac{\begin{vmatrix} 3 & -2 \\ 5 & 4 \end{vmatrix}}{-11} = \frac{12 - (-10)}{-11} = \frac{22}{-11} = -2$$

$(2, -2)$

43. $\begin{bmatrix} 20 & 8 & \vdots & 11 \\ 12 & -24 & \vdots & 21 \end{bmatrix}$

$$D = \begin{vmatrix} 20 & 8 \\ 12 & -24 \end{vmatrix} = -480 - 96 = -576$$

$$x = \frac{D_x}{D} = \frac{\begin{vmatrix} 11 & 8 \\ 21 & -24 \end{vmatrix}}{-576} = \frac{-264 - 168}{-576} = \frac{-432}{-576} = \frac{3}{4}$$

$$y = \frac{D_y}{D} = \frac{\begin{vmatrix} 20 & 11 \\ 12 & 21 \end{vmatrix}}{-576} = \frac{420 - 132}{-576} = \frac{288}{-576} = -\frac{1}{2}$$

$\left(\frac{3}{4}, -\frac{1}{2} \right)$

45. $\begin{bmatrix} -0.4 & 0.8 & \vdots & 1.6 \\ 2 & -4 & \vdots & 5 \end{bmatrix}$

$$D = \begin{vmatrix} -0.4 & 0.8 \\ 2 & -4 \end{vmatrix} = 1.6 - 1.6 = 0$$

Cannot be solved by Cramer's Rule because $D = 0$.

Solve by elimination.

$$-4x + 8y = 16 \implies -4x + 8y = 16$$
$$2x - 4y = 5 \implies \underline{4x - 8y = 10}$$
$$0 \neq 26$$

Inconsistent; no solution

47. $\begin{bmatrix} 3 & 6 & \vdots & 5 \\ 6 & 14 & \vdots & 11 \end{bmatrix}$

$$D = \begin{vmatrix} 3 & 6 \\ 6 & 14 \end{vmatrix} = 42 - 36 = 6$$

$$x = \frac{D_x}{D} = \frac{\begin{vmatrix} 5 & 6 \\ 11 & 14 \end{vmatrix}}{6} = \frac{70 - 66}{6} = \frac{4}{6} = \frac{2}{3}$$

$$y = \frac{D_y}{D} = \frac{\begin{vmatrix} 3 & 5 \\ 6 & 11 \end{vmatrix}}{6} = \frac{33 - 30}{6} = \frac{3}{6} = \frac{1}{2}$$

$\left(\frac{2}{3}, \frac{1}{2} \right)$

49. $\begin{bmatrix} 4 & -1 & 1 & \vdots & -5 \\ 2 & 2 & 3 & \vdots & 10 \\ 5 & -2 & 6 & \vdots & 1 \end{bmatrix}$

$$D = \begin{vmatrix} 4 & -1 & 1 \\ 2 & 2 & 3 \\ 5 & -2 & 6 \end{vmatrix} = (1)\begin{vmatrix} 2 & 2 \\ 5 & -2 \end{vmatrix} - (3)\begin{vmatrix} 4 & -1 \\ 5 & -2 \end{vmatrix} + (6)\begin{vmatrix} 4 & -1 \\ 2 & 2 \end{vmatrix}$$

$$= (1)(-14) + (-3)(-3) + (6)(10)$$

$$= -14 + 9 + 60 = 55$$

—CONTINUED—

49. —CONTINUED—

$$x = \frac{\begin{vmatrix} -5 & -1 & 1 \\ 10 & 2 & 3 \\ 1 & -2 & 6 \end{vmatrix}}{55} = \frac{(1)\begin{vmatrix} 10 & 2 \\ 1 & -2 \end{vmatrix} - (3)\begin{vmatrix} -5 & -1 \\ 1 & -2 \end{vmatrix} + (6)\begin{vmatrix} -5 & -1 \\ 10 & 2 \end{vmatrix}}{55}$$

$$= \frac{(1)(-22) + (-3)(11) + (6)(0)}{55}$$

$$= \frac{-22 - 33}{55} = \frac{-55}{55} = -1$$

$$y = \frac{\begin{vmatrix} 4 & -5 & 1 \\ 2 & 10 & 3 \\ 5 & 1 & 6 \end{vmatrix}}{55} = \frac{(1)\begin{vmatrix} 2 & 10 \\ 5 & 1 \end{vmatrix} - (3)\begin{vmatrix} 4 & -5 \\ 5 & 1 \end{vmatrix} + (6)\begin{vmatrix} 4 & -5 \\ 2 & 10 \end{vmatrix}}{55}$$

$$= \frac{(1)(-48) + (-3)(29) + (6)(50)}{55}$$

$$= \frac{-48 - 87 + 300}{55} = \frac{165}{55} = 3$$

$$z = \frac{\begin{vmatrix} 4 & -1 & -5 \\ 2 & 2 & 10 \\ 5 & -2 & 1 \end{vmatrix}}{55} = \frac{(5)\begin{vmatrix} -1 & -5 \\ 2 & 10 \end{vmatrix} - (-2)\begin{vmatrix} 4 & -5 \\ 2 & 10 \end{vmatrix} + (1)\begin{vmatrix} 4 & -1 \\ 2 & 2 \end{vmatrix}}{55}$$

$$= \frac{(5)(0) + (2)(50) + (1)(10)}{55}$$

$$= \frac{0 + 100 + 10}{55} = \frac{110}{55} = 2$$

$(-1, 3, 2)$

51. $\begin{bmatrix} 3 & 4 & 4 & \vdots & 11 \\ 4 & -4 & 6 & \vdots & 11 \\ 6 & -6 & 0 & \vdots & 3 \end{bmatrix}$

$$D = \begin{vmatrix} 3 & 4 & 4 \\ 4 & -4 & 3 \\ 6 & -6 & 0 \end{vmatrix} = (4)\begin{vmatrix} 4 & -4 \\ 6 & -6 \end{vmatrix} - (6)\begin{vmatrix} 3 & 4 \\ 6 & -6 \end{vmatrix} + 0$$

$$= (4)(0) - (6)(-42) + 0$$

$$= 252$$

$$x = \frac{\begin{vmatrix} 11 & 4 & 4 \\ 11 & -4 & 6 \\ 3 & -6 & 0 \end{vmatrix}}{252} = \frac{(4)\begin{vmatrix} 11 & -4 \\ 3 & -6 \end{vmatrix} - (6)\begin{vmatrix} 11 & 4 \\ 3 & -6 \end{vmatrix} + 0}{252}$$

$$= \frac{(4)(-54) - (6)(-78)}{252}$$

$$= \frac{-216 - 468}{252} = \frac{252}{252} = 1$$

—CONTINUED—

51. —CONTINUED—

$$y = \frac{\begin{vmatrix} 3 & 11 & 4 \\ 4 & 11 & 6 \\ 6 & 3 & 0 \end{vmatrix}}{252} = \frac{(4)\begin{vmatrix} 4 & 11 \\ 6 & 3 \end{vmatrix} - (6)\begin{vmatrix} 3 & 11 \\ 6 & 3 \end{vmatrix} + 0}{252}$$

$$= \frac{(4)(-54) - (6)(-57)}{252}$$

$$= \frac{-216 + 342}{252} = \frac{126}{252} = \frac{1}{2}$$

$$z = \frac{\begin{vmatrix} 3 & 4 & 11 \\ 4 & -4 & 11 \\ 6 & -6 & 3 \end{vmatrix}}{252} = \frac{(3)\begin{vmatrix} -4 & 11 \\ -6 & 3 \end{vmatrix} - (4)\begin{vmatrix} 4 & 11 \\ -6 & 3 \end{vmatrix} + (6)\begin{vmatrix} 4 & 11 \\ -4 & 11 \end{vmatrix}}{252}$$

$$= \frac{(3)(54) - (4)(78) + (6)(88)}{252}$$

$$= \frac{162 - 312 + 528}{252} = \frac{378}{252} = \frac{3}{2}$$

$$\left(1, \frac{1}{2}, \frac{3}{2}\right)$$

53. $\begin{bmatrix} 3 & 3 & 4 & \vdots & 1 \\ 3 & 5 & 9 & \vdots & 2 \\ 5 & 9 & 14 & \vdots & 4 \end{bmatrix}$

$$D = \begin{vmatrix} 3 & 3 & 4 \\ 3 & 5 & 9 \\ 5 & 9 & 17 \end{vmatrix} = (3)\begin{vmatrix} 5 & 9 \\ 9 & 17 \end{vmatrix} - (3)\begin{vmatrix} 3 & 4 \\ 9 & 17 \end{vmatrix} + (5)\begin{vmatrix} 3 & 4 \\ 5 & 9 \end{vmatrix}$$

$$= (3)(4) - (3)(15) + (5)(7)$$

$$= 12 - 45 + 35$$

$$= 2$$

$$a = \frac{\begin{vmatrix} 1 & 3 & 4 \\ 2 & 5 & 9 \\ 4 & 9 & 17 \end{vmatrix}}{2} = \frac{(1)\begin{vmatrix} 5 & 9 \\ 9 & 17 \end{vmatrix} - (2)\begin{vmatrix} 3 & 4 \\ 9 & 17 \end{vmatrix} + (4)\begin{vmatrix} 3 & 4 \\ 5 & 9 \end{vmatrix}}{2}$$

$$= \frac{(1)(4) - (2)(15) + (4)(7)}{2}$$

$$= \frac{4 - 30 + 28}{2} = \frac{2}{2} = 1$$

$$b = \frac{\begin{vmatrix} 3 & 1 & 4 \\ 3 & 2 & 9 \\ 5 & 4 & 17 \end{vmatrix}}{2} = \frac{(3)\begin{vmatrix} 2 & 9 \\ 4 & 17 \end{vmatrix} - (2)\begin{vmatrix} 3 & 9 \\ 5 & 17 \end{vmatrix} + (4)\begin{vmatrix} 3 & 2 \\ 5 & 4 \end{vmatrix}}{2}$$

$$= \frac{(3)(-2) - (1)(6) + (4)(2)}{2}$$

$$= \frac{-6 - 6 + 8}{2} = \frac{-4}{2} = -2$$

—CONTINUED—

53. —CONTINUED—

$$c = \frac{\begin{vmatrix} 3 & 3 & 1 \\ 3 & 5 & 2 \\ 5 & 9 & 4 \end{vmatrix}}{2} = \frac{(1)\begin{vmatrix} 3 & 5 \\ 5 & 9 \end{vmatrix} - (2)\begin{vmatrix} 3 & 3 \\ 5 & 9 \end{vmatrix} + (4)\begin{vmatrix} 3 & 3 \\ 3 & 5 \end{vmatrix}}{2}$$

$$= \frac{(1)(2) - (2)(12) + (4)(6)}{2}$$

$$= \frac{2 - 24 + 24}{2} = \frac{2}{2} = 1$$

$(1, -2, 1)$

55. $\begin{bmatrix} 5 & -3 & 2 & \vdots & 2 \\ 2 & 2 & -3 & \vdots & 3 \\ 1 & -7 & 8 & \vdots & -4 \end{bmatrix}$

$$D = \begin{vmatrix} 5 & -3 & 2 \\ 2 & 2 & -3 \\ 1 & 7 & 8 \end{vmatrix} = (5)\begin{vmatrix} 2 & -3 \\ -7 & 8 \end{vmatrix} - (2)\begin{vmatrix} -3 & 2 \\ -7 & 8 \end{vmatrix} + (1)\begin{vmatrix} -3 & 2 \\ 2 & -3 \end{vmatrix}$$

$$= (5)(-5) - (2)(-10) + (1)(5)$$

$$= -25 + 20 + 5$$

$$= 0$$

Cannot be solved by Cramer's Rule because $D = 0$.

57. $\begin{bmatrix} -3 & 10 & \vdots & 22 \\ 9 & -3 & \vdots & 0 \end{bmatrix}$

$$D = \begin{vmatrix} -3 & 10 \\ 9 & -3 \end{vmatrix} = -81$$

$$x = \frac{D_x}{D} = \frac{\begin{vmatrix} 22 & 10 \\ 0 & -3 \end{vmatrix}}{-81} = \frac{-66}{-81} = \frac{22}{27}$$

$$y = \frac{D_y}{D} = \frac{\begin{vmatrix} -3 & 22 \\ 9 & 0 \end{vmatrix}}{-81} = \frac{-198}{-81} = \frac{22}{9}$$

Keystrokes:

det D

[MATRX] [EDIT 1] 2 [ENTER] 2 [ENTER]

Enter each number in matrix followed by [ENTER]

[QUIT]

[MATRX] [MATH 1] [MATRX 1] [ENTER]

det D_x

[MATRX] [EDIT 2] 2 [ENTER] 2 [ENTER]

Enter each number in matrix followed by [ENTER]
[QUIT]

[MATRX] [MATH 1] [MATRX 2] [ENTER]

det D_y

[MATRX] [EDIT 3] 2 [ENTER] 2 [ENTER]

Enter each number in matrix followed by [ENTER]
[QUIT]

[MATRX] [MATH 1] [MATRX 3] [ENTER]

59. $D = \begin{vmatrix} 3 & -2 & 3 \\ 1 & 3 & 6 \\ 1 & 2 & 9 \end{vmatrix} = 48$

$x = \dfrac{D_x}{D} = \dfrac{\begin{vmatrix} 8 & -2 & 3 \\ -3 & 3 & 6 \\ -5 & 2 & 9 \end{vmatrix}}{48} = \dfrac{153}{48} = \dfrac{51}{16}$

$y = \dfrac{D_y}{D} = \dfrac{\begin{vmatrix} 3 & 8 & 3 \\ 1 & -3 & 6 \\ 1 & -5 & 9 \end{vmatrix}}{48} = \dfrac{-21}{48} = \dfrac{-7}{16}$

$z = \dfrac{D_z}{D} = \dfrac{\begin{vmatrix} 3 & -2 & 8 \\ 1 & 3 & -3 \\ 1 & 2 & -5 \end{vmatrix}}{48} = \dfrac{-39}{48} = -\dfrac{13}{16}$

$\left(\dfrac{51}{16}, -\dfrac{7}{16}, -\dfrac{13}{16} \right)$

Keystrokes:

det D [MATRX] [EDIT 1] 3 [ENTER] 3 [ENTER]

     Enter each number in matrix followed by [ENTER].

     [QUIT]

     [MATRX] [MATH 1] [MATRX 1] [ENTER]

det D_x [MATRX] [EDIT 2] 3 [ENTER] 3 [ENTER]

     Enter each number in matrix followed by [ENTER].

     [QUIT]

     [MATRX] [MATH 1] [MATRX 2] [ENTER]

det D_y [MATRX] [EDIT 3] 3 [ENTER] 3 [ENTER]

     Enter each number in matrix followed by [ENTER].

     [QUIT]

     [MATRX] [MATH 1] [MATRX 3] [ENTER]

det D_z [MATRX] [EDIT 4] 3 [ENTER] 3 [ENTER]

     Enter each number in matrix followed by [ENTER].

     [QUIT]

     [MATRX] [MATH 1] [MATRX 4] [ENTER]

61. $(5 - x)(2 - x) - 4 = 0$

$10 - 7x + x^2 - 4 = 0$

$x^2 - 7x + 6 = 0$

$(x - 6)(x - 1) = 0$

$x = 6 \quad x = 1$

63. $(x_1, y_1) = (0, 3), (x_2, y_2) = (4, 0), (x_3, y_3) = (8, 5)$

$\begin{vmatrix} x_1 & y_1 & 1 \\ x_2 & y_2 & 1 \\ x_3 & y_3 & 1 \end{vmatrix} = \begin{vmatrix} 0 & 3 & 1 \\ 4 & 0 & 1 \\ 8 & 5 & 1 \end{vmatrix} = 32$

Area $= +\dfrac{1}{2}(32) = 16$

65. $(x_1, y_1) = (0, 0), (x_2, y_2) = (3, 1), (x_3, y_3) = (1, 5)$

$\begin{vmatrix} x_1 & y_1 & 1 \\ x_2 & y_2 & 1 \\ x_3 & y_3 & 1 \end{vmatrix} = \begin{vmatrix} 0 & 0 & 1 \\ 3 & 1 & 1 \\ 1 & 5 & 1 \end{vmatrix} = (1)\begin{vmatrix} 3 & 1 \\ 1 & 5 \end{vmatrix}$

$= (1)(14) = 14$

Area $= +\dfrac{1}{2}(14) = 7$

67. $(x_1, y_1) = (-2, 1), (x_2, y_2) = (3, -1), (x_3, y_3) = (1, 6)$

$\begin{vmatrix} x_1 & y_1 & 1 \\ x_2 & y_2 & 1 \\ x_3 & y_3 & 1 \end{vmatrix} = \begin{vmatrix} -2 & 1 & 1 \\ 3 & -1 & 1 \\ 1 & 6 & 1 \end{vmatrix} = (1)\begin{vmatrix} 3 & -1 \\ 1 & 6 \end{vmatrix} - (1)\begin{vmatrix} -2 & 1 \\ 1 & 6 \end{vmatrix} + (1)\begin{vmatrix} -2 & 1 \\ 3 & -1 \end{vmatrix}$

$= (1)(19) - (1)(-13) + (1)(-1)$

$= 19 + 13 - 1$

$= 31$

Area $= +\dfrac{1}{2}(31) = \dfrac{31}{2}$ or $15\dfrac{1}{2}$

69. $(x_1, y_1) = \left(0, \frac{1}{2}\right)$ $(x_2, y_2) = \left(\frac{5}{2}, 0\right)$ $(x_3, y_3) = (4, 3)$

$$\begin{vmatrix} x_1 & y_1 & 1 \\ x_2 & y_2 & 1 \\ x_3 & y_3 & 1 \end{vmatrix} = \begin{vmatrix} 0 & \frac{1}{2} & 1 \\ \frac{5}{2} & 0 & 1 \\ 4 & 3 & 1 \end{vmatrix} = 0\begin{vmatrix} 0 & 1 \\ 3 & 1 \end{vmatrix} - \frac{1}{2}\begin{vmatrix} \frac{5}{2} & 1 \\ 4 & 1 \end{vmatrix} + 1\begin{vmatrix} \frac{5}{2} & 0 \\ 4 & 3 \end{vmatrix}$$

$$= 0 - \frac{1}{2}\left(\frac{5}{2} - 4\right) + 1\left(\frac{15}{2} - 0\right)$$

$$= -\frac{1}{2}\left(-\frac{3}{2}\right) + 1\left(\frac{15}{2}\right)$$

$$= \frac{3}{4} + \frac{15}{2}$$

$$= \frac{3}{4} + \frac{30}{4}$$

$$= \frac{33}{4}$$

Area $= \frac{1}{2}\left(\frac{33}{4}\right) = \frac{33}{8}$

71. Verbal model:

| Area of Shaded Region | = | Area of Triangle 1 | + | Area of Triangle 2 |

Equation: $A = 11.5 + 4.5$

$= 16$

Let $(x_1, y_1) = (-1, 2)$ $(x_2, y_2) = (4, 0)$ $(x_3, y_3) = (3, 5)$

$$\begin{vmatrix} x_1 & y_1 & 1 \\ x_2 & y_2 & 1 \\ x_3 & y_3 & 1 \end{vmatrix} = \begin{vmatrix} -1 & 2 & 1 \\ 4 & 0 & 1 \\ 3 & 5 & 1 \end{vmatrix} = -4\begin{vmatrix} 2 & 1 \\ 5 & 1 \end{vmatrix} + 0 - 1\begin{vmatrix} -1 & 2 \\ 3 & 5 \end{vmatrix}$$

$$= -4(-3) - 1(-11) = 12 + 11 = 23$$

Area $= \frac{1}{2}(23) = 11.5$

Let $(x_1, y_1) = (3, 5)$ $(x_2, y_2) = (4, 0)$ $(x_3, y_3) = (5, 4)$

$$\begin{vmatrix} x_1 & y_1 & 1 \\ x_2 & y_2 & 1 \\ x_3 & y_3 & 1 \end{vmatrix} = \begin{vmatrix} 3 & 5 & 1 \\ 4 & 0 & 1 \\ 5 & 4 & 1 \end{vmatrix} = -4\begin{vmatrix} 5 & 1 \\ 4 & 1 \end{vmatrix} + 0 - 1\begin{vmatrix} 3 & 5 \\ 5 & 4 \end{vmatrix}$$

$$= -4(1) - 1(-13) = -4 + 13 = 9$$

Area $= \frac{1}{2}(9) = 4.5$

73. *Verbal Model:*

| Area of Shaded Region | = | Area of Rectangle | − | Area of Triangle |

Equation: $A = (9)(4) - 9.5$

$= 36 - 9.5$

$= 26.5$

Let $(x_1, y_1) = (-3, -1)$, $(x_2, y_2) = (2, -2)$, $(x_3, y_3) = (1, 2)$

$$\begin{vmatrix} x_1 & y_1 & 1 \\ x_2 & y_2 & 1 \\ x_3 & y_3 & 1 \end{vmatrix} = \begin{vmatrix} -3 & -1 & 1 \\ 2 & -2 & 1 \\ 1 & 2 & 1 \end{vmatrix} = 19$$

Area $= \frac{1}{2}(19) = 9.5$

75.

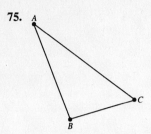

From diagram the coordinates of A, B, C are determined to be $A(0, 20)$, $B(10, -5)$ and $C(28, 0)$.

$$\begin{vmatrix} x_1 & y_1 & 1 \\ x_2 & y_2 & 1 \\ x_3 & y_3 & 1 \end{vmatrix} = \begin{vmatrix} 0 & 20 & 1 \\ 10 & -5 & 1 \\ 28 & 0 & 1 \end{vmatrix} = -500$$

Area $= -\frac{1}{2}(-500) = 250$ mi^2

77. Let $(x_1, y_1) = (-1, 11)$, $(x_2, y_2) = (0, 8)$, $(x_3, y_3) = (2, 2)$

$$\begin{vmatrix} x_1 & y_1 & 1 \\ x_2 & y_2 & 1 \\ x_3 & y_3 & 1 \end{vmatrix} = \begin{vmatrix} -1 & 11 & 1 \\ 0 & 8 & 1 \\ 2 & 2 & 1 \end{vmatrix} = (-1)\begin{vmatrix} 8 & 1 \\ 2 & 1 \end{vmatrix} + 0 + (2)\begin{vmatrix} 11 & 1 \\ 8 & 1 \end{vmatrix}$$

$$= (-1)(6) + (2)(3)$$

$$= -6 + 6$$

$$= 0$$

The three points are collinear.

79. $(x_1, y_1) = (-1, -5)$, $(x_2, y_2) = (1, -1)$, $(x_3, y_3) = (4, 5)$

$$\begin{vmatrix} x_1 & y_1 & 1 \\ x_2 & y_2 & 1 \\ x_3 & y_3 & 1 \end{vmatrix} = \begin{vmatrix} -1 & -5 & 1 \\ 1 & -1 & 1 \\ 4 & 5 & 1 \end{vmatrix} = (1)\begin{vmatrix} 1 & -1 \\ 4 & 5 \end{vmatrix} - (1)\begin{vmatrix} -1 & -5 \\ 4 & 5 \end{vmatrix} + (1)\begin{vmatrix} -1 & -5 \\ 1 & -1 \end{vmatrix}$$

$$= (1)(9) - (1)(15) + (1)(6)$$

$$= 9 - 15 + 6$$

$$= 0$$

The three points are collinear.

81. Let $(x_1, y_1) = \left(-2, \frac{1}{3}\right)$, $(x_2, y_2) = (2, 1)$, $(x_3, y_3) = \left(3, \frac{1}{5}\right)$

$$\begin{vmatrix} x_1 & y_1 & 1 \\ x_2 & y_2 & 1 \\ x_3 & y_3 & 1 \end{vmatrix} = \begin{vmatrix} -2 & \frac{1}{3} & 1 \\ 2 & 1 & 1 \\ 3 & \frac{1}{5} & 1 \end{vmatrix} = (1)\begin{vmatrix} 2 & 1 \\ 3 & \frac{1}{5} \end{vmatrix} - (1)\begin{vmatrix} -2 & \frac{1}{3} \\ 3 & \frac{1}{5} \end{vmatrix} + (1)\begin{vmatrix} -2 & \frac{1}{3} \\ 2 & 1 \end{vmatrix}$$

$$= (1)\left(-\frac{13}{5}\right) - (1)\left(-\frac{7}{5}\right) + (1)\left(-\frac{8}{3}\right)$$

$$= -\frac{13}{5} + \frac{7}{5} - \frac{8}{3}$$

$$= -\frac{18}{15} - \frac{40}{15}$$

$$= -\frac{58}{15}$$

The three points are not collinear.

83. $(x_1, y_1) = (0, 0), (x_2, y_2) = (5, 3)$

$$\begin{vmatrix} x & y & 1 \\ 0 & 0 & 1 \\ 5 & 3 & 1 \end{vmatrix} = 0$$

$(1) \begin{vmatrix} x & y \\ 5 & 3 \end{vmatrix} = 0$

$(1)(3x - 5y) = 0$

$3x - 5y = 0$

85. $(x_1, y_1) = (10, 7), (x_2, y_2) = (-2, -7)$

$$\begin{vmatrix} x & y & 1 \\ 10 & 7 & 1 \\ -2 & -7 & 1 \end{vmatrix} = 0$$

$(1) \begin{vmatrix} 10 & 7 \\ -2 & -7 \end{vmatrix} - (1) \begin{vmatrix} x & y \\ -2 & -7 \end{vmatrix} + (1) \begin{vmatrix} x & y \\ 10 & 7 \end{vmatrix} = 0$

$(1)(-56) - (-7x + 2y) + (1)(7x - 10y) = 0$

$-56 + 7x - 2y + 7x - 10y = 0$

$14x - 12y - 56 = 0$

$7x - 6y - 28 = 0$

87. $(x_1, y_1) = \left(-2, \frac{3}{2}\right), (x_2, y_2) = (3, -3)$

$$\begin{vmatrix} x & y & 1 \\ -2 & \frac{3}{2} & 1 \\ 3 & -3 & 1 \end{vmatrix} = 0$$

$x \begin{vmatrix} \frac{3}{2} & 1 \\ -3 & 1 \end{vmatrix} - y \begin{vmatrix} -2 & 1 \\ 3 & 1 \end{vmatrix} + 1 \begin{vmatrix} -2 & \frac{3}{2} \\ 3 & -3 \end{vmatrix} = 0$

$\frac{9}{2}x + 5y + \frac{3}{2} = 0$

$9x + 10y + 3 = 0$

89. $(x_1, y_1) = (2, 3.6) \quad (x_2, y_2) = (8, 10)$

$$\begin{vmatrix} x & y & 1 \\ 2 & 3.6 & 1 \\ 8 & 10 & 1 \end{vmatrix} = 0$$

$x \begin{vmatrix} 3.6 & 1 \\ 10 & 1 \end{vmatrix} - y \begin{vmatrix} 2 & 1 \\ 8 & 1 \end{vmatrix} + 1 \begin{vmatrix} 2 & 3.6 \\ 8 & 10 \end{vmatrix} = 0$

$x(3.6 - 10) - y(2 - 8) + 1(20 - 28.8) = 0$

$-6.4x + 6y - 8.8 = 0$

$-3.2x + 3y - 4.4 = 0$

$32x - 30y + 44 = 0$

91. $1 = a(0)^2 + b(0) + c \implies 1 = c$

$-3 = a(1)^2 + b(1) + c \implies -3 = a + b + c$

$21 = a(-2)^2 + b(-2) + c \implies 21 = 4a + 2b + c$

$$\begin{bmatrix} 0 & 0 & 1 & \vdots & 1 \\ 1 & 1 & 1 & \vdots & -3 \\ 4 & -2 & 1 & \vdots & 21 \end{bmatrix}$$

$D = \begin{vmatrix} 0 & 0 & 1 \\ 1 & 1 & 1 \\ 4 & -2 & 1 \end{vmatrix} = (1) \begin{vmatrix} 1 & 1 \\ 4 & -2 \end{vmatrix} = (1)(-6) = -6$

$a = \dfrac{\begin{vmatrix} 1 & 0 & 1 \\ -3 & 1 & 1 \\ 21 & -2 & 1 \end{vmatrix}}{-6} = \dfrac{(1)\begin{vmatrix} 1 & 1 \\ -2 & 1 \end{vmatrix} - 0 + (1)\begin{vmatrix} -3 & 1 \\ 21 & -2 \end{vmatrix}}{-6}$

$\qquad = \dfrac{(1)(3) + (1)(-15)}{-6} = \dfrac{-12}{-6} = 2$

$b = \dfrac{\begin{vmatrix} 0 & 1 & 1 \\ 1 & -3 & 1 \\ 4 & 21 & 1 \end{vmatrix}}{-6} = \dfrac{-(1)\begin{vmatrix} 1 & 1 \\ 4 & 1 \end{vmatrix} + (1)\begin{vmatrix} 1 & -3 \\ 4 & 21 \end{vmatrix}}{-6}$

$\qquad = \dfrac{(-1)(3) + (1)(33)}{-6} = \dfrac{36}{-6} = -6$

$c = \dfrac{\begin{vmatrix} 0 & 0 & 1 \\ 1 & 1 & -3 \\ 4 & -2 & 21 \end{vmatrix}}{-6} = \dfrac{(1)\begin{vmatrix} 1 & 1 \\ 4 & -2 \end{vmatrix}}{-6} = \dfrac{(1)(-6)}{-6} = 1$

$y = 2x^2 - 6x + 1$

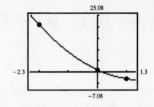

93. $6 = a(-2)^2 + b(-2) + c \implies 6 = 4a - 2b + c$

$-2 = a\ (2)^2 + b\ (2) + c \implies -2 = 4a + 2b + c$

$0 = a\ (4)^2 + b\ (4) + c \implies 6 = 16a + 4b + c$

$$\begin{bmatrix} 4 & -2 & 1 & \vdots & 6 \\ 4 & 2 & 1 & \vdots & -2 \\ 16 & 4 & 1 & \vdots & 0 \end{bmatrix} \quad D = \begin{vmatrix} 4 & -2 & 1 \\ 4 & 2 & 1 \\ 16 & 4 & 1 \end{vmatrix} = -48$$

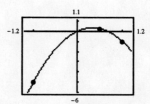

$$a = \frac{D_a}{D} = \frac{\begin{vmatrix} 6 & -2 & 1 \\ -2 & 2 & 1 \\ 0 & 4 & 1 \end{vmatrix}}{-48} = \frac{-24}{-48} = \frac{1}{2}$$

$$b = \frac{D_b}{D} = \frac{\begin{vmatrix} 4 & 6 & 1 \\ 4 & -2 & 1 \\ 16 & 0 & 1 \end{vmatrix}}{-48} = \frac{96}{-48} = -2$$

$$c = \frac{D_c}{D} = \frac{\begin{vmatrix} 4 & -2 & 6 \\ 4 & 2 & -2 \\ 16 & 4 & 0 \end{vmatrix}}{-48} = \frac{0}{-48} = 0$$

$$y = \frac{1}{2}x^2 - 2x$$

95. $-1 = a(1)^2 + b(1) + c \implies -1 = a + b + c$

$-5 = a(-1)^2 + b(-1) + c \implies -5 = a - b + c$

$\frac{1}{4} = a\left(\frac{1}{2}\right)^2 + b\left(\frac{1}{2}\right) + c \implies \frac{1}{4} = \frac{1}{4}a + \frac{1}{2}b + c$ or $1 = a + 2b + 4c$

$$\begin{bmatrix} 1 & 1 & 1 & \vdots & -1 \\ 1 & -1 & 1 & \vdots & -5 \\ 1 & 2 & 4 & \vdots & 1 \end{bmatrix}$$

$$D = \begin{vmatrix} 1 & 1 & 1 \\ 1 & -1 & 1 \\ 1 & 2 & 4 \end{vmatrix} = (1)\begin{vmatrix} -1 & 1 \\ 2 & 4 \end{vmatrix} - (1)\begin{vmatrix} 1 & 1 \\ 2 & 4 \end{vmatrix} + (1)\begin{vmatrix} 1 & 1 \\ -1 & 1 \end{vmatrix}$$

$$= (1)(-6) - (1)(2) + (1)(2)$$

$$= -6 - 2 + 2$$

$$= -6$$

$$a = \frac{\begin{vmatrix} -1 & 1 & 1 \\ -5 & -1 & 1 \\ 1 & 2 & 4 \end{vmatrix}}{-6} = \frac{(-1)\begin{vmatrix} -1 & 1 \\ 2 & 4 \end{vmatrix} - (1)\begin{vmatrix} -5 & 1 \\ 1 & 4 \end{vmatrix} + (1)\begin{vmatrix} -5 & -1 \\ 1 & 2 \end{vmatrix}}{-6}$$

$$= \frac{(-1)(-6) - (1)(-21) + (1)(-9)}{-6}$$

$$= \frac{6 + 21 - 9}{-6} = \frac{18}{-6} = -3$$

—CONTINUED—

95. —CONTINUED—

$$b = \frac{\begin{vmatrix} 1 & -1 & 1 \\ 1 & -5 & 1 \\ 1 & 1 & 4 \end{vmatrix}}{-6} = \frac{(1)\begin{vmatrix} -5 & 1 \\ 1 & 4 \end{vmatrix} - (1)\begin{vmatrix} -1 & 1 \\ 1 & 4 \end{vmatrix} + (1)\begin{vmatrix} -1 & 1 \\ -5 & 1 \end{vmatrix}}{-6}$$

$$= \frac{(1)(-21) - (1)(-5) + (1)(4)}{-6}$$

$$= \frac{-21 + 5 + 4}{-6} = \frac{-12}{-6} = 2$$

$$c = \frac{\begin{vmatrix} 1 & 1 & -5 \\ 1 & -1 & -5 \\ 1 & 2 & 1 \end{vmatrix}}{-6} = \frac{(1)\begin{vmatrix} -1 & -5 \\ 2 & 1 \end{vmatrix} - (1)\begin{vmatrix} 1 & -1 \\ 2 & 1 \end{vmatrix} + (1)\begin{vmatrix} 1 & -1 \\ -1 & -5 \end{vmatrix}}{-6}$$

$$= \frac{(1)(9) - (1)(3) + (1)(-5)}{-6} = \frac{9 - 3 - 6}{-6} = 0$$

$$y = -3x^2 + 2x$$

97. (a) $(5, 584.7)\ (6, 624.8)\ (7, 689.2)$

$a(5)^2 + b(5) + c = 584.7 \implies 25a + 5b + c = 584.7$

$a(6)^2 + b(6) + c = 624.8 \implies 36a + 6b + c = 624.8$

$a(7)^2 + b(7) + c = 689.2 \implies 49a + 7b + c = 689.2$

$$\det(A) = \begin{vmatrix} 25 & 5 & 1 \\ 36 & 6 & 1 \\ 49 & 7 & 1 \end{vmatrix} = -2$$

$$a_1 = \frac{\begin{vmatrix} 584.7 & 5 & 1 \\ 624.8 & 6 & 1 \\ 689.2 & 7 & 1 \end{vmatrix}}{-2} = \frac{-24.3}{-2} \qquad b_1 = \frac{\begin{vmatrix} 25 & 584.7 & 1 \\ 36 & 624.8 & 1 \\ 49 & 689.2 & 1 \end{vmatrix}}{-2} = \frac{187.1}{-2}$$

$$= 12.15 \qquad\qquad\qquad = -93.55$$

$$c_1 = \frac{\begin{vmatrix} 25 & 5 & 584.7 \\ 36 & 6 & 624.8 \\ 49 & 7 & 689.2 \end{vmatrix}}{-2} = \frac{-1497.4}{-2} = 748.7$$

$$y_1 = 12.15t^2 - 93.55t + 748.7$$

(b) $(5, 743.4)\ (6, 791.4)\ (7, 870.7)$

$a(5)^2 + b(5) + c = 743.4 \implies 25a^2 + 5b + c = 743.4$

$a(6)^2 + b(6) + c = 791.4 \implies 36a^2 + 6b + c = 791.4$

$a(7)^2 + b(7) + c = 870.7 \implies 49a^2 + 7b + c = 870.7$

$$\det(A) = \begin{vmatrix} 25 & 5 & 1 \\ 36 & 6 & 1 \\ 49 & 7 & 1 \end{vmatrix} = -2 \qquad a_2 = \frac{\begin{vmatrix} 743.4 & 5 & 1 \\ 791.4 & 6 & 1 \\ 870.7 & 7 & 1 \end{vmatrix}}{-2} = \frac{-31.3}{-2} = 15.65$$

—CONTINUED—

97. —CONTINUED—

$$b_2 = \frac{\begin{vmatrix} 25 & 743.4 & 1 \\ 36 & 791.4 & 1 \\ 49 & 870.7 & 1 \end{vmatrix}}{-2} = \frac{248.3}{-2}$$

$$= -124.15$$

$$y_2 = 15.65t^2 - 124.15t + 972.9$$

$$c_2 = \frac{\begin{vmatrix} 25 & 5 & 743.4 \\ 36 & 6 & 791.4 \\ 49 & 7 & 870.7 \end{vmatrix}}{-2} = \frac{-1945.8}{-2}$$

$$= 972.9$$

(c)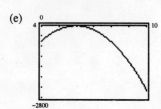

(d) $y_1 - y_2 = (12.15t^2 - 93.55t + 748.7) - (15.65t^2 - 124.15t + 972.9)$

$$= -3.5t^2 + 30.6t - 224.2$$

(e)

The trade deficit is increasing.

99. (a) $\begin{bmatrix} k & 1-k & \vdots & 1 \\ 1-k & k & \vdots & 3 \end{bmatrix}$

$$D = \begin{vmatrix} k & 1-k \\ 1-k & k \end{vmatrix} = k^2 - (1-k)^2 = k^2 - (1 - 2k + k^2) = k^2 - 1 + 2k - k^2 = 2k - 1$$

$$x = \frac{D_x}{D} = \frac{\begin{vmatrix} 1 & 1-k \\ 3 & k \end{vmatrix}}{2k-1} = \frac{k - 3(1-k)}{2k-1} = \frac{k - 3 + 3k}{2k-1} = \frac{4k-3}{2k-1}$$

$$y = \frac{D_y}{D} = \frac{\begin{vmatrix} k & 1 \\ 1-k & 3 \end{vmatrix}}{2k-1} = \frac{3k - 1(1-k)}{2k-1} = \frac{3k - 1 + k}{2k-1} = \frac{4k-1}{2k-1}$$

$$\left(\frac{4k-3}{2k-1}, \frac{4k-1}{2k-1} \right)$$

(b) $2k - 1 = 0$

$$2k = 1$$

$$k = \frac{1}{2}$$

101. A determinant is a real number associated with a square matrix.

103. The minor of an entry of a square matrix is the determinant of the matrix that remains after deleting the row and column in which the entry occurs.

Review Exercises for Chapter 7

1. (a) $(3, 4)$

$$3(3) + 7(4) \overset{?}{=} 2$$

$$37 \neq 2$$

Not a solution

(b) $(3, -1)$

$$3(3) + 7(-1) \overset{?}{=} 2 \qquad 5(3) + 6(-1) \overset{?}{=} 9$$

$$2 = 2 \qquad\qquad 9 = 9$$

Solution

3. (a) $(4, -5)$

$$4^2 + (-5)^2 \overset{?}{=} 41$$

$$41 = 41$$

$$20(4) + 10(-5) \overset{?}{=} 30$$

$$80 - 50 \overset{?}{=} 30$$

$$30 = 30$$

Solution

(b) $(7, 12)$

$$7^2 + 12^2 \overset{?}{=} 41$$

$$193 \neq 41$$

Not a solution

5. Solve each equation for y.

$$x + y = 2 \qquad x - y = 0$$

$$y = -x + 2 \qquad -y = -x$$

$$y = x$$

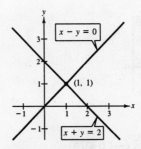

Point of intersection is $(1, 1)$.

7. Solve each equation for y.

$$x - y = 3 \qquad -x + y = 1$$

$$-y = -x + 3 \qquad y = x + 1$$

$$y = x - 3$$

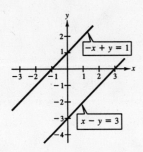

No solution

9. Solve each equation for y.

$$2x - y = 0 \qquad -x + y = 4$$

$$-y = -2x \qquad y = x + 4$$

$$y = 2x$$

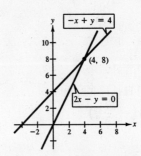

Point of intersection is $(4, 8)$.

11. Solve each equation for y.

$$2x + y = 4 \qquad -4x - 2y = -8$$

$$y = -2x + 4 \qquad -2y = 4x - 8$$

$$y = -2x + 4$$

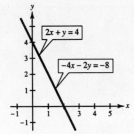

Infinite number of solutions.

13. Solve each equation for y.

$$3x - 2y = -2 \qquad -5x + 2y = 2$$

$$-2y = -3x - 2 \qquad 2y = 5x + 2$$

$$y = \tfrac{3}{2}x + 1 \qquad y = \tfrac{5}{2}x + 1$$

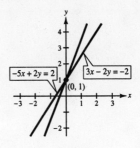

Point of intersection is $(0, 1)$.

15. Solve each equation for y.

$$5x - 3y = 3 \qquad\qquad 2x + 2y = 14$$
$$-3y = -5x + 3 \qquad\qquad 2y = -2x + 14$$
$$y = \tfrac{5}{3}x - 1 \qquad\qquad y = -1x + 7$$

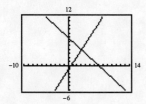

Keystrokes:

y_1 [Y=] [(] 5 [÷] 3 [)] [X,T,θ] [−] 1 [ENTER]

y_2 [(−)] [X,T,θ] [+] 7 [GRAPH]

Solution is $(3, 4)$

17. Solve each equation for y.

$$y = x^2 - 4 \qquad 2x - 3y = 11$$
$$-3y = -2x + 11$$
$$y = \tfrac{2}{3}x - \tfrac{11}{3}$$

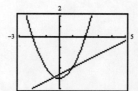

Keystrokes:

y_1 [Y=] [X,T,θ] [x²] [−] 4 [ENTER]

y_2 [(] 2 [÷] 3 [)] [X,T,θ] [−] 11 [÷] 3 [GRAPH]

Solutions are $(1, -3)$ and $\left(-\tfrac{1}{3}, -\tfrac{35}{9}\right)$

19. $x = -2 - 4y$

$$2(-2 - 4y) + 3y = 1$$
$$-4 - 8y + 3y = 1$$
$$-5y = 5$$
$$y = -1$$
$$x = -2 - 4(-1)$$
$$x = 2$$

$(2, -1)$

21. $2y = 5x + 4$

$$y = \frac{5x + 4}{2}$$
$$10x - 4\left(\frac{5x + 4}{2}\right) = 7$$
$$10x - 2(5x + 4) = 7$$
$$10x - 10x - 8 = 7$$
$$-8 \neq 7$$

No solution

23. $3x = 7y + 5$

$$x = \frac{7y + 5}{3}$$

$$5\left(\frac{7y + 5}{3}\right) - 9y = -5$$

$$5(7y + 5) - 27y = -15$$

$$35y + 25 - 27y = -15$$

$$8y = -40$$

$$y = -5$$

$$x = \frac{7(-5) + 5}{3}$$

$$x = -10$$

$(-10, -5)$

25. $y = 5x^2$

$$y = -15x - 10$$

$$5x^2 = -15x - 10$$

$$5x^2 + 15x + 10 = 0$$

$$x^2 + 3x + 2 = 0$$

$$(x + 2)(x + 1) = 0$$

$x = -2$	$x = -1$
$y = 5(-2)^2$	$y = 5(-1)^2$
$= 20$	$= 5$

$(-2, 20), (-1, 5)$

27. $x^2 + y^2 = 1$

$$x + y = -1$$

$$y = -1 - x$$

$$x^2 + (-1 - x)^2 = 1$$

$$x^2 + 1 + 2x + x^2 = 1$$

$$2x^2 + 2x = 0$$

$$2x(x + 1) = 0$$

$x = 0$	$x = -1$
$y = -1$	$y = 0$

$(0, -1), (-1, 0)$

29. $\begin{aligned} x + y = 0 &\implies x + y = 0 \\ 2x + y = 0 &\implies \underline{-2x - y = 0} \end{aligned}$

$$-x = 0$$

$$x = 0$$

$$0 + y = 0$$

$$y = 0$$

$(0, 0)$

31. $\begin{aligned} 2x - y = 2 &\implies 16x - 8y = 16 \\ 6x + 8y = 39 &\implies \underline{6x + 8y = 39} \end{aligned}$

$$22x = 55$$

$$x = \frac{55}{22} = \frac{5}{2}$$

$$2\left(\tfrac{5}{2}\right) - y = 2$$

$$5 - y = 2$$

$$-y = -3$$

$$y = 3$$

$\left(\frac{5}{2}, 3\right)$

33. $0.2x + 0.3y = 0.14 \Rightarrow 2x + 3y = 1.4 \Rightarrow -4x - 6y = -2.8$
 $0.4x + 0.5y = 0.20 \Rightarrow 4x + 5y = 2 \quad \Rightarrow \underline{ 4x + 5y = 2 }$

$$-y = -0.8$$
$$y = 0.8$$

$$2x + 3(0.8) = 1.4$$
$$2x + 2.4 = 1.4$$
$$2x = -1$$
$$x = -\tfrac{1}{2} = -0.5$$

$(-0.5, 0.8)$

35.
$$x - y - 2z = -1$$
$$2x + 3y + z = -2$$
$$5x + 4y + 2z = 4$$

$$x - y - 2z = -1$$
$$5y + 5z = 0$$
$$9y + 12z = 9$$

$$x - y - 2z = -1$$
$$y + z = 0$$
$$9y + 12z = 9$$

$$x - y - 2z = -1$$
$$y + z = 0$$
$$3z = 9$$

$$x - y - 2z = -1$$
$$y + z = 0$$
$$z = 3$$

$$y + 3 = 0$$
$$y = -3$$
$$x - (-3) - 2(3) = -1$$
$$x + 3 - 6 = -1$$
$$x = 2$$

$(2, -3, 3)$

37.
$$x - y - z = 1$$
$$-2x + y + 3z = -5$$
$$3x + 4y - z = 6$$

$$x - y - z = 1$$
$$-y + z = -3$$
$$7y + 2z = 3$$

$$x - y - z = 1$$
$$-y + z = -3$$
$$9z = -18$$
$$z = -2$$

$$-y + (-2) = -3$$
$$-y = -1$$
$$y = 1$$
$$x - 1 - (-2) = 1$$
$$x = 0$$

$(0, 1, -2)$

39.
$$x - 4z = 17$$
$$-2x + 4y + 3z = -14$$
$$5x - y + 2z = -3$$

$$x - 4z = 17$$
$$18x + 11z = -26$$
$$5x - y + 2z = -3$$

$$x - 4z = 17$$
$$83z = -332$$
$$5x - y + 2z = -3$$

$$z = -4$$
$$x - 4(-4) = 17$$
$$x = 1$$
$$5(1) - y + 2(-4) = -3$$
$$-y = 0$$
$$y = 0$$

$(1, 0, -4)$

41.
$$\begin{bmatrix} 5 & 4 & \vdots & 2 \\ -1 & 1 & \vdots & -22 \end{bmatrix}$$

$$\begin{matrix} R_1 \\ R_2 \end{matrix} \begin{bmatrix} -1 & 1 & \vdots & -22 \\ 5 & 4 & \vdots & 2 \end{bmatrix}$$

$$-R_1 \begin{bmatrix} 1 & -1 & \vdots & 22 \\ 5 & 4 & \vdots & 2 \end{bmatrix}$$

$$-5R_1 + R_2 \begin{bmatrix} 1 & -1 & \vdots & 22 \\ 0 & 9 & \vdots & -108 \end{bmatrix}$$

$$\tfrac{1}{9}R_2 \begin{bmatrix} 1 & -1 & \vdots & 22 \\ 0 & 1 & \vdots & -12 \end{bmatrix}$$

$y = -12 \qquad x - (-12) = 22 \qquad (10, -12)$

$\qquad\qquad\qquad\qquad x = 10$

43.
$$\begin{bmatrix} .2 & -.1 & \vdots & .07 \\ .4 & -.5 & \vdots & -.01 \end{bmatrix}$$

$$\begin{matrix} 10R_1 \\ 10R_2 \end{matrix} \begin{bmatrix} 2 & -1 & \vdots & .7 \\ 4 & -5 & \vdots & -.1 \end{bmatrix}$$

$$\tfrac{1}{2}R_1 \begin{bmatrix} 1 & -\tfrac{1}{2} & \vdots & .35 \\ 4 & -5 & \vdots & -.1 \end{bmatrix}$$

$$-4R_1 + R_2 \begin{bmatrix} 1 & -\tfrac{1}{2} & \vdots & .35 \\ 0 & -3 & \vdots & -1.5 \end{bmatrix}$$

$$-\tfrac{1}{3}R_2 \begin{bmatrix} 1 & -\tfrac{1}{2} & \vdots & .35 \\ 0 & 1 & \vdots & 5 \end{bmatrix}$$

$y = .5 \qquad x - \tfrac{1}{2}(.5) = .35 \qquad (0.6, 0.5)$

$\qquad\qquad\qquad\qquad x = .6$

45.
$$\begin{bmatrix} 1 & 2 & 6 & \vdots & 4 \\ -3 & 2 & -1 & \vdots & -4 \\ 4 & 0 & 2 & \vdots & 16 \end{bmatrix}$$

$$\begin{matrix} 3R_1 + R_2 \\ -4R_1 + R_3 \end{matrix} \begin{bmatrix} 1 & 2 & 6 & \vdots & 4 \\ 0 & 8 & 17 & \vdots & 8 \\ 0 & -8 & -22 & \vdots & 0 \end{bmatrix}$$

$$\tfrac{1}{8}R_2 \begin{bmatrix} 1 & 2 & 6 & \vdots & 4 \\ 0 & 1 & \tfrac{17}{8} & \vdots & 1 \\ 0 & -8 & -22 & \vdots & 0 \end{bmatrix}$$

$$8R_2 + R_3 \begin{bmatrix} 1 & 2 & 6 & \vdots & 4 \\ 0 & 1 & \tfrac{17}{8} & \vdots & 1 \\ 0 & 0 & -5 & \vdots & 8 \end{bmatrix}$$

$$-\tfrac{1}{5}R_3 \begin{bmatrix} 1 & 2 & 6 & \vdots & 4 \\ 0 & 1 & \tfrac{17}{8} & \vdots & 1 \\ 0 & 0 & 1 & \vdots & -\tfrac{8}{5} \end{bmatrix}$$

$z = -\tfrac{8}{5} \qquad y + \tfrac{17}{8}\left(-\tfrac{8}{5}\right) = 1 \qquad x + 2\left(\tfrac{22}{5}\right) + 6\left(-\tfrac{8}{5}\right) = 4$

$\qquad\qquad\qquad\qquad y = \tfrac{22}{5} \qquad\qquad\qquad\qquad x = \tfrac{24}{5}$

$\left(\tfrac{24}{5}, \tfrac{22}{5}, -\tfrac{8}{5}\right)$

47.
$$\begin{bmatrix} 2 & 3 & 3 & \vdots & 3 \\ 6 & 6 & 12 & \vdots & 13 \\ 12 & 9 & -1 & \vdots & 2 \end{bmatrix}$$

$$\tfrac{1}{2}R_1 \begin{bmatrix} 1 & \tfrac{3}{2} & \tfrac{3}{2} & \vdots & \tfrac{3}{2} \\ 6 & 6 & 12 & \vdots & 13 \\ 12 & 9 & -1 & \vdots & 2 \end{bmatrix}$$

$$\begin{matrix} -6R_1 + R_2 \\ -12R_1 + R_3 \end{matrix} \begin{bmatrix} 1 & \tfrac{3}{2} & \tfrac{3}{2} & \vdots & \tfrac{3}{2} \\ 0 & -3 & 3 & \vdots & 4 \\ 0 & -9 & -19 & \vdots & -16 \end{bmatrix}$$

$$-\tfrac{1}{3}R_2 \begin{bmatrix} 1 & \tfrac{3}{2} & \tfrac{3}{2} & \vdots & \tfrac{3}{2} \\ 0 & 1 & -1 & \vdots & -\tfrac{4}{3} \\ 0 & -9 & -19 & \vdots & -16 \end{bmatrix}$$

$$9R_2 + R_3 \begin{bmatrix} 1 & \tfrac{3}{2} & \tfrac{3}{2} & \vdots & \tfrac{3}{2} \\ 0 & 1 & -1 & \vdots & -\tfrac{4}{3} \\ 0 & 0 & -28 & \vdots & -28 \end{bmatrix}$$

$$-\tfrac{1}{28}R_3 \begin{bmatrix} 1 & \tfrac{3}{2} & \tfrac{3}{2} & \vdots & \tfrac{3}{4} \\ 0 & 1 & -1 & \vdots & -\tfrac{4}{5} \\ 0 & 0 & 0 & \vdots & 1 \end{bmatrix}$$

$x_3 = 1$

$x_2 - 1 = -\tfrac{4}{3}$

$x_2 = -\tfrac{1}{3}$

$x_1 + \tfrac{3}{2}\left(\tfrac{-1}{3}\right) + \tfrac{3}{2}(1) = \tfrac{3}{2}$

$x_1 - \tfrac{1}{2} + \tfrac{3}{2} = \tfrac{3}{2}$

$x_1 + 1 = \tfrac{3}{2}$

$x_1 = \tfrac{1}{2}$

$\left(\tfrac{1}{2}, \tfrac{-1}{3}, 1\right)$

49. $\det(A) = \begin{vmatrix} 7 & 10 \\ 10 & 15 \end{vmatrix} = (7)(15) - (10)(10) = 105 - 100 = 5$

51. $\det(A) = \begin{vmatrix} 8 & 6 & 3 \\ 6 & 3 & 0 \\ 3 & 0 & 2 \end{vmatrix}$

$= (3)\begin{vmatrix} 6 & 3 \\ 3 & 0 \end{vmatrix} - 0\begin{vmatrix} 8 & 3 \\ 6 & 0 \end{vmatrix} + 2\begin{vmatrix} 8 & 6 \\ 6 & 3 \end{vmatrix}$ (third row)

$= (3)(-9) - 0 + (2)(-12)$

$= -27 - 24$

$= -51$

53. $\det(A) = \begin{vmatrix} 8 & 3 & 2 \\ 1 & -2 & 4 \\ 6 & 0 & 5 \end{vmatrix}$

$= 6\begin{vmatrix} 3 & 2 \\ -2 & 4 \end{vmatrix} - 0 + 5\begin{vmatrix} 8 & 3 \\ 1 & -2 \end{vmatrix}$ (third row)

$= (6)(16) + (5)(-19)$

$= 1$

55. $\begin{bmatrix} 7 & 12 & \vdots & 63 \\ 2 & 3 & \vdots & 15 \end{bmatrix}$

$D = \begin{vmatrix} 7 & 12 \\ 2 & 3 \end{vmatrix} = 21 - 24 = -3$

$x = \dfrac{D_x}{D} = \dfrac{\begin{vmatrix} 63 & 12 \\ 15 & 3 \end{vmatrix}}{-3} = \dfrac{189 - 180}{-3} = \dfrac{9}{-3} = -3$

$y = \dfrac{D_y}{D} = \dfrac{\begin{vmatrix} 7 & 63 \\ 2 & 15 \end{vmatrix}}{-3} = \dfrac{105 - 126}{-3} = \dfrac{-21}{-3} = 7$

$(-3, 7)$

57. $\begin{bmatrix} 3 & -2 & \vdots & 16 \\ 12 & -8 & \vdots & -5 \end{bmatrix}$

$D = \begin{vmatrix} 3 & -2 \\ 12 & -8 \end{vmatrix} = -24 + 24 = 0$

Cannot be solved by Cramer's Rule because $D = 0$. Solve by elimination.

$-12x + 8y = -64$

$\underline{12x - 8y = -5}$

$\,0 \neq -69$

Inconsistent; no solution

59. $\begin{bmatrix} -1 & 1 & 2 & \vdots & 1 \\ 2 & 3 & 1 & \vdots & -2 \\ 5 & 4 & 2 & \vdots & 4 \end{bmatrix}$

$D = \begin{vmatrix} -1 & 1 & 2 \\ 2 & 3 & 1 \\ 5 & 4 & 2 \end{vmatrix} = (-1)\begin{vmatrix} 3 & 1 \\ 4 & 2 \end{vmatrix} - (1)\begin{vmatrix} 2 & 1 \\ 5 & 2 \end{vmatrix} + (2)\begin{vmatrix} 2 & 3 \\ 5 & 4 \end{vmatrix}$

$= (-1)(2) - (1)(-1) + (2)(-7)$

$= -2 + 1 - 14 = -15$

$x = \dfrac{\begin{vmatrix} 1 & 1 & 2 \\ -2 & 3 & 1 \\ 4 & 4 & 2 \end{vmatrix}}{-15} = \dfrac{(1)\begin{vmatrix} 3 & 1 \\ 4 & 2 \end{vmatrix} - (1)\begin{vmatrix} -2 & 1 \\ 4 & 2 \end{vmatrix} + (2)\begin{vmatrix} -2 & 3 \\ 4 & 4 \end{vmatrix}}{-15}$

$= \dfrac{(1)(2) - (1)(-8) + (2)(-20)}{-15}$

$= \dfrac{2 + 8 - 40}{-15} = \dfrac{-30}{-15} = 2$

$y = \dfrac{\begin{vmatrix} -1 & 1 & 2 \\ 2 & -2 & 1 \\ 5 & 4 & 2 \end{vmatrix}}{-15} = \dfrac{(-1)\begin{vmatrix} -2 & 1 \\ 4 & 2 \end{vmatrix} - (1)\begin{vmatrix} 2 & 1 \\ 5 & 2 \end{vmatrix} + (2)\begin{vmatrix} 2 & -2 \\ 5 & 4 \end{vmatrix}}{-15}$

$= \dfrac{(-1)(-8) - (1)(-1) + (2)(18)}{-15}$

$= \dfrac{8 + 1 + 36}{-15} = \dfrac{45}{-15} = -3$

—CONTINUED—

59. —CONTINUED—

$$z = \frac{\begin{vmatrix} -1 & 1 & 1 \\ 2 & 2 & -2 \\ 5 & 4 & 4 \end{vmatrix}}{-15} = \frac{(-1)\begin{vmatrix} 3 & -2 \\ 4 & 4 \end{vmatrix} - (1)\begin{vmatrix} 2 & -2 \\ 5 & 4 \end{vmatrix} + (1)\begin{vmatrix} 2 & 3 \\ 5 & 4 \end{vmatrix}}{-15}$$

$$= \frac{(-1)(20) - (1)(18) + (1)(-7)}{-15}$$

$$= \frac{-20 - 18 - 7}{-15} = \frac{-45}{-15} = 3 \qquad (2, -3, 3)$$

61. $\quad 3x - y = 6$

$\quad -3x + 2y = -10$

There are many other correct solutions. Write equations so that $\left(\frac{2}{3}, -4\right)$ satisfies each equation.

63. *Verbal Model:*

$$\boxed{\text{Total Cost}} = \boxed{\text{Cost per unit}} \cdot \boxed{\text{Number of units}} + \boxed{\text{Initial cost}}$$

$$\boxed{\text{Total Revenue}} = \boxed{\text{Price per unit}} \cdot \boxed{\text{Number of units}}$$

Labels: Total cost $= C$

Cost per unit $= 3.75$

Number of units $= x$

Initial cost $= 25,000$

Total revenue $= R$

Price per unit $= 5.25$

System of equations: $\quad C = 3.75x + 25,000$

$\quad R = 5.25x$

$\quad R = C$

$5.25x = 3.75x + 25,000$

$1.50x = 25,000$

$x = 16,666.\overline{6} \approx 16,667$ items

65. *Verbal Model:*

$$\boxed{\text{Gallons Solution 1}} + \boxed{\text{Gallons Solution 2}} = \boxed{100}$$

$$\boxed{\text{Value Solution 1}} + \boxed{\text{Value Solution 2}} = \boxed{0.60(100)}$$

Labels: Gallons Solution 1 $= x$

Gallons Solution 2 $= y$

System of equations:

$\quad x + \quad y = \quad 100$

$0.75x + 0.50y = 0.60(100)$

$x = 100 - y$

$75(100 - y) + 50y = 60(100)$

$7500 - 75y + 50y = 6000$

$-25y = -1500$

$y = 60$ gallons at 50% solution

$x = 100 - 60 = 40$ gallons at 75% solution

67. *Verbal model:*

$$2 \cdot \boxed{\text{Length}} + 2 \cdot \boxed{\text{Width}} = \boxed{\text{Perimeter}}$$

$$\boxed{\text{Length}} = 1.50 \cdot \boxed{\text{Width}}$$

Labels: Length $= x$

Width $= y$

System of equations:

$\quad 2x + 2y = 480$

$\quad x = 1.50y$

$2(1.50y) + 2y = 480$

$3y + 2y = 480$

$5y = 480$

$y = 96$ meters in width

$x = 1.50(96) = 144$ meters in length

69. *Verbal Model:*

$$\boxed{\begin{matrix}\text{Number}\\\text{Tapes 1}\end{matrix}} + \boxed{\begin{matrix}\text{Number}\\\text{Tapes 2}\end{matrix}} = 650$$

$$\boxed{\begin{matrix}\text{Receipts}\\\text{Tapes 1}\end{matrix}} + \boxed{\begin{matrix}\text{Receipts}\\\text{Tapes 2}\end{matrix}} = 7717.50$$

Labels: Number Tapes 1 $= x$

Number Tapes 2 $= y$

System of equations:

$$x + y = 650$$
$$9.95x + 14.95y = 7717.50$$

$$y = 650 - x$$

$$9.95x + 14.95(650 - x) = 7717.50$$

$$9.95x + 9717.50 - 14.95x = 7717.50$$

$$-5x = -2000$$

$$x = 400 \text{ tapes at } \$9.95$$

$$y = 650 - 400 = 250 \text{ tapes at } \$14.95$$

71. *Verbal Model:*

$$\boxed{\begin{matrix}\text{Speed}\\\text{Plane 1}\end{matrix}} \cdot \boxed{\text{Time}} + \boxed{\begin{matrix}\text{Speed}\\\text{Plane 2}\end{matrix}} \cdot \boxed{\text{Time}} = \boxed{\text{Distance}}$$

$$\boxed{\begin{matrix}\text{Speed}\\\text{Plane 2}\end{matrix}} = \boxed{\begin{matrix}\text{Speed}\\\text{Plane 1}\end{matrix}} + 40$$

Labels: Speed Plane 1 $= x$

Speed Plane 2 $= y$

Time $= \frac{50}{60} = \frac{5}{6}$ hr

Distance $= 450$ miles

System of equations:

$$x \cdot \tfrac{5}{6} + y \cdot \tfrac{5}{6} = 450$$
$$y = x + 40$$

$$\tfrac{5}{6}x + \tfrac{5}{6}(x + 40) = 450$$

$$\tfrac{5}{6}(2x + 40) = 450$$

$$2x + 40 = 540$$

$$2x = 500$$

$$x = 250 \text{ mph}$$

$$y = 250 + 40 = 290 \text{ mph}$$

73. *Verbal model:*

$$\boxed{\begin{array}{c}\text{Number}\\1\end{array}} + \boxed{\begin{array}{c}\text{Number}\\2\end{array}} + \boxed{\begin{array}{c}\text{Number}\\3\end{array}} = 68$$

$$\boxed{\begin{array}{c}\text{Number}\\2\end{array}} = 4 + \boxed{\begin{array}{c}\text{Number}\\1\end{array}}$$

$$\boxed{\begin{array}{c}\text{Number}\\3\end{array}} = 2 \cdot \boxed{\begin{array}{c}\text{Number}\\1\end{array}}$$

Labels: Number 1 $= x$ Number 2 $= y$ Number 3 $= z$

System of equations:

$$\begin{aligned} x + y + z &= 68 \\ y &= 4 + x \\ z &= 2x \end{aligned}$$

$$\begin{aligned} x + y + z &= 68 & x + y + z &= 68 \\ -x + y \phantom{{}+z} &= 4 & y + \tfrac{1}{2}z &= 36 \\ -2x \phantom{{}+y} + z &= 0 & 2z &= 64 \end{aligned}$$

$$\begin{aligned} x + y + z &= 68 & z &= 32 \\ 2y + z &= 72 & y + \tfrac{1}{2}(32) &= 36 \\ 2y + 3z &= 136 & y &= 20 \end{aligned}$$

$$\begin{aligned} x + y + z &= 68 \\ y + \tfrac{1}{2}z &= 36 & x + 20 + 32 &= 68 \\ 2y + 3z &= 136 & x &= 16 \end{aligned}$$

$(16, 20, 32)$

75.

$$\begin{aligned} 0 &= a(-5)^2 + b(-5) + c &&\Longrightarrow & 0 &= 25a - 5b + c \\ -6 &= a(1)^2 + b(1) + c &&\Longrightarrow & -6 &= a + b + c \\ 14 &= a(2)^2 + b(2) + c &&\Longrightarrow & 14 &= 4a + 2b + c \end{aligned}$$

$$\begin{bmatrix} 25 & -5 & 1 & \vdots & 0 \\ 1 & 1 & 1 & \vdots & -6 \\ 4 & 2 & 1 & \vdots & 14 \end{bmatrix}$$

$$D = \begin{vmatrix} 25 & -5 & 1 \\ 1 & 1 & 1 \\ 4 & 2 & 1 \end{vmatrix} = -42$$

$$a = \frac{D_a}{D} = \frac{\begin{vmatrix} 0 & -5 & 1 \\ -6 & 1 & 1 \\ 14 & 2 & 1 \end{vmatrix}}{-42} = \frac{-126}{-42} = 3$$

$$b = \frac{D_b}{D} = \frac{\begin{vmatrix} 25 & 0 & 1 \\ 1 & -6 & 1 \\ 4 & 14 & 1 \end{vmatrix}}{-42} = \frac{-462}{-42} = 11$$

$$c = \frac{\begin{vmatrix} 25 & -5 & 0 \\ 1 & 1 & -6 \\ 4 & 2 & 14 \end{vmatrix}}{-42} = \frac{840}{-42} = -20 \qquad y = 3x^2 + 11x - 20$$

77. $(x_1, y_1) = (1, 0)$, $(x_2, y_2) = (5, 0)$, $(x_3, y_3) = (5, 8)$

$$\begin{vmatrix} x_1 & y_1 & 1 \\ x_2 & y_2 & 1 \\ x_3 & y_3 & 1 \end{vmatrix} = \begin{vmatrix} 1 & 0 & 1 \\ 5 & 0 & 1 \\ 5 & 8 & 1 \end{vmatrix} = -0 + 0 - (8)\begin{vmatrix} 1 & 1 \\ 5 & 1 \end{vmatrix} = (-8)(-4) = 32$$

Area $= +\dfrac{1}{2}(32) = 16$

79. $(x_1, y_1) = (1, 2)$, $(x_2, y_2) = (4, -5)$, $(x_3, y_3) = (3, 2)$

$$\begin{vmatrix} x_1 & y_1 & 1 \\ x_2 & y_2 & 1 \\ x_3 & y_3 & 1 \end{vmatrix} = \begin{vmatrix} 1 & 2 & 1 \\ 4 & -5 & 1 \\ 3 & 2 & 1 \end{vmatrix} = (1)\begin{vmatrix} 4 & -5 \\ 3 & 2 \end{vmatrix} - (1)\begin{vmatrix} 1 & 2 \\ 3 & 2 \end{vmatrix} + (1)\begin{vmatrix} 1 & 2 \\ 4 & -5 \end{vmatrix}$$

$$= (1)(23) - (1)(-4) + (1)(-13)$$

$$= 23 + 4 - 13$$

$$= 14$$

Area $= +\dfrac{1}{2}(14) = 7$

81. $\begin{vmatrix} x & y & 1 \\ -4 & 0 & 1 \\ 4 & 4 & 1 \end{vmatrix} = 0$

$$-(-4)\begin{vmatrix} y & 1 \\ 4 & 1 \end{vmatrix} + 0 - (1)\begin{vmatrix} x & y \\ 4 & 4 \end{vmatrix} = 0$$

$$(4)(y - 4) - (1)(4x - 4y) = 0$$

$$4y - 16 - 4x + 4y = 0$$

$$-4x + 8y - 16 = 0$$

$$x - 2y + 4 = 0$$

83. $\begin{vmatrix} x & y & 1 \\ -\frac{5}{2} & 3 & 1 \\ \frac{7}{2} & 1 & 1 \end{vmatrix} = 0$

$$(1)\begin{vmatrix} -\frac{5}{2} & 3 \\ \frac{7}{2} & 1 \end{vmatrix} - (1)\begin{vmatrix} x & y \\ \frac{7}{2} & 1 \end{vmatrix} + (1)\begin{vmatrix} x & y \\ -\frac{5}{2} & 3 \end{vmatrix} = 0$$

$$(1)(-13) - (1)\left(x - \tfrac{7}{2}y\right) + (1)\left(3x + \tfrac{5}{2}y\right) = 0$$

$$-13 - x + \tfrac{7}{2}y + 3x + \tfrac{5}{2}y = 0$$

$$2x + 6y - 13 = 0$$

Chapter Test for Chapter 7

1. (a) $(3, -4)$

$$2(3) - 2(-4) \overset{?}{=} 1$$

$$6 + 8 \neq 1$$

Not a solution

(b) $\left(1, \tfrac{1}{2}\right)$

$$2(1) - 2\left(\tfrac{1}{2}\right) \overset{?}{=} 1 \qquad -1 + 2\left(\tfrac{1}{2}\right) \overset{?}{=} 0$$

$$2 - 1 = 1 \qquad\qquad -1 + 1 = 0$$

$$1 = 1 \qquad\qquad\quad 0 = 0$$

Solution

2.

$$5x - y = 6$$
$$4x - 3y = -4$$
$$-y = -5x + 6$$
$$y = 5x - 6$$

$$y = 5(2) - 6$$
$$y = 4$$
$$(2, 4)$$

$$4x - 3(5x - 6) = -4$$
$$4x - 15x + 18 = -4$$
$$-11x = -22$$
$$x = 2$$

3.

$x + y = 8$	$x + (10 - x^2) = 8$	$y = 10 - 2^2$	$y = 10 - (-1)^2$
$x^2 + y = 10$	$0 = x^2 - x - 2$	$y = 6$	$y = 9$
$y = 10 - x^2$	$0 = (x - 2)(x + 1)$		
$(2, 6), (-1, 9)$	$x = 2 \quad x = -1$		

4.

$(3, 2)$

5.

$$3x - 4y = -14$$
$$\underline{-3x + \ y = \ \ 8}$$

$-3y = -6$	$3x - 4(2) = -14$
$y = \ \ \ 2$	$3x \ \ \ \ \ = \ -6$
	$x \ \ \ \ \ = \ -2$

$(-2, 2)$

6.

$8x + 3y = \ \ \ 3 \implies 16x + 6y = \ \ \ 6$		$8\left(\frac{1}{4}\right) + 3y = 3$
$4x - 6y = -1 \implies \underline{\ \ 4x - 6y = -1}$		$3y = 1$
$20x \ \ \ \ \ \ = \ \ \ 5$		$y = \frac{1}{3}$
$x \ \ \ \ \ \ = \ \ \ \frac{1}{4}$		

$\left(\frac{1}{4}, \frac{1}{3}\right)$

7.

$x + 2y - \ 4z = \ \ 0$
$3x + \ y - \ 2z = \ \ 5$
$3x - \ y + \ 2z = \ \ 7$

$x + 2y - \ 4z = \ \ 0$
$\ \ \ \ \ \ -5y + 10z = \ \ 5$
$\ \ \ \ \ \ -7y + 14z = \ \ 7$

$x + 2y - \ 4z = \ \ 0$
$\ \ \ \ \ \ \ \ \ y - \ 2z = -1$
$\ \ \ \ -7y + 14z = \ \ 7$

$x + 2y - \ 4z = \ \ 0$
$\ \ \ \ \ \ \ \ \ y - \ 2z = -1$
$\ \ \ \ \ \ \ \ \ \ \ \ \ \ 0 = \ \ 0$

let $a = z$ (a is any real number)

$y = \ \ 2z - 1$
$x = -2y + 4z$
$\ \ = -2(2z - 1) + 4z$
$\ \ = -4z + 2 + 4z$
$x = \ \ 2$

$(2, 2a - 1, a)$

8.

$$\begin{bmatrix} 1 & 0 & -3 & \vdots & -10 \\ 0 & -2 & 2 & \vdots & 0 \\ 1 & -2 & 0 & \vdots & -7 \end{bmatrix}$$

$$\begin{matrix} -\frac{1}{2}R_2 \\ -R_1 + R_3 \end{matrix} \begin{bmatrix} 1 & 0 & -3 & \vdots & -10 \\ 0 & 1 & -1 & \vdots & 0 \\ 0 & -2 & 3 & \vdots & 3 \end{bmatrix}$$

$$2R_2 + R_3 \begin{bmatrix} 1 & 0 & -3 & \vdots & -10 \\ 0 & 1 & -1 & \vdots & 0 \\ 0 & 0 & 1 & \vdots & 3 \end{bmatrix}$$

$z = 3$	$y - 3 = 0$	$x - 3(3) = -10$
	$y = 3$	$x = -1$

$(-1, 3, 3)$

9.

$$\begin{bmatrix} 1 & -3 & 1 & \vdots & -3 \\ 3 & 2 & -5 & \vdots & 18 \\ 0 & 1 & 1 & \vdots & -1 \end{bmatrix}$$

$$-3R_1 + R_2 \begin{bmatrix} 1 & -3 & 1 & \vdots & -3 \\ 0 & 11 & -8 & \vdots & 27 \\ 0 & 1 & 1 & \vdots & -1 \end{bmatrix}$$

$$\begin{matrix} R_2 \\ R_3 \end{matrix} \begin{bmatrix} 1 & -3 & 1 & \vdots & -3 \\ 0 & 1 & 1 & \vdots & -1 \\ 0 & 11 & -8 & \vdots & 27 \end{bmatrix}$$

$$\begin{matrix} 3R_2 + R_1 \\ \\ -11R_2 + R_3 \end{matrix} \begin{bmatrix} 1 & 0 & 4 & \vdots & -6 \\ 0 & 1 & 1 & \vdots & -1 \\ 0 & 0 & -19 & \vdots & 38 \end{bmatrix}$$

$$\begin{matrix} \\ \\ -\frac{1}{19}R_3 \end{matrix} \begin{bmatrix} 1 & 0 & 4 & \vdots & -6 \\ 0 & 1 & 1 & \vdots & -1 \\ 0 & 0 & 1 & \vdots & -2 \end{bmatrix}$$

$$z = -2 \qquad y + (-2) = -1 \qquad x + 4(-2) = -6$$
$$y = 1 \qquad\qquad x = 2$$

$$(2, 1, -2)$$

10. $\begin{bmatrix} 2 & -7 & \vdots & 7 \\ 3 & 7 & \vdots & 13 \end{bmatrix}$

$$D = \begin{vmatrix} 2 & -7 \\ 3 & 7 \end{vmatrix} = 14 + 21 = 35$$

$$x = \frac{D_x}{D} = \frac{\begin{vmatrix} 7 & -7 \\ 13 & 7 \end{vmatrix}}{35} = \frac{49 + 91}{35} = \frac{140}{35} = 4$$

$$y = \frac{D_y}{D} = \frac{\begin{vmatrix} 2 & 7 \\ 3 & 13 \end{vmatrix}}{35} = \frac{26 - 21}{35} = \frac{5}{35} = \frac{1}{7}$$

$$\left(4, \tfrac{1}{7}\right)$$

11.

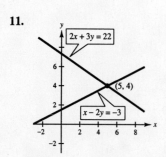

Solution: $(5, 4)$

12. $\begin{bmatrix} 3 & -2 & 1 & \vdots & 12 \\ 1 & -3 & 0 & \vdots & 2 \\ -3 & 0 & -9 & \vdots & -6 \end{bmatrix}$

$$\begin{matrix} R_1 \\ \\ -\frac{1}{3}R_3 \end{matrix} \begin{bmatrix} 1 & 0 & 3 & \vdots & 2 \\ 1 & -3 & 0 & \vdots & 2 \\ 3 & -2 & 1 & \vdots & 12 \end{bmatrix}$$

$$\begin{matrix} -1R_1 + R_2 \\ -3R_1 + R_3 \end{matrix} \begin{bmatrix} 1 & 0 & 3 & \vdots & 2 \\ 0 & -3 & -3 & \vdots & 0 \\ 0 & -2 & -8 & \vdots & 6 \end{bmatrix}$$

$$\begin{matrix} -\frac{1}{3}R_2 \\ -\frac{1}{2}R_3 \end{matrix} \begin{bmatrix} 1 & 0 & 3 & \vdots & 2 \\ 0 & 1 & 1 & \vdots & 0 \\ 0 & 1 & 4 & \vdots & -3 \end{bmatrix}$$

$$\begin{matrix} \\ \\ -R_2 + R_3 \end{matrix} \begin{bmatrix} 1 & 0 & 3 & \vdots & 2 \\ 0 & 1 & 1 & \vdots & 0 \\ 0 & 0 & 3 & \vdots & -3 \end{bmatrix}$$

$$\begin{matrix} \\ \\ \frac{1}{3}R_3 \end{matrix} \begin{bmatrix} 1 & 0 & 3 & \vdots & 2 \\ 0 & 1 & 1 & \vdots & 0 \\ 0 & 0 & 1 & \vdots & -1 \end{bmatrix}$$

$$\begin{matrix} -3R_3 + R_1 \\ -R_3 + R_2 \end{matrix} \begin{bmatrix} 1 & 0 & 0 & \vdots & 5 \\ 0 & 1 & 0 & \vdots & 1 \\ 0 & 0 & 1 & \vdots & -1 \end{bmatrix}$$

$$(5, 1, -1)$$

13.

$$\begin{bmatrix} 4 & 1 & 2 & \vdots & -4 \\ 0 & 3 & 1 & \vdots & 8 \\ -3 & 1 & -3 & \vdots & 5 \end{bmatrix}$$

$$\frac{1}{4}R_1 \begin{bmatrix} 1 & \frac{1}{4} & \frac{1}{2} & \vdots & -1 \\ 0 & 3 & 1 & \vdots & 8 \\ -3 & 1 & -3 & \vdots & 5 \end{bmatrix}$$

$$\begin{matrix} \\ \frac{1}{3}R_2 \\ 3R_1 + R_3 \end{matrix} \begin{bmatrix} 1 & \frac{1}{4} & \frac{1}{2} & \vdots & -1 \\ 0 & 1 & \frac{1}{3} & \vdots & \frac{8}{3} \\ 0 & \frac{7}{4} & -\frac{3}{2} & \vdots & 2 \end{bmatrix}$$

$$\begin{matrix} -\frac{1}{4}R_2 + R_1 \\ \\ -\frac{7}{4}R_2 + R_3 \end{matrix} \begin{bmatrix} 1 & 0 & \frac{5}{12} & \vdots & -\frac{5}{3} \\ 0 & 1 & \frac{1}{3} & \vdots & \frac{8}{3} \\ 0 & 0 & -\frac{25}{12} & \vdots & -\frac{8}{3} \end{bmatrix}$$

$$\begin{matrix} \\ \\ -\frac{12}{25}R_3 \end{matrix} \begin{bmatrix} 1 & 0 & \frac{5}{12} & \vdots & -\frac{5}{3} \\ 0 & 1 & \frac{1}{3} & \vdots & \frac{8}{3} \\ 0 & 0 & 1 & \vdots & \frac{32}{25} \end{bmatrix}$$

$$\begin{matrix} -\frac{5}{12}R_3 + R_1 \\ -\frac{1}{3}R_3 + R_2 \\ \\ \end{matrix} \begin{bmatrix} 1 & 0 & 0 & \vdots & -\frac{11}{5} \\ 0 & 1 & 0 & \vdots & \frac{56}{25} \\ 0 & 0 & 1 & \vdots & \frac{32}{25} \end{bmatrix}$$

$$\left(-\frac{11}{5}, \frac{56}{25}, \frac{32}{25}\right)$$

14. Inconsistent: no solutions

Consistent: one solution or infinitely many solutions

15.
$$\begin{vmatrix} 3 & -2 & 0 \\ -1 & 5 & 3 \\ 2 & 7 & 1 \end{vmatrix} = 0 - (3)\begin{vmatrix} 3 & -2 \\ 2 & 7 \end{vmatrix} + (1)\begin{vmatrix} 3 & -2 \\ -1 & 5 \end{vmatrix}$$

$$= (-3)(25) + (1)(13)$$

$$= -75 + 13$$

$$= -62$$

16.
$$\begin{vmatrix} 5 & -8 \\ 3 & a \end{vmatrix} = 0$$

$$5a + 24 = 0$$

$$5a = -24$$

$$a = -\frac{24}{5}$$

17. $x + y = 2$

$2x - y = 13$

There are many correct answers.

Write equations so that $(5, -3)$ satisfies each equation.

18. *Verbal model:*

$$\boxed{\begin{matrix}\text{Distance}\\\text{Person 1}\\\text{Drives}\end{matrix}} + \boxed{\begin{matrix}\text{Distance}\\\text{Person 2}\\\text{Drives}\end{matrix}} = \boxed{\begin{matrix}\text{Total}\\\text{Distance}\end{matrix}}$$

$$\boxed{\begin{matrix}\text{Distance}\\\text{Person 1}\\\text{Drives}\end{matrix}} = 4 \cdot \boxed{\begin{matrix}\text{Distance}\\\text{Person 2}\\\text{Drives}\end{matrix}}$$

Labels: Distance Person 1 Drives $= x$

Distance Person 2 Drives $= y$

System of equations: $\quad x + y = 200$
$\qquad\qquad\qquad x = 4y$

By substitution

$4y + y = 200$

$5y = 200$

$y = 40$ miles

$x = 4(40) = 160$ miles

19. $4 = a(0)^2 - b(0) - c \implies 4 = \quad\quad -c$

$3 = a(1)^2 - b(1) - c \implies 3 = \quad a - \quad b - c$

$6 = a(2)^2 - b(2) - c \implies 6 = 4a - 2b - c$

$$\begin{bmatrix} 0 & 0 & 1 & \vdots & 4 \\ 1 & 1 & 1 & \vdots & 3 \\ 4 & 2 & 1 & \vdots & 6 \end{bmatrix}$$

$$\begin{bmatrix} 1 & 1 & 1 & \vdots & 3 \\ 4 & 2 & 1 & \vdots & 6 \\ 0 & 0 & 1 & \vdots & 4 \end{bmatrix} = \begin{bmatrix} 1 & 1 & 1 & \vdots & 3 \\ 0 & -2 & -3 & \vdots & -6 \\ 0 & 0 & 1 & \vdots & 4 \end{bmatrix} = \begin{bmatrix} 1 & 1 & 1 & \vdots & 3 \\ 0 & 1 & \frac{3}{2} & \vdots & 3 \\ 0 & 0 & 1 & \vdots & 4 \end{bmatrix}$$

$c = 4 \quad\quad b - \frac{3}{2}(4) = \quad 3 \quad\quad\quad a - (-3) - 4 = 3$

$\quad\quad\quad\quad\quad\quad\quad\quad b = -3 \quad\quad\quad\quad\quad\quad a = 2$

$y = 2x^2 - 3x - 4$

20. *Verbal model:* $\boxed{\begin{array}{c}\text{Investment}\\1\end{array}} - \boxed{\begin{array}{c}\text{Investment}\\2\end{array}} - \boxed{\begin{array}{c}\text{Investment}\\3\end{array}} = \$25{,}000$

$4.5\% \cdot \boxed{\begin{array}{c}\text{Investment}\\1\end{array}} - 5\% \cdot \boxed{\begin{array}{c}\text{Investment}\\2\end{array}} - 8\% \cdot \boxed{\begin{array}{c}\text{Investment}\\3\end{array}} = \$1275$

$\boxed{\begin{array}{c}\text{Investment}\\1\end{array}} - 4000 = \boxed{\begin{array}{c}\text{Investment}\\3\end{array}} - 10{,}000$

Labels: Investment 1 $= x$

             Investment 2 $= y$

             Investment 3 $= z$

System of equations:
$$x - \quad y - \quad z = 25{,}000$$
$$0.045x - 0.05y - 0.08z = 1275$$

$$y - 4000 = z - 10{,}000 \implies y - z = 6000$$

Using determinants and Cramer's Rule:

$$\begin{bmatrix} 1 & 1 & 1 & 25{,}000 \\ .045 & .05 & .08 & 1275 \\ 0 & 1 & -1 & 6000 \end{bmatrix} \qquad D = \begin{vmatrix} 1 & 1 & 1 \\ .045 & .05 & .08 \\ 0 & 1 & -1 \end{vmatrix} = -.04$$

$$x = \frac{D_x}{D} = \frac{\begin{vmatrix} 25{,}000 & 1 & 1 \\ 1275 & .05 & .08 \\ 6000 & 1 & -1 \end{vmatrix}}{-.04} = \frac{-520}{-0.4} = \$13{,}000$$

$$y = \frac{D_y}{D} = \frac{\begin{vmatrix} 1 & 25{,}000 & 1 \\ .045 & 1275 & .08 \\ 0 & 6000 & -1 \end{vmatrix}}{-.04} = \frac{-360}{-.04} = \$9{,}000$$

$$z = \frac{D_z}{D} = \frac{\begin{vmatrix} 1 & 1 & 25{,}000 \\ .045 & .05 & 1275 \\ 0 & 1 & 6000 \end{vmatrix}}{-.04} = \frac{-120}{-.04} = \$3{,}000$$

Use your graphing calculator to find each determinant.

$\$13{,}000$ at 4.5% $\$9{,}000$ at 5% $\$3{,}000$ at 8%

21. $(x_1, y_1) = (0, 0)$, $(x_2, y_2) = (5, 4)$, $(x_3, y_3) = (6, 0)$

$$\begin{vmatrix} x_1 & y_1 & 1 \\ x_2 & y_2 & 1 \\ x_3 & y_3 & 1 \end{vmatrix} = \begin{vmatrix} 0 & 0 & 1 \\ 5 & 4 & 1 \\ 6 & 0 & 1 \end{vmatrix} = (1)\begin{vmatrix} 5 & 4 \\ 6 & 0 \end{vmatrix} = (1)(-24) = -24$$

Area $= -\frac{1}{2}(-24) = 12$

C H A P T E R 8
Rational Expressions, Equations, and Functions

CHAPTER 8
Rational Expressions, Equations, and Functions

Section 8.1   Rational Expressions and Functions

Solutions to Odd-Numbered Exercises

1. $x - 8 \neq 0$

$x \neq 8$

$D = (-\infty, 8) \cup (8, \infty)$

3. $x + 4 \neq 0$

$x \neq -4$

$D = (-\infty, -4) \cup (-4, \infty)$

5. $4 \neq 0$

$D = (-\infty, \infty)$

7. $D = (-\infty, \infty)$

9. $x^2 + 4 \neq 0$

$D = (-\infty, \infty)$

11. $y(y + 3) \neq 0$

$y \neq 0 \quad y \neq -3$

$D = (-\infty, -3) \cup (-3, 0) \cup (0, \infty)$

13. $\quad t^2 - 16 \neq 0$

$(t - 4)(t + 4) \neq 0$

$t \neq 4 \quad t \neq -4$

$D = (-\infty, -4) \cup (-4, 4) \cup (4, \infty)$

15. $\quad y^2 - 3y \neq 0$

$y(y - 3) \neq 0$

$y \neq 0 \quad y \neq 3$

$D = (-\infty, 0) \cup (0, 3) \cup (3, \infty)$

17. $\quad x^2 - 5x + 6 \neq 0$

$(x - 3)(x - 2) \neq 0$

$x \neq 3 \quad x \neq 2$

$D = (-\infty, 2) \cup (2, 3) \cup (3, \infty)$

19. $\quad 3u^2 - 2u - 5 \neq 0$

$(3u - 5)(u + 1) \neq 0$

$u \neq \frac{5}{3} \quad u \neq -1$

$D = (-\infty, -1) \cup \left(-1, \frac{5}{3}\right) \cup \left(\frac{5}{3}, \infty\right)$

21. (a) $f(1) = \dfrac{4(1)}{1 + 3} = \dfrac{4}{4} = 1$

(c) $f(-3) = \dfrac{4(-3)}{-3 + 3} = \dfrac{-12}{0}$

$ = \text{not possible; undefined}$

(b) $f(-2) = \dfrac{4(-2)}{-2 + 3} = \dfrac{-8}{1} = -8$

(d) $f(0) = \dfrac{4(0)}{0 + 3} = \dfrac{0}{3} = 0$

23. (a) $g(0) = \dfrac{0^2 - 4(0)}{0^2 - 9} = 0$

(c) $g(3) = \dfrac{3^2 - 4(3)}{3^2 - 9} = \dfrac{9 - 12}{9 - 9} = \dfrac{-3}{0}$

$ = \text{not possible; undefined}$

(b) $g(4) = \dfrac{4^2 - 4(4)}{4^2 - 9} = \dfrac{16 - 16}{16 - 9} = \dfrac{0}{7} = 0$

(d) $g(-3) = \dfrac{(-3)^2 - 4(-3)}{(-3)^2 - 9} = \dfrac{9 + 12}{9 - 9} = \dfrac{21}{0}$

$ = \text{not possible; undefined}$

25. (a) $h(10) = \dfrac{10^2}{10^2 - 10 - 2} = \dfrac{100}{88} = \dfrac{25}{22}$

(b) $h(0) = \dfrac{0^2}{0^2 - 0 - 2} = \dfrac{0}{-2} = 0$

(c) $h(-1) = \dfrac{(-1)^2}{(-1)^2 - (-1) - 2} = \dfrac{1}{1 + 1 - 2} = \dfrac{1}{0}$

$= $ not possible; undefined

(d) $h(2) = \dfrac{2^2}{2^2 - 2 - 2} = \dfrac{4}{4 - 2 - 2} = \dfrac{4}{0}$

$= $ not possible; undefined

27. Since length must be positive, $x \geq 0$. Since $\dfrac{500}{x}$ must be defined, $x \neq 0$. Therefore, the domain is $x > 0$ or $(0, \infty)$.

29. $x = $ units of a product

$D = \{1, 2, 3, 4, \ldots\}$

31. Since p is the percent of air pollutants in the stack emission of a utility, $0 \leq p \leq 100$. Since

$$\dfrac{80{,}000p}{100 - p}$$

must be defined, $p \neq 100$. Therefore, the domain is $[0, 100)$.

33. $\dfrac{5}{6} = \dfrac{5(x + 3)}{6(x + 3)}, \quad x \neq -3$

35. $\dfrac{x}{2} = \dfrac{3x(x + 16)^2}{2(3(x + 16)^2)}, \quad x \neq -16$

37. $\dfrac{x + 5}{3x} = \dfrac{(x + 5)(x(x - 2))}{3x^2(x - 2)}, \quad x \neq 2$

39. $\dfrac{8x}{x - 5} = \dfrac{8x(x + 2)}{x^2 - 3x - 10}, \quad x \neq -2$

41. $\dfrac{5x}{25} = \dfrac{5x}{5 \cdot 5} = \dfrac{x}{5}$

43. $\dfrac{12y^2}{2y} = \dfrac{2 \cdot 6 \cdot y \cdot y}{2 \cdot y}$

$= 6y, \quad y \neq 0$

45. $\dfrac{18x^2y}{15xy^4} = \dfrac{3 \cdot 6 \cdot x \cdot x \cdot y}{3 \cdot 5 \cdot x \cdot y \cdot y^3}$

$= \dfrac{6x}{5y^3}, \quad x \neq 0$

47. $\dfrac{3x^2 - 9x}{12x^2} = \dfrac{3x(x - 3)}{12x^2} = \dfrac{(x - 3)}{4x}$

49. $\dfrac{x^2(x - 8)}{x(x - 8)} = \dfrac{x \cdot x(x - 8)}{x(x - 8)}$

$= x, \quad x \neq 0, x \neq 8$

51. $\dfrac{2x - 3}{4x - 6} = \dfrac{2x - 3}{2(2x - 3)} = \dfrac{1}{2}, \quad x \neq \dfrac{3}{2}$

53. $\dfrac{5 - x}{3x - 15} = \dfrac{-1(x - 5)}{3(x - 5)}$

$= -\dfrac{1}{3}, \quad x \neq 5$

55. $\dfrac{a + 3}{a^2 + 6a + 9} = \dfrac{a + 3}{(a + 3)(a + 3)}$

$= \dfrac{1}{a + 3}$

57. $\dfrac{x^2 - 7x}{x^2 - 14x + 49} = \dfrac{x(x - 7)}{(x - 7)(x - 7)}$

$= \dfrac{x}{x - 7}$

59. $\dfrac{y^3 - 4y}{y^2 + 4y - 12} = \dfrac{y(y^2 - 4)}{(y + 6)(y - 2)}$

$= \dfrac{y(y - 2)(y + 2)}{(y + 6)(y - 2)}$

$= \dfrac{y(y + 2)}{y + 6}, \quad y \neq 2$

61. $\dfrac{x^3 - 4x}{x^2 - 5x + 6} = \dfrac{x(x^2 - 4)}{(x - 3)(x - 2)}$

$= \dfrac{x(x - 2)(x + 2)}{(x - 3)(x - 2)}$

$= \dfrac{x(x + 2)}{x - 3}, \quad x \neq 2$

63. $\dfrac{3x^2 - 7x - 20}{12 + x - x^2} = \dfrac{(3x + 5)(x - 4)}{-1(x^2 - x - 12)}$

$= \dfrac{(3x + 5)(x - 4)}{-1(x - 4)(x + 3)}$

$= -\dfrac{3x + 5}{x + 3}, \quad x \neq 4$

65. $\dfrac{2x^2 + 19x + 24}{2x^2 - 3x - 9} = \dfrac{(2x + 3)(x + 8)}{(2x + 3)(x - 3)}$

$= \dfrac{x + 8}{x - 3}, \quad x \neq -\dfrac{3}{2}$

67. $\dfrac{15x^2 + 7x - 4}{25x^2 - 16} = \dfrac{(5x + 4)(3x - 1)}{(5x + 4)(5x - 4)}$

$= \dfrac{3x - 1}{5x - 4}, \quad x \neq -\dfrac{4}{5}$

69. $\dfrac{3xy^2}{xy^2 + x} = \dfrac{3xy^2}{x(y^2 + 1)}$

$= \dfrac{3y^2}{y^2 + 1}, \quad x \neq 0$

71. $\dfrac{y^2 - 64x^2}{5(3y + 24x)} = \dfrac{(y - 8x)(y + 8x)}{15(y + 8x)}$

$\qquad = \dfrac{y - 8x}{15}, \quad y \ne -8x$

73. $\dfrac{5xy + 3x^2y^2}{xy^3} = \dfrac{xy(5 + 3xy)}{xy \cdot y^2}$

$\qquad = \dfrac{5 + 3xy}{y^2}, \quad x \ne 0$

75. $\dfrac{u^2 - 4v^2}{u^2 + uv - 2v^2} = \dfrac{(u - 2v)(u + 2v)}{(u - v)(u + 2v)}$

$\qquad = \dfrac{u - 2v}{u - v}, \quad u \ne -2v$

77. $\dfrac{3m^2 - 12n^2}{m^2 + 4mn + 4n^2} = \dfrac{3(m^2 - 4n^2)}{(m + 2n)(m + 2n)}$

$\qquad = \dfrac{3(m - 2n)(m + 2n)}{(m + 2n)(m + 2n)}$

$\qquad = \dfrac{3(m - 2n)}{m + 2n}$

79. $\dfrac{x - 4}{4} \ne x - 1$

$\dfrac{10 - 4}{4} \ne 10 - 1$ Choose a value such as 10 for x and evaluate both sides.

$\dfrac{6}{4} \ne 9$

81. $\dfrac{3x + 2}{4x + 2} \ne \dfrac{3}{4}$

$\dfrac{3(0) + 2}{4(0) + 2} \ne \dfrac{3}{4}$ Choose a value such as 0 for x and evaluate both sides.

$1 \ne \dfrac{3}{4}$

83.

x	-2	-1	0	1	2	3	4
$\dfrac{x^2 - x - 2}{x - 2}$	-1	0	1	2	Undefined	4	5
$x + 1$	-1	0	1	2	3	4	5

Domain of $\dfrac{x^2 - x - 2}{x - 2}$ is $(-\infty, 2) \cup (2, \infty)$.

Domain of $x + 1$ is $(-\infty, \infty)$.

The two expressions are equal for all replacements of the variable x except 2.

85. $\dfrac{\text{Area of shaded portion}}{\text{Area of total figure}} = \dfrac{x(x + 1)}{(x + 1)(x + 3)} = \dfrac{x}{x + 3}, \quad x > 0$

87. (a) *Verbal Model:* $\boxed{\begin{array}{c}\text{Total}\\\text{cost}\end{array}} = \boxed{\begin{array}{c}\text{Number}\\\text{of units}\end{array}} \cdot \boxed{\begin{array}{c}\text{Cost per}\\\text{unit}\end{array}} + \boxed{\begin{array}{c}\text{Initial}\\\text{cost}\end{array}}$

Labels: Total cost $= C$

Number of units $= x$

Equation: $2500 + 9.25x = C$

(b) *Verbal Model:* $\boxed{\begin{array}{c}\text{Average}\\\text{cost}\end{array}} = \boxed{\begin{array}{c}\text{Total}\\\text{cost}\end{array}} \div \boxed{\begin{array}{c}\text{Number}\\\text{of units}\end{array}}$

Label: Average cost $= \overline{C}$

Equation: $\overline{C} = \dfrac{2500 + 9.25x}{x}$

(c) Domain $= \{1, 2, 3, 4, \ldots\}$

(d) $\dfrac{2500 + 9.25(100)}{100} = \$34.25$

89. (a) *Verbal Model:* $\boxed{\text{Distance}} = \boxed{\text{Rate}} \cdot \boxed{\text{Time}}$

Van: $45(t + 3)$

Car: $60t$

(b) Distance between van and car $= d$

$$= 45(t + 3) - 60t$$
$$= 45t + 135 - 60t$$
$$= 135 - 15t$$
$$= 15(9 - t)$$

(c) $\dfrac{\text{Distance of car}}{\text{Distance of van}} = \dfrac{60t}{45(t + 3)} = \dfrac{4t}{3(t + 3)}$

91. $\dfrac{\text{Circular pool volume}}{\text{Rectangular pool volume}} = \dfrac{\pi(3d)^2(d + 2)}{d(3d)(3d + 6)}$

$$= \dfrac{\pi(3d)^2(d + 2)}{3d^2 \cdot 3(d + 2)}$$

$$= \dfrac{\pi(3d)^2(d + 2)}{(3d)^2(d + 2)}$$

$$= \pi$$

93. Average cost of Medicare per person $= \dfrac{107.30 + 15.09t \text{ billion}}{34.26 + 0.65t \text{ million}} = \dfrac{(10,730 + 1509t)1000}{3426 + 65t}$

95. Let u and v be polynomials. The algebraic expression u/v is a rational expression.

97. The rational expression is in simplified form if the numerator and denominator have no factors in common (other than ± 1).

99. You can cancel only common factors.

Section 8.2 Multiplying and Dividing Rational Expressions

1. (a) $x = 10$: $\dfrac{10 - 10}{4(10)} = \dfrac{0}{40} = 0$

(b) $x = 0$: $\dfrac{0 - 10}{4(0)} = \dfrac{-10}{0} = $ undefined

(c) $x = -2$: $\dfrac{-2 - 10}{4(-2)} = \dfrac{-12}{-8} = \dfrac{3}{2}$

(d) $x = 12$: $\dfrac{12 - 10}{4(12)} = \dfrac{2}{48} = \dfrac{1}{24}$

3. $\dfrac{7}{3y} = \dfrac{7x^2}{3y(x^2)}, \quad x \neq 0$

5. $\dfrac{3x}{x - 4} = \dfrac{3x(x + 2)^2}{(x - 4)(x + 2)^2}, \quad x \neq -2$

7. $\dfrac{3u}{7v} = \dfrac{3u(u + 1)}{7v(u + 1)}, \quad u \neq -1$

9. $\dfrac{13x}{x - 2} = \dfrac{13x((-1)(2 + x))}{4 - x^2}, \quad x \neq -2$

11. $\dfrac{45}{28} \cdot \dfrac{77}{60} = \dfrac{9 \cdot 5 \cdot 7 \cdot 11}{7 \cdot 4 \cdot 6 \cdot 10} = \dfrac{33}{16}$

13. $7x \cdot \dfrac{9}{14x} = \dfrac{7x \cdot 3 \cdot 3}{7 \cdot 2 \cdot x} = \dfrac{9}{2}$

15. $\dfrac{8s^3}{9s} \cdot \dfrac{6s^2}{32s} = \dfrac{8s^3 \cdot 3 \cdot 2s \cdot s}{3 \cdot 3 \cdot s \cdot 8 \cdot 2 \cdot 2 \cdot s} = \dfrac{s^3}{6}, \quad s \neq 0$

17. $16u^4 \cdot \dfrac{12}{8u^2} = \dfrac{8 \cdot 2 \cdot u^2 \cdot u^2 \cdot 12}{8 \cdot u^2} = 24u^2, \quad u \neq 0$

19. $\dfrac{8}{3 + 4x} \cdot (9 + 12x) = \dfrac{8 \cdot 3(3 + 4x)}{3 + 4x}$

$$= 24, \quad x \neq -\dfrac{3}{4}$$

21. $\dfrac{8u^2v}{3u + v} \cdot \dfrac{u + v}{12u} = \dfrac{4 \cdot 2 \cdot u \cdot u \cdot v(u + v)}{(3u + v) \cdot 4 \cdot 3 \cdot u}$

$$= \dfrac{2uv(u + v)}{3(3u + v)}, \quad u \neq 0$$

23. $\dfrac{12 - r}{3} \cdot \dfrac{3}{r - 12} = \dfrac{-1(r - 12) \cdot 3}{3(r - 12)} = -1, \quad r \neq 12$

25. $\dfrac{(2x - 3)(x + 8)}{x^3} \cdot \dfrac{x}{3 - 2x} = \dfrac{(2x - 3)(x + 8)x}{x \cdot x^2 \cdot -1(2x - 3)}$

$$= \dfrac{x + 8}{-x^2}, \quad x \neq \dfrac{3}{2}$$

27. $\dfrac{4r - 12}{r - 2} \cdot \dfrac{r^2 - 4}{r - 3} = \dfrac{4(r - 3)(r - 2)(r + 2)}{(r - 2) \cdot (r - 3)} = 4(r + 2), \quad r \neq 3, r \neq 2$

29. $\dfrac{2t^2 - t - 15}{t + 2} \cdot \dfrac{t^2 - t - 6}{t^2 - 6t + 9} = \dfrac{(2t + 5)(t - 3)(t - 3)(t + 2)}{(t + 2)(t - 3)(t - 3)} = 2t + 5, \quad t \neq 3, t \neq -2$

31. $(x^2 - 4y^2) \cdot \dfrac{xy}{(x - 2y)^2} = (x - 2y)(x + 2y) \cdot \dfrac{xy}{(x - 2y)^2} = \dfrac{(x + 2y)xy}{x - 2y}$

33. $\dfrac{x^2 + 2xy - 3y^2}{(x + y)^2} \cdot \dfrac{x^2 - y^2}{x + 3y} = \dfrac{(x + 3y)(x - y)}{(x + y)^2} \cdot \dfrac{(x - y)(x + y)}{x + 3y} = \dfrac{(x - y)^2}{x + y}, \quad x \neq -3y$

35. $\dfrac{x + 5}{x - 5} \cdot \dfrac{2x^2 - 9x - 5}{3x^2 + x - 2} \cdot \dfrac{x^2 - 1}{x^2 + 7x + 10} = \dfrac{x + 5}{x - 5} \cdot \dfrac{(2x + 1)(x - 5)}{(3x - 2)(x + 1)} \cdot \dfrac{(x - 1)(x + 1)}{(x + 5)(x + 2)}$

$$= \dfrac{(x + 5)(2x + 1)(x - 5)(x - 1)(x + 1)}{(x - 5)(3x - 2)(x + 1)(x + 5)(x + 2)}$$

$$= \dfrac{(2x + 1)(x - 1)}{(3x - 2)(x + 2)}, \quad x \neq \pm 5, -1$$

37. $\dfrac{9 - x^2}{2x + 3} \cdot \dfrac{4x^2 + 8x - 5}{4x^2 - 8x + 3} \cdot \dfrac{6x^4 - 2x^3}{8x^2 + 4x} = \dfrac{(3 - x)(3 + x)}{2x + 3} \cdot \dfrac{(2x + 5)(2x - 1)}{(2x - 3)(2x - 1)} \cdot \dfrac{2x^3(3x - 1)}{4x(2x + 1)}$

$$= \dfrac{-1(x - 3)(x + 3)(2x + 5)x^2(3x - 1)}{(2x + 3)(2x - 3)2(2x + 1)}$$

$$= \dfrac{(x^2 - 9)(2x + 5)x^2(3x - 1)}{(2x + 3)(3 - 2x)2(2x + 1)}, \quad x \neq 0, \dfrac{1}{2}$$

39. $\dfrac{x^3 + 3x^2 - 4x - 12}{x^3 - 3x^2 - 4x + 12} \cdot \dfrac{x^2 - 9}{x} = \dfrac{x^2(x + 3) - 4(x + 3)}{x^2(x - 3) - 4(x - 3)} \cdot \dfrac{(x + 3)(x - 3)}{x}$

$$= \dfrac{(x + 3)(x^2 - 4)(x + 3)(x - 3)}{(x - 3)(x^2 - 4) \cdot x}$$

$$= \dfrac{(x + 3)^2}{x}, \quad x \neq -2, 2, 3$$

41. $\dfrac{-5}{12} \div \dfrac{45}{32} = \dfrac{-5}{12} \cdot \dfrac{32}{45} = \dfrac{-5 \cdot 8 \cdot 4}{4 \cdot 3 \cdot 5 \cdot 9} = \dfrac{-8}{27}$

43. $x^2 \div \dfrac{3x}{4} = x^2 \cdot \dfrac{4}{3x} = \dfrac{4x}{3}, \quad x \neq 0$

45. $\dfrac{7xy^2}{10u^2v} \div \dfrac{21x^3}{45uv} = \dfrac{7xy^2}{10u^2v} \cdot \dfrac{45uv}{21x^3}$

$$= \dfrac{7xy^2 \cdot 3 \cdot 3 \cdot 5 \cdot u \cdot v}{5 \cdot 2 \cdot u \cdot u \cdot v \cdot 7 \cdot 3x \cdot x^2}$$

$$= \dfrac{3y^2}{2ux^2}, \quad v \neq 0$$

47. $\dfrac{3(a + b)}{4} \div \dfrac{(a + b)^2}{2} = \dfrac{3(a + b)}{4} \cdot \dfrac{2}{(a + b)^2}$

$$= \dfrac{3(a + b) \cdot 2}{2 \cdot 2 \cdot (a + b)(a + b)}$$

$$= \dfrac{3}{2(a + b)}$$

49. $\dfrac{(x^3y)^2}{(x+2y)^2} \div \dfrac{x^2y}{(x+2y)^3} = \dfrac{(x^3y)^2}{(x+2y)^2} \cdot \dfrac{(x+2y)^3}{x^2y}$

$$= \dfrac{(x^3y)(x^3y)(x+2y)^2(x+2y)}{(x+2y)^2 x^2 y}$$

$$= \dfrac{(x^3y)(x^2 \cdot xy)(x+2y)}{x^2 y}$$

$$= x^4 y(x+2y), \quad x \neq 0, y \neq 0, x \neq -2y$$

51. $\dfrac{\left(\dfrac{x^2}{12}\right)}{\left(\dfrac{5x}{18}\right)} = \dfrac{x^2}{12} \div \dfrac{5x}{18}$

$$= \dfrac{x^2}{12} \cdot \dfrac{18}{5x}$$

$$= \dfrac{x^2 \cdot 3 \cdot 3 \cdot 2}{2 \cdot 2 \cdot 3 \cdot 5 \cdot x}$$

$$= \dfrac{3x}{10}, \quad x \neq 0$$

53. $\dfrac{\left(\dfrac{25x^2}{x-5}\right)}{\left(\dfrac{10x}{5+4x-x^2}\right)} = \dfrac{25x^2}{x-5} \div \dfrac{10x}{5+4x-x^2}$

$$= \dfrac{25x^2}{x-5} \cdot \dfrac{5+4x-x^2}{10x}$$

$$= \dfrac{5 \cdot 5 \cdot x \cdot x \cdot (-1)(x^2-4x-5)}{(x-5) \cdot 5 \cdot 2 \cdot x}$$

$$= \dfrac{5 \cdot x \cdot (-1)(x-5)(x+1)}{(x-5)2}$$

$$= \dfrac{-5x(x+1)}{2}, \quad x \neq 0, 5, -1$$

55. $\dfrac{16x^2+8x+1}{3x^2+8x-3} \div \dfrac{4x^2-3x-1}{x^2+6x+9} = \dfrac{16x^2+8x+1}{3x^2+8x-3} \cdot \dfrac{x^2+6x+9}{4x^2-3x-1}$

$$= \dfrac{(4x+1)(4x+1)}{(3x-1)(x+3)} \cdot \dfrac{(x+3)(x+3)}{(4x+1)(x-1)}$$

$$= \dfrac{(4x+1)(4x+1)(x+3)(x+3)}{(3x-1)(x+3)(4x+1)(x-1)}$$

$$= \dfrac{(4x+1)(x+3)}{(3x-1)(x-1)}, \quad x \neq -3, -\dfrac{1}{4}$$

57. $\dfrac{x^2+3x-2x-6}{x^2-4} \div \dfrac{x+3}{x^2+4x+4} = \dfrac{x(x+3)-2(x+3)}{x^2-4} \cdot \dfrac{x^2+4x+4}{x+3}$

$$= \dfrac{(x+3)(x-2)}{(x-2)(x+2)} \cdot \dfrac{(x+2)(x+2)}{x+3}$$

$$= \dfrac{(x+3)(x-2)(x+2)(x+2)}{(x-2)(x+2)(x+3)}$$

$$= (x+2), \quad x \neq \pm 2, -3$$

59. $\dfrac{\left(\dfrac{x^2-3x-10}{x^2-4x+4}\right)}{\left(\dfrac{21+4x-x^2}{x^2-5x-14}\right)} = \dfrac{x^2-3x-10}{x^2-4x+4} \div \dfrac{21+4x-x^2}{x^2-5x-14}$

$$= \dfrac{x^2-3x-10}{x^2-4x+4} \cdot \dfrac{x^2-5x-14}{1(x^2-4x-21)}$$

$$= \dfrac{(x-5)(x+2)}{(x-2)(x-2)} \cdot \dfrac{(x-7)(x+2)}{-1(x-7)(x+3)}$$

$$= -\dfrac{(x^2-3x-10)(x+2)}{(x^2-4x+4)(x+3)}, \quad x \neq \pm 2, 7$$

61. $\left[\dfrac{x^2}{9} \cdot \dfrac{3(x+4)}{x^2+2x}\right] \div \dfrac{x}{x+2} = \dfrac{x^2}{9} \cdot \dfrac{3(x+4)}{x(x+2)} \cdot \dfrac{x+2}{x}$

$$= \dfrac{x+4}{3}, \quad x \neq -2, 0$$

63. $\left[\dfrac{xy + y}{4x} \div (3x + 3)\right] \div \dfrac{y}{3x} = \dfrac{y(x + 1)}{4x} \cdot \dfrac{1}{3(x + 1)} \cdot \dfrac{3x}{y} = \dfrac{1}{4}, \quad x \neq -1, 0, y \neq 0$

65. $\dfrac{2x^2 + 5x - 25}{3x^2 + 5x + 2} \cdot \dfrac{3x^2 + 2x}{x + 5} \div \left(\dfrac{x}{x + 1}\right)^2 = \dfrac{(2x - 5)(x + 5)}{(3x + 2)(x + 1)} \cdot \dfrac{x(3x + 2)}{x + 5} \cdot \left(\dfrac{x + 1}{x}\right)^2$

$$= \dfrac{(2x - 5)(x + 5)x(3x + 2)(x + 1)(x + 1)}{(3x + 2)(x + 1)(x + 5)x \cdot x}$$

$$= \dfrac{(2x - 5)(x + 1)}{x}, x \neq -1, -5, -\dfrac{2}{3}$$

67. $x^3 \cdot \dfrac{x^{2n} - 9}{x^{2n} + 4x^n + 3} \div \dfrac{x^{2n} - 2x^n - 3}{x} = x^3 \cdot \dfrac{(x^n - 3)(x^n + 3)}{(x^n + 3)(x^n + 1)} \cdot \dfrac{x}{(x^n - 3)(x^n + 1)}$

$$= \dfrac{x^4}{(x^n + 1)^2}, \quad x^n \neq -3, 3, 0$$

69. *Keystrokes:*

y_1 `Y=` `(` `(` `(` `3` `X,T,θ` `+` `2` `)` `÷` `X,T,θ` `)` `×`
`(` `X,T,θ` `x²` `÷` `(` `9` `X,T,θ` `x²` `−` `4` `)` `)` `)` `ENTER`
y_2 `X,T,θ` `÷` `(` `3` `X,T,θ` `−` `2` `)` `GRAPH`

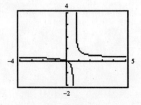

71. *Keystrokes:*

y_1 `Y=` `(` `(` `(` `3` `X,T,θ` `+` `15` `)` `÷` `X,T,θ` `^` `4` `)` `÷`
`(` `(` `X,T,θ` `+` `5` `)` `÷` `X,T,θ` `x²` `)` `)` `ENTER`
y_2 `3` `÷` `X,T,θ` `x²` `GRAPH`

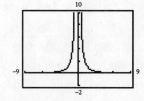

73. Area $= \left(\dfrac{2w + 3}{3}\right)\left(\dfrac{w}{2}\right) = \dfrac{(2w + 3)w}{6}$

75. $\dfrac{\text{Unshaded Area}}{\text{Total Area}} = \dfrac{\dfrac{x}{2} \cdot \dfrac{x}{2}}{x(2x + 1)} = \left[\dfrac{x}{2} \cdot \dfrac{x}{2}\right] \div [x(2x + 1)]$

$$= \dfrac{x}{2} \cdot \dfrac{x}{2} \cdot \dfrac{1}{x(2x + 1)}$$

$$= \dfrac{x}{4(2x + 1)}$$

77. $\dfrac{\text{Unshaded Area}}{\text{Total Area}} = \dfrac{x \cdot \dfrac{x}{2}}{x(2x + 1)} = \dfrac{x}{2(2x + 1)}$

79. (a) $\dfrac{20 \text{ pages}}{1 \text{ minute}} = \dfrac{20 \text{ pages}}{60 \text{ seconds}} = \dfrac{1 \text{ page}}{3 \text{ seconds}}$,

$t = 3$ seconds or $\dfrac{1}{20}$ minutes

(b) $\dfrac{3 \text{ seconds}}{1 \text{ page}} \cdot x$ pages $= 3x$ seconds or $\dfrac{x}{20}$ minutes

(c) $\dfrac{3 \text{ seconds}}{1 \text{ page}} \cdot 35$ pages $= 3 \cdot 35$ seconds

$= 105$ seconds or $\dfrac{7}{4}$ minutes

81. (a) *Keystrokes:*

y_1 [Y=] 6357 [+] 1070 [X,T,θ] [x^2] [ENTER]

y_2 6115.2 [+] 590.7 [X,T,θ] [x^2] [GRAPH]

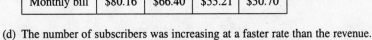

(b) *Verbal Model:*

$$\boxed{\text{Average monthly bill}} = \boxed{\dfrac{\text{Revenue}}{\text{Number of subscribers}}}$$

Equation: $AMB = \dfrac{(6115.2 + 590.7t^2)1000}{(6357 + 1070t^2)12} = \dfrac{6115200 + 590700t^2}{(6357 + 1070t^2)12}$

(c)

Year, t	0	2	4	6
Monthly bill	$80.16	$66.40	$55.21	$50.70

(d) The number of subscribers was increasing at a faster rate than the revenue.

83. Divide a rational expression by a polynomial by multiplying the rational expression by the reciprocal of the polynomial.

85. Invert the divisor, not the dividend.

Section 8.3   Adding and Subtracting Rational Expressions

1. $\dfrac{5}{8} + \dfrac{7}{8} = \dfrac{5+7}{8} = \dfrac{12}{8} = \dfrac{3}{2}$

3. $\dfrac{5x}{8} - \dfrac{7x}{8} = \dfrac{-2x}{8} = \dfrac{-x}{4}$

5. $\dfrac{2}{3a} - \dfrac{11}{3a} = \dfrac{2-11}{3a} = \dfrac{-9}{3a} = \dfrac{-3}{a}$

7. $\dfrac{x}{9} - \dfrac{x+2}{9} = \dfrac{x-(x+2)}{9} = \dfrac{x-x-2}{9} = -\dfrac{2}{9}$

9. $\dfrac{z^2}{3} + \dfrac{z^2-2}{3} = \dfrac{z^2+z^2-2}{3} = \dfrac{2z^2-2}{3}$

11. $\dfrac{2x+5}{3} + \dfrac{1-x}{3} = \dfrac{2x+5+1-x}{3} = \dfrac{x+6}{3}$

13. $\dfrac{3y}{3} - \dfrac{3y-3}{3} - \dfrac{7}{3} = \dfrac{3y-(3y-3)-7}{3}$

$\qquad = \dfrac{3y-3y+3-7}{3}$

$\qquad = -\dfrac{4}{3}$

15. $\dfrac{3y-22}{y-6} - \dfrac{2y-16}{y-6} = \dfrac{3y-22-(2y-16)}{y-6}$

$\qquad = \dfrac{3y-22-2y+16}{y-6}$

$\qquad = \dfrac{y-6}{y-6}$

$\qquad = 1, y \neq 6$

17. $\dfrac{2x-1}{x(x-3)} + \dfrac{1-x}{x(x-3)} = \dfrac{2x-1+1-x}{x(x-3)}$

$\qquad = \dfrac{x}{x(x-3)}$

$\qquad = \dfrac{1}{x-3}, \quad x \neq 0$

19. $5x^2 = 5 \cdot x \cdot x$

$20x^3 = 5 \cdot 2 \cdot 2 \cdot x \cdot x \cdot x$

$LCM = 20x^3$

21. $9y^3 = 3 \cdot 3 \cdot y \cdot y \cdot y$

$12y = 2 \cdot 2 \cdot 3 \cdot y$

$LCM = 3 \cdot 3 \cdot 2 \cdot 2 \cdot y \cdot y \cdot y = 36y^3$

23. $15x^2 = 5 \cdot 3 \cdot x \cdot x$

$3(x+5) = 3 \cdot (x+5)$

$LCM = 15x^2(x+5)$

25. $63z^2(z + 1) = 7 \cdot 9 \cdot z \cdot z(z + 1)$

$14(z + 1)^4 = 7 \cdot 2 \cdot (z + 1)^4$

$\text{LCM} = 126z^2(z + 1)^4$

27. $8t(t + 2) = 2 \cdot 2 \cdot 2 \cdot t \cdot (t + 2)$

$14(t^2 - 4) = 2 \cdot 7 \cdot (t + 2)(t - 2)$

$\text{LCM} = 2 \cdot 2 \cdot 2 \cdot 7 \cdot t \cdot (t + 2)(t - 2) = 56t(t^2 - 4)$

29. $6(x^2 - 4) = 6(x - 2)(x + 2)$

$2x(x + 2) = 2 \cdot x \cdot (x + 2)$

$\text{LCM} = 6x(x - 2)(x + 2)$

31. $\dfrac{7x^2}{4a(x^2)} = \dfrac{7}{4a}, \quad x \neq 0$

33. $\dfrac{5r(u + 1)}{3v(u + 1)} = \dfrac{5r}{3v}, \quad u \neq -1$

35. $\dfrac{7y(-1(x + 2))}{4 - x^2} = \dfrac{7y}{x - 2}, \quad x \neq -2$

$4 - x^2 = (2 - x)(2 + x)$

$= -1(x - 2)(2 + x)$

37. $\dfrac{n + 8}{3n - 12} = \dfrac{n + 8}{3(n - 4)} = \dfrac{n + 8(2n^2)}{3(n - 4)(2n^2)} = \dfrac{2n^2(n + 8)}{6n^2(n - 4)}$

$\dfrac{10}{6n^2} = \dfrac{10}{3 \cdot 2n^2} = \dfrac{10(n - 4)}{3 \cdot 2n^2(n - 4)} = \dfrac{10(n - 4)}{6n^2(n - 4)}$

$\text{LCD} = 6n^2(n - 4)$

39. $\dfrac{2}{x^2(x - 3)} = \dfrac{2(x + 3)}{x^2(x - 3)(x + 3)}$

$\dfrac{5}{x(x + 3)} = \dfrac{5x(x - 3)}{x^2(x + 3)(x - 3)}$

$\text{LCD} = x^2(x - 3)(x + 3) = x^2(x^2 - 9)$

41. $\dfrac{v}{2v^2 + 2v} = \dfrac{v}{2v(v + 1)} = \dfrac{v(3v)}{2v(v + 1)(3v)} = \dfrac{3v^2}{6v^2(v + 1)}$

$\dfrac{4}{3v^2} = \dfrac{4(2(v + 1))}{3v^2(2(v + 1))} = \dfrac{8v + 8}{6v^2(v + 1)}$

$\text{LCD} = 6v^2(v + 1)$

43. $\dfrac{x - 8}{x^2 - 25} = \dfrac{x - 8}{(x - 5)(x + 5)}$

$= \dfrac{(x - 8)(x - 5)}{(x - 5)(x + 5)(x - 5)} = \dfrac{(x - 8)(x - 5)}{(x - 5)^2(x + 5)}$

$\dfrac{9x}{x^2 - 10x + 25} = \dfrac{9x}{(x - 5)^2}$

$= \dfrac{9x(x + 5)}{(x - 5)^2(x + 5)} = \dfrac{9x(x + 5)}{(x - 5)^2(x + 5)}$

$\text{LCD} = (x - 5)^2(x + 5)$

45. $\dfrac{5}{4x} - \dfrac{3}{5} = \dfrac{5(5)}{4x(5)} - \dfrac{3(4x)}{5(4x)} = \dfrac{25}{20x} - \dfrac{12x}{20x} = \dfrac{25 - 12x}{20x}$

47. $\dfrac{7}{a} + \dfrac{14}{a^2} = \dfrac{7(a)}{a(a)} + \dfrac{14(1)}{a^2(1)} = \dfrac{7a}{a^2} + \dfrac{14}{a^2} = \dfrac{7a + 14}{a^2}$

49. $\dfrac{20}{x - 4} + \dfrac{20}{4 - x} = \dfrac{20(1)}{(x - 4)(1)} + \dfrac{20(-1)}{(4 - x)(-1)}$

$= \dfrac{20}{x - 4} - \dfrac{20}{x - 4}$

$= \dfrac{20 - 20}{x - 4} = 0, \quad x \neq 4$

51. $\dfrac{3x}{x - 8} - \dfrac{6}{8 - x} = \dfrac{3x(1)}{(x - 8)(1)} - \dfrac{6(-1)}{(8 - x)(-1)}$

$= \dfrac{3x}{x - 8} + \dfrac{6}{x - 8}$

$= \dfrac{3x + 6}{x - 8}$

53. $25 + \dfrac{10}{x + 4} = \dfrac{25(x + 4)}{1(x + 4)} + \dfrac{10(1)}{(x + 4)(1)}$

$= \dfrac{25(x + 4)}{x + 4} + \dfrac{10}{x + 4}$

$= \dfrac{25x + 100 + 10}{x + 4} = \dfrac{25x + 110}{x + 4}$

55. $\dfrac{3x}{3x - 2} + \dfrac{2}{2 - 3x} = \dfrac{3x(1)}{3x - 2(1)} + \dfrac{2(-1)}{(2 - 3x)(-1)}$

$= \dfrac{3x}{3x - 2} + \dfrac{-2}{3x - 2}$

$= \dfrac{3x - 2}{3x - 2} = 1, \quad x \neq \dfrac{2}{3}$

57. $-\dfrac{1}{6x} + \dfrac{1}{6(x-3)} = \dfrac{-1(x-3)}{6x(x-3)} + \dfrac{1(x)}{6(x-3)x}$

$$= \dfrac{-(x-3)}{6x(x-3)} + \dfrac{x}{6x(x-3)}$$

$$= \dfrac{-x+3+x}{6x(x-3)}$$

$$= \dfrac{3}{6x(x-3)}$$

$$= \dfrac{1}{2x(x-3)}$$

59. $\dfrac{x}{x+3} - \dfrac{5}{x-2} = \dfrac{x(x-2)}{(x+3)(x-2)} - \dfrac{5(x+3)}{(x-2)(x+3)}$

$$= \dfrac{x(x-2)}{(x+3)(x-2)} - \dfrac{5(x+3)}{(x-2)(x+3)}$$

$$= \dfrac{x^2-2x-5x-15}{(x+3)(x-2)}$$

$$= \dfrac{x^2-7x-15}{(x+3)(x-2)}$$

61. $\dfrac{3}{x+1} - \dfrac{2}{x} = \dfrac{3x}{(x+1)x} - \dfrac{2(x+1)}{x(x+1)}$

$$= \dfrac{3x}{x(x+1)} - \dfrac{2(x+1)}{x(x+1)}$$

$$= \dfrac{3x-2x-2}{x(x+1)}$$

$$= \dfrac{x-2}{x(x+1)}$$

63. $\dfrac{3}{x-5} + \dfrac{2}{x+5} = \dfrac{3(x+5)}{(x-5)(x+5)} + \dfrac{2(x-5)}{(x+5)(x-5)}$

$$= \dfrac{3(x+5)}{(x-5)(x+5)} + \dfrac{2(x-5)}{(x+5)(x-5)}$$

$$= \dfrac{3x+15+2x-10}{(x-5)(x+5)}$$

$$= \dfrac{5x+5}{(x-5)(x+5)}$$

65. $\dfrac{4}{x^2} - \dfrac{4}{x^2+1} = \dfrac{4(x^2+1)}{x^2(x^2+1)} - \dfrac{4x^2}{(x^2+1)x^2}$

$$= \dfrac{4(x^2+1)}{x^2(x^2+1)} - \dfrac{4x^2}{x^2(x^2+1)}$$

$$= \dfrac{4x^2+4-4x^2}{x^2(x^2+1)}$$

$$= \dfrac{4}{x^2(x^2+1)}$$

67. $\dfrac{x}{x^2-9} + \dfrac{3}{x^2-5x+6} = \dfrac{x}{(x-3)(x+3)} + \dfrac{3}{(x-3)(x-2)}$

$$= \dfrac{x(x-2)}{(x-2)(x-3)(x+3)} + \dfrac{3(x+3)}{(x-2)(x-3)(x+3)}$$

$$= \dfrac{x^2-2x+3x+9}{(x-2)(x-3)(x+3)}$$

$$= \dfrac{x^2+x+9}{(x-2)(x-3)(x+3)}$$

69. $\dfrac{4}{x-4} + \dfrac{16}{(x-4)^2} = \dfrac{4(x-4)}{(x-4)(x-4)} + \dfrac{16(1)}{(x-4)^2(1)}$

$$= \dfrac{4x-16}{(x-4)^2} + \dfrac{16}{(x-4)^2}$$

$$= \dfrac{4x-16+16}{(x-4)^2}$$

$$= \dfrac{4x}{(x-4)^2}$$

71. $\dfrac{y}{x^2+xy} - \dfrac{x}{xy+y^2} = \dfrac{y}{x(x+y)} - \dfrac{x}{y(x+y)}$

$$= \dfrac{y(y)}{x(x+y)(y)} - \dfrac{x(x)}{y(x+y)(x)}$$

$$= \dfrac{y^2}{xy(x+y)} - \dfrac{x^2}{xy(x+y)}$$

$$= \dfrac{y^2-x^2}{xy(x+y)}$$

$$= \dfrac{(y-x)(y+x)}{xy(x+y)} = \dfrac{y-x}{xy}, \quad x \neq -y$$

73. $\dfrac{4}{x} - \dfrac{2}{x^2} + \dfrac{4}{x+3} = \dfrac{4x(x+3)}{x(x)(x+3)} - \dfrac{2(x+3)}{x^2(x+3)} + \dfrac{4(x^2)}{(x+3)x^2}$

$\qquad = \dfrac{4x^2 + 12x}{x^2(x+3)} - \dfrac{2x+6}{x^2(x+3)} + \dfrac{4x^2}{x^2(x+3)}$

$\qquad = \dfrac{4x^2 + 12x - 2x - 6 + 4x^2}{x^2(x+3)}$

$\qquad = \dfrac{8x^2 + 10x - 6}{x^2(x+3)}$

$\qquad = \dfrac{2(4x^2 + 5x - 3)}{x^2(x+3)}$

75. $\dfrac{3u}{u^2 - 2uv + v^2} + \dfrac{2}{u-v} - \dfrac{u}{u-v}$

$\qquad = \dfrac{3u}{(u-v)^2} + \dfrac{2-u}{u-v}$

$\qquad = \dfrac{3u(1)}{(u-v)^2(1)} + \dfrac{(2-u)(u-v)}{(u-v)(u-v)}$

$\qquad = \dfrac{3u}{(u-v)^2} + \dfrac{2u - 2v - u^2 + uv}{(u-v)^2}$

$\qquad = \dfrac{3u + 2u - 2v - u^2 + uv}{(u-v)^2}$

$\qquad = \dfrac{5u - 2v - u^2 + uv}{(u-v)^2}$

$\qquad = -\dfrac{u^2 - uv - 5u + 2v}{(u-v)^2}$

77. $\dfrac{x+2}{x-1} - \dfrac{2}{x+6} - \dfrac{14}{x^2 + 5x - 6} = \dfrac{(x+2)(x+6)}{(x-1)(x+6)} - \dfrac{2(x-1)}{(x+6)(x-1)} - \dfrac{14(1)}{(x+6)(x-1)(1)}$

$\qquad = \dfrac{x^2 + 8x + 12}{(x-1)(x+6)} - \dfrac{2x-2}{(x+6)(x-1)} - \dfrac{14}{(x+6)(x-1)}$

$\qquad = \dfrac{x^2 + 8x + 12 - 2x + 2 - 14}{(x-1)(x+6)}$

$\qquad = \dfrac{x^2 + 6x}{(x-1)(x+6)}$

$\qquad = \dfrac{x(x+6)}{(x-1)(x+6)}$

$\qquad = \dfrac{x}{x-1}, \quad x \neq -6$

79. *Keystrokes:*

y_1 [Y=] [(] [2] [÷] [X,T,θ] [)] [+] [(] [4] [÷]
[(] [X,T,θ] [−] [2] [)] [)] [ENTER]

y_2 [(] [6] [X,T,θ] [−] [4] [)] [÷] [(] [X,T,θ] [(] [X,T,θ] [−] [2] [)] [)] [GRAPH]

$\dfrac{2}{x} + \dfrac{4}{(x-2)} = \dfrac{2(x-2)}{x(x-2)} + \dfrac{4x}{x(x-2)} = \dfrac{2x - 4 + 4x}{x(x-2)} = \dfrac{6x - 4}{x(x-2)}$

$y_1 = y_2$

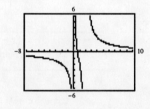

81. $\dfrac{\frac{1}{2}}{\left(3 + \frac{1}{x}\right)} = \dfrac{\frac{1}{2}}{\left(3 + \frac{1}{x}\right)} \cdot \dfrac{2x}{2x}$

$\qquad = \dfrac{\frac{1}{2} \cdot 2x}{3(2x) + \frac{1}{x}(2x)}$

$\qquad = \dfrac{x}{6x + 2}$

$\qquad = \dfrac{x}{2(3x+1)}, \quad x \neq 0$

83. $\dfrac{\left(\frac{4}{x} + 3\right)}{\left(\frac{4}{x} - 3\right)} = \dfrac{\left(\frac{4}{x} + 3\right)}{\left(\frac{4}{x} - 3\right)} \cdot \dfrac{x}{x}$

$\qquad = \dfrac{4 + 3x}{4 - 3x}, \quad x \neq 0$

85. $\dfrac{\left(16x - \dfrac{1}{x}\right)}{\left(\dfrac{1}{x} - 4\right)} = \dfrac{\left(16x - \dfrac{1}{x}\right)}{\left(\dfrac{1}{x} - 4\right)} \cdot \dfrac{x}{x}$

$= \dfrac{16x(x) - \dfrac{1}{x}(x)}{\dfrac{1}{x}(x) - 4(x)}$

$= \dfrac{16x^2 - 1}{1 - 4x}$

$= \dfrac{(4x - 1)(4x + 1)}{-1(4x - 1)}$

$= \dfrac{4x + 1}{-1}$

$= -4x - 1, \quad x \neq 0, \dfrac{1}{4}$

87. $\dfrac{\left(3 + \dfrac{9}{x - 3}\right)}{\left(4 + \dfrac{12}{x - 3}\right)} = \dfrac{\left(3 + \dfrac{9}{x - 3}\right)}{\left(4 + \dfrac{12}{x - 3}\right)} \cdot \dfrac{x - 3}{x - 3}$

$= \dfrac{3(x - 3) + \dfrac{9}{x - 3}(x - 3)}{4(x - 3) + \dfrac{12}{x - 3}(x - 3)}$

$= \dfrac{3x - 9 + 9}{4x - 12 + 12}$

$= \dfrac{3x}{4x} = \dfrac{3}{4}, \quad x \neq 0, 3$

89. $\dfrac{\left(\dfrac{3}{x^2} + \dfrac{1}{x}\right)}{\left(2 - \dfrac{4}{5x}\right)} = \dfrac{\left(\dfrac{3}{x^2} + \dfrac{1}{x}\right)}{\left(2 - \dfrac{4}{5x}\right)} \cdot \dfrac{5x^2}{5x^2}$

$= \dfrac{15 + 5x}{10x^2 - 4x}$

$= \dfrac{5(3 + x)}{2x(5x - 2)}$

91. $\dfrac{\left(\dfrac{y}{x} - \dfrac{x}{y}\right)}{\left(\dfrac{x + y}{xy}\right)} = \dfrac{\left(\dfrac{y}{x} - \dfrac{x}{y}\right)}{\left(\dfrac{x + y}{xy}\right)} \cdot \dfrac{xy}{xy}$

$= \dfrac{\dfrac{y}{x}(xy) - \dfrac{x}{y}(xy)}{\left(\dfrac{x + y}{xy}\right)xy}$

$= \dfrac{y^2 - x^2}{x + y}$

$= \dfrac{(y - x)(y + x)}{x + y}$

$= y - x, \quad x \neq 0, y \neq 0, x \neq -y$

93. $\dfrac{\left(1 - \dfrac{1}{y}\right)}{\left(\dfrac{1 - 4y}{y - 3}\right)} = \dfrac{\left(1 - \dfrac{1}{y}\right)}{\left(\dfrac{1 - 4y}{y - 3}\right)} \cdot \dfrac{y(y - 3)}{y(y - 3)}$

$= \dfrac{y(y - 3) - (y - 3)}{y(1 - 4y)}$

$= \dfrac{y^2 - 3y - y + 3}{y - 4y^2}$

$= \dfrac{y^2 - 4y + 3}{-y(-1 + 4y)}$

$= -\dfrac{(y - 3)(y - 1)}{y(4y - 1)}, \quad y \neq 3$

95. $\dfrac{\left(\dfrac{x}{x - 3} - \dfrac{2}{3}\right)}{\left(\dfrac{10}{3x} + \dfrac{x^2}{x - 3}\right)} = \dfrac{\left(\dfrac{x}{x - 3} - \dfrac{2}{3}\right)}{\left(\dfrac{10}{3x} + \dfrac{x^2}{x - 3}\right)} \cdot \dfrac{3x(x - 3)}{3x(x - 3)}$

$= \dfrac{3x^2 - 2x(x - 3)}{10(x - 3) + 3x^2}$

$= \dfrac{3x^2 - 2x^2 + 6x}{10x - 30 + 3x^3}$

$= \dfrac{x^2 + 6x}{3x^3 + 10x - 30}, \quad x \neq 0, \quad x \neq 3$

97. $\dfrac{f(2 + h) - f(2)}{h} = \dfrac{\dfrac{1}{2 + h} - \dfrac{1}{2}}{h}$

$= \dfrac{\dfrac{1}{2 + h} - \dfrac{1}{2}}{h} \cdot \dfrac{2(2 + h)}{2(2 + h)}$

$= \dfrac{2 - (2 + h)}{2h(2 + h)}$

$= \dfrac{2 - 2 - h}{2h(2 + h)}$

$= \dfrac{-h}{2h(2 + h)}$

$= \dfrac{-1}{2(2 + h)}$

99.

x	-3	-2	-1	0	1	2	3
$\dfrac{\left(1 - \dfrac{1}{x}\right)}{\left(1 - \dfrac{1}{x^2}\right)}$	$\dfrac{3}{2}$	2	Undef.	Undef.	Undef.	$\dfrac{2}{3}$	$\dfrac{3}{4}$
$\dfrac{x}{x + 1}$	$\dfrac{3}{2}$	2	Undef.	0	$\dfrac{1}{2}$	$\dfrac{2}{3}$	$\dfrac{3}{4}$

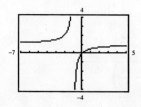

Keystrokes:

y_1 [Y=] [(] 1 [−] 1 [÷] [X,T,θ] [)] [÷]

[(] 1 [−] 1 [÷] [X,T,θ] [x²] [)] [ENTER]

y_2 [X,T,θ] [÷] [(] [X,T,θ] [+] 1 [)] [GRAPH]

Zero and one are not in the domain of

$$\dfrac{1 - \dfrac{1}{x}}{1 - \dfrac{1}{x^2}}$$

but are in the domain of $\dfrac{x}{x + 1}$. The two expressions are equivalent except at $x = 0$ and $x = 1$.

101. $\dfrac{t}{4} + \dfrac{t}{6} = \dfrac{t(3)}{4(3)} + \dfrac{t(2)}{6(2)}$

$= \dfrac{3t}{12} + \dfrac{2t}{12}$

$= \dfrac{5t}{12}$

103. $\dfrac{\dfrac{x}{4} + \dfrac{x}{6}}{2} = \dfrac{\left(\dfrac{x}{4} + \dfrac{x}{6}\right)}{2} \cdot \dfrac{12}{12}$

$= \dfrac{\dfrac{x}{4}(12) + \dfrac{x}{6}(12)}{2(12)}$

$= \dfrac{3x + 2x}{24}$

$= \dfrac{5x}{24}$

105. $\dfrac{\dfrac{x}{3} - \dfrac{x}{5}}{3} \cdot \dfrac{15}{15} = 5x - 3x = \dfrac{2x}{45}$

Thus,

$x_1 = \dfrac{x}{5} + \dfrac{2x}{45} = \dfrac{9x}{45} + \dfrac{2x}{45} = \dfrac{11x}{45}$

$x_2 = \dfrac{11x}{45} + \dfrac{2x}{45} = \dfrac{13x}{45}.$

107. $\dfrac{1}{\left(\dfrac{1}{R_1} + \dfrac{1}{R_2}\right)} = \dfrac{1}{\left(\dfrac{1}{R_1} + \dfrac{1}{R_2}\right)} \cdot \dfrac{R_1 R_2}{R_1 R_2}$

$\quad = \dfrac{R_1 R_2}{\dfrac{1}{R_1}(R_1 R_2) + \dfrac{1}{R_2}(R_1 R_2)}$

$\quad = \dfrac{R_1 R_2}{R_2 + R_1}$

109. (a) *Verbal Model:* $\boxed{\text{Distance}} = \boxed{\text{Rate}} \cdot \boxed{\text{Time}}$

$\qquad \dfrac{\text{Distance}}{\text{Rate}} = \text{Time}$

$\qquad$ *Equation:* $\quad$ Upstream $t = \dfrac{10}{5 - x}$

$\qquad\qquad\qquad$ Downstream $t = \dfrac{10}{5 + x}$

(b) Total time $= t(x) = \dfrac{10}{5 - x} + \dfrac{10}{5 + x}$

(c) $t(x) = \dfrac{10(5 + x)}{(5 - x)(5 + x)} + \dfrac{10(5 - x)}{(5 - x)(5 + x)}$

$\qquad = \dfrac{50 + 10x + 50 - 10x}{(5 - x)(5 + x)}$

$\qquad = \dfrac{100}{(5 - x)(5 + x)}$

111. Rewrite each fraction in terms of the lowest common denominator, combine the numerators, and place the result over the lowest common denominator.

113. $\dfrac{x - 1}{x + 4} - \dfrac{4x - 11}{x + 4} = \dfrac{(x - 1) - (4x - 11)}{x + 4}$

$\qquad = \dfrac{x - 1 - 4x + 11}{x + 4}$

$\qquad = \dfrac{-3x + 10}{x + 4}$

The subtraction must be distributed to both terms of the numerator of the second fraction.

Mid-Chapter Quiz for Chapter 8

1. $\qquad y(y - 4) \neq 0$

$y \neq 0 \quad y - 4 \neq 0$

$\qquad\qquad y \neq 4$

$D = (-\infty, 0) \cup (0, 4) \cup (4, \infty)$

2. (a) $h(-3) = \dfrac{(-3)^2 - 9}{(-3)^2 - (-3) - 2} = \dfrac{9 - 9}{9 + 3 - 2}$

$\qquad = \dfrac{0}{10} = 0$

(c) $h(-1) = \dfrac{(-1)^2 - 9}{(-1)^2 - (-1) - 2} = \dfrac{1 - 9}{1 + 1 - 2}$

$\qquad = \dfrac{-8}{0} = \text{undefined}$

(b) $h(0) = \dfrac{0^2 - 9}{0^2 - 0 - 2} = \dfrac{-9}{-2}$

$\qquad = \dfrac{9}{2}$

(d) $h(5) = \dfrac{5^2 - 9}{5^2 - 5 - 2} = \dfrac{25 - 9}{25 - 5 - 2}$

$\qquad = \dfrac{16}{18} = \dfrac{8}{9}$

3. $\dfrac{9y^2}{6y} = \dfrac{3y}{2}$

4. $\dfrac{8u^3 v^2}{36uv^3} = \dfrac{2u^2}{9v}$

5. $\dfrac{4x^2 - 1}{x - 2x^2} = \dfrac{(2x - 1)(2x + 1)}{x(1 - 2x)}$

$\qquad = \dfrac{(2x - 1)(2x + 1)}{-x(2x - 1)}$

$\qquad = \dfrac{2x + 1}{-x}$

6. $\dfrac{(z + 3)^2}{2z^2 + 5z - 3} = \dfrac{(z + 3)(z + 3)}{(2z - 1)(z + 3)}$

$\qquad = \dfrac{z + 3}{2z - 1}$

7. $\dfrac{7ab + 3a^2b^2}{a^2b} = \dfrac{ab(7 + 3ab)}{a^2b} = \dfrac{7 + 3ab}{a}$

8. $\dfrac{2mn^2 - n^3}{2m^2 + mn - n^2} = \dfrac{n^2(2m - n)}{(2m - n)(m + n)} = \dfrac{n^2}{m + n}$

9. $\dfrac{11t^2}{6} \cdot \dfrac{9}{33t} = \dfrac{11t^2(9)}{6(33t)} = \dfrac{t}{2}$

10. $(x^2 + 2x) \cdot \dfrac{5}{x^2 - 4} = \dfrac{x(x + 2)5}{(x - 2)(x + 2)} = \dfrac{5x}{x - 2}$

11. $\dfrac{4}{3(x - 1)} \cdot \dfrac{12x}{6(x^2 + 2x - 3)} = \dfrac{4(12x)}{3(x - 1)6(x + 3)(x - 1)} = \dfrac{8x}{3(x - 1)^2(x + 3)}$

12. $\dfrac{5u}{3(u + v)} \cdot \dfrac{2(u^2 - v^2)}{3v} \div \dfrac{25u^2}{18(u - v)} = \dfrac{5u \cdot 2(u - v)(u + v) \cdot 18(u - v)}{3(u + v)(3v)(25u^2)} = \dfrac{4(u - v)^2}{5uv}$

13. $\dfrac{\dfrac{9t^2}{3 - t}}{\dfrac{6t}{t - 3}} \cdot \dfrac{t - 3}{t - 3} = \dfrac{-9t^2}{6t} = -\dfrac{3t}{2}$

14. $\dfrac{\dfrac{10}{x^2 + 2x}}{\dfrac{15}{x^2 + 3x + 2}} = \dfrac{\dfrac{10}{x(x + 2)}}{\dfrac{15}{(x + 2)(x + 1)}} \cdot \dfrac{x(x + 2)(x + 1)}{x(x + 2)(x + 1)}$

$\qquad = \dfrac{10(x + 1)}{15x} = \dfrac{2(x + 1)}{3x}$

15. $\dfrac{4x}{x + 5} - \dfrac{3x}{4} = \dfrac{4x(4)}{4(x + 5)} - \dfrac{3x(x + 5)}{4(x + 5)}$

$\qquad = \dfrac{16x - 3x^2 - 15x}{4(x + 5)}$

$\qquad = \dfrac{-3x^2 + x}{4(x + 5)}$

$\qquad = \dfrac{x(1 - 3x)}{4(x + 5)}$

16. $4 + \dfrac{x}{x^2 - 4} - \dfrac{2}{x^2} = \dfrac{4x^2(x^2 - 4)}{x^2(x^2 - 4)} + \dfrac{x(x^2)}{x^2(x^2 - 4)} - \dfrac{2(x^2 - 4)}{x^2(x^2 - 4)}$

$\qquad = \dfrac{4x^4 - 16x^2 + x^3 - 2x^2 + 8}{x^2(x^2 - 4)}$

$\qquad = \dfrac{4x^4 + x^3 - 18x^2 + 8}{x^2(x^2 - 4)}$

17. $\dfrac{\left(1 - \dfrac{2}{x}\right)}{\left(\dfrac{3}{x} - \dfrac{4}{5}\right)} = \dfrac{\left(1 - \dfrac{2}{x}\right)}{\left(\dfrac{3}{x} - \dfrac{4}{5}\right)} \cdot \dfrac{5x}{5x}$

$\qquad = \dfrac{5x - 10}{15 - 4x}$

18. $\dfrac{\left(\dfrac{3}{x} + \dfrac{x}{3}\right)}{\left(\dfrac{x + 3}{6x}\right)} = \dfrac{\dfrac{3}{x} + \dfrac{x}{3}}{\dfrac{x + 3}{6x}} \cdot \dfrac{6x}{6x}$

$\qquad = \dfrac{18 + 2x^2}{x + 3} = \dfrac{2(9 + x^2)}{x + 3}$

19. (a) *Verbal Model:* $\boxed{\text{Average cost}} = \boxed{\text{Total cost}} \div \boxed{\text{Number of units}}$

Equation: Average cost $= \dfrac{6000 + 10.50x}{x}$

(b) Average cost when $x = 500$ units are produced $= \dfrac{6000 + 10.50(500)}{500} = \$22.50$

20. $\dfrac{\text{Shaded Portion}}{\text{Total Area}} = \dfrac{\left[\frac{1}{2}(x+4)(x+2)\right] - \left[\frac{1}{2}\left(\frac{x(x+2)}{x+4}\right)x\right]}{\left[\frac{1}{2}(x+4)(x+2)\right]}$

$$= \frac{\frac{1}{2}(x+2)\left[(x+4) - \dfrac{x^2}{x+4}\right]}{\frac{1}{2}(x+4)(x+2)}$$

$$= \frac{\dfrac{(x+4)^2 - x^2}{(x+4)}}{x+4}$$

$$= \frac{x^2 + 8x + 16 - x^2}{(x+4)} \cdot \frac{1}{x+4}$$

$$= \frac{8x + 16}{(x+4)^2}$$

$$= \frac{8(x+2)}{(x+4)^2}$$

Section 8.4   Solving Rational Equations

1. (a) $x = 0$

$\dfrac{0}{3} - \dfrac{0}{5} \overset{?}{=} \dfrac{4}{3}$

$0 \neq \dfrac{4}{3}$

Not a solution

(b) $x = -1$

$\dfrac{-1}{3} - \dfrac{-1}{5} \overset{?}{=} \dfrac{4}{3}$

$\dfrac{-5}{15} - \dfrac{-3}{15} \overset{?}{=} \dfrac{20}{15}$

$\dfrac{-5}{15} + \dfrac{3}{15} \overset{?}{=} \dfrac{20}{15}$

$\dfrac{-2}{15} \neq \dfrac{20}{15}$

Not a solution

(c) $x = \dfrac{1}{8}$

$\dfrac{1/8}{3} - \dfrac{1/8}{5} \overset{?}{=} \dfrac{4}{3}$

$\dfrac{1}{24} - \dfrac{1}{40} \overset{?}{=} \dfrac{4}{3}$

$\dfrac{5}{120} - \dfrac{3}{120} \overset{?}{=} \dfrac{160}{120}$

$\dfrac{2}{120} \neq \dfrac{160}{120}$

Not a solution

(d) $x = 10$

$\dfrac{10}{3} - \dfrac{10}{5} \overset{?}{=} \dfrac{4}{3}$

$\dfrac{50}{15} - \dfrac{30}{15} \overset{?}{=} \dfrac{20}{15}$

$\dfrac{20}{15} = \dfrac{20}{15}$

Solution

3. (a) $x = -1$

$\dfrac{-1}{4} + \dfrac{3}{4(-1)} \overset{?}{=} 1$

$\dfrac{-1}{4} + \dfrac{-3}{4} \overset{?}{=} 1$

$-1 \neq 1$

Not a solution

(b) $x = 1$

$\dfrac{1}{4} + \dfrac{3}{4(1)} \overset{?}{=} 1$

$\dfrac{1}{4} + \dfrac{3}{4} \overset{?}{=} 1$

$1 = 1$

Solution

(c) $x = 3$

$\dfrac{3}{4} + \dfrac{3}{4(3)} \overset{?}{=} 1$

$\dfrac{3}{4} + \dfrac{3}{12} \overset{?}{=} 1$

$\dfrac{3}{4} + \dfrac{1}{4} \overset{?}{=} 1$

$1 = 1$

Solution

(d) $x = 2$

$\dfrac{2}{4} + \dfrac{3}{4(2)} \overset{?}{=} 1$

$\dfrac{4}{8} + \dfrac{3}{8} \overset{?}{=} 1$

$\dfrac{7}{8} \neq 1$

Not a solution

5. $\dfrac{x}{6} - 1 = \dfrac{2}{3}$

$6\left(\dfrac{x}{6} - 1\right) = \left(\dfrac{2}{3}\right)6$

$x - 6 = 4$

$x = 10$

Check: $\dfrac{10}{6} - 1 \overset{?}{=} \dfrac{2}{3}$

$\dfrac{5}{3} - \dfrac{3}{3} \overset{?}{=} \dfrac{2}{3}$

$\dfrac{2}{3} = \dfrac{2}{3}$

7. $\dfrac{z + 2}{3} = 4 - \dfrac{z}{12}$

$12\left(\dfrac{z + 2}{3}\right) = \left(4 - \dfrac{z}{12}\right)12$

$4(z + 2) = 48 - z$

$4z + 8 = 48 - z$

$5z = 40$

$z = 8$

Check: $\dfrac{8 + 2}{3} \overset{?}{=} 4 - \dfrac{8}{12}$

$\dfrac{10}{3} \overset{?}{=} \dfrac{12}{3} - \dfrac{2}{3}$

$\dfrac{10}{3} = \dfrac{10}{3}$

9. $\dfrac{2y - 9}{6} = 3y - \dfrac{3}{4}$

$(12)\left(\dfrac{2y - 9}{6}\right) = \left(3y - \dfrac{3}{4}\right)(12)$

$2(2y - 9) = 36y - 9$

$4y - 18 = 36y - 9$

$-9 = 32y$

$-\dfrac{9}{32} = y$

Check: $\dfrac{2\left(-\frac{9}{32}\right) - 9}{6} \overset{?}{=} 3\left(-\dfrac{9}{32}\right) - \dfrac{3}{4}$

$\dfrac{1}{6}\left(-\dfrac{18}{32} - \dfrac{288}{32}\right) \overset{?}{=} -\dfrac{27}{32} - \dfrac{24}{32}$

$\dfrac{1}{6}\left(-\dfrac{306}{32}\right) \overset{?}{=} -\dfrac{51}{32}$

$-\dfrac{51}{32} = -\dfrac{51}{32}$

11. $\dfrac{4t}{3} = 15 - \dfrac{t}{6}$

$6\left(\dfrac{4t}{3}\right) = \left(15 - \dfrac{t}{6}\right)6$

$8t = 90 - t$

$9t = 90$

$t = 10$

Check: $\dfrac{4(10)}{3} \overset{?}{=} 15 - \dfrac{10}{6}$

$\dfrac{40}{3} \overset{?}{=} \dfrac{45}{3} - \dfrac{5}{3}$

$\dfrac{40}{3} = \dfrac{40}{3}$

13. $\dfrac{5y - 1}{12} + \dfrac{y}{3} = -\dfrac{1}{4}$

$12\left(\dfrac{5y - 1}{12} + \dfrac{y}{3}\right) = \left(-\dfrac{1}{4}\right)12$

$5y - 1 + 4y = -3$

$9y = -2$

$y = -\dfrac{2}{9}$

15. $\dfrac{h + 2}{5} - \dfrac{h - 1}{9} = \dfrac{2}{3}$

$45\left(\dfrac{h + 2}{5} - \dfrac{h - 1}{9}\right) = \left(\dfrac{2}{3}\right)45$

$9(h + 2) - 5(h - 1) = 30$

$9h + 18 - 5h + 5 = 30$

$4h + 23 = 30$

$4h = 7$

$h = \dfrac{7}{4}$

Check: $\dfrac{\frac{7}{4} + 2}{5} - \dfrac{\frac{7}{4} - 1}{9} \overset{?}{=} \dfrac{2}{3}$

$\dfrac{1}{5}\left(\dfrac{7}{4} + \dfrac{8}{4}\right) - \dfrac{1}{9}\left(\dfrac{7}{4} - \dfrac{4}{4}\right) \overset{?}{=} \dfrac{2}{3}$

$\dfrac{1}{5}\left(\dfrac{15}{4}\right) - \dfrac{1}{9}\left(\dfrac{3}{4}\right) \overset{?}{=} \dfrac{2}{3}$

$\dfrac{3}{4} - \dfrac{1}{12} \overset{?}{=} \dfrac{2}{3}$

$\dfrac{9}{12} - \dfrac{1}{12} \overset{?}{=} \dfrac{2}{3}$

$\dfrac{8}{12} \overset{?}{=} \dfrac{2}{3}$

$\dfrac{2}{3} = \dfrac{2}{3}$

17. $\dfrac{x+5}{4} - \dfrac{3x-8}{3} = \dfrac{4-4}{12}$

$12\left(\dfrac{x+5}{4} - \dfrac{3x-8}{3}\right) = \left(\dfrac{4-x}{12}\right)12$

$3(x+5) - 4(3x-8) = 4 - x$

$3x + 15 - 12x + 32 = 4 - x$

$-9x + 47 = 4 - x$

$-8x = -43$

$x = \dfrac{43}{8}$

Check: $\dfrac{\frac{43}{8}+5}{4} - \dfrac{3\left(\frac{43}{8}\right)-8}{3} \overset{?}{=} \dfrac{4-\left(\frac{43}{8}\right)}{12}$

$\dfrac{1}{4}\left(\dfrac{43}{8} + \dfrac{40}{8}\right) - \dfrac{1}{3}\left(\dfrac{129}{8} - \dfrac{64}{8}\right) \overset{?}{=} \dfrac{1}{12}\left(\dfrac{32}{8} - \dfrac{43}{8}\right)$

$\dfrac{1}{4}\left(\dfrac{83}{8}\right) - \dfrac{1}{3}\left(\dfrac{65}{8}\right) \overset{?}{=} \dfrac{1}{12}\left(-\dfrac{11}{8}\right)$

$\dfrac{1}{8}\left(\dfrac{83}{4} - \dfrac{65}{3}\right) \overset{?}{=} \dfrac{1}{8}\left(-\dfrac{11}{12}\right)$

$\dfrac{1}{8}\left(\dfrac{249}{12} - \dfrac{260}{12}\right) \overset{?}{=} \dfrac{1}{8}\left(-\dfrac{11}{12}\right)$

$\dfrac{1}{8}\left(-\dfrac{11}{12}\right) = \dfrac{1}{8}\left(-\dfrac{11}{12}\right)$

19. $\dfrac{9}{25-y} = -\dfrac{1}{4}$

$4(25-y)\left(\dfrac{9}{25-y}\right) = \left(-\dfrac{1}{4}\right)4(25-y)$

$36 = -(25-y)$

$36 = -25 + y$

$61 = y$

Check: $\dfrac{9}{25-61} \overset{?}{=} -\dfrac{1}{4}$

$-\dfrac{9}{36} = -\dfrac{1}{4}$

$-\dfrac{1}{4} = -\dfrac{1}{4}$

21. $5 - \dfrac{12}{a} = \dfrac{5}{3}$

$3a\left(5 - \dfrac{12}{a}\right) = \left(\dfrac{5}{3}\right)3a$

$15a - 36 = 5a$

$10a = 36$

$a = \dfrac{36}{10}$

$a = \dfrac{18}{5}$

Check: $5 - \dfrac{12}{\dfrac{18}{5}} \overset{?}{=} \dfrac{5}{3}$

$5 - \dfrac{60}{18} = \dfrac{5}{3}$

$\dfrac{15}{3} - \dfrac{10}{3} = \dfrac{5}{3}$

$\dfrac{5}{3} = \dfrac{5}{3}$

23. $\dfrac{4}{x} - \dfrac{7}{5x} = -\dfrac{1}{2}$

$10x\left(\dfrac{4}{x} - \dfrac{7}{5x}\right) = \left(-\dfrac{1}{2}\right)10x$

$40 - 14 = -5x$

$26 = -5x$

$\dfrac{26}{-5} = x$

Check: $\dfrac{4}{\dfrac{26}{-5}} - \dfrac{7}{5\left(\dfrac{26}{-5}\right)} \overset{?}{=} -\dfrac{1}{2}$

$-\dfrac{20}{26} + \dfrac{7}{26} \overset{?}{=} -\dfrac{1}{2}$

$-\dfrac{13}{26} \overset{?}{=} -\dfrac{1}{2}$

$-\dfrac{1}{2} = -\dfrac{1}{2}$

25. $\dfrac{12}{y+5} + \dfrac{1}{2} = 2$

$2(y+5)\left(\dfrac{12}{y+5} + \dfrac{1}{2}\right) = (2)2(y+5)$

$24 + y + 5 = 4(y+5)$

$y + 29 = 4y + 20$

$9 = 3y$

$3 = y$

Check: $\dfrac{12}{3+5} + \dfrac{1}{2} \overset{?}{=} 2$

$\dfrac{3}{2} + \dfrac{1}{2} = 2$

$\dfrac{4}{2} = 2$

$2 = 2$

27.
$$\frac{5}{x} = \frac{25}{3(x+2)}$$

$$3x(x+2)\left(\frac{5}{x}\right) = \left(\frac{25}{3(x+2)}\right)3x(x+2)$$

$$15(x+2) = 25x$$

$$15x + 30 = 25x$$

$$30 = 10x$$

$$3 = x$$

Check: $\dfrac{5}{3} \overset{?}{=} \dfrac{25}{3(3+2)}$

$$\frac{5}{3} = \frac{25}{15}$$

$$\frac{5}{3} = \frac{5}{3}$$

29.
$$\frac{8}{3x+5} = \frac{1}{x+2}$$

$$(3x+5)(x+2)\left(\frac{8}{3x+5}\right) = \left(\frac{1}{x+2}\right)(3x+5)(x+2)$$

$$8(x+2) = 3x+5$$

$$8x + 16 = 3x + 5$$

$$5x = -11$$

$$x = -\frac{11}{5}$$

Check: $\dfrac{8}{3\left(-\frac{11}{5}\right)+5} \overset{?}{=} \dfrac{1}{-\frac{11}{5}+2}$

$$\frac{8}{-\frac{33}{5}+\frac{25}{5}} = \frac{1}{-\frac{11}{5}+\frac{10}{5}}$$

$$\frac{8}{-\frac{8}{5}} = \frac{1}{-\frac{1}{5}}$$

$$-5 = -5$$

31.
$$\frac{3}{x+2} - \frac{1}{x} = \frac{1}{5x}$$

$$5x(x+2)\left(\frac{3}{x+2} - \frac{1}{x}\right) = \left(\frac{1}{5x}\right)5x(x+2)$$

$$15x - 5(x+2) = x+2$$

$$15x - 5x - 10 = x+2$$

$$10x - 10 = x+2$$

$$9x = 12$$

$$x = \frac{12}{9}$$

$$x = \frac{4}{3}$$

Check: $\dfrac{1}{\frac{4}{3}+2} - \dfrac{1}{\frac{4}{3}} \overset{?}{=} \dfrac{1}{5\left(\frac{4}{3}\right)}$

$$\frac{1}{\frac{10}{3}} - \frac{1}{\frac{4}{3}} = \frac{1}{\frac{20}{3}}$$

$$\frac{9}{10} - \frac{3}{4} = \frac{3}{20}$$

$$\frac{18}{20} - \frac{15}{20} = \frac{3}{20}$$

$$\frac{3}{20} = \frac{3}{20}$$

33.
$$\frac{1}{2} = \frac{18}{x^2}$$

$$2x^2\left(\frac{1}{2}\right) = \left(\frac{18}{x^2}\right)2x^2$$

$$x^2 = 36$$

$$x^2 - 36 = 0$$

$$(x-6)(x+6) = 0$$

$$x = 6 \quad x = -6$$

Check: $\dfrac{1}{2} \overset{?}{=} \dfrac{18}{6^2}$

$$\frac{1}{2} = \frac{18}{36}$$

$$\frac{1}{2} = \frac{1}{2}$$

$\dfrac{1}{2} \overset{?}{=} \dfrac{18}{(-6)^2}$

$$\frac{1}{2} = \frac{18}{36}$$

$$\frac{1}{2} = \frac{1}{2}$$

35. $\dfrac{32}{t} = 2t$ **Check:** $\dfrac{32}{4} \overset{?}{=} 2(4)$ $\dfrac{32}{-4} \overset{?}{=} 2(-4)$

$\quad\quad t\left(\dfrac{32}{t}\right) = (2t)t$ $8 = 8$ $-8 = -8$

$\quad\quad\quad 32 = 2t^2$

$\quad\quad\quad 16 = t^2$

$\quad\quad\quad\quad 0 = t^2 - 16$

$\quad\quad\quad\quad 0 = (t - 4)(t + 4)$

$\quad t = 4 \quad t = -4$

37. $x + 1 = \dfrac{72}{x}$ **Check:** $-9 + 1 \overset{?}{=} \dfrac{72}{-9}$ $8 + 1 \overset{?}{=} \dfrac{72}{8}$

$\quad\quad x(x + 1) = \left(\dfrac{72}{x}\right)x$ $-8 = -8$ $9 = 9$

$\quad\quad\quad x^2 + x = 72$

$\quad\quad x^2 + x - 72 = 0$

$\quad (x + 9)(x - 8) = 0$

$\quad\quad x = -9 \quad x = 8$

39. $1 = \dfrac{16}{y} - \dfrac{39}{y^2}$ **Check:** $1 \overset{?}{=} \dfrac{16}{13} - \dfrac{39}{13^2}$ $1 \overset{?}{=} \dfrac{16}{3} - \dfrac{39}{3^2}$

$\quad\quad y^2(1) = \left[\dfrac{16}{y} - \dfrac{39}{y^2}\right]y^2$ $1 = \dfrac{16}{13} - \dfrac{3}{13}$ $1 = \dfrac{16}{3} - \dfrac{13}{3}$

$\quad\quad\quad y^2 = 16y - 39$ $1 = 1$ $1 = 1$

$\quad y^2 - 16y + 39 = 0$

$\quad (y - 13)(y - 3) = 0$

$\quad\quad y = 13 \quad y = 3$

41. $\dfrac{2x}{3x - 10} - \dfrac{5}{x} = 0$

$\quad x(3x - 10)\left(\dfrac{2x}{3x - 10} - \dfrac{5}{x}\right) = (0)x(3x - 10)$

$\quad\quad 2x^2 - 5(3x - 10) = 0$

$\quad\quad 2x^2 - 15x + 50 = 0$

No real solution

43. $5x\left(\dfrac{2x}{5}\right) = \left(\dfrac{x^2 - 5x}{5x}\right)5x$ **Check:** $x = 0$ $x = -5$

$\quad\quad 2x^2 = x^2 - 5x$ $\dfrac{2(0)}{5} \overset{?}{=} \dfrac{0^2 - 5(0)}{5(0)}$ $\dfrac{2(-5)}{5} \overset{?}{=} \dfrac{(-5)^2 - 5(-5)}{5(-5)}$

$\quad\quad x^2 + 5x = 0$ $0 \neq$ undefined $\dfrac{-10}{5} = \dfrac{25 + 25}{-25}$

$\quad\quad x(x + 5) = 0$ so $x = 0$ is extraneous. $-2 = -2$

$\quad x = 0 \quad x + 5 = 0$

$\quad\quad\quad\quad x = -5$

45.
$$\frac{2}{6q + 5} - \frac{3}{4(6q + 5)} = \frac{1}{28}$$

$$28(6q + 5)\left(\frac{2}{6q + 5} - \frac{3}{4(6q + 5)}\right) = \left(\frac{1}{28}\right)28(6q + 5)$$

$$28(2) - 7(3) = 6q + 5$$

$$56 - 21 = 6q + 5$$

$$35 = 6q + 5$$

$$30 = 6q$$

$$5 = q$$

Check: $\dfrac{2}{6(5) + 5} - \dfrac{3}{4[6(5) + 5]} \stackrel{?}{=} \dfrac{1}{28}$

$$\frac{2}{30 + 5} - \frac{3}{4(30 + 5)} \stackrel{?}{=} \frac{1}{28}$$

$$\frac{2}{35} - \frac{3}{4(35)} \stackrel{?}{=} \frac{1}{28}$$

$$\frac{8}{140} - \frac{3}{140} \stackrel{?}{=} \frac{1}{28}$$

$$\frac{5}{140} \stackrel{?}{=} \frac{1}{28}$$

$$\frac{1}{28} = \frac{1}{28}$$

47.
$$\frac{4}{2x + 3} + \frac{17}{5x - 3} = 3$$

$$(5x - 3)(2x + 3)\left(\frac{4}{2x + 3} + \frac{17}{5x - 3}\right) = (3)(2x + 3)(5x - 3)$$

$$4(5x - 3) + 17(2x + 3) = 3(10x^2 + 9x - 9)$$

$$20x - 12 + 34x + 51 = 30x^2 + 27x - 27$$

$$0 = 30x^2 - 27x - 66$$

$$0 = 10x^2 - 9x - 22$$

$$0 = (10x + 11)(x - 2)$$

$$x = -\frac{11}{10} \quad x = 2$$

Check: $\dfrac{4}{2\left(-\frac{11}{10}\right) + 3} + \dfrac{17}{5\left(-\frac{11}{10}\right) - 3} \stackrel{?}{=} 3$

$$\frac{4}{-\frac{22}{10} + \frac{30}{10}} + \frac{17}{-\frac{55}{10} - \frac{30}{10}} \stackrel{?}{=} 3$$

$$\frac{4}{\frac{8}{10}} + \frac{17}{-\frac{85}{10}} \stackrel{?}{=} 3$$

$$\frac{4}{\frac{4}{5}} + \frac{17}{-\frac{17}{2}} \stackrel{?}{=} 3$$

$$5 + -2 \stackrel{?}{=} 3$$

$$3 = 3$$

Check: $\dfrac{4}{2(2) + 3} + \dfrac{17}{5(2) - 3} \stackrel{?}{=} 3$

$$\frac{4}{7} + \frac{17}{7} \stackrel{?}{=} 3$$

$$3 = 3$$

49.
$$\frac{2}{x - 10} - \frac{3}{x - 2} = \frac{6}{x^2 - 12x + 20}$$

$$\frac{2}{x - 10} - \frac{3}{x - 2} = \frac{6}{(x - 10)(x - 2)}$$

$$(x - 10)(x - 2)\left(\frac{2}{x - 10} - \frac{3}{x - 2}\right) = \left(\frac{6}{(x - 10)(x - 2)}\right)(x - 10)(x - 2)$$

$$2(x - 2) - 3(x - 10) = 6$$

$$2x - 4 - 3x + 30 = 6$$

$$-x + 26 = 6$$

$$-x = -20$$

$$x = 20$$

Check:

$$\frac{2}{20 - 10} - \frac{3}{20 - 2} \stackrel{?}{=} \frac{6}{(20)^2 - 12(20) + 20}$$

$$\frac{2}{10} - \frac{3}{18} \stackrel{?}{=} \frac{6}{400 - 240 + 20}$$

$$\frac{1}{5} - \frac{1}{6} \stackrel{?}{=} \frac{6}{180}$$

$$\frac{6}{30} - \frac{5}{30} \stackrel{?}{=} \frac{1}{30}$$

$$\frac{1}{30} = \frac{1}{30}$$

51.　$\dfrac{x+3}{x^2-9}+\dfrac{4}{3-x}-2=0$

$\dfrac{x+3}{(x-3)(x+3)}-\dfrac{4}{x-3}-2=0$

$(x-3)\left(\dfrac{1}{x-3}-\dfrac{4}{x-3}-2\right)=0(x-3)$

$1-4-2(x-3)=0$

$-3-2x+6=0$

$-2x=-3$

$x=\dfrac{3}{2}$

Check: $\dfrac{\frac{3}{2}+3}{\left(\frac{3}{2}\right)^2-9}+\dfrac{4}{3-\frac{3}{2}}-2\overset{?}{=}0$

$\dfrac{\frac{3}{2}+\frac{6}{2}}{\frac{9}{4}-\frac{36}{4}}+\dfrac{4}{\frac{6}{2}-\frac{3}{2}}-2\overset{?}{=}0$

$\dfrac{\frac{9}{2}}{-\frac{27}{4}}+\dfrac{4}{\frac{3}{2}}-2\overset{?}{=}0$

$-\dfrac{2}{3}+\dfrac{8}{3}-\dfrac{6}{3}\overset{?}{=}0$

$0=0$

53.　$\dfrac{x}{x-2}+\dfrac{3x}{x-4}=\dfrac{-2(x-6)}{x^2-6x+8}$

$(x-2)(x-4)\left(\dfrac{x}{x-2}+\dfrac{3x}{x-4}\right)=\left(\dfrac{-2(x-6)}{(x-4)(x-2)}\right)(x-2)(x-4)$

$x(x-4)+3x(x-2)=-2(x-6)$

$x^2-4x+3x^2-6x=-2x+12$

$4x^2-8x-12=0$

$x^2-2x-3=0$

$(x-3)(x+1)=0$

$x=3\quad x=-1$

Check: $\dfrac{3}{3-2}+\dfrac{3(3)}{3-4}\overset{?}{=}\dfrac{-2(3-6)}{3^2-6(3)+8}$

$3+\dfrac{9}{-1}\overset{?}{=}\dfrac{6}{-1}$

$-6=-6$

Check:

$\dfrac{-1}{-1-2}+\dfrac{3(-1)}{-1-4}\overset{?}{=}\dfrac{-2(-1-6)}{(-1)^2-6(-1)+8}$

$\dfrac{1}{3}+\dfrac{3}{5}\overset{?}{=}\dfrac{14}{15}$

$\dfrac{5}{15}+\dfrac{9}{15}\overset{?}{=}\dfrac{14}{15}$

$\dfrac{14}{15}=\dfrac{14}{15}$

55.　$\dfrac{2(x+7)}{x+4}-2=\dfrac{2x+20}{2x+8}$

$2(x+4)\left(\dfrac{2(x+7)}{x+4}-2\right)=\left(\dfrac{2x+20}{2(x+4)}\right)2(x+4)$

$4(x+7)-2\cdot2(x+4)=2x+20$

$4x+28-4x-16=2x+20$

$12=2x+20$

$-8=2x$

$-4=x$

Check: $\dfrac{2[-4+7]}{-4+4}-2\overset{?}{=}\dfrac{2(-4)+20}{2(-4)+8}$

$\dfrac{6}{0}-2\neq\dfrac{12}{0}$

Division by zero is undefined. Solution is extraneous, so equation has no solution.

57. $\dfrac{x}{2} = \dfrac{2 - \dfrac{3}{x}}{1 - \dfrac{1}{x}}$

$\dfrac{x}{2} = \dfrac{2 - \dfrac{3}{x}}{1 - \dfrac{1}{x}} \cdot \dfrac{x}{x}$

$2(x - 1)\left(\dfrac{x}{2}\right) = \left(\dfrac{2x - 3}{x - 1}\right)2(x - 1)$

$x(x - 1) = (2x - 3)2$

$x^2 - x = 4x - 6$

$x^2 - 5x + 6 = 0$

$(x - 3)(x - 2) = 0$

$x = 3 \quad x = 2$

Check: $\dfrac{3}{2} \overset{?}{=} \dfrac{2 - \dfrac{3}{3}}{1 - \dfrac{1}{3}}$

$\dfrac{3}{2} = \dfrac{1}{\dfrac{2}{3}}$

$\dfrac{3}{2} = \dfrac{3}{2}$

$\dfrac{2}{2} \overset{?}{=} \dfrac{2 - \dfrac{3}{2}}{1 - \dfrac{1}{2}}$

$1 = \dfrac{\dfrac{1}{2}}{\dfrac{1}{2}}$

$1 = 1$

59. x-intercept: $(-2, 0)$

$0 = \dfrac{x + 2}{x - 2}$

$(x - 2)(0) = \left(\dfrac{x + 2}{x - 2}\right)(x - 2)$

$0 = x + 2$

$-2 = x$

61. x-intercepts: $(-1, 0)$ and $(1, 0)$

$0 = x - \dfrac{1}{x}$

$x(0) = \left(x - \dfrac{1}{x}\right)x$

$0 = x^2 - 1$

$0 = (x - 1)(x + 1)$

$x - 1 = 0 \quad x + 1 = 0$

$x = 1 \qquad x = -1$

63. (a) *Keystrokes:*

$\boxed{\text{Y=}}\ \boxed{(}\ \boxed{\text{X,T,}\theta}\ \boxed{-}\ 4\ \boxed{)}\ \boxed{\div}\ \boxed{(}\ \boxed{\text{X,T,}\theta}\ \boxed{+}\ 5\ \boxed{)}\ \boxed{\text{GRAPH}}$

x-intercept: $(4, 0)$

(b) $0 = \dfrac{x - 4}{x + 5}$

$0 = x - 4$

$4 = x$

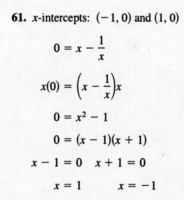

65. *Keystrokes:*

$\boxed{\text{Y=}}\ 1\ \boxed{\div}\ \boxed{\text{X,T,}\theta}\ \boxed{+}\ 4\ \boxed{\div}\ \boxed{(}\ \boxed{\text{X,T,}\theta}\ \boxed{-}\ 5\ \boxed{)}\ \boxed{\text{GRAPH}}$

x-intercept: $(1, 0)$

$0 = \dfrac{1}{x} + \dfrac{4}{x - 5}$

$x(x - 5)(0) = \left(\dfrac{1}{x} + \dfrac{4}{x - 5}\right)x(x - 5)$

$0 = x - 5 + 4x$

$5 = 5x$

$1 = x$

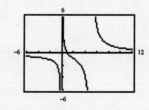

67. *Keystrokes:*

$\boxed{Y=}$ $\boxed{(}$ $\boxed{X,T,\theta}$ $\boxed{+}$ $\boxed{1}$ $\boxed{)}$ $\boxed{-}$ $\boxed{6}$ $\boxed{\div}$ $\boxed{X,T,\theta}$ $\boxed{GRAPH}$

x-intercepts: $(-3, 0)$ and $(2, 0)$

$$0 = (x + 1) - \frac{6}{x}$$

$$x(0) = \left[(x + 1) - \frac{6}{x}\right]x$$

$$0 = x^2 + x - 6$$

$$0 = (x + 3)(x - 2)$$

$$x + 3 = 0 \qquad x - 2 = 0$$

$$x = -3 \qquad x = 2$$

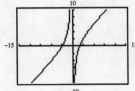

69. (a) *Keystrokes:*

$\boxed{Y=}$ $\boxed{(}$ $\boxed{X,T,\theta}$ $\boxed{-}$ $\boxed{1}$ $\boxed{)}$ $\boxed{-}$ $\boxed{12}$ $\boxed{\div}$ $\boxed{X,T,\theta}$ $\boxed{GRAPH}$

x-intercepts: $(-3, 0)$ and $(4, 0)$

(b) $$0 = (x - 1) - \frac{12}{x}$$

$$0 = x^2 - x - 12$$

$$0 = (x - 4)(x + 3)$$

$$x - 4 = 0 \qquad x + 3 = 0$$

$$x = 4 \qquad x = -3$$

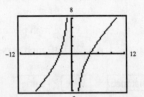

71. *Verbal Model:* $\boxed{\text{Number}} + \boxed{\text{Reciprocal}} = \boxed{\dfrac{65}{8}}$

Labels: Number $= x$

Reciprocal $= \dfrac{1}{x}$

Equation: $$x + \frac{1}{x} = \frac{65}{8}$$

$$8x\left(x + \frac{1}{x}\right) = \left(\frac{65}{8}\right)8x$$

$$8x^2 + 8 = 65x$$

$$8x^2 - 65x + 8 = 0$$

$$(8x - 1)(x - 8) = 0$$

$$x = \frac{1}{8} \quad x = 8$$

73. *Verbal Model:* $\boxed{\text{Distance}} \div \boxed{\text{Rate}} = \boxed{\text{Time}}$

$\dfrac{\boxed{\text{Distance}}}{\boxed{\text{Rate}}} = \boxed{\text{Time}}$

$\boxed{\begin{array}{c}\text{Time traveled} \\ \text{with wind}\end{array}} = \boxed{\begin{array}{c}\text{Time traveled} \\ \text{without wind}\end{array}}$

Labels: Speed of the wind $= x$

Equation: $$\frac{680}{300 + x} = \frac{520}{300 - x}$$

$$(300 + x)(300 - x)\left(\frac{680}{300 + x}\right) = \left(\frac{520}{300 - x}\right)(300 + x)(300 - x)$$

$$680(300 - x) = 520(300 + x)$$

$$204,000 - 680x = 156,000 + 520x$$

$$-1200x = -48,000$$

$$x = 40 \text{ miles per hour}$$

75. *Verbal Model:* $\boxed{\text{Distance}} = \boxed{\text{Rate}} \cdot \boxed{\text{Time}}$

$$\frac{\boxed{\text{Distance person 1}}}{\boxed{\text{Rate person 1}}} = \frac{\boxed{\text{Distance person 2}}}{\boxed{\text{Rate person 2}}}$$

Labels: Rate person 1 $= x + 2$

Rate person 2 $= x$

Equation:
$$\frac{5}{x + 2} = \frac{4}{x}$$

$$x(x + 2)\left(\frac{5}{x + 2}\right) = \left(\frac{4}{x}\right)x(x + 2)$$

$$5x = 4(x + 2)$$

$$5x = 4x + 8$$

$$x = 8 \text{ mph person 2}$$

$$x + 2 = 10 \text{ mph person 1}$$

77. *Verbal Model:* $\boxed{\text{Distance}} = \boxed{\text{Rate}} \cdot \boxed{\text{Time}}$

$$\frac{\text{Distance}}{\text{Rate}} = \text{Time}$$

$$\boxed{\begin{array}{c}\text{Time traveled}\\\text{upstream}\end{array}} + \boxed{\begin{array}{c}\text{Time traveled}\\\text{downstream}\end{array}} = \boxed{\begin{array}{c}\text{Total}\\\text{time}\end{array}}$$

Labels: Speed of the current $= x$

Equation:
$$\frac{48}{20 - x} + \frac{48}{20 + x} = 5$$

$$(20 - x)(20 + x)\left(\frac{48}{20 - x} + \frac{48}{20 + x}\right) = (5)(20 - x)(20 + x)$$

$$48(20 + x) + 48(20 - x) = 5(400 - x^2)$$

$$960 + 48x + 960 - 48x = 2000 - 5x^2$$

$$1920 = 2000 - 5x^2$$

$$5x^2 = 80$$

$$x^2 = 16$$

$$x = 4 \text{ mph}$$

79. *Verbal Model:* $\boxed{\begin{array}{c}\text{Cost per person}\\\text{current group}\end{array}} - \boxed{\begin{array}{c}\text{Cost per person}\\\text{new group}\end{array}} = \boxed{4000}$

Labels: Persons in current group $= x$

Persons in new group $= x + 2$

Equation:
$$\frac{240,000}{x} - \frac{240,000}{x + 2} = 4000$$

$$x(x + 2)\left(\frac{240,000}{x} - \frac{240,000}{x + 2}\right) = (4000)x(x + 2)$$

$$240,000(x + 2) - 240,000x = 4000(x^2 + 2x)$$

$$240,000x + 480,000 - 240,000x = 4000x^2 + 8000x$$

$$0 = 4000x^2 + 8000x - 480,000$$

$$0 = x^2 + 2x - 120$$

$$0 = (x + 12)(x - 10)$$

$$x + 12 = 0 \qquad x - 10 = 0$$

$$x = -12 \qquad x = 10 \text{ people}$$

81. *Verbal Model:* $\boxed{\begin{array}{c}\text{Cost per person}\\\text{original group}\end{array}} - \boxed{\begin{array}{c}\text{Cost per person}\\\text{new group}\end{array}} = \boxed{1300}$

Labels: Persons in current group $= x$

Persons in new group $= x + 3$

Equation:

$$\frac{78{,}000}{x} - \frac{78{,}000}{x+3} = 1300$$

$$x(x+3)\left(\frac{78{,}000}{x} - \frac{78{,}000}{x+3}\right) = (1300)x(x+3)$$

$$78{,}000(x+3) - 78{,}000x = 1300x(x+3)$$

$$78{,}000x + 234{,}000 - 78{,}000x = 1300x^2 + 3900x$$

$$0 = 1300x^2 + 3900x - 234{,}000$$

$$0 = x^2 + 3x - 180$$

$$0 = (x+15)(x-12)$$

$$x = -15 \quad x = 12 \text{ persons}$$

83. (a) *Keystrokes:*

$\boxed{Y=}\ \boxed{(}\ 120{,}000\ \boxed{X,T,\theta}\ \boxed{)}\ \boxed{\div}\ \boxed{(}\ 100\ \boxed{-}\ \boxed{X,T,\theta}\ \boxed{)}\ \boxed{GRAPH}$

(b) *Verbal Model:* $\boxed{\text{Cost}} = \dfrac{120{,}000p}{100 - p}$

Equation:

$$680{,}000 = \frac{120{,}000p}{100 - p}$$

$$(100 - p)(680{,}000) = \left(\frac{120{,}000p}{100 - p}\right)(100 - p)$$

$$68{,}000{,}000 - 680{,}000p = 120{,}000p$$

$$68{,}000{,}000 = 800{,}000p$$

$$85\% = p$$

85.

$\dfrac{1}{6} + \dfrac{1}{6} = \dfrac{1}{t}$

$t + t = 6$

$2t = 6$

$t = 3 \text{ hours}$

$\dfrac{1}{3} + \dfrac{1}{5} = \dfrac{1}{t}$

$5t + 3t = 15$

$8t = 15$

$t = \dfrac{15}{8} \text{ minutes}$

$\dfrac{1}{5} + \dfrac{1}{2\frac{1}{2}} = \dfrac{1}{t}$

$\dfrac{1}{5} + \dfrac{1}{\frac{5}{2}} = \dfrac{1}{t}$

$\dfrac{1}{5} + \dfrac{2}{5} = \dfrac{1}{t}$

$t + 2t = 5$

$3t = 5$

$t = \dfrac{5}{3} \text{ hours}$

Person #1	Person #2	Together
6 hours	6 hours	3 hours
3 minutes	5 minutes	$\frac{15}{8}$ minutes
5 hours	$2\frac{1}{2}$ hours	$\frac{5}{3}$ hours

87. *Verbal Model:* $\boxed{\text{Rate Person 1}} + \boxed{\text{Rate Person 2}} = \boxed{\text{Rate Together}}$

Labels: Second landscaper's time $= x$

First landscaper's time $= \dfrac{3}{2}x$

Equation:

$$\frac{1}{x} + \frac{1}{\frac{3}{2}x} = \frac{1}{9}$$

$$9x\left(\frac{1}{x} + \frac{2}{3x}\right) = \left(\frac{1}{9}\right)9x$$

$$9 + 6 = x$$

$$15 \text{ hours} = x$$

$$22\frac{1}{2} \text{ hours} = \frac{45}{2} = \frac{3}{2}x$$

89. *Verbal Model:* $\boxed{\text{Rate Pipe 1}} + \boxed{\text{Rate Pipe 2}} = \boxed{\text{Rate Together}}$

Labels: Second pipe's time $= x$

First pipe's time $= \dfrac{5}{4}x$

Equation:

$$\frac{1}{x} + \frac{1}{\frac{5}{4}x} = \frac{1}{5}$$

$$5x\left(\frac{1}{x} + \frac{4}{5x}\right) = \left(\frac{1}{5}\right)5x$$

$$5 + 4 = x$$

$$9 \text{ hours} = x$$

$$11\frac{1}{4} \text{ hours} = \frac{45}{4} = \frac{5}{4}x$$

91. $y = \dfrac{87{,}709 - 1236(0)}{1000 - 93(0)} \approx 87.7$ $y = \dfrac{87{,}709 - 1236(3)}{1000 - 93(3)} \approx 116.5$

$y = \dfrac{87{,}709 - 1236(1)}{1000 - 93(1)} \approx 95.3$ $y = \dfrac{87{,}709 - 1236(4)}{1000 - 93(4)} \approx 131.8$

$y = \dfrac{87{,}709 - 1236(2)}{1000 - 93(2)} \approx 104.7$ $y = \dfrac{87{,}709 - 1236(5)}{1000 - 93(5)} \approx 152.4$

93. (a) Domain $= \{4, 6, 8, 10, \ldots\}$ (c) $135 = 43.4 + \dfrac{9353}{x^2}$

(b)

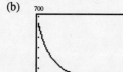

$$91.6 = \frac{9353}{x^2}$$

$$x^2 = \frac{9353}{91.6}$$

$$x \approx 10d$$

95. (d)
$$t(x) = \frac{10}{5 - x} + \frac{10}{5 + x} \quad \text{(from Section 8.3, Exercise 109)}$$

$$6\tfrac{1}{4} = \frac{10}{5 - x} + \frac{10}{5 + x}$$

$$4(5 - x)(5 + x)\left(\frac{25}{4}\right) = \left(\frac{10}{5 - x} + \frac{10}{5 + x}\right)4(5 - x)(5 + x)$$

$$25(25 - x^2) = 40(5 + x) + 40(5 - x)$$

$$625 - 25x^2 = 200 + 40x + 200 - 40x$$

$$0 = 25x^2 - 225$$

$$0 = x^2 - 9$$

$$x = 3 \text{ miles per hour}$$

(e) $t(x) = \dfrac{10}{5 - 4} + \dfrac{10}{5 + 4}$

$t(x) = 10 + \dfrac{10}{9} = \dfrac{90}{9} + \dfrac{10}{9} = \dfrac{100}{9}$

$t(x) = 11\tfrac{1}{9}$ or 11.1 hours

Yes

97. Solve a rational equation by multiplying both sides of the equation by the lowest common denominator. Then solve the resulting equation, checking for any extraneous solutions.

99. (a) Simplify each side by removing symbols of grouping, combining like terms, and reducing fractions on one or both sides.

(b) Add (or subtract) the same quantity to (from) both sides of the equation.

(c) Multiply (or divide) both sides of the equation by the same nonzero real number.

(d) Interchange the two sides of the equation.

101. When the equation involves only two fractions, one on each side of the equation, the equation can be solved by cross-multiplication.

Section 8.5   Graphs of Rational Functions

1. (a)

x	0	0.5	0.9	0.99	0.999
y	-4	-8	-40	-400	-4000

x	2	1.5	1.1	1.01	1.001
y	4	8	40	400	4000

x	2	5	10	100	1000
y	4	1	0.44444	0.0404	0.004

(b)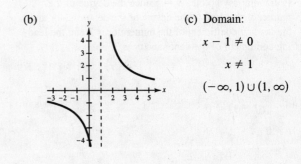

(c) Domain:

$x - 1 \ne 0$

$x \ne 1$

$(-\infty, 1) \cup (1, \infty)$

3. (a)

x	2	2.5	2.9	2.99	2.999
y	1	0	-8	-98	-998

x	4	3.5	3.1	3.01	3.001
y	3	4	12	102	1002

x	4	5	10	100	1000
y	3	2.5	2.143	2.010	2.001

(b)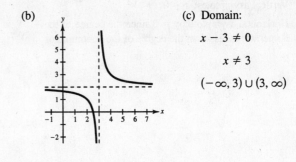

(c) Domain:

$x - 3 \ne 0$

$x \ne 3$

$(-\infty, 3) \cup (3, \infty)$

5. (a)

x	2	2.5	2.9	2.99	2.999
y	-1.2	-2.727	-14.75	-149.7	-1500

x	4	3.5	3.1	3.01	3.001
y	1.714	3.231	15.246	150.25	1500.2

x	4	5	10	100	1000
y	1.714	0.938	0.330	0.030	0.003

(b)

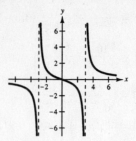

(c) Domain:
$$x^2 - 9 \neq 0$$
$$(x - 3)(x + 3) \neq 0$$
$$x \neq 3 \quad x \neq -3$$
$$(-\infty, -3) \cup (-3, 3) \cup (3, \infty)$$

7. Domain: $x^2 \neq 0$
$$x \neq 0$$
$$(-\infty, 0) \cup (0, \infty)$$

Vertical asymptote: $x = 0$

Horizontal asymptote: $y = 0$ since the degree of the numerator is less than the degree of the denominator.

9. Domain: $x + 8 \neq 0$
$$x \neq -8$$
$$(-\infty, -8) \cup (-8, \infty)$$

Vertical asymptote: $x = -8$

Horizontal asymptote: $y = 1$ since the degree of the numerator is equal to the degree of the denominator and the leading coefficients are 1.

11. Domain: $3t - 9 \neq 0$
$$t \neq 3$$
$$(-\infty, 3) \cup (3, \infty)$$

Vertical asymptote: $t = 3$

Horizontal asymptote: $y = \frac{2}{3}$ since the degree of the numerator is equal to the degree of the denominator and the leading coefficient of the numerator is 2 and the leading coefficient of the denominator is 3.

13. Domain: $1 - 3x \neq 0$
$$1 \neq 3x$$
$$\frac{1}{3} \neq x$$
$$\left(-\infty, \frac{1}{3}\right) \cup \left(\frac{1}{3}, \infty\right)$$

Vertical asymptote: $x = \frac{1}{3}$

Horizontal asymptote: $y = \frac{5}{3}$ since the degree of the numerator is equal to the degree of the denominator and the leading coefficient of the numerator is -5 and the leading coefficient of the denominator is -3.

15. $t(t - 1) \neq 0$
$$t \neq 0 \quad t - 1 \neq 0$$
$$t \neq 1$$
$$(-\infty, 0) \cup (0, 1) \cup (1, \infty)$$

Vertical asymptotes: $t = 0, t = 1$

Horizontal asymptote: $y = 0$ since the degree of the numerator is less than the degree of the denominator.

17. Domain: $x^2 + 1 \neq 0$
$$(-\infty, \infty)$$
no real solution

Vertical asymptote: none

Horizontal asymptote: $y = 2$ since the degree of the numerator is equal to the degree of the denominator and the leading coefficient of the numerator is 2 and the leading coefficient of the denominator is 1.

19. Domain: $x^2 - 1 \neq 0$

$$(x - 1)(x + 1) \neq 0$$

$$x \neq 1 \quad x \neq -1$$

$$(-\infty, -1) \cup (-1, 1) \cup (1, \infty)$$

Vertical asymptotes: $x = 1, x = -1$

Horizontal asymptote: $y = 1$ since the degree of the numerator is equal to the degree of the denominator and the leading coefficient of the numerator is 1 and the leading coefficient of the denominator is 1.

21. $g(z) = \dfrac{z}{z} \cdot \dfrac{1}{1} - \dfrac{2}{z} = \dfrac{z - 2}{z}$

Domain: $z \neq 0$

$$(-\infty, 0) \cup (0, \infty)$$

Vertical asymptote: $z = 0$

Horizontal asymptote: $y = 1$ since the degree of the numerator is equal to the degree of the denominator and the leading coefficients are 1.

23. $g(x) = 2x + \dfrac{4}{x} = \dfrac{x}{x} \cdot \dfrac{2x}{1} + \dfrac{4}{x} = \dfrac{2x^2 + 4}{x}$

Domain: $x \neq 0$

$$(-\infty, 0) \cup (0, \infty)$$

Vertical asymptote: $x = 0$

Horizontal asymptote: none since the degree of the numerator is greater than the degree of the denominator.

25. $f(x) = \dfrac{2}{x + 1}$ matches with graph (d).

Vertical asymptote: $x + 1 = 0$

$$x = -1$$

Horizontal asymptote: $y = 0$

27. $f(x) = \dfrac{x - 2}{x - 1}$ matches with graph (b).

Vertical asymptote: $x - 1 = 0$

$$x = 1$$

Horizontal asymptote: $y = 1$

29. (d)

31. (a)

33. y-intercept: $g(0) = \dfrac{5}{0} =$ undefined, none

x-intercept: none, numerator is never zero.

Vertical asymptote: $x = 0$

Horizontal asymptote: $y = 0$ since the degree of the numerator is less than the degree of the denominator.

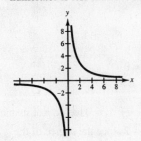

35. y-intercept: $g(0) = \dfrac{5}{0 - 4} = -\dfrac{5}{4}$

x-intercept: none, numerator is never zero.

Vertical asymptote: $x - 4 = 0$

$$x = 4$$

Horizontal asymptote: $y = 0$ since the degree of the numerator is less than the degree of the denominator.

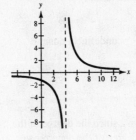

37. y-intercept: $f(0) = \dfrac{1}{0 - 2} = -\dfrac{1}{2}$

x-intercept: none, numerator is never zero.

Vertical asymptote: $x - 2 = 0$

$$x = 2$$

Horizontal asymptote: $y = 0$ since the degree of the numerator is less than the degree of the denominator.

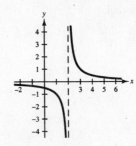

39. y-intercept: $g(0) = \dfrac{1}{2 - 0} = \dfrac{1}{2}$

x-intercept: none, numerator is never zero

Vertical asymptote: $2 - x = 0$

$$x = 2$$

Horizontal asymptote: $y = 0$ since the degree of the numerator is less than the degree of the denominator.

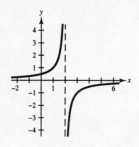

41. y-intercept: $y = \dfrac{3(0)}{0^2 + 4(0)}$ = undefined, none

x-intercept: $0 = \dfrac{3x}{x^2 + 4x} = \dfrac{3x}{x(x + 4)}$

$$0 = \dfrac{3}{x + 4}; \text{ none}$$

Vertical asymptote: $x^2 + 4x = 0$

$$x(x + 4) = 0$$

$$x = -4$$

Horizontal asymptote: $y = 0$ since the degree of the numerator is less than the degree of the denominator.

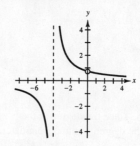

43. y-intercept: $h(0) = \dfrac{3(0)^2}{0^2 - 3(0)}$ = undefined, none

x-intercept: $0 = \dfrac{3u^2}{u^2 - 3u} = \dfrac{3u^2}{u(u - 3)}$

$$0 = \dfrac{3u}{u - 3}$$

$$0 = 3u$$

$$0 = u, \text{ none, since } h(0) \text{ is undefined.}$$

Vertical asymptote: $u^2 - 3u = 0$

$$u(u - 3) = 0$$

$$u = 3$$

Horizontal asymptote: $y = 3$ since the degrees are equal and the leading coefficient of the numerator is 3 and the leading coefficient of the denominator is 1.

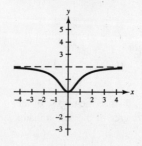

45. y-intercept: $y = \dfrac{2(0) + 4}{0}$ = undefined, none.

x-intercept: $2x + 4 = 0$

$$x = -2$$

Vertical asymptote: $x = 0$

Horizontal asymptote: $y = 2$ since the degree of the numerator is equal to the degree of the denominator and the leading coefficient of the numerator is 2 and the leading coefficient of the denominator is 1.

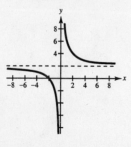

47. y-intercept: $y = \dfrac{2(0)^2}{0^2 + 1} = 0$

x-intercept: $x = 0$

Vertical asymptote: none, $x^2 + 1 = 0$ has no real solutions.

Horizontal asymptote: $y = 2$ since the degree of the numerator is equal to the degree of the denominator and the leading coefficient of the numerator is 2 and the leading coefficient of the denominator is 1.

49. *y*-intercept: $y = \dfrac{4}{0^2 + 1} = 4$

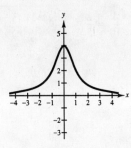

x-intercept: none, numerator is never zero.

Vertical asymptote: none, $x^2 + 1 \neq 0$

no real solution

Horizontal asymptote: $y = 0$ since the degree of the numerator is less than the degree of the denominator.

51. *y*-intercept: $g(0) = 3 - \dfrac{2}{0} = $ undefined, none

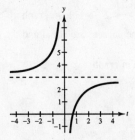

x-intercept: $0 = 3 - \dfrac{2}{t}$

$0 = 3t - 2$

$2 = 3t$

$\dfrac{2}{3} = t$

Vertical asymptote: $t = 0$

Horizontal asymptote: $y = 3$

53. *y*-intercept: $y = \dfrac{-0}{0^2 - 4} = 0$

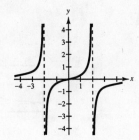

x-intercept: $0 = -\dfrac{x}{x^2 - 4}$

$0 = -x$

$0 = x$

Vertical asymptote: $x = 2, x = -2$

$x^2 - 4 = 0$

$(x - 2)(x + 2) = 0$

$x = 2 \quad x = -2$

Horizontal asymptote: $y = 0$ since the degree of the numerator is less than the degree of the denominator.

55. *y*-intercept: $y = \dfrac{3(0)^2}{0^2 - 0 - 2} = \dfrac{0}{-2} = 0$

x-intercept: $3x^2 = 0$

$x^2 = 0$

$x = 0$

Vertical asymptotes: $\quad x^2 - x - 2 = 0$

$(x - 2)(x + 1) = 0$

$x - 2 = 0 \qquad x - 2 = 0$

$x = 2 \qquad\quad x = 2$

Vertical asymptote: none

Horizontal asymptote: $y = 3$ since the degree of the numerator is equal to the degree of the denominator and the leading coefficient of the numerator is 3 and the leading coefficient of the denominator is 1.

57. y-intercept: $f(0) = \dfrac{0^2 - 4}{0^2 - 3(0) - 10} = \dfrac{4}{10} = \dfrac{2}{5}$

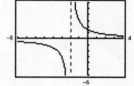

x-intercept: $0 = \dfrac{x^2 - 4}{x^2 - 3x - 10}$

$$0 = x^2 - 4$$

$$0 = (x - 2)(x + 2)$$

$x = 2$ undefined at $x = -2$

Vertical asymptotes: $x^2 - 3x - 10 = 0$

$$(x - 5)(x + 2) = 0$$

$x = 5 \quad \cancel{x} \cancel{2} \to$ hole in graph

Horizontal asymptote: $y = 1$ since the degrees are equal and the leading coefficients are 1.

$f(x) = \dfrac{(x - 2)\cancel{(x + 2)}}{(x - 5)\cancel{(x + 2)}} \Big\}$ gives a hole in graph at $x = -2$

59. Domain: $x + 2 \neq 0$

$$x \neq -2$$

$$(-\infty, -2) \cup (-2, \infty)$$

Vertical asymptote: $x = -2$

Horizontal asymptote: $y = 0$

Keystrokes:

y_1 [Y=] 3 [÷] [(] [X,T,θ] [+] 2 [)] [GRAPH]

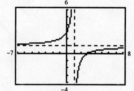

61. Domain: $x - 1 \neq 0$

$$x \neq 1$$

$$(-\infty, 1) \cup (1, \infty)$$

Vertical asymptote: $x = 1$

Horizontal asymptote: $y = 1$

Keystrokes:

[Y=] [(] [X,T,θ] [−] 3 [)] [÷] [(] [X,T,θ] [−] 1 [GRAPH]

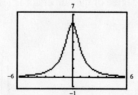

63. Domain: $t^2 + 1 \neq 0$

$$(-\infty, \infty)$$

Vertical asymptote: none

Horizontal asymptote: $y = 0$

Keystrokes:

[Y=] 6 [÷] [(] [X,T,θ] [x²] [+] 1 [)] [GRAPH]

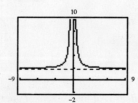

65. Domain: $x^2 \neq 0$

$$x \neq 0$$

$$(-\infty, 0) \cup (0, \infty)$$

Vertical asymptote: $x = 0$

Horizontal asymptote: $y = 2$

Keystrokes:

[Y=] [(] 2 [(] [X,T,θ] [x²] [+] 1 [)] [)] [÷] [X,T,θ] [x²] [GRAPH]

67. $y = \dfrac{3}{x} + \dfrac{1}{x-2} = \dfrac{4x-6}{x^2-2x}$

Domain: $x \neq 0 \quad x - 2 \neq 0$

$\qquad\qquad\qquad\quad x \neq 2$

$(-\infty, 0) \cup (0, 2) \cup (2, \infty)$

Vertical asymptotes: $x^2 - 2x = 0, x(x-2) = 0, x = 0, x = 2$

Horizontal asymptote: $y = 0$ since the degree of the numerator is less than the degree of the denominator.

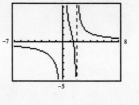

Keystrokes: [Y=] 3 [÷] [X,T,θ] [+] 1 [÷] [(] [(] [X,T,θ] [−] 2 [)] [GRAPH] or

[Y=] ([4 [X,T,θ] [−] 6) ÷ ([X,T,θ] [x²] [−] 2 [X,T,θ]) [GRAPH]

69. Reduce $g(x)$ to lowest terms.

$$g(x) = \frac{4 - 2x}{x - 2} = \frac{2(2 - x)}{x - 2} = -2$$

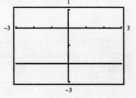

Keystrokes: [Y=] [(] 4 [−] 2 [X,T,θ] [)] [÷] [(] [X,T,θ] [−] 2 [)] [GRAPH]

There is no vertical asymptote because the fraction is not reduced to lowest terms.

71. (a) Average cost $= \dfrac{\text{Cost}}{\text{Number of units}}$

$$\overline{C} = \frac{2500 + 0.50x}{x}, \quad 0 < x$$

(b) $\overline{C} = \dfrac{2500 + 0.50(1000)}{1000} = \$3$

$\overline{C} = \dfrac{2500 + 0.50(10,000)}{10,000} = \$0.75$

(c) *Keystrokes:*

[Y=] [(] 2500 [+] .5 [X,T,θ] [)] [÷] [X,T,θ] [GRAPH]

Horizontal asymptote

$\overline{C} = \$0.50$ since the degree of the numerator is equal to the degree of the denominator and the leading coefficient of the numerator is 0.50 and the leading coefficient of the denominator is 1. As the number of units produced increases, the average cost is approximately $0.50.

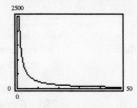

73. (a) $C = 0$ is the horizontal asymptote, since the degree of the numerator is less than the degree of the denominator. The meaning in the context of the problem is that the chemical is eliminated from the body.

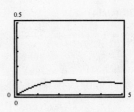

(b) *Keystrokes:* [Y=] 2 [X,T,θ] [÷] [(] 4 [X,T,θ] [x²] [+] 25 [)] [GRAPH]

Maximum occurs when $t \approx 2.5$.

75. (a) answers will vary.

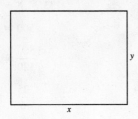

(c) Domain: $x > 0$ or $(0, \infty)$

(d) Minimum perimeter: 20 units × 20 units

Keystrokes:

[Y=] 2 [(] [X,T,θ] [+] 400 [÷] [X,T,θ] [)] [GRAPH]

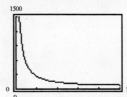

(b) $\quad A = x \cdot y \qquad P = 2l + 2w$

$\quad 400 = x \cdot y \qquad P = 2(l + w)$

$\quad \dfrac{400}{x} = y \qquad P = 2\left(x + \dfrac{400}{x}\right)$

77. $y = \dfrac{2(x + 1)}{x - 3}$

79. $y = \dfrac{x - 6}{(x - 4)(x + 2)}$

81. (f)

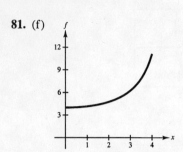

Since $t(x) \le 12$, domain is $[0, 4]$.

(g) No x-intercept

An x-intercept would mean that the time necessary for the trip would be zero for some value of x.

(h) $x = 5$. As x approaches 5, the speed for paddling upstream approaches 0.

83. An asymptote of a graph is a line to which the graph becomes arbitrarily close as $|x|$ or $|y|$ increases without bound.

85. No, not when the domain is all reals. For example,

$$f(x) = \frac{1}{x^2 + 1} \text{ has no vertical asymptote.}$$

Section 8.6   Variation

1. $I = kV$

3. $V = kt$

5. $u = kv^2$

7. $p = \dfrac{k}{d}$

9. $P = \dfrac{k}{\sqrt{1 + r}}$

11. $A = klw$

13. $P = \dfrac{k}{V}$

15. The area of a triangle varies jointly as the base and height.

17. The area of a rectangle varies jointly as the length and the width.

19. The volume of a right circular cylinder varies jointly as the square of the radius and the height.

21. The average speed varies directly as the distance and inversely as the time.

23. $s = kt$

$20 = k(4)$

$5 = k$

$s = 5t$

25. $F = kx^2$

$500 = k(40)^2$

$\frac{500}{1600} = k$

$\frac{5}{16} = k$

$F = \frac{5}{16}x^2$

27. $H = ku$

$100 = k(40)$

$\frac{100}{40} = k$

$\frac{5}{2} = k$

$H = \frac{5}{2}u$

29. $n = \dfrac{k}{m}$

$32 = \dfrac{k}{1.5}$

$48 = k$

$n = \dfrac{48}{m}$

31. $g = \dfrac{k}{\sqrt{z}}$

$\dfrac{4}{5} = \dfrac{k}{\sqrt{25}}$

$4 = k$

$g = \dfrac{4}{\sqrt{z}}$

33. $F = kxy$

$500 = k(15)(8)$

$\dfrac{500}{120} = k$

$\dfrac{25}{6} = k$

$F = \dfrac{25}{6}xy$

35. $d = k\left(\dfrac{x^2}{r}\right)$

$3000 = k\left(\dfrac{10^2}{4}\right)$

$3000 = k(25)$

$120 = k$

$d = \dfrac{120x^2}{r}$

37. (a) $R = kx$ (b) Price per unit

$3875 = k(500)$

$7.75 = k$

$R = 7.75x$

$R = 7.75(635)$

$R = \$4921.25$

39. (a) $d = kF$ $d = \dfrac{1}{10}F$

$5 = k(50)$

$\dfrac{5}{50} = k$ $d = \dfrac{1}{10}(20)$

$\dfrac{1}{10} = k$ $d = 2$ inches

(b) $d = \dfrac{1}{10}F$

$1.5 = \dfrac{1}{10}F$

15 pounds $= F$

41. $d = kF$

$7 = k(10.5)$

$\dfrac{7}{10.5} = k$

$\dfrac{70}{105} = k$

$\dfrac{2}{3} = k$

$12 = \dfrac{2}{3}F$

18 pounds $= F$

43. $v = kt$

$96 = k(3)$

$32 = k$

acceleration $= 32$ ft/sec^2

45. $d = ks^2$

$75 = k(30)^2$

$\dfrac{75}{900} = k$

$\dfrac{1}{12} = k$

$d = \dfrac{1}{12}(50)^2$

$d = 208.\overline{3}$ feet

47. $F = ks^2$

$F = k(2s)^2$

$F = 4ks^2$

$F = 4(ks^2)$

F will change by a factor of 4.

49. $p = kA$

9-inch: $6.78 = k(\pi)(4.5)^2$ 12-inch: $9.78 = k(\pi)(6)^2$ 15-inch: $12.18 = k(\pi)(7.5)^2$

$6.78 = 20.25\pi k$ $9.78 = 36\pi k$ $12.18 = 56.25\pi k$

$\dfrac{6.78}{20.25\pi} = k$ $\dfrac{9.78}{36\pi} = k$ $\dfrac{12.18}{56.25\pi} = k$

$0.106575 \approx k$ $0.0864745 \approx k$ $0.068923 \approx k$

No, the price of the pizza is not directly proportional to its area. The 15-inch pizza at \$12.18 is the best buy.

51. $x = \dfrac{k}{p}$

$800 = \dfrac{k}{5}$

$4000 = k$

$x = \dfrac{4000}{6}$

$x = 666.\overline{6} \approx 667$ units

53. $W_m = k \cdot W_e$ $W_m = \dfrac{1}{6} \cdot W_e$

$60 = k \cdot 360$ $54 = \dfrac{1}{6} \cdot x$

$k = \dfrac{1}{6}$ $x = 324$ pounds

55. $I = \dfrac{k}{d^2}$

$I = \dfrac{k}{18^2}$ $I = \dfrac{k}{36^2}$

$I = \dfrac{k}{324}$ $I = \dfrac{k}{1296}$

I will change by a factor of $\frac{324}{1296}$ or $\frac{1}{4}$.

57. $p = \dfrac{k}{t}$

$38 = \dfrac{k}{3}$

$114 = k$

So, $p = \dfrac{114}{t}$.

$p = \dfrac{114}{6.5}$

$p = 17.5\%$

59. (a) $I = Krt$

$202.50 = K(0.09)(3)$ $I = 750rt$

$202.50 = K(0.27)$ $I = 750(0.09)(4)$

$750 = K$ $I = \$270$

   (b) K is the principal or the amount of investment.

61.

x	2	4	6	8	10
$y = kx^2$	4	16	36	64	100

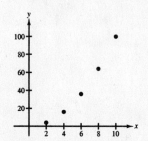

63.

x	2	4	6	8	10
$y = kx^2$	2	8	18	32	50

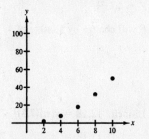

65.

x	2	4	6	8	10
$y = \dfrac{k}{x^2}$	$\dfrac{1}{2}$	$\dfrac{1}{8}$	$\dfrac{1}{18}$	$\dfrac{1}{32}$	$\dfrac{1}{50}$

67.

x	2	4	6	8	10
$y = \dfrac{k}{x^2}$	$\dfrac{5}{2}$	$\dfrac{5}{8}$	$\dfrac{5}{18}$	$\dfrac{5}{32}$	$\dfrac{1}{10}$

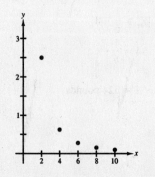

69.

x	10	20	30	40	50
y	$\dfrac{2}{5}$	$\dfrac{1}{5}$	$\dfrac{2}{15}$	$\dfrac{1}{10}$	$\dfrac{2}{25}$

$\dfrac{2}{5} = \dfrac{k}{10}$ $\dfrac{1}{5} = \dfrac{k}{20}$

$4 = k$ $4 = k$

Using any two pairs of numbers, k is 4.

71.

x	10	20	30	40	50
y	-3	-6	-9	-12	-15

$$-3 = k \cdot 10 \qquad -6 = k \cdot 20$$

$$-\tfrac{3}{10} = k \qquad -\tfrac{6}{20} = k$$

$$-\tfrac{3}{10} = k$$

Using any two pairs of numbers, k is $-\tfrac{3}{10}$.

75. $y = kx^2$

$y = k(2x)^2$

$y = k(4x^2)$

$y = 4kx^2$

y will quadruple.

73. Increase. $y = kx$ and $k > 0$ so if one variable increases the other also increases.

Review Exercises for Chapter 8

1. $y - 8 \neq 0$

$y \neq 8$

$D = (-\infty, 8) \cup (8, \infty)$

3. $u^2 - 7u + 6 \neq 0$

$(u - 6)(u - 1) \neq 0$

$u \neq 6 \quad u \neq 1$

$D = (-\infty, 1) \cup (1, 6) \cup (6, \infty)$

5. $\dfrac{6x^4y^2}{15xy^2} = \dfrac{2 \cdot 3x \cdot x^3 \cdot y^2}{5 \cdot 3x \cdot y^2} = \dfrac{2x^3}{5}, \quad x \neq 0, y \neq 0$

7. $\dfrac{5b - 15}{30b - 120} = \dfrac{5(b - 3)}{30(b - 4)}$

$\qquad = \dfrac{5(b - 3)}{5 \cdot 6(b - 4)}$

$\qquad = \dfrac{b - 3}{6(b - 4)}$

9. $\dfrac{9x - 9y}{y - x} = \dfrac{9(x - y)}{-1(x - y)} = -9, \quad x \neq y$

11. $\dfrac{x^2 - 5x}{2x^2 - 50} = \dfrac{x(x - 5)}{2(x^2 - 25)}$

$\qquad = \dfrac{x(x - 5)}{2(x - 5)(x + 5)}$

$\qquad = \dfrac{x}{2(x + 5)}, \quad x \neq 5$

13. $3x(x^2y)^2 = 3x(x^4y^2) = 3x^5y^2$

15. $\dfrac{24x^4}{15x} = \dfrac{8x^3}{5}, \quad x \neq 0$

17. $\dfrac{7}{8} \cdot \dfrac{2x}{y} \cdot \dfrac{y^2}{14x^2} = \dfrac{7 \cdot 2 \cdot x \cdot y \cdot y}{2 \cdot 2 \cdot 2 \cdot y \cdot 7 \cdot 2 \cdot x \cdot x}$

$\qquad = \dfrac{y}{8x}, \quad y \neq 0$

19. $\dfrac{60z}{z + 6} \cdot \dfrac{z^2 - 36}{5} = \dfrac{5 \cdot 12z(z - 6)(z + 6)}{(z + 6)5}$

$\qquad = 12z(z - 6), \quad z \neq -6$

21. $\dfrac{u}{u-3} \cdot \dfrac{3u-u^2}{4u^2} = \dfrac{u}{u-3} \cdot \dfrac{-u(u-3)}{4u^2}$

$$= -\dfrac{1}{4}, \quad u \neq 0, u \neq 3$$

23. $\dfrac{6/x}{2/x^3} = \dfrac{6}{x} \div \dfrac{2}{x^3} = \dfrac{3 \cdot 2}{x} \cdot \dfrac{x \cdot x^2}{2} = 3x^2, \quad x \neq 0$

25. $25y^2 \div \dfrac{xy}{5} = 25y \cdot y \cdot \dfrac{5}{xy} = \dfrac{125y}{x}, \quad y \neq 0$

27. $\dfrac{x^2-7x}{x+1} \div \dfrac{x^2-14x+49}{x^2-1} = \dfrac{x(x-7)}{x+1} \cdot \dfrac{(x-1)(x+1)}{(x-7)(x-7)}$

$$= \dfrac{x(x-1)}{x-7}, \quad x \neq -1, x \neq 1$$

29. $\dfrac{\left(\dfrac{6x^2}{x^2+2x-35}\right)}{\left(\dfrac{x^3}{x^2-25}\right)} = \dfrac{\dfrac{6x^2}{(x+7)(x-5)}}{\dfrac{x^3}{(x-5)(x+5)}} \cdot \dfrac{(x+7)(x-5)(x+5)}{(x+7)(x-5)(x+5)}$

$$= \dfrac{6x^2(x+5)}{x^3(x+7)}$$

$$= \dfrac{6(x+5)}{x(x+7)}, \quad x \neq 5, x \neq -5$$

31. $\dfrac{4}{9} - \dfrac{11}{9} = \dfrac{4-11}{9} = -\dfrac{7}{9}$

33. $\dfrac{15}{16} - \dfrac{5}{24} - 1 = \dfrac{15(3)}{16(3)} - \dfrac{5(2)}{24(2)} - \dfrac{1(48)}{1(48)}$

$$= \dfrac{45 - 10 - 48}{48}$$

$$= -\dfrac{13}{48}$$

35. $\dfrac{1}{x+5} + \dfrac{3}{x-12} = \dfrac{1}{x+5}\left(\dfrac{x-12}{x-12}\right) + \dfrac{3}{x-12}\left(\dfrac{x+5}{x+5}\right)$

$$= \dfrac{x-12}{(x+5)(x-12)} + \dfrac{3(x+5)}{(x-12)(x+5)}$$

$$= \dfrac{x-12+3x+15}{(x+5)(x-12)}$$

$$= \dfrac{4x+3}{(x+5)(x-12)}$$

37. $5x + \dfrac{2}{x-3} - \dfrac{3}{x+2} = \dfrac{5x(x-3)(x+2)}{(x-3)(x+2)} + \dfrac{2}{(x-3)}\left(\dfrac{x+2}{x+2}\right) - \dfrac{3}{(x+2)}\left(\dfrac{x-3}{x-3}\right)$

$$= \dfrac{5x^3 - 5x^2 - 30x + 2x + 4 - 3x + 9}{(x-3)(x+2)}$$

$$= \dfrac{5x^3 - 5x^2 - 31x + 13}{(x-3)(x+2)}$$

39. $\dfrac{6}{x} - \dfrac{6x-1}{x^2+4} = \dfrac{6(x^2+4)}{x(x^2+4)} - \dfrac{6x-1(x)}{x^2+4(x)} = \dfrac{6x^2+24-6x^2+x}{x(x^2+4)} = \dfrac{24+x}{x(x^2+4)}$

41. $\dfrac{5}{x + 3} - \dfrac{4x}{(x + 3)^2} - \dfrac{1}{x - 3} = \dfrac{5}{x + 3}\left(\dfrac{(x + 3)(x - 3)}{(x + 3)(x - 3)}\right) - \dfrac{4x}{(x + 3)^2}\left(\dfrac{x - 3}{x - 3}\right) - \dfrac{1}{x - 3}\left(\dfrac{(x + 3)^2}{(x + 3)^2}\right)$

$$= \dfrac{5x^2 - 45 - 4x^2 + 12x - x^2 - 6x - 9}{(x + 3)^2(x - 3)}$$

$$= \dfrac{6x - 54}{(x + 3)^2(x - 3)}$$

43. $\dfrac{3t}{\left(5 - \dfrac{2}{t}\right)} \cdot \dfrac{t}{t} = \dfrac{3t^2}{5t - 2}, \quad t \neq 0$

45. $\dfrac{\left(\dfrac{1}{a^2 - 16} - \dfrac{1}{a}\right)}{\left(\dfrac{1}{a^2 + 4a} + 4\right)} \cdot \dfrac{a(a - 4)(a + 4)}{a(a - 4)(a + 4)} = \dfrac{a - (a - 4)(a + 4)}{a - 4 + 4a(a - 4)(a + 4)}$

$$= \dfrac{a - (a^2 - 16)}{a - 4 + 4a(a^2 - 16)}$$

$$= \dfrac{a - a^2 + 16}{a - 4 + 4a^3 - 64a}$$

$$= \dfrac{-a^2 + a + 16}{4a^3 - 63a - 4}, a \neq 0, a \neq -4$$

47. *Keystrokes:*

y_1 $\boxed{\text{Y=}}$ $\boxed{(}$ $\boxed{(}$ $\boxed{(}$ $\boxed{\text{X,T,}\theta}$ $\boxed{x^2}$ $\boxed{+}$ 6 $\boxed{\text{X,T,}\theta}$ $\boxed{+}$ 9 $\boxed{)}$ $\boxed{\div}$ $\boxed{\text{X,T,}\theta}$ $\boxed{x^2}$ $\boxed{)}$

$\boxed{\times}$ $\boxed{(}$ $\boxed{(}$ $\boxed{\text{X,T,}\theta}$ $\boxed{x^2}$ $\boxed{-}$ 3 $\boxed{\text{X,T,}\theta}$ $\boxed{)}$ $\boxed{\div}$ $\boxed{(}$ $\boxed{\text{X,T,}\theta}$ $\boxed{+}$ 3 $\boxed{)}$ $\boxed{)}$ $\boxed{\text{ENTER}}$

y_2 $\boxed{(}$ $\boxed{\text{X,T,}\theta}$ $\boxed{x^2}$ $\boxed{-}$ 9 $\boxed{)}$ $\boxed{\div}$ $\boxed{\text{X,T,}\theta}$ $\boxed{\text{GRAPH}}$

$\dfrac{x^2 + 6x + 9}{x^2} \cdot \dfrac{x^2 - 3x}{x + 3} = \dfrac{(x + 3)(x + 3)x(x - 3)}{x^2(x + 3)} = \dfrac{x^2 - 9}{x}$

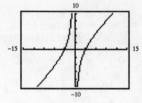

49. *Keystrokes:*

y_1 $\boxed{\text{Y=}}$ $\boxed{(}$ $\boxed{(}$ 1 $\boxed{\div}$ $\boxed{\text{X,T,}\theta}$ $\boxed{)}$ $\boxed{-}$ $\boxed{(}$ 1 $\boxed{\div}$ 2 $\boxed{)}$ $\boxed{)}$ $\boxed{\div}$ 2 $\boxed{\text{X,T,}\theta}$ $\boxed{\text{ENTER}}$

y_2 $\boxed{\text{Y=}}$ $\boxed{(}$ 2 $\boxed{-}$ $\boxed{\text{X,T,}\theta}$ $\boxed{)}$ $\boxed{\div}$ 4 $\boxed{\text{X,T,}\theta}$ $\boxed{x^2}$ $\boxed{\text{GRAPH}}$

$\dfrac{\left(\dfrac{1}{x} - \dfrac{1}{2}\right)}{2x} \cdot \dfrac{2x}{2x} = \dfrac{2 - x}{4x^2}$

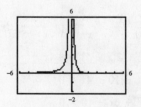

51. $\quad \dfrac{3x}{8} = -15 + \dfrac{x}{4}$

$8\left(\dfrac{3}{8}x\right) = (-15)8 + \left(\dfrac{x}{4}\right)8$

$\quad 3x = -120 + 2x$

$\quad x = -120$

Check: $\dfrac{3(-120)}{8} \stackrel{?}{=} -15 + -\dfrac{120}{4}$

$-\dfrac{360}{8} \stackrel{?}{=} -15 + -30$

$-45 = -45$

53. $(3t)\left(8 - \dfrac{12}{t}\right) = \dfrac{1}{3}(3t)$

$24t - 36 = t$

$23t = 36$

$t = \dfrac{36}{23}$

Check: $8 - \dfrac{12}{\left(\frac{36}{23}\right)} \stackrel{?}{=} \dfrac{1}{3}$

$8 - \dfrac{23}{3} \stackrel{?}{=} \dfrac{1}{3}$

$\dfrac{24}{3} - \dfrac{23}{3} \stackrel{?}{=} \dfrac{1}{3}$

$\dfrac{1}{3} = \dfrac{1}{3}$

55. $\dfrac{2}{y} - \dfrac{1}{3y} = \dfrac{1}{3}$ **Check:** $\dfrac{2}{5} - \dfrac{1}{3(5)} \overset{?}{=} \dfrac{1}{3}$

$3y\left(\dfrac{2}{y} - \dfrac{1}{3y}\right) = \left(\dfrac{1}{3}\right)3y$ $\dfrac{2}{5} - \dfrac{1}{15} \overset{?}{=} \dfrac{1}{3}$

$\qquad\qquad 6 - 1 = y$ $\dfrac{6}{15} - \dfrac{1}{15} \overset{?}{=} \dfrac{1}{3}$

$\qquad\qquad\quad 5 = y$ $\dfrac{5}{15} \overset{?}{=} \dfrac{1}{3}$

$\dfrac{1}{3} = \dfrac{1}{3}$

57. $r = 2 + \dfrac{24}{r}$ **Check:** $6 \overset{?}{=} 2 + \dfrac{24}{6}$

$r(r) = \left(2 + \dfrac{24}{r}\right)r$ $6 \overset{?}{=} 2 + 4$

$r^2 = 2r + 24$ $6 = 6$

$r^2 - 2r - 24 = 0$ **Check:**

$(r - 6)(r + 4) = 0$ $-4 \overset{?}{=} 2 + \dfrac{24}{-4}$

$r = 6 \quad r = -4$ $-4 \overset{?}{=} 2 - 6$

$-4 = -4$

59. $8\left(\dfrac{6}{x} - \dfrac{1}{x + 5}\right) = 15$

$\left(\dfrac{6}{x} - \dfrac{1}{x + 5}\right) = \dfrac{15}{8}$

$8x(x + 5)\left(\dfrac{6}{x} - \dfrac{1}{x + 5}\right) = \left(\dfrac{15}{8}\right)8x(x + 5)$

$48(x + 5) - 8x = 15x(x + 5)$

$48x + 240 - 8x = 15x^2 + 75x$

$240 + 40x = 15x^2 + 75x$

$0 = 15x^2 + 35x - 240$

$0 = 5(3x^2 + 7x - 48)$

$0 = 5(3x + 16)(x - 3)$

$3x + 16 = 0 \qquad\qquad x - 3 = 0$

$3x = -16 \qquad\qquad x = 3$

$x = -\dfrac{16}{3}$

Check: $8\left(\dfrac{6}{-\frac{16}{3}} - \dfrac{1}{-\frac{16}{3} + 5}\right) \overset{?}{=} 15$

$8\left(-\dfrac{18}{16} - \dfrac{1}{-\frac{16}{3} + \frac{15}{3}}\right) \overset{?}{=} 15$

$8\left(-\dfrac{9}{8} - \dfrac{1}{-\frac{1}{3}}\right) \overset{?}{=} 15$

$8\left(-\dfrac{9}{8} + 3\right) \overset{?}{=} 15$

$8\left(-\dfrac{9}{8} + \dfrac{24}{8}\right) \overset{?}{=} 15$

$8\left(\dfrac{15}{8}\right) \overset{?}{=} 15$

$15 = 15$

Check: $8\left(\dfrac{6}{3} - \dfrac{1}{3 + 5}\right) \overset{?}{=} 15$

$8\left(2 - \dfrac{1}{8}\right) \overset{?}{=} 15$

$8\left(\dfrac{16}{8} - \dfrac{1}{8}\right) \overset{?}{=} 15$

$8\left(\dfrac{15}{8}\right) \overset{?}{=} 15$

$15 = 15$

61. $(x - 5)\left(\dfrac{4x}{x - 5} + \dfrac{2}{x}\right) = \left(-\dfrac{4}{x - 5}\right)x(x - 5)$

$4x^2 + 2(x - 5) = -4x$

$4x^2 + 2x - 10 + 4x = 0$

$4x^2 + 6x - 10 = 0$

$2x^2 + 3x - 5 = 0$

$(2x + 5)(x - 1) = 0$

$x = -\dfrac{5}{2} \quad x = 1$

Check: $\dfrac{4\left(-\frac{5}{2}\right)}{\left(-\frac{5}{2}\right) - 5} + \dfrac{2}{\left(-\frac{5}{2}\right)} \overset{?}{=} -\dfrac{4}{\left(-\frac{5}{2}\right) - 5}$

$\dfrac{-10}{-\frac{15}{2}} - \dfrac{4}{5} \overset{?}{=} -\dfrac{4}{-\frac{15}{2}}$

$\dfrac{20}{15} - \dfrac{12}{15} \overset{?}{=} \dfrac{8}{15}$

$\dfrac{8}{15} = \dfrac{8}{15}$

Check: $\dfrac{4(1)}{1 - 5} + \dfrac{2}{1} \overset{?}{=} -\dfrac{4}{1 - 5}$

$\dfrac{4}{-4} + 2 \overset{?}{=} -\dfrac{4}{-4}$

$1 = 1$

63.
$$\frac{12}{x^2 + x - 12} - \frac{1}{x - 3} = -1$$

$$(x - 3)(x + 4)\left(\frac{12}{x^2 + x - 12} - \frac{1}{x - 3}\right) = (-1)(x - 3)(x + 4)$$

$$12 - (x + 4) = -1(x^2 + x - 12)$$

$$12 - x - 4 = -x^2 - x + 12$$

$$(x^2 - 4) = 0$$

$$(x - 2)(x + 2) = 0$$

$$x - 2 = 0 \qquad x + 2 = 0$$

$$x = 2 \qquad x = -2$$

65.
$$\frac{5}{x^2 - 4} - \frac{6}{x - 2} = -5$$

$$(x - 2)(x + 2)\left(\frac{5}{x^2 - 4} - \frac{6}{x - 2}\right) = (-5)(x - 2)(x + 2)$$

$$5 - 6(x + 2) = -5(x^2 - 4)$$

$$5 - 6x - 12 = -5x^2 + 20$$

$$5x^2 - 6x - 27 = 0$$

$$(5x + 9)(x - 3) = 0$$

$$5x + 9 = 0 \qquad x - 3 = 0$$

$$x = -\frac{9}{5} \qquad x = 3$$

67. *Keystrokes:*

x-intercept: $(-3, 0)$

$$0 = \frac{1}{x} - \frac{1}{2x - 3}$$

$$0 = 2x + 3 - x$$

$$0 = x + 3$$

$$-3 = x$$

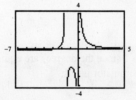

69. (b)

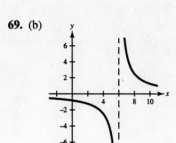

71. (a)

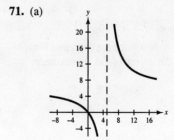

73. *y*-intercept: $f(0) = \dfrac{-5}{0^2} = $ undefined; none

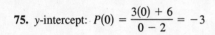

x-intercept: $0 = \dfrac{-5}{x^2}$

$0 = -5$; none

Vertical asymptote: $x^2 = 0$

$x = 0$

Horizontal asymptote: $y = 0$ since the degree of the numerator is less than the degree of the denominator.

75. *y*-intercept: $P(0) = \dfrac{3(0) + 6}{0 - 2} = -3$

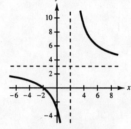

x-intercept: $0 = \dfrac{3x + 6}{x - 2}$

$0 = 3x + 6$

$-2 = x$

Vertical asymptote: $x = 2$

Horizontal asymptote: $y = 3$ since the degrees are equal and the leading coefficient of the numerator is 3 and the leading coefficient of the denominator is 1.

77. *y*-intercept: $g(0) = \dfrac{2 + 0}{1 - 0} = 2$

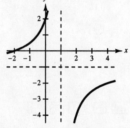

x-intercept: $0 = \dfrac{2 + x}{1 - x}$

$0 = 2 + x$

$-2 = x$

Vertical asymptote: $x = 1$

Horizontal asymptote: $y = -1$ since the degrees are equal and the leading coefficient of the numerator is 1 and the leading coefficient of the denominator is -1.

79. *y*-intercept: $f(0) = \dfrac{0}{0^2 + 1} = 0$

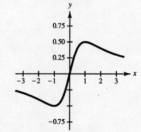

x-intercept: $0 = \dfrac{x}{x^2 + 1}$

$0 = x$

Vertical asymptote: none; $x^2 + 1 = 0$

Horizontal asymptote: $y = 0$ since the degree of the numerator is less than the degree of the denominator.

81. *y*-intercept: $h(0) = \dfrac{4}{(0 - 1)^2} = 4$

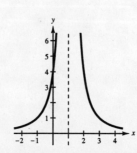

x-intercept: $0 = \dfrac{4}{(x - 1)^2}$

$0 = 4$; none

Vertical asymptote: $x = 1$

Horizontal asymptote: $y = 0$ since the degree of the numerator is less than the degree of the denominator.

83. *y*-intercept: $y = \dfrac{0}{0^2 - 1} = 0$

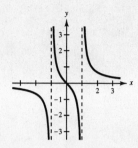

x-intercept: $0 = \dfrac{x}{x^2 - 1}$

$0 = x$

Vertical asymptotes: $x^2 - 1 = 0$

$$x = 1, \quad x = -1$$

Horizontal asymptote: $y = 0$ since the degree of the numerator is less than the degree of the denominator.

$$y = \frac{2x^2}{x^2 - 4x} = \frac{2x^2}{x(x - 4)} = \frac{2x}{x - 4}$$

85. *y*-intercept: $y = \dfrac{2(0)^2}{0^2 - 4(0)} =$ undefined, none

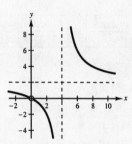

x-intercept: $0 = \dfrac{2x^2}{x^2 - 4x}$

$0 = 2x^2$

$0 = x$, none, hole at $x = 0$

Vertical asymptote: $x - 4 = 0$

$$x = 4$$

Horizontal asymptote: $y = 2$ since the degrees are equal and the leading coefficient of the numerator is 2 and the leading coefficient of the denominator is 1.

87. $y = \dfrac{x - 4}{x^2 - 3x - 4} = \dfrac{x - 4}{(x - 4)(x + 1)} = \dfrac{1}{x + 1}$

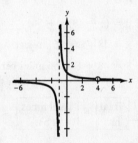

y-intercept: $y = \dfrac{0 - 4}{0^2 - 3(0) - 4} = \dfrac{-4}{-4} = 1$

x-intercept: $0 = \dfrac{1}{x + 1}$

$0 = 1$; none

Vertical asymptote: $x = -1$

Horizontal asymptote: $y = 0$ since the degree of the numerator is less than the degree of the denominator.

Since the function is undefined for $x = 4$, and $\dfrac{(x - 4)}{(x - 4)(x + 1)} = \dfrac{1}{x + 1}$, there is a hole at $x = 4$.

89. $y = \dfrac{3x}{x - 4}$

91. $y = k\sqrt[3]{x}$

$12 = k\sqrt[3]{8}$

$\dfrac{12}{2} = k$

$6 = k$

$y = 6\sqrt[3]{x}$

93. $T = krs^2$

$5000 = k(0.09)(1000)^2$

$\dfrac{5000}{90{,}000} = k$

$\dfrac{1}{18} = k$

$T = \dfrac{1}{18}rs^2$

95. Domain: $(0, 6]$

$P = 2\left(w + \dfrac{36}{w}\right), \quad w \neq 0$

97. *Verbal Model:*

$$\boxed{\text{Distance}} = \boxed{\text{Rate}} \cdot \boxed{\text{Time}}$$

$$\boxed{\dfrac{\text{Distance}}{\text{Rate}}} = \boxed{\text{Time}}$$

$$\boxed{\dfrac{\text{Original trip distance}}{\text{Rate}}} = \boxed{\dfrac{\text{Round trip distance}}{\text{Rate}}} + \dfrac{1}{6}$$

Labels: Rate original trip $= x$

Rate return trip $= x + 8$

Equation:

$$\dfrac{56}{x} = \dfrac{56}{x + 8} + \dfrac{1}{6}$$

$$6x(x + 8)\left(\dfrac{56}{x}\right) = \left(\dfrac{56}{x + 8} + \dfrac{1}{6}\right)6x(x + 8)$$

$$336(x + 8) = 336x + x(x + 8)$$

$$336x + 2688 = 336x + x^2 + 8x$$

$$0 = x^2 + 8x - 2688$$

$$0 = (x - 48)(x + 56)$$

$$x = 48 \quad \cancel{x = -56} \text{ reject}$$

$$x + 8 = 48 + 8 = 56 \text{ miles per hour}$$

99. *Verbal Model:*

$$\boxed{\genfrac{}{}{0pt}{}{\text{Batting}}{\text{Average}}} = \boxed{\genfrac{}{}{0pt}{}{\text{Total}}{\text{hits}}} \div \boxed{\genfrac{}{}{0pt}{}{\text{Total times}}{\text{at bat}}} = 0.400$$

Labels: Current times at bat $= 150$

Current hits $= 45$

Additional consecutive times $= x$

Equation:

$$\dfrac{45 + x}{150 + x} = 0.400$$

$$(150 + x)\left(\dfrac{45 + x}{150 + x}\right) = 0.4(150 + x)$$

$$45 + x = 0.4(150 + x)$$

$$45 + x = 60 + 0.4x$$

$$45 + 0.6x = 60$$

$$0.6x = 15$$

$$x = \dfrac{15}{0.6}$$

$$x = 25$$

Thus, the player must hit safely 25 consecutive times to obtain a batting average of 0.400.

101. *Verbal Model:* $\boxed{\begin{array}{c}\text{Share per}\\\text{person now}\end{array}} = \boxed{\begin{array}{c}\text{Share per}\\\text{person later}\end{array}} + \boxed{5000}$

Labels: People presently in group $= x$

People in new group $= x + 2$

Equation: $\dfrac{60{,}000}{x} = \dfrac{60{,}000}{x + 2} + 5000$

$$x(x + 2)\left(\dfrac{60{,}000}{x}\right) = \left(\dfrac{60{,}000}{x + 2} + 5000\right)x(x + 2)$$

$$60{,}000(x + 2) = 60{,}000 + 5000x(x + 2)$$

$$60{,}000x + 120{,}000 = 60{,}000x + 5000x^2 + 10{,}000x$$

$$0 = 5000x^2 + 10{,}000x - 120{,}000$$

$$0 = x^2 + 2x - 24$$

$$0 = (x + 6)(x - 4)$$

$$x = -6 \quad x = 4 \text{ people}$$

103. *Verbal Model:* $\boxed{\text{Rate person 1}} + \boxed{\text{Rate person 2}} = \boxed{\text{Rate together}}$

Labels: Supervisor's time $= 12$

Your time $= 15$

Time together $= x$

Equation: $\dfrac{1}{12} + \dfrac{1}{15} = \dfrac{1}{x}$

$$60x\left(\dfrac{1}{12} + \dfrac{1}{15}\right) = \left(\dfrac{1}{x}\right)60x$$

$$5x + 4x = 60$$

$$9x = 60$$

$$x = \dfrac{60}{9}$$

$$x = \dfrac{20}{3} = 6\dfrac{2}{3} \text{ min or 6 min 40 sec}$$

105. (a) $N = \dfrac{20[4 + 3(5)]}{1 + 0.05(5)} = 304$

$N = \dfrac{20[4 + 3(10)]}{1 + 0.05(10)} \approx 453$

$N = \dfrac{20[4 + 3(25)]}{1 + 0.05(25)} \approx 702$

(b) $752 = \dfrac{20(4 + 3t)}{1 + 0.05t}$

$$752(1 + 0.05t) = 20(4 + 3t)$$

$$752 + 37.6t = 80 + 60t$$

$$672 = 22.4t$$

$$30 \text{ years} = t$$

107. $d = ks^2$

$d = k(2s)^2$

$d = 4ks^2 = 4(ks^2)$

The stopping distance increases by a factor of 4.

109. $F = \dfrac{k}{r^2}$

$$200 = \dfrac{k}{(4000)^2}$$

$$3{,}200{,}000{,}000 = k$$

$$F = \dfrac{3{,}200{,}000{,}000}{r^2} = \dfrac{3{,}200{,}000{,}000}{(4500)^2}$$

$$F \approx 158.02 \text{ pounds}$$

Chapter Test for Chapter 8

1.
$$y^2 - 25 \neq 0$$
$$(y - 5)(y + 5) \neq 0$$
$$y \neq 5, -5$$
$$D = (-\infty, -5) \cup (-5, 5) \cup (5, \infty)$$

2. LCD $= x^3(x - 3)(x + 3)$

3. (a) $\dfrac{2 - x}{3x - 6} = \dfrac{2 - x}{-3(-x + 2)} = -\dfrac{1}{3}, \quad x \neq 2$

(b) $\dfrac{2a^2 - 5a - 12}{5a - 20} = \dfrac{(2a + 3)(a - 4)}{5(a - 4)} = \dfrac{2a + 3}{5}, \quad a \neq 4$

4. $\dfrac{4z^3}{5} \cdot \dfrac{25}{12z^2} = \dfrac{4 \cdot z^2 \cdot z \cdot 5 \cdot 5}{5 \cdot 4 \cdot 3 \cdot z^2} = \dfrac{5z}{3}, \quad z \neq 0$

5. $\dfrac{y^2 + 8x + 16}{2(y - 2)} \cdot \dfrac{8y - 16}{(y + 4)^3} = \dfrac{(y + 4)^2 \cdot 8(y - 2)}{2(y - 2)(y + 4)^2(y + 4)}$

$$= \dfrac{4}{y + 4}, \quad y \neq 2$$

6. $(4x^2 - 9) \cdot \dfrac{2x + 3}{2x^2 - x - 3} = \dfrac{(2x - 3)(2x + 3)(2x + 3)}{(2x - 3)(x + 1)}$

$$= \dfrac{(2x + 3)^2}{x + 1}, \quad x \neq \dfrac{3}{2}$$

7. $\dfrac{(2xy^2)^3}{15} \div \dfrac{12x^3}{21} = \dfrac{(2xy^2)^3}{15} \cdot \dfrac{21}{12x^3}$

$$= \dfrac{8x^3y^6 \cdot 7 \cdot 3}{5 \cdot 3 \cdot 4 \cdot 3x^3}$$

$$= \dfrac{14y^6}{15}, \quad x \neq 0$$

8. $\dfrac{\left(\dfrac{3x}{x + 2}\right)}{\left(\dfrac{12}{x^3 + 2x^2}\right)} = \dfrac{3x}{x + 2} \div \dfrac{12}{x^3 + 12x^2}$

$$= \dfrac{3x}{x + 2} \cdot \dfrac{x^2(x + 2)}{12}$$

$$= \dfrac{x^3}{4}, \quad x \neq 0, -2$$

9. $\dfrac{\left(9x - \dfrac{1}{x}\right)}{\left(\dfrac{1}{x} - 3\right)} = \dfrac{\left(9x - \dfrac{1}{x}\right)}{\left(\dfrac{1}{x} - 3\right)} \cdot \dfrac{x}{x}$

$$= \dfrac{9x(x) - \dfrac{1}{x}(x)}{\dfrac{1}{x}(x) - 3(x)}$$

$$= \dfrac{9x^2 - 1}{1 - 3x}$$

$$= \dfrac{(3x - 1)(3x + 1)}{-1(-1 + 3x)}$$

$$= -(3x + 1), \quad x \neq 0, \dfrac{1}{3}$$

10. $2x + \dfrac{1 - 4x^2}{x + 1} = 2x\left(\dfrac{x + 1}{x + 1}\right) + \dfrac{1 - 4x^2}{x + 1}$

$$= \dfrac{2x^2 + 2x}{x + 1} + \dfrac{1 - 4x^2}{x + 1}$$

$$= \dfrac{-2x^2 + 2x + 1}{x + 1}$$

11. $\dfrac{5x}{x + 2} - \dfrac{2}{x^2 - x - 6} = \dfrac{5x}{x + 2} - \dfrac{2}{(x - 3)(x + 2)}$

$$= \dfrac{5x}{x + 2}\left(\dfrac{x - 3}{x - 3}\right) - \dfrac{2}{(x - 3)(x + 2)}$$

$$= \dfrac{5x^2 - 15x - 2}{(x + 2)(x - 3)}$$

12. $\dfrac{3}{x} - \dfrac{5}{x^2} + \dfrac{2x}{x^2 + 2x + 1} = \dfrac{3}{x} - \dfrac{5}{x^2} + \dfrac{2x}{(x + 1)^2}$

$$= \frac{3}{x}\!\left(\frac{x(x+1)^2}{x(x+1)^2}\right) - \frac{5}{x^2}\!\left(\frac{x(x+1)^2}{x(x+1)^2}\right) + \frac{2x}{(x+1)^2}\!\left(\frac{x^2}{x^2}\right)$$

$$= \frac{3x(x^2 + 2x + 1) - 5(x^2 + 2x + 1) + 2x^3}{x^2(x + 1)^2}$$

$$= \frac{3x^3 + 6x^2 + 3x - 5x^2 - 10x - 5 + 2x^3}{x^2(x + 1)^2}$$

$$= \frac{5x^3 + x^2 - 7x - 5}{x^2(x + 1)^2}$$

13. (a) $f(x) = \dfrac{3}{x - 3}$

y-intercept: $f(0) = \dfrac{3}{0 - 3} = -1$

x-intercept: none, numerator is never zero.

Vertical asymptote: $x - 3 = 0$

$\qquad\qquad\qquad\quad x = 3$

Horizontal asymptote: $y = 0$ since the degree of the numerator is less than the degree of the denominator.

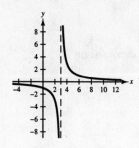

(b) $g(x) = \dfrac{3x}{x - 3}$

y-intercept: $y = 0$

x-intercept: $x = 0$

Vertical asymptote: $x - 3 = 0$

$\qquad\qquad\qquad\quad x = 3$

Horizontal asymptote: $y = 3$ since the degree of the numerator is equal to the degree of the denominator and the leading coefficient of the numerator is 3 and the leading coefficient of the denominator is 1.

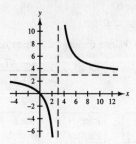

14.
$$\frac{3}{h + 2} = \frac{1}{8}$$

$$8(h + 2)\!\left(\frac{3}{h + 2}\right) = \left(\frac{1}{8}\right)8(h + 2)$$

$$3(8) = h + 2$$

$$24 = h + 2$$

$$22 = h$$

Check: $\dfrac{3}{22 + 2} \overset{?}{=} \dfrac{1}{8}$

$$\frac{3}{24} \overset{?}{=} \frac{1}{8}$$

$$\frac{1}{8} = \frac{1}{8}$$

15.
$$\frac{2}{x+5} - \frac{3}{x+3} = \frac{1}{x}$$

$$x(x+3)(x+5)\left(\frac{2}{x+5} - \frac{3}{x+3}\right) = \left(\frac{1}{x}\right)x(x+3)(x+5)$$

$$2x(x+3) - 3x(x+5) = (x+5)(x+3)$$

$$2x^2 + 6x - 3x^2 - 15x = x^2 + 3x + 5x + 15$$

$$-2x^2 - 17x - 15 = 0$$

$$(-2x - 15)(x+1) = 0$$

$$-2x - 15 = 0 \qquad x + 1 = 0$$

$$-2x = 15 \qquad\qquad x = -1$$

$$x = -\frac{15}{2}$$

Check:

$$\frac{2}{-\frac{15}{2}+5} - \frac{3}{-\frac{15}{2}+3} \overset{?}{=} \frac{1}{-\frac{15}{2}}$$

$$-\frac{12}{15} + \frac{10}{15} \overset{?}{=} -\frac{2}{15}$$

$$-\frac{2}{15} = -\frac{2}{15}$$

Check:

$$\frac{2}{-1+5} - \frac{3}{-1+3} \overset{?}{=} -\frac{1}{1}$$

$$\frac{2}{4} - \frac{3}{2} \overset{?}{=} -1$$

$$\frac{1}{2} - \frac{3}{2} \overset{?}{=} -1$$

$$-1 = -1$$

16. $f(x) = \dfrac{x-4}{x^2 - x - 12} = \dfrac{x-4}{(x-4)(x+3)} = \dfrac{1}{x+3}$

y-intercept: $f(0) = \dfrac{1}{0+3} = \dfrac{1}{3}$

x-intercept: $0 = \dfrac{1}{x+3}$

$0 = 1$; none

Vertical asymptote: $x + 3 = 0$

$$x = -3$$

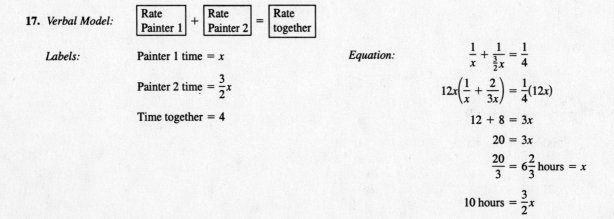

Horizontal asymptote: $y = 0$ since the degree of the numerator is less than the degree of the denominator.

Since $f(x)$ is undefined at $x = 4$, and since $\dfrac{\cancel{(x-4)}}{\cancel{(x-4)}(x+3)} = \dfrac{1}{x+3}$, there is a hole at $x = 4$.

17. *Verbal Model:* | Rate Painter 1 | + | Rate Painter 2 | = | Rate together |

Labels: Painter 1 time $= x$

Painter 2 time $= \dfrac{3}{2}x$

Time together $= 4$

Equation:

$$\frac{1}{x} + \frac{1}{\frac{3}{2}x} = \frac{1}{4}$$

$$12x\left(\frac{1}{x} + \frac{2}{3x}\right) = \frac{1}{4}(12x)$$

$$12 + 8 = 3x$$

$$20 = 3x$$

$$\frac{20}{3} = 6\frac{2}{3} \text{ hours} = x$$

$$10 \text{ hours} = \frac{3}{2}x$$

18. $S = \dfrac{kx^2}{y}$

19. $v = k\sqrt{u}$

$$\frac{3}{2} = k\sqrt{36}$$

$$\frac{1}{4} = k$$

$$v = \frac{1}{4}\sqrt{u}$$

20. $P = \dfrac{k}{v}$ $P = \dfrac{180}{v}$ $0.75v = 180$

$1 = \dfrac{k}{180}$ $0.75 = \dfrac{180}{v}$ $v = \dfrac{180}{0.75}$

$180 = k$ $v = 240$ cubic meters

C H A P T E R 9
Radicals and Complex Numbers

CHAPTER 9
Radicals and Complex Numbers

Section 9.1 Radicals and Rational Exponents

Solutions to Odd-Numbered Exercises

1. $\sqrt{64} = 8$ because $8 \cdot 8 = 64$

3. $-\sqrt{49} = -7$ because $7 \cdot 7 = 49$

5. $\sqrt[3]{-8} = -2$ because $-2 \cdot -2 \cdot -2 = -8$

7. $\sqrt{-1}$ is not a real number because no real number multiplied by itself yields -1.

9. Because $7^2 = 49$, 7 is a square root of 49.

11. Because $4.2^3 = 74.088$, 4.2 is a cube root of 74.088.

13. Because $45^2 = 2025$, 45 is called the square root of 2025.

15. $\sqrt{8^2} = |8| = 8$

    (index is even)

17. $\sqrt{(-10)^2} = |-10| = 10$

    (index is even)

19. $\sqrt{-9^2} = $ not a real number

    (even root of a negative number)

21. $-\sqrt{\left(\frac{2}{3}\right)^2} = -\frac{2}{3}$

    (index is even)

23. $\sqrt{-\left(\frac{3}{10}\right)^2} = $ not a real number

    (even root of a negative number)

25. $\left(\sqrt{5}\right)^2 = 5$

    (inverse property of powers and roots)

27. $-\left(\sqrt{23}\right)^2 = -23$

    (inverse property of powers and roots)

29. $\sqrt[3]{(5)^3} = 5$

    (index is odd)

31. $\sqrt[3]{10^3} = 10$

    (index is odd)

33. $-\sqrt[3]{(-6)^3} = 6$

    (index is odd)

35. $\sqrt[3]{\left(-\frac{1}{4}\right)^3} = -\frac{1}{4}$

    (index is odd)

37. $\left(\sqrt[3]{11}\right)^3 = 11$

    (inverse property of powers and roots)

39. $\left(-\sqrt[3]{24}\right)^3 = -24$

    (inverse property of powers and roots)

41. $\sqrt[4]{3^4} = 3$

    (inverse property of powers and roots)

43. $-\sqrt[4]{-5^4} = $ not a real number

    (even root of a negative number)

45. $\sqrt{6}$ is not rational because 6 is not a perfect square.

47. $\sqrt{900}$ is rational because $30 \cdot 30 = 900$, a perfect square.

49. *Radical Form*   *Rational Exponent Form*

    $\sqrt{16} = 4$ $16^{1/2} = 4$

51. *Radical Form*   *Rational Exponent Form*

    $\sqrt[3]{27^2} = 9$ $27^{2/3} = 9$

53. *Radical Form*   *Rational Exponent Form*

    $\sqrt[4]{256^3} = 64$ $256^{3/4} = 64$

55. $25^{1/2} = \sqrt{25} = 5$

    Root is 2. Power is 1.

57. $-36^{1/2} = -\sqrt{36} = -6$

    Root is 2. Power is 1.

59. $-(16)^{3/4} = -\left(\sqrt[4]{16}\right)^3 = -8$

    Root is 4. Power is 3.

61. $32^{-2/5} = \dfrac{1}{\left(\sqrt[5]{32}\right)^2} = \dfrac{1}{2^2} = \dfrac{1}{4}$

Root is 5. Power is 2.

63. $(-27)^{-2/3} = \dfrac{1}{(-27)^{2/3}} = \dfrac{1}{\left(\sqrt[3]{-27}\right)^2}$

$= \dfrac{1}{9}$

Root is 3. Power is 2.

65. $\left(\dfrac{8}{27}\right)^{2/3} = \left(\sqrt[3]{\dfrac{8}{27}}\right)^2 = \left(\dfrac{2}{3}\right)^2 = \dfrac{4}{9}$

Root is 3. Power is 2.

67. $\left(\dfrac{121}{9}\right)^{-1/2} = \left(\dfrac{9}{121}\right)^{1/2} = \sqrt{\dfrac{9}{121}} = \dfrac{3}{11}$

Root is 2. Power is 1.

69. $(3^3)^{2/3} = 3^{3 \cdot 2/3} = 3^2 = 9$

Root is 3. Power is 2.

71. $-(4^4)^{3/4} = -4^{4 \cdot 3/4} = -4^3 = -64$

Root is 4. Power is 3.

73. $\left(\dfrac{1}{5^3}\right)^{-2/3} = (5^3)^{2/3} = 5^{3 \cdot 2/3} = 5^2 = 25$

Root is 3. Power is 2.

75. $\sqrt{t} = t^{1/2}$

Root is 2. Power is 1.

77. $x\sqrt[4]{x^3} = x \cdot x^{3/4} = x^{1+3/4} = x^{7/4}$

Root is 4. Power is 3.

79. $u^2\sqrt[3]{u} = u^2 \cdot u^{1/3} = u^{2+1/3} = u^{7/3}$

Root is 3. Power is 1.

81. $s\sqrt[4]{s^5} = s^4 \cdot s^{5/2} = s^{4+5/2} = s^{13/2}$

Root is 2. Power is 5.

83. $\dfrac{\sqrt{x}}{\sqrt{x^3}} = \dfrac{x^{1/2}}{x^{3/2}} = x^{1/2-3/2} = x^{-1} = \dfrac{1}{x}$

85. $\dfrac{\sqrt[4]{t}}{\sqrt{t^5}} = \dfrac{t^{1/4}}{t^{5/2}} = t^{1/4-5/2}$

$= t^{1/4-10/4}$

$= t^{-9/4}$

$= \dfrac{1}{t^{9/4}}$

87. $\sqrt[3]{x^2} \cdot \sqrt[3]{x^7} = x^{2/3} \cdot x^{7/3} = x^{2/3+7/3} = x^{9/3}$

$= x^3$

89. $\sqrt[4]{y^3} \cdot \sqrt[3]{y} = y^{3/4} \cdot y^{1/3} = y^{3/4+1/3}$

$= y^{9/12+4/12} = y^{13/12}$

91. $\sqrt[4]{x^3 y} = (x^3 y)^{1/4}$

$= x^{3/4} y^{1/4}$

93. $z^2\sqrt{y^5 z^4} = z^2 \cdot (y^5 z^4)^{1/2} = z^2 y^{5/2} z^2$

$= z^{2+2} y^{5/2} = z^4 y^{5/2}$

95. $3^{1/4} \cdot 3^{3/4} = 3^{1/4+3/4}$

$= 3^{4/4}$

$= 3^1$

97. $(2^{1/2})^{2/3} = 2^{1/2 \cdot 2/3}$

$= 2^{1/3}$

$= \sqrt[3]{2}$

99. $\dfrac{2^{1/5}}{2^{6/5}} = 2^{1/5-6/5}$

$= 2^{-5/5}$

$= 2^{-1} = \dfrac{1}{2}$

101. $(c^{3/2})^{1/3} = c^{3/2 \cdot 1/3}$

$= c^{1/2}$

$= \sqrt{c}$

103. $\dfrac{18y^{4/3}z^{-1/3}}{24y^{-2/3}z} = \dfrac{6 \cdot 3y^{4/3-(-2/3)}z^{-1/3-1}}{6 \cdot 4}$

$\qquad = \dfrac{3y^{6/3}z^{-4/3}}{4}$

$\qquad = \dfrac{3y^2}{4z^{4/3}}$

105. $(3x^{-1/3}y^{3/4})^2 = 3^2x^{-2/3}y^{3/2}$

$\qquad = \dfrac{9y^{3/2}}{x^{2/3}}$

107. $\left(\dfrac{x^{1/4}}{x^{1/6}}\right)^3 = (x^{1/4-1/6})^3$

$\qquad = (x^{3/12-2/12})^3$

$\qquad = (x^{1/12})^3$

$\qquad = x^{3/12}$

$\qquad = x^{1/4}$

109. $\sqrt{\sqrt[4]{y}} = (y^{1/4})^{1/2}$

$\qquad = y^{1/4 \cdot 1/2}$

$\qquad = y^{1/8}$

$\qquad = \sqrt[8]{y}$

111. $\sqrt[4]{\sqrt{x^3}} = \sqrt[4]{x^{3/2}}$

$\qquad = (x^{3/2})^{1/4}$

$\qquad = x^{3/2 \cdot 1/4}$

$\qquad = x^{3/8}$

113. $\dfrac{(x+y)^{3/4}}{\sqrt[4]{x+y}} = \dfrac{(x+y)^{3/4}}{(x+y)^{1/4}}$

$\qquad = (x+y)^{3/4-1/4}$

$\qquad = (x+y)^{2/4}$

$\qquad = (x+y)^{1/2}$

$\qquad = \sqrt{x+y}$

115. $\dfrac{(3u-2v)^{2/3}}{\sqrt{(3u-2v)^3}} = \dfrac{(3u-2v)^{2/3}}{(3u-2v)^{3/2}} = (3u-2v)^{2/3-3/2}$

$\qquad = (3u-2v)^{4/6-9/6}$

$\qquad = (3u-2v)^{-5/6}$

$\qquad = \dfrac{1}{(3u-2v)^{5/6}}$

117. $\sqrt{73} \approx 8.5440$

Scientific: 73 $\boxed{\sqrt{}}$

Graphing: $\boxed{\sqrt{}}$ 73 $\boxed{\text{ENTER}}$

119. $315^{2/5} = \left(\sqrt[5]{315}\right)^2 \approx 9.9845$

Scientific: 315 $\boxed{y^x}$ $\boxed{(}$ 2 $\boxed{\div}$ 5 $\boxed{)}$ $\boxed{=}$

Graphing: 315 $\boxed{\wedge}$ $\boxed{(}$ 2 $\boxed{\div}$ 5 $\boxed{)}$ $\boxed{\text{ENTER}}$

121. $1698^{-3/4} \approx 0.0038$

Scientific: 1698 $\boxed{y^x}$ $\boxed{(}$ 3 $\boxed{\div}$ 4 $\boxed{+/-}$ $\boxed{)}$ $\boxed{=}$

Graphing: 1698 $\boxed{\wedge}$ $\boxed{(}$ $\boxed{(-)}$ 3 $\boxed{\div}$ 4 $\boxed{)}$ $\boxed{\text{ENTER}}$

123. $\sqrt[4]{342} \approx 4.3004$

$\qquad \sqrt[4]{342} = 342^{1/4}$

Scientific: 342 $\boxed{y^x}$ $\boxed{(}$ 1 $\boxed{\div}$ 4 $\boxed{)}$ $\boxed{=}$

Graphing: 342 $\boxed{\wedge}$ $\boxed{(}$ 1 $\boxed{\div}$ 4 $\boxed{)}$ $\boxed{\text{ENTER}}$

125. $\sqrt[3]{545^2} \approx 66.7213$

$\qquad \sqrt[3]{545^2} = 545^{2/3}$

Scientific: 545 $\boxed{y^x}$ $\boxed{(}$ 2 $\boxed{\div}$ 3 $\boxed{)}$ $\boxed{=}$

Graphing: 545 $\boxed{\wedge}$ $\boxed{(}$ 2 $\boxed{\div}$ 3 $\boxed{)}$ $\boxed{\text{ENTER}}$

127. $\dfrac{8-\sqrt{35}}{2} \approx 1.0420$

Scientific: $\boxed{(}$ 8 $\boxed{-}$ 35 $\boxed{\sqrt{}}$ $\boxed{)}$ $\boxed{\div}$ 2 $\boxed{=}$

Graphing: $\boxed{(}$ 8 $\boxed{-}$ $\boxed{\sqrt{}}$ 35 $\boxed{)}$ $\boxed{\div}$ 2 $\boxed{\text{ENTER}}$

129. $\dfrac{3+\sqrt{17}}{9} \approx 0.7915$

Scientific: $\boxed{(}$ 3 $\boxed{+}$ 17 $\boxed{\sqrt{}}$ $\boxed{)}$ $\boxed{\div}$ 9 $\boxed{=}$

Graphing: $\boxed{(}$ 3 $\boxed{+}$ $\boxed{\sqrt{}}$ 17 $\boxed{)}$ $\boxed{\div}$ 9 $\boxed{\text{ENTER}}$

131. $f(x) = 3\sqrt{x}, \; x \geq 0,$

Domain $= [0, \infty)$

133. The domain of $g(x) = \dfrac{2}{\sqrt[4]{x}}$ is the set of all

nonnegative real numbers or $(0, \infty)$.

135. $f(x) = \sqrt{-x}, \qquad -x \geq 0$

$x \leq 0$

Domain $= (-\infty, 0]$

137. *Keystrokes:*

$\boxed{\text{Y=}}\; 5 \; \boxed{\div} \; 4 \; \boxed{\text{MATH 5}} \; \boxed{\text{X,T,}\theta} \; \boxed{\text{MATH 3}} \; \boxed{\text{GRAPH}}$

Domain is $(0, \infty)$ so graphing utility did complete the graph.

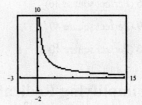

139. *Keystrokes:*

$\boxed{\text{Y=}}\; 2 \; \boxed{\text{X,T,}\theta} \; \boxed{\wedge} \; \boxed{(} \; 3 \; \boxed{\div} \; 5 \; \boxed{)} \; \boxed{\text{GRAPH}}$

Domain is $(-\infty, \infty)$ so graphing utility did complete the graph.

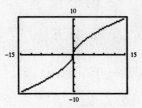

141. $x^{1/2}(2x - 3) = 2x^{3/2} - 3x^{1/2}$

143. $y^{-1/3}(y^{1/3} + 5y^{4/3}) = y^0 + 5y^{3/3}$

$\qquad\qquad\qquad\qquad = 1 + 5y$

145. $r = 1 - \left(\dfrac{25{,}000}{75{,}000}\right)^{1/8}$

$\quad = 1 - \left(\dfrac{1}{3}\right)^{1/8}$

$\quad \approx 0.128 \approx 12.8\%$

147. *Verbal Model:* $\boxed{\text{Area}} = \boxed{\text{side}} \cdot \boxed{\text{side}}$

Labels: Area $= 529$

side $= x$

Equation: $529 = x \cdot x$

$529 = x^2$

$\sqrt{529} = x$

$23 = x$

23 feet $\times$ 23 feet

149. $d = \sqrt{l^2 + w^2 + h^2}$

$\quad = \sqrt{9^2 + 5^2 + 2^2}$

$\quad = \sqrt{81 + 25 + 4}$

$\quad = \sqrt{110}$

$\quad \approx 10.49$ cm

151. (a) $r = \sqrt{\dfrac{2v}{\pi l}}$

$\quad\; r = \sqrt{\dfrac{2(35{,}350)}{\pi(100)}}$

$\quad\; r \approx 15$ feet

(b) $h = \sqrt{r^2 - \left(\dfrac{a}{2}\right)^2}$

$\quad\; h = \sqrt{15^2 - \left(\dfrac{a}{2}\right)^2}$

(c) $h = \sqrt{15^2 - \left(\dfrac{25}{2}\right)^2}$

$\quad\; h = \sqrt{225 - 156.25}$

$\quad\; h = \sqrt{68.75}$

$\quad\; h \approx 8.29$ feet

(d) $a = 2\sqrt{r^2 - h^2}$

$\quad\; a = 2\sqrt{15^2 - 8^2}$

$\quad\;\; = 2\sqrt{225 - 64}$

$\quad\;\; = 2\sqrt{161}$

$\quad\;\; \approx 25.38$ feet

153. Given $\sqrt[n]{a}$, a is the radicand and n is the index.

155. No. $\sqrt{2}$ is an irrational number. Its decimal representation is a nonterminating, nonrepeating decimal.

157. (a) "Last digits:" 1 (Perfect square 81)

4 (Perfect square 64)

5 (Perfect square 25)

6 (Perfect square 36)

9 (Perfect square 49)

0 (Perfect square 100)

(b) Yes, 4,322,788,986 ends in a 6, but it is not a perfect square.

Section 9.2   Simplifying Radical Expressions

1. $\sqrt{20} = \sqrt{4 \cdot 5} = \sqrt{2^2 \cdot 5} = 2\sqrt{5}$

3. $\sqrt{50} = \sqrt{25 \cdot 2} = \sqrt{5^2 \cdot 2} = 5\sqrt{2}$

5. $\sqrt{96} = \sqrt{16 \cdot 6} = \sqrt{4^2 \cdot 6} = 4\sqrt{6}$

7. $\sqrt{216} = \sqrt{36 \cdot 6} = \sqrt{6^2 \cdot 6} = 6\sqrt{6}$

9. $\sqrt{1183} = \sqrt{169 \cdot 7} = \sqrt{13^2 \cdot 7} = 13\sqrt{7}$

11. $\sqrt{0.04} = \sqrt{4 \cdot 0.01} = \sqrt{4}\sqrt{0.01} = 2 \cdot 0.1 = 0.2$

13. $\sqrt{0.0072} = \sqrt{36 \cdot 2 \cdot 0.0001}$
$= \sqrt{36} \cdot \sqrt{2} \cdot \sqrt{0.0001}$
$= 6 \cdot 0.01\sqrt{2}$
$= 0.06\sqrt{2}$

15. $\sqrt{2.42} = \sqrt{121 \cdot 2 \cdot 0.01}$
$= \sqrt{121} \cdot \sqrt{2} \cdot \sqrt{0.01}$
$= 11 \cdot 0.1\sqrt{2}$
$= 1.1\sqrt{2}$

17. $\sqrt{\dfrac{15}{4}} = \dfrac{\sqrt{15}}{2}$

19. $\sqrt{\dfrac{13}{25}} = \dfrac{\sqrt{13}}{5}$

21. $\sqrt{9x^5} = \sqrt{3^2 x^4 \cdot x}$
$= 3 \cdot x^2 \cdot \sqrt{x}$
$= 3x^2\sqrt{x}$

23. $\sqrt{48y^4} = \sqrt{16 \cdot 3 \cdot y^4} = 4y^2\sqrt{3}$

25. $\sqrt{117y^6} = \sqrt{9 \cdot 13 \cdot y^6} = 3|y^3|\sqrt{13}$

27. $\sqrt{120x^2y^3} = \sqrt{4 \cdot 30 \cdot x^2 \cdot y^2 \cdot y} = 2|x|y\sqrt{30y}$

29. $\sqrt{192a^5b^7} = \sqrt{64 \cdot 3 \cdot a^4 \cdot a \cdot b^6 \cdot b}$
$= 8a^2b^3\sqrt{3ab}$

31. $\sqrt[3]{48} = \sqrt[3]{16 \cdot 3} = \sqrt[3]{2^4 \cdot 3} = 2\sqrt[3]{3 \cdot 2} = 2\sqrt[3]{6}$

33. $\sqrt[3]{112} = \sqrt[3]{8 \cdot 14}$
$= \sqrt[3]{8} \cdot \sqrt[3]{14}$
$= 2\sqrt[3]{14}$

35. $\sqrt[3]{40x^5} = \sqrt[3]{8 \cdot 5 \cdot x^3 \cdot x^2}$
$= 2x\sqrt[3]{5x^2}$

37. $\sqrt[4]{324y^6} = \sqrt[4]{81 \cdot 4 \cdot y^4 \cdot y^2}$
$= 3|y|\sqrt[4]{4y^2}$
$= 3|y|\sqrt[4]{2^2y^2}$
$= 3|y|\sqrt{2y}$

39. $\sqrt[3]{x^4y^3} = \sqrt[3]{x^3 \cdot x \cdot y^3} = xy\sqrt[3]{x}$

41. $\sqrt[4]{3x^4y^2} = \sqrt[4]{x^4} \cdot \sqrt[4]{3y^2}$
$= |x|\sqrt[4]{3y^2}$

43. $\sqrt[5]{32x^5y^6} = \sqrt[5]{2^5 \cdot x^5 \cdot y^5 \cdot y} = 2xy\sqrt[5]{y}$

45. $\sqrt[3]{\dfrac{35}{64}} = \dfrac{\sqrt[3]{35}}{4}$

47. $\sqrt[5]{\dfrac{15}{243}} = \dfrac{\sqrt[5]{15}}{3}$

49. $\sqrt[5]{\dfrac{32x^2}{y^5}} = \sqrt[5]{\dfrac{2^5 x^2}{y^5}}$

$\phantom{\sqrt[5]{\dfrac{32x^2}{y^5}}} = \dfrac{2}{y}\sqrt[5]{x^2}$

51. $\sqrt[3]{\dfrac{54a^4}{b^9}} = \sqrt[3]{\dfrac{3^3 \cdot 2 \cdot a^3 \cdot a}{b^9}} = \dfrac{3a}{b^3}\sqrt[3]{2a}$

53. $\sqrt{\dfrac{32a^4}{b^2}} = \dfrac{\sqrt{16 \cdot 2 \cdot a^4}}{\sqrt{b^2}} = \dfrac{4a^2\sqrt{2}}{|b|}$

55. $\sqrt[4]{(3x^2)^4} = 3x^2$

57. $\sqrt{\dfrac{1}{3}} = \dfrac{1}{\sqrt{3}} \cdot \dfrac{\sqrt{3}}{\sqrt{3}} = \dfrac{\sqrt{3}}{3}$

59. $\dfrac{1}{\sqrt{7}} = \dfrac{1}{\sqrt{7}} \cdot \dfrac{\sqrt{7}}{\sqrt{7}} = \dfrac{\sqrt{7}}{7}$

61. $\dfrac{12}{\sqrt{3}} = \dfrac{12}{\sqrt{3}} \cdot \dfrac{\sqrt{3}}{\sqrt{3}} = \dfrac{12\sqrt{3}}{3} = 4\sqrt{3}$

63. $\sqrt[4]{\dfrac{5}{4}} = \dfrac{\sqrt[4]{5}}{\sqrt[4]{2^2}} \cdot \dfrac{\sqrt[4]{2^2}}{\sqrt[4]{2^2}} = \dfrac{\sqrt[4]{5 \cdot 2^2}}{\sqrt[4]{2^4}} = \dfrac{\sqrt[4]{20}}{2}$

65. $\dfrac{6}{\sqrt[3]{32}} = \dfrac{6}{\sqrt[3]{2^3 \cdot 2^2}} = \dfrac{6}{2\sqrt[3]{2^2}} \cdot \dfrac{\sqrt[3]{2}}{\sqrt[3]{2}} = \dfrac{6\sqrt[3]{2}}{2\sqrt[3]{2^3}} = \dfrac{6\sqrt[3]{2}}{4} = \dfrac{3\sqrt[3]{2}}{2}$

67. $\dfrac{1}{\sqrt{y}} = \dfrac{1}{\sqrt{y}} \cdot \dfrac{\sqrt{y}}{\sqrt{y}} = \dfrac{\sqrt{y}}{\sqrt{y^2}} = \dfrac{\sqrt{y}}{y}$

69. $\sqrt{\dfrac{4}{x}} = \dfrac{\sqrt{4}}{\sqrt{x}} = \dfrac{2}{\sqrt{x}} \cdot \dfrac{\sqrt{x}}{\sqrt{x}} = \dfrac{2\sqrt{x}}{x}$

71. $\dfrac{1}{\sqrt{2x}} = \dfrac{1}{\sqrt{2x}} \cdot \dfrac{\sqrt{2x}}{\sqrt{2x}} = \dfrac{\sqrt{2x}}{2x}$

73. $\dfrac{6}{\sqrt{3b^3}} = \dfrac{6}{b\sqrt{3b}} \cdot \dfrac{\sqrt{3b}}{\sqrt{3b}} = \dfrac{6\sqrt{3b}}{3b^2} = \dfrac{2\sqrt{3b}}{b^2}$

75. $\sqrt[3]{\dfrac{2x}{3y}} = \dfrac{\sqrt[3]{2x}}{\sqrt[3]{3y}} \cdot \dfrac{\sqrt[3]{3^2 y^2}}{\sqrt[3]{3^2 y^2}} = \dfrac{\sqrt[3]{2x \cdot 3^2 y^2}}{\sqrt[3]{3^3 y^3}} = \dfrac{\sqrt[3]{18xy^2}}{3y}$

77. $\dfrac{a^3}{\sqrt[3]{ab^2}} = \dfrac{a^3}{\sqrt[3]{ab^2}} \cdot \dfrac{\sqrt[3]{a^2 b}}{\sqrt[3]{a^2 b}} = \dfrac{a^3 \sqrt[3]{a^2 b}}{\sqrt[3]{a^3 b^3}} = \dfrac{a^3 \sqrt[3]{a^2 b}}{ab} = \dfrac{a^2 \sqrt[3]{a^2 b}}{b}$

79. $3\sqrt{2} - \sqrt{2} = 2\sqrt{2}$

81. $12\sqrt{8} - 3\sqrt[3]{8} = 12\sqrt{4 \cdot 2} - 3 \cdot 2 = 24\sqrt{2} - 6$

83. $\sqrt[4]{3} - 5\sqrt[4]{7} - 12\sqrt[4]{3} = -11\sqrt[4]{3} - 5\sqrt[4]{7}$

85. $2\sqrt[3]{54} + 12\sqrt[3]{16} = 2\sqrt[3]{27 \cdot 2} + 12\sqrt[3]{8 \cdot 2}$

$\phantom{2\sqrt[3]{54} + 12\sqrt[3]{16}} = 6\sqrt[3]{2} + 24\sqrt[3]{2} = 30\sqrt[3]{2}$

87. $5\sqrt{9x} - 3\sqrt{x} = 15\sqrt{x} - 3\sqrt{x} = 12\sqrt{x}$

89. $\sqrt{25y} + \sqrt{64y} = 5\sqrt{y} + 8\sqrt{y} = 13\sqrt{y}$

91. $10\sqrt[3]{z} - \sqrt[3]{z^4} = 10\sqrt[3]{z} - \sqrt[3]{z^3 \cdot z} = 10\sqrt[3]{z} - z\sqrt[3]{z}$

$\phantom{10\sqrt[3]{z} - \sqrt[3]{z^4}} = (10 - z)\sqrt[3]{z}$

93. $\sqrt{5} - \dfrac{3}{\sqrt{5}} = \sqrt{5} - \left(\dfrac{3}{\sqrt{5}} \cdot \dfrac{\sqrt{5}}{\sqrt{5}}\right) = \sqrt{5} - \dfrac{3\sqrt{5}}{5}$

$\phantom{\sqrt{5} - \dfrac{3}{\sqrt{5}}} = \left(1 - \dfrac{3}{5}\right)\sqrt{5}$

$\phantom{\sqrt{5} - \dfrac{3}{\sqrt{5}}} = \dfrac{2}{5}\sqrt{5}$

95. $\sqrt{20} - \sqrt{\dfrac{1}{5}} = \sqrt{4 \cdot 5} - \sqrt{\dfrac{1}{5}} \cdot \dfrac{\sqrt{5}}{\sqrt{5}} = 2\sqrt{5} - \dfrac{\sqrt{5}}{5}$

$\phantom{\sqrt{20} - \sqrt{\dfrac{1}{5}}} = \left(2 - \dfrac{1}{5}\right)\sqrt{5}$

$\phantom{\sqrt{20} - \sqrt{\dfrac{1}{5}}} = \dfrac{9}{5}\sqrt{5}$

97. $\sqrt{2x} - \dfrac{3}{\sqrt{2x}} = \sqrt{2x} - \left(\dfrac{3}{\sqrt{2x}} \cdot \dfrac{\sqrt{2x}}{\sqrt{2x}}\right) = \sqrt{2x} - \dfrac{3\sqrt{2x}}{2x}$

$\phantom{\sqrt{2x} - \dfrac{3}{\sqrt{2x}}} = \dfrac{2x\sqrt{2x}}{2x} - \dfrac{3\sqrt{2x}}{2x}$

$\phantom{\sqrt{2x} - \dfrac{3}{\sqrt{2x}}} = \dfrac{2x\sqrt{2x} - 3\sqrt{2x}}{2x}$

$\phantom{\sqrt{2x} - \dfrac{3}{\sqrt{2x}}} = \dfrac{\sqrt{2x}(2x - 3)}{2x}$

99. $\sqrt{7} + \sqrt{18} > \sqrt{7 + 18}$

101. $5 > \sqrt{3^2 + 2^2}$

103. $c = \sqrt{a^2 + b^2}$
$\quad = \sqrt{6^2 + 3^2}$
$\quad = \sqrt{36 + 9} = \sqrt{45} = \sqrt{9 \cdot 5} = 3\sqrt{5}$

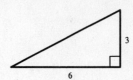

105. $c = \sqrt{a^2 + b^2}$
$\quad = \sqrt{9^2 + 6^2}$
$\quad = \sqrt{81 + 36}$
$\quad = \sqrt{117}$
$\quad = 3\sqrt{13}$

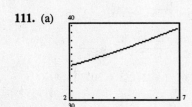

107. (a) $c = \sqrt{a^2 + b^2}$
$\quad c = \sqrt{(15)^2 + (5)^2}$
$\quad c = \sqrt{225 + 25}$
$\quad c = \sqrt{250}$
$\quad c = \sqrt{25 \cdot 10}$
$\quad c = 5\sqrt{10}$

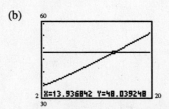

(b) Area of roof $= 2 \cdot$ Length $\cdot$ Width
$$A = 2 \cdot 40 \cdot 5\sqrt{10}$$
$$A = 400\sqrt{10}$$

Thus, the total area of the roof is $400\sqrt{10} \approx 1264.9$ square feet.

109. $f = \dfrac{1}{100}\sqrt{\dfrac{400 \times 10^6}{5}} \approx 8.9443 \times 10^1 \approx 89.443 \approx 89.44$ cycles per second

111. (a)

(b)

The average salary will reach \$48,000 after 14 years so $1990 + 14 = 2004$.

113. $\left(\dfrac{5}{\sqrt{3}}\right)^2 = \dfrac{5}{\sqrt{3}} \cdot \dfrac{5}{\sqrt{3}} = \dfrac{25}{3}$

No. When you rationalize the denominator, the value of the number is not changed.

$$\dfrac{5}{\sqrt{3}} = \dfrac{5}{\sqrt{3}} \cdot \dfrac{\sqrt{3}}{\sqrt{3}} = \dfrac{5\sqrt{3}}{3} \neq \dfrac{25}{3}$$

115. Example: $\sqrt{6} \cdot \sqrt{15} = \sqrt{6 \cdot 15} = \sqrt{3 \cdot 2 \cdot 3 \cdot 5} = 3\sqrt{10}$

117. $\sqrt{2} + \sqrt{18}$ is not in simplest form because $\sqrt{18}$ can be simplified to $3\sqrt{2}$ and then added to $\sqrt{2}$.
$\quad \sqrt{2} + \sqrt{18} = \sqrt{2} + 3\sqrt{2} = 4\sqrt{2}$

119. $\sqrt{x^2} \neq x$ for all negative values of x.
$\quad$ Example: $\sqrt{(-8)^2} = \sqrt{64} = 8$

Section 9.3   Multiplying and Dividing Radical Expressions

1. $\sqrt{2} \cdot \sqrt{8} = \sqrt{2 \cdot 8}$
$\qquad = \sqrt{16}$
$\qquad = 4$

3. $\sqrt{3} \cdot \sqrt{6} = \sqrt{3 \cdot 6}$
$\qquad = \sqrt{18}$
$\qquad = \sqrt{9 \cdot 2}$
$\qquad = 3\sqrt{2}$

5. $\sqrt[3]{12} \cdot \sqrt[3]{6} = \sqrt[3]{12 \cdot 6}$
$\qquad = \sqrt[3]{8 \cdot 9}$
$\qquad = 2\sqrt[3]{9}$

7. $\sqrt[4]{8} \cdot \sqrt[4]{6} = \sqrt[4]{8 \cdot 6} = \sqrt[4]{2^4 \cdot 3}$
$\qquad = 2\sqrt[4]{3}$

9. $\sqrt{5}(2 - \sqrt{3}) = 2\sqrt{5} - \sqrt{5}\sqrt{3} = 2\sqrt{5} - \sqrt{15}$

11. $\sqrt{2}(\sqrt{20} + 8) = \sqrt{2}\sqrt{20} + 8\sqrt{2}$
$\qquad = \sqrt{40} + 8\sqrt{2}$
$\qquad = 2\sqrt{10} + 8\sqrt{2}$

13. $\sqrt{6}(\sqrt{12} - \sqrt{3}) = \sqrt{6}\sqrt{12} - \sqrt{6}\sqrt{3}$
$\qquad = \sqrt{72} - \sqrt{18}$
$\qquad = \sqrt{36 \cdot 2} - \sqrt{9 \cdot 2}$
$\qquad = 6\sqrt{2} - 3\sqrt{2}$
$\qquad = 3\sqrt{2}$

15. $\sqrt{2}(\sqrt{18} - \sqrt{10}) = \sqrt{2}\sqrt{18} - \sqrt{2}\sqrt{10}$
$\qquad = \sqrt{36} - \sqrt{20}$
$\qquad = 6 - \sqrt{4 \cdot 5}$
$\qquad = 6 - 2\sqrt{5}$

17. $\sqrt{y}(\sqrt{y} + 4) = (\sqrt{y})^2 + 4\sqrt{y}$
$\qquad = y + 4\sqrt{y}$

19. $\sqrt{a}(4 - \sqrt{a}) = \sqrt{a} \cdot 4 - \sqrt{a}\sqrt{a}$
$\qquad = 4\sqrt{a} - a$

21. $\sqrt[3]{4}(\sqrt[3]{2} - 7) = \sqrt[3]{4}\sqrt[3]{2} - 7\sqrt[3]{4}$
$\qquad = \sqrt[3]{8} - 7\sqrt[3]{4}$
$\qquad = 2 - 7\sqrt[3]{4}$

23. $(\sqrt{3} + 2)(\sqrt{3} - 2) = (\sqrt{3})^2 - 2^2$
$\qquad = 3 - 4$
$\qquad = -1$

25. $(\sqrt{5} + 3)(\sqrt{3} - 5) = \sqrt{15} - 5\sqrt{5} + 3\sqrt{3} - 15$

27. $(\sqrt{20} + 2)^2 = (\sqrt{20})^2 + 2 \cdot \sqrt{20} \cdot 2 + 2^2$
$\qquad = 20 + 4\sqrt{20} + 4$
$\qquad = 24 + 4\sqrt{4 \cdot 5}$
$\qquad = 24 + 8\sqrt{5}$

29. $(\sqrt[3]{6} - 3)(\sqrt[3]{4} + 3) = \sqrt[3]{6}\sqrt[3]{4} + 3\sqrt[3]{6} - 3\sqrt[3]{4} - 9$
$\qquad = \sqrt[3]{24} + 3\sqrt[3]{6} - 3\sqrt[3]{4} - 9$
$\qquad = \sqrt[3]{8 \cdot 3} + 3\sqrt[3]{6} - 3\sqrt[3]{4} - 9$
$\qquad = 2\sqrt[3]{3} + 3\sqrt[3]{6} - 3\sqrt[3]{4} - 9$

31. $(10 + \sqrt{2x})^2 = 10^2 + 2 \cdot 10 \cdot \sqrt{2x} + (\sqrt{2x})^2$
$\qquad = 100 + 20\sqrt{2x} + 2x$

33. $(9\sqrt{x} + 2)(5\sqrt{x} - 3) = (9\sqrt{x})(5\sqrt{x}) - 27\sqrt{x} + 10\sqrt{x} - 6$
$\qquad = 45x - 17\sqrt{x} - 6$

35. $(3\sqrt{x} - 5)(3\sqrt{x} + 5) = (3\sqrt{x})^2 - 5^2$
$\qquad = 9x - 25$

37. $(\sqrt[3]{2x} + 5)^2 = (\sqrt[3]{2x})^2 + 2 \cdot 5\sqrt[3]{2x} + 5^2$
$\qquad = \sqrt[3]{(2x)^2} + 10\sqrt[3]{2x} + 25$
$\qquad = \sqrt[3]{4x^2} + 10\sqrt[3]{2x} + 25$

39. $\left(\sqrt[3]{y} + 2\right)\left(\sqrt[3]{y^2} - 5\right) = \sqrt[3]{y} \cdot \sqrt[3]{y^2} - 5\sqrt[3]{y} + 2\sqrt[3]{y^2} - 10$

$$= \sqrt[3]{y^3} - 5\sqrt[3]{y} + 2\sqrt[3]{y^2} - 10$$

$$= y - 5\sqrt[3]{y} + 2\sqrt[3]{y^2} - 10$$

41. $\left(\sqrt[3]{t} + 1\right)\left(\sqrt[3]{t^2} + 4\sqrt[3]{t} - 3\right) = \sqrt[3]{t}\sqrt[3]{t^2} + \sqrt[3]{t} \cdot 4\sqrt[3]{t} - 3\sqrt[3]{t} + \sqrt[3]{t^2} + 4\sqrt[3]{t} - 3$

$$= \sqrt[3]{t^3} + 4\sqrt[3]{t^2} - 3\sqrt[3]{t} + \sqrt[3]{t^2} + 4\sqrt[3]{t} - 3$$

$$= t + 5\sqrt[3]{t^2} + \sqrt[3]{t} - 3$$

43. $5x\sqrt{3} + 15\sqrt{3} = 5\sqrt{3}(x + 3)$

45. $4\sqrt{12} - 2x\sqrt{27} = 4\sqrt{4 \cdot 3} - 2x\sqrt{9 \cdot 3}$

$$= 8\sqrt{3} - 6x\sqrt{3}$$

$$= 2\sqrt{3}(4 - 3x)$$

47. $6u^2 + \sqrt{18u^3} = 6u^2 + \sqrt{9 \cdot 2u^2 \cdot u}$

$$= 6u^2 + 3u\sqrt{2u}$$

$$= 3u\left(2u + \sqrt{2u}\right)$$

49. $2 + \sqrt{5}$, conjugate $= 2 - \sqrt{5}$

$$\text{product} = \left(2 + \sqrt{5}\right)\left(2 - \sqrt{5}\right)$$

$$= 2^2 - \left(\sqrt{5}\right)^2$$

$$= 4 - 5 = -1$$

51. $\sqrt{11} - \sqrt{3}$, conjugate $= \sqrt{11} + \sqrt{3}$

$$\text{product} = \left(\sqrt{11} - \sqrt{3}\right)\left(\sqrt{11} + \sqrt{3}\right)$$

$$= \left(\sqrt{11}\right)^2 - \left(\sqrt{3}\right)^2$$

$$= 11 - 3 = 8$$

53. $\sqrt{15} + 3$,

$$\text{conjugate} = \sqrt{15} - 3$$

$$\text{product} = \left(\sqrt{15} + 3\right)\left(\sqrt{15} - 3\right)$$

$$= \sqrt{15} \cdot \sqrt{15} - 3\sqrt{15} + 3\sqrt{15} - 9$$

$$= 15 - 9 = 6$$

55. $\sqrt{x} - 3$, conjugate $= \sqrt{x} + 3$

$$\text{product} = \left(\sqrt{x} - 3\right)\left(\sqrt{x} + 3\right)$$

$$= \left(\sqrt{x}\right)^2 - 3^2$$

$$= x - 9$$

57. $\sqrt{2u} - \sqrt{3}$, conjugate $= \sqrt{2u} + \sqrt{3}$

$$\text{product} = \left(\sqrt{2u} - \sqrt{3}\right)\left(\sqrt{2u} + \sqrt{3}\right)$$

$$= \left(\sqrt{2u}\right)^2 - \left(\sqrt{3}\right)^2$$

$$= 2u - 3$$

59. $2\sqrt{2} + \sqrt{4}$, conjugate $= 2\sqrt{2} - \sqrt{4}$

$$\text{product} = \left(2\sqrt{2} + \sqrt{4}\right)\left(2\sqrt{2} - \sqrt{4}\right)$$

$$= \left(2\sqrt{2}\right)^2 - \left(\sqrt{4}\right)^2$$

$$= 4 \cdot 2 - 4$$

$$= 8 - 4 = 4$$

61. $\sqrt{x} + \sqrt{y}$, conjugate $= \sqrt{x} - \sqrt{y}$

$$\text{product} = \left(\sqrt{x} + \sqrt{y}\right)\left(\sqrt{x} - \sqrt{y}\right)$$

$$= \left(\sqrt{x}\right)^2 - \left(\sqrt{y}\right)^2$$

$$= x - y$$

63. $\dfrac{4 - 8\sqrt{x}}{12} = \dfrac{4\left(1 - 2\sqrt{x}\right)}{12}$

$$= \dfrac{1 - 2\sqrt{x}}{3}$$

65. $\dfrac{-2y + \sqrt{12y^3}}{8y} = \dfrac{-2y + 2y\sqrt{3y}}{8y}$

$$= \dfrac{2y\left(-1 + \sqrt{3y}\right)}{8y}$$

$$= \dfrac{-1 + \sqrt{3y}}{4}$$

67. (a) $f(2 - \sqrt{3}) = (2 - \sqrt{3})^2 - 6(2 - \sqrt{3}) + 1$

$= 4 - 4\sqrt{3} + 3 - 12 + 6\sqrt{3} + 1$

$= 2\sqrt{3} - 4$

(b) $f(3 - 2\sqrt{2}) = (3 - 2\sqrt{2})^2 - 6(3 - 2\sqrt{2}) + 1$

$= 9 - 12\sqrt{2} + 8 - 18 + 12\sqrt{2} + 1$

$= 0$

69. (a) $f(1 + \sqrt{2}) = (1 + \sqrt{2})^2 - 2(1 + \sqrt{2}) - 1$

$= 1 + 2\sqrt{2} + 2 - 2 - 2\sqrt{2} - 1$

$= 0$

(b) $f(\sqrt{4}) = (\sqrt{4})^2 - 2\sqrt{4} - 1$

$= 4 - 4 - 1$

$= -1$

71. $\dfrac{6}{\sqrt{2} - 2} = \dfrac{6}{\sqrt{2} - 2} \cdot \dfrac{\sqrt{2} + 2}{\sqrt{2} + 2} = \dfrac{6(\sqrt{2} + 2)}{(\sqrt{2})^2 - 2^2} = \dfrac{6(\sqrt{2} + 2)}{2 - 4} = \dfrac{6(\sqrt{2} + 2)}{-2} = -3(\sqrt{2} + 2)$

73. $\dfrac{7}{\sqrt{3} + 5} = \dfrac{7}{\sqrt{3} + 5} \cdot \dfrac{\sqrt{3} - 5}{\sqrt{3} - 5} = \dfrac{7(\sqrt{3} - 5)}{(\sqrt{3})^2 - 5^2} = \dfrac{7(\sqrt{3} - 5)}{3 - 25} = \dfrac{7(\sqrt{3} - 5)}{-22} = \dfrac{7(5 - \sqrt{3})}{22}$

75. $\dfrac{3}{2\sqrt{10} - 5} = \dfrac{3}{2\sqrt{10} - 5} \cdot \dfrac{2\sqrt{10} + 5}{2\sqrt{10} + 5}$

$= \dfrac{3(2\sqrt{10} + 5)}{(2\sqrt{10})^2 - 5^2}$

$= \dfrac{3(2\sqrt{10} + 5)}{40 - 25}$

$= \dfrac{3(2\sqrt{10} + 5)}{15}$

$= \dfrac{2\sqrt{10} + 5}{5}$

77. $\dfrac{2}{\sqrt{6} + \sqrt{2}} = \dfrac{2}{\sqrt{6} + \sqrt{2}} \cdot \dfrac{\sqrt{6} - \sqrt{2}}{\sqrt{6} - \sqrt{2}} = \dfrac{2(\sqrt{6} - \sqrt{2})}{6 - 2}$

$= \dfrac{2(\sqrt{6} - \sqrt{2})}{4}$

$= \dfrac{\sqrt{6} - \sqrt{2}}{2}$

79. $\dfrac{9}{\sqrt{3} - \sqrt{7}} = \dfrac{9}{\sqrt{3} - \sqrt{7}} \cdot \dfrac{\sqrt{3} + \sqrt{7}}{\sqrt{3} + \sqrt{7}} = \dfrac{9(\sqrt{3} + \sqrt{7})}{(\sqrt{3})^2 - (\sqrt{7})^2} = \dfrac{9(\sqrt{3} + \sqrt{7})}{3 - 7} = \dfrac{9(\sqrt{3} + \sqrt{7})}{-4} = \dfrac{-9(\sqrt{3} + \sqrt{7})}{4}$

81. $(\sqrt{7} + 2) \div (\sqrt{7} - 2) = \dfrac{\sqrt{7} + 2}{\sqrt{7} - 2} \cdot \dfrac{\sqrt{7} + 2}{\sqrt{7} + 2} = \dfrac{(\sqrt{7})^2 + 2\sqrt{7} + 2\sqrt{7} + 4}{(\sqrt{7})^2 - 2^2} = \dfrac{7 + 4\sqrt{7} + 4}{7 - 4} = \dfrac{11 + 4\sqrt{7}}{3}$

83. $(\sqrt{x} - 5) \div (2\sqrt{x} - 1) = \dfrac{\sqrt{x} - 5}{2\sqrt{x} - 1} \cdot \dfrac{2\sqrt{x} + 1}{2\sqrt{x} + 1} = \dfrac{2x + \sqrt{x} - 10\sqrt{x} - 5}{(2\sqrt{x})^2 - 1^2} = \dfrac{2x - 9\sqrt{x} - 5}{4x - 1}$

85. $\dfrac{3x}{\sqrt{15} - \sqrt{3}} = \dfrac{3x}{\sqrt{15} - \sqrt{3}} \cdot \dfrac{\sqrt{15} + \sqrt{3}}{\sqrt{15} + \sqrt{3}} = \dfrac{3x(\sqrt{15} + \sqrt{3})}{(\sqrt{15})^2 - (\sqrt{3})^2} = \dfrac{3x(\sqrt{15} + \sqrt{3})}{15 - 3}$

$= \dfrac{3x(\sqrt{15} + \sqrt{3})}{12} = \dfrac{x\sqrt{15} + x\sqrt{3}}{4}$

87. $\dfrac{2t^2}{\sqrt{5} - \sqrt{t}} = \dfrac{2t^2}{\sqrt{5} - \sqrt{t}} \cdot \dfrac{\sqrt{5} + \sqrt{t}}{\sqrt{5} + \sqrt{t}} = \dfrac{2t^2(\sqrt{5} + \sqrt{t})}{(\sqrt{5})^2 - (\sqrt{t})^2} = \dfrac{2t^2(\sqrt{5} + \sqrt{t})}{5 - t} = \dfrac{2t^2(\sqrt{5} + \sqrt{t})}{5 - t}$

89. $\dfrac{8a}{\sqrt{3a} + \sqrt{a}} = \dfrac{8a}{\sqrt{3a} + \sqrt{a}} \cdot \dfrac{\sqrt{3a} - \sqrt{a}}{\sqrt{3a} - \sqrt{a}} = \dfrac{8a(\sqrt{3a} - \sqrt{a})}{(\sqrt{3a})^2 - (\sqrt{a})^2} = \dfrac{8a(\sqrt{3a} - \sqrt{a})}{3a - a} = \dfrac{8a(\sqrt{3a} - \sqrt{a})}{2a} = 4(\sqrt{3a} - \sqrt{a})$

91. $\dfrac{3(x-4)}{x^2-\sqrt{x}} = \dfrac{3(x-4)}{x^2-\sqrt{x}} \cdot \dfrac{x^2+\sqrt{x}}{x^2+\sqrt{x}} = \dfrac{3(x-4)(x^2+\sqrt{x})}{(x^2)^2-(\sqrt{x})^2} = \dfrac{3(x-4)(x^2+\sqrt{x})}{x^4-x}$

$= \dfrac{3(x-4)(x^2+\sqrt{x})}{x(x^3-1)} = \dfrac{3(x-4)(x^2+\sqrt{x})}{x(x-1)(x^2+x+1)}$

93. $\dfrac{\sqrt{u+v}}{\sqrt{u-v}-\sqrt{u}} = \dfrac{\sqrt{u+v}}{\sqrt{u-v}-\sqrt{u}} \cdot \dfrac{\sqrt{u-v}+\sqrt{u}}{\sqrt{u-v}+\sqrt{u}} = \dfrac{\sqrt{u+v}(\sqrt{u-v}+\sqrt{u})}{u-v-u}$

$= \dfrac{\sqrt{u+v}(\sqrt{u-v}+\sqrt{u})}{-v} = \dfrac{-\sqrt{u+v}(\sqrt{u-v}+\sqrt{u})}{v}$

95. *Keystrokes:*

y_1 [Y=] 10 [÷] [(] [√] [X,T,θ] [+] 1 [)] [ENTER]

y_2 [(] 10 [(] [√] [X,T,θ] [−] 1 [)] [)] [÷] [(] [X,T,θ] [−] 1 [)] [GRAPH]

$y_1 = y_2$, except at $x = 1$

$\dfrac{10}{\sqrt{x}+1} = \dfrac{10}{\sqrt{x}+1} \cdot \dfrac{\sqrt{x}-1}{\sqrt{x}-1} = \dfrac{10(\sqrt{x}-1)}{x-1}$, $x \neq 1$

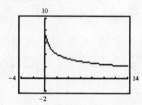

97. *Keystrokes:*

y_1 [Y=] 2 [√] [X,T,θ] [÷] [(] 2 [−] [√] [X,T,θ] [)] [ENTER]

y_2 2 [(] 2 [√] [X,T,θ] [+] [X,T,θ] [)] [÷] [(] 4 [−] [X,T,θ] [)] [GRAPH]

$y_1 = y_2$

$\dfrac{2\sqrt{x}}{2-\sqrt{x}} = \dfrac{2\sqrt{x}}{2-\sqrt{x}} \cdot \dfrac{2+\sqrt{x}}{2+\sqrt{x}}$

$= \dfrac{2\sqrt{x}(2+\sqrt{x})}{2^2-(\sqrt{x})^2}$

$= \dfrac{2\sqrt{x}(2+\sqrt{x})}{4-x}$

$= \dfrac{2(2\sqrt{x}+x)}{4-x}$

99. $\dfrac{\sqrt{2}}{7} = \dfrac{\sqrt{2}}{7} \cdot \dfrac{\sqrt{2}}{\sqrt{2}} = \dfrac{2}{7\sqrt{2}}$

101. $\dfrac{\sqrt{7}+\sqrt{3}}{5} = \dfrac{\sqrt{7}+\sqrt{3}}{5} \cdot \dfrac{\sqrt{7}-\sqrt{3}}{\sqrt{7}-\sqrt{3}}$

$= \dfrac{(\sqrt{7})^2-(\sqrt{3})^2}{5(\sqrt{7}-\sqrt{3})}$

$= \dfrac{7-3}{5(\sqrt{7}-\sqrt{3})}$

$= \dfrac{4}{5(\sqrt{7}-\sqrt{3})}$

103. Area $= h \cdot w$

$= \sqrt{24^2-(8\sqrt{3})^2} \cdot 8\sqrt{3}$

$= \sqrt{576-192} \cdot 8\sqrt{3}$

$= \sqrt{384} \cdot 8\sqrt{3}$

$= 8\sqrt{1152}$

$= 8\sqrt{2^7 \cdot 3^2}$

$= 8 \cdot 2^3 \cdot 3\sqrt{2}$

$= 192\sqrt{2}$ square inches

105. $\dfrac{500k}{\dfrac{1}{\sqrt{k^2+1}}+\dfrac{k^2}{\sqrt{k^2+1}}} = \dfrac{500k}{\dfrac{1+k^2}{\sqrt{k^2+1}}}$

$\qquad\qquad = \dfrac{500k\sqrt{k^2+1}}{1+k^2}$

107. $\sqrt{3}\left(1 - \sqrt{6}\right) = \sqrt{3} - \sqrt{3} \cdot \sqrt{6}$

      Distributive Property

$\qquad = \sqrt{3} - \sqrt{9 \cdot 2}$

      Multiplication Property of Radicals

$\qquad = \sqrt{3} - 3\sqrt{2}$

      Simplify radicals.

109. $\left(3 - \sqrt{2}\right)\left(3 + \sqrt{2}\right) = 9 - 2 = 7$

    Multilpying the number by its conjugate yields the difference of two squares. Squaring a square root eliminates the radical.

Mid-Chapter Quiz for Chapter 9

1. $\sqrt{225} = 15$ because $15 \cdot 15 = 225$

2. $\sqrt[4]{\dfrac{81}{16}} = \dfrac{3}{2}$ because $\dfrac{3}{2} \cdot \dfrac{3}{2} \cdot \dfrac{3}{2} \cdot \dfrac{3}{2} = \dfrac{81}{16}$

3. $64^{1/2} = \sqrt{64} = 8$ because $8 \cdot 8 = 64$

4. $(-27)^{2/3} = \sqrt[3]{(-27)^2} = \left(\sqrt[3]{-27}\right)^2 = (-3)^2 = 9$

5. $\sqrt{27x^2} = \sqrt{9 \cdot 3 \cdot x^2} = 3|x|\sqrt{3}$

6. $\sqrt[4]{81x^6} = \sqrt[4]{81 \cdot x^4 \cdot x^2} = 3|x|\sqrt[4]{x^2}$

$\qquad\qquad = 3|x|\sqrt{x}$

7. $\sqrt{\dfrac{4u^3}{9}} = \dfrac{\sqrt{4 \cdot u^2 \cdot u}}{\sqrt{9}} = \dfrac{2|u|\sqrt{u}}{3}$

8. $\sqrt[3]{\dfrac{16}{u^6}} = \dfrac{\sqrt[3]{16}}{\sqrt[3]{u^6}} = \dfrac{\sqrt[3]{16}}{u^2} = \dfrac{\sqrt[3]{8 \cdot 2}}{u^2} = \dfrac{2\sqrt[3]{2}}{u^2}$

9. $\sqrt{200y} - 3\sqrt{8y} = \sqrt{100 \cdot 2y} - 3\sqrt{4 \cdot 2y}$

$\qquad\qquad = 10\sqrt{2y} - 6\sqrt{2y}$

$\qquad\qquad = 4\sqrt{2y}$

10. $6x\sqrt[3]{5x^2} + 2\sqrt[3]{40x^4} = 6x\sqrt[3]{5x^2} + 2\sqrt[3]{8 \cdot 5 \cdot x^3 \cdot x}$

$\qquad\qquad = 6x\sqrt[3]{5x^2} + 4x\sqrt[3]{5x}$

11. $\sqrt{8}\left(3 + \sqrt{32}\right) = 3\sqrt{8} + \sqrt{256}$

$\qquad\qquad = 3\sqrt{4 \cdot 2} + \sqrt{2^8}$

$\qquad\qquad = 6\sqrt{2} + 2^4$

$\qquad\qquad = 6\sqrt{2} + 16$

12. $\left(\sqrt{50} - 4\right)\sqrt{2} = \sqrt{100} - 4\sqrt{2}$

$\qquad\qquad = \sqrt{10^2} - 4\sqrt{2}$

$\qquad\qquad = 10 - 4\sqrt{2}$

13. $\left(\sqrt{6} + 3\right)\left(4\sqrt{6} - 7\right) = \sqrt{6} \cdot 4\sqrt{6} - 7\sqrt{6} + 12\sqrt{6} - 21$

$\qquad\qquad\qquad = 24 + 5\sqrt{6} - 21$

$\qquad\qquad\qquad = 3 + 5\sqrt{6}$

14. $\left(9 + 2\sqrt{3}\right)\left(2 + 7\sqrt{3}\right) = 18 + 63\sqrt{3} + 4\sqrt{3} + 2\sqrt{3} \cdot 7\sqrt{3}$

$\qquad\qquad\qquad = 18 + 67\sqrt{3} + 14(3)$

$\qquad\qquad\qquad = 18 + 67\sqrt{3} + 42$

$\qquad\qquad\qquad = 60 + 67\sqrt{3}$

15. $\dfrac{\sqrt{7}}{1+\sqrt{3}} \cdot \dfrac{1-\sqrt{3}}{1-\sqrt{3}} = \dfrac{\sqrt{7}(1-\sqrt{3})}{1-(\sqrt{3})^2} = \dfrac{\sqrt{7}(1-\sqrt{3})}{1-3} = \dfrac{\sqrt{7}(1-\sqrt{3})}{-2}$

$$= \dfrac{\sqrt{7}-\sqrt{21}}{-2} = \dfrac{\sqrt{21}-\sqrt{7}}{2}$$

16. $\dfrac{6\sqrt{2}}{2\sqrt{2}-4} \cdot \dfrac{2\sqrt{2}+4}{2\sqrt{2}+4} = \dfrac{6\sqrt{2}(2\sqrt{2}+4)}{(2\sqrt{2})^2 - 4^2} = \dfrac{12(\sqrt{2})^2 + 24\sqrt{2}}{8-16}$

$$= \dfrac{24+24\sqrt{2}}{-8} = \dfrac{24(1+\sqrt{2})}{-8} = -3(1+\sqrt{2})$$

$$= -3 - 3\sqrt{2}$$

17. $4 \div (\sqrt{6}+3) = \dfrac{4}{\sqrt{6}+3} \cdot \dfrac{\sqrt{6}-3}{\sqrt{6}-3} = \dfrac{4(\sqrt{6}-3)}{(\sqrt{6})^2 - 3^2} = \dfrac{4(\sqrt{6}-3)}{6-9}$

$$= \dfrac{4(\sqrt{6}-3)}{-3}$$

$$= \dfrac{4}{3}(3-\sqrt{6})$$

18. $(4\sqrt{2} - 2\sqrt{3}) \div (\sqrt{2}+\sqrt{6}) = \dfrac{4\sqrt{2}-2\sqrt{3}}{\sqrt{2}+\sqrt{6}} \cdot \dfrac{\sqrt{2}-\sqrt{6}}{\sqrt{2}-\sqrt{6}} = \dfrac{4\sqrt{2}\cdot\sqrt{2} - 4\sqrt{2}\cdot\sqrt{6} - 2\sqrt{3}\cdot\sqrt{2} + 2\sqrt{3}\cdot\sqrt{6}}{(\sqrt{2})^2 - (\sqrt{6})^2}$

$$= \dfrac{4(2) - 4\sqrt{12} - 2\sqrt{6} + 2\sqrt{18}}{2-6} = \dfrac{8 - 4\sqrt{4\cdot3} - 2\sqrt{6} + 2\sqrt{9\cdot2}}{-4} = \dfrac{8 - 8\sqrt{3} - 2\sqrt{6} + 6\sqrt{2}}{-4}$$

$$= \dfrac{2(4 - 4\sqrt{3} - \sqrt{6} + 3\sqrt{2})}{-4} = \dfrac{4 - 4\sqrt{3} - \sqrt{6} + 3\sqrt{2}}{-2} = \dfrac{1}{2}(4\sqrt{3} + \sqrt{6} - 3\sqrt{2} - 4)$$

19. $1 + \sqrt{4}$, conjugate $= 1 - \sqrt{4}$

product $= (1+\sqrt{4})(1-\sqrt{4})$

$\qquad = 1^2 - (\sqrt{4})^2$

$\qquad = 1 - 4$

$\qquad = -3$

20. $\sqrt{10} - 5$, conjugate $= \sqrt{10} + 5$

product $= (\sqrt{10}-5)(\sqrt{10}+5)$

$\qquad = (\sqrt{10})^2 - 5^2$

$\qquad = 10 - 25$

$\qquad = -15$

21. $\sqrt{5^2 + 12^2} = \sqrt{25 + 144}$

$\qquad\qquad = \sqrt{169}$

$\qquad\qquad = 13$

$\qquad\qquad 13 \neq 17$

$\sqrt{5^2 + 12^2} \neq 17$

22. $C = \sqrt{2^2 + 2^2}$

$\qquad = \sqrt{4+4} = \sqrt{8}$

Equation:

$P = 2(7) + 2(4\frac{1}{2}) + 4(\sqrt{8})$

$\qquad = 14 + 9 + 8\sqrt{2}$

$\qquad = 23 + 8\sqrt{2}$ inches

Section 9.4 Solving Radical Equations

1. (a) $x = -4$ $\sqrt{-4} - 10 \neq 0$ Not a solution

    (b) $x = -100$ $\sqrt{-100} - 10 \neq 0$ Not a solution

    (c) $x = \sqrt{10}$ $\sqrt{\sqrt{10}} - 10 \neq 0$ Not a solution

    (d) $x = 100$ $\sqrt{100} - 10 = 0$ A solution

3. (a) $x = -60$ $\sqrt[3]{-60-4} \neq 4$ Not a solution

    (b) $x = 68$ $\sqrt[3]{68-4} = 4$ A solution

    (c) $x = 20$ $\sqrt[3]{20-4} \neq 4$ Not a solution

    (d) $x = 0$ $\sqrt[3]{0-4} \neq 4$ Not a solution

5. $\sqrt{x} = 20$ **Check:** $\sqrt{400} \overset{?}{=} 20$

$\left(\sqrt{x}\right)^2 = 20^2$ $20 = 20$

$x = 400$

7. $\sqrt{x} = 3$ **Check:** $\sqrt{9} \overset{?}{=} 3$

$\left(\sqrt{x}\right)^2 = 3^2$ $3 = 3$

$x = 9$

9. $\sqrt[3]{z} = 3$ **Check:** $\sqrt[3]{27} \overset{?}{=} 3$

$\left(\sqrt[3]{z}\right)^3 = 3^3$ $3 = 3$

$z = 27$

11. $\sqrt{y} - 7 = 0$ **Check:** $\sqrt{49} - 7 \overset{?}{=} 0$

$\sqrt{y} = 7$ $7 - 7 \overset{?}{=} 0$

$\left(\sqrt{y}\right)^2 = 7^2$ $0 = 0$

$y = 49$

13. $\sqrt{u} + 13 = 0$ **Check:** $\sqrt{169} + 13 \overset{?}{=} 0$

$\sqrt{u} = -13$ $13 + 13 \neq 0$

$\left(\sqrt{u}\right)^2 = (-13)^2$ No solution

$u = 169$

15. $\sqrt{x} - 8 = 0$ **Check:** $\sqrt{64} - 8 \overset{?}{=} 0$

$\sqrt{x} = 8$ $8 - 8 \overset{?}{=} 0$

$\left(\sqrt{x}\right)^2 = 8^2$ $0 = 0$

$x = 64$

17. $\sqrt{10x} = 30$ **Check:** $\sqrt{10 \cdot 90} \overset{?}{=} 30$

$\left(\sqrt{10x}\right)^2 = 30^2$ $\sqrt{900} \overset{?}{=} 30$

$10x = 900$ $30 = 30$

$x = 90$

19. $\sqrt{-3x} = 9$ **Check:** $\sqrt{-3(-27)} \overset{?}{=} 9$

$\left(\sqrt{-3x}\right)^2 = 9^2$ $\sqrt{81} \overset{?}{=} 9$

$-3x = 81$ $9 = 9$

$x = -27$

21. $\sqrt{5t} - 2 = 0$ **Check:** $\sqrt{5\left(\frac{4}{5}\right)} - 2 \overset{?}{=} 0$

$\sqrt{5t} = 2$ $\sqrt{4} - 2 \overset{?}{=} 0$

$\left(\sqrt{5t}\right)^2 = 2^2$ $2 - 2 \overset{?}{=} 0$

$5t = 4$ $0 = 0$

$t = \frac{4}{5}$

23. $\sqrt{3y + 1} = 4$ **Check:** $\sqrt{3(5) + 1} \overset{?}{=} 4$

$\left(\sqrt{3y + 1}\right)^2 = 4^2$ $\sqrt{16} \overset{?}{=} 4$

$3y + 1 = 16$ $4 = 4$

$3y = 15$

$y = 5$

25. $\sqrt{4 - 5x} = -3$ **Check:** $\sqrt{4 - 5(-1)} \overset{?}{=} -3$

$\left(\sqrt{4 - 5x}\right)^2 = (-3)^2$ $\sqrt{9} \overset{?}{=} -3$

$4 - 5x = 9$ $3 \neq -3$

$-5x = 5$ No solution

$x = -1$

27. $\sqrt{3y + 5} - 3 = 4$ **Check:** $\sqrt{3\left(\frac{44}{3}\right) + 5} - 3 \overset{?}{=} 4$

$\sqrt{3y + 5} = 7$ $\sqrt{49} - 3 \overset{?}{=} 4$

$\left(\sqrt{3y + 5}\right)^2 = 7^2$ $7 - 3 \overset{?}{=} 4$

$3y + 5 = 49$ $4 = 4$

$3y = 44$

$y = \frac{44}{3}$

29. $5\sqrt{x + 2} = 8$ **Check:** $5\sqrt{\frac{14}{25} + 2} \overset{?}{=} 8$

$\left(5\sqrt{x + 2}\right)^2 = 8^2$ $5\sqrt{\frac{64}{25}} \overset{?}{=} 8$

$25(x + 2) = 64$ $5 \cdot \frac{8}{5} \overset{?}{=} 8$

$25x + 50 = 64$ $8 = 8$

$25x = 14$

$x = \frac{14}{25}$

31. $\sqrt{3x + 2} + 5 = 0$ **Check:** $\sqrt{3\left(\frac{23}{3}\right) + 2} + 5 \overset{?}{=} 0$

$\sqrt{3x + 2} = -5$ $\sqrt{23 + 2} + 5 \overset{?}{=} 0$

$\left(\sqrt{3x + 2}\right)^2 = (-5)^2$ $\sqrt{25} + 5 \overset{?}{=} 0$

$3x + 2 = 25$ $5 + 5 \overset{?}{=} 0$

$3x = 23$ $10 \neq 0$

$x = \frac{23}{3}$ No solution

33. $\sqrt{x + 3} = \sqrt{2x - 1}$

$\left(\sqrt{x + 3}\right)^2 = \left(\sqrt{2x - 1}\right)^2$

$x + 3 = 2x - 1$

$4 = x$

Check: $\sqrt{4 + 3} \overset{?}{=} \sqrt{2(4) - 1}$

$\sqrt{7} = \sqrt{7}$

35. $\sqrt{3y - 5} - 3\sqrt{y} = 0$

$\sqrt{3y - 5} = 3\sqrt{y}$

$\left(\sqrt{3y - 5}\right)^2 = \left(3\sqrt{y}\right)^2$

$3y - 5 = 9y$

$-5 = 6y$

$-\frac{5}{6} = y$

Check: $\sqrt{3\left(-\frac{5}{6}\right) - 5} - 3\sqrt{-\frac{5}{6}} \overset{?}{=} 0$

No solution

37. $\sqrt[3]{3x - 4} = \sqrt[3]{x + 10}$

$\left(\sqrt[3]{3x - 4}\right)^3 = \left(\sqrt[3]{x + 10}\right)^3$

$3x - 4 = x + 10$

$2x = 14$

$x = 7$

Check: $\sqrt[3]{3(7) - 4} \overset{?}{=} \sqrt[3]{7 + 10}$

$\sqrt[3]{17} = \sqrt[3]{17}$

39. $\sqrt[3]{2x + 15} - \sqrt[3]{x} = 0$

$\sqrt[3]{2x + 15} = \sqrt[3]{x}$

$\left(\sqrt[3]{2x + 15}\right)^3 = \left(\sqrt[3]{x}\right)^3$

$2x + 15 = x$

$x = -15$

Check: $\sqrt[3]{2(-15) + 15} - \sqrt[3]{-15} \overset{?}{=} 0$

$\sqrt[3]{-15} - \sqrt[3]{-15} \overset{?}{=} 0$

$0 = 0$

41. $\sqrt{x^2 + 5} = x + 3$

$\left(\sqrt{x^2 + 5}\right)^2 = (x + 3)^2$

$x^2 + 5 = x^2 + 6x + 9$

$-4 = 6x$

$-\frac{4}{6} = x$

$-\frac{2}{3} = x$

Check: $\sqrt{\left(-\frac{2}{3}\right)^2 + 5} \overset{?}{=} -\frac{2}{3} + 3$

$\sqrt{\frac{4}{9} + \frac{45}{9}} \overset{?}{=} -\frac{2}{3} + \frac{9}{3}$

$\sqrt{\frac{49}{9}} \overset{?}{=} \frac{7}{3}$

$\frac{7}{3} = \frac{7}{3}$

43. $\sqrt{2x} = x - 4$

$\left(\sqrt{2x}\right)^2 = (x - 4)^2$

$2x = x^2 - 8x + 16$

$0 = x^2 - 10x + 16$

$0 = (x - 8)(x - 2)$

$8 = x, x = 2$

Not a solution

Check: $\sqrt{2(8)} \overset{?}{=} 8 - 4$

$\sqrt{16} \overset{?}{=} 4$

$4 = 4$

$\sqrt{2(2)} \overset{?}{=} 2 - 4$

$\sqrt{4} \overset{?}{=} -2$

$2 \neq -2$

45. $\sqrt{8x + 1} = x + 2$

$\left(\sqrt{8x + 1}\right)^2 = (x + 2)^2$

$8x + 1 = x^2 + 4x + 4$

$0 = x^2 - 4x + 3$

$0 = (x - 3)(x - 1)$

$3 = x, \qquad x = 1$

Check: $\sqrt{8(3) + 1} \stackrel{?}{=} 3 + 2$

$\sqrt{25} \stackrel{?}{=} 5$

$5 = 5$

$\sqrt{8(1) + 1} \stackrel{?}{=} 1 + 2$

$\sqrt{9} \stackrel{?}{=} 3$

$3 = 3$

47. $\sqrt{z + 2} = 1 + \sqrt{2}$

$\left(\sqrt{z + 2}\right)^2 = \left(1 + \sqrt{z}\right)^2$

$z + 2 = 1 + 2\sqrt{z} + z$

$1 = 2\sqrt{z}$

$1^2 = \left(2\sqrt{z}\right)^2$

$1 = 4z$

$\frac{1}{4} = z$

Check: $\sqrt{\frac{1}{4} + 2} \stackrel{?}{=} 1 + \sqrt{\frac{1}{4}}$

$\sqrt{\frac{9}{4}} \stackrel{?}{=} 1 + \frac{1}{2}$

$\frac{3}{2} = \frac{3}{2}$

49. $\sqrt{2t + 3} = 3 - \sqrt{2t}$

$\left(\sqrt{2t + 3}\right)^2 = \left(3 - \sqrt{2t}\right)^2$

$2t + 3 = 9 - 6\sqrt{2t} + 2t$

$-6 = -6\sqrt{2t}$

$1 = \sqrt{2t}$

$1^2 = \left(\sqrt{2t}\right)^2$

$1 = 2t$

$\frac{1}{2} = t$

Check: $\sqrt{2\left(\frac{1}{2}\right) + 3} \stackrel{?}{=} 3 - \sqrt{2\left(\frac{1}{2}\right)}$

$\sqrt{1 + 3} \stackrel{?}{=} 3 - \sqrt{1}$

$\sqrt{4} \stackrel{?}{=} 3 - 1$

$2 = 2$

51. $\sqrt{x + 5} - \sqrt{x} = 1$

$\sqrt{x + 5} = 1 + \sqrt{x}$

$\left(\sqrt{x + 5}\right)^2 = \left(1 + \sqrt{x}\right)^2$

$x + 5 = 1 + 2\sqrt{x} + x$

$4 = 2\sqrt{x}$

$2 = \sqrt{x}$

$2^2 = \left(\sqrt{x}\right)^2$

$4 = x$

Check: $\sqrt{4 + 5} - \sqrt{4} \stackrel{?}{=} 1$

$\sqrt{9} - \sqrt{4} \stackrel{?}{=} 1$

$3 - 2 \stackrel{?}{=} 1$

$1 = 1$

53. $t^{3/2} = 8$

$\sqrt{t^3} = 8$

$\left(\sqrt{t^3}\right)^2 = 8^2$

$t^3 = 64$

$t = 4$

Check: $4^{3/2} \stackrel{?}{=} 8$

$\left(\sqrt{4}\right)^3 \stackrel{?}{=} 8$

$2^3 \stackrel{?}{=} 8$

$8 = 8$

55. $3y^{1/3} = 18$

$y^{1/3} = 6$

$\sqrt[3]{y} = 6$

$\left(\sqrt[3]{y}\right)^3 = 6^3$

$y = 216$

Check: $3(216)^{1/3} \overset{?}{=} 18$

$3\sqrt[3]{216} \overset{?}{=} 18$

$3 \cdot 6 \overset{?}{=} 18$

$18 = 18$

57. $(x + 4)^{2/3} = 4$

$\sqrt[3]{(x + 4)^2} = 4$

$\left(\sqrt[3]{(x + 4)^2}\right)^3 = (4)^3$

$(x + 4)^2 = 64$

$x + 4 = \pm\sqrt{64}$

$x = -4 \pm 8$

$= 4, -12$

Check: $(4 + 4)^{2/3} \overset{?}{=} 4$

$8^{2/3} \overset{?}{=} 4$

$2^2 = 4$

$(-12 + 4)^{2/3} \overset{?}{=} 4$

$(-8)^{2/3} \overset{?}{=} 4$

$(-2)^2 = 4$

59. $(2x + 5)^{1/3} + 3 = 0$

$\sqrt[3]{(2x + 5)} = -3$

$\left(\sqrt[3]{2x + 5}\right)^3 = (-3)^3$

$2x + 5 = -27$

$2x = -32$

$x = -16$

Check: $(2(-16) + 5)^{1/3} + 3 \overset{?}{=} 0$

$(-32 + 5)^{1/3} + 3 \overset{?}{=} 0$

$(-27)^{1/3} + 3 \overset{?}{=} 0$

$-3 + 3 \overset{?}{=} 0$

$0 = 0$

61. *Keystrokes:*

y_1 [Y=] [√] [X,T,θ] [ENTER]

y_2 2 [(] 2 [−] [X,T,θ] [)] [GRAPH]

Approximate solution: $x \approx 1.407$

Check algebraically: $\sqrt{1.407} \overset{?}{=} 2(2 - 1.407)$

$1.186 = 1.186$

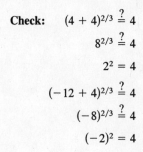

63. *Keystrokes:*

y_1 [Y=] [√] [(] [X,T,θ] [x²] [+] 1 [)] [ENTER]

y_2 5 [−] 2 [X,T,θ] [GRAPH]

Approximate solution: $x \approx 1.569$

Check algebraically: $\sqrt{1.569^2 + 1} \overset{?}{=} 5 - 2(1.569)$

$1.86 = 1.86$

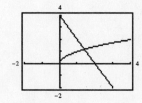

65. *Keystrokes:*

y_1 [Y=] [√] [(] [X,T,θ] [+] 3 [)] [ENTER]

y_2 5 [−] [√] [X,T,θ] [GRAPH]

Approximate solution: $x \approx 4.840$

Check algebraically: $\sqrt{4.840 + 3} \overset{?}{=} 5 - \sqrt{4.840}$

$2.8 = 2.8$

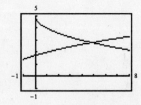

67. *Keystrokes:*

y_1 [Y=] 4 [MATH 4] [X,T,θ] [ENTER]

y_2 7 [−] [X,T,θ] [GRAPH]

Approximate solution: $x \approx 1.978$

Check algebraically: $4\sqrt[3]{1.978} \stackrel{?}{=} 7 - 1.978$

$$5.02 = 5.02$$

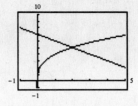

69. *Keystrokes:*

y_1 [Y=] [√] [(] 15 [−] 4 [X,T,θ] [)] [ENTER]

y_2 2 [X,T,θ] [GRAPH]

Solution: $x = 1.5$

Check algebraically: $\sqrt{15 - 4(1.5)} \stackrel{?}{=} 2(1.5)$

$$\sqrt{9} \stackrel{?}{=} 3$$

$$3 = 3$$

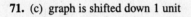

71. (c) graph is shifted down 1 unit

73. (d) graph is shifted left 3 units and upward 1 unit

75. (f) graph is shifted down 1 unit

77.
$$15^2 = x^2 + 12^2$$
$$225 = x^2 + 144$$
$$81 = x^2$$
$$\sqrt{81} = x^2$$
$$9 = x$$

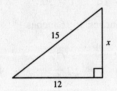

79.
$$c^2 = a^2 + b^2$$
$$13^2 = x^2 + 5^2$$
$$169 = x^2 + 25$$
$$144 = x^2$$
$$\sqrt{144} = x$$
$$12 = x$$

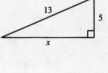

81.
$$13.75^2 = 8.25^2 + x^2$$
$$x^2 = 13.75^2 - 8.25^2$$
$$x = \sqrt{121}$$
$$x = 11 \text{ inches}$$

83. $c = \sqrt{32^2 + 26^2}$

$\quad = \sqrt{1024 + 676}$

$\quad = \sqrt{1700}$

$\quad = 10\sqrt{17}$

$\quad \approx 41.23$ feet

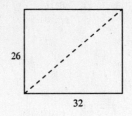

85. $17^2 = x^2 + 8^2$

$\quad x^2 = 17^2 - 8^2$

$\quad x = \sqrt{289 - 64}$

$\quad x = \sqrt{225}$

$\quad x = 15$ feet

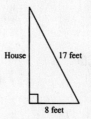

87. $\quad P = 2l + 2w$

$\quad 92 = 2l + 2w$

$\quad 46 = l + w$

$\quad 46 - w = l$

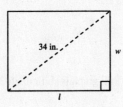

$34^2 = w^2 + (46 - w)^2$

$1156 = w^2 + 2116 - 92w + w^2$

$0 = 2w^2 - 92w + 960$

$0 = w^2 - 46w + 480$

$0 = (w - 30)(w - 16)$

$w = 30 \qquad w = 16$

$l = 16 \qquad l = 30$

30 inches $\times$ 16 inches

89. $\qquad S = \pi r \sqrt{r^2 + h^2}$

$\qquad \dfrac{S}{\pi r} = \sqrt{r^2 + h^2}$

$\qquad \left(\dfrac{S}{\pi r}\right)^2 = \left(\sqrt{r^2 + h^2}\right)^2$

$\qquad \dfrac{S^2}{\pi^2 r^2} = r^2 + h^2$

$\qquad \dfrac{S^2}{\pi^2 r^2} - r^2 = h^2$

$\qquad \dfrac{S^2 - \pi^2 r^4}{\pi^2 r^2} = h^2$

$\qquad \sqrt{\dfrac{S^2 - \pi^2 r^4}{\pi^2 r^2}} = h$

$\qquad \dfrac{\sqrt{S^2 - \pi^2 r^4}}{\pi r} = h$

91. $\qquad 2 = \sqrt{\dfrac{d}{16}}$

$\qquad 2^2 = \left(\sqrt{\dfrac{d}{16}}\right)^2$

$\qquad 4 = \dfrac{d}{16}$

$\qquad 64 \text{ feet} = d$

93. $v = \sqrt{2(32)50}$

$\quad v = \sqrt{3200}$

$\quad v = 40\sqrt{2}$

$\quad v \approx 56.57$ feet per second

95. $\qquad 60 = \sqrt{2(32)h}$

$\qquad 60^2 = \left(\sqrt{64h}\right)^2$

$\qquad 3600 = 64h$

$\qquad \dfrac{3600}{64} = h$

$\qquad 56.25 \text{ feet} = h$

97.
$$1.5 = 2\pi\sqrt{\frac{L}{32}}$$

$$\left(\frac{1.5}{2\pi}\right)^2 = \left(\sqrt{\frac{L}{32}}\right)^2$$

$$\frac{2.25}{4\pi^2} = \frac{L}{32}$$

$$\frac{2.25}{4\pi^2}(32) = L$$

$$1.82 \text{ feet} \approx L$$

99.
$$30.02 = 50 - \sqrt{0.8(x-1)}$$

$$\sqrt{0.8(x-1)} = 19.98$$

$$\left(\sqrt{0.8(x-1)}\right)^2 = (19.98)^2$$

$$0.8(x-1) = 399.2004$$

$$0.8x - 0.8 = 399.2004$$

$$0.8x = 400.0004$$

$$x = 500.0005$$

$$\approx 500 \text{ units}$$

101. (a) *Keystrokes:*

y_1 [Y=] 133.5 [+] 9.3 [X,T,θ] [+] 18 [√] [X,T,θ] [GRAPH]

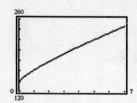

(b)

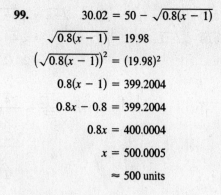

$\approx$ 5 years from 1990 $\approx$ 1995

103.
$$R = \left(\frac{A}{P}\right)^{1/n} - 1$$

$$0.07 = \left(\frac{25{,}000}{P}\right)^{1/10} - 1$$

$$1.07 = \left(\frac{25{,}000}{P}\right)^{1/10}$$

$$1.07 = \sqrt[10]{\frac{25{,}000}{P}}$$

$$1.07^{10} = \left(\sqrt[10]{\frac{25{,}000}{P}}\right)^{10}$$

$$1.07^{10} = \frac{25{,}000}{P}$$

$$P = \frac{25{,}000}{1.07^{10}}$$

$$P \approx \$12{,}708.73$$

105. No. It is not an operation that necessarily yields an equivalent equation. There may be extraneous solutions.

107. $\left(\sqrt{x} + \sqrt{6}\right)^2 \neq \left(\sqrt{x}\right)^2 + \left(\sqrt{6}\right)^2$

$\left(\sqrt{x} + \sqrt{6}\right)^2$ must be multiplied by FOIL.

Section 9.5   Complex Numbers

1. $\sqrt{-4} = \sqrt{-1 \cdot 4}$
$= \sqrt{-1} \cdot \sqrt{4}$
$= 2i$

3. $-\sqrt{-144} = -\sqrt{144 \cdot -1}$
$= -\sqrt{144} \cdot \sqrt{-1}$
$= -12i$

5. $\sqrt{\frac{-4}{25}} = \sqrt{\frac{4}{25} \cdot -1}$
$= \sqrt{\frac{4}{25}} \cdot \sqrt{-1}$
$= \frac{2}{5}i$

7. $\sqrt{-0.09} = \sqrt{0.09 \cdot -1}$
$= \sqrt{0.09} \cdot \sqrt{-1}$
$= 0.3i$

9. $\sqrt{-8} = \sqrt{4 \cdot 2 \cdot -1}$
$= \sqrt{4} \cdot \sqrt{2} \cdot \sqrt{-1}$
$= 2i\sqrt{2}$

11. $\sqrt{-27} = \sqrt{-1 \cdot 9 \cdot 3}$
$= \sqrt{-1} \cdot \sqrt{9} \cdot \sqrt{3}$
$= 3i\sqrt{3}$

13. $\sqrt{-7} = \sqrt{7 \cdot -1}$
$= \sqrt{7} \cdot \sqrt{-1}$
$= i\sqrt{7}$

15. $\dfrac{\sqrt{-12}}{\sqrt{-3}} = \dfrac{\sqrt{4 \cdot 3 \cdot -1}}{\sqrt{3 \cdot -1}}$
$= \dfrac{\sqrt{4} \cdot \sqrt{3} \cdot \sqrt{-1}}{\sqrt{3} \cdot \sqrt{-1}}$
$= \sqrt{4} = 2$

or

$\dfrac{\sqrt{-12}}{\sqrt{-3}} = \sqrt{\dfrac{-12}{-3}} = \sqrt{4} = 2$

17. $\dfrac{\sqrt{-20}}{\sqrt{4}} = \sqrt{\dfrac{-20}{4}}$
$= \sqrt{-5}$
$= \sqrt{5} \cdot \sqrt{-1}$
$= i\sqrt{5}$

19. $\sqrt{\dfrac{-18}{64}} = \sqrt{\dfrac{-1 \cdot 9 \cdot 2}{64}}$
$= \dfrac{3i}{8}\sqrt{2}$

21. $\sqrt{-16} + \sqrt{-36} = 4i + 6i$
$= (4 + 6)i$
$= 10i$

23. $\sqrt{-50} - \sqrt{-8} = 5i\sqrt{2} - 2i\sqrt{2}$
$= (5\sqrt{2} - 2\sqrt{2})i$
$= 3\sqrt{2}i$

25. $\sqrt{-48} + \sqrt{-12} - \sqrt{-27} = \sqrt{16 \cdot 3 \cdot -1} + \sqrt{4 \cdot 3 \cdot -1} - \sqrt{9 \cdot 3 \cdot -1}$
$= 4i\sqrt{3} + 2i\sqrt{3} - 3i\sqrt{3}$
$= (4i + 2i - 3i)\sqrt{3}$
$= 3i\sqrt{3}$

27. $\sqrt{-8}\sqrt{-2} = (2i\sqrt{2})(i\sqrt{2})$
$= 2 \cdot 2 \cdot i^2$
$= 4(-1) = -4$

29. $\sqrt{-18}\sqrt{-3} = (3i\sqrt{2})(i\sqrt{3})$
$= 3\sqrt{6} \cdot i^2$
$= -3\sqrt{6}$

31. $\sqrt{-0.16}\sqrt{-1.21} = (0.4i)(1.1i) = 0.44i^2 = -0.44$

33. $\sqrt{-3}(\sqrt{-3} + \sqrt{-4}) = i\sqrt{3}(i\sqrt{3} + 2i)$
$= (i\sqrt{3})^2 + 2\sqrt{3}i^2$
$= -3 - 2\sqrt{3}$

35. $\sqrt{-5}(\sqrt{-16} - \sqrt{-10}) = i\sqrt{5}(4i - i\sqrt{10})$
$= i^2 4\sqrt{5} - i^2\sqrt{50}$
$= -4\sqrt{5} + 5\sqrt{2}$
$= 5\sqrt{2} - 4\sqrt{5}$

37. $\sqrt{-2}(3 - \sqrt{-8}) = i\sqrt{2}(3 - 2i\sqrt{2})$
$= 3\sqrt{2}i - 2i^2(2)$
$= 3\sqrt{2}i + 4$

39. $(\sqrt{-16})^2 = (4i)^2$
$= 16i^2$
$= -16$

41. $(\sqrt{-4})^3 = (2i)^3$
$= 8i^3$
$= -8i$

43. $3 - 4i = a + bi$
$a = 3 \quad b = -4$

45. $5 - 4i = (a + 3) + (b - 1)i$

$a + 3 = 5 \qquad b - 1 = -4$

$a = 2 \qquad\quad b = -3$

47. $-4 - \sqrt{-8} = a + bi$

$-4 - 2i\sqrt{2} = a + bi$

$-4 = a \qquad -2i\sqrt{2} = bi$

$-2\sqrt{2} = b$

49. $(a + 5) + (b - 1)i = 7 - 3i$

$a + 5 = 7 \qquad b - 1 = -3$

$a = 2 \qquad\quad b = -2$

51. $(4 - 3i) + (6 + 7i) = (4 + 6) + (-3 + 7)i$

$= 10 + 4i$

53. $(-4 - 7i) + (-10 - 33i) = (-4 - 10) + (-7 - 33)i$

$= -14 - 40i$

55. $13i - (14 - 7i) = (-14) + (13 + 7)i$

$= -14 + 20i$

57. $(30 - i) - (18 + 6i) + 3i^2 = 30 - i - 18 - 6i - 3$

$= 9 - 7i$

59. $6 - (3 - 4i) + 2i = 6 - 3 + 4i + 2i$

$= 3 + 6i$

61. $\left(\frac{4}{3} + \frac{1}{3}i\right) + \left(\frac{5}{6} + \frac{7}{6}i\right) = \left(\frac{4}{3} + \frac{5}{6}\right) + \left(\frac{1}{3} + \frac{7}{6}\right)i$

$= \left(\frac{8}{6} + \frac{5}{6}\right) + \left(\frac{2}{6} + \frac{7}{6}\right)i$

$= \frac{13}{6} + \frac{9}{6}i$

$= \frac{13}{6} + \frac{3}{2}i$

63. $15i - (3 - 25i) + \sqrt{-81} = 15i - 3 + 25i + 9i$

$= -3 + (15 + 25 + 9)i$

$= -3 + 49i$

65. $8 - \left(5 - \sqrt{-63}\right) + (4 - 5i) = 8 - 5 + 3i\sqrt{7} + 4 - 5i$

$= 7 + \left(3\sqrt{7} - 5\right)i$

67. $(3i)(12i) = 36i^2$

$= -36$

69. $(3i)(-8i) = -24i^2$

$= -24(-1)$

$= 24$

71. $(-6i)(-i)(6i) = 36i^3 = -36i$

73. $(-3i)^3 = -27i^3 = 27i$

75. $(-3i)^2 = 9i^2$

$= 9(-1)$

$= -9$

77. $-5(13 + 2i) = -65 - 10i$

79. $4i(-3 - 5i) = -12i - 20i^2 = 20 - 12i$

81. $(9 - 2i)\left(\sqrt{-4}\right) = (9 - 2i)(2i)$

$= 18i - 4i^2$

$= 18i + 4$

$= 4 + 18i$

83. $\sqrt{-20}\left(6 + 2\sqrt{5}i\right) = 2i\sqrt{5}\left(6 + 2\sqrt{5}i\right)$

$= 12i\sqrt{5} + 4i^2(5)$

$= -20 + 12i\sqrt{5}$

85. $(4 + 3i)(-7 + 4i) = -28 + 16i - 21i + 12i^2$

$= -28 - 12 - 5i$

$= -40 - 5i$

87. $(-7 + 7i)(4 - 2i) = -28 + 14i + 28i - 14i^2$

$$= -28 + 42i + 14$$

$$= -14 + 42i$$

89. $\left(-2 + \sqrt{-5}\right)\left(-2 - \sqrt{-5}\right) = \left(-2 + i\sqrt{5}\right)\left(-2 - i\sqrt{5}\right)$

$$= 4 + 2i\sqrt{5} - 2i\sqrt{5} - 5i^2$$

$$= 4 + 5$$

$$= 9$$

91. $(3 - 4i)^2 = 3^2 - 2(3)(4i) + (4i)^2$

$$= 9 - 24i + 16i^2$$

$$= 9 - 16 - 24i$$

$$= -7 - 24i$$

93. $(2 + 5i)^2 = 2^2 + 2(2)(5i) + (5i)^2$

$$= 4 + 20i + 25i^2$$

$$= 4 - 25 + 20i$$

$$= -21 + 20i$$

95. $(2 + i)^3 = (2 + i)(2 + i)(2 + i)$

$$= (4 + 2(2)i + i^2)(2 + i)$$

$$= (4 + 4i - 1)(2 + i)$$

$$= (3 + 4i)(2 + i)$$

$$= 6 + 3i + 8i + 4i^2$$

$$= 6 + 3i + 8i - 4$$

$$= 2 + 11i$$

97. $2 + i$, conjugate $= 2 - i$

$$\text{product} = (2 + i)(2 - i)$$

$$= 2^2 - i^2$$

$$= 4 + 1$$

$$= 5$$

99. $-2 - 8i$, conjugate $= -2 + 8i$

$$\text{product} = (-2 - 8i)(-2 + 8i)$$

$$= (-2)^2 - (8i)^2$$

$$= 4 - 64i^2 = 4 + 64 = 68$$

101. $5 - \sqrt{6}i$, conjugate $= 5 + \sqrt{6}i$

$$\text{product} = \left(5 - \sqrt{6}i\right)\left(5 + \sqrt{6}i\right)$$

$$= 5^2 - \left(\sqrt{6}i\right)^2$$

$$= 25 - 6i^2 = 25 + 6 = 31$$

103. $10i$, conjugate $= -10i$

$$\text{product} = (10i)(-10i)$$

$$= -(10i)^2$$

$$= -100i^2$$

$$= 100$$

105. $1 + \sqrt{-3} = 1 + i\sqrt{3}$, conjugate $= 1 - i\sqrt{3}$

$$\text{product} = \left(1 + i\sqrt{3}\right)\left(1 - i\sqrt{3}\right)$$

$$= 1^2 - \left(i\sqrt{3}\right)^2$$

$$= 1 - 3i^2$$

$$= 1 + 3$$

$$= 4$$

107. $1.5 + \sqrt{-0.25}$, conjugate $= 1.5 - \sqrt{-0.25}$

$$\text{product} = (1.5 + 0.5i)(1.5 - 0.5i)$$
$$= 1.5^2 - (0.5i)^2$$
$$= 2.25 + 0.25$$
$$= 2.5$$

109. $\dfrac{20}{2i} = \dfrac{10}{i} \cdot \dfrac{-i}{-i} = \dfrac{-10i}{1}$

$$= 0 - 10i$$

111. $\dfrac{4}{1-i} = \dfrac{4}{1-i} \cdot \dfrac{1+i}{1+i} = \dfrac{4(1+i)}{1+1}$

$$= \dfrac{4(1+i)}{2}$$
$$= 2(1+i)$$
$$= 2 + 2i$$

113. $\dfrac{-12}{2+7i} = \dfrac{-12}{2+7i} \cdot \dfrac{2-7i}{2-7i} = \dfrac{-12(2-7i)}{4+49}$

$$= \dfrac{-12(2-7i)}{53}$$
$$= \dfrac{-24 + 84i}{53}$$
$$= \dfrac{-24}{53} + \dfrac{84}{53}i$$

115. $\dfrac{4i}{1-3i} = \dfrac{4i}{1-3i} \cdot \dfrac{1+3i}{1+3i} = \dfrac{4i(1+3i)}{1+9}$

$$= \dfrac{4i + 12i^2}{10}$$
$$= \dfrac{-12 + 4i}{10} = \dfrac{4(-3+i)}{10}$$
$$= \dfrac{2(-3+i)}{5} = \dfrac{-6+2i}{5}$$
$$= -\dfrac{6}{5} + \dfrac{2}{5}i$$

117. $\dfrac{2+3i}{1+2i} = \dfrac{2+3i}{1+2i} \cdot \dfrac{1-2i}{1-2i} = \dfrac{2 - 4i + 3i - 6i^2}{1+4}$

$$= \dfrac{2 + 6 - i}{5}$$
$$= \dfrac{8-i}{5}$$
$$= \dfrac{8}{5} - \dfrac{1}{5}i$$

119. $\dfrac{1}{1-2i} + \dfrac{4}{1+2i} = \dfrac{1}{1-2i} \cdot \dfrac{1+2i}{1+2i} + \dfrac{4}{1+2i} \cdot \dfrac{1-2i}{1-2i}$

$$= \dfrac{1+2i}{1+4} + \dfrac{4-8i}{1+4} = \dfrac{1+2i}{5} + \dfrac{4-8i}{5}$$
$$= \dfrac{(1+4) + (2-8)i}{5} = \dfrac{5-6i}{5}$$
$$= 1 - \dfrac{6}{5}i$$

121. $\dfrac{i}{4-3i} - \dfrac{5}{2+i} = \dfrac{i}{4-3i} \cdot \dfrac{4+3i}{4+3i} - \dfrac{5}{2+i} \cdot \dfrac{2-i}{2-i}$

$$= \dfrac{4i + 3i^2}{16+9} - \dfrac{10-5i}{4+1} = \dfrac{-3+4i}{25} - \dfrac{10-5i}{5} \cdot \dfrac{5}{5}$$
$$= \dfrac{-3+4i}{25} - \dfrac{50-25i}{25} = \dfrac{(-3-50) + (4+25)i}{25}$$
$$= \dfrac{-53 + 29i}{25} = \dfrac{-53}{25} + \dfrac{29}{25}i$$

123. (a) $x = -1 + 2i$

$$(-1 + 2i)^2 + 2(-1 + 2i) + 5 \overset{?}{=} 0$$

$$1 - 4i + 4i^2 - 2 + 4i + 5 \overset{?}{=} 0$$

$$1 - 4 - 2 + 5 \overset{?}{=} 0$$

$$0 = 0 \text{ Solution}$$

(b) $x = -1 - 2i$

$$(-1 - 2i)^2 + 2(-1 - 2i) + 5 \overset{?}{=} 0$$

$$1 + 4i + 4i^2 - 2 - 4i + 5 \overset{?}{=} 0$$

$$1 - 4 - 2 + 5 \overset{?}{=} 0$$

$$0 = 0 \text{ Solution}$$

125. (a) $x = -4$

$$(-4)^3 + 4(-4)^2 + 9(-4) + 36 \overset{?}{=} 0$$

$$-64 + 64 - 36 + 36 \overset{?}{=} 0$$

$$0 = 0 \text{ Solution}$$

(b) $x = -3i$

$$(-3i)^3 + 4(-3i)^2 + 9(-3i) + 36 \overset{?}{=} 0$$

$$-27i^3 + 36i^2 - 27i + 36 \overset{?}{=} 0$$

$$27i - 36 - 27i + 36 \overset{?}{=} 0$$

$$0 = 0 \text{ Solution}$$

127. (a) $\left(\dfrac{-5 + 5\sqrt{3}i}{2} \right)^3 = \left(\dfrac{-5}{2} + \dfrac{5}{2}\sqrt{3}i \right)^2 \left(\dfrac{-5}{2} + \dfrac{5}{2}\sqrt{3}i \right)$

$$= \left(\dfrac{25}{4} - \dfrac{25}{2}\sqrt{3}i + \dfrac{25}{4}(3)i^2 \right) \left(\dfrac{-5}{2} + \dfrac{5}{2}\sqrt{3}i \right)$$

$$= \left(\dfrac{25}{4} - \dfrac{25}{2}\sqrt{3}i - \dfrac{75}{4} \right) \left(\dfrac{-5}{2} + \dfrac{5}{2}\sqrt{3}i \right)$$

$$= \left(\dfrac{-50}{4} - \dfrac{25}{2}\sqrt{3}i \right) \left(\dfrac{-5}{2} + \dfrac{5}{2}\sqrt{3}i \right)$$

$$= \left(\dfrac{-25}{2} - \dfrac{25}{2}\sqrt{3}i \right) \left(\dfrac{-5}{2} + \dfrac{5}{2}\sqrt{3}i \right)$$

$$= \dfrac{125}{4} - \dfrac{125}{4}\sqrt{3}i + \dfrac{125}{4}\sqrt{3}i - \dfrac{125}{4}(3)i^2$$

$$= \dfrac{125}{4} + \dfrac{375}{4}$$

$$= \dfrac{500}{4} = 125$$

(b) use same method as part (a)

129. (a) $1, \dfrac{-1 + \sqrt{3}i}{2}, \dfrac{-1 - \sqrt{3}i}{2}$

(b) $2, \dfrac{-2 + 2\sqrt{3}i}{2} = -1 + \sqrt{3}i, \dfrac{-2 - 2\sqrt{3}i}{2} = -1 - \sqrt{3}i$

(c) $4, \dfrac{-4 + 4\sqrt{3}i}{2} = -2 + 2\sqrt{3}i, \dfrac{-4 - 4\sqrt{3}i}{2} = -2 - 2\sqrt{3}i$

131. $(a + bi) + (a - bi) = (a + a) + (b - b)i$

$$= 2a + 0i$$

133. $(a + bi) - (a - bi) = (a - a) + (b + b)i$

$$= 0 + 2bi$$

135. $i = \sqrt{-1}$

137. $\sqrt{-3}\sqrt{-3} = \sqrt{(-3)(-3)} = \sqrt{9} = 3$

The product rule for radicals does not hold if both radicands are negative.

$\sqrt{-3}\sqrt{-3} = i\sqrt{3} \cdot i\sqrt{3} = i^2(3) = -3$

139. $3 - 2i$, conjugate $= 3 + 2i$

$$\text{product} = (3 - 2i)(3 + 2i)$$
$$= 3^2 - (2i)^2$$
$$= 9 + 4$$
$$= 13$$

Review Exercises for Chapter 9

1. $\sqrt{49} = 7$ because $7 \cdot 7 = 49$

3. $-\sqrt{81} = -9$ because $9 \cdot 9 = 81$

5. $\sqrt[3]{-8} = -2$ because $-2 \cdot -2 \cdot -2 = -8$

7. $-\sqrt[3]{64} = -4$ because $4 \cdot 4 \cdot 4 = 64$

9. $\sqrt{(1.2)^2} = 1.2$ (inverse property of powers and roots)

11. $\sqrt{\left(\frac{5}{6}\right)^2} = \frac{5}{6}$ (inverse property of powers and roots)

13. $\sqrt[3]{-\left(\frac{1}{5}\right)^3} = -\frac{1}{5}$ (inverse property of powers and roots)

15. $\sqrt{-2^2} = 2i$

17. $49^{1/2} = 7$

19. $\sqrt[3]{216} = 6$

21. $27^{4/3} = \left(\sqrt[3]{27}\right)^4 = 3^4 = 81$

23. $-(5^2)^{3/2} = -\left(\sqrt{25}\right)^3 = -5^3 = -125$

25. $8^{-4/3} = \dfrac{1}{8^{4/3}} = \dfrac{1}{\left(\sqrt[3]{8}\right)^4} = \dfrac{1}{2^4} = \dfrac{1}{16}$

27. $-\left(\frac{27}{64}\right)^{2/3} = -\left(\sqrt[3]{\frac{27}{64}}\right)^2$
$= -\left(\frac{3}{4}\right)^2$
$= -\frac{9}{16}$

29. $x^{3/4} \cdot x^{-1/6} = x^{3/4 + (-1/6)}$
$= x^{9/12 + (-2)/12}$
$= x^{7/12}$

31. $z\sqrt[3]{z^2} = z \cdot z^{2/3}$
$= z^{1 + 2/3}$
$= z^{5/3}$

33. $\dfrac{\sqrt[4]{x^3}}{\sqrt{x^4}} = \dfrac{x^{3/4}}{x^{4/2}} = x^{3/4 - 2} = x^{3/4 - 8/4} = x^{-5/4} = \dfrac{1}{x^{5/4}}$

35. $\sqrt[3]{a^3 b^2} = a\sqrt[3]{b^2}$

37. $\sqrt[4]{\sqrt{x}} = \sqrt[4]{x^{1/2}} = (x^{1/2})^{1/4} = x^{1/8} = \sqrt[8]{x}$

39. $\dfrac{(3x + 2)^{2/3}}{\sqrt[3]{3x + 2}} = \dfrac{(3x + 2)^{2/3}}{(3x + 2)^{1/3}}$
$= (3x + 2)^{2/3 - 1/3}$
$= (3x + 2)^{1/3}$
$= \sqrt[3]{3x + 2}$

41. $75^{-3/4} = 0.0392377 \approx 0.04$

43. $\sqrt{13^2 - 4(2)(7)} = 10.630146 \approx 10.63$

45. *Keystrokes:*

$\boxed{\text{Y=}}\ 3\ \boxed{\sqrt[3]{\ }}\ 2\ \boxed{\text{X,T,}\theta}\ \boxed{\text{GRAPH}}$

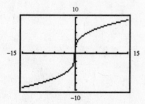

Domain $= (-\infty, \infty)$

47. *Keystrokes:*

$\boxed{\text{Y=}}\ 4\ \boxed{\text{X,T,}\theta}\ \boxed{\wedge}\ .75\ \boxed{\text{GRAPH}}$

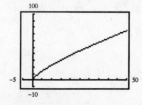

Domain $= [0, \infty)$

49. $\sqrt{360} = \sqrt{36 \cdot 10}$
$= 6\sqrt{10}$

51. $\sqrt{75u^5v^4} = \sqrt{25 \cdot 3 \cdot u^4 \cdot u \cdot v^4}$
$= 5u^2v^2\sqrt{3u}$

53. $\sqrt{0.25x^4y} = \sqrt{25 \times 10^{-2}x^4y}$
$= 5 \times 10^{-1}x^2\sqrt{y}$
$= 0.5x^2\sqrt{y}$

55. $\sqrt[4]{64a^2b^5} = \sqrt[4]{16 \cdot 4 \cdot a^2 \cdot b^4 \cdot b}$
$= 2b\sqrt[4]{4a^2b}$

57. $\sqrt[3]{48a^3b^4} = \sqrt[3]{8 \cdot 6a^3b^3b}$
$= 2ab\sqrt[3]{6b}$

59. $\sqrt{\dfrac{5}{6}} = \sqrt{\dfrac{5}{6}} \cdot \dfrac{\sqrt{6}}{\sqrt{6}} = \dfrac{\sqrt{30}}{6}$

61. $\dfrac{3}{\sqrt{12x}} = \dfrac{3}{\sqrt{4 \cdot 3x}} = \dfrac{3}{2\sqrt{3x}} \cdot \dfrac{\sqrt{3x}}{\sqrt{3x}} = \dfrac{3\sqrt{3x}}{6x} = \dfrac{\sqrt{3x}}{2x}$

63. $\dfrac{2}{\sqrt[3]{2x}} = \dfrac{2}{\sqrt[3]{2x}} \cdot \dfrac{\sqrt[3]{2^2x^2}}{\sqrt[3]{2^2x^2}} = \dfrac{2\sqrt[3]{4x^2}}{\sqrt[3]{8x^3}} = \dfrac{2\sqrt[3]{4x^2}}{2x} = \dfrac{\sqrt[3]{4x^2}}{x}$

65. $2\sqrt{7} - 5\sqrt{7} + 4\sqrt{7} = \sqrt{7}(2 - 5 + 4)$
$= \sqrt{7}$

67. $3\sqrt{40} - 10\sqrt{90} = 3\sqrt{4 \cdot 10} - 10\sqrt{9 \cdot 10}$
$= 6\sqrt{10} - 30\sqrt{10}$
$= -24\sqrt{10}$

69. $5\sqrt{x} - \sqrt[3]{x} + 9\sqrt{x} - 8\sqrt[3]{x} = 5\sqrt{x} + 9\sqrt{x} - \sqrt[3]{x} - 8\sqrt[3]{x}$
$= (5 + 9)\sqrt{x} + (-1 - 8)\sqrt[3]{x}$
$= 14\sqrt{x} - 9\sqrt[3]{x}$

71. $10\sqrt[4]{y + 3} - 3\sqrt[4]{y + 3} = (10 - 3)\sqrt[4]{y + 3}$
$= 7\sqrt[4]{y + 3}$

73. $\sqrt{25x} + \sqrt{49x} - \sqrt[3]{8x} = 5\sqrt{x} + 7\sqrt{x} - 2\sqrt[3]{x}$
$= 12\sqrt{x} - 2\sqrt[3]{x}$

75. $\sqrt{5} - \dfrac{3}{\sqrt{5}} = \sqrt{5} - \dfrac{3}{\sqrt{5}} \cdot \dfrac{\sqrt{5}}{\sqrt{5}}$
$= \sqrt{5} - \dfrac{3\sqrt{5}}{5}$
$= \sqrt{5} \cdot \dfrac{5}{5} - \dfrac{3\sqrt{5}}{5}$
$= \dfrac{5\sqrt{5}}{5} - \dfrac{3\sqrt{5}}{5}$
$= \dfrac{2\sqrt{5}}{5}$

77. $\sqrt{15} \cdot \sqrt{20} = \sqrt{15 \cdot 20}$
$= \sqrt{300}$
$= \sqrt{100 \cdot 3}$
$= 10\sqrt{3}$

79. $\sqrt{5}(\sqrt{10} + 3) = \sqrt{5}\sqrt{10} + \sqrt{5} \cdot 3$
$= \sqrt{50} + 3\sqrt{5}$
$= \sqrt{25 \cdot 2} + 3\sqrt{5}$
$= 5\sqrt{2} + 3\sqrt{5}$

81. $\sqrt{10}(\sqrt{2} + \sqrt{5}) = \sqrt{10}\sqrt{2} + \sqrt{10}\sqrt{5}$
$= \sqrt{20} + \sqrt{50}$
$= \sqrt{4 \cdot 5} + \sqrt{25 \cdot 2}$
$= 2\sqrt{5} + 5\sqrt{2}$

83. $(2\sqrt{3} + 7)(\sqrt{6} - 2) = 2\sqrt{3}\sqrt{6} - 4\sqrt{3} + 7\sqrt{6} - 14$
$= 2\sqrt{18} - 4\sqrt{3} + 7\sqrt{6} - 14$
$= 6\sqrt{2} - 4\sqrt{3} + 7\sqrt{6} - 14$

85. $(\sqrt{5} + 6)^2 = (\sqrt{5})^2 + 2(6)\sqrt{5} + 6^2 = 5 + 12\sqrt{5} + 36 = 41 + 12\sqrt{5}$

87. $(\sqrt{3} - \sqrt{x})(\sqrt{3} + \sqrt{x}) = 3 - \sqrt{3x} + \sqrt{3x} - x = 3 - x$

89. $\dfrac{3}{1-\sqrt{2}} \cdot \dfrac{1+\sqrt{2}}{1+\sqrt{2}} = \dfrac{3(1+\sqrt{2})}{1^2-(\sqrt{2})^2}$

$\phantom{\dfrac{3}{1-\sqrt{2}} \cdot \dfrac{1+\sqrt{2}}{1+\sqrt{2}}} = \dfrac{3(1+\sqrt{2})}{1-2}$

$\phantom{\dfrac{3}{1-\sqrt{2}} \cdot \dfrac{1+\sqrt{2}}{1+\sqrt{2}}} = \dfrac{3(1+\sqrt{2})}{-1}$

$\phantom{\dfrac{3}{1-\sqrt{2}} \cdot \dfrac{1+\sqrt{2}}{1+\sqrt{2}}} = -3(1+\sqrt{2})$

91. $\dfrac{3\sqrt{8}}{2\sqrt{2}+\sqrt{3}} \cdot \dfrac{2\sqrt{2}-\sqrt{3}}{2\sqrt{2}-\sqrt{3}} = \dfrac{6\sqrt{16}-3\sqrt{24}}{(2\sqrt{2})^2-(\sqrt{3})^2}$

$\phantom{\dfrac{3\sqrt{8}}{2\sqrt{2}+\sqrt{3}} \cdot \dfrac{2\sqrt{2}-\sqrt{3}}{2\sqrt{2}-\sqrt{3}}} = \dfrac{24-6\sqrt{6}}{8-3}$

$\phantom{\dfrac{3\sqrt{8}}{2\sqrt{2}+\sqrt{3}} \cdot \dfrac{2\sqrt{2}-\sqrt{3}}{2\sqrt{2}-\sqrt{3}}} = \dfrac{24-6\sqrt{6}}{5}$

93. $\dfrac{\sqrt{2}-1}{\sqrt{3}-4} = \dfrac{\sqrt{2}-1}{\sqrt{3}-4} \cdot \dfrac{\sqrt{3}+4}{\sqrt{3}+4}$

$\phantom{\dfrac{\sqrt{2}-1}{\sqrt{3}-4}} = \dfrac{\sqrt{6}+4\sqrt{2}-\sqrt{3}-4}{(\sqrt{3})^2-4^2}$

$\phantom{\dfrac{\sqrt{2}-1}{\sqrt{3}-4}} = \dfrac{\sqrt{6}+4\sqrt{2}-\sqrt{3}-4}{3-16}$

$\phantom{\dfrac{\sqrt{2}-1}{\sqrt{3}-4}} = \dfrac{\sqrt{6}+4\sqrt{2}-\sqrt{3}-4}{-13}$

$\phantom{\dfrac{\sqrt{2}-1}{\sqrt{3}-4}} = -\dfrac{\sqrt{6}+4\sqrt{2}-\sqrt{3}-4}{13}$

95. $\dfrac{\sqrt{x}+10}{\sqrt{x}-10} = \dfrac{\sqrt{x}+10}{\sqrt{x}-10} \cdot \dfrac{\sqrt{x}+10}{\sqrt{x}+10}$

$\phantom{\dfrac{\sqrt{x}+10}{\sqrt{x}-10}} = \dfrac{x+10\sqrt{x}+10\sqrt{x}+100}{(\sqrt{x})^2-10^2}$

$\phantom{\dfrac{\sqrt{x}+10}{\sqrt{x}-10}} = \dfrac{x+20\sqrt{x}+100}{x-100}$

97. *Keystrokes:*

y_1 [Y=] [√] [(] 5 [÷] [(] 2 [X,T,θ] [)] [)] [)] [ENTER]

y_2 [√] [(] 10 [X,T,θ] [)] [÷] [(] 2 [X,T,θ] [)] [GRAPH]

$\sqrt{\dfrac{5}{2x}} = \sqrt{\dfrac{5}{2x}} \cdot \dfrac{\sqrt{2x}}{\sqrt{2x}} = \dfrac{\sqrt{10x}}{2x}$

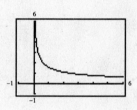

99. *Keystrokes:*

y_1 [Y=] 5 [√] [X,T,θ] [−] 2 [√] [X,T,θ] [ENTER]

y_2 3 [√] [X,T,θ] [GRAPH]

$5\sqrt{x} - 2\sqrt{x} = (5-2)\sqrt{x} = 3\sqrt{x}$

101. $\sqrt{y} = 15$ **Check:** $\sqrt{225} \overset{?}{=} 15$

$(\sqrt{y})^2 = (15)^2$ $\phantom{\text{Check: }} 15 = 15$

$y = 225$

103. $\sqrt{3x} + 9 = 0$ **Check:** $\sqrt{3 \cdot 27} + 9 \overset{?}{=} 0$

$\sqrt{3} = -9$ $\phantom{\text{Check: }} \sqrt{81} + 9 \overset{?}{=} 0$

$(\sqrt{3x})^2 = (-9)^2$ $\phantom{\text{Check: }} 9 + 9 \overset{?}{=} 0$

$3x = 81$ $\phantom{\text{Check: }} 18 \neq 0$

$x = \dfrac{81}{3}$ $\phantom{\text{Check: }}$ No real solution

$x = 27$

105. $\sqrt{2(a-7)} = 14$ **Check:** $\sqrt{2(105-7)} \overset{?}{=} 14$

$(\sqrt{2(a-7)})^2 = (14)^2$ $\phantom{\text{Check: }} \sqrt{196} \overset{?}{=} 14$

$2(a-7) = 196$ $\phantom{\text{Check: }} 14 = 14$

$2a - 14 = 196$

$2a = 210$

$a = 105$

107. $\sqrt[3]{5x-7} - 3 = -1$ **Check:** $\sqrt[3]{5(3)-7} - 3 \overset{?}{=} -1$

$\sqrt[3]{5x-7} = 2$ $\phantom{\text{Check: }} \sqrt[3]{8} - 3 \overset{?}{=} -1$

$(\sqrt[3]{5x-7})^3 = 2^3$ $\phantom{\text{Check: }} 2 - 3 \overset{?}{=} -1$

$5x - 7 = 8$ $\phantom{\text{Check: }} -1 = -1$

$5x = 15$

$x = 3$

109. $\sqrt[3]{5x + 2} - \sqrt[3]{7x - 8} = 0$

$\qquad \sqrt[3]{5x + 2} = \sqrt[3]{7x - 8}$

$\qquad \left(\sqrt[3]{5x + 2}\right)^3 = \left(\sqrt[3]{7x - 8}\right)^3$

$\qquad 5x + 2 = 7x - 8$

$\qquad 10 = 2x$

$\qquad 5 = x$

Check: $\sqrt[3]{5(5) + 2} - \sqrt[3]{7(5) - 8} \overset{?}{=} 0$

$\qquad \sqrt[3]{27} - \sqrt[3]{27} \overset{?}{=} 0$

$\qquad 0 = 0$

111. $\sqrt{2(x + 5)} = x + 5$

$\qquad \left(\sqrt{2(x + 5)}\right)^2 = (x + 5)^2$

$\qquad 2(x + 5) = x^2 + 10x + 25$

$\qquad 2x + 10 = x^2 + 10x + 25$

$\qquad 0 = x^2 + 8x + 15$

$\qquad 0 = (x + 5)(x + 3)$

$\qquad -5 = x, \quad x = -3$

Check: $\sqrt{2(-5 + 5)} \overset{?}{=} -5 + 5$

$\qquad \sqrt{0} \overset{?}{=} 0$

$\qquad 0 = 0$

$\qquad \sqrt{2(-3 + 5)} \overset{?}{=} -3 + 5$

$\qquad \sqrt{4} \overset{?}{=} 2$

$\qquad 2 = 2$

113. $\sqrt{v - 6} = 6 - v$

$\qquad \left(\sqrt{v - 6}\right)^2 = (6 - v)^2$

$\qquad v - 6 = 36 - 12v + v^2$

$\qquad 0 = v^2 - 13v + 42$

$\qquad 0 = (v - 6)(v - 7)$

$\qquad v = 6, \quad\quad v = 7$

Check: $\sqrt{6 - 6} \overset{?}{=} 6 - 6$

$\qquad 0 = 0$

$\qquad \sqrt{7 - 6} \overset{?}{=} 6 - 7$

$\qquad 1 \neq -1$

not a solution

115. $\sqrt{1 + 6x} = 2 - \sqrt{6x}$

$\qquad \left(\sqrt{1 + 6x}\right)^2 = \left(2 - \sqrt{6x}\right)^2$

$\qquad 1 + 6x = 4 - 4\sqrt{6x} + 6x$

$\qquad 1 = 4 - 4\sqrt{6x}$

$\qquad -3 = -4\sqrt{6x}$

$\qquad (3)^2 = \left(4\sqrt{6x}\right)^2$

$\qquad 9 = 16(6x)$

$\qquad \dfrac{9}{96} = x$

$\qquad \dfrac{3}{32} = x$

Check: $\sqrt{1 + 6\left(\dfrac{3}{32}\right)} \overset{?}{=} 2 - \sqrt{6\left(\dfrac{3}{32}\right)}$

$\qquad \sqrt{\dfrac{32}{32} + \dfrac{18}{32}} \overset{?}{=} 2 - \sqrt{\dfrac{18}{32}}$

$\qquad \sqrt{\dfrac{50}{32}} \overset{?}{=} 2 - \sqrt{\dfrac{9 \cdot 2}{16 \cdot 2}}$

$\qquad \sqrt{\dfrac{25 \cdot 2}{16 \cdot 2}} \overset{?}{=} 2 - \sqrt{\dfrac{9 \cdot 2}{16 \cdot 2}}$

$\qquad \sqrt{\dfrac{25}{16}} \overset{?}{=} 2 - \sqrt{\dfrac{9}{16}}$

$\qquad \dfrac{5}{4} \overset{?}{=} 2 - \dfrac{3}{4}$

$\qquad \dfrac{5}{4} \overset{?}{=} \dfrac{8}{4} - \dfrac{3}{4}$

$\qquad \dfrac{5}{4} = \dfrac{5}{4}$

117. $\sqrt{-48} = \sqrt{16 \cdot 3 \cdot -1} = 4i\sqrt{3}$

119. $10 - 3\sqrt{-27} = 10 - 3\sqrt{-1 \cdot 9 \cdot 3}$

$\qquad = 10 - 3\sqrt{-1} \cdot \sqrt{9} \cdot \sqrt{3}$

$\qquad = 10 - 9i\sqrt{3}$

121. $\frac{3}{4} - 5\sqrt{-\frac{3}{25}} = \frac{3}{4} - 5\sqrt{\frac{3}{25} \cdot -1}$

$= \frac{3}{4} - \frac{5}{5}i\sqrt{3}$

$= \frac{3}{4} - i\sqrt{3}$

123. $\sqrt{-81} + \sqrt{-36} = 9i + 6i$

$= 15i$

125. $\sqrt{-121} - \sqrt{-84} = 11i - 2i\sqrt{21}$

127. $\sqrt{-5}\sqrt{-5} = i\sqrt{5} \cdot i\sqrt{5} = i^2 \cdot 5 = -5$

129. $\sqrt{-10}\left(\sqrt{-4} - \sqrt{-7}\right) = i\sqrt{10}\left(2i - i\sqrt{7}\right)$

$= 2i^2\sqrt{10} - i^2\sqrt{70}$

$= -2\sqrt{10} + \sqrt{70}$

131. $4x - \sqrt{-36} = 8 - 2yi$

$4x - 6i = 8 - 2yi$

$4x = 8 \qquad -6 = -2y$

$x = 2 \qquad 3 = y$

133. $24 + \sqrt{-5y} = 6x + 25i$

$24 + i\sqrt{5y} = 6x + 25i$

$24 = 6x \qquad \sqrt{5y} = 25$

$4 = x \qquad 5y = 625$

$y = 125$

135. $(-4 + 5i) - (-12 + 8i) = (-4 + 12) + (5 - 8)i$

$= 8 - 3i$

137. $(3 - 8i) + (5 + 12i) = 3 - 8i + 5 + 12i$

$= (3 + 5) + (-8 + 12)i$

$= 8 + 4i$

139. $(4 - 3i)(4 + 3i) = 4^2 - (3i)^2$

$= 16 + 9$

$= 25$

141. $(6 - 5i)^2 = 6^2 - 2(6)(5i) + (5i)^2$

$= 36 - 60i - 25$

$= 11 - 60i$

143. $\frac{7}{3i} = \frac{7}{3i} \cdot \frac{-i}{-i} = \frac{-7i}{-3i^2} = \frac{-7i}{3}$

145. $\frac{4i}{2 - 8i} = \frac{4i}{2 - 8i} \cdot \frac{2 + 8i}{2 + 8i}$

$= \frac{8i + 32i^2}{2^2 - (8i)^2}$

$= \frac{8i - 32}{4 + 64}$

$= \frac{8i - 32}{68}$

$= \frac{-8 + 2i}{17}$

$= \frac{-8}{17} + \frac{2}{17}i$

147. $\frac{3 - 5i}{6 + i} = \frac{3 - 5i}{6 + i} \cdot \frac{6 - i}{6 - i}$

$= \frac{18 - 3i - 30i + 5i^2}{6^2 - i^2}$

$= \frac{18 - 33i - 5}{36 + 1}$

$= \frac{13 - 33i}{37}$

$= \frac{13}{37} - \frac{33}{37}i$

149. $c = \sqrt{3^2 + 3^2} = \sqrt{9 + 9} = \sqrt{18}$

Equation: $P = 2(8) + 2\left(2\frac{1}{2}\right) + 4\left(\sqrt{18}\right)$

$\qquad = 16 + 5 + 12\sqrt{2}$

$\qquad = 21 + 12\sqrt{2}$ inches

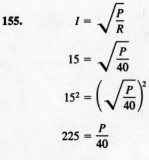

151.
$$1.3 = 2\pi\sqrt{\frac{L}{32}}$$
$$\frac{1.3}{2\pi} = \sqrt{\frac{L}{32}}$$
$$\left(\frac{1.3}{2\pi}\right)^2 = \left(\sqrt{\frac{L}{32}}\right)^2$$
$$\frac{1.69}{4\pi^2} = \frac{L}{32}$$
$$\frac{1.69}{4\pi^2}(32) = L$$
$$1.3698624 = L \approx 1.37 \text{ feet}$$

153.
$$I = \sqrt{\frac{P}{R}}$$
$$5 = \sqrt{\frac{P}{20}}$$
$$5^2 = \left(\sqrt{\frac{P}{20}}\right)^2$$
$$25 = \frac{P}{20}$$
$$500 \text{ watts} = P$$

155.
$$I = \sqrt{\frac{P}{R}}$$
$$15 = \sqrt{\frac{P}{40}}$$
$$15^2 = \left(\sqrt{\frac{P}{40}}\right)^2$$
$$225 = \frac{P}{40}$$
$$9000 \text{ watts} = P$$

157.
$$v = \sqrt{2gh}$$
$$25 = \sqrt{2(32)h}$$
$$25^2 = \left(\sqrt{2(32)h}\right)^2$$
$$625 = 2(32)h$$
$$9.77 \text{ feet} \approx h$$

Chapter Test for Chapter 9

1. (a) $16^{3/2} = \left(\sqrt{16}\right)^3$

$\qquad = 4^3$

$\qquad = 64$

(b) $\sqrt{5}\sqrt{20} = \sqrt{5 \cdot 20}$

$\qquad = \sqrt{100}$

$\qquad = 10$

2. (a) $27^{-2/3} = \dfrac{1}{27^{2/3}}$

$\qquad = \dfrac{1}{9}$

(b) $\sqrt{2}\sqrt{18} = \sqrt{2 \cdot 18}$

$\qquad = \sqrt{36}$

$\qquad = 6$

3. (a) $\left(\dfrac{x^{1/2}}{x^{1/3}}\right)^2 = \dfrac{x}{x^{2/3}}$

$\qquad = x^{1 - 2/3} = x^{1/3}$

(b) $5^{1/4} \cdot 5^{7/4} = 5^{1/4 + 7/4}$

$\qquad = 5^{8/4} = 5^2 = 25$

4. (a) $\sqrt{\dfrac{32}{9}} = \sqrt{\dfrac{16 \cdot 2}{9}} = \dfrac{4}{3}\sqrt{2}$

(b) $\sqrt[3]{24} = \sqrt[3]{8 \cdot 3} = 2\sqrt[3]{3}$

5. (a) $\sqrt{24x^3} = \sqrt{4 \cdot 6 \cdot x^2 \cdot x}$

$\qquad = 2x\sqrt{6x}$

(b) $\sqrt[4]{16x^5y^8} = \sqrt[4]{16x^4xy^8}$

$\qquad = 2xy^2\sqrt[4]{x}$

6. $\dfrac{3}{\sqrt{6}} = \dfrac{3}{\sqrt{6}} \cdot \dfrac{\sqrt{6}}{\sqrt{6}} = \dfrac{3\sqrt{6}}{6} = \dfrac{\sqrt{6}}{2}$

Multiply the numerator and denominator of a fraction by a factor such that no radical contains a fraction and no denominator of a fraction contains a radical.

7. $5\sqrt{3x} - 3\sqrt{75x} = 5\sqrt{3x} - 3\sqrt{25 \cdot 3x}$

$\qquad = 5\sqrt{3x} - 15\sqrt{3x}$

$\qquad = -10\sqrt{3x}$

8. $\sqrt{5}\left(\sqrt{15x} + 3\right) = \sqrt{75x} + 3\sqrt{5}$

$\qquad = \sqrt{25 \cdot 3x} + 3\sqrt{5}$

$\qquad = 5\sqrt{3x} + 3\sqrt{5}$

9. $\left(4 - \sqrt{2x}\right)^2 = 16 - 8\sqrt{2x} + 2x$

10. $7\sqrt{27} + 14y\sqrt{12} = 7\sqrt{9 \cdot 3} + 14y\sqrt{4 \cdot 3}$

$\qquad = 21\sqrt{3} + 28y\sqrt{3}$

$\qquad = 7\sqrt{3}(3 + 4y)$

11. $\sqrt{3y} - 6 = 3$

$\sqrt{3y} = 9$

$\left(\sqrt{3y}\right)^2 = 9^2$

$3y = 81$

$y = 27$

Check: $\sqrt{3(27)} - 6 \overset{?}{=} 3$

$\sqrt{81} - 6 \overset{?}{=} 3$

$9 - 6 \overset{?}{=} 3$

$3 = 3$

12. $\sqrt{x^2 - 1} = x - 2$

$\left(\sqrt{x^2 - 1}\right)^2 = (x - 2)^2$

$x^2 - 1 = x^2 - 4x + 4$

$4x = 5$

$x = \tfrac{5}{4}$

No solution

Check: $\sqrt{\left(\tfrac{5}{4}\right)^2 - 1} \overset{?}{=} \tfrac{5}{4} - 2$

$\sqrt{\tfrac{25}{16} - \tfrac{16}{16}} \overset{?}{=} \tfrac{5}{4} - \tfrac{8}{4}$

$\sqrt{\tfrac{9}{16}} \overset{?}{=} -\tfrac{3}{4}$

$\tfrac{3}{4} \neq -\tfrac{3}{4}$

13. $\sqrt{x} - x + 6 = 0$

$\left(\sqrt{x}\right)^2 = (x - 6)^2$

$x = x^2 - 12x + 36$

$0 = x^2 - 13x + 36$

$0 = (x - 9)(x - 4)$

$0 = x - 9 \qquad 0 = x - 4$

$9 = x \qquad\quad 4 = x$

$\qquad\qquad$ Not a solution

Check: $\sqrt{9} - 9 + 6 \overset{?}{=} 0$

$3 - 9 + 6 \overset{?}{=} 0$

$0 = 0$

$\sqrt{4} - 4 + 6 \overset{?}{=} 0$

$2 - 4 + 6 \overset{?}{=} 0$

$4 \neq 0$

14. $3x + \sqrt{-4y} = 12 + 40i$

$3x + 2\sqrt{y}i = 12 + 40i$

$3x = 12 \qquad 2\sqrt{y} = 40$

$x = 4 \qquad\quad \sqrt{y} = 20$

$\qquad\qquad\quad y = 400$

15. $27 - \sqrt{-16y} = 9x - 4i$

$27 - 4\sqrt{y}i = 9x - 4i$

$27 = 9x \qquad -4\sqrt{y} = -4$

$3 = x \qquad\quad \sqrt{y} = 1$

$\qquad\qquad\quad y = 1$

16. $(2 + 3i) - \sqrt{-25} = 2 + 3i - 5i = 2 - 2i$

17. $(2 - 3i)^2 = (2 - 3i)(2 - 3i)$

$= 4 - 6i - 6i + 9i^2$

$= 4 - 12i - 9$

$= -5 - 12i$

18. $\sqrt{-16}(1 + \sqrt{4}) = 4i(1 + 2i)$

$= 4i + 8i^2$

$= -8 + 4i$

19. $(3 - 2i)(1 + 5i) = 3 + 13i - 10i^2$

$= 3 + 13i + 10$

$= 13 + 13i$

20. $\dfrac{5 - 2i}{i} = \dfrac{5 - 2i}{i} \cdot \dfrac{-i}{-i} = \dfrac{-5i + 2i^2}{-i^2} = -2 - 5i$

21.

$v = \sqrt{2gh}$

$80 = \sqrt{2(32)h}$

$80 = \sqrt{64h}$

$80^2 = \left(\sqrt{64h}\right)^2$

$6400 = 64h$

$100 \text{ feet} = h$

Cumulative Test for Chapters 7–9

1. $x + y = 1$

$2x - y = -1$

Solution c

2. $4x + 3y = 16$

$-8x - 6y = -32$

Solution e

3. $-3x + y = 10$

$x + 3y = 10$

Solution d

4. $5x - 5y = 10$

$-x + y = 5$

Solution f

5. $4x - 2y = 2$

$2x + 2y = 10$

Solution a

6. $-x + y = 0$

$3x - 2y = -1$

Solution b

7. $x - y = 1$

$-y = -x + 1$

$y = x - 1$

Keystrokes:

y_1 [Y=] [X,T,θ] [−] 1 [ENTER]

y_2 [(−)] 2 [X,T,θ] [+] 5 [GRAPH]

Solution = (2, 1)

$2x + y = 5$

$y = -2x + 5$

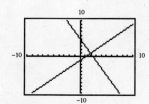

8. $x = 5y + 13$

$4(5y + 13) + 2y = 8$

$20y + 52 + 2y = 8$

$22y = -44$

$y = -2$

$x = 5(-2) + 13$

$= 3$

$(3, -2)$

9. $\quad 4x - 3y = 8 \implies \quad 4x - 3y = 8$

$\qquad -2x + y = -6 \implies \underline{-6x + 3y = -18}$

$\qquad\qquad\qquad\qquad\qquad -2x = -10 \qquad 4(5) - 3y = 8$

$\qquad\qquad\qquad\qquad\qquad\quad x = 5 \qquad\qquad -3y = -12$

$\qquad\qquad\qquad\qquad\qquad\qquad\qquad\qquad\qquad\qquad\qquad y = 4$

$(5, 4)$

10. $D = \begin{vmatrix} 2 & -1 \\ 3 & 1 \end{vmatrix} = (2) - (-3) = 5$

$x = \dfrac{\begin{vmatrix} 4 & -1 \\ -5 & 1 \end{vmatrix}}{5} = -\dfrac{1}{5}$

$y = \dfrac{\begin{vmatrix} 2 & 4 \\ 3 & -5 \end{vmatrix}}{5} = -\dfrac{22}{5}$

$\left(-\dfrac{1}{5}, -\dfrac{22}{5}\right)$

11. $\dfrac{x^2 + 8x + 16}{18x^2} \cdot \dfrac{2x^4 + 4x^3}{x^2 - 16} = \dfrac{(x + 4)^2}{18x^2} \cdot \dfrac{2x^3(x + 2)}{(x - 4)(x + 4)} = \dfrac{x(x + 4)(x + 2)}{9(x - 4)}$

12. $\dfrac{2}{x} - \dfrac{x}{x^3 + 3x^2} + \dfrac{1}{x + 3} = \dfrac{2}{x} - \dfrac{x}{x^2(x + 3)} + \dfrac{1}{x + 3}$

$\qquad\qquad\qquad\qquad\qquad = \dfrac{2}{x} - \dfrac{1}{x(x + 3)} + \dfrac{1}{x + 3}$

$\qquad\qquad\qquad\qquad\qquad = \dfrac{2}{x}\left(\dfrac{x + 3}{x + 3}\right) - \dfrac{1}{x(x + 3)}\left(\dfrac{1}{1}\right) + \dfrac{1}{x + 3}\left(\dfrac{x}{x}\right)$

$\qquad\qquad\qquad\qquad\qquad = \dfrac{2x + 6}{x(x + 3)} - \dfrac{1}{x(x + 3)} + \dfrac{x}{x(x + 3)}$

$\qquad\qquad\qquad\qquad\qquad = \dfrac{2x + 6 - 1 + x}{x(x + 3)}$

$\qquad\qquad\qquad\qquad\qquad = \dfrac{3x + 5}{x(x + 3)}$

13. $\dfrac{\left(\dfrac{x}{y} - \dfrac{y}{x}\right)}{\left(\dfrac{x - y}{xy}\right)} = \dfrac{\left(\dfrac{x}{y} - \dfrac{y}{x}\right)}{\left(\dfrac{x - y}{xy}\right)} \cdot \dfrac{xy}{xy}$

$\qquad\qquad = \dfrac{x^2 - y^2}{x - y}$

$\qquad\qquad = \dfrac{(x - y)(x + y)}{x - y}$

$\qquad\qquad = x + y$

14. $\sqrt{-2}\left(\sqrt{-8} + 3\right) = i\sqrt{2}\left(2i\sqrt{2} + 3\right)$

$\qquad\qquad\qquad\qquad\quad = 2i^2 \cdot 2 + 3i\sqrt{2}$

$\qquad\qquad\qquad\qquad\quad = -4 + 3i\sqrt{2}$

15. $(3 - 4i)^2 = 3^2 + 2(3)(-4i) + (4i)^2$

$\qquad\qquad\quad = 9 - 24i + 16i^2$

$\qquad\qquad\quad = 9 - 16 - 24i$

$\qquad\qquad\quad = -7 - 24i$

16. $\left(\dfrac{t^{1/2}}{t^{1/4}}\right)^2 = \dfrac{t}{t^{1/2}} = t^{1 - 1/2}$

$\qquad\qquad\quad = t^{1/2}$

17. $10\sqrt{20x} + 3\sqrt{125x} = 10\sqrt{4 \cdot 5x} + 3\sqrt{25 \cdot 5x}$

$$= 20\sqrt{5x} + 15\sqrt{5x}$$

$$= 35\sqrt{5x}$$

18. $\left(\sqrt{2x} - 3\right)^2 = 2x - 6\sqrt{2x} + 9$

19. $\dfrac{6}{\sqrt{10} - 2} = \dfrac{6}{\sqrt{10} - 2} \cdot \dfrac{\sqrt{10} + 2}{\sqrt{10} + 2}$

$$= \dfrac{6\left(\sqrt{10} + 2\right)}{10 - 4}$$

$$= \dfrac{6\left(\sqrt{10} + 2\right)}{6}$$

$$= \sqrt{10} + 2$$

20. $\dfrac{1 - 2i}{4 + i} = \dfrac{1 - 2i}{4 + i} \cdot \dfrac{4 - i}{4 - i}$

$$= \dfrac{4 - i - 8i + 2i^2}{16 - i^2} = \dfrac{4 - 9i - 2}{16 + 1}$$

$$= \dfrac{2 - 9i}{17} = \dfrac{2}{17} - \dfrac{9}{17}i$$

21. $\dfrac{1}{x} + \dfrac{4}{10 - x} = 1$

$$x(10 - x)\left(\dfrac{1}{x} + \dfrac{4}{10 - x}\right) = (1)x(10 - x)$$

$$10 - x + 4x = 10x - x^2$$

$$x^2 - 7x + 10 = 0$$

$$(x - 5)(x - 2) = 0$$

$$x = 5 \qquad x = 2$$

Check: $\dfrac{1}{5} + \dfrac{4}{10 - 5} \overset{?}{=} 1$

$$\dfrac{1}{5} + \dfrac{4}{5} \overset{?}{=} 1$$

$$\dfrac{5}{5} \overset{?}{=} 1$$

$$1 = 1$$

$\dfrac{1}{2} + \dfrac{4}{10 - 2} \overset{?}{=} 1$

$$\dfrac{1}{2} + \dfrac{4}{8} \overset{?}{=} 1$$

$$\dfrac{1}{2} + \dfrac{1}{2} \overset{?}{=} 1$$

$$1 = 1$$

22. $\dfrac{x - 3}{x} + 1 = \dfrac{x - 4}{x - 6}$

$$x(x - 6)\left(\dfrac{x - 3}{x} + 1\right) = \left(\dfrac{x - 4}{x - 6}\right)x(x - 6)$$

$$(x - 6)(x - 3) + x(x - 6) = x(x - 4)$$

$$x^2 - 9x + 18 + x^2 - 6x = x^2 - 4x$$

$$x^2 - 11x + 18 = 0$$

$$(x - 9)(x - 2) = 0$$

$$x = 9 \qquad x = 2$$

Check: $\dfrac{9 - 3}{9} + 1 \overset{?}{=} \dfrac{9 - 4}{9 - 6}$

$$\dfrac{6}{9} + 1 \overset{?}{=} \dfrac{5}{3}$$

$$\dfrac{2}{3} + \dfrac{3}{3} \overset{?}{=} \dfrac{5}{3}$$

$$\dfrac{5}{3} = \dfrac{5}{3}$$

$\dfrac{2 - 3}{2} + 1 \overset{?}{=} \dfrac{2 - 4}{2 - 6}$

$$\dfrac{-1}{2} + \dfrac{2}{2} \overset{?}{=} \dfrac{-2}{-4}$$

$$\dfrac{1}{2} = \dfrac{1}{2}$$

23. $\sqrt{x} - x + 12 = 0$

$$\sqrt{x} = x - 12$$

$$\left(\sqrt{x}\right)^2 = (x - 12)^2$$

$$x = x^2 - 24x + 144$$

$$0 = x^2 - 25x + 144$$

$$0 = (x - 16)(x - 9)$$

$$x = 16 \qquad x = 9$$

Not a solution

Check: $\sqrt{16} - 16 + 12 \overset{?}{=} 0$

$$4 - 16 + 12 \overset{?}{=} 0$$

$$0 = 0$$

$\sqrt{9} - 9 + 12 \overset{?}{=} 0$

$$3 - 9 + 12 \overset{?}{=} 0$$

$$6 \neq 0$$

24. $\sqrt{5-x} + 10 = 11$ **Check:** $\sqrt{5-4} + 10 \overset{?}{=} 11$

$\sqrt{5-x} = 1$ $\sqrt{1} + 10 \overset{?}{=} 11$

$\left(\sqrt{5-x}\right)^2 = 1^2$ $11 = 11$

$5 - x = 1$

$-x = -4$

$x = 4$

25. $f(x) = \dfrac{2}{x-2}$ **26.** $f(x) = \dfrac{2x}{x-2}$ **27.** $f(x) = \dfrac{2x^2}{x-2}$ **28.** $f(x) = \dfrac{2}{x^2-2}$

Solution b   Solution d   Solution a   Solution c

29. $d = k \cdot s^2$ $d = \dfrac{2}{25} \cdot 40^2$ **30.** $N = \dfrac{k}{t+1}$ $N = \dfrac{300}{5+1}$

$50 = k \cdot 25^2$ $N = 50$ prey

$50 = k \cdot 625$ $d = \dfrac{2}{25} \cdot 1600$ $300 = \dfrac{k}{0+1}$

$\dfrac{50}{625} = k$ $d = 128$ feet $300 = k$

$\dfrac{2}{25} = k$

31. $\pi r_2{}^2(5) = \pi r_1{}^2(3)$ **32.** $c = \sqrt{4^2 + 4^2}$

$r_2{}^2 = \dfrac{3}{5} r_1{}^2$ $= \sqrt{32}$

$r_2 = \sqrt{\dfrac{3r_1{}^2}{5}}$ $P = 4(4) + 4\left(\sqrt{32}\right)$

$r_2 = \dfrac{\sqrt{15 r_1{}^2}}{5}$ $P = 16 + 16\sqrt{2}$ inches

$r_2 = r_1 \dfrac{\sqrt{15}}{5}$ $P \approx 38.6$ inches

C H A P T E R   1 0
Quadratic Equations and Inequalities

CHAPTER 10
Quadratic Equations and Inequalities

Section 10.1   Factoring and Extracting Square Roots

Solutions to Odd-Numbered Exercises

1. $x^2 - 12x + 35 = 0$

$(x - 5)(x - 7) = 0$

$x = 5 \quad x = 7$

3. $x^2 + x - 72 = 0$

$(x + 9)(x - 8) = 0$

$x = -9 \quad x = 8$

5. $x^2 + 4x = 45$

$x^2 + 4x - 45 = 0$

$(x + 9)(x - 5) = 0$

$x = -9 \quad x = 5$

7. $x^2 - 12x + 36 = 0$

$(x - 6)(x - 6) = 0$

$x - 6 = 0 \quad x - 6 = 0$

$x = 6 \quad x = 6$

9. $9x^2 + 24x + 16 = 0$

$(3x + 4)(3x + 4) = 0$

$3x + 4 = 0 \quad 3x + 4 = 0$

$3x = -4 \quad 3x = -4$

$x = -\frac{4}{3} \quad x = -\frac{4}{3}$

11. $4x^2 - 12x = 0$

$4x(x - 3) = 0$

$4x = 0 \quad x - 3 = 0$

$x = 0 \quad x = 3$

13. $u(u - 9) - 12(u - 9) = 0$

$(u - 9)(u - 12) = 0$

$u - 9 = 0 \quad u - 12 = 0$

$u = 9 \quad u = 12$

15. $3x(x - 6) - 5(x - 6) = 0$

$(x - 6)(3x - 5) = 0$

$x - 6 = 0 \quad 3x - 5 = 0$

$x = 6 \quad x = \frac{5}{3}$

17. $(y - 4)(y - 3) = 6$

$y^2 - 7y + 12 - 6 = 0$

$y^2 - 7y + 6 = 0$

$(y - 6)(y - 1) = 0$

$y - 6 = 0 \quad y - 1 = 0$

$y = 6 \quad y = 1$

19. $2x(3x + 2) = 5 - 6x^2$

$6x^2 + 4x = 5 - 6x^2$

$12x^2 + 4x - 5 = 0$

$(6x + 5)(2x - 1) = 0$

$6x + 5 = 0 \quad 2x - 1 = 0$

$x = -\frac{5}{6} \quad x = \frac{1}{2}$

21. $x^2 = 64$

$x = \pm\sqrt{64}$

$x = \pm 8$

23. $6x^2 = 54$

$x^2 = 9$

$x = \pm\sqrt{9}$

$x = \pm 3$

25. $25x^2 = 16$

$x^2 = \frac{16}{25}$

$x = \pm\sqrt{\frac{16}{25}}$

$x = \pm\frac{4}{5}$

27. $\frac{1}{2}y^2 = 32$

$y^2 = 64$

$y = \pm\sqrt{64}$

$y = \pm 8$

29. $4x^2 - 25 = 0$

$4x^2 = 25$

$x^2 = \frac{25}{4}$

$x = \pm\sqrt{\frac{25}{4}}$

$x = \pm\frac{5}{2}$

31. $4u^2 - 225 = 0$

$u^2 = \frac{225}{4}$

$u = \pm\sqrt{\frac{225}{4}}$

$u = \pm\frac{15}{2}$

33. $(x + 4)^2 = 169$

$x + 4 = \pm\sqrt{169}$

$x = -4 \pm 13$

$x = 9, -17$

35. $(x - 3)^2 = 0.25$

$x - 3 = \pm\sqrt{0.25}$

$x = 3 \pm 0.5$

$x = 3.5, 2.5$

37. $(x - 2)^2 = 7$

$$x - 2 = \pm\sqrt{7}$$
$$x = 2 \pm \sqrt{7}$$

39. $(2x + 1)^2 = 50$

$$2x + 1 = \pm\sqrt{50}$$
$$2x = -1 \pm 5\sqrt{2}$$
$$x = \frac{-1 \pm 5\sqrt{2}}{2}$$

41. $(4x - 3)^2 - 98 = 0$

$$(4x - 3)^2 = 98$$
$$4x - 3 = \pm\sqrt{98}$$
$$4x = 3 \pm 7\sqrt{2}$$
$$x = \frac{3 \pm 7\sqrt{2}}{4}$$

43. $z^2 = -36$

$$z = \pm\sqrt{-36}$$
$$z = \pm 6i$$

45. $x^2 + 4 = 0$

$$x^2 = -4$$
$$x = \pm\sqrt{-4}$$
$$x = \pm 2i$$

47. $9u^2 + 17 = 0$

$$9u^2 = -17$$
$$u = \pm\sqrt{-\frac{17}{9}}$$
$$= \pm i\frac{\sqrt{17}}{3}$$

49. $(t - 3)^2 = -25$

$$t - 3 = \pm\sqrt{-25}$$
$$t = 3 \pm 5i$$

51. $(3z + 4)^2 + 144 = 0$

$$(3z + 4)^2 = -144$$
$$3z + 4 = \pm\sqrt{-144}$$
$$3z + 4 = \pm 12i$$
$$3z = -4 \pm 12i$$
$$z = \frac{-4 \pm 12i}{3}$$
$$z = -\frac{4}{3} \pm 4i$$

53. $(2x + 3)^2 = -54$

$$2x + 3 = \pm\sqrt{-54}$$
$$2x = -3 \pm 3i\sqrt{6}$$
$$x = -\frac{3}{2} \pm \frac{3i\sqrt{6}}{2}$$

55. $9(x + 6)^2 = -121$

$$(x + 6)^2 = \frac{-121}{9}$$
$$x + 6 = \pm\sqrt{\frac{-121}{9}}$$
$$x = -6 \pm \frac{11}{3}i$$

57. $(x - 1)^2 = -27$

$$x - 1 = \pm\sqrt{-27}$$
$$x = 1 \pm 3i\sqrt{3}$$

59. $(x + 1)^2 + 0.04 = 0$

$$(x + 1)^2 = -0.04$$
$$x + 1 = \pm\sqrt{-0.04}$$
$$x = -1 \pm 0.2i$$

61. $\left(c - \dfrac{2}{3}\right)^2 + \dfrac{1}{9} = 0$

$$\left(c - \frac{2}{3}\right)^2 = -\frac{1}{9}$$
$$c - \frac{2}{3} = \pm\sqrt{-\frac{1}{9}}$$
$$c = \frac{2}{3} \pm \frac{1}{3}i$$

63. $\left(x + \dfrac{7}{3}\right)^2 = -\dfrac{38}{9}$

$$x + \frac{7}{3} = \pm\sqrt{-\frac{38}{9}}$$
$$x = -\frac{7}{3} \pm \frac{i}{3}\sqrt{38}$$

65. $2x^2 - 5x = 0$

$$x(2x - 5) = 0$$
$$x = 0 \qquad 2x - 5 = 0$$
$$x = \frac{5}{2}$$

67. $2x^2 + 5x - 12 = 0$

$$(2x - 3)(x + 4) = 0$$
$$x = \frac{3}{2} \qquad x = -4$$

69. $x^2 - 900 = 0$

$$x^2 = 900$$
$$x = \pm 30$$

71. $x^2 + 900 = 0$

$$x^2 = -900$$
$$x = \pm\sqrt{-900}$$
$$x = \pm 30i$$

73. $\frac{2}{3}x^2 = 6$

$\frac{3}{2} \cdot \frac{2}{3}x^2 = 6 \cdot \frac{3}{2}$

$x^2 = 9$

$x = \pm 3$

75. $(x - 5)^2 - 100 = 0$

$(x - 5)^2 = 100$

$x - 5 = \pm 10$

$x = 15, -5$

77. $(x - 5)^2 + 100 = 0$

let $u = (x - 5)$

$u^2 + 100 = 0$

$(u + 10i)(u - 10i) = 0$

$u + 10i = 0 \qquad u - 10i = 0$

$u = -10i \qquad u = 10i$

$x - 5 = -10i \qquad x - 5 = 10i$

$x = 5 - 10i \qquad x = 5 + 10i$

79. $(x + 2)^2 + 18 = 0$

$(x + 2)^2 = -18$

$x + 2 = \pm\sqrt{-18}$

$x = -2 \pm 3i\sqrt{2}$

81. *Keystrokes:*

$\boxed{Y=}\ \boxed{X,T,\theta}\ \boxed{x^2}\ \boxed{-}\ 9\ \boxed{GRAPH}$

x-intercepts are -3 and 3.

$0 = x^2 - 9$

$= (x - 3)(x + 3)$

$x - 3 = 0 \qquad x + 3 = 0$

$x = 3 \qquad x = -3$

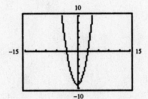

83. *Keystrokes:*

$\boxed{Y=}\ \boxed{X,T,\theta}\ \boxed{x^2}\ \boxed{-}\ 2\ \boxed{X,T,\theta}\ \boxed{-}\ 15\ \boxed{GRAPH}$

x-intercepts are -3 and 5.

$0 = x^2 - 2x - 15$

$0 = (x - 5)(x + 3)$

$x - 5 = 0 \qquad x + 3 = 0$

$x = 5 \qquad x = -3$

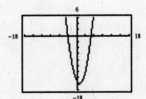

85. *Keystrokes:*

$\boxed{Y=}\ 4\ \boxed{-}\ \boxed{(}\ \boxed{X,T,\theta}\ \boxed{-}\ 3\ \boxed{)}\ \boxed{x^2}\ \boxed{GRAPH}$

x-intercepts are 1 and 5.

$0 = 4 - (x - 3)^2$

$(x - 3)^2 = 4$

$x - 3 = \pm 2$

$x = 5, 1$

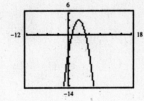

87. *Keystrokes:*

$\boxed{\text{Y=}}$ 2 $\boxed{\text{X,T,}\theta}$ $\boxed{x^2}$ $\boxed{-}$ $\boxed{\text{X,T,}\theta}$ $\boxed{-}$ 6 $\boxed{\text{GRAPH}}$

x-intercepts are $-\frac{3}{2}$ and 2.

$0 = 2x^2 - x - 6$

$0 = (2x + 3)(x - 2)$

$x = -\frac{3}{2} \qquad x = 2$

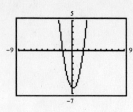

89. *Keystrokes:*

$\boxed{\text{Y=}}$ 3 $\boxed{\text{X,T,}\theta}$ $\boxed{x^2}$ $\boxed{-}$ 8 $\boxed{\text{X,T,}\theta}$ $\boxed{-}$ 16 $\boxed{\text{GRAPH}}$

x-intercepts are $-\frac{4}{3}$ and 4.

$0 = 3x^2 - 8x - 16$

$0 = (3x + 4)(x - 14)$

$x = -\frac{4}{3} \qquad x = 4$

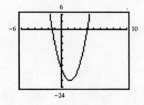

91. *Keystrokes:*

$\boxed{\text{Y=}}$ $\boxed{\text{X,T,}\theta}$ $\boxed{x^2}$ $\boxed{+}$ 7 $\boxed{\text{GRAPH}}$

$0 = x^2 + 7$

$-7 = x^2$

$\pm\sqrt{-7} = x$

$\pm i\sqrt{7} = x$

The equation has complex roots.

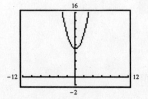

93. *Keystrokes:*

$\boxed{\text{Y=}}$ $\boxed{(}$ $\boxed{\text{X,T,}\theta}$ $\boxed{-}$ 1 $\boxed{)}$ $\boxed{x^2}$ $\boxed{+}$ 1 $\boxed{\text{GRAPH}}$

$0 = (x - 1)^2 + 1$

$-1 = (x - 1)^2$

$\pm i = x - 1$

$1 \pm i = x$

The equation has complex roots.

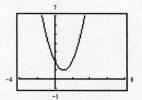

95. *Keystrokes:*

$\boxed{\text{Y=}}$ $\boxed{(}$ $\boxed{\text{X,T,}\theta}$ $\boxed{+}$ 3 $\boxed{)}$ $\boxed{x^2}$ $\boxed{+}$ 5 $\boxed{\text{GRAPH}}$

$0 = (x + 3)^2 + 5$

$-5 = (x + 3)^2$

$\pm\sqrt{-5} = x + 3$

$\pm\sqrt{5}i = x + 3$

$-3 \pm \sqrt{5}i = x$

The equation has complex roots.

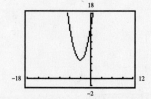

97. $x^2 + y^2 = 4$

$y^2 = 4 - x^2$

$y = \pm\sqrt{4 - x^2}$

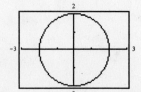

99. $x^2 + 4y^2 = 4$

$$4y^2 = 4 - x^2$$

$$y^2 = \frac{4 - x^2}{4}$$

$$y = \pm\sqrt{\frac{4 - x^2}{4}} = \pm\frac{\sqrt{4 - x^2}}{2}$$

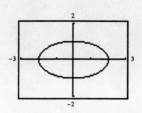

101. $\quad\quad\quad x^4 - 5x^2 + 4 = 0$

$$(x^2 - 4)(x^2 - 1) = 0$$

$$(x - 2)(x + 2)(x - 1)(x + 1) = 0$$

$$x - 2 = 0 \quad\quad x + 2 = 0 \quad\quad x - 1 = 0 \quad\quad x + 1 = 0$$

$$x = 2 \quad\quad\quad x = -2 \quad\quad\quad x = 1 \quad\quad\quad x = -1$$

103. $\quad x^4 - 5x^2 + 6 = 0$

$$(x^2 - 3)(x^2 - 2) = 0$$

$$x^2 - 3 = 0 \quad\quad x^2 - 2 = 0$$

$$x^2 = 3 \quad\quad\quad x^2 = 2$$

$$x = \pm\sqrt{3} \quad\quad x = \pm\sqrt{2}$$

105. $\quad x^4 - 3x^2 - 4 = 0$

$$(x^2 - 4)(x^2 + 1) = 0$$

$$x^2 - 4 = 0 \quad\quad x^2 + 1 = 0$$

$$x^2 = 4 \quad\quad\quad x^2 = -1$$

$$x = \pm 2 \quad\quad\quad x = \pm i$$

107. $\quad (x^2 - 4)^2 + 2(x^2 - 4) - 3 = 0$

$$[(x^2 - 4) + 3][(x^2 - 4) - 1] = 0$$

$$(x^2 - 1)(x^2 - 5) = 0$$

$$x^2 - 1 = 0 \quad\quad x^2 - 5 = 0$$

$$x^2 = 1 \quad\quad\quad x^2 = 5$$

$$x = \pm 1 \quad\quad\quad x = \pm\sqrt{5}$$

109. $\quad\quad x - 7\sqrt{x} + 10 = 0$

$$\text{let } u = \sqrt{x}$$

$$\left(\sqrt{x}\right)^2 - 7\left(\sqrt{x}\right) + 10 = 0$$

$$u^2 - 7u + 10 = 0$$

$$(u - 5)(u - 2) = 0$$

$$u = 5 \quad\quad u = 2$$

$$\sqrt{x} = 5 \quad\quad \sqrt{x} = 2$$

$$x = 25 \quad\quad x = 4$$

Check: $25 - 7\sqrt{25} + 10 \overset{?}{=} 0$

$$25 - 35 + 10 \overset{?}{=} 0$$

$$0 = 0$$

Check: $\quad 4 - 7\sqrt{4} + 10 \overset{?}{=} 0$

$$4 - 14 + 10 \overset{?}{=} 0$$

$$0 = 0$$

111. $\quad x^{2/3} - x^{1/3} - 6 = 0$

$$(x^{1/3} - 3)(x^{1/3} + 2) = 0$$

$$x^{1/3} - 3 = 0 \quad\quad x^{1/3} + 2 = 0$$

$$x^{1/3} = 3 \quad\quad\quad x^{1/3} = -2$$

$$x = 27 \quad\quad\quad x = -8$$

113. $2x^{2/3} - 7x^{1/3} + 5 = 0$

let $u = x^{1/3}$

$2(x^{1/3})^2 - 7x^{1/3} + 5 = 0$

$2u^2 - 7u + 5 = 0$

$(2u - 5)(u - 1) = 0$

$2u - 5 = 0 \qquad u - 1 = 0$

$2u = 5 \qquad\qquad u = 1$

$u = \frac{5}{2} \qquad\qquad x^{1/3} = 1$

$x^{1/3} = \frac{5}{2} \qquad (x^{1/3}) = 1^3$

$(x^{1/3}) = \left(\frac{5}{2}\right)^3 \qquad x = 1$

$x = \frac{125}{8}$

115. $x^{2/5} - 3x^{1/5} + 2 = 0$

$(x^{1/5} - 2)(x^{1/5} - 1) = 0$

$x^{1/5} = 2 \qquad x^{1/5} = 1$

$x = 2^5 \qquad\quad x = 1^5$

$x = 32 \qquad\quad x = 1$

117. $2x^{2/5} - 7x^{1/5} + 3 = 0$

$(2x^{1/5} - 1)(x^{1/5} - 3) = 0$

$x^{1/5} = \frac{1}{2} \qquad x^{1/5} = 3$

$x = \left(\frac{1}{2}\right)^5 \qquad x = 3^5$

$x = \frac{1}{32} \qquad\quad x = 243$

119. $\dfrac{1}{x^2} - \dfrac{3}{x} + 2 = 0$

$1 - 3x + 2x^2 = 0$

$2x^2 - 3x + 1 = 0$

$(2x - 1)(x - 1) = 0$

$2x - 1 = 0 \qquad x - 1 = 0$

$x = \frac{1}{2} \qquad\qquad x = 1$

121. $(x - 5)(x - (-2)) = 0$

$(x - 5)(x + 2) = 0$

$x^2 - 5x + 2x - 10 = 0$

$x^2 - 3x - 10 = 0$

123. $\left[x - \left(1 + \sqrt{2}\right)\right]\left[x - \left(1 - \sqrt{2}\right)\right] = 0$

$\left[(x - 1) - \sqrt{2}\right]\left[(x - 1) + \sqrt{2}\right] = 0$

$(x - 1)^2 - \left(\sqrt{2}\right)^2 = 0$

$x^2 - 2x + 1 - 2 = 0$

$x^2 - 2x - 1 = 0$

125. $(x - 5i)(x - (-5i)) = 0$

$(x - 5i)(x + 5i) = 0$

$x^2 - 25i^2 = 0$

$x^2 + 25 = 0$

127. $0 - -16t^2 + 256$

$16t^2 = 256$

$t^2 = 16$

$t = 4 \text{ seconds}$

129. $0 = -16t^2 + 128$

$16t^2 = 128$

$t^2 = 8$

$t = \pm\sqrt{8}$

$t = \pm 2\sqrt{2}$

$t = 2\sqrt{2} \approx 2.828 \text{ seconds}$

131. $0 = 144 + 128 - 16^2$

$0 = -16t^2 + 128t + 144$

$0 = -16(t^2 - 8t - 9)$

$0 = -16(t - 9)(t + 1)$

$t - 9 = 0 \qquad t + 1 = 0$

$t = 9 \text{ seconds} \qquad \bowtie$

133. $1685.40 = 1500(1 + r)^2$

$1.1236 = (1 + r)^2$

$1.06 = 1 + r$

$.06 = r$

$6\% = r$

135.
$$892 = (26.6 + t)^2$$
$$\sqrt{892} = 26.6 + t$$
$$\sqrt{892} - 26.6 = t$$
$$3 \approx t$$

Year 1993

137. (a) $h_0 = 100$ feet $v_0 = 0$ feet/sec $h = 0$
$$0 = 16t^2 + 0 \cdot t + 100$$
$$16t^2 = 100$$
$$t^2 = 6.25$$
$$t = \sqrt{6.25}$$
$$t = 2.5 \text{ seconds}$$

Extracting the roots method was used because the quadratic equation did not have a linear term.

(b) $h_0 = 100$ feet $v_0 = 32$ feet/sec $h = 100$ feet
$$100 = -16t^2 + 32t + 100$$
$$0 = -16t^2 + 32t$$
$$0 = -16t(t - 2)$$
$$-16t = 0 \qquad t - 2 = 0$$
$$t = 0 \text{ seconds} \qquad t = 2 \text{ seconds}$$

Factoring method was used because the quadratic equation did not have a constant term.

139. Factoring and the Zero-Factor Property allow you to solve a quadratic equation by converting it into two linear equations that you already know how to solve.

143. To solve an equation of quadratic form, determine an algebraic expression u such that substitution yields the quadratic equation $au^2 + bu + c = 0$. Solve this quadratic equation for u and then, through back-substitution, find the solution of the original equation.

141. False. The solutions are $x = 5$ and $x = -5$.

Section 10.2   Completing the Square

1. $x^2 + 8x + 16 \quad \left[16 = \left(\dfrac{8}{2} \right)^2 \right]$

3. $y^2 - 20y + 100 \quad \left[100 = \left(-\dfrac{20}{2} \right)^2 \right]$

5. $x^2 - 16x + 64 \quad \left[64 = \left(-\dfrac{16}{2} \right)^2 \right]$

7. $t^2 + 5t + \dfrac{25}{4} \quad \left[\dfrac{25}{4} = \left(\dfrac{5}{2} \right)^2 \right]$

9. $x^2 - 9x + \dfrac{81}{4} \quad \left[\dfrac{81}{4} = \left(-\dfrac{9}{2} \right)^2 \right]$

11. $a^2 - \dfrac{1}{3}a + \dfrac{1}{36} \quad \left[\dfrac{1}{36} = \left[\left(-\dfrac{1}{3} \right) \left(\dfrac{1}{2} \right) \right]^2 \right]$

13. $y^2 - \dfrac{3}{5}y + \dfrac{9}{100} \quad \left[\dfrac{9}{100} = \left[\left(-\dfrac{3}{5} \right) \left(\dfrac{1}{2} \right) \right]^2 \right]$

15. $r^2 - 0.4r + 0.04 \quad \left[0.04 = \left(-\dfrac{0.4}{2} \right)^2 \right]$

17. (a) $x^2 - 20x + 100 = 100$
$$(x - 10)^2 = 100$$
$$x - 10 = \pm 10$$
$$x = 10 \pm 10$$
$$x = 20, 0$$

(b) $x^2 - 20x = 0$
$$x(x - 20) = 0$$
$$x = 0 \qquad x = 20$$

19. (a) $x^2 + 6x + 9 = 0 + 9$
$$(x + 3)^2 = 9$$
$$x + 3 = \pm 3$$
$$x = -3 \pm 3$$
$$x = -6, 0$$

(b) $x^2 + 6x = 0$
$$x(x + 6) = 0$$
$$x = 0 \qquad x + 6 = 0$$
$$x = -6, 0$$

21. (a) $y^2 - 5y = 0$

$y^2 - 5y + \frac{25}{4} = \frac{25}{4}$

$\left(y - \frac{5}{2}\right)^2 = \frac{25}{4}$

$y - \frac{5}{2} = \pm\frac{5}{2}$

$y = \frac{5}{2} \pm \frac{5}{2}$

$= 0, 5$

(b) $y^2 - 5y = 0$

$y(y - 5) = 0$

$y = 0 \qquad y - 5 = 0$

$y = 5$

23. (a) $t^2 - 8t + 16 = -7 + 16$

$(t - 4)^2 = 9$

$t - 4 = \pm 3$

$t = 4 \pm 3$

$t = 7, 1$

(b) $t^2 - 8t + 7 = 0$

$(t - 7)(t - 1) = 0$

$t = 7 \qquad t = 1$

25. (a) $x^2 + 2x + 1 = 24 + 1$

$(x + 1)^2 = 25$

$x + 1 = \pm 5$

$x = -1 \pm 5$

$x = 4, -6$

(b) $x^2 + 2x - 24 = 0$

$(x + 6)(x - 4) = 0$

$x = -6 \qquad x = 4$

27. (a) $x^2 + 7x + \frac{49}{4} = -12 + \frac{49}{4}$

$\left(x + \frac{7}{2}\right)^2 = \frac{1}{4}$

$x + \frac{7}{2} = \pm\frac{1}{2}$

$x = -\frac{7}{2} \pm \frac{1}{2}$

$x = -\frac{6}{2}, -\frac{8}{2}$

$x = -3, -4$

(b) $x^2 + 7x + 12 = 0$

$(x + 4)(x + 3) = 0$

$x = -4 \qquad x = -3$

29. (a) $x^2 - 3x + \frac{9}{4} = 18 + \frac{9}{4}$

$\left(x - \frac{3}{2}\right)^2 = \frac{81}{4}$

$x - \frac{3}{2} = \pm\frac{9}{2}$

$x = \frac{3}{2} \pm \frac{9}{2}$

$x = \frac{12}{2}, -\frac{6}{2}$

$x = 6, -3$

(b) $x^2 - 3x - 18 = 0$

$(x - 6)(x + 3) = 0$

$x = 6 \qquad x = -3$

31. (a) $2x^2 - 14x + 12 = 0$

$x^2 - 7x + 6 = 0$

$x^2 - 7x = -6$

$x^2 - 7x + \frac{49}{4} = -6 + \frac{49}{4}$

$\left(x - \frac{7}{2}\right)^2 = -\frac{24}{4} + \frac{49}{4}$

$\left(x - \frac{7}{2}\right)^2 = \frac{25}{4}$

$x - \frac{7}{2} = \pm\frac{5}{2}$

$x = \frac{7}{2} \pm \frac{5}{2}$

$x = \frac{12}{2}, \frac{2}{2}$

$x = 6, 1$

(b) $2x^2 - 14x + 12 = 0$

$x^2 - 7x + 6 = 0$

$(x - 6)(x - 1) = 0$

$x = 6 \qquad x = 1$

33. (a) $4x^2 + 4x - 15 = 0$

$$x^2 + x - \frac{15}{4} = 0$$

$$x^2 + x = \frac{15}{4}$$

$$x^2 + x + \frac{1}{4} = \frac{15}{4} + \frac{1}{4}$$

$$\left(x + \frac{1}{2}\right)^2 = \frac{16}{4}$$

$$x + \frac{1}{2} = \pm\sqrt{4}$$

$$x = -\frac{1}{2} \pm 2$$

$$x = \frac{3}{2}, -\frac{5}{2}$$

(b) $\quad 4x^2 + 4x - 15 = 0$

$$(2x - 3)(2x + 5) = 0$$

$$x = \frac{3}{2} \qquad x = -\frac{5}{2}$$

35. $x^2 - 4x - 3 = 0$

$$x^2 - 4x + 4 = 3 + 4$$

$$(x - 2)^2 = 7$$

$$x - 2 = \pm\sqrt{7}$$

$$x = 2 \pm \sqrt{7}$$

$$x \approx 4.65, -0.65$$

37. $x^2 + 4x - 3 = 0$

$$x^2 + 4x + 4 = 3 + 4$$

$$(x + 2)^2 = 7$$

$$x + 2 = \pm\sqrt{7}$$

$$x = -2 \pm \sqrt{7}$$

$$x \approx 0.65, -4.65$$

39. $u^2 - 4u + 1 = 0$

$$u^2 - 4u + 4 = -1 + 4$$

$$(u - 2)^2 = 3$$

$$u - 2 = \pm\sqrt{3}$$

$$u = 2 \pm \sqrt{3}$$

$$u \approx 3.73, 0.27$$

41. $x^2 + 2x + 3 = 0$

$$x^2 + 2x + 1 = -3 + 1$$

$$(x + 1)^2 = -2$$

$$x + 1 = \pm\sqrt{-2}$$

$$x = -1 \pm i\sqrt{2}$$

$$x \approx -1 + 1.41i$$

$$x \approx -1 - 1.41i$$

43. $x^2 - 10x - 2 = 0$

$$x^2 - 10x + 25 = 2 + 25$$

$$(x - 5)^2 = 27$$

$$x - 5 = \pm\sqrt{27}$$

$$x = 5 \pm 3\sqrt{3}$$

$$x \approx 10.20, -0.20$$

45. $y^2 + 20y + 10 = 0$

$$y^2 + 20y + 100 = -10 + 100$$

$$(y + 10)^2 = 90$$

$$y + 10 = \pm\sqrt{90}$$

$$y = -10 \pm 3\sqrt{10}$$

$$y \approx -0.51, -19.49$$

47. $t^2 + 5t + 3 = 0$

$$t^2 + 5t + \frac{25}{4} = -3 + \frac{25}{4}$$

$$\left(t + \frac{5}{2}\right)^2 = \frac{13}{4}$$

$$t + \frac{5}{2} = \pm\sqrt{\frac{13}{4}}$$

$$t = -\frac{5}{2} \pm \frac{\sqrt{13}}{2}$$

$$t = \frac{-5 \pm \sqrt{13}}{2}$$

$$t \approx -0.70, -4.30$$

49. $v^2 + 3v - 2 = 0$

$$v^2 + 3v + \frac{9}{4} = 2 + \frac{9}{4}$$

$$\left(v + \frac{3}{2}\right)^2 = \frac{17}{4}$$

$$v + \frac{3}{2} = \pm\sqrt{\frac{17}{4}}$$

$$v = -\frac{3}{2} \pm \sqrt{\frac{17}{4}}$$

$$v = -\frac{3}{2} \pm \frac{\sqrt{17}}{2}$$

$$v = \frac{-3 \pm \sqrt{17}}{2}$$

$$v \approx 0.56, -3.56$$

51. $-x^2 + x - 1 = 0$

$$x^2 - x + 1 = 0$$

$$x^2 - x + \frac{1}{4} = -1 + \frac{1}{4}$$

$$\left(x - \frac{1}{2}\right)^2 = -\frac{3}{4}$$

$$x - \frac{1}{2} = \pm\sqrt{-\frac{3}{4}}$$

$$x = \frac{1}{2} \pm \frac{i\sqrt{3}}{2}$$

$$x = \frac{1 \pm i\sqrt{3}}{2}$$

$$x \approx 0.5 + 0.87i$$

$$x \approx 0.5 - 0.87i$$

53. $x^2 - 7x + 12 = 0$

$$x^2 - 7x + \frac{49}{4} = -12 + \frac{49}{4}$$

$$\left(x - \frac{7}{2}\right)^2 = \frac{-48}{4} + \frac{49}{4}$$

$$\left(x - \frac{7}{2}\right)^2 = \frac{1}{4}$$

$$x - \frac{7}{2} = \pm\sqrt{\frac{1}{4}}$$

$$x = \frac{7}{2} \pm \frac{1}{2}$$

$$x = 4, 3$$

55. $x^2 - \frac{2}{3}x - 3 = 0$

$$x^2 - \frac{2}{3}x + \frac{1}{9} = 3 + \frac{1}{9}$$

$$\left(x - \frac{1}{3}\right)^2 = \frac{28}{9}$$

$$x - \frac{1}{3} = \pm\sqrt{\frac{28}{9}}$$

$$x = \frac{1}{3} \pm \frac{2}{3}\sqrt{7}$$

$$x = \frac{1 \pm 2\sqrt{7}}{3}$$

$$x \approx 2.10, -1.43$$

57. $v^2 + \frac{3}{4}v - 2 = 0$

$$v^2 + \frac{3}{4}v + \frac{9}{64} = 2 + \frac{9}{64}$$

$$\left(v + \frac{3}{8}\right)^2 = \frac{128}{64} + \frac{9}{64}$$

$$\left(v + \frac{3}{8}\right)^2 = \frac{137}{64}$$

$$v + \frac{3}{8} = \pm\sqrt{\frac{137}{64}}$$

$$v = -\frac{3}{8} \pm \frac{\sqrt{137}}{8}$$

$$v \approx 1.09, -1.84$$

59. $2x^2 + 8x + 3 = 0$

$$x^2 + 4x + 4 = -\frac{3}{2} + 4$$

$$(x + 2)^2 = \frac{5}{2}$$

$$x + 2 = \pm\sqrt{\frac{5}{2}} \cdot \frac{\sqrt{2}}{\sqrt{2}}$$

$$x = -2 \pm \frac{\sqrt{10}}{2}$$

$$x \approx -0.42, -3.58$$

61. $3x^2 + 9x + 5 = 0$

$$x^2 + 3x + \frac{9}{4} = -\frac{5}{3} + \frac{9}{4}$$

$$\left(x + \frac{3}{2}\right)^2 = \frac{-20 + 27}{12}$$

$$\left(x + \frac{3}{2}\right)^2 = \frac{7}{12}$$

$$x + \frac{3}{2} = \pm\sqrt{\frac{7}{12}} \cdot \frac{\sqrt{3}}{\sqrt{3}}$$

$$x = -\frac{3}{2} \pm \frac{\sqrt{21}}{6}$$

$$x = \frac{-9 \pm \sqrt{21}}{6}$$

$$x \approx -0.74, -2.26$$

63. $4y^2 + 4y - 9 = 0$

$$y^2 + y + \frac{1}{4} = \frac{9}{4} + \frac{1}{4}$$

$$\left(y + \frac{1}{2}\right)^2 = \frac{10}{4}$$

$$y + \frac{1}{2} = \pm\sqrt{\frac{10}{4}}$$

$$y = -\frac{1}{2} \pm \frac{\sqrt{10}}{2}$$

$$y = \frac{-1 \pm \sqrt{10}}{2}$$

$$y \approx 1.08, -2.08$$

65. $5x^2 - 3x + 10 = 0$

$$x^2 - \frac{3}{5}x = -2$$

$$x^2 - \frac{3}{5}x + \frac{9}{100} = -2 + \frac{9}{100}$$

$$\left(x - \frac{3}{10}\right)^2 = \frac{-200}{100} + \frac{9}{100}$$

$$\left(x - \frac{3}{10}\right)^2 = -\frac{191}{100}$$

$$x - \frac{3}{10} = \pm\sqrt{-\frac{191}{100}}$$

$$x = \frac{3}{10} \pm \frac{\sqrt{191}}{10}i$$

$$x \approx 0.30 + 1.38i, 0.30 - 1.38i$$

67. $x(x - 7) = 2$

$$x^2 - 7x + \frac{49}{4} = 2 + \frac{49}{4}$$

$$\left(x - \frac{7}{2}\right)^2 = \frac{8 + 49}{4}$$

$$\left(x - \frac{7}{2}\right)^2 = \frac{57}{4}$$

$$x - \frac{7}{2} = \pm\sqrt{\frac{57}{4}}$$

$$x = \frac{7}{2} \pm \frac{\sqrt{57}}{2}$$

$$x = \frac{7 \pm \sqrt{57}}{2}$$

$$x \approx 7.27, -0.27$$

69. $0.5t^2 + t + 2 = 0$

$$t^2 + 2t = -4$$

$$t^2 + 2t + 1 = -4 + 1$$

$$(t + 1)^2 = -3$$

$$t + 1 = \pm\sqrt{-3}$$

$$t + 1 = \pm\sqrt{3}i$$

$$t = -1 \pm \sqrt{3}i$$

$$t \approx -1 + 1.73i, -1 - 1.73i$$

71. $0.1x^2 + 0.2x + 0.5 = 0$

$$x^2 + 2x + 5 = 0$$

$$x^2 + 2x + 1 = -5 + 1$$

$$(x + 1)^2 = -4$$

$$x + 1 = \pm\sqrt{-4}$$

$$x = -1 \pm 2i$$

73. $\dfrac{x}{2} - \dfrac{1}{x} = 1$

$$2x\left(\frac{x}{2} - \frac{1}{x}\right) = (1)2x$$

$$x^2 - 2 = 2x$$

$$x^2 - 2x + 1 = 2 + 1$$

$$(x - 1)^2 = 3$$

$$x - 1 = \pm\sqrt{3}$$

$$x = 1 \pm \sqrt{3}$$

75.

$$\frac{x^2}{4} = \frac{x+1}{2}$$

$$2x^2 = 4x + 4$$

$$2x^2 - 4x - 4 = 0$$

$$x^2 - 2x - 2 = 0$$

$$x^2 - 2x + 1 = 2 + 1$$

$$(x - 1)^2 = 3$$

$$x - 1 = \pm\sqrt{3}$$

$$x = 1 \pm \sqrt{3}$$

77.

$$\sqrt{2x+1} = x - 3$$

$$\left(\sqrt{2x+1}\right)^2 = (x - 3)^2$$

$$2x + 1 = x^2 - 6x + 9$$

$$0 = x^2 - 8x + 8$$

$$+16 - 8 = x^2 - 8x + 16$$

$$8 = (x - 4)^2$$

$$\pm\sqrt{8} = x - 4$$

$$4 \pm \sqrt{8} = x$$

$$4 \pm 2\sqrt{2} = x$$

79. *Keystrokes:*

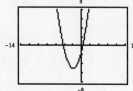

$$0 = x^2 + 4x - 1$$

$$1 = x^2 + 4x$$

$$1 + 4 = x^2 + 4x + 4$$

$$5 = (x + 2)^2$$

$$\pm\sqrt{5} = x + 2$$

$$-2 \pm \sqrt{5} = x$$

$$x \approx .236$$

$$x \approx -4.236$$

81. *Keystrokes:*

$$\boxed{Y=}\ \boxed{X,T,\theta}\ \boxed{x^2}\ \boxed{-}\ 2\ \boxed{X,T,\theta}\ \boxed{-}\ 5\ \boxed{GRAPH}$$

$$0 = x^2 - 2x - 5$$

$$5 = x^2 - 2x$$

$$1 + 5 = x^2 - 2x + 1$$

$$6 = (x - 1)^2$$

$$\pm\sqrt{6} = x - 1$$

$$1 \pm \sqrt{6} = x$$

$$x \approx 3.449$$

$$x \approx -1.449$$

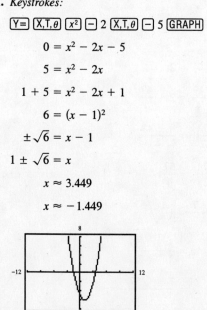

83. *Keystrokes:*

$$\boxed{Y=}\ \boxed{(}\ 1\ \boxed{\div}\ 3\ \boxed{)}\ \boxed{X,T,\theta}\ \boxed{x^2}\ \boxed{+}\ 2\ \boxed{X,T,\theta}\ \boxed{-}\ 6\ \boxed{GRAPH}$$

$$0 = \tfrac{1}{3}x^2 + 2x - 6$$

$$0 = x^2 + 6x - 18$$

$$18 = x^2 + 6x$$

$$9 + 18 = x^2 + 6x + 9$$

$$27 = (x + 3)^2$$

$$\pm\sqrt{27} = x + 3$$

$$-3 \pm 3\sqrt{3} = x$$

$$x \approx 2.20$$

$$x \approx -8.20$$

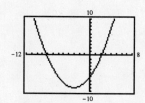

85. *Keystrokes:*

$\boxed{Y=}$ $\boxed{(-)}$ $\boxed{X,T,\theta}$ $\boxed{x^2}$ $\boxed{-}$ $\boxed{X,T,\theta}$ $\boxed{+}$ 3 $\boxed{\text{GRAPH}}$

$-x^2 - x + 3 = 0$

$x^2 + x - 3 = 0$

$x^2 + x = 3$

$x^2 + x + \dfrac{1}{4} = 3 + \dfrac{1}{4}$

$\left(x + \dfrac{1}{2}\right)^2 = \dfrac{13}{4}$

$x + \dfrac{1}{2} = \pm\dfrac{\sqrt{13}}{2}$

$x = -\dfrac{1}{2} \pm \dfrac{\sqrt{13}}{2}$

$x \approx 1.30, -2.30$

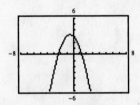

87. (a) Area of square $= x \cdot x = x^2$

Area of vertical rectangle $= 4 \cdot x = 4x$

Area of horizontal rectangle $= 4 \cdot x = 4x$

Total area $= x^2 + 4x + 4x = x^2 + 8x$

(b) Area of small square $= 4 \cdot 4 = 16$

Total area $= x^2 + 8x + 16$

(c) $(x + 4)(x + 4) = x^2 + 8x + 16$

89. *Verbal model:* $\boxed{\text{Area}} = \frac{1}{2} \cdot \boxed{\text{Base}} \cdot \boxed{\text{Height}}$

Labels: Base $= x$

Height $= x + 2$

Equation: $12 = \frac{1}{2}x(x + 2)$

$24 = x^2 + 2x$

$0 = x^2 + 2x - 24$

$0 = (x + 6)(x - 4)$

$x = -6$ $x = 4$ cm base

not a solution $x + 2 = 6$ cm height

91. *Verbal model:* $\boxed{\text{Area}} = \boxed{\text{Length}} \cdot \boxed{\text{Width}}$

Labels: Length $= x$

Width $= \dfrac{200 - 4x}{3}$

Equation: $1400 = 2\left[x \cdot \left(\dfrac{200 - 4x}{3}\right)\right]$

$1400 = 2\left[\dfrac{200}{3}x - \dfrac{4x^2}{3}\right]$

$1400 = \dfrac{400x}{3} - \dfrac{8x^2}{3}$

$4200 = 400x - 8x^2$

$8x^2 - 400x + 4200 = 0$

$x^2 - 50x + 525 = 0$

$(x - 35)(x - 15) = 0$

$x - 35 = 0$ $x - 15 = 0$

$x = 35$ ft. $x = 15$ ft.

$\dfrac{200 - 4x}{3} = 20$ ft. $\dfrac{200 - 4x}{3} = 46\dfrac{2}{3}$ ft.

93. *Verbal model:* $\boxed{\text{side 1}}^2 + \boxed{\text{side 2}}^2 = \boxed{\text{Hypotenuse}}^2$

Labels: side 1 = x

side 2 = $400 - x$

Equation: $x^2 + (400 - x)^2 = 300^2$

$x^2 + 160,000 - 2(400)x + x^2 = 90,000$

$x^2 + 160,000 - 800x + x^2 = 90,000$

$2x^2 - 800x + 70,000 = 0$

$x^2 - 400x + 35,000 = 0$

$x^2 - 400x = -35,000$

$x^2 - 400x + 40,000 = -35,000 + 40,000$

$(x - 200)^2 = 5000$

$x - 200 = \pm\sqrt{5000}$

$x - 200 = \pm 50\sqrt{2}$

$x = 200 \pm 50\sqrt{2}$

$x = 200 \pm 50\sqrt{2}$ meters

$x \approx 270.71$ meters and 129.29 meters

95. *Equation:* $12,000 = x\left(100 - \frac{1}{10}x\right)$

$12,000 = 100x - \frac{1}{10}x^2$

$120,000 = 1000x - x^2$

$x^2 - 1000x + 120,000 = 0$

$x^2 - 1000x = -120,000$

$x^2 - 1000x + 250,000 = -120,000 + 250,000$

$(x - 500)^2 = 130,000$

$x - 500 = \pm\sqrt{130,000}$

$x - 500 = \pm 100\sqrt{13}$

$x = 500 \pm 100\sqrt{13}$

$x \approx 860.56, \ 139.44$

Thus, 139 or 861 units must be sold.

97. $\frac{25}{4}$. Divide the coefficient of the first-degree term by 2, and square the result to obtain $\left(\frac{5}{2}\right)^2 = \frac{25}{4}$.

99. Yes. $x^2 + 1 = 0$

101. True. Given the solutions $x = r_1$ and $x = r_2$, the quadratic equation can be written as $(x - r_1)(x - r_2) = 0$.

Section 10.3   The Quadratic Formula

1. $2x^2 = 7 - 2x$

   $2x^2 + 2x - 7 = 0$

3. $x(10 - x) = 5$

   $10x - x^2 = 5$

   $-x^2 + 10x - 5 = 0$

   $x^2 - 10x + 5 = 0$

5. (a) $x = \dfrac{11 \pm \sqrt{11^2 - 4(1)(28)}}{2(1)}$

   $x = \dfrac{11 \pm \sqrt{121 - 112}}{2}$

   $x = \dfrac{11 \pm \sqrt{9}}{2}$

   $x = \dfrac{11 \pm 3}{2}$

   $x = 7, 4$

   (b) $(x - 7)(x - 4) = 0$

   $x - 7 = 0 \qquad x - 4 = 0$

   $\qquad x = 7 \qquad\qquad x = 4$

7. (a) $x = \dfrac{-6 \pm \sqrt{6^2 - 4(1)(8)}}{2(1)}$

   $x = \dfrac{6 \pm \sqrt{36 - 32}}{2}$

   $x = \dfrac{-6 \pm \sqrt{4}}{2}$

   $x = \dfrac{-6 \pm 2}{2} \qquad x = -2, -4$

   (b) $(x + 4)(x + 2) = 0$

   $x + 4 = 0 \qquad\qquad x + 2 = 0$

   $\quad x = -4 \qquad\qquad\quad x = -2$

9. (a) $x = \dfrac{-4 \pm \sqrt{4^2 - 4(4)(1)}}{2(4)}$

   $x = \dfrac{-4 \pm \sqrt{16 - 16}}{8}$

   $x = \dfrac{-4}{8} = \dfrac{-1}{2}$

   (b) $(2x + 1)(2x + 1) = 0$

   $2x + 1 = 0 \qquad\qquad 2x + 1 = 0$

   $\qquad x = -\dfrac{1}{2} \qquad\qquad x = -\dfrac{1}{2}$

11. (a) $x = \dfrac{-12 \pm \sqrt{12^2 - 4(4)(9)}}{2(4)}$

   $x = \dfrac{-12 \pm \sqrt{144 - 144}}{8}$

   $x = \dfrac{-12 \pm 0}{8}$

   $x = -\dfrac{12}{8} = -\dfrac{3}{2}$

   (b) $(2x + 3)(2x + 3) = 0$

   $2x + 3 = 0 \quad 2x + 3 = 0$

   $\qquad x = -\dfrac{3}{2} \qquad\qquad x = -\dfrac{3}{2}$

13. (a) $x = \dfrac{1 \pm \sqrt{(-1)^2 - 4(6)(-2)}}{2(6)}$

   $x = \dfrac{1 \pm \sqrt{1 + 48}}{12}$

   $x = \dfrac{1 \pm \sqrt{49}}{12}$

   $x = \dfrac{1 \pm 7}{12}$

   $x = \dfrac{8}{12}, -\dfrac{6}{12} = \dfrac{2}{3}, -\dfrac{1}{2}$

   (b) $(3x - 2)(2x + 1) = 0$

   $3x - 2 = 0 \qquad\qquad 2x + 1 = 0$

   $\qquad x = \dfrac{2}{3} \qquad\qquad\quad x = -\dfrac{1}{2}$

15. (a) $x = \dfrac{-(-5) \pm \sqrt{(-5)^2 - 4(1)(-300)}}{2(1)}$

$x = \dfrac{5 \pm \sqrt{25 + 1200}}{2}$

$x = \dfrac{5 \pm \sqrt{1225}}{2}$

$x = \dfrac{5 \pm 35}{2}$

$x = 20, -15$

(b) $(x - 20)(x + 15) = 0$

$x - 20 = 0 \qquad x + 15 = 0$

$x = 20 \qquad\quad x = -15$

17. $x = \dfrac{-(-2) \pm \sqrt{(-2)^2 - 4(1)(-4)}}{2(1)}$

$x = \dfrac{2 \pm \sqrt{4 + 16}}{2}$

$x = \dfrac{2 \pm \sqrt{20}}{2}$

$x = \dfrac{2 \pm 2\sqrt{5}}{2}$

$x = \dfrac{2(1 \pm \sqrt{5})}{2}$

$x = 1 \pm \sqrt{5}$

19. $t = \dfrac{-4 \pm \sqrt{4^2 - 4(1)(1)}}{2(1)}$

$t = \dfrac{-4 \pm \sqrt{16 - 4}}{2}$

$t = \dfrac{4 \pm \sqrt{12}}{2}$

$t = \dfrac{-4 \pm 2\sqrt{3}}{2}$

$t = \dfrac{2(-2 \pm \sqrt{3})}{2}$

$t = -2 \pm \sqrt{3}$

21. $x = \dfrac{-6 \pm \sqrt{6^2 - 4(1)(-3)}}{2(1)}$

$x = \dfrac{-6 \pm \sqrt{36 + 12}}{2}$

$x = \dfrac{-6 \pm \sqrt{48}}{2}$

$x = \dfrac{-6 \pm 4\sqrt{3}}{2}$

$x = \dfrac{2(-3 \pm 2\sqrt{3})}{2}$

$x = -3 \pm 2\sqrt{3}$

23. $x = \dfrac{-(-10) \pm \sqrt{(-10)^2 - 4(1)(23)}}{2(1)}$

$x = \dfrac{10 \pm \sqrt{100 - 92}}{2}$

$x = \dfrac{10 \pm \sqrt{8}}{2}$

$x = \dfrac{10 \pm 2\sqrt{2}}{2}$

$x = \dfrac{2(5 \pm \sqrt{2})}{2}$

$x = 5 \pm \sqrt{2}$

25. $x = \dfrac{-3 \pm \sqrt{3^2 - 4(2)(3)}}{2(2)}$

$x = \dfrac{-3 \pm \sqrt{9 - 24}}{4}$

$x = \dfrac{-3 \pm \sqrt{-15}}{4}$

$x = \dfrac{-3 \pm i\sqrt{15}}{4}$

$x = \dfrac{-3}{4} \pm \dfrac{\sqrt{15}}{4}i$

27. $v = \dfrac{-(-2) \pm \sqrt{(-2)^2 - 4(3)(-1)}}{2(3)}$

$v = \dfrac{2 \pm \sqrt{4 + 12}}{6}$

$v = \dfrac{2 \pm \sqrt{16}}{6}$

$v = \dfrac{2 \pm 4}{6}$

$v = \dfrac{6}{6}, \dfrac{-2}{6}$

$v = 1, -\dfrac{1}{3}$

29. $x = \dfrac{-4 \pm \sqrt{4^2 - 4(2)(-3)}}{2(2)}$

$x = \dfrac{-4 \pm \sqrt{16 + 24}}{4}$

$x = \dfrac{-4 \pm \sqrt{40}}{4}$

$x = \dfrac{-4 \pm 2\sqrt{10}}{4}$

$x = \dfrac{2(-2 \pm \sqrt{10})}{4}$

$x = \dfrac{-2 \pm \sqrt{10}}{2}$

31. $z = \dfrac{-6 \pm \sqrt{6^2 - 4(9)(-4)}}{2(9)}$

$z = \dfrac{-6 \pm \sqrt{36 + 144}}{18}$

$z = \dfrac{-6 \pm \sqrt{180}}{18}$

$z = \dfrac{-6 \pm 6\sqrt{5}}{18}$

$z = \dfrac{6(-1 \pm \sqrt{5})}{18}$

$z = \dfrac{-1 \pm \sqrt{5}}{3}$

33. $x = \dfrac{-(-6) \pm \sqrt{(-6)^2 - 4(-4)(3)}}{2(-4)}$

$x = \dfrac{6 \pm \sqrt{36 + 48}}{-8}$

$x = \dfrac{6 \pm \sqrt{84}}{-8}$

$x = \dfrac{6 \pm 2\sqrt{21}}{-8}$

$x = \dfrac{-3 \pm \sqrt{21}}{4}$

35. $4x^2 - 3x + 1 = 0$

$x = \dfrac{-(-3) \pm \sqrt{(-3)^2 - 4(4)(1)}}{2(4)}$

$x = \dfrac{3 \pm \sqrt{9 - 16}}{8}$

$x = \dfrac{3 \pm \sqrt{-7}}{8} = \dfrac{3}{8} \pm \dfrac{\sqrt{7}}{8}i$

37. $2x^2 - 5x - 6 = 0$

$x = \dfrac{-(-5) \pm \sqrt{(-5)^2 - 4(2)(-6)}}{2(2)}$

$x = \dfrac{5 \pm \sqrt{25 + 48}}{4}$

$x = \dfrac{5 \pm \sqrt{73}}{4}$

39. $\qquad 9x^2 = 1 + 9x$

$9x^2 - 9x - 1 = 0$

$x = \dfrac{-(-9) \pm \sqrt{(-9)^2 - 4(9)(-1)}}{2(9)}$

$x = \dfrac{9 \pm \sqrt{81 + 36}}{18}$

$x = \dfrac{9 \pm \sqrt{117}}{18}$

$x = \dfrac{9}{18} \pm \dfrac{3\sqrt{13}}{18}$

$x = \dfrac{1}{2} \pm \dfrac{\sqrt{13}}{6}$ or $\dfrac{3 \pm \sqrt{13}}{6}$

41. $3x - 2x^2 - 4 + 5x^2 = 0$

$\qquad 3x^2 + 3x - 4 = 0$

$x = \dfrac{-3 \pm \sqrt{3^2 - 4(3)(-4)}}{2(3)}$

$x = \dfrac{-3 \pm \sqrt{9 + 48}}{6}$

$x = \dfrac{-3 \pm \sqrt{57}}{6}$

43. $x = \dfrac{-(-0.4) \pm \sqrt{(-0.4)^2 - 4(1)(-0.16)}}{2(1)}$

$x = \dfrac{0.4 \pm \sqrt{0.16 + 0.64}}{2}$

$x = \dfrac{0.4 \pm \sqrt{0.80}}{2}$

$x = \dfrac{0.4 \pm 2\sqrt{0.2}}{2}$

$x = 0.2 \pm \sqrt{0.2}$ or $\dfrac{1 \pm \sqrt{5}}{5}$

45. $x = \dfrac{-1 \pm \sqrt{1^2 - 4(2.5)(-0.9)}}{2(2.5)}$

$x = \dfrac{-1 \pm \sqrt{1 + 9}}{5}$

$x = \dfrac{-1 \pm \sqrt{10}}{5}$

47. $b^2 - 4ac = 1^2 - 4(1)(1)$

$= 1 - 4$

$= -3$

2 distinct imaginary solutions

49. $b^2 - 4ac = (-5)^2 - 4(2)(-4)$

$= 25 + 32$

$= 57$

2 distinct irrational solutions

51. $b^2 - 4ac = 7^2 - 4(5)(3)$

$= 49 - 60$

$= -11$

2 distinct imaginary solutions

53. $b^2 - 4ac = (-12)^2 - 4(4)(9)$

$= 144 - 144$

$= 0$

1 rational repeated solution

55. $b^2 - 4ac = (-1)^2 - 4(3)(2)$

$= 1 - 24$

$= -23$

2 distinct imaginary solutions

57. $z^2 - 169 = 0$

$z^2 = 169$

$z = \pm 13$

59. $5y^2 + 15y = 0$

$5y(y + 3) = 0$

$5y = 0 \qquad y + 3 = 0$

$y = 0 \qquad\quad y = -3$

61. $25(x - 3)^2 - 36 = 0$

$(x - 3)^2 = \dfrac{36}{25}$

$x - 3 = \pm\sqrt{\dfrac{36}{25}}$

$x = 3 \pm \dfrac{6}{5}$

$x = \dfrac{15}{5} \pm \dfrac{6}{5}$

$x = \dfrac{21}{5}, \dfrac{9}{5}$

63. $2y(y - 18) + 3(y - 18) = 0$

$(y - 18)(2y + 3) = 0$

$y - 18 = 0 \qquad 2y + 3 = 0$

$y = 18 \qquad\qquad 2y = -3$

$y = -\dfrac{3}{2}$

65. $x^2 + 8x + 25 = 0$

$x^2 + 8x + 16 = -25 + 16$

$(x + 4)^2 = -9$

$x + 4 = \pm\sqrt{-9}$

$x = -4 \pm 3i$

67. $x^2 - 24x + 128 = 0$

$x^2 - 24x + 144 = -128 + 144$

$(x - 12)^2 = 16$

$x - 12 = \pm\sqrt{16}$

$x = 12 \pm 4$

$x = 16, 8$

69. $x = \dfrac{-(-13) \pm \sqrt{(-13)^2 - 4(3)(169)}}{2(3)}$

$x = \dfrac{13 \pm \sqrt{169 - 2028}}{6}$

$x = \dfrac{13 \pm \sqrt{-1859}}{6}$

$x = \dfrac{13}{6} \pm \dfrac{13\sqrt{11}}{6}i$

71. $x = \dfrac{-15 \pm \sqrt{15^2 - 4(18)(-50)}}{2(18)}$

$x = \dfrac{-15 \pm \sqrt{225 + 3600}}{36}$

$x = \dfrac{-15 \pm \sqrt{3825}}{36}$

$x = \dfrac{-15 \pm 15\sqrt{17}}{36}$

$x = \dfrac{-5 \pm 5\sqrt{17}}{12}$

73. $1.2x^2 - 0.8x - 5.5 = 0$

$12x^2 - 8x - 55 = 0$

$(6x + 11)(2x - 5) = 0$

$6x + 11 = 0 \qquad 2x - 5 = 0$

$6x = -11 \qquad 2x = 5$

$x = -\frac{11}{6} \qquad x = \frac{5}{2}$

75. *Keystrokes:*

$\boxed{Y=}$ 3 $\boxed{X,T,\theta}$ $\boxed{x^2}$ $\boxed{-}$ 6 $\boxed{X,T,\theta}$ $\boxed{+}$ 1 $\boxed{GRAPH}$

$0 = 3x^2 - 6x + 1$

$x = \dfrac{-(-6) \pm \sqrt{(-6)^2 - 4(3)(1)}}{2(3)}$

$x = \dfrac{6 \pm \sqrt{36 - 12}}{6}$

$x = \dfrac{6 \pm \sqrt{24}}{6}$

$x \approx 1.82, 0.18$

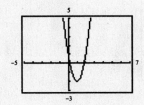

77. *Keystrokes:*

$\boxed{Y=}$ $\boxed{(-)}$ $\boxed{(}$ 4 $\boxed{X,T,\theta}$ $\boxed{x^2}$ $\boxed{-}$ 20 $\boxed{X,T,\theta}$ $\boxed{+}$ 25 $\boxed{)}$ $\boxed{GRAPH}$

$0 = -(4x^2 - 20x + 25)$

$= 4x^2 - 20x + 25$

$x = \dfrac{-(-20) \pm \sqrt{(-20)^2 - 4(4)(25)}}{2(4)}$

$x = \dfrac{20 \pm \sqrt{400 - 400}}{8}$

$x = \dfrac{20}{8} = \dfrac{5}{2} = 2.50$

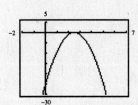

79. *Keystrokes:*

$\boxed{Y=}$ 5 $\boxed{X,T,\theta}$ $\boxed{x^2}$ $\boxed{-}$ 18 $\boxed{X,T,\theta}$ $\boxed{+}$ 6 $\boxed{GRAPH}$

$$x = \frac{-(-18) \pm \sqrt{(-18)^2 - 4(5)(6)}}{2(5)}$$

$$x = \frac{18 \pm \sqrt{324 - 120}}{10}$$

$$x = \frac{18 \pm \sqrt{204}}{10}$$

$$x = \frac{18 \pm 2\sqrt{51}}{10}$$

$$x = \frac{2(9 \pm \sqrt{51})}{10}$$

$$x = \frac{9 \pm \sqrt{51}}{5}$$

$$x \approx 3.23, 0.37$$

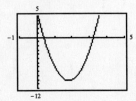

81. *Keystrokes:*

$\boxed{Y=}$ $\boxed{(-)}$ .04 $\boxed{X,T,\theta}$ $\boxed{x^2}$ $\boxed{+}$ 4 $\boxed{X,T,\theta}$ $\boxed{-}$ .8 $\boxed{GRAPH}$

$$x = \frac{-4 \pm \sqrt{4^2 - 4(-0.04)(-0.8)}}{2(-0.04)}$$

$$x = \frac{-4 \pm \sqrt{16 - 0.128}}{-0.08}$$

$$x = \frac{-4 \pm \sqrt{15.872}}{-0.08}$$

$$x = \frac{-4 \pm 16\sqrt{0.062}}{-0.08}$$

$$x = \frac{4(-1 \pm 4\sqrt{0.062})}{-0.08}$$

$$x = \frac{-1 \pm 4\sqrt{0.062}}{-0.02}$$

$$x \approx 0.20, 99.80$$

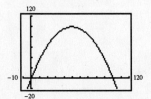

83. *Keystrokes:*

$\boxed{Y=}$ 2 $\boxed{X,T,\theta}$ $\boxed{x^2}$ $\boxed{-}$ 5 $\boxed{X,T,\theta}$ $\boxed{+}$ 5 $\boxed{GRAPH}$

$$b^2 - 4ac = (-5)^2 - 4(2)(5)$$
$$= 25 - 40$$
$$= -15$$

No real solutions

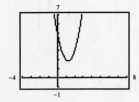

85. *Keystrokes:*

$\boxed{Y=}$ $\boxed{X,T,\theta}$ $\boxed{x^2}$ $\boxed{+}$ 6 $\boxed{X,T,\theta}$ $\boxed{-}$ 40 $\boxed{GRAPH}$

$$b^2 - 4ac = 6^2 - 4(1)(-40)$$
$$= 36 + 160$$
$$= 196$$

Two real solutions

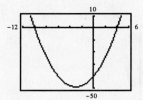

87. $\dfrac{2x^2}{5} - \dfrac{x}{2} = 1$

$10\left(\dfrac{2x^2}{5} - \dfrac{x}{2}\right) = (1)10$

$4x^2 - 5x = 10$

$4x^2 - 5x - 10 = 0$

$x = \dfrac{-(-5) \pm \sqrt{(-5)^2 - 4(4)(-10)}}{2(4)}$

$x = \dfrac{5 \pm \sqrt{25 + 160}}{8}$

$x = \dfrac{5 \pm \sqrt{185}}{8}$

89. $\sqrt{x + 3} = x - 1$

$\left(\sqrt{x + 3}\right)^2 = (x - 1)^2$

$x + 3 = x^2 - 2x + 1$

$0 = x^2 - 3x - 2$

$x = \dfrac{-(-3) \pm \sqrt{(-3)^2 - 4(1)(-2)}}{2(1)}$

$x = \dfrac{3 \pm \sqrt{9 + 8}}{2}$

$x = \dfrac{3 \pm \sqrt{17}}{2}$

$x = \dfrac{3 + \sqrt{17}}{2}$

$x = \dfrac{3 - \sqrt{17}}{2}$ does not check.

91. (a) $b^2 - 4ac > 0$

$(-6)^2 - 4(1)c > 0$

$36 - 4c > 0$

$-4c > -36$

$c < 9$

(b) $b^2 - 4ac = 0$

$(-6)^2 - 4(1)c = 0$

$36 - 4c = 0$

$-4c = -36$

$c = 9$

(c) $b^2 - 4ac < 0$

$(-6)^2 - 4(1)c < 0$

$36 - 4c < 0$

$-4c < -36$

$c > 9$

93. (a) $b^2 - 4ac > 0$

$8^2 - 4(1)c > 0$

$64 - 4c > 0$

$-4c > -64$

$c < 16$

(b) $b^2 - 4ac = 0$

$8^2 - 4(1)c = 0$

$64 - 4c = 0$

$-4c = -64$

$c = 16$

(c) $b^2 - 4ac < 0$

$8^2 - 4(1)c < 0$

$64 - 4c < 0$

$-4c < -64$

$c > 16$

95. *Verbal model:* $\boxed{\text{Area}} = \boxed{\text{Length}} \cdot \boxed{\text{Width}}$

Labels: Length $= x + 6.3$

Width $= x$

Equation: $58.14 = (x + 6.3) \cdot x$

$58.14 = x^2 + 6.3x$

$0 = x^2 + 6.3x - 58.14$

$x = \dfrac{-6.3 \pm \sqrt{6.3^2 - 4(1)(-58.14)}}{2(1)}$

$x = \dfrac{-6.3 \pm \sqrt{39.69 + 232.56}}{2}$

$x = \dfrac{-6.3 \pm \sqrt{272.25}}{2}$

$x \approx 5.1$ inches

$x + 6.3 \approx 11.4$ inches

97. (a) $50 = -16t^2 + 40t + 50$

$0 = -16t^2 + 40t$

$0 = -8(2t^2 - 5t)$

$0 = 2t^2 - 5t$

$0 = t(2t - 5)$

$0 = t \quad 2t - 5 = 0$

$t = \dfrac{5}{2} = 2.5$ seconds

(b) $0 = -16t^2 + 40t + 50$

$0 = -2(8t^2 - 20t - 25)$

$t = \dfrac{-(-20) \pm \sqrt{(-20)^2 - 4(8)(-25)}}{2(8)}$

$t = \dfrac{20 \pm \sqrt{400 + 800}}{16}$

$t = \dfrac{20 \pm \sqrt{1200}}{16}$

$t = \dfrac{20 \pm 20\sqrt{3}}{16}$

$t = \dfrac{4(5 \pm 5\sqrt{3})}{16}$

$t = \dfrac{5 + 5\sqrt{3}}{4}, \dfrac{5 - 5\sqrt{3}}{4}$ reject

$t \approx 3.4$ seconds

99. (a) *Keystrokes:*

$\boxed{Y=}$ 831.3 $\boxed{-}$ 85.71 $\boxed{X,T,\theta}$ $\boxed{+}$ 3.452 $\boxed{X,T,\theta}$ $\boxed{x^2}$ $\boxed{\text{GRAPH}}$

(b) $750 = 831.3 - 85.71t + 3.452t^2$

$0 = 3.452t^2 - 85.71t + 81.3$

$t = \dfrac{-(-85.71) \pm \sqrt{(-85.71)^2 - 4(3.452)(81.3)}}{2(3.452)}$

$t = \dfrac{85.71 \pm \sqrt{6223.6137}}{6.904} = \dfrac{85.71 \pm 78.89}{6.904}$

$t \approx .9879$ year 1991

(c) $y = 831.3 - 85.71(7) + 3.452(7)^2$

$y = 400,500$

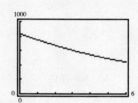

101.

	x_1, x_2	$x_1 + x_2$	$x_1 \cdot x_2$
(a) $x^2 - x - 6 = 0$	$3, -2$	1	-6
$(x - 3)(x + 2) = 0$			
$x = 3 \quad x = -2$			
(b) $2x^2 + 5x - 3 = 0$	$\frac{1}{2}, -3$	$-\frac{5}{2}$	$-\frac{3}{2}$
$(2x - 1)(x + 3) = 0$			
$x = \frac{1}{2} \quad x = -3$			
(c) $4x^2 - 9 = 0$	$\frac{3}{2}, -\frac{3}{2}$	0	$-\frac{9}{4}$
$(2x - 3)(2x + 3)$			
$x = \frac{3}{2} \quad x = -\frac{3}{2}$			
(d) $x^2 - 10x + 34 = 0$	$5 + 3i, 5 - 3i$	10	34

(d) $x^2 - 10x + 25 = -34 + 25$

$(x - 5)^2 = -9$

$x - 5 = \pm\sqrt{-9}$

$x = 5 \pm 3i$

103. (c) $h = -16t^2 + v_0 t + h_0$ $v_0 = 32$ feet/sec

$50 = -16t^2 + 32t + 100$ $h = 50$ feet

$0 = -16t^2 + 32t + 50$

$0 = 8t^2 - 16t - 25$ (Divide by -2)

$t = \dfrac{-(-16) \pm \sqrt{(-16)^2 - 4(8)(-25)}}{2(8)}$

$t = \dfrac{16 \pm \sqrt{256 + 800}}{16}$

$t = \dfrac{16 \pm \sqrt{1056}}{16} = \dfrac{16 \pm 4\sqrt{66}}{16} = \dfrac{4 \pm \sqrt{66}}{4} \approx 3.0$ seconds

Quadratic formula method was used because the numbers were large and equation would not factor.

(d) $0 = -16t^2 + 32t + 100$ $0 = -16t^2 + 32t + 84$

$0 = 4t^2 - 8t - 25$ $0 = 4t^2 - 8t - 21$

$t = \dfrac{-(-8) \pm \sqrt{(-8)^2 - 4(4)(-25)}}{2(4)}$ $t = \dfrac{-(-8) \pm \sqrt{(-8)^2 - 4(4)(-21)}}{2(4)}$

$t = \dfrac{8 \pm \sqrt{64 + 400}}{8}$ $t = \dfrac{8 \pm \sqrt{64 + 336}}{8}$

$t = \dfrac{8 \pm \sqrt{464}}{8}$ $t = \dfrac{8 \pm \sqrt{400}}{8}$

$t \approx 3.7$ seconds $t \approx 3.5$ seconds

105. $b^2 - 4ac$. If the discriminant is positive, the quadratic equation has two real solutions; if it is zero, the equation has one (repeated) real solution; and if it is negative, the equation has no real solutions.

107. The four methods are factoring, extracting square roots, completing the square, and the Quadratic Formula.

Mid-Chapter Quiz for Chapter 10

1. $2x^2 - 72 = 0$

$2(x^2 - 36) = 0$

$2(x - 6)(x + 6) = 0$

$x - 6 = 0 \qquad x + 6 = 0$

$x = 6 \qquad\quad x = -6$

2. $2x^2 + 3x - 20 = 0$

$(2x - 5)(x + 4) = 0$

$2x - 5 = 0 \qquad x + 4 = 0$

$x = \frac{5}{2} \qquad\quad x = -4$

3. $t^2 = 12$

$t = \pm\sqrt{12}$

$t = \pm 2\sqrt{3}$

4. $(u - 3)^2 - 16 = 0$

$(u - 3)^2 = 16$

$u - 3 = \pm 4$

$u = 3 \pm 4 = 7, -1$

5. $s^2 + 10s + 1 = 0$

$s^2 + 10s = -1$

$s^2 + 10s + 25 = -1 + 25$

$(s + 5)^2 = 24$

$s + 5 = \pm\sqrt{24}$

$s = -5 \pm 2\sqrt{6}$

6. $2y^2 + 6y - 5 = 0$

$y^2 + 3y = \dfrac{5}{2}$

$y^2 + 3y + \dfrac{9}{4} = \dfrac{5}{2} + \dfrac{9}{4}$

$\left(y + \dfrac{3}{2}\right)^2 = \dfrac{10}{4} + \dfrac{9}{4}$

$\left(y + \dfrac{3}{2}\right)^2 = \dfrac{19}{4}$

$y + \dfrac{3}{2} = \pm\dfrac{\sqrt{19}}{2}$

$y = -\dfrac{3}{2} \pm \dfrac{\sqrt{19}}{2}$

7. $x = \dfrac{-4 \pm \sqrt{4^2 - 4(1)(-6)}}{2(1)}$

$x = \dfrac{-4 \pm \sqrt{16 + 24}}{2}$

$x = \dfrac{-4 \pm \sqrt{40}}{2}$

$x = \dfrac{-4 \pm 2\sqrt{10}}{2} = -2 \pm \sqrt{10}$

8. $v = \dfrac{-(-3) \pm \sqrt{(-3)^2 - 4(6)(-4)}}{2(6)}$

$v = \dfrac{3 \pm \sqrt{9 + 96}}{12}$

$v = \dfrac{3 \pm \sqrt{105}}{12}$

9. $x = \dfrac{-5 \pm \sqrt{5^2 - 4(1)(7)}}{2(1)}$

$x = \dfrac{-5 \pm \sqrt{25 - 28}}{2}$

$x = \dfrac{-5 \pm \sqrt{-3}}{2}$

$x = \dfrac{-5 \pm i\sqrt{3}}{2} = -\dfrac{5}{2} \pm \dfrac{\sqrt{3}}{2}i$

10. $\quad 36 = (t - 4)^2$

$\quad \pm 6 = t - 4$

$\quad 4 \pm 6 = t$

$\quad 10, -2 = t$

11. $(x - 10)(x + 3) = 0$

$\quad (x - 10) = 0 \qquad x + 3 = 0$

$\qquad\qquad x = 10 \qquad\quad x = -3$

12. $x^2 - 3x - 10 = 0$

$\quad (x - 5)(x + 2) = 0$

$\quad x - 5 = 0 \qquad x + 2 = 0$

$\qquad x = 5 \qquad\quad x = -2$

13. $(2b - 3)(2b - 3) = 0$

$\quad 2b - 3 = 0 \qquad 2b - 3 = 0$

$\qquad b = \dfrac{3}{2} \qquad\quad b = \dfrac{3}{2}$

14. $m = \dfrac{-10 \pm \sqrt{10^2 - 4(3)(5)}}{2(3)}$

$m = \dfrac{-10 \pm \sqrt{100 - 60}}{6}$

$m = \dfrac{-10 \pm \sqrt{40}}{6}$

$m = \dfrac{-10 \pm 2\sqrt{10}}{6}$

$m = \dfrac{-5 \pm \sqrt{10}}{3}$

15. $\quad x - 2\sqrt{x} - 24 = 0$

$\qquad\qquad\quad$ let $u = \sqrt{x}$

$\left(\sqrt{x}\right)^2 - 2\sqrt{x} - 24 = 0$

$\qquad u^2 - 2u - 24 = 0$

$\qquad (u - 6)(u + 4) = 0$

$\quad u = 6 \qquad\qquad u = -4$

$\quad \sqrt{x} = 6 \qquad\quad \sqrt{x} = -4$

$\quad x = 6^2 \qquad\qquad x = (-4)^2$

$\quad x = 36 \qquad\qquad x = 16$

Check: $\qquad\qquad\qquad$ Not a solution

$36 - 2\sqrt{36} - 24 \overset{?}{=} 0 \qquad 16 - 2\sqrt{16} - 24 \overset{?}{=} 0$

$\quad 36 - 12 - 24 \overset{?}{=} 0 \qquad\quad 16 - 8 - 24 \overset{?}{=} 0$

$\qquad\qquad\qquad 0 = 0 \qquad\qquad\qquad -16 \neq 0$

16. $\quad x^4 + 7x^2 + 12 = 0$

$\quad (x^2 + 4)(x^2 + 3) = 0$

$\quad x^2 = -4 \qquad\quad x^2 = -3$

$\quad x = \pm\sqrt{-4} \qquad x = \pm\sqrt{-3}$

$\quad x = \pm 2i \qquad\quad x = \pm\sqrt{3}i$

17. *Keystrokes:*

$\boxed{Y=}$ .5 $\boxed{X,T,\theta}$ $\boxed{x^2}$ $\boxed{-}$ 3 $\boxed{X,T,\theta}$ $\boxed{-}$ 1 $\boxed{GRAPH}$

$0 = .5x^2 - 3x - 1$

$0 = x^2 - 6x - 2$

$x = \dfrac{-(-6) \pm \sqrt{(-6)^2 - 4(1)(-2)}}{2(1)}$

$x = \dfrac{6 \pm \sqrt{36 + 8}}{2}$

$x = \dfrac{6 \pm \sqrt{44}}{2}$

$x = \dfrac{6 \pm 2\sqrt{11}}{2}$

$x = 3 \pm \sqrt{11}$

$x \approx 6.32$ and -0.32

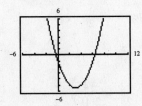

18. *Keystrokes:*

$\boxed{Y=}$ $\boxed{X,T,\theta}$ $\boxed{x^2}$ $\boxed{+}$ .045 $\boxed{X,T,\theta}$ $\boxed{-}$ 4 $\boxed{GRAPH}$

$0 = x^2 + 0.45x - 4$

$x = \dfrac{-0.45 \pm \sqrt{(0.45)^2 - 4(1)(-4)}}{2(1)}$

$x = \dfrac{-0.45 \pm \sqrt{0.2025 + 16}}{2}$

$x = \dfrac{-0.45 \pm \sqrt{16.2025}}{2}$

$x \approx 1.79$ and -2.24

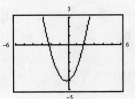

19.

$500 = x(20 - 0.2x)$

$\qquad = 20x - 0.2x^2$

$0.2x^2 - 20x + 500 = 0$

$x^2 - 100x + 2500 = 0$

$x = \dfrac{-(-100) \pm \sqrt{(-100)^2 - 4(1)(2500)}}{2(1)}$

$x = \dfrac{100 \pm \sqrt{10,000 - 10,000}}{2}$

$x = \dfrac{100 \pm \sqrt{0}}{2} = 50$ units

20. *Verbal model:* $\boxed{\text{Area}} = \boxed{\text{Length}} \cdot \boxed{\text{Width}}$

Equation:

$2275 = x \cdot (100 - x)$

$2275 = 100x - x^2$

$0 = x^2 - 100x + 2275$

$0 = (x - 35)(x - 65)$

$x - 35 = 0 \qquad\qquad x - 65 = 0$

$x = 35$ meters $\qquad x = 65$ meters

35 meters $\times$ 65 meters

Section 10.4 Graphs of Quadratic Functions

1. $y = 4 - 2x$ (e)

3. $y = x^2 - 3$ (b)

5. $y = (x - 2)^2$ (d)

7. $y = x^2 + 2 = (x - 0)^2 + 2$
vertex $(0, 2)$

9. $y = x^2 - 4x + 7$
$= (x^2 - 4x + 4) + 7 - 4$
$= (x - 2)^2 + 3$
vertex $= (2, 3)$

11. $y = x^2 + 6x + 5$
$y = (x^2 + 6x + 9) + 5 - 9$
$y = (x + 3)^2 - 4$
vertex $= (-3, -4)$

13. $y = -x^2 + 6x - 10$
$y = -1(x^2 - 6x) - 10$
$y = -1(x^2 - 6x + 9) - 10 + 9$
$y = -1(x - 3)^2 - 1$
vertex $(3, -1)$

15. $y = -x^2 + 2x - 7$
$= -1(x^2 - 2x + 1) - 7 + 1$
$= -1(x - 1)^2 - 6$
vertex $= (1, -6)$

17. $y = 2x^2 + 6x + 2$
$= 2\left(x^2 + 3x + \dfrac{9}{4}\right) + 2 - \dfrac{9}{2}$
$= 2\left(x + \dfrac{3}{2}\right)^2 - \dfrac{5}{2}$
vertex $= \left(-\dfrac{3}{2}, -\dfrac{5}{2}\right)$

19. $a = 1$ $b = -8$
$x = \dfrac{-b}{2a} = \dfrac{-(-8)}{2(1)} = 4$
$f\left(-\dfrac{b}{2a}\right) = 4^2 - 8(4) + 15$
$= 16 - 32 + 15$
$= -1$
vertex $= (4, -1)$

21. $a = -1$ $b = -2$
$x = \dfrac{-b}{2a} = \dfrac{-(-2)}{2(-1)} = -1$
$g\left(\dfrac{-b}{2a}\right) = -(-1)^2 - 2(-1) + 1$
$= -1 + 2 + 1$
$= 2$
vertex $= (-1, 2)$

23. $a = 4$ $b = 4$
$x = \dfrac{-b}{2a} = \dfrac{-4}{2(4)} = \dfrac{-1}{2}$
$y = 4\left(-\dfrac{1}{2}\right)^2 + 4\left(-\dfrac{1}{2}\right) + 4$
$= 4\left(\dfrac{1}{4}\right) - 2 + 4$
$= 1 - 2 + 4$
$= 3$
vertex $= \left(-\dfrac{1}{2}, 3\right)$

25. $2 > 0$ opens up
vertex $= (0, 2)$

27. $-1 < 0$ opens down
vertex $= (10, 4)$

29. $1 > 0$ opens up
vertex $= (0, -6)$

31. $-1 < 0$ opens down
vertex $= (3, 0)$

33. $y = 25 - x^2$
$0 = 25 - x^2$
$x^2 = 25$
$x = \pm 5$
$(5, 0), (-5, 0)$
$y = 25 - x^2$
$y = 25 - 0^2$
$y = 25$
$(0, 25)$

35. $y = x^2 - 9x$
$0 = x^2 - 9x$
$0 = x(x - 9)$
$(0, 0), (9, 0)$
$y = x^2 - 9x$
$y = 0^2 - 9(0)$
$y = 0$
$(0, 0)$

37. $y = 4x^2 - 12x + 9$

$0 = 4x^2 - 12x + 9$

$0 = (2x - 3)^2$

$0 = 2x - 3$

$\dfrac{3}{2} = x$

$\left(\dfrac{3}{2}, 0\right)$

$y = 4x^2 - 12x + 9$

$y = 4(0)^2 - 12(0) + 9$

$y = 9$

$(0, 9)$

39. $y = x^2 - 3x + 3$

$0 = x^2 - 3x + 3$

$x = \dfrac{3 \pm \sqrt{9 - 12}}{2}$

$= \dfrac{3 \pm \sqrt{-3}}{2}$

no x-intercepts

$y = x^2 - 3x + 3$

$y = 0^2 - 3(0) + 3$

$y = 3$

$(0, 3)$

41. x-intercepts

$0 = x^2 - 4$

$0 = (x - 2)(x + 2)$

$x = 2 \quad x = -2$

vertex

$g(x) = (x - 0)^2 - 4$

$(0, -4)$

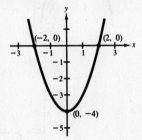

43. x-intercepts

$0 = -x^2 + 4$

$x^2 = 4$

$x = \pm 2$

vertex

$f(x) = -(x - 0)^2 + 4$

$(0, 4)$

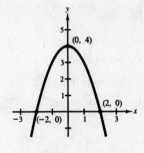

45. x-intercepts

$0 = x^2 - 3x$

$0 = x(x - 3)$

$0 = x \quad x = 3$

vertex

$f(x) = \left(x^2 - 3x + \dfrac{9}{4}\right) - \dfrac{9}{4}$

$f(x) = \left(x - \dfrac{3}{2}\right)^2 - \dfrac{9}{4}$

$\left(\dfrac{3}{2}, -\dfrac{9}{4}\right)$

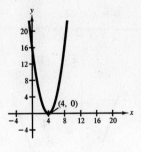

47. x-intercepts

$0 = -x^2 + 3x$

$0 = -x(x - 3)$

$0 = x \quad x = 3$

vertex

$y = -1\left(x^2 - 3x + \dfrac{9}{4}\right) + \dfrac{9}{4}$

$= -1\left(x - \dfrac{3}{2}\right)^2 + \dfrac{9}{4}$

$\left(\dfrac{3}{2}, \dfrac{9}{4}\right)$

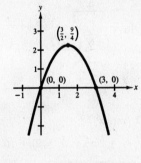

49. x-intercepts

$0 = (x - 4)^2$

$0 = x - 4$

$4 = x$

vertex

$y = (x - 4)^2 + 0$

$(4, 0)$

51. x-intercepts

$0 = x^2 - 8x + 15$

$0 = (x - 5)(x - 3)$

$5 = x \quad x = 3$

vertex

$y = (x^2 - 8x + 16) + 15 - 16$

$\quad = (x - 4)^2 - 1$

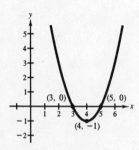

53. x-intercepts

$0 = x^2 + 6x + 5$

$0 = (x + 5)(x + 1)$

$-5 = x \quad x = -1$

vertex

$y = -(x^2 + 6x + 9) - 5 + 9$

$y = -(x + 3)^2 + 4$

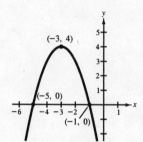

55. x-intercepts

$0 = -x^2 + 6x - 7$

$x = \dfrac{-6 \pm \sqrt{6^2 - 4(-1)(-7)}}{2(-1)}$

$x = \dfrac{-6 \pm \sqrt{36 - 28}}{-2}$

$x = \dfrac{-6 \pm \sqrt{8}}{-2}$

$x = \dfrac{-6 \pm 2\sqrt{2}}{2}$

$x = -3 \pm \sqrt{2}$

vertex

$q(x) = -x^2 + 6x - 7$

$q(x) = -(x^2 - 6x) - 7$

$q(x) = -(x^2 - 6x + 9 - 9) - 7$

$q(x) = -(x^2 - 6x + 9) + 9 - 7$

$q(x) = -(x - 3)^2 + 2$

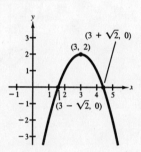

57. vertex

$y = 2(x^2 + 6x + 9) + 16 - 18$

$y = 2(x + 3)^2 - 2$

x-intercepts

$0 = x^2 + 6x + 8$

$0 = (x + 4)(x + 2)$

$-4 = x \quad x = -2$

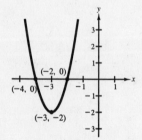

59. vertex

$y = \frac{1}{2}(x^2 - 2x + 1) - \frac{3}{2} - \frac{1}{2}$

$y = \frac{1}{2}(x - 1)^2 - 2$

x-intercepts

$0 = x^2 - 2x - 3$

$0 = (x - 3)(x + 1)$

$3 = x \quad x = -1$

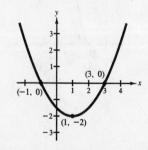

61. $y = \frac{3}{5}(x^2 - 8x + 16) + \frac{38}{5} - \frac{48}{5}$

$y = \frac{3}{5}(x - 4)^2 - 2$

$0 = 3x^2 - 24x + 38$

$x = \frac{24 \pm \sqrt{576 - 456}}{6}$

$x = \frac{24 \pm \sqrt{120}}{6} = \frac{12 \pm \sqrt{30}}{3}$

$\approx 5.83, 2.17$

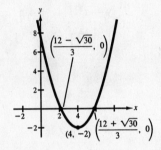

63. $f(x) = -\frac{1}{3}x^2 + 5$

$f(x) = -\frac{1}{3}(x - 0)^2 + 5$

$0 = -\frac{1}{3}x^2 + 5$

$\frac{1}{3}x^2 = 5$

$x^2 = 15$

$x = \pm\sqrt{15}$

$x \approx 3.87, -3.87$

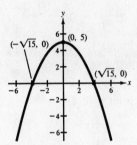

65.

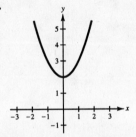

Vertical shift 2 units up.

67.

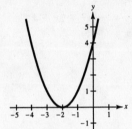

Horizontal shift 2 units left.

69.

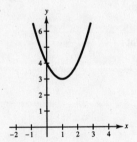

Horizontal shift 1 unit right.
Vertical shift 3 units up.

71.

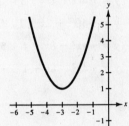

Horizontal shift 3 units left.
Vertical shift 1 unit up.

73. *Keystrokes:*

$\boxed{\text{Y=}}\ \boxed{(}\ \boxed{(}\ 1\ \boxed{\div}\ 6\ \boxed{)}\ \boxed{)}\ \boxed{(}\ 2\ \boxed{\text{X,T,}\theta}\ \boxed{x^2}\ \boxed{-}\ 8\ \boxed{\text{X,T,}\theta}\ \boxed{+}\ 11\ \boxed{)}\ \boxed{\text{GRAPH}}$

vertex $= (2, 0.5)$

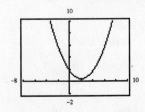

75. *Keystrokes:*

$\boxed{Y=}$ $\boxed{(-)}$ .7 $\boxed{X,T,\theta}$ $\boxed{x^2}$ $\boxed{(-)}$ 2.7 $\boxed{X,T,\theta}$ $\boxed{+}$ 2.3 $\boxed{\text{GRAPH}}$

vertex $= (-1.9, 4.9)$

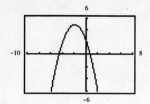

77. vertex $= (0, 4)$ point $= (-2, 0)$

$y = a(x - 0)^2 + 4 \qquad y = -1(x - 0)^2 + 4$

$0 = a(-2 - 0)^2 + 4 \qquad y = -x^2 + 4$

$0 = 4a + 4$

$-4 = 4a$

$-1 = a$

79. vertex $= (-2, 2)$ point $= (0, 2)$

$y = a(x - (-2))^2 + (-2) \qquad y = 1(x + 2)^2 - 2$

$y = a(x + 2)^2 - 2 \qquad\qquad y = (x + 2)^2 - 2$

$2 = a(0 + 2)^2 - 2 \qquad\qquad y = x^2 + 4x + 4 - 2$

$2 = 4a - 2 \qquad\qquad\qquad y = x^2 + 4x + 2$

$4 = 4a$

$1 = a$

81. vertex $= (2, 6)$ point $= (0, 4)$

$y = a(x - 2)^2 + 6 \qquad y = -\frac{1}{2}(x - 2)^2 + 6$

$4 = a(0 - 2)^2 + 6 \qquad y = -\frac{1}{2}(x^2 - 4x + 4) + 6$

$4 = a(4) + 6 \qquad\qquad y = -\frac{1}{2}x^2 + 2x - 2 + 6$

$-2 = a(4) \qquad\qquad\quad y = -\frac{1}{2}x^2 + 2x + 4$

$-\frac{2}{4} = a$

83. $y = 1(x - 2)^2 + 1 = x^2 - 4x + 5$

85. $0 = a(0 - 2)^2 - 4$

$4 = a(4)$

$1 = a$

$y = 1(x - 2)^2 - 4 = x^2 - 4x$

87. $4 = a(1 - 3)^2 + 2$

$2 = a(4)$

$\frac{1}{2} = a$

$y = \frac{1}{2}(x - 3)^2 + 2 = \frac{1}{2}x^2 - 3x + \frac{13}{2}$

89. $1 = a(0 - (-1))^2 + 5 \qquad y = -4(x + 1)^2 + 5$

$1 = a(1) + 5 \qquad\qquad\quad y = -4(x^2 + 2x + 1) + 5$

$-4 = a \qquad\qquad\qquad\quad y = -4x^2 - 8x - 4 + 5$

$\qquad\qquad\qquad\qquad\qquad y = -4x^2 - 8x + 1$

91. Horizontal shift 3 units right

93. Horizontal shift 2 units right

Vertical shift 3 units down

95. (a) $y = -\dfrac{1}{12}(0)^2 + 2(0) + 4$

$\qquad y = 4$ feet

(b) $y = -\dfrac{1}{12}x^2 + 2x + 4$

$\qquad y = -\dfrac{1}{12}(x^2 - 24x + 144) + 4 + 12$

$\qquad y = -\dfrac{1}{12}(x - 12)^2 + 16$

Maximum height $= 16$ feet

(c) $0 = -\dfrac{1}{12}x^2 + 2x + 4$

$\qquad 0 = x^2 - 24x - 48$

$\qquad x = \dfrac{24 \pm \sqrt{576 + 192}}{2}$

$\qquad\quad \approx 25.86$ feet

97. $y = -\frac{4}{9}x^2 + \frac{24}{9}x + 10$

$y = -\frac{4}{9}(x^2 - 6x) + 10$

$y = -\frac{4}{9}(x^2 - 6x + 9 - 9) + 10$

$y = -\frac{4}{9}(x^2 - 6x + 9) + 4 + 10$

$y = -\frac{4}{9}(x - 3)^2 + 14$

The maximum height of the diver is 14 ft.

99. (a)

(b) vertex = $(3.65, 110, 810)$ 1993, 110,800 reserves

101. (a) $P = (100 + x)[90 - x(0.15)] - (100 + x)60$

$P = (100 + x)[90 - x(0.15) - 60]$

$P = (100 + x)(30 - 0.15x)$

$P = 3000 - 15x + 30x = 0.15x^2$

$P = 3000 + 15x - 0.15x^2$

$P = 3000 + 15x - \frac{3}{20}x^2$

(b) $P = -\frac{3}{20}x^2 + 15x + 3000$

$P = -\frac{3}{20}(x^2 - 100x + 2500) + 3000 + 375$

$P = -\frac{3}{20}(x - 50)^2 + 3375$

vertex = $(50, 3375)$

order size for maximum profit

$P = 100 + 50 = 150$ radios

(c) Recommend pricing scheme if price reductions are restricted to orders between 100 and 150 orders.

103. *Keystrokes:*

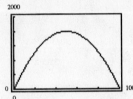

$x \approx 50$ when A is maximum

105. $100 = a(500 - 0)^2 + 0$

$100 = a(250,000)$

$\dfrac{100}{250,000} = a$

$\dfrac{1}{2500} = a$

$y = \dfrac{1}{2500}(x - 0)^2 + 0$

$y = \dfrac{1}{2500}x^2$

107. The graph of the quadratic function $f(x) = ax^2 + bx + c$ is a parabola.

109. To find any x-intercepts, set $y = 0$ and solve the resulting equation for x.

To find the y-intercept, set $x = 0$ and solve the resulting equation for y.

111. The discriminant of a quadratic function tells how many x-intercepts the parabola has. If positive, there are 2 x-intercepts; if zero, 1 x-intercept; and if negative, no x-intercepts.

113. Find the y-coordinate of the vertex. This is the maximum (or minimum) value of a quadratic function.

Section 10.5   Applications of Quadratic Equations

1. *Verbal model:* $\boxed{\begin{array}{c}\text{Selling price} \\ \text{per doz eggs}\end{array}} = \boxed{\begin{array}{c}\text{Cost per} \\ \text{doz eggs}\end{array}} + \boxed{\begin{array}{c}\text{Profit per} \\ \text{doz eggs}\end{array}}$

Equation: $\dfrac{21.60}{x} = \dfrac{21.60}{x+6} + 0.30$

Labels: Number eggs sold $= x$
Number eggs purchased $= x + 6$

$$21.60(x+6) = 21.60x + 0.30x(x+6)$$

$$21.6x + 129.6 = 21.6x + 0.3x^2 + 1.8x$$

$$0 = 0.3x^2 + 1.8x - 129.6$$

$$0 = 3x^2 + 18x - 1296$$

$$0 = x^2 + 6x - 432$$

$$0 = (x+24)(x-18)$$

$$x = -24 \quad x = 18 \text{ dozen}$$

Selling price $= \dfrac{21.60}{18} = \$1.20$ per dozen

3. *Verbal model:* $\boxed{\begin{array}{c}\text{Selling price} \\ \text{per video}\end{array}} = \boxed{\begin{array}{c}\text{Cost per} \\ \text{video}\end{array}} + \boxed{\begin{array}{c}\text{Profit per} \\ \text{video}\end{array}}$

Labels: Number videos sold $= x$
Number videos purchased $= x + 8$

Equation: $\dfrac{480}{x} = \dfrac{480}{x+8} + 10$

$$480(x+8) = 480x + 10x(x+8)$$

$$480x + 3840 = 480x + 10x^2 + 80x$$

$$0 = 10x^2 + 80x - 3840$$

$$0 = x^2 + 8x - 384$$

$$0 = (x+24)(x-16)$$

$$x = -24 \quad x = 16 \text{ videos}$$

Selling price $= \dfrac{480}{16} = \$30$

5. *Verbal model:* $2\boxed{\text{Length}} + 2\boxed{\text{Width}} = \boxed{\text{Perimeter}}$

Labels: Length $= l$
Width $= 0.75\,l$

Equation: $2l + 2(0.75l) = 42$

$$2l + 1.5l = 42$$

$$3.5l = 42$$

$$l = 12 \text{ inches}$$

$$w = 0.75\,l = 9 \text{ inches}$$

Verbal model: $\boxed{\text{Length}} \cdot \boxed{\text{Width}} = \boxed{\text{Area}}$

Equation: $12 \cdot 9 = A$

108 square inches $= A$

7. *Verbal model:* $\boxed{\text{Area}} = \boxed{\text{Length}} \cdot \boxed{\text{Width}}$

Labels: Length $= 2.5w$
Width $= w$

Equation: $250 = 2.5w \cdot w$

$$250 = 2.5w^2$$

$$100 = w^2$$

$$10 = w$$

$$25 = 2.5w$$

Verbal model: $2\boxed{\text{Length}} + 2\boxed{\text{Width}} = \boxed{\text{Perimeter}}$

Equation: $2(25) + 2(10) = P$

70 feet $= P$

9. *Verbal model:* $\boxed{\text{Length}}$ · $\boxed{\text{Width}}$ = $\boxed{\text{Area}}$

Labels: Length = l

Width = $\frac{1}{3}l$

Equation: $l \cdot \frac{1}{3}l = 192$

$\frac{1}{3}l^2 = 192$

$l^2 = 576$

$l = 24$ inches

$w = \frac{1}{3}l = 8$ inches

Verbal model: 2 $\boxed{\text{Length}}$ + 2 $\boxed{\text{Width}}$ = $\boxed{\text{Perimeter}}$

Equation: $2(24) + 2(8) = P$

$48 + 16 = P$

64 inches $= P$

11. *Verbal model:* 2 $\boxed{\text{Length}}$ + 2 $\boxed{\text{Width}}$ = $\boxed{\text{Perimeter}}$

Labels: Length = $w + 3$

Width = w

Equation: $2(w + 3) + 2w = 54$

$2w + 6 + 2w = 54$

$4w = 48$

$w = 12$ km

$l = w + 3 = 15$ km

Verbal model: $\boxed{\text{Length}}$ · $\boxed{\text{Width}}$ = $\boxed{\text{Area}}$

Equation: $15 \cdot 12 = 180$ square kilometers $= A$

13. *Verbal model:* $\boxed{\text{Length}}$ · $\boxed{\text{Width}}$ = $\boxed{\text{Area}}$

Labels: Length = l

Width = $l - 20$

Equation: $l \cdot (l - 20) = 12{,}000$

$l^2 - 20l = 12{,}000$

$l^2 - 20l + 100 = 12{,}000 + 100$

$(l - 10)^2 = 12{,}100$

$l - 10 = \pm\sqrt{12{,}100}$

$l = 10 + 110 = 120$ meters

$w = l - 20 = 100$ meters

Verbal model: 2 $\boxed{\text{Length}}$ + 2 $\boxed{\text{Width}}$ = $\boxed{\text{Perimeter}}$

Equation: $2(120) + 2(100) = 440$ meters $= P$

15. *Verbal model:* $\boxed{\text{Area}}$ = $\boxed{\text{Length}}$ · $\boxed{\text{Width}}$

Labels: Length = $x + 4$

Width = x

Equation: $192 = (x + 4)x$

$192 = x^2 + 4x$

$0 = x^2 + 4x - 192$

$0 = (x + 16)(x - 12)$

$x = -16$ $x = 12$ inches

$x + 4 = 16$ inches

17. *Verbal model:* $\boxed{\text{Area}}$ = $\frac{1}{2}$ · $\boxed{\text{Height}}$ · $\boxed{\text{Base}}$

Labels: Height = $x - 8$

Base = x

Equation: $192 = \frac{1}{2}(x - 8)x$

$384 = x^2 - 8x$

$0 = x^2 - 8x - 384$

$0 = (x - 24)(x + 16)$

$x = 24$ inches   reject $x = -16$

$x - 8 = 16$ inches

19. *Verbal model:* $\boxed{\text{Length}}$ · $\boxed{\text{Width}}$ = $\boxed{\text{Area}}$

Labels: Length = $350 - 2x$

Width = x

Equation: $(350 - 2x) \cdot x = 12{,}500$

$350x - 2x^2 = 12{,}500$

$2x^2 - 350x + 12{,}500 = 0$

$x^2 - 175x + 6{,}250 = 0$

$x = \dfrac{175 \pm \sqrt{175^2 - 4(1)(6{,}250)}}{2(1)}$

$x = \dfrac{175 \pm \sqrt{5625}}{2} = \dfrac{175 \pm 75}{2}$

$x = 125, 50$

$350 - 2x = 100, 250$

100 ft × 125 ft. or 50 ft × 250 ft.

21. *Verbal model:* $\boxed{\text{Side 1}}$ + $\boxed{\text{Side 2}}$ + $\boxed{\text{Side 3}}$ = 550

Equation: $x + x + b = 550$

$$2x + b = 550$$

$$b = 550 - 2x$$

Verbal model: $\boxed{\frac{1}{2}}$ · $\boxed{\text{Height}}$ $\boxed{(}$ $\boxed{\text{Base 1}}$ + $\boxed{\text{Base 2}}$ $\boxed{)}$ = $\boxed{\text{Area}}$

Labels: Height = x

Base 1 = x

Base 2 = 6

Equation: $\frac{1}{2}x(x + b) = 43{,}560$

$$\frac{1}{2}x(x + 550 - 2x) = 43{,}560$$

$$\frac{1}{2}x(-x + 550) = 43{,}560$$

$$-\frac{1}{2}x^2 + 275x = 43{,}560$$

$$-x^2 + 550x = 87{,}120$$

$$0 = x^2 - 550x + 87{,}120$$

This has no real solution, so it would be impossible to have an area of 43,560 square feet.

23. *Verbal model:* $\boxed{\text{Height}}$ · $\boxed{\text{Width}}$ = $\boxed{\text{Area}}$

Labels: Height = x

Width = $48 - 2x$

Equation: $x \cdot (48 - 2x) = 288$

$$2x^2 - 48x + 288 = 0$$

$$x^2 - 24x + 144 = 0$$

$$(x - 12)(x - 12) = 0$$

$$x = 12$$

height = 12 inches

width = $48 - 2(12)$

$$= 48 - 24 = 24 \text{ inches}$$

25. $A = P(1 + r)^2$

$3499.20 = 3000(1 + r)^2$

$1.1664 = (1 + r)^2$

$1.08 = 1 + r$

$0.08 = r$ or 8%

27. $A = P(1 + r)^2$

$280.90 = 250.00(1 + r)^2$

$\dfrac{280.90}{250.00} = (1 + r)^2$

$1.1236 = (1 + r)^2$

$1.06 = 1 + r$

$.06 = r$

$6\% = r$

29. $A = P(1 + r)^2$

$8420.20 = 8000.00(1 + r)^2$

$1.052525 = (1 + r)^2$

$1.0259 \approx 1 + r$

$.0259 \approx r$ or 2.59%

31. *Verbal model:*

$$\boxed{\begin{array}{c}\text{Cost per}\\\text{member}\end{array}} \cdot \boxed{\begin{array}{c}\text{Number of}\\\text{members}\end{array}} = \boxed{\$240}$$

Labels: Number of members $= x$

Number going to game $= x + 8$

Equation:

$$\left(\frac{240}{x} - 1\right) \cdot (x + 8) = 240$$

$$\left(\frac{240 - x}{x}\right)(x + 8) = 240$$

$$(240 - x)(x + 8) = 240x$$

$$240x + 1920 - x^2 - 8x = 240x$$

$$-x^2 - 8x + 1920 = 0$$

$$x^2 + 8x - 1920 = 0$$

$$(x + 48)(x - 40) = 0$$

$$x = -48 \quad x = 40$$

$$x + 8 = 48$$

33. *Verbal model:*

$$\boxed{\begin{array}{c}\text{Investment per}\\\text{person; current group}\end{array}} - \boxed{\begin{array}{c}\text{Investment per}\\\text{person; new group}\end{array}} = \boxed{6000}$$

Labels: Number in current group $= x$

Number in new group $= x + 3$

Equation:

$$\frac{80,000}{x} - \frac{80,000}{x + 3} = 6000$$

$$x(x + 3)\left(\frac{80,000}{x} - \frac{80,000}{x + 3}\right) = (6000)x(x + 3)$$

$$80,000(x + 3) - 80,000x = 6000(x^2 + 3x)$$

$$80,000x + 240,000 - 80,000x = 6000x^2 + 18,000x$$

$$0 = 6000x^2 + 18,000x - 240,000$$

$$0 = x^2 + 3x - 40$$

$$0 = (x + 8)(x - 5)$$

$$x + 8 = 0 \qquad x - 5 = 0$$

$$\cancel{x = -8} \qquad\qquad x = 5 \text{ investors}$$

35. *Common Formula:* $a^2 + b^2 = c^2$

Equation:

$$x^2 + (18 - x)^2 = 16^2$$

$$x^2 + 324 - 36x + x^2 = 256$$

$$2x^2 - 36x + 68 = 0$$

$$x^2 - 18x + 34 = 0$$

$$x = \frac{18 \pm \sqrt{18^2 - 4(1)(34)}}{2(1)}$$

$$x = \frac{18 \pm \sqrt{324 - 136}}{2} = \frac{18 \pm \sqrt{188}}{2}$$

$$x = 15.855655, \quad \text{reject } 2.1443454$$

$$\approx 15.86 \text{ miles}$$

37. (a) $d = \sqrt{(3 + x)^2 + (4 + x)^2}$

Keystrokes:

$\boxed{Y=}\ \boxed{\sqrt{\ }}\ \boxed{(}\ \boxed{(}\ 3\ \boxed{+}\ \boxed{X,T,\theta}\ \boxed{)}\ \boxed{x^2}\ \boxed{+}\ \boxed{(}\ 4\ \boxed{+}\ \boxed{X,T,\theta}\ \boxed{)}\ \boxed{x^2}\ \boxed{)}\ \boxed{GRAPH}$

Approximate value of $x \approx 3.55$ when $d = 10$.

(b) $10 = \sqrt{(3 + x)^2 + (4 + x)^2}$

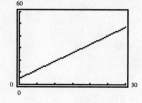

$100 = (3 + x)^2 + (4 + x)^2$

$ = 9 + 6x + x^2 + 16 + 8x + x^2$

$0 = 2x^2 + 14x - 75$

$x = \dfrac{-14 \pm \sqrt{14^2 - 4(2)(-75)}}{2(2)}$

$x = \dfrac{-14 \pm \sqrt{196 + 600}}{4}$

$x = \dfrac{-14 \pm \sqrt{796}}{4}$

$x = \dfrac{14 \pm 2\sqrt{199}}{4}$

$x = \dfrac{-7 \pm \sqrt{199}}{2} \approx 3.55$ meters

39. *Verbal model:* $\boxed{\begin{array}{c}\text{Work done by}\\ \text{Person 1}\end{array}} + \boxed{\begin{array}{c}\text{Work done by}\\ \text{Person 2}\end{array}} = \boxed{\text{One complete job}}$

Labels: Time to do job by Person 1 $= x$

Time to do job by Person 2 $= x + 2$

Equation: $\dfrac{1}{x}(5) + \dfrac{1}{x + 2}(5) = 1$

$x(x + 2)\left[(5)\left(\dfrac{1}{x} + \dfrac{1}{x + 2}\right) = 1\right]x(x + 2)$

$5(x + 2) + 5x = x(x + 2)$

$5x + 10 + 5x = x^2 + 2x$

$-x^2 + 8x + 10 = 0$

$x^2 - 8x - 10 = 0$

$x = \dfrac{8 \pm \sqrt{(-8)^2 - 4(1)(-10)}}{2(1)}$

$x = \dfrac{8 \pm \sqrt{64 + 40}}{2}$

$x = \dfrac{8 \pm \sqrt{104}}{2}$

$x \approx 9.1$ hours, reject -1.1

$x + 2 \approx 11.1$ hours

41. *Verbal model:* | Rate Company A | + | Rate Company B | = | Rate together |

Labels: Time Company A $= x + 3$

Time Company B $= x$

Equation: $$\frac{1}{x + 3} + \frac{1}{x} = \frac{1}{4}$$

$$4x(x + 3)\left(\frac{1}{x + 3} + \frac{1}{x}\right) = \left(\frac{1}{4}\right)4x(x + 3)$$

$$4x + 4(x + 3) = x(x + 3)$$

$$4x + 4x + 12 = x^2 + 3x$$

$$0 = x^2 - 5x - 12$$

$$x = \frac{-(-5) \pm \sqrt{(-5)^2 - 4(1)(-12)}}{2(1)}$$

$$x = \frac{5 \pm \sqrt{25 + 48}}{2}$$

$$x = \frac{5 \pm \sqrt{73}}{2}$$

$$x \approx 6.8 \text{ days} \quad \cancel{x = 1.8}$$

$$x + 3 \approx 9.8$$

43. $h = h_0 - 16t^2$

$0 = 144 - 16t^2$

$16t^2 = 144$

$t^2 = 9$

$t = 3 \text{ seconds}$

45. $h = h_0 - 16t^2$

$0 = 1454 - 16t^2$

$16t^2 = 1454$

$t^2 = 90.875$

$t = 9.532838 \text{ seconds} \approx 9.5 \text{ seconds}$

47. $h = 3 + 75t - 16t^2$

$0 = 3 + 75t - 16t^2$

$0 = 16t^2 - 75t - 3$

$t = \dfrac{75 \pm \sqrt{(-75)^2 - 4(16)(-3)}}{2(16)}$

$t = \dfrac{75 \pm \sqrt{5625 + 192}}{32}$

$t = \dfrac{75 \pm \sqrt{5817}}{32}$

$t = \dfrac{75 \pm 76.26926}{32}$

$t = 4.7271644, \quad \text{reject } -0.0396644$

$\approx 4.7 \text{ seconds}$

49. (a) $336 = -16t^2 + 160t$

$0 = -16t^2 + 160t - 336$

$0 = t^2 - 10t + 21$

$0 = (t - 7)(t - 3)$

at 3 seconds and at 7 seconds

(b) $0 = -16t^2 + 160t$

$0 = -16t(t - 10)$

$t = 0, 10$

after 10 seconds.

51. *Verbal model:* $\boxed{\text{Integer}} \cdot \boxed{\text{Integer}} = \boxed{\text{Product}}$

Labels: First integer $= n$

Second integer $= n + 1$

Equation:
$$n \cdot (n + 1) = 240$$
$$n^2 + n + \frac{1}{4} = 240 + \frac{1}{4}$$
$$\left(n + \frac{1}{2}\right)^2 = \frac{960 + 1}{4}$$
$$n + \frac{1}{2} = \pm\sqrt{\frac{961}{4}}$$
$$n = -\frac{1}{2} \pm \frac{\sqrt{961}}{2}$$
$$n = \frac{-1 \pm 31}{2}$$

$n = 15 \qquad n = -16$

$n + 1 = 16 \qquad n + 1 = 2 - 15 \Big\}$ reject

53. *Verbal model:* $\boxed{\begin{array}{c}\text{Even}\\\text{integer}\end{array}} \cdot \boxed{\begin{array}{c}\text{Even}\\\text{integer}\end{array}} = \boxed{\text{Product}}$

Labels: First even integer $= 2n$

Second even integer $= 2n + 2$

Equation:
$$2n \cdot (2n + 2) = 224$$
$$4n^2 + 4n = 224$$
$$n^2 + n = 56$$
$$n^2 + n - 56 = 0$$
$$(n + 8)(n - 7) = 0$$

$n + 8 = 0 \qquad\qquad n - 7 = 0$

$n = -8 \qquad\qquad\quad n = 7$

reject$\begin{cases} 2n = -16 & 2n = 14 \\ 2n + 2 = -14 & 2n + 2 = 16 \end{cases}$

55. *Verbal model:* $\boxed{\begin{array}{c}\text{Odd}\\\text{integer}\end{array}} \cdot \boxed{\begin{array}{c}\text{Odd}\\\text{integer}\end{array}} = \boxed{\text{Product}}$

Labels: First odd integer $= 2n + 1$

Second odd integer $= 2n + 3$

Equation:
$$(2n + 1) \cdot (2n + 3) = 483$$
$$4n^2 + 8n + 3 = 483$$
$$4n^2 + 8n - 480 = 0$$
$$n^2 + 2n - 120 = 0$$
$$(n + 12)(n - 10) = 0$$

$n + 12 = 0 \qquad n - 10 = 0$

~~$n = -12$~~ $\qquad\quad n = 10$

$\qquad\qquad\qquad 2n + 1 = 21$

$\qquad\qquad\qquad 2n + 3 = 23$

57. *Verbal model:* $\boxed{\text{Original time}} = \boxed{\text{New time}} + \boxed{\frac{1}{5}}$

Labels: Speed $= x$

Increased speed $= x + 40$

Equation:
$$\frac{720}{x} = \frac{720}{x + 40} + \frac{1}{5}$$
$$720(5)(x + 40) = 720(5x) + x(x + 40)$$
$$3600x + 144{,}000 = 3600x + x^2 + 40x$$
$$0 = x^2 + 40x - 144{,}000$$
$$x = \frac{-40 \pm \sqrt{40^2 - 4(1)(-144{,}000)}}{2(1)}$$
$$x = \frac{40 \pm \sqrt{1600 + 576{,}000}}{2}$$
$$x = \frac{-40 \pm 760}{2}$$
$$x = 360, \; \cancel{-400}$$
$$x + 40 = 400 \text{ miles per hour}$$

59. *Verbal model:* | Total Cost | = | Wage Cost | + | Fuel Cost |

Label: Time $= x$

Equation: $20.39 = 5x + x\left[\dfrac{\left(\dfrac{110}{x}\right)^2}{600}\right]$

$$20.39 = 5x + \dfrac{121}{6x}$$

$$122.34x = 30x^2 + 121$$

$$0 = 30x^2 - 122.34x + 121$$

$$x = \dfrac{-(-122.34) \pm \sqrt{(-122.34)^2 - 4(30)(121)}}{2(30)}$$

$$x = \dfrac{122.34 \pm \sqrt{477.0756}}{60}$$

$$x \approx 2.39,\ 1.67$$

$$v = \dfrac{110}{2.39} \approx 46\ \text{mi/hr}$$

or

$$v = \dfrac{110}{1.67} \approx 65\ \text{mi/hr}$$

61. (a) $a + b = 20$ $A = \pi a b$

          $b = 20 - a$ $A = \pi a(20 - a)$

(b)

a	4	7	10	13	16
A	201.1	285.9	314.2	285.9	201.1

$A = \pi(4)(20 - 4)$ $A = \pi(7)(20 - 7)$ $A = \pi(10)(20 - 10)$

   $= \pi(4)(16)$ $= \pi(7)(13)$ $= \pi(10)(10)$

   $= 64\pi$ $= 91\pi$ $= 100\pi$

   ≈ 201.1 ≈ 285.9 ≈ 314.2

$A = \pi(13)(20 - 13)$ $A = \pi(16)(20 - 16)$

   $= \pi(13)(7)$ $= \pi(16)(4)$

   $= 91\pi$ $= 64\pi$

   ≈ 285.9 ≈ 201.1

(c) $300 = \pi a(20 - a)$ (d) $A = \pi a(20 - a)$

     $0 = 20\pi a - \pi a^2 - 300$ *Keystrokes:*

     $0 = \pi a^2 - 20\pi a + 300$

     $a = \dfrac{-(-20\pi) \pm \sqrt{(-20\pi)^2 - 4(\pi)(300)}}{2(\pi)}$

     $a = \dfrac{20\pi \pm \sqrt{177.9305761}}{2\pi}$

     $a \approx 12.1,\ 7.9$

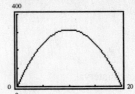

63. Guidelines for solving word problems:

   (a) Write a verbal model that will describe what you need to know.

   (b) Assign labels to each part of the verbal model—numbers to the known quantities and letters to the variable quantities.

   (c) Use the labels to write an algebraic model based on the verbal model.

   (d) Solve the resulting algebraic equation and check your solution.

65. Unit Analysis

$$\frac{9 \text{ dollars}}{\text{hour}} \cdot (20 \text{ hours}) = 180 \text{ dollars}$$

67. An example of a quadratic equation that has only one repeated solution is $(x + 4)^2 = 0$. Any equation of the form $(x - c)^2 = 0$, where c is a constant will have only one repeated solution.

Section 10.6   Quadratic and Rational Inequalities

1. $x(2x - 5) = 0$

$x = 0 \quad 2x - 5 = 0$

$x = \frac{5}{2}$

Critical numbers $= 0, \frac{5}{2}$

3. $4x^2 - 81 = 0$

$x^2 = \frac{81}{4}$

$x = \pm\frac{9}{2}$

Critical numbers: $\frac{9}{2}, -\frac{9}{2}$

5. $x(x + 3) - 5(x + 3) = 0$

$(x - 5)(x + 3) = 0$

$x = 5 \quad x = -3$

Critical numbers: $5, -3$

7. $x^2 - 4x + 3 = 0$

$(x - 3)(x - 1) = 0$

$x = 3 \quad x = 1$

Critical numbers $= 3, 1$

9. $4x^2 - 20x + 25 = 0$

$(2x - 5)^2 = 0$

$2x - 5 = 0$

$x = \frac{5}{2}$

Critical number: $\frac{5}{2}$

11. Negative: $(-\infty, 4)$

Positive: $(4, \infty)$

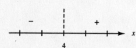

Choose a test value from each interval.

$(-\infty, 4) \Rightarrow x = 0 \Rightarrow 0 - 4 = -4 < 0$

$(4, \infty) \Rightarrow x = 5 \Rightarrow 5 - 4 = 1 > 0$

13. Negative: $(6, \infty)$

Positive: $(-\infty, 6)$

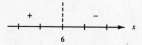

Choose a test value from each interval.

$(-\infty, 6) \Rightarrow x = 0 \Rightarrow 3 - \frac{1}{2}(0) = 3 > 0$

$(6, \infty) \Rightarrow x = 8 \Rightarrow 3 - \frac{1}{2}(8) = -1 < 0$

15. Positive: $(-\infty, 0)$

Negative: $(0, 4)$

Positive: $(4, \infty)$

Choose a test value from each interval.

$(-\infty, 0) \Rightarrow x = -1 \Rightarrow 2(-1)(-1 - 4) = 10 > 0$

$(0, 4) \Rightarrow x = 1 \Rightarrow 2(1)(1 - 4) = -6 < 0$

$(4, \infty) \Rightarrow x = 5 \Rightarrow 2(5)(5 - 4) = 10 > 0$

17. $4 - x^2 = (2 - x)(2 + x)$

Negative: $(-\infty, -2) \cup (2, \infty)$

Positive: $(-2, 2)$

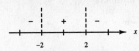

Choose a test value from each interval.

$(-\infty, -2) \Rightarrow x = -3 \Rightarrow (2-3)(2 + -3) = -5 < 0$

$(-2, 2) \Rightarrow x = 0 \Rightarrow (2 - 0)(2 + 0) = 4 > 0$

$(2, \infty) \Rightarrow x = 3 \Rightarrow (2 - 3)(2 + 3) = -5 < 0$

19. $(x - 5)(x + 1)$

Positive: $(-\infty, -1)$

Negative: $(-1, 5)$

Positive: $(5, \infty)$

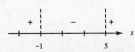

Choose a test value from each interval.

$(-\infty, -1) \Rightarrow x = -2 \Rightarrow (-2 - 5)(-2 + 1) = 7 > 0$

$(-1, 5) \Rightarrow x = 0 \Rightarrow (0 - 5)(0 + 1) = -5 < 0$

$(5, \infty) \Rightarrow x = 6 \Rightarrow (6 - 5)(6 + 1) = 7 > 0$

21. $2(x + 3) \geq 0$

Critical number: $x = -3$

Test intervals:

Negative: $(-\infty, -3]$

Positive: $[-3, \infty)$

Solution: $[-3, \infty)$

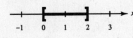

23. $-\dfrac{3}{4}x + 6 < 0$

Critical number: $x = 8$

Test intervals:

Negative: $(8, \infty)$

Positive: $(-\infty, 8)$

Solution: $(8, \infty)$

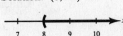

25. $3x(x - 2) < 0$

Critical number: $x = 0, 2$

Test intervals:

Positive: $(-\infty, 0)$

Negative: $(0, 2)$

Positive: $(2, \infty)$

Solution: $(0, 2)$

27. $3x(2 - x) \geq 0$

Critical numbers: $x = 0, 2$

Test intervals:

Negative: $(-\infty, 0]$

Positive: $[0, 2]$

Negative: $[2, \infty)$

Solution: $[0, 2]$

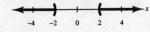

29. $x^2 > 4$

$x^2 - 4 > 0$

$(x - 2)(x + 2) > 0$

Critical numbers: $x = 2, -2$

Test intervals:

Positive: $(-\infty, 2)$

Negative: $(-2, 2)$

Positive: $(2, \infty)$

Solution: $(-\infty, -2) \cup (2, \infty)$

31. $x^2 + 3x - 10 \leq 0$

$(x + 5)(x - 2) \leq 0$

Critical number: $x = -5, 2$

Test intervals:

Positive: $(-\infty, -5]$

Negative: $[-5, 2]$

Positive: $[2, \infty)$

Solution: $[-5, 2]$

33. $u^2 + 2u - 2 > 1$

$u^2 + 2u - 3 > 0$

$(u + 3)(u - 1) > 0$

Critical numbers: $u = -3, 1$

Test intervals:

Positive: $(-\infty, -3)$

Negative: $(-3, 1)$

Positive: $(1, \infty)$

Solution: $(-\infty, -3) \cup (1, \infty)$

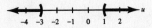

35. $x^2 + 4x + 5 < 0$

$$x = \frac{-4 \pm \sqrt{16 - 20}}{2}$$

No critical numbers

$x^2 + 4x + 5$ is not less than zero for any value of x.

Solution: none

37. $(x + 1)^2 \geq 0$

$(x + 1)^2 \geq 0$ for all real numbers

Solution: $(-\infty, \infty)$

39. $x^2 - 4x + 2 > 0$

$$x = \frac{4 \pm \sqrt{16 - 8}}{2}$$

$$= \frac{4 \pm \sqrt{8}}{2} = \frac{4 \pm 2\sqrt{2}}{2}$$

$$= 2 \pm \sqrt{2}$$

Critical numbers: $x = 2 + \sqrt{2}, 2 - \sqrt{2}$

Test intervals:

Positive: $\left(-\infty, 2 - \sqrt{2}\right)$

Negative: $\left(2 - \sqrt{2}, 2 + \sqrt{2}\right)$

Positive: $\left(2 + \sqrt{2}, \infty\right)$

Solution: $\left(-\infty, 2 - \sqrt{2}\right) \cup \left(2 + \sqrt{2}, \infty\right)$

41. $x^2 - 6x + 9 \geq 0$

$$(x - 3)^2 \geq 0$$

$(x - 3)^2 \geq 0$ for all real numbers

43. $u^2 - 10u + 25 < 0$

$$(u - 5)(u - 5) < 0$$

Critical number: $u = 5$

Test intervals:

Positive: $(-\infty, 5)$

Positive: $(5, \infty)$

Solution: none

45. $3x^2 + 2x - 8 \leq 0$

$$(3x - 4)(x + 2) \leq 0$$

Critical numbers: $x = \frac{4}{3}, -2$

Test intervals:

Positive: $(-\infty, -2]$

Negative: $\left[-2, \frac{4}{3}\right]$

Positive: $\left[\frac{4}{3}, \infty\right)$

Solution: $\left[-2, \frac{4}{3}\right]$

47. $-6u^2 + 19u - 10 > 0$

$$6u^2 - 19u + 10 < 0 \quad \text{(Multiply by } -1\text{)}$$

$$(3u - 2)(2u - 5) < 0$$

Critical numbers: $u = \frac{2}{3}, \frac{5}{2}$

Test intervals:

Positive: $\left(-\infty, \frac{2}{3}\right)$

Negative: $\left(\frac{2}{3}, \frac{5}{2}\right)$

Positive: $\left(\frac{5}{2}, \infty\right)$

Solution: $\left(\frac{2}{3}, \frac{5}{2}\right)$

49. $2u^2 - 7u - 4 > 0$

$$(2u + 1)(u - 4) > 0$$

Critical numbers: $u = -\frac{1}{2}, 4$

Test intervals:

Positive: $\left(-\infty, -\frac{1}{2}\right)$

Negative: $\left(-\frac{1}{2}, 4\right)$

Positive: $(4, \infty)$

Solution: $\left(-\infty, -\frac{1}{2}\right) \cup (4, \infty)$

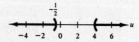

51. $4x^2 + 28x + 49 \leq 0$

$(2x + 7)(2x + 7) \leq 0$

Critical number: $x = \frac{-7}{2}$

Test intervals:

Positive: $\left(-\infty, -\frac{7}{2}\right)$

Positive: $\left(-\frac{7}{2}, \infty\right)$

Solution: $-\frac{7}{2}$

$$\underset{\substack{ \\ \text{-5 \quad -4 \quad -3 \quad -2 \quad -1 \quad 0}}}{\overset{-\frac{7}{2}}{\bullet}} \quad x$$

53. $(x - 5)^2 > 0$ for all real numbers except 5.

Solution: none

55. $6 - (x^2 - 10x + 25) < 0$

$6 - x^2 + 10x - 25 < 0$

$x^2 - 10x + 19 > 0$

$x = \dfrac{10 \pm \sqrt{100 - 76}}{2}$

$= \dfrac{10 \pm \sqrt{24}}{2} = \dfrac{10 \pm 2\sqrt{6}}{2}$

$= 5 \pm \sqrt{6}$

Critical numbers: $x = 5 + \sqrt{6}, 5 - \sqrt{6}$

Test intervals:

Positive: $\left(-\infty, 5 - \sqrt{6}\right)$

Negative: $\left(5 - \sqrt{6}, 5 + \sqrt{6}\right)$

Positive: $\left(5 + \sqrt{6}, \infty\right)$

Solution: $\left(-\infty, 5 - \sqrt{6}\right) \cup \left(5 + \sqrt{6}, \infty\right)$

57. $16 \leq (u + 5)^2$

$(u + 5)^2 \geq 16$

$u^2 + 10u + 25 - 16 \geq 0$

$u^2 + 10u + 9 \geq 0$

$(u + 9)(u + 1) \geq 0$

Critical numbers: $x = -9, -1$

Test intervals:

Positive: $(-\infty, -9]$

Negative: $(-9, -1]$

Positive: $[-1, \infty)$

Solution: $(-\infty, -9] \cup [-1, \infty)$

59. $x(x - 2)(x + 2) > 0$

Critical numbers: $x = 0, 2, -2$

Test intervals:

Negative: $(-\infty, -2)$

Positive: $(-2, 0)$

Negative: $(0, 2)$

Positive: $(2, \infty)$

Solution: $(-2, 0) \cup (2, \infty)$

61. *Keystrokes:*

$\boxed{\text{Y=}}\ \boxed{\text{X,T,}\theta}\ \boxed{x^2}\ \boxed{-}\ 6\ \boxed{\text{X,T,}\theta}\ \boxed{\text{GRAPH}}$

$(0, 6)$

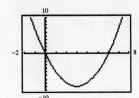

63. *Keystrokes:*

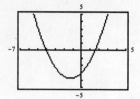

$(-\infty, -4) \cup \left(\frac{3}{2}, \infty\right)$

65. *Keystrokes:*

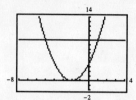

$(-\infty, -5] \cup [1, \infty)$

67. *Keystrokes:*

y_1 Y= 9 − 0.2 (X,T,θ − 2) x^2

y_2 4 GRAPH

$(-\infty, -3) \cup (7, \infty)$

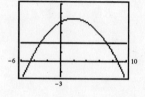

69. Critical number: $x = 3$

71. Critical numbers: $x = 0, -5$

73. $\dfrac{5}{x - 3} > 0$

Critical number: $x = 3$

Test intervals:

Negative: $(-\infty, 3)$

Positive: $(3, \infty)$

Solution: $(3, \infty)$

75. $\dfrac{-5}{x - 3} > 0$

Critical number: $x = 3$

Test intervals:

Positive: $(-\infty, 3)$

Negative: $(3, \infty)$

Solution: $(-\infty, 3)$

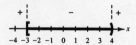

77. $\dfrac{x}{x - 3} < 0$

Critical numbers: $x = 0, 3$

Test intervals:

Positive: $(-\infty, 0)$

Negative: $(0, 3)$

Positive: $(3, \infty)$

Solution: $(0, 3)$

79. $\dfrac{x + 3}{x - 4} \le 0$

Critical numbers: $x = -3, 4$

Test intervals:

Positive: $(-\infty, -3]$

Negative: $[-3, 4)$

Positive: $(4, \infty)$

Solution: $[-3, 4)$

81. $\dfrac{y-4}{y+6} < 0$

Critical numbers: $y = 4, -6$

Test intervals:

Positive: $(-\infty, -6)$

Negative: $(-6, 4)$

Positive: $(4, \infty)$

Solution: $(-6, 4)$

83. $\dfrac{y-3}{y-11} \ge 0$

Critical numbers: $y = 3, \dfrac{11}{2}$

Test intervals:

Positive: $(-\infty, 3]$

Negative: $\left[3, \dfrac{11}{2}\right)$

Positive: $\left(\dfrac{11}{2}, \infty\right)$

Solution: $(-\infty, 3] \cup \left(\dfrac{11}{2}, \infty\right)$

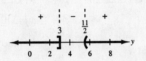

85. $\dfrac{x+2}{4x+6} \le 0$

Critical numbers: $x = -2, -\dfrac{3}{2}$

Test intervals:

Positive: $(-\infty, -2]$

Negative: $\left[-2, -\dfrac{3}{2}\right)$

Positive: $\left(-\dfrac{3}{2}, \infty\right)$

Solution: $\left[-2, -\dfrac{3}{2}\right)$

87. $\dfrac{3(u-3)}{u+1} < 0$

Critical numbers: $u = 3, -1$

Test intervals:

Positive: $(-\infty, -1)$

Negative: $(-1, 3)$

Positive: $(3, \infty)$

Solution: $(-1, 3)$

89. $\dfrac{6}{x-4} > 2$

$\dfrac{6}{x-4} - 2 > 0$

$\dfrac{6 - 2(x-4)}{x-4} > 0$

$\dfrac{6 - 2x + 8}{x-4} > 0$

$\dfrac{14 - 2x}{x-4} > 0$

$\dfrac{-2(-7 + x)}{x-4} > 0$

Critical numbers: $x = 7, 4$

Test intervals:

Negative: $(-\infty, 4)$

Positive: $(4, 7)$

Negative: $(7, \infty)$

Solution: $(4, 7)$

91. $\dfrac{4x}{x+2} < -1$

$\dfrac{4x}{x+2} + 1 < 0$

$\dfrac{4x + (x+2)}{x+2} < 0$

$\dfrac{5x+2}{x+2} < 0$

Critical numbers: $x = -\dfrac{2}{5}, -2$

Test intervals:

Positive: $(-\infty, -2)$

Negative: $\left(-2, -\dfrac{2}{5}\right)$

Positive: $\left(-\dfrac{2}{5}, \infty\right)$

Solution: $\left(-2, -\dfrac{2}{5}\right)$

93. $\dfrac{x-1}{x-3} \le 2$

$\dfrac{x-1}{x-3} - 2 \le 0$

$\dfrac{x-1-2(x-3)}{x-3} \le 0$

$\dfrac{x-1-2x+6}{x-3} \le 0$

$\dfrac{-x+5}{x-3} \le 0$

Critical numbers: $x = 5, 3$

Test intervals:

Negative: $(-\infty, 3)$

Positive: $(3, 5]$

Negative: $[5, \infty)$

Solution: $(-\infty, 3) \cup [5, \infty)$

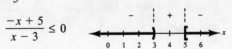

95. *Keystrokes:*

[Y=] 1 [÷] [X,T,θ] [−] [X,T,θ] [GRAPH]

Solution: $(-\infty, -1) \cup (0, 1)$

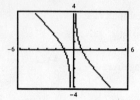

97. *Keystrokes:*

[Y=] [(] [X,T,θ] [+] 6 [)] [÷] [(] [X,T,θ] [+] 1 [)] [−] 2 [GRAPH]

Solution: $(-\infty, -1) \cup (4, \infty)$

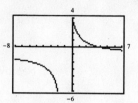

99. *Keystrokes:*

y_1 [Y=] [(] 6 [X,T,θ] [−] 3 [)] [÷] [(] [X,T,θ] [+] 5 [)] [ENTER]

y_2 2 [GRAPH]

Solution: $(-5, 3.25)$

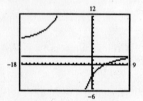

101. *Keystrokes:*

y_1 [Y=] [X,T,θ] [+] 1 [÷] [X,T,θ] [ENTER]

y_2 3 [GRAPH]

Solution: $(0, 0.382) \cup (2.618, \infty)$

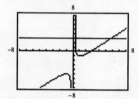

103. *Keystrokes:*

[Y=] 3 [X,T,θ] [÷] [(] [X,T,θ] [−] 2 [GRAPH]

(a) Solution $[0, 2)$

   (Look at *x*-axis and vertical asymptote $x = 2$)

(b) $(2, 4]$

   (Graph $y = 6$ as y_2 and find the intersection.)

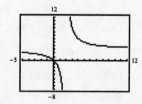

105. *Keystrokes:*

[Y=] 2 [X,T,θ] [x²] [÷] [(] [X,T,θ] [x²] [+] 4 [)] [GRAPH]

(a) Solution: $(-\infty, -2] \cup [2, \infty)$

   (Graph $y = 1$ as y_2 and find the intersection.)

(b) Solution $(-\infty, \infty)$

   (Notice graph stays below line $y = 2$.)

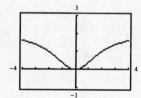

107.
$$\text{height} > 240$$
$$-16t^2 + 128t > 240$$
$$-16t^2 + 128t - 240 > 0$$
$$t^2 - 8t + 15 < 0$$
$$(t - 3)(t - 5) < 0$$

Critical numbers: $x = 3, 5$

Test intervals:

Positive: $(-\infty, 3)$

Negative: $(3, 5)$

Positive: $(5, \infty)$

Solution: $(3, 5)$

109.
$$1000(1 + r)^2 > 1150$$
$$1000(1 + 2r + r^2) > 1150$$
$$1000 + 2000r + 1000r^2 > 1150$$
$$1000r^2 + 2000r - 150 > 0$$
$$20r^2 + 40r - 3 > 0$$

Critical numbers: $r = \dfrac{-40 + \sqrt{1840}}{40}, \dfrac{-40 - \sqrt{1840}}{40}$

r cannot be negative.

Test intervals:

Negative: $\left(0, \dfrac{-40 + \sqrt{1840}}{40}\right)$

Positive: $\left(\dfrac{-40 + \sqrt{1840}}{40}, \infty\right)$

Solution: $\left(\dfrac{-40 + \sqrt{1840}}{40}, \infty\right)$

$(0.0724, \infty), \quad r > 7.24\%$

111. *Verbal model:* $\boxed{\text{Profit}} > 1,650,000$

$\boxed{\text{Revenue}} - \boxed{\text{Cost}} = \text{Profit} > 1,650,000$

$$x(50 - 0.0002x) - [12x + 150,000] > 1,650,000$$
$$50x - 0.0002x^2 - 12x - 150,000 > 1,650,000$$
$$-0.0002x^2 + 38x - 150,000 > 1,650,000$$
$$0 > 0.0002x^2 - 38x + 1,800,000$$
$$0 > (0.0002x - 20)(x - 90,000)$$

Critical numbers: 90,000, 100,000

Test intervals:

Positive: $(0, 90,000)$

Negative: $(90,000, 100,000)$

Positive: $(100,000, \infty)$

Solution: $(90,000, 100,000)$

$90,000 \le x \le 100,000$ units

113.
$$\text{Area} > 240$$
$$l(32 - l) > 240$$
$$32l - l^2 > 240$$
$$-l^2 + 32l - 240 > 0$$
$$l^2 - 32l + 240 < 0$$
$$(l - 20)(l - 12) < 0$$

Critical numbers: $l = 20, 12$

Test intervals:

Positive: $(-\infty, 12)$

Negative: $(12, 20)$

Positive: $(20, \infty)$

Solution: $(12, 20)$

115. (a) *Keystrokes:*

$\boxed{Y=}$ $\boxed{(}$ 244.20 $\boxed{-}$ 13.23 $\boxed{X,T,\theta}$ $\boxed{)}$ $\boxed{\div}$ $\boxed{(}$ 1 $\boxed{-}$ .13

$\boxed{X,T,\theta}$ $\boxed{+}$ .005 $\boxed{X,T,\theta}$ $\boxed{x^2}$ $\boxed{)}$ $\boxed{GRAPH}$

(b) Let $y_2 = 400$ and find the intersection of the graphs.

Solution: $[5.7, 13.7], \; 5.7 \le t \le 13 \cdot 7$

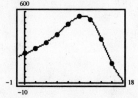

117. The direction of the inequality is reversed, when both sides are multiplied by a negative real number.

119. A polynomial can change signs only at the x-values that make the polynomial zero. The zeros of the polynomial are called the ciritical numbers, and they are used ro determine the test intervals in solving polynomial inequalities.

121. $x^2 + 1 < 0$ is one example of a quadratic inequality that has no real solution. Any inequality of the form $x^2 + c < 0$, c any positive constant or $-x^2 - c > 0$, c any positive constant will not have a real solution.

Review Exercises for Chapter 10

1. $x^2 + 12x = 0$

$x(x + 12) = 0$

$x = 0 \qquad x + 12 = 0$

$x = 0 \qquad\qquad x = -12$

3. $\qquad 4y^2 - 1 = 0$

$(2y - 1)(2y + 1) = 0$

$2y - 1 = 0 \qquad 2y + 1 = 0$

$y = \frac{1}{2} \qquad\qquad y = -\frac{1}{2}$

5. $4y^2 + 20y + 25 = 0$

$(2y + 5)(2y + 5) = 0$

$2y + 5 = 0 \qquad 2y + 5 = 0$

$2y = -5 \qquad\qquad 2y = -5$

$y = -\frac{5}{2} \qquad\qquad y = -\frac{5}{2}$

7. $2x^2 - 2x - 180 = 0$

$2(x^2 - x - 90) = 0$

$2(x - 10)(x + 9) = 0$

$x - 10 = 0 \qquad x + 9 = 0$

$x = 10 \qquad\qquad x = -9$

9. $\qquad 6x^2 - 12x = 4x^2 - 3x + 18$

$2x^2 - 9x - 18 = 0$

$(2x + 3)(x - 6) = 0$

$x = -\frac{3}{2} \qquad x = 6$

11. $4x^2 = 10,000$

$x^2 = 2500$

$x = \pm\sqrt{2500}$

$x = \pm 50$

13. $y^2 - 12 = 0$

$y^2 = 12$

$y = \pm\sqrt{12}$

$y = \pm 2\sqrt{3}$

15. $(x - 16)^2 = 400$

$x - 16 = \pm\sqrt{400}$

$x = 16 \pm 20$

$x = 36, -4$

17. $z^2 = -121$

$z = \pm\sqrt{-121}$

$z = \pm 11i$

19. $y^2 + 50 = 0$

$y^2 = -50$

$y = \pm\sqrt{-50}$

$y = \pm 5\sqrt{2}i$

21. $(y + 4)^2 + 18 = 0$

$(y + 4)^2 = -18$

$y + 4 = \pm\sqrt{-18}$

$y = -4 \pm 3\sqrt{2}i$

23. $\qquad x^4 - 4x^2 - 5 = 0$

$(x^2 - 5)(x^2 + 1) = 0$

$\qquad\qquad x^2 + 1 = 0$

$x^2 - 5 = 0 \qquad x^2 = -1$

$x^2 = 5 \qquad\quad x = \pm\sqrt{-1}$

$x = \pm\sqrt{5} \qquad x = \pm i$

25. $\qquad x - 4\sqrt{x} + 3 = 0$

$(\sqrt{x} - 3)(\sqrt{x} - 1) = 0$

$(\sqrt{x} - 3) = 0 \qquad (\sqrt{x} - 1) = 0$

$\sqrt{x} = 3 \qquad\qquad \sqrt{x} = 1$

$(\sqrt{x})^2 = 3^2 \qquad (\sqrt{x})^2 = 1^2$

$x = 9 \qquad\qquad x = 1$

Check: $\qquad\qquad$ Check:

$9 - 4\sqrt{9} + 3 \stackrel{?}{=} 0 \qquad 1 - 4\sqrt{1} + 3 \stackrel{?}{=} 0$

$9 - 12 + 3 \stackrel{?}{=} 0 \qquad 1 - 4 + 3 \stackrel{?}{=} 0$

$0 = 0 \qquad\qquad\qquad 0 = 0$

27. $(x^2 - 2x)^2 - 4(x^2 - 2x) - 5 = 0$

$[(x^2 - 2x) - 5][(x^2 - 2x) + 1] = 0$

$(x^2 - 2x - 5)(x^2 - 2x + 1) = 0$

$x = \dfrac{-(-2) \pm \sqrt{(-2)^2 - 4(1)(-5)}}{2(1)}$

$x = \dfrac{2 \pm \sqrt{4 + 20}}{2}$

$x = \dfrac{2 \pm \sqrt{24}}{2}$

$x = \dfrac{2 \pm 2\sqrt{6}}{2}$ $\qquad (x - 1)^2 = 0$

$x = 1 \pm \sqrt{6}$ $\qquad\qquad x = 1$

29. $x^{2/3} + 3x^{1/3} - 28 = 0$

$(x^{1/3} + 7)(x^{1/3} - 4) = 0$

$x^{1/3} + 7 = 0 \qquad\qquad x^{1/3} - 4 = 0$

$x^{1/3} = -7 \qquad\qquad\quad x^{1/3} = 4$

$\sqrt[3]{x} = -7 \qquad\qquad\quad \sqrt[3]{x} = 4$

$\left(\sqrt[3]{x}\right)^3 = (-7)^3 \qquad \left(\sqrt[3]{x}\right)^3 = 4^3$

$x = -343 \qquad\qquad\quad x = 64$

31. $x^2 - 6x - 3 = 0$

$x^2 - 6x + 9 = 3 + 9$

$(x - 3)^2 = 12$

$x - 3 = \pm\sqrt{12}$

$x = 3 \pm 2\sqrt{3}$

33. $x^2 - 3x + 3 = 0$

$x^2 - 3x + \dfrac{9}{4} = -3 + \dfrac{9}{4}$

$\left(x - \dfrac{3}{2}\right)^2 = \dfrac{-12 + 9}{4}$

$\left(x - \dfrac{3}{2}\right)^2 = -\dfrac{3}{4}$

$x - \dfrac{3}{2} = \pm\sqrt{-\dfrac{3}{4}}$

$x = \dfrac{3}{2} \pm \dfrac{i\sqrt{3}}{2}$

35. $y^2 - \dfrac{2}{3}y + 2 = 0$

$y^2 - \dfrac{2}{3}y = -2$

$y^2 - \dfrac{2}{3}y + \dfrac{1}{9} = -2 + \dfrac{1}{9}$

$\left(y - \dfrac{1}{3}\right)^2 = \dfrac{-17}{9}$

$y - \dfrac{1}{3} = \pm\sqrt{\dfrac{-17}{9}}$

$y = \dfrac{1}{3} \pm \dfrac{\sqrt{17}i}{3}$

37. $2y^2 + 10y + 3 = 0$

$y^2 + 5y + \dfrac{25}{4} = -\dfrac{3}{2} + \dfrac{25}{4}$

$\left(y + \dfrac{5}{2}\right)^2 = \dfrac{-6 + 25}{4}$

$\left(y + \dfrac{5}{2}\right)^2 = \dfrac{19}{4}$

$y + \dfrac{5}{2} = \pm\sqrt{\dfrac{19}{4}}$

$y = -\dfrac{5}{2} \pm \dfrac{\sqrt{19}}{2}$

39. $y^2 + y - 30 = 0$

$$y = \frac{-1 \pm \sqrt{1^2 - 4(1)(-30)}}{2(1)}$$

$$y = \frac{-1 \pm \sqrt{1 + 120}}{2}$$

$$y = \frac{-1 \pm \sqrt{121}}{2}$$

$$y = \frac{-1 \pm 11}{2}$$

$$y = 5, -6$$

41. $2y^2 + y - 21 = 0$

$$y = \frac{-1 \pm \sqrt{1^2 - 4(2)(-21)}}{2(2)}$$

$$y = \frac{-1 \pm \sqrt{1 + 168}}{4}$$

$$y = \frac{-1 \pm \sqrt{169}}{4}$$

$$y = \frac{-1 \pm 13}{4}$$

$$y = 3, -\frac{7}{2}$$

43. $5x^2 - 16x + 2 = 0$

$$x = \frac{-(-16) \pm \sqrt{(-16)^2 - 4(5)(2)}}{2(5)}$$

$$x = \frac{16 \pm \sqrt{256 - 40}}{10}$$

$$x = \frac{16 \pm \sqrt{216}}{10}$$

$$x = \frac{16 \pm 6\sqrt{6}}{10}$$

$$x = \frac{8 \pm 3\sqrt{6}}{5}$$

45. $0.3t^2 - 2t + 5 = 0$

$$t = \frac{-(-2) \pm \sqrt{(-2)^2 - 4(0.3)(5)}}{2(0.3)}$$

$$t = \frac{2 \pm \sqrt{4 - 6}}{0.6}$$

$$t = \frac{2 \pm \sqrt{-2}}{0.6}$$

$$t = \frac{2 \pm i\sqrt{2}}{0.6}$$

$$t = \frac{20 \pm 10\sqrt{2}i}{6} = \frac{10}{3} \pm \frac{5\sqrt{2}i}{3}$$

47. $b^2 - 4ac = 4^2 - 4(1)(4)$

$$= 16 - 16$$

$$= 0$$

One repeated rational solution.

49. $b^2 - 4ac = (-1)^2 - 4(1)(-20)$

$$= 1 + 80$$

$$= 81$$

Two distinct rational solutions.

51. $b^2 - 4ac = 17^2 - 4(3)(10)$

$$= 289 - 120$$

$$= 169$$

Two distinct rational solutions.

53. $b^2 - 4ac = (-6)^2 - 4(1)(21)$

$$= 36 - 84$$

$$= -48$$

Two distinct imaginary solutions.

55. $f(x) = x^2 - 8x + 3$

$$= (x^2 - 8x + 16) + 3 - 16$$

$$= (x - 4)^2 - 13$$

vertex $= (4, -13)$

57. $h(u) = 2u^2 - u + 3$

$$= 2\left(u^2 - \tfrac{1}{2}u\right) + 3$$

$$= 2\left(u^2 - \tfrac{1}{2}u + \tfrac{1}{16}\right) + 3 - \tfrac{1}{8}$$

$$= 2\left(u - \tfrac{1}{4}\right)^2 + \tfrac{23}{8}$$

vertex $= \left(\tfrac{1}{4}, \tfrac{23}{8}\right)$

59. x-intercepts   vertex

$0 = x^2 + 8x$ $y = x^2 + 8x + 16 - 16$

$0 = x(x + 8)$ $y = (x + 4)^2 - 16$

$x = 0$ $x = -8$ $(-4, -16)$

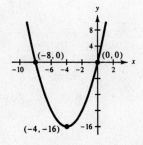

61. x-intercepts   vertex

$0 = x^2 - 6x + 5$ $y = (x^2 - 6x + 9) + 5 - 9$

$0 = (x - 5)(x - 1)$ $y = (x - 3)^2 - 4$

$x = 5$ $x = 1$ $(3, -4)$

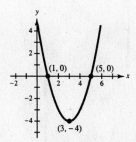

63.

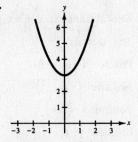

Vertical shift 3 units up

65.

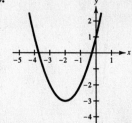

Horizontal shift 2 units left

Vertical shift 3 units down

67. $y = a(x - h)^2 + k$

$y = -2(x - 3)^2 + 5$

69. $y = a(h - x)^2 + k$

$y = a(x - 2)^2 - 5$

$3 = a(0 - 2)^2 - 5$

$3 = a(4) - 5$

$8 = a(4)$

$2 = a$

$y = 2(x - 2)^2 - 5$ or $y = 2x^2 - 8x + 3$

71. $y = a(x - h)^2 + k$

$1 = a(1 - 5)^2 + 0$

$1 = a(16)$

$\frac{1}{16} = a$

$y = \frac{1}{16}(x - 5)^2 + 0$ or $y = \frac{1}{16}x^2 - \frac{5}{8}x + \frac{25}{16}$

73. Critical numbers: $x = 0, 7$

Test intervals:

Negative: $(-\infty, 0)$

Positive: $(0, 7)$

Negative: $(7, \infty)$

Solution: $(0, 7)$

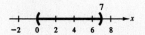

75.
$$16 - (x - 2)^2 \leq 0$$
$$(4 - x + 2)(4 + x - 2) \leq 0$$
$$(6 - x)(2 + x) \leq 0$$

Critical numbers: $x = -2, 6$

Test intervals:

Negative: $(-\infty, 2]$

Positive: $[-2, 6]$

Negative: $[6, \infty)$

Solution: $(-\infty, -2] \cup [6, \infty)$

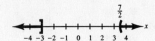

77.
$$2x^2 + 3x - 20 < 0$$
$$(2x - 5)(x + 4) < 0$$

Critical numbers: $x = -4, \frac{5}{2}$

Test intervals:

Positive: $(-\infty, -4)$

Negative: $\left(-4, \frac{5}{2}\right)$

Positive: $\left(\frac{5}{2}, \infty\right)$

Solution: $\left(-4, \frac{5}{2}\right)$

79. Critical numbers: $x = -3, \frac{7}{2}$

Test intervals:

Positive: $(-\infty, -3]$

Negative: $\left[-3, \frac{7}{2}\right]$

Positive: $\left(\frac{7}{2}, \infty\right)$

Solution: $[-\infty, -3] \cup \left(\frac{7}{2}, \infty\right)$

81.
$$\frac{2x - 2}{x + 6} + 2 < 0$$
$$\frac{2x - 2 + 2(x + 6)}{x + 6} < 0$$
$$\frac{2x - 2 + 2x + 12}{x + 6} < 0$$
$$\frac{4x + 10}{x + 6} < 0$$

Critical numbers: $x = -6, -\frac{5}{2}$

Test intervals:

Positive: $(-\infty, -6)$

Negative: $\left(-6, -\frac{5}{2}\right)$

Positive: $\left(-\frac{5}{2}, \infty\right)$

Solution: $\left(-6, -\frac{5}{2}\right)$

83. *Verbal model:*

Selling price per car	$=$	Cost per car	$+$	Profit per car

Labels: Number cars sold $= x$

Number cars purchased $= x + 4$

Equation:
$$\frac{80,000}{x} = \frac{80,000}{x + 4} + 1,000$$
$$x(x + 4)\left(\frac{80,000}{x}\right) = \left(\frac{80,000}{x + 4} + 1,000\right)x(x + 4)$$
$$80,000(x + 4) = 80,000x + 1,000x(x + 4)$$
$$80,000x + 320,000 = 80,000x + 1,000x^2 + 4,000x$$
$$0 = 1,000x^2 + 4,000x - 320,000$$
$$0 = x^2 + 4x - 320$$
$$0 = (x + 20)(x - 16)$$

reject $x = -20$ $x = 16$ cars

Average price per car $= \dfrac{80,000}{16} = \$5,000$

85. *Verbal model:* $\boxed{\text{Area}} = \boxed{\text{Length}} \cdot \boxed{\text{Width}}$

Labels: Width $= x$

Length $= x + 12$

Equation: $108 = (x + 12)x$

$0 = x^2 + 12x - 108$

$0 = (x + 18)(x - 6)$

reject $x = -18$ $x = 6$ inches

$x + 12 = 18$ inches

87. *Formula:* $A = P(1 + r)^2$

$21,424.50 = 20,000(1 + r)^2$

$1.071225 = (1 + r)^2$

$1.035 = 1 + r$

$.035 = r$ or 3.5%

89. *Verbal model:* $\boxed{\begin{array}{c}\text{Cost per person}\\ \text{Current Group}\end{array}} - \boxed{\begin{array}{c}\text{Cost per person}\\ \text{New Group}\end{array}} = \boxed{\$1.50}$

Labels: Number in Current Group $= x$

Number in New Group $= x + 8$

Equation: $\dfrac{360}{x} - \dfrac{360}{x + 8} = 1.50$

$[x(x + 8)]\left(\dfrac{360}{x} - \dfrac{360}{x + 8}\right) = (1.50)[x(x + 8)]$

$360(x + 8) - 360x = 1.50(x^2 + 8x)$

$360x + 2880 - 360x = 1.50x^2 + 12x$

$0 = 1.5x^2 + 12x - 2880$

$0 = x^2 + 8x - 1920$

$0 = (x + 48)(x - 40)$

$x + 48 = 0$ $x - 40 = 0$

$\cancel{x = -48}$ $x = 40$

$x + 8 = 48$

91. *Verbal model:* $\boxed{\begin{array}{c}\text{Cost per}\\ \text{ticket}\end{array}} \cdot \boxed{\begin{array}{c}\text{Number of}\\ \text{tickets}\end{array}} = \boxed{\$96}$

Labels: Number in team $= x$

Number going to game $= x + 3$

Equation: $\left(\dfrac{96}{x} - 1.60\right)(x + 3) = 96$

$\left(\dfrac{96 - 1.60x}{x}\right)(x + 3) = 96$

$(96 - 1.6x)(x + 3) = 96x$

$96x - 1.6x^2 - 4.8x + 288 = 96x$

$1.6x^2 + 4.8x - 288 = 0$

$x^2 + 3x - 180 = 0$

$(x - 12)(x + 15) = 0$

$x - 12 = 0$ $x + 15 = 0$

$x = 12$ $x = -15$ reject

$x + 3 = 15$

93. *Formula:* $c^2 = a^2 + b^2$ $a + b = 140$

Labels: $c = 100$ $x + b = 140$

$a = x$ $b = 140 - x$

$b = 140 - x$

Equation: $100^2 = x^2 + (140 - x)^2$

$10,000 = x^2 + 19,600 - 280x + x^2$

$0 = 2x^2 - 280x + 9,600$

$0 = x^2 - 140x + 4800$

$0 = (x - 60)(x - 80)$

$x = 60$ $x = 80$

$140 - x = 80$ $140 - x = 60$

60 feet and 80 feet

95. *Verbal model:* | Work done by Person 1 | + | Work done by Person 2 | = | One complete job |

Labels: Time Person 1 $= x$

Labels: Time Person 2 $= x + 2$

Equation:

$$\frac{1}{x}(10) + \frac{1}{x+2}(10) = 1$$

$$x(x+2)\left[10\left(\frac{1}{x} + \frac{1}{x+2}\right)\right] = [1]x(x+2)$$

$$10(x+2) + 10x = x(x+2)$$

$$10x + 20 + 10x = x^2 + 2x$$

$$0 = x^2 - 18x - 20$$

$$x = \frac{-(-18) \pm \sqrt{(-18)^2 - 4(1)(-20)}}{2(1)}$$

$$x = \frac{18 \pm \sqrt{324 + 80}}{2}$$

$$x = \frac{18 \pm \sqrt{404}}{2}$$

$$x = \frac{18 \pm 2\sqrt{101}}{2}$$

$$x = 9 \pm \sqrt{101}$$

$$x \approx 19 \qquad \bowtie$$

$$x + 2 \approx 21$$

19 hours, 21 hours

97. (a) $256 = -16t^2 + 64t + 192$

$0 = -16t^2 + 64t - 64$

$0 = t^2 - 4t + 4$

$0 = (t - 2)^2$

$t = 2$ seconds

(b) $0 = -16t^2 + 64t + 192$

$0 = -16(t^2 - 4t - 12)$

$0 = -16(t + 2)(t - 6)$

$t + 2 = 0 \qquad t - 6 = 0$

discard $t = -2 \qquad t = 6$ seconds

99. (a) *Keystrokes:*

$\boxed{Y=}\ \boxed{(-)}\ \boxed{X,T,\theta}\ \boxed{x^2}\ \boxed{\div}\ 10\ \boxed{+}\ 3\ \boxed{X,T,\theta}\ \boxed{+}\ 6\ \boxed{GRAPH}$

(b) $y = -\dfrac{1}{10}(0)^2 + 3(0) + 6$

$y = 0 + 0 + 6$

$y = 6$ feet

(c) $x = -\dfrac{b}{2a}$

$= -\dfrac{3}{2\left(-\frac{1}{10}\right)}$

$= \dfrac{-3}{-\frac{1}{5}}$

$= 15$

$y = \dfrac{1}{10}(15)^2 + 3(15) + 6$

$= -\dfrac{1}{10}(225) + 45 + 6$

$= -22.5 + 45 + 6$

$= 28.5$ feet

(d) $0 = \dfrac{-1}{10}x^2 + 3x + 6$

$x = \dfrac{-3 \pm \sqrt{3^2 - 4\left(-\frac{1}{10}\right)(6)}}{2\left(-\frac{1}{10}\right)}$

$x = \dfrac{-3 \pm \sqrt{9 + 2.4}}{-\frac{1}{5}}$

$x = \dfrac{-3 \pm \sqrt{11.4}}{-\frac{1}{5}}$

$x = -5\left(-3 \pm \sqrt{11.4}\right) = 15 \pm 5\sqrt{11.4} \approx 31.88$ or $\cancel{-1.88}$

The ball is 31.88 ft from the child when it hits the ground.

101. (a) *Keystrokes:*

$\boxed{Y=}$ 653.50 $\boxed{-}$ 379.89 $\boxed{X,T,\theta}$ $\boxed{+}$ 62.75 $\boxed{X,T,\theta}$ $\boxed{x^2}$ $\boxed{GRAPH}$

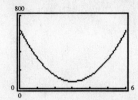

(b) $x = -\dfrac{b}{2a} = \dfrac{-(379.89)}{2(62.75)} \approx 3$ year 1993

(c) $N = 653.50 - 379.89(10) + 6275(10)^2 \approx 3130$ orders

103. $\overline{C} = \dfrac{C}{x} = \dfrac{50{,}000 + 1.2x}{x} = \dfrac{50{,}000}{x} + 1.2$

$\overline{C} < 5$

$\dfrac{50{,}000}{x} + 1.2 < 5$

$\dfrac{50{,}000}{x} - 3.8 < 0$

$\dfrac{50{,}000 - 3.8x}{x} < 0$

Critical numbers: $x = 0,\ 13158$

Test intervals:

x must be positive

Positive: $(0, 13{,}158)$

Negative: $(13{,}158, \infty)$

Solution: $(13{,}158, \infty)$

105. $h = -16t^2 + 312t$

$-16t^2 + 312t > 1200$

$-16t^2 + 312t - 1200 > 0$ (Divide by -16)

$t^2 - 19.5t + 75 < 0$

$t = \dfrac{-(-19.5) \pm \sqrt{(-19.5)^2 - 4(1)(75)}}{2(1)}$

$t = \dfrac{19.5 \pm \sqrt{80.25}}{2}$

$t \approx 14.2,\ 5.3$

Critical numbers: $t = 14.2,\ 5.3$

Test intervals:

Positive: $(-\infty, 5.3)$

Negative: $(5.3, 14.2)$

Positive: $(14.2, \infty)$

Solution: $(5.3, 14.2)$

$5.3 < t < 14.2$

Chapter Test for Chapter 10

1. $x(x + 5) - 10(x + 5) = 0$

$(x + 5)(x - 10) = 0$

$x + 5 = 0 \qquad x - 10 = 0$

$x = -5 \qquad\quad x = 10$

2. $8x^2 - 21x - 9 = 0$

$(8x + 3)(x - 3) = 0$

$8x + 3 = 0 \qquad x - 3 = 0$

$x = -\tfrac{3}{8} \qquad\quad x = 3$

3. $(x - 2)^2 = 0.09$

$x - 2 = \pm 0.3$

$x = 2 \pm 0.3$

$x = 2.3,\ 1.7$

4. $(x + 3)^2 + 81 = 0$

$(x + 3)^2 = -81$

$x + 3 = \pm\sqrt{-81}$

$x = -3 \pm 9i$

5. $2x^2 - 6x + 3 = 0$

$$x^2 - 3x + \frac{9}{4} = -\frac{3}{2} + \frac{9}{4}$$

$$\left(x - \frac{3}{2}\right)^2 = \frac{-6 + 9}{4}$$

$$\left(x - \frac{3}{2}\right)^2 = \frac{3}{4}$$

$$x - \frac{3}{2} = \pm\sqrt{\frac{3}{4}}$$

$$x = \frac{3}{2} \pm \frac{\sqrt{3}}{2}$$

6. $2y(y - 2) = 7$

$$2y^2 - 4y - 7 = 0$$

$$y = \frac{-(-4) \pm \sqrt{(-4)^2 - 4(2)(-7)}}{2(2)}$$

$$y = \frac{4 \pm \sqrt{16 + 56}}{4}$$

$$y = \frac{4 \pm \sqrt{72}}{4}$$

$$y = \frac{4 \pm 6\sqrt{2}}{4}$$

$$y = \frac{2 \pm 3\sqrt{2}}{2} \approx 7.41 \text{ and } -0.41$$

7. $x - 5\sqrt{x} + 4 = 0$

$$\left(\sqrt{x} - 4\right)\left(\sqrt{x} - 1\right) = 0$$

$$\sqrt{x} - 4 = 0 \qquad \sqrt{x} - 1 = 0$$

$$\sqrt{x} = 4 \qquad \sqrt{x} = 1$$

$$\left(\sqrt{x}\right)^2 = 4^2 \qquad \left(\sqrt{x}\right)^2 = 1^2$$

$$x = 16 \qquad x = 1$$

Check: **Check:**

$$16 - 5\sqrt{16} + 4 \overset{?}{=} 0 \qquad 1 - 5\sqrt{1} + 4 \overset{?}{=} 0$$

$$16 - 20 + 4 \overset{?}{=} 0 \qquad 1 - 5 + 4 \overset{?}{=} 0$$

$$0 = 0 \qquad\qquad 0 = 0$$

8. $x^4 + 6x^2 - 16 = 0$

$$(x^2 + 8)(x^2 - 2) = 0$$

$$x^2 + 8 = 0 \qquad x^2 - 2 = 0$$

$$x^2 = -8 \qquad x^2 = 2$$

$$x = \pm\sqrt{-8} \qquad x = \pm\sqrt{2}$$

$$x = \pm 2\sqrt{2}i$$

9. $b^2 - 4ac = (-12)^2 - 4(5)(10)$

$$= 144 - 200$$

$$= -56$$

2 imaginary solutions.

10. $(x - (-4))(x - 5) = 0$

$$(x + 4)(x - 5) = 0$$

$$x^2 - x - 20 = 0$$

11. $y = -2(x - 2)^2 + 8$

vertex $= (2, 8)$

$$y = -2(0 - 2)^2 + 8$$

$$y = -2(4) + 8$$

$$y = 0 \quad (0, 0); x \text{ \& } y\text{-intercept}$$

$$0 = -2(x - 2)^2 + 8$$

$$-8 = -2(x - 2)^2$$

$$4 = (x - 2)^2$$

$$\pm 2 = x - 2$$

$$2 \pm 2 = x$$

$$0, 4 = x \quad (0, 0), (4, 0); x\text{-intercepts}$$

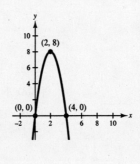

12. $y = a(x - h)^2 + k$

$4 = a(0 - 3)^2 - 2$

$6 = a(9)$

$\frac{6}{9} = a$

$\frac{2}{3} = a$

$y = \frac{2}{3}(x - 3)^2 - 2$ or $\frac{2}{3}x^2 - 4x + 4$

13. $y = a(x - h)^2 + k$

$y = a(x - 3)^2 - 2$

$6 = a(-1 - 3)^2 - 2$

$6 = a(16) - 2$

$8 = a(16)$

$\frac{1}{2} = \frac{8}{16} = a$

$y = \frac{1}{2}(x - 3)^2 - 2$

14. $16 \le (x - 2)^2$

$(x - 2)^2 \ge 16$

$x^2 - 4x + 4 \ge 16$

$x^2 - 4x - 12 \ge 0$

$(x - 6)(x + 2) \ge 0$

Critical numbers: $x = -2, 6$

Test intervals:

Positive: $(-\infty, -2]$

Negative: $[-2, 6]$

Positive: $[6, \infty)$

Solution: $(-\infty, -2] \cup [6, \infty)$

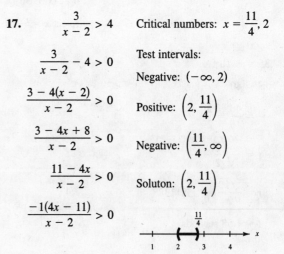

15. $2x(x - 3) < 0$

Critical numbers: $x = 0, 3$

Test intervals:

Positive: $(-\infty, 0)$

Negative: $(0, 3)$

Positive: $(3, \infty)$

Solution: $(0, 3)$

16. $\dfrac{3u + 2}{u - 3} \le 2$

$\dfrac{3u + 2}{u - 3} - \dfrac{2(u - 3)}{u - 3} \le 0$

$\dfrac{3u + 2 - 2u + 6}{u - 3} \le 0$

$\dfrac{u + 8}{u - 3} \le 0$

Critical numbers: $u = -8, 3$

Test intervals:

Positive: $(-\infty, -8]$

Negative: $[-8, 3)$

Positive: $(3, \infty)$

Soluton: $[-8, 3)$

17. $\dfrac{3}{x - 2} > 4$

$\dfrac{3}{x - 2} - 4 > 0$

$\dfrac{3 - 4(x - 2)}{x - 2} > 0$

$\dfrac{3 - 4x + 8}{x - 2} > 0$

$\dfrac{11 - 4x}{x - 2} > 0$

$\dfrac{-1(4x - 11)}{x - 2} > 0$

Critical numbers: $x = \dfrac{11}{4}, 2$

Test intervals:

Negative: $(-\infty, 2)$

Positive: $\left(2, \dfrac{11}{4}\right)$

Negative: $\left(\dfrac{11}{4}, \infty\right)$

Soluton: $\left(2, \dfrac{11}{4}\right)$

18. *Verbal model:* $\boxed{\text{Area}} = \boxed{\text{Length}} \cdot \boxed{\text{Width}}$

Labels: Length $= l$

Width $= l - 8$

Equation: $240 = l \cdot (l - 8)$

$0 = l^2 - 8l - 240$

$0 = (l - 20)(l + 12)$

$0 = l - 20 \qquad 0 = l + 12$

$20 \text{ feet} = l \qquad -12 = l \text{ reject}$

$12 \text{ feet} = l - 8$

19. *Verbal model:* $\boxed{\begin{array}{c}\text{Cost per person}\\\text{Current Group}\end{array}} - \boxed{\begin{array}{c}\text{Cost per person}\\\text{New Group}\end{array}} = 6.25$

Labels: Number Current Group $= x$

Number New Group $= x$

Equation: $\dfrac{1250}{x} - \dfrac{1250}{x + 10} = 6.25$

$x(x + 10)\left(\dfrac{1250}{x} - \dfrac{1250}{x + 10}\right) = (6.25)x(x + 10)$

$1250(x + 10) - 1250x = 6.25x(x + 10)$

$1250x + 12500 - 1250x = 6.25x^2 + 62.5x$

$0 = 6.25x^2 + 62.5x - 12500$

$0 = x^2 + 10x - 2000$

$0 = (x + 50)(x - 40)$

$\text{reject } x = -50 \qquad x = 40 \text{ club members}$

20. $35 = -16t^2 + 75$

$16t^2 = 40$

$t^2 = \dfrac{40}{16} = \dfrac{5}{2}$

$t = \sqrt{\dfrac{5}{2}}$

$t = \dfrac{\sqrt{10}}{2} \approx 1.5811388$

$t \approx 1.58 \text{ seconds}$

21. $R = -\dfrac{1}{20}(n^2 - 240n), \ 80 \le n \le 160$

$R = -\dfrac{1}{20}(n^2 - 240n + 14{,}400) + 720$

$R = -\dfrac{1}{20}(n - 120)^2 + 720$

$n = 120 \text{ passengers will produce a maximum revenue}$

22. $h = -16t^2 + 288t$

$-16t^2 + 288t > 1040$

$-16t^2 + 288t - 1040 > 0 \qquad \text{(Divide by } -16)$

$t^2 - 18t + 65 < 0$

$(t - 5)(t - 13) < 0$

Critical numbers: $t = 5, 13$

Test intervals:

t must be positive

Positive: $(0, 5)$

Negative: $(5, 13)$

Positive: $(13, \infty)$

Solution: $(5, 13)$

$5 < t < 13 \text{ seconds}$

CHAPTER 11
Exponential and Logarithmic Functions

CHAPTER 11
Exponential and Logarithmic Functions

Section 11.1   Exponential Functions

Solutions to Odd-Numbered Exercises

1. $2^x \cdot 2^{x-1} = 2^{x+(x-1)} = 2^{2x-1}$

3. $\dfrac{e^{x+2}}{e^x} = e^{x+2-x} = e^2$

5. $(2e^x)^3 = 8e^{3x}$

7. $\sqrt[3]{-8e^{3x}} = -2e^x$ because
$-2 \cdot -2 \cdot -2 \cdot e^x \cdot e^x \cdot e^x = -8e^{3x}.$

9. $4^{\sqrt{3}} \approx 11.036$

Keystrokes:

Scientific: 4 $\boxed{y^x}$ 3 $\boxed{\sqrt{}}$ $\boxed{=}$

Graphing: 4 $\boxed{\wedge}$ $\boxed{\sqrt{}}$3 $\boxed{\text{ENTER}}$

11. $e^{1/3} \approx 1.396$

Keystrokes:

Scientific: $\boxed{(}$1$\boxed{\div}$3$\boxed{)}$ $\boxed{\text{Inv}}$ $\boxed{\ln x}$ $\boxed{=}$

Graphing: $\boxed{e^x}$ $\boxed{(}$1 $\boxed{\div}$ 3 $\boxed{)}$ $\boxed{\text{ENTER}}$

13. $4(3e^4)^{1/2} = 4 \cdot 3^{1/2} \cdot e^2 \approx 51.193$

Keystrokes:

Scientific: 4 $\boxed{\times}$ 3 $\boxed{y^x}$ 0.5 $\boxed{\times}$ 2 $\boxed{\text{Inv}}$ $\boxed{\ln x}$ $\boxed{=}$

Graphing: 4 $\boxed{\times}$ 3 $\boxed{\wedge}$ 0.5 $\boxed{\times}$ $\boxed{e^x}$ 2 $\boxed{\text{ENTER}}$

15. $\dfrac{4e^3}{12e^2} = \dfrac{e}{3} \approx 0.906$

Keystrokes:

Scientific: 1 $\boxed{\text{Inv}}$ $\boxed{\ln x}$ $\boxed{\div}$ 3 $\boxed{=}$

Graphing: $\boxed{e}$ $\boxed{\div}$ 3 $\boxed{\text{ENTER}}$

17. (a) $f(-2) = 3^{-2} = \dfrac{1}{9}$

(b) $f(0) = 3^0 = 1$

(c) $f(1) = 3^1 = 3$

19. (a) $g(-1) = 1.07^{-1} \approx 0.935$

(b) $g(3) = 1.07^3 \approx 1.225$

(c) $g(\sqrt{5}) = 1.07^{\sqrt{5}} \approx 1.163$

21. (a) $f(0) = 500(\tfrac{1}{2})^0 = 500$

(b) $f(1) = 500(\tfrac{1}{2})^1 = 250$

(c) $f(\pi) = 500(\tfrac{1}{2})^\pi = 56.657$

23. (a) $f(0) = 1000(1.05)^{(2)(0)} = 1000$

(b) $f(5) = 1000(1.05)^{2(5)} = 1628.895$

(c) $f(10) = 1000(1.05)^{2(10)} = 2653.298$

25. (a) $h(5) = \dfrac{5000}{(1.06)^{8(5)}} \approx 486.11$

(b) $h(10) = \dfrac{5000}{(1.06)^{8(10)}} \approx 47.261$

(c) $h(20) = \dfrac{5000}{(1.06)^{8(20)}} \approx 0.447$

27. (a) $g(-4) = 10e^{-0.5(-4)} = 10e^2 \approx 73.891$

(b) $g(4) = 10e^{-0.5(4)} = 10e^{-2} \approx 1.353$

(c) $g(8) = 10e^{-0.5(8)} = 10e^{-4} \approx 0.183$

29. (a) $g(0) = \dfrac{1000}{2 + e^{-0.12(0)}} \approx 333.333$

(b) $g(10) = \dfrac{1000}{2 + e^{-0.12(10)}} \approx 434.557$

(c) $g(50) = \dfrac{1000}{2 + e^{-0.12(50)}} \approx 499.381$

31.

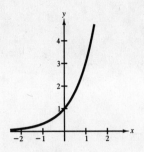

Table of values:

x	-2	-1	0	1	2
$f(x)$	0.1	0.3	1	3	9

33.

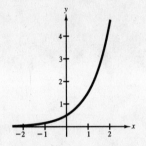

Table of values:

x	-2	-1	0	1	2
$h(x)$	0.1	0.2	0.5	1.5	4.5

35.

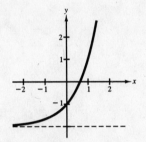

Table of values:

x	-2	-1	0	1	2
$g(x)$	-1.9	-1.7	-1	1	7

37.

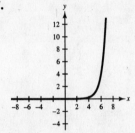

Table of values:

x	-1	0	1	5	6
$f(x)$	2.4×10^{-4}	9.8×10^{-4}	0.004	1	4

39.

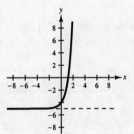

Table of values:

x	-2	-1	0	1	2
$f(x)$	-4.9	-4.8	-4	-1	11

41.

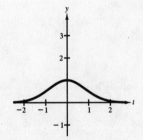

Table of values:

t	-2	-1	0	1	2
$f(t)$	0.1	0.5	1	0.5	0.1

43.

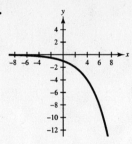

Table of values:

x	-2	-1	0	1	2
$f(x)$	-5	-0.7	-1	-1.4	-2

45.

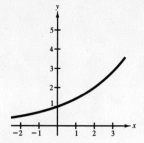

Table of values:

x	-2	-1	0	1	2
$h(x)$	0.5	0.7	1	1.4	2

47.

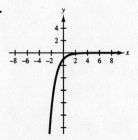

Table of values:

x	-2	-1	0	1	2
$f(x)$	-9	-3	-1	-0.3	-0.1

49.

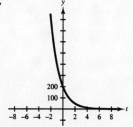

Table of values:

t	-2	-1	0	1	2
$g(t)$	800	400	200	100	50

51. $f(x) = 2^x$

(b) Basic graph

53. $f(x) = 2^{-x}$

(e) Basic graph reflected in the y-axis

55. $f(x) = 2^{x-1}$

(f) Basic graph shifted 1 unit right

57. $f(x) = \left(\frac{1}{2}\right)^x - 2$

(h) Basic graph reflected in y-axis and shifted 2 units down

59. $y = 5^{x/3}$

Keystrokes:

Y= 5 ^ ((X,T,θ ÷ 3) GRAPH

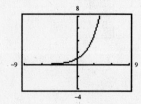

61. $y = 5^{(x-2)/3}$

Keystrokes:

Y= 5 ^ (((X,T,θ − 2) ÷ 3) GRAPH

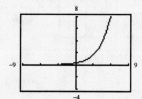

63. $y = 500(1.06)^t$

Keystrokes:

[Y=] 500 [(] 1.06 [)] [^] [X,T,θ] [GRAPH]

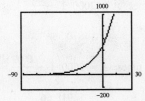

65. $y = 3e^{0.2x}$

Keystrokes:

[Y=] 3 [eˣ] 0.2 [X,T,θ] [GRAPH]

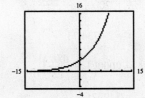

67. $P(t) = 100e^{-0.1t}$

Keystrokes:

[Y=] 100 [eˣ] [(−)] 0.1 [X,T,θ] [GRAPH]

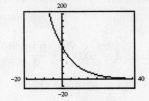

69. $y = 6e^{-x^2/3}$

Keystrokes:

[Y=] 6 [eˣ] [(−)] [(] [X,T,θ] [x²] [÷] 3 [)] [GRAPH]

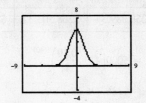

71. Vertical shift 1 unit down

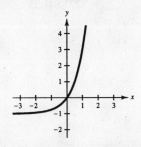

73. Horizontal shift 2 units left

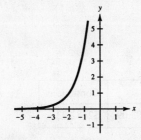

75. Reflection in the x-axis

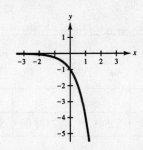

77. (a) $f(x) = 2x$ Algebraic (Linear)

     (b) $f(x) = \sqrt{2x}$ Algebraic (Radical)

     (c) $f(x) = 2^x$ Exponential

     (d) $f(x) = 2x^2$ Algebraic (Quadratic)

79. $y = 16\left(\frac{1}{2}\right)^{80/30} = 2.520$ grams

Keystrokes:

16 [×] 0.5 [yˣ] [(] 8 [÷] 3 [)] [=]   Scientific

16 [×] 0.5 [^] [(] 8 [÷] 3 [)] [ENTER]   Graphing

81.

n	1	4	12	365	Continuous
A	\$466.10	\$487.54	\$492.68	\$495.22	\$495.30

Compounded 1 time:
$$A = 100\left(1 + \frac{0.08}{1}\right)^{1(20)}$$
$$= \$466.10$$

Compounded 4 times:
$$A = 100\left(1 + \frac{0.08}{4}\right)^{4(20)}$$
$$= \$487.54$$

Compounded 12 times:
$$A = 100\left(1 + \frac{0.08}{12}\right)^{12(20)}$$
$$= \$492.68$$

Compounded 365 times:
$$A = 100\left(1 + \frac{0.08}{365}\right)^{365(20)}$$
$$= \$495.22$$

Compounded continuously:
$$A = Pe^{rt}$$
$$= 100e^{0.08(20)}$$
$$= 495.30$$

83.

n	1	4	12	365	Continuous
A	\$4734.73	\$4870.38	\$4902.71	\$4918.66	\$4919.21

Compounded 1 time: $A = 2000\left(1 + \dfrac{0.09}{1}\right)^{1(10)}$

$\qquad\qquad\qquad = \$4734.73$

Compounded 12 times: $A = 2000\left(1 + \dfrac{0.09}{12}\right)^{12(10)}$

$\qquad\qquad\qquad = \$4902.71$

Compounded continuously: $A = 2000e^{0.09(10)}$

$\qquad\qquad\qquad = \$4919.21$

Compounded 4 times: $A = 2000\left(1 + \dfrac{0.09}{4}\right)^{4(10)}$

$\qquad\qquad\qquad = \$4870.38$

Compounded 365 times: $A = 2000\left(1 + \dfrac{0.09}{365}\right)^{365(10)}$

$\qquad\qquad\qquad = \$4918.66$

85.

n	1	4	12	365	Continuous
A	\$226,296.28	\$259,889.34	\$268,503.32	\$272,841.23	\$272,990.75

Compounded 1 time: $A = 5000\left(1 + \dfrac{0.10}{1}\right)^{1(40)}$

$\qquad\qquad\qquad = \$226,296.28$

Compounded 12 times: $A = 5000\left(1 + \dfrac{0.10}{12}\right)^{12(40)}$

$\qquad\qquad\qquad = \$268,503.32$

Compounded continuously: $A = 5000^{0.10(40)}$

$\qquad\qquad\qquad = \$272,990.75$

Compounded 4 times: $A = 5000\left(1 + \dfrac{0.10}{4}\right)^{4(40)}$

$\qquad\qquad\qquad = \$259,889.34$

Compounded 365 times: $A = 5000\left(1 + \dfrac{0.10}{365}\right)^{365(40)}$

$\qquad\qquad\qquad = \$272,841.23$

87.

n	1	4	12	365	Continuous
P	\$2541.75	\$2498.00	\$2487.98	\$2483.09	\$2482.93

Compounded 1 time: $\qquad 5000 = P\left(1 + \dfrac{0.07}{1}\right)^{1(10)}$

$\qquad\qquad \dfrac{5000}{(1.07)^{10}} = P$

$\qquad\qquad \$2541.75 = P$

Compounded 12 times: $\qquad 5000 = P\left(1 + \dfrac{0.07}{12}\right)^{12(10)}$

$\qquad\qquad \dfrac{5000}{(1.00583)^{120}} = P$

$\qquad\qquad \$2487.98 = P$

Compounded Continuously: $\qquad 5000 = Pe^{0.07(10)}$

$\qquad\qquad \dfrac{5000}{e^{0.7}} = P$

$\qquad\qquad \$2482.93 = P$

Compounded 4 times: $\qquad 5000 = \left(1 + \dfrac{0.07}{4}\right)^{4(10)}$

$\qquad\qquad \dfrac{5000}{(1.0175)^{40}} = P$

$\qquad\qquad \$2498.00 = P$

Compounded 365 times:

$\qquad\qquad 5000 = P\left(1 + \dfrac{0.07}{365}\right)^{365(10)}$

$\qquad\qquad \dfrac{5000}{(1.0001918)^{3.650}} = P$

$\qquad\qquad \$2483.09 = P$

89.

n	1	4	12	365	Continuous
P	\$18,429.30	\$15,830.43	\$15,272.04	\$15,004.64	\$14,995.58

Compounded 1 time: $1{,}000{,}000 = P\left(1 + \dfrac{0.105}{1}\right)^{1(40)}$

$$\frac{1{,}000{,}000}{(1.105)^{40}} = P$$

$$\$18{,}429.30 = P$$

Compounded 4 times: $1{,}000{,}000 = P\left(1 + \dfrac{0.105}{4}\right)^{4(40)}$

$$\frac{1{,}000{,}000}{(1.02625)^{160}} = P$$

$$\$15{,}830.43 = P$$

Compounded 12 times: $1{,}000{,}000 = P\left(1 + \dfrac{0.105}{12}\right)^{12(40)}$

$$\frac{1{,}000{,}000}{(1.00875)^{480}} = P$$

$$\$15{,}272.04 = P$$

Compounded 365 times: $1{,}000{,}000 = P\left(1 + \dfrac{0.105}{365}\right)^{365(40)}$

$$\frac{1{,}000{,}000}{(1.002877)^{14{,}600}} = P$$

$$\$15{,}004.64 = P$$

Compounded continuously: $1{,}000{,}000 = Pe^{0.105(40)}$

$$\frac{1{,}000{,}000}{e^{4.2}} = P$$

$$\$14{,}995.58 = P$$

91. (a) $p = 25 - 0.4e^{0.02(100)}$

$\qquad = 25 - 0.4e^{2}$

$\qquad \approx \$22.04$

(b) $p = 25 - 0.4e^{0.02(125)}$

$\qquad = 25 - 0.4e^{2.5}$

$\qquad \approx \$20.13$

93. (a) $v(5) = 64{,}000(2)^{5/15}$

$\qquad = 64{,}000(2)^{1/3}$

$\qquad \approx \$80{,}634.95$

(b) $v(20) = 64{,}000(2)^{20/15}$

$\qquad = 64{,}000(2)^{4/3}$

$\qquad \approx \$161{,}269.89$

95. (a) $V(t) = 16{,}000\left(\frac{3}{4}\right)^{t}$

(b)

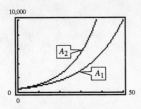

$$V(2) = 16{,}000\left(\tfrac{3}{4}\right)^{2} = 9000$$

97. (a) The balances in the accounts after t years are modeled by $A_1 = 500e^{0.06t}$ and $A_2 = 500e^{0.08t}$.

(b) *Keystrokes:*

y_1 [Y=] 500 [e^x] 0.06 [X,T,θ] [ENTER]

y_2 500 [e^x] 0.08 [X,T,θ] [GRAPH]

(c) $A_2 - A_1 = 500e^{0.08t} - 500e^{0.06t}$

$\qquad\qquad = 500(e^{0.08t} - e^{0.06t})$

Keystrokes:

y_1 [Y=] 500 [(] [e^x] 0.08 [X,T,θ] [−] [e^x]

0.06 [X,T,θ] [)] [GRAPH]

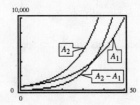

(d) The difference between the functions increases at an increasing rate.

99. (a) *Keystrokes:*

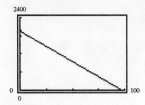

(b) $h = 1950 + 50e^{-1.6(0)} - 20(0) = 2000$ feet

$h = 1950 + 50e^{-1.6(25)} - 20(25) = 1450$ feet

$h = 1950 + 50e^{-1.6(50)} - 20(50) = 950$ feet

$h = 1950 + 50e^{-1.6(75)} - 20(75) = 450$ feet

(c) The parachutist will reach the ground at 97.5 seconds.

101. (a) *Graph model:*

Plot data:

Keystrokes: [STAT] [EDIT 1]

Enter each x entry in L 1 followed by [ENTER].

Enter each y entry in L 2 followed by [ENTER].

[STAT PLOT] [ENTER] [ENTER] [ZOOM 9]   or set window.

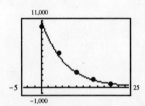

(b) *Keystrokes:* [Y=] 10,958 [e^x] [(-)] .15 [X,T,θ]
The model fits the data.

(c)

h	0	5	10	15	20
P	10,332	5583	2376	1240	517
Approx.	10,958	5176	2445	1155	546

(d) At an altitude of 8 kilometers, P is 3300 kilograms per square meter. Use table.

Keystrokes: [TABLE] 8 [ENTER]

(e) If P is 2000 kilograms per square meter, altitude is 11.3 kilometers. Graph $y_2 = 2000$ and find the intersection point.

103. (a)

x	1	10	100	1000	10,000
$\left(1 + \dfrac{1}{x}\right)^x$	2	2.5937	2.7048	2.7169	2.7181

(c) As x gets larger and larger, $\left(1 + \dfrac{1}{x}\right)^x$ approaches e.

(b) *Keystrokes:*

[Y=] [(] 1 [+] 1 [÷] [X,T,θ] [)] [^] [X,T,θ] [GRAPH]

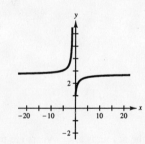

Yes, the graph is approaching a horizontal asymptote.

105. Polynomial functions have terms with variable bases and constant exponents. Exponential functions have terms with constant bases and variable exponents.

107. $f(x) = 3^x$ is an increasing function and $g(x) = \left(\frac{1}{3}\right)^x$ is a decreasing function.

109. False. e is an irrational number.

$\dfrac{271,801}{99,990}$ is rational because its equivalent decimal form is a repeating decimal.

Section 11.2   Inverse Functions

1. (a) $(f \circ g)(x) = (2x - 4) - 3 = 2x - 7$

  (b) $(g \circ f)(x) = 2(x - 3) - 4 = 2x - 6 - 4 = 2x - 10$

  (c) $(f \circ g)(4) = 2(4) - 7 = 1$

  (d) $(g \circ f)(7) = 2(7) - 10 = 4$

3. (a) $(f \circ g)(x) = (2x^2 - 6) + 5 = 2x^2 - 1$

  (b) $(g \circ f)(x) = 2(x + 5)^2 - 6 = 2(x^2 + 10x + 25) - 6$
  $$= 2x^2 + 20x + 50 - 6$$
  $$= 2x^2 + 20x + 44$$

  (c) $(f \circ g)(2) = 2(2)^2 - 1 = 2(4) - 1 = 7$

  (d) $(g \circ f)(-3) = 2(-3)^2 + 20(-3) + 44$
  $$= 2(9) - 60 + 44$$
  $$= 2$$

5. (a) $(f \circ g)(x) = |3x - 3|$

  (b) $(g \circ f)(x) = 3|x - 3|$

  (c) $(f \circ g)(1) = |3 - 3| = 0$

  (d) $(g \circ f)(2) = 3|2 - 3| = 3$

7. (a) $(f \circ g)(x) = \sqrt{x + 5 - 4} = \sqrt{x + 1}$

  (b) $(g \circ f)(x) = \sqrt{x - 4} + 5$

  (c) $(f \circ g)(3) = \sqrt{3 + 1} = 2$

  (d) $(g \circ f)(8) = \sqrt{8 - 4} + 5 = 2 + 5 = 7$

9. (a) $(f \circ g)(x) = \dfrac{1}{\dfrac{2}{x^2} - 3} \cdot \dfrac{x^2}{x^2} = \dfrac{x^2}{2 - 3x^2}$

  (b) $(g \circ f)(x) = \dfrac{2}{\left(\dfrac{1}{x - 3}\right)^2} = 2(x - 3)^2$

  (c) $(f \circ g)(-1) = \dfrac{(-1)^2}{2 - 3(-1)^2} = \dfrac{1}{2 - 3} = \dfrac{1}{-1} = -1$

  (d) $(g \circ f)(2) = 2(2 - 3)^2 = 2(-1)^2 = 2$

11. (a) $f(1) = -1$

  (b) $g(-1) = -2$

  (c) $(g \circ f)(1) = g[f(1)]$
  $$= g[-1]$$
  $$= -2$$

13. (a) $(f \circ g)(-3) = f[g(-3)] = f[1] = -1$

  (b) $(g \circ f)(-2) = g[f(-2)] = g[3] = 1$

15. (a) $f(3) = 10$

  (b) $g(10) = 1$

  (c) $(g \circ f)(3) = g[f(3)] = g[10] = 1$

17. (a) $(g \circ f)(4) = g[f(4)] = g[17] = 0$

  (b) $(f \circ g)(2) = f[g(2)] = f[3] = 10$

19. $f(x) = x + 1, \quad g(x) = 2x - 5$

  (a) $f \circ g = (2x - 5) + 1 = 2x - 4$
  Domain: $(-\infty, \infty)$

  (b) $g \circ f = 2(x + 1) - 5 = 2x + 2 - 5 = 2x - 3$
  Domain: $(-\infty, \infty)$

21. $f(x) = \sqrt{x}, \quad g(x) = x - 2$

  (a) $f \circ g = \sqrt{x - 2}$ Domain: $[2, \infty)$

  (b) $g \circ f = \sqrt{x} - 2$ Domain: $[0, \infty)$

23. $f(x) = x^2 - 1, \quad g(x) = \sqrt{x + 3}$

  (a) $f \circ g = \left(\sqrt{x + 3}\right)^2 - 1 = x + 3 - 1 = x + 2$
  Domain: $[-3, \infty)$

  (b) $g \circ f = \sqrt{(x^2 - 1) + 3} = \sqrt{x^2 + 2}$
  Domain: $(-\infty, \infty)$

25. $f(x) = \dfrac{x}{x + 5},$ $g(x) = \sqrt{x - 1}$

(a) $f \circ g = \dfrac{\sqrt{x - 1}}{\sqrt{x - 1} + 5}$ Domain: $[1, \infty)$

(b) $g \circ f = \sqrt{\dfrac{x}{x + 5} - 1}$ Domain: $(-\infty, -5)$

27. $f(x) = x^2 - 2$

No, it does not have an inverse because it is possible to find a horizontal line that intersects the graph of f at more than one point.

29. $f(x) = x^2,\ x \geq 0$

Yes, it does have an inverse because no horizontal line intersects the graph of f at more than one point.

31. $g(x) = \sqrt{25 - x^2}$

No, it does not have an inverse because it is possible to find a horizontal line that intersects the graph of g at more than one point.

33. *Keystrokes:*

[Y=] [X,T,θ] [^] 3 [−] 1 [GRAPH]

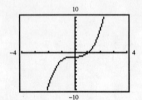

One-to-one

35. *Keystrokes:*

[Y=] [MATH] 4 [(] 5 [−] [X,T,θ] [)] [GRAPH]

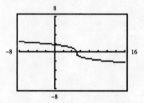

One-to-one

37. *Keystrokes:*

[Y=] [X,T,θ] [^] 4 [−] 6 [GRAPH]

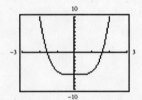

Not one-to-one

39. *Keystrokes:*

[Y=] 5 [÷] [X,T,θ] [GRAPH]

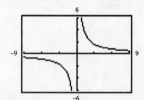

One-to-one

41. *Keystrokes:*

[Y=] 4 [÷] [(] [X,T,θ] [x²] [+] 1 [)] [GRAPH]

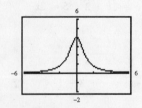

Not one-to-one

43. $f(g(x)) = 10\left(\dfrac{1}{10}x\right) = x$

$g(f(x)) = \dfrac{10x}{10} = x$

45. $f(g(x)) = (x - 15) + 15 = x$

$g(f(x)) = (x + 15) - 15 = x$

47. $f(g(x)) = 1 - 2\left[\frac{1}{2}(1 - x)\right]$

$\qquad = 1 - (1 - x) = 1 - 1 + x = x$

$g(f(x)) = \frac{1}{2}[1 - (1 - 2x)]$

$\qquad = \frac{1}{2}[1 - 1 + 2x] = \frac{1}{2}[2x] = x$

49. $f(g(x)) = 2 - 3\left[\frac{1}{3}(2 - x)\right] = 2 - (2 - x) = x$

  $g(f(x)) = \frac{1}{3}[2 - (2 - 3x)] = \frac{1}{3}[3x] = x$

51. $f(g(x)) = \sqrt[3]{x^3 - 1 + 1} = \sqrt[3]{x^3} = x$

  $g(f(x)) = \left(\sqrt[3]{x + 1}\right)^3 - 1 = x + 1 - 1 = x$

53. $f(g(x)) = \dfrac{1}{\dfrac{1}{x}} = x$

  $g(f(x)) = \dfrac{1}{\dfrac{1}{x}} = x$

55. $f^{-1}(x) = \dfrac{x}{5}$

  $f(f^{-1}(x)) = f\left(\dfrac{x}{5}\right) = 5\left(\dfrac{x}{5}\right) = x$

  $f^{-1}(f(x)) = f^{-1}(5x) = \dfrac{5x}{5} = x$

57. $f^{-1}(x) = 2x$

  $f(f^{-1}(x)) = f(2x) = \frac{1}{2}(2x) = x$

  $f^{-1}(f(x)) = f^{-1}\left(\frac{1}{2}x\right) = 2\left(\frac{1}{2}x\right) = x$

59. $f^{-1}(x) = x - 10$

  $f(f^{-1}(x)) = f(x - 10) = x - 10 + 10 = x$

  $f^{-1}(f(x)) = f^{-1}(x + 10) = x + 10 - 10 = x$

61. $f^{-1}(x) = 3 - x$

  $f(f^{-1}(x)) = f(3 - x) = 3 - (3 - x) = 3 - 3 + x = x$

  $f^{-1}(f(x)) = f^{-1}(3 - x) = 3 - (3 - x) = 3 - 3 + x = x$

63. $f^{-1}(x) = \sqrt[7]{x}$

  $f(f^{-1}(x)) = f(\sqrt[7]{x}) = \left(\sqrt[7]{x}\right)^7 = x$

  $f^{-1}(f(x)) = f^{-1}(x^7) = \sqrt[7]{x^7} = x$

65. $f^{-1}(x) = x^3$

  $f(f^{-1}(x)) = f(x^3) = \sqrt[3]{x^3} = x$

  $f^{-1}(f(x)) = f^{-1}(\sqrt[3]{x}) = \left(\sqrt[3]{x}\right)^3 = x$

67. $f(x) = 8x$

   $y = 8x$

   $x = 8y$

   $\dfrac{x}{8} = y$

   $f^{-1}(x) = \dfrac{x}{8}$

69. $g(x) = x + 25$

   $y = x + 25$

   $x = y + 25$

   $x - 25 = y$

   $g^{-1}(x) = x - 25$

71. $g(x) = 3 - 4x$

   $y = 3 - 4x$

   $x = 3 - 4y$

   $x - 3 = -4y$

   $\dfrac{x - 3}{-4} = y$

   $\dfrac{3 - x}{4}$ or $\dfrac{x - 3}{-4} = g^{-1}(x)$

73. $g(t) = \frac{1}{4}t + 2$

   $y = \frac{1}{4}t + 2$

   $t = \frac{1}{4}y + 2$

   $t - 2 = \frac{1}{4}y$

   $4(t - 2) = y$

   $4t - 8 = g^{-1}(t)$

75. $h(x) = \sqrt{x}$

   $y = \sqrt{x}$

   $x = \sqrt{y}$

   $x^2 = y$

   $x^2 = h^{-1}(x),\quad x \geq 0$

77. $f(t) = t^3 - 1$

   $y = t^3 - 1$

   $t = y^3 - 1$

   $t + 1 = y^3$

   $\sqrt[3]{t + 1} = y$

   $\sqrt[3]{t + 1} = f^{-1}(t)$

79. $g(s) = \dfrac{5}{s + 4}$

   $y = \dfrac{5}{s + 4}$

   $s = \dfrac{5}{y + 4}$

   $y + 4 = \dfrac{5}{s}$

   $g^{-1}(s) = \dfrac{5}{s} - 4,\quad s \neq 0$

81. $f(x) = \sqrt{x + 3}$

   $y = \sqrt{x + 3}$

   $x = \sqrt{y + 3}$

   $x^2 = y + 3$

   $x^2 - 3 = y$

   $x^2 - 3 = f^{-1}(x),\quad x \geq 0$

83. $f(x) = x + 4, \quad f^{-1}(x) = x - 4$

$(0, 4) \qquad\qquad (4, 0)$

$(-4, 0) \qquad\quad (0, -4)$

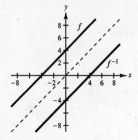

85. $f(x) = 3x - 1, \quad f^{-1}(x) = \frac{1}{3}(x + 1)$

$(0, -1) \qquad\qquad (-1, 0)$

$\left(\frac{1}{3}, 0\right) \qquad\qquad \left(0, \frac{1}{3}\right)$

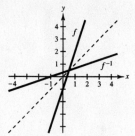

87. $f(x) = x^2 - 1, \quad f^{-1}(x) = \sqrt{x + 1}$

$(0, -1) \qquad\qquad (-1, 0)$

$(1, 0) \qquad\qquad (0, 1)$

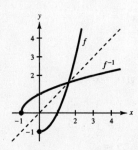

89. (b)

91. (d)

93. *Keystrokes:*

y_1 Y= (1 ÷ 3) X,T,θ ENTER

y_2 3 X,T,θ GRAPH

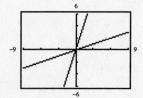

95. *Keystrokes:*

y_1 Y= √ (X,T,θ + 1) ENTER

y_2 X,T,θ x^2 − 1 ÷ (X,T,θ TEST 4 0) GRAPH

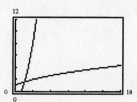

97. *Keystrokes:*

y_1 Y= (1 ÷ 8) X,T,θ MATH 3 ENTER

y_2 2 MATH 4 X,T,θ GRAPH

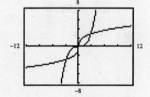

99. *Keystrokes:*

y_1 Y= 3 X,T,θ + 4 ENTER

y_2 (X,T,θ − 4) ÷ 3 GRAPH

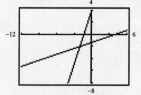

101. $f(x) = (x - 2)^2,\ \ x \geq 2$

$y = (x - 2)^2$

$x = (y - 2)^2$

$\sqrt{x} = y - 2$

$\sqrt{x} + 2 = y$

$\sqrt{x} + 2 = f^{-1}(x),\ \ x \geq 0$

103. $f(x) = |x| + 1,\ \ x \geq 0$

$y = |x| + 1$

$x = |y| + 1$

$x - 1 = |y|$

$x - 1 = y$

$x - 1 = f^{-1}(x),\ \ x \geq 1$

105.

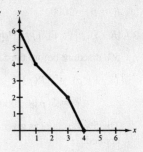

x	0	1	3	4
f^{-1}	6	4	2	0

107.

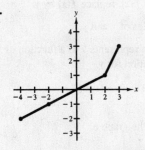

x	-4	-2	2	3
f^{-1}	-2	-1	1	3

109. (a) $y = 3 - 2x$

$x = 3 - 2y$

$2y = 3 - x$

$y = \dfrac{3 - x}{2}$

$f^{-1}(x) = \dfrac{3 - x}{2}$

(b) $y = \dfrac{3 - x}{2}$

$x = \dfrac{3 - y}{2}$

$2x = 3 - y$

$y = 3 - 2x$

$(f^{-1})^{-1}(x) = 3 - 2x$

111. (a) Total cost = Cost of \$0.50 commodity + Cost of \$0.75 commodity

$y = 0.50x + 0.75(100 - x)$

(b) $y = 0.50x + 0.75(100 - x)$

$y = 0.50x + 75 - 0.75x$

$y = -0.25x + 75$

$x = -0.25y + 75$

$x - 75 = -0.25y$

$\dfrac{x - 75}{-0.25} = y$

$-4(x - 75) = y$

$4(75 - x) = y$

x: total cost

y: number of pounds at \$0.50 per pound

(c) $50 \leq x \leq 75$

If you buy only the cheaper commodity, your cost will be \$50. If you buy only the more expensive commodity, your cost will be \$75. Any combination will lie between \$50 and \$75.

(d) $y = 4(75 - 60)$

$y = 4(15)$

$y = 60$

Thus, 60 pounds of the \$0.50 per pound commodity is purchased.

113. (a) $f(g(x)) = 0.02x - 200{,}000$

(b) $g(f(x)) = 0.02(x - 200{,}000),\ \ x > 200{,}000$

This part represents the bonus because it gives 2% of sales over \$200,000.

115. (a) $R = p - 2000$

(c) $(R \circ S)(p) = R[S(p)] = R(0.95p) = 0.95p - 2000$

5% discount before the $2000 rebate is given.

$(S \circ R)(p) = S[R(p)]$

$\qquad = S(p - 2000) = 0.95(p - 2000)$

The 5% discount is given after the $2000 rebate is applied.

(b) $S = p - 0.05p$

$\quad S = 0.95p$

(d) $(R \circ S)(26,000) = 0.95(26,000) - 2000 = \$22,700$

$(S \circ R)(26,000) = 0.95(26,000 - 2000) = \$22,800$

$R \circ S$ yields the smaller cost because the dealer discount is based on a larger amount.

117. True, the x-coordinate of a point on the graph of f becomes the y-coordinate of a point on the graph of f^{-1}.

119. False: $\quad f(x) = \sqrt{x - 1}$ Domain $[1, \infty)$

$f^{-1}(x) = x^2 + 1$ Domain $[0, \infty)$

121. If $f(x) = 2x$ and $g(x) = x^2$, then $(f \circ g)(x) = 2x^2$ and $(g \circ f)(x) = 4x^2$.

123. (a) In the equation for $f(x)$, replace $f(x)$ by y.

(b) Interchange the roles of x and y.

(c) If the new equation represents y as a function of x, solve the new equation for y.

(d) Replace y by $f^{-1}(x)$.

125. Graphically, a function f has an inverse function if and only if no horizontal line intersects the graph of f at more than one point. This is equivalent to saying that the function f is one-to-one.

Section 11.3   Logarithmic Functions

1. $\log_5 25 = 2$

$\quad 5^2 = 25$

3. $\log_4 \frac{1}{16} = -2$

$\quad 4^{-2} = \frac{1}{16}$

5. $\log_3 \frac{1}{243} = -5$

$\quad 3^{-5} = \frac{1}{243}$

7. $\log_{36} 6 = \frac{1}{2}$

$\quad 36^{1/2} = 6$

9. $\log_8 4 = \frac{2}{3}$

$\quad 8^{2/3} = 4$

11. $\log_2 2.462 \approx 1.3$

$\quad 2^{1.3} \approx 2.462$

13. $\quad 7^2 = 49$

$\log_7 49 = 2$

15. $\quad 3^{-2} = \frac{1}{9}$

$\log_3 \frac{1}{9} = -2$

17. $\quad 8^{2/3} = 4$

$\log_8 4 = \frac{2}{3}$

19. $25^{-1/2} = \frac{1}{5}$

$\log_{25} \frac{1}{5} = -\frac{1}{2}$

21. $\quad 4^0 = 1$

$\log_4 1 = 0$

23. $\quad 5^{1.4} \approx 9.518$

$\log_5 9.518 \approx 1.4$

25. $\log_2 8 = 3$ because $2^3 = 8$.

27. $\log_{10} 10 = 1$ because $10^1 = 10$.

29. $\log_{10} 1000 = 3$ because $10^3 = 1000$.

31. $\log_2 \frac{1}{4} = -2$ because $2^{-2} = \frac{1}{4}$.

33. $\log_4 \frac{1}{64} = -3$ because $4^{-3} = \frac{1}{64}$.

35. $\log_{10} \dfrac{1}{10,000} = -4$ because $10^{-4} = \dfrac{1}{10,000}$.

37. $\log_2(-3)$ is not possible because there is no power to which 2 can be raised to obtain -3.

39. $\log_4 1 = 0$ because $4^0 = 1$.

41. $\log_5 (-6)$ is not possible because there is no power to which 5 can be raised to obtain -6.

43. $\log_9 3 = \frac{1}{2}$ because $9^{1/2} = 3$.

45. $\log_{16} 8 = \frac{3}{4}$ because $16^{3/4} = 8$.

47. $\log_7 7^4 = 4$ because $7^4 = 7^4$.

49. $\log_{10} 31 \approx 1.4914$

51. $\log_{10} 0.85 \approx -0.0706$

53. $\log_{10}(\sqrt{2} + 4) \approx 0.7335$

55.

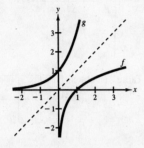

f and g are inverse functions.

57.

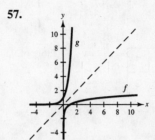

f and g are inverse functions.

59. f and g are inverse functions.

61. f and g are inverse functions.

63. $h(x) = 3 + \log_2 x$

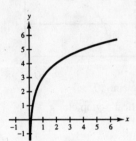

Vertical shift 3 units up

65. $h(x) = \log_2(x - 2)$

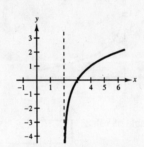

Horizontal shift 2 units right

67. $h(x) = \log_2(-x)$

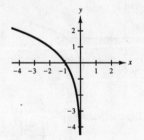

Reflection in the x-axis

69. $f(x) = 4 + \log_3 x$
matches graph (e)

71. $f(x) = -\log_3 x$
matches graph (d)

73. $f(x) = \log_3(x - 4)$
matches graph (a)

75. $f(x) = \log_5 x$

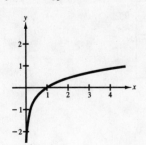

Table of values:

x	1	5
y	0	1

77. $f(x) = -\log_2 t$

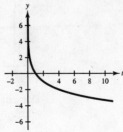

Table of values:

x	1	2
y	0	-1

79. $f(x) = 3 + \log_2 x$

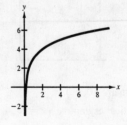

Table of values:

x	1	2
y	3	4

81. $g(x) = \log_2(x - 3)$

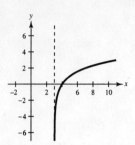

Table of values:

x	4	7
y	0	2

83. $f(x) = \log_{10}(10x)$

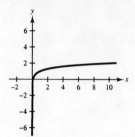

Table of values:

x	1	10
y	1	2

85. $f(x) = \log_4 x$

Domain: $(0, \infty)$

Vertical asymptote: $x = 0$

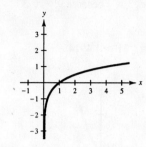

Table of values:

x	1	4
y	0	1

87. $h(x) = \log_4(x - 3)$

Domain: $(3, \infty)$

Vertical asymptote: $x = 3$

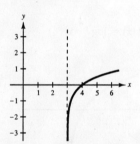

Table of values:

x	4	13
y	0	1

89. $y = -\log_3 x + 2$

Domain: $(0, \infty)$

Vertical asymptote: $x = 0$

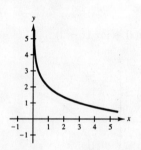

Table of values:

x	1	3
y	2	1

91. $y = 5 \log_{10} x$

Keystrokes:

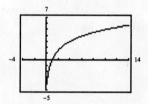

Domain: $(0, \infty)$

Vertical asymptote: $x = 0$

93. $y = -3 + 5 \log_{10} x$

Keystrokes: $\boxed{Y=}\ \boxed{(-)}\ 3\ \boxed{+}\ 5\ \boxed{LOG}\ \boxed{X,T,\theta}\ \boxed{GRAPH}$

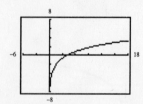

Domain: $(0, \infty)$

Vertical asymptote: $x = 0$

95. $y = \log_{10}\left(\dfrac{x}{5}\right)$

Keystrokes: $\boxed{Y=}\ \boxed{LOG}\ \boxed{X,T,\theta}\ \boxed{\div}\ 5\ \boxed{GRAPH}$

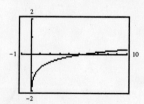

Domain: $(0, \infty)$

Vertical asymptote: $x = 0$

97. $\ln 25 \approx 3.2189$

99. $\ln 0.75 \approx -0.2877$

101. $\ln\left(\dfrac{1 + \sqrt{5}}{3}\right) \approx 0.0757$

103. (b) Basic graph shifted 1 unit left

105. (d) Basic graph shifted $\frac{3}{2}$ unit right

107. (f) Basic graph multiplied by 10

109. $f(x) = -\ln x$

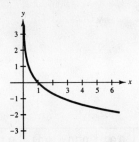

Table of values:

x	1	e
y	0	-1

111. $f(x) = 3 \ln x$

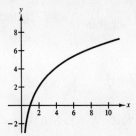

Table of values:

x	1	e
y	0	3

113. $f(x) = 1 + \ln x$

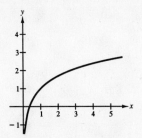

Table of values:

x	1	e
y	1	2

115. $g(t) = 2 \ln(t - 4)$

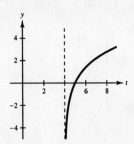

Table of values:

x	5	6
y	0	1.4

117. *Keystrokes:*

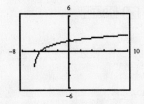

Domain: $(-6, \infty)$

Vertical asymptote: $x = -6$

119. *Keystrokes:*

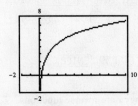

Domain: $(0, \infty)$

Vertical asymptote: $t = 0$

121. $\log_8 132 = \dfrac{\log 132}{\log 8} \approx 2.3481$

$ = \dfrac{\ln 132}{\ln 8} \approx 2.3481$

123. $\log_3 7 = \dfrac{\log 7}{\log 3} \approx 1.7712$

$ = \dfrac{\ln 7}{\ln 3} \approx 1.7712$

125. $\log_2 0.72 = \dfrac{\log 0.72}{\log 2} \approx -0.4739$

$ = \dfrac{\ln 0.72}{\ln 2} \approx -0.4739$

127. $\log_{15} 1250 = \dfrac{\log 1250}{\log 15} \approx 2.6332$

$\phantom{\log_{15} 1250} = \dfrac{\ln 1250}{\ln 15} \approx 2.6332$

129. $\log_{(1/2)} 4 = \dfrac{\log 4}{\log 0.5} = -2$

$\phantom{\log_{(1/2)} 4} = \dfrac{\ln 4}{\ln 0.5} = -2$

131. $\log_4 \sqrt{42} = \dfrac{\log \sqrt{42}}{\log 4} \approx 1.3481$

$\phantom{\log_4 \sqrt{42}} = \dfrac{\ln \sqrt{42}}{\ln 4} \approx 1.3481$

133. $\log_2(1 + e) = \dfrac{\log(1 + e)}{\log 2} \approx 1.8946$

$= \dfrac{\ln(1 + e)}{\ln 2} \approx 1.8946$

135. $h = 116 \log_{10}(55 + 40) - 176$

$= 116 \log_{10}(95) - 176$

≈ 53.4 inches

137. r of 0.07: $t = \dfrac{\ln 2}{0.07} \approx 9.9021$ r of 0.08: $t = \dfrac{\ln 2}{0.08} \approx 8.6643$

r of 0.09: $t = \dfrac{\ln 2}{0.09} \approx 7.7016$ r of 0.10: $t = \dfrac{\ln 2}{0.10} \approx 6.9315$

r of 0.11: $t = \dfrac{\ln 2}{0.11} \approx 6.3013$ r of 0.12: $t = \dfrac{\ln 2}{0.12} \approx 5.7762$

r	0.07	0.08	0.09	0.10	0.11	0.12
t	9.9	8.7	7.7	6.9	6.3	5.8

139. (a) *Keystrokes:*

[Y=] 10 [LN] [(] [(] 10 [+] [√] [(] 100 [−] [X,T,θ] [x²] [)] [)] [÷] [X,T,θ] [)] [−] [√] [(] 100 [−] [X,T,θ] [x²] [)] [GRAPH]

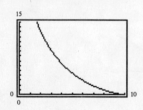

Domain: (0, 10]

(b) Vertical asymptote: $x = 0$

(c) $y = 13.126$ when $x = 2$. Trace to $x = 2$.

141. (a) *Graph model:*

Plot data:

Keystrokes: [STAT] [EDIT 1]

Enter each x entry in L1 followed by [ENTER].

Enter each y entry in L2 followed by [ENTER].

[STAT PLOT] [ENTER] [ENTER] [ZOOM 9] or set window.

Keystrokes: [Y=] 435.33 [−] 527.72 [X,T,θ] [+] 396.68 [LN] [X,T,θ] [+] 88.05 [X,T,θ] [x²] [GRAPH]

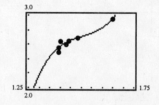

(b) $2.81

Keystrokes: [TABLE] 1.59 [ENTER]

143. $f(x) = 2^x$ and $g(x) = \log_2 x$ are inverse functions.

145. $\log_a a^x = x$ because $a^x = a^x$.

147. $\log_b x = \dfrac{\log x}{\log b} = \dfrac{\ln x}{\ln b}$

149. $f(x) = \log_{10} x$

     $f^{-1}(x) = 10^x$

151. If $f(x)$ is negative, then $0 < x < 1$.

153. $f(a) = \log_{10} a$ $f(b) = \log_{10} b$

     $10^{f(a)} = a$ $10^{f(b)} = b$

     $\dfrac{a}{b} = \dfrac{10^{f(a)}}{10^{f(b)}} = \dfrac{10^{3f(b)}}{10^{f(b)}} = 10^{2f(b)} = 10^{2\log_{10} b}$

                        $= 10^{\log_{10} b^2} = b^2$

Mid-Chapter Quiz for Chapter 11

1. (a) $f(2) = \left(\dfrac{4}{3}\right)^2 = \dfrac{16}{9}$

(b) $f(0) = \left(\dfrac{4}{3}\right)^0 = 1$

(c) $f(-1) = \left(\dfrac{4}{3}\right)^{-1} = \dfrac{3}{4}$

(d) $f(1.5) = \left(\dfrac{4}{3}\right)^{1.5} \approx 1.54$

$\qquad = \dfrac{8\sqrt{3}}{9}$

2. $g(x) = 2^{-0.5x}$

Domain: $(-\infty, \infty)$

Range: $(0, \infty)$

3.

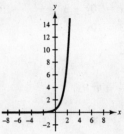

4.

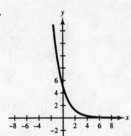

5.

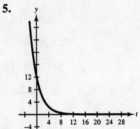

6.

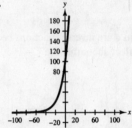

7. Compounded 1 time per year: $A = 750\left(1 + \dfrac{0.075}{1}\right)^{1(20)}$

$\approx \$3185.89$

Compounded 12 times per year: $A = 750\left(1 + \dfrac{0.075}{12}\right)^{12(20)}$

$\approx \$3345.61$

Compounded 4 times per year: $A = 750\left(1 + \dfrac{0.075}{4}\right)^{4(20)}$

$\approx \$3314.90$

Compounded 365 times per year: $A = 750\left(1 + \dfrac{0.075}{365}\right)^{365(20)}$

$\approx \$3360.75$

Compounded continuously: $A = Pe^{rt}$

$= 750e^{0.075(20)}$

$\approx \$3361.27$

8. $A = 2.23e^{(0.04)(5)} = \$2.72$

9. (a) $(f \circ g)(x) = f[g(x)] = 2x^3 - 3$

(b) $(g \circ f)(x) = g[f(x)] = (2x - 3)^3$

(c) $(fg)(-2) = f[g(-2)] = f[-8] = 2(-8) - 3 = -19$

(d) $(g \circ f)(4) = g[f(4)] = g[5] = 5^3 = 125$

10. $f[g(x)] = 3 - 5\left[\dfrac{1}{5}(3 - x)\right] = 3 - 1(3 - x) = 3 - 3 + x = x$

$g[f(x)] = \dfrac{1}{5}[3 - (3 - 5x)] = \dfrac{1}{5}[3 - 3 + 5x] = \dfrac{1}{5}[5x] = x$

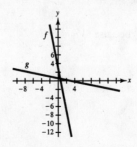

11. $h(x) = 10x + 3$

$y = 10x + 3$

$x = 10y + 3$

$x - 3 = 10y$

$\dfrac{x - 3}{10} = y$

$\dfrac{x - 3}{10} = h^{-1}(x)$

12. $g(t) = \dfrac{1}{2}t^3 + 2$

$y = \dfrac{1}{2}t^3 + 2$

$t = \dfrac{1}{2}y^3 + 2$

$t - 2 = \dfrac{1}{2}y^3$

$2t - 4 = y^3$

$\sqrt[3]{2t - 4} = y$

$\sqrt[3]{2t - 4} = g^{-1}(t)$

13. $\log_4\left(\dfrac{1}{16}\right) = -2$

$4^{-2} = \dfrac{1}{16}$

14. $3^4 = 81$

$\log_3 81 = 4$

15. $\log_5 125 = 3$ because $5^3 = 125.$

16. f and g are inverse functions because the graphs of f and g reflect about the line $y = x$.

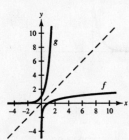

17. *Keystrokes:*

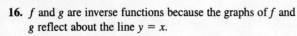

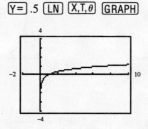

18. *Keystrokes:*

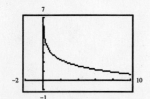

19. $f(x) = \log_5(x - 2) + 1$

The graph of $f(x) = \log_5 x$ has been shifted 3 units right and 1 unit up, so $h = 2, k = 1$.

20. $\log_6 450 = \dfrac{\log 450}{\log 6} \approx 3.4096$

Section 11.4   Properties of Logarithms

1. $\log_5 5^2 = 2 \cdot \log_5 5 = 2 \cdot 1 = 2$

3. $\log_2\left(\dfrac{1}{8}\right)^3 = \log_2(2^{-3})^3 = \log_2 2^{-9}$

$= -9 \cdot \log_2 2 = -9 \cdot 1 = -9$

5. $\log_6 \sqrt{6} = \log_6 6^{1/2} = \dfrac{1}{2}$ because $6^{1/2} = 6^{1/2}.$

7. $\ln 8^0 = 0 \cdot \ln 8 = 0$ *or* $\ln 8^0 = \ln 1 = 0$

9. $\ln e^4 = 4 \ln e = 4(1) = 4$

11. $\log_4 2 + \log_4 8 = \log_4 2 \cdot 8 = \log_4 16 = 2$ because $4^2 = 16.$

13. $\log_8 4 + \log_8 16 = \log_8 4 \cdot 16 = \log_8 64 = 2$ because $8^2 = 64$.

15. $\log_4 8 - \log_4 2 = \log_4 \frac{8}{2} = \log_4 4 = 1$ because $4^1 = 4$.

17. $\log_6 72 - \log_6 2 = \log_6 \frac{72}{2} = \log_6 36 = 2$ because $6^2 = 36$.

19. $\log_2 5 - \log_2 40 = \log_2 \frac{5}{40} = \log_2 \frac{1}{8} = \log_2 2^{-3} = -3$ because $2^{-3} = 2^{-3}$.

21. $\ln e^8 + \ln e^4 = \ln e^8 \cdot e^4 = \ln e^{12}$

$\qquad = 12 \ln e = 12 \cdot 1 = 12$

23. $\ln \frac{e^3}{e^2} = \ln e = 1$

25. $\log_4 4 = \log_4 2 + \log_4 2 = 0.5000 + 0.5000 = 1$

27. $\log_4 6 = \log_4 2 \cdot 3 = \log_4 2 + \log_4 3$

$\qquad = 0.5000 + 0.7925 \approx 1.2925$

29. $\log_4 \frac{3}{2} = \log_4 3 - \log_4 2 = 0.7925 - 0.5000 \approx 0.2925$

31. $\log_4 \sqrt{2} = \frac{1}{2} \log_4 2 = \frac{1}{2}(0.5000) = 0.25$

33. $\log_4(3 \cdot 2^4) = \log_4 3 + 4 \log_4 2$

$\qquad = 0.7925 + 4(0.5000) \approx 2.7925$

35. $\log_4 3^0 = \log_4 1 = 0$

37. $\log_{10} 9 = \log_{10} 3^2 = 2 \log_{10} 3 \approx 2(0.477) \approx 0.954$

39. $\log_{10} 36 = \log_{10}(3 \cdot 12) = \log_{10} 3 + \log_{10} 12$

$\qquad \approx 0.477 + 1.079$

$\qquad \approx 1.556$

41. $\log_{10} \sqrt{36} = \log_{10} 36^{1/2} = \frac{1}{2} \log_{10} 36$

$\qquad \approx \frac{1}{2}(1.556)$

$\qquad \approx 0.778$

43. $\log_3 11x = \log_3 11 + \log_3 x$

45. $\log_7 x^2 = 2 \log_7 x$

47. $\log_5 x^{-2} = -2 \log_5 x$

49. $\log_4 \sqrt{3x} = \log_4(3x)^{1/2} = \frac{1}{2} \log_4(3x)$

$\qquad = \frac{1}{2}(\log_4 3 + \log_4 x)$

51. $\ln 3y = \ln 3 + \ln y$

53. $\log_2 \frac{z}{17} = \log_2 z - \log_2 17$

55. $\ln \frac{5}{x-2} = \ln 5 - \ln(x-2)$

57. $\ln x^2(y-2) = \ln x^2 + \ln(y-2)$

$\qquad = 2 \ln x + \ln(y-2)$

59. $\log_4[x^6(x-7)^2] = \log_4 x^6 + \log_4(x-7)^2$

$\qquad = 6 \log_4 x + 2 \log_4(x-7)$

61. $\log_3 \sqrt[3]{x+1} = \frac{1}{3} \log_3(x+1)$

63. $\ln \sqrt{x(x+2)} = \frac{1}{2}[\ln x + \ln(x+2)]$

65. $\ln\left(\frac{x+1}{x-1}\right)^2 = 2 \ln\left(\frac{x+1}{x-1}\right)$

$\qquad = 2[\ln(x+1) - \ln(x-1)]$

67. $\ln \sqrt[3]{\frac{x^2}{x+1}} = \ln\left(\frac{x^2}{x+1}\right)^{1/3} = \frac{1}{3} \ln\left(\frac{x^2}{x+1}\right)$

$\qquad = \frac{1}{3}[\ln x^2 - \ln(x+1)]$

$\qquad = \frac{1}{3}[2 \ln x - \ln(x+1)]$

69. $\ln \dfrac{a^3(b-4)}{c^2} = \ln a^3 + \ln(b-4) - \ln c^2$

$\qquad = 3 \ln a + \ln(b-4) - 2 \ln c$

71. $\ln \dfrac{x\sqrt[3]{y}}{(wz)^4} = \ln x + \ln \sqrt[3]{y} - \ln(wz)^4$

$\qquad = \ln x + \ln y^{1/3} - 4 \ln(wz)$

$\qquad = \ln x + \dfrac{1}{3} \ln y - 4(\ln w + \ln z)$

73. $\log_6[a\sqrt{b}(c-d)^3] = \log_6 a + \log_6 \sqrt{b} + \log_6(c-d)^3$

$\qquad = \log_6 a + \log_6 b^{1/2} + 3 \log_6(c-d)$

$\qquad = \log_6 a + \dfrac{1}{2} \log_6 b + 3 \log_6(c-d)$

75. $\ln\left[(x+y)\dfrac{\sqrt[5]{w+2}}{3t}\right] = \ln(x+y) + \ln \sqrt[5]{w+2} - \ln(3t)$

$\qquad = \ln(x+y) + \ln(w+2)^{1/5} - (\ln 3 + \ln t)$

$\qquad = \ln(x+y) + \dfrac{1}{5}\ln(w+2) - (\ln 3 + \ln t)$

77. $\log_{12} x - \log_{12} 3 = \log_{12} \dfrac{x}{3}$

79. $\log_2 3 + \log_2 x = \log_2 3x$

81. $\log_{10} 4 - \log_{10} x = \log_{10} \dfrac{4}{x}$

83. $4 \ln b = \ln b^4, \quad b > 0$

85. $-2 \log_5 2x = \log_5(2x)^{-2}$

$\qquad = \log_5 \dfrac{1}{4x^2}, \; x > 0$

87. $\dfrac{1}{3}\ln(2x+1) = \ln \sqrt[3]{2x+1}$

89. $\log_3 2 + \dfrac{1}{2}\log_3 y = \log_3 2 + \log_3 \sqrt{y}$

$\qquad = \log_3 2\sqrt{y}$

91. $2 \ln x + 3 \ln y - \ln z = \ln \dfrac{x^2 y^3}{z}, \quad x > 0, \, y > 0, \, z > 0$

93. $5 \ln 2 - \ln x + 3 \ln y = \ln 2^5 - \ln x + \ln y^3$

$\qquad = \ln 32 - \ln x + \ln y^3$

$\qquad = \ln \dfrac{32y^3}{x}, \; x > 0, \, y > 0$

95. $4(\ln x + \ln y) = \ln(xy)^4 \quad \text{or} \quad \ln x^4 y^4, \; x > 0, \, y > 0$

97. $2[\ln x - \ln(x+1)] = 2 \ln \dfrac{x}{x+1} = \ln\left(\dfrac{x}{x+1}\right)^2$

$\qquad = \ln \dfrac{x^2}{(x+1)^2}, \; x > 0$

99. $\log_4(x+8) - 3 \log_4 x = \log_4(x+8) - \log_4 x^3$

$\qquad = \log_4 \dfrac{(x+8)}{x^3}, \; x > 0$

101. $\dfrac{1}{2}\log_5(x+2) - \log_5(x-3) = \log_5(x+2)^{1/2} - \log(x-3)$

$\qquad = \log_5 \dfrac{\sqrt{x+2}}{x-3}$

103. $5 \log_6(c+d) - \dfrac{1}{2}\log_6(m-n) = \log_6(c+d)^5 - \log_6(m-n)^{1/2}$

$\qquad = \log_6 \dfrac{(c+d)^5}{\sqrt{m-n}}$

105. $\dfrac{1}{5}(3 \log_2 x - 4 \log_2 y) = \dfrac{1}{5}(\log_2 x^3 - \log_2 y^4)$

$$= \dfrac{1}{5}\left(\log_2 \dfrac{x^3}{y^4}\right)$$

$$= \log_2 \sqrt[5]{\dfrac{x^3}{y^4}}, \quad y > 0$$

107. $\dfrac{1}{5} \log_6(x - 3) - 2 \log_6 x - 3 \log(x + 1) = \log_6(x - 3)^{1/5} - \log_6 x^2 - \log_6(x + 1)^3$

$$= \log_6 \dfrac{\sqrt[5]{x - 3}}{x^2(x + 1)^3}, \quad x > 3$$

109. $\ln 3e^2 = \ln 3 + \ln e^2$
$\quad = \ln 3 + 2 \ln e$
$\quad = \ln 3 + 2$

111. $\log_5 \sqrt{50} = \dfrac{1}{2}[\log_5(5^2 \cdot 2)]$

$$= \dfrac{1}{2}[2 \log_5 5 + \log_5 2]$$

$$= \dfrac{1}{2}[2 + \log_5 2]$$

$$= 1 + \dfrac{1}{2} \log_5 2$$

113. $\log_4 \dfrac{4}{x^2} = \log_4 4 - \log_4 x^2$

$$= 1 - \log_4 x^2$$

$$= 1 - 2 \log_4 x$$

115.

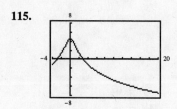

Keystrokes:

y_1 [Y=] [(] 10 [÷] [(] [X,T,θ] [x²] [+] 1 [)] [)] [x²] [ENTER]

y_2 [LN] [y_1] [ENTER]

y^3 2 [(] [LN] 10 [−] [LN] [(] [X,T,θ] [x²] [+] 1 [)] [)] [GRAPH]

Graph y_2 and y_3.

117.

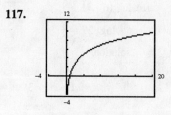

Keystrokes:

y_1 [Y=] [LN] [(] [X,T,θ] [x²] [(] [X,T,θ] [+] 2 [)] [)] [ENTER]

y_2 2 [LN] [X,T,θ] [+] [LN] [(] [X,T,θ] [+] 2 [)] [GRAPH]

119. Choose two values for x and y, such as $x = 3$ and $y = 5$, and show the two expressions are not equal.

$\dfrac{\ln 3}{\ln 5} \neq \ln \dfrac{3}{5} = \ln 3 - \ln 5$ 　　　　　 or 　　　 $\dfrac{\ln e}{\ln e} \neq \ln \dfrac{e}{e}$

$0.6826062 \neq -0.5108256 = -0.5108256$ 　　　　　　　　　 $1 \neq \ln 1$

$1 \neq 0$

121. $B = 10 \log_{10}\left(\dfrac{I}{10^{-16}}\right)$ 　　 or 　　 $B = 10[\log_{10} 10^{-10} + 16]$

$\quad = 10[\log_{10} I - \log_{10} 10^{-16}]$ 　　　　　 $= 10[-10 + 16]$

$\quad = 10[\log_{10} I - (-16)]$ 　　　　　　　 $= 60$ decibels

$\quad = 10[\log_{10} I + 16]$

123. $E = 1.4(\log_{10} C_2 - \log_{10} C_1) = 1.4\left(\log_{10} \dfrac{C_2}{C_1}\right)$ 　　　　 **125.** True, $\ln e^{2-x} = (2 - x) \ln e$

$$= (2 - x)(1) = 2 - x$$

127. True, $\log_8 4 + \log_8 16 = \log_8 4 \cdot 16$

$$= \log_8 64$$
$$= 2$$

129. False, $\log_3(u \cdot v) = \log_3 u + \log_3 v$

131. True, $f(ax) = \log_a ax = \log_a a + \log_a x$

$$= 1 + \log_a x$$
$$= 1 + f(x)$$

133. False; 0 is not in the domain of f.

135. False; $f(x - 3) = \ln(x - 3) \neq \ln x - \ln 3$.

137. False; if $v = u^2$, then $f(v) = \ln u^2 = 2 \ln u = 2f(u)$.

Section 11.5   Solving Exponential and Logarithmic Equations

1. (a) $3^{2(1)-5} \overset{?}{=} 27$

$3^{-3} \neq 27$

not a solution

(b) $3^{2(4)-5} \overset{?}{=} 27$

$3^3 = 27$

solution

3. (a) $e^{-5+\ln 45 +5} \overset{?}{=} 45$

$e^{\ln 45} \overset{?}{=} 45$

$45 = 45$

solution

(b) $e^{-5+e^{45}+5} \overset{?}{=} 45$

$e^{e^{45}} \neq 45$

not a solution

5. (a) $\log_9(6 \cdot 27) \overset{?}{=} \frac{3}{2}$

$\log_9 162 \neq \frac{3}{2}$

not a solution

(b) $\log_9\left(6 \cdot \frac{9}{2}\right) \overset{?}{=} \frac{3}{2}$

$\log_9 27 = \frac{3}{2}$

solution

7. $2^x = 2^5$

so $x = 5$

9. $3^{x+4} = 3^{12}$

so $x + 4 = 12$

$x = 8$

11. $3^{x-1} = 3^7$

so $x - 1 = 7$

$x = 8$

13. $4^{3x} = 16$

$4^{3x} = 4^2$

so $3x = 2$

$x = \frac{2}{3}$

15. $6^{2x-1} = 216$

$6^{2x-1} = 6^3$

so $2x - 1 = 3$

$2x = 4$

$x = 2$

17. $5^x = \frac{1}{125}$

$5^x = 5^{-3}$

so $x = -3$

19. $2^{x+2} = \frac{1}{16}$

$2^{x+2} = 2^{-4}$

so $x + 2 = -4$

$x = -6$

21. $4^{x+3} = 32^x$

$(2^2)^{x+3} = (2^5)^x$

so $2(x + 3) = 5x$

$2x + 6 = 5x$

$6 = 3x$

$2 = x$

23. $\ln 5x = \ln 22$

so $5x = 22$

$x = \frac{22}{5}$

25. $\log_6 3x = \log_6 18$

so $3x = 18$

$x = 6$

27. $\ln(2x - 3) = \ln 15$

so $2x - 3 = 15$

$2x = 18$

$x = 9$

29. $\log_2(x + 3) = \log_2 7$

so $x + 3 = 7$

$x = 4$

31. $\log_5(2x - 3) = \log_5(4x - 5)$

so $2x - 3 = 4x - 5$

$2 = 2x$

$1 = x$

No solution since expressions on either side are undefined for $x = 1$.

33. $\log_3(2 - x) = 2$
$2 - x = 3^2$
$-x = 7$
$x = -7$

35. $\ln e^{2x-1} = (2x - 1)\ln e$
$= (2x - 1)(1)$
$= 2x - 1$

37. $10^{\log_{10} 2x} = 2x, \quad x > 0$

39. $2^x = 45$
$\log_2 2^x = \log_2 45$
$x = \dfrac{\log 45}{\log 2}$
$x \approx 5.49$

41. $3^x = 3.6$
$\log_3 3^x = \log_3 3.6$
$x = \dfrac{\log 3.6}{\log 3} \approx 1.17$

43. $10^{2y} = 52$
$\log 10^{2y} = \log 52$
$2y = \log 52$
$y = \dfrac{\log 52}{2}$
$y \approx 0.86$

45. $7^{3y} = 126$
$\log_7 7^{3y} = \log_7 126$
$3y = \log_7 126$
$y = \dfrac{\log_7 126}{3}$
$y = \dfrac{\log 126}{3 \log 7}$
$y \approx 0.83$

47. $3^{x+4} = 6$
$\log_3 3^{x+4} = \log_3 6$
$x + 4 = \log_3 6$
$x = \dfrac{\log 6}{\log 3} - 4$
$x \approx -2.37$

49. $10^{x+6} = 250$
$\log 10^{x+6} = \log 250$
$x + 6 = \log 250$
$x = \log 250 - 6$
$x \approx -3.60$

51. $3e^x = 42$
$e^x = 14$
$\ln e^x = \ln 14$
$x = \ln 14$
$x \approx 2.64$

53. $\dfrac{1}{4}e^x = 5$
$e^x = 20$
$\ln e^x = \ln 20$
$x = \ln 20$
$x \approx 3.00$

55. $\dfrac{1}{2}e^{3x} = 20$
$e^{3x} = 40$
$\ln e^{3x} = \ln 40$
$3x = \ln 40$
$x = \dfrac{\ln 40}{3} \approx 1.23$

57. $250(1.04)^x = 1000$
$(1.04)^x = 4$
$\log_{1.04} 1.04^x = \log_{1.04} 4$
$x = \log_{1.04} 4$
$x = \dfrac{\log 4}{\log 1.04}$
$x \approx 35.35$

59. $300e^{x/2} = 9000$
$e^{x/2} = 30$
$\ln e^{x/2} = \ln 30$
$\dfrac{x}{2} = \ln 30$
$x = 2\ln 30$
$x \approx 6.80$

61. $1000^{0.12x} = 25{,}000$
$\log_{1000} 1000^{0.12x} = \log_{1000} 25{,}000$
$0.12x = \log_{1000} 25{,}000$
$x = \dfrac{\log_{1000} 25{,}000}{0.12}$
$x = \dfrac{\log 25{,}000}{0.12 \log 1000}$
$x \approx 12.22$

63. $\dfrac{1}{5}4^{x+2} = 300$
$4^{x+2} = 1500$
$\log_4 4^{x+2} = \log_4 1500$
$x + 2 = \dfrac{\log 1500}{\log 4}$
$x = \dfrac{\log 1500}{\log 4} - 2$
$x \approx 3.28$

65. $6 + 2^{x-1} = 1$
$2^{x-1} = -5$
$\log_2 2^{x-1} = \log_2(-5)$
No solution
$\log_2(-5)$ is not possible.

67. $7 + e^{2-x} = 28$
$e^{2-x} = 21$
$\ln e^{2-x} = \ln 21$
$2 - x = \ln 21$
$-x = \ln 21 - 2$
$x = 2 - \ln 21$
$x \approx -1.04$

69. $8 - 12e^{-x} = 7$

$-12e^{-x} = -1$

$e^{-x} = \dfrac{1}{12}$

$\ln e^{-x} = \ln \dfrac{1}{12}$

$-x = \ln \dfrac{1}{12}$

$x = -\ln \dfrac{1}{12} \approx 2.48$

71. $4 + e^{2x} = 10$

$e^{2x} = 6$

$\ln e^{2x} = \ln 6$

$2x = \ln 6$

$x = \dfrac{\ln 6}{2}$

$x \approx 0.90$

73. $32 + e^{7x} = 46$

$e^{7x} = 14$

$\ln e^{7x} = \ln 14$

$7x = \ln 14$

$x = \dfrac{\ln 14}{7}$

$x \approx 0.38$

75. $23 - 5e^{x+1} = 3$

$-5e^{x+1} = -20$

$e^{x+1} = 4$

$\ln e^{x+1} = \ln 4$

$x + 1 = \ln 4$

$x = \ln 4 - 1$

$x \approx 0.39$

77. $4(1 + e^{x/3}) = 84$

$1 + e^{x/3} = 21$

$e^{x/3} = 20$

$\ln e^{x/3} = \ln 20$

$\dfrac{x}{3} = \ln 20$

$x = 3 \ln 20$

$x \approx 8.99$

79. $\dfrac{8000}{(1.03)^t} = 6000$

$\dfrac{8000}{6000} = (1.03)^t$

$\dfrac{4}{3} = (1.03)^t$

$\log_{1.03} \dfrac{4}{3} = \log_{1.03} 1.03^t$

$\log_{1.03} \dfrac{4}{3} = t$

$9.73 \approx t$

81. $\dfrac{300}{2 - e^{-0.15t}} = 200$

$\dfrac{300}{200} = 2 - e^{-0.15t}$

$\dfrac{3}{2} - 2 = -e^{-0.15t}$

$-\dfrac{1}{2} = -e^{-0.15t}$

$\ln\left(\dfrac{1}{2}\right) = \ln e^{-0.15t}$

$\ln\left(\dfrac{1}{2}\right) = -0.15t$

$\dfrac{\ln\left(\frac{1}{2}\right)}{-0.15} = t \approx 4.62$

83. $\log_{10} x = 3$

$10^{\log_{10} x} = 10^3$

$x = 1000.00$

85. $\log_2 x = 4.5$

$2^{\log_2 x} = 2^{4.5}$

$x = 2^{4.5}$

$x = 22.63$

87. $4 \log_3 x = 28$

$\log_3 x = 7$

$3^{\log_3 x} = 3^7$

$x = 3^7$

$x = 2187.00$

89. $16 \ln x = 30$

$\ln x = \dfrac{30}{16}$

$e^{\ln x} = e^{15/8}$

$x = e^{15/8}$

$x \approx 6.52$

91. $\log_{10} 4x = 2$

$10^{\log_{10} 4x} = 10^2$

$4x = 10^2$

$x = \dfrac{10^2}{4}$

$x = \dfrac{100}{4} = 25.00$

93. $\ln 2x = 3$

$\quad e^{\ln 2x} = e^3$

$\quad 2x = e^3$

$\quad x = \dfrac{e^3}{2}$

$\quad x \approx 10.04$

95. $\ln x^2 = 6$

$\quad e^{\ln x^2} = e^6$

$\quad x^2 = e^6$

$\quad x = \pm\sqrt{e^6}$

$\quad x \approx \pm 20.09$

97. $2\log_4(x + 5) = 3$

$\quad \log_4(x + 5) = \dfrac{3}{2}$

$\quad 4^{\log_4(x+5)} = 4^{1.5}$

$\quad x + 5 = 4^{1.5}$

$\quad x = 4^{1.5} - 5$

$\quad x = 3.00$

99. $2\log_8(x + 3) = 3$

$\quad \log_8(x + 3) = \dfrac{3}{2}$

$\quad 8^{\log_8(x+3)} = 8^{3/2}$

$\quad x + 3 = 8^{1.5}$

$\quad x = 8^{1.5} - 3$

$\quad x \approx 19.63$

101. $1 - 2\ln x = -4$

$\quad -2\ln x = -5$

$\quad \ln x = \dfrac{5}{2}$

$\quad e^{\ln x} = e^{2.5}$

$\quad x = e^{2.5}$

$\quad x \approx 12.18$

103. $-1 + 3\log_{10}\dfrac{x}{2} = 8$

$\quad 3\log_{10}\dfrac{x}{2} = 9$

$\quad \log_{10}\dfrac{x}{2} = 3$

$\quad 10^{\log_{10}(x/2)} = 10^3$

$\quad \dfrac{x}{2} = 10^3$

$\quad x = 2(10)^3$

$\quad x = 2000.00$

105. $\log_4 x + \log_4 5 = 2$

$\quad \log_4 x(5) = 2$

$\quad 4^{\log_4 5x} = 4^2$

$\quad 5x = 16$

$\quad x = \dfrac{16}{5}$

$\quad x = 3.20$

107. $\log_6(x + 8) + \log_6 3 = 2$

$\quad \log_6(x + 8)(3) = 2$

$\quad 6^{\log_6 3(x+8)} = 6^2$

$\quad 3x + 24 = 36$

$\quad 3x = 12$

$\quad x = 4.00$

109. $\log_5(x + 3) - \log_5 x = 1$

$\quad \log_5\left(\dfrac{x + 3}{x}\right) = 1$

$\quad 5^{\log_5[(x+3)/x]} = 5^1$

$\quad \dfrac{x + 3}{x} = 5$

$\quad x + 3 = 5x$

$\quad 3 = 4x$

$\quad \dfrac{3}{4} = x$

$\quad 0.75 = x$

111. $\log_{10} x + \log_{10}(x - 3) = 1$

$\quad \log_{10} x(x - 3) = 1$

$\quad 10^{\log_{10} x(x-3)} = 10^1$

$\quad x(x - 3) = 10$

$\quad x^2 - 3x - 10 = 0$

$\quad (x - 5)(x + 2) = 0$

$\quad x = 5, \ x = -2 \text{ (which is extraneous)}$

113. $\log_2(x - 1) + \log_2(x + 3) = 3$

$\quad \log_2(x - 1)(x + 3) = 3$

$\quad x^2 + 2x - 3 = 2^3$

$\quad x^2 + 2x - 11 = 0$

$\quad x = \dfrac{-2 \pm \sqrt{4 - 4(1)(-11)}}{2(1)} = \dfrac{-2 \pm \sqrt{4 + 44}}{2}$

$\qquad = \dfrac{-2 \pm \sqrt{48}}{2}$

$\quad x \approx 2.46 \text{ and } -4.46 \text{ (which is extraneous)}$

115. $\log_4 3x + \log_4(x - 2) = \dfrac{1}{2}$

$$\log_4 3x(x - 2) = \frac{1}{2}$$

$$4^{\log_4 3x(x-2)} = 4^{1/2}$$

$$3x(x - 2) = 2$$

$$3x^2 - 6x = 2$$

$$3x^2 - 6x - 2 = 0$$

$$x = \frac{-(-6) \pm \sqrt{(-6)^2 - 4(3)(-2)}}{2(3)}$$

$$= \frac{6 \pm \sqrt{36 + 24}}{6} = \frac{6 \pm \sqrt{60}}{6}$$

$x \approx 2.29$ and -0.29 (which is extraneous)

117. $\log_2 x + \log_2(x + 2) - \log_2 3 = 4$

$$\log_2 \frac{x(x + 2)}{3} = 4$$

$$2^{\log_2(x^2 + 2x/3)} = 2^4$$

$$\frac{x^2 + 2x}{3} = 16$$

$$x^2 + 2x = 48$$

$$x^2 + 2x - 48 = 0$$

$$(x + 8)(x - 6) = 0$$

$x = -8$ (which is extraneous)

$x = 6.00$

119. *Keystrokes:*

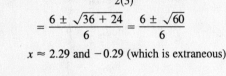

x-intercept

$1.3974 \approx 1.40$

$(1.40, 0)$

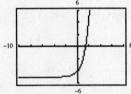

121. *Keystrokes:*

Y= 6 LN (.4 X,T,θ) − 13 GRAPH

x-intercept

$21.822846 \approx 21.82$

$(21.82, 0)$

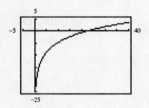

123. *Keystrokes:*

y_1 Y= 2 ENTER

y_2 e^x X,T,θ GRAPH

Point of intersection: $(0.69, 2)$

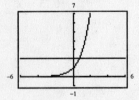

125. *Keystrokes:*

y_1 Y= 3 ENTER

y_2 2 LN (X,T,θ + 3) GRAPH

Point of intersection: $(1.48, 3)$

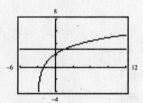

127. *Formula:* $A = Pe^{rt}$

Labels: Principal $= P = \$10,000$

Amount $= A = \$11,972,17$

Time $= t = 2$ years

Annual interest rate $= r$

Equation: $11,972.17 = 10,000e^{r(2)}$

$$\frac{11,972.17}{10,000} = e^{2r}$$

$$1.197217 = e^{2r}$$

$$\ln 1.197217 = \ln(e^{2r})$$

$$0.18 \approx 2r$$

$$0.09 \approx r \approx 9\%$$

129. $5000 = 2500e^{0.09t}$

$$\frac{5000}{2500} = e^{0.09t}$$

$$2 = e^{0.09t}$$

$$\ln 2 = \ln(e^{0.09t})$$

$$\ln 2 = 0.09t$$

$$\frac{\ln 2}{0.09} = t$$

7.70 years $\approx t$

131.
$$B = 10 \log_{10}\left(\frac{I}{10^{-16}}\right)$$
$$75 = 10 \log_{10}\left(\frac{I}{10^{-16}}\right)$$
$$7.5 = \log_{10}\left(\frac{I}{10^{-16}}\right)$$
$$10^{7.5} = 10^{\log_{10}(I/10^{-16})}$$
$$10^{7.5} = \frac{I}{10^{-16}}$$
$$(10^{7.5})(10^{-16}) = I$$
$$10^{-8.5} = I$$
$$3.1623 \times 10^{-9} = I \text{ watts per square centimeter}$$

133.
$$2.5 = 15.7 - 2.48 \ln m$$
$$-13.2 = -2.48 \ln m$$
$$5.322580645 = \ln m$$
$$e^{5.322580645} = m$$
$$205 \approx m$$

135. (a)
$$72 = 80 - \log_{10}(t + 1)^{12}$$
$$-8 = -\log_{10}(t + 1)^{12}$$
$$8 = \log_{10}(t + 1)^{12}$$
$$8 = 12 \log_{10}(t + 1)$$
$$\frac{8}{12} = \log_{10}(t + 1)$$
$$\frac{2}{3} = \log_{10}(t + 1)$$
$$t + 1 = 10^{2/3}$$
$$t = 10^{2/3} - 1$$
$$t \approx 3.64 \text{ months}$$

(b) *Keystrokes:*

y_1 [Y=] 80 [−] 12 [LOG] [(] [X,T,θ] [+] 1 [)] [ENTER]

y_2 72 [GRAPH]

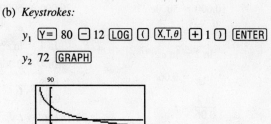

(c) Answers will vary.

137. (a)
$$Kt = \ln \frac{T - S}{T_0 - S}$$
$$K(4) = \ln \frac{32° - 0°}{60° - 0°}$$
$$K(4) = \ln \frac{32°}{60°}$$
$$K = \frac{1}{4} \ln \frac{8}{15}$$
$$K \approx -0.1572$$

(b)
$$t = \frac{1}{K} \ln \frac{T - S}{T_0 - S}$$
$$t = \frac{1}{-0.1572} \ln \frac{32° - (-10°)}{60° - (-10°)}$$
$$t = \frac{1}{-0.1572} \ln \frac{42°}{70°}$$
$$t \approx 3.25 \text{ hours}$$

(c)
$$t = \frac{1}{K} \ln \frac{T - S}{T_0 - S}$$
$$t = \frac{1}{-0.1572} \ln \frac{32° - 0°}{50° - 0°}$$
$$t = \frac{1}{-0.1572} \ln \frac{32°}{50°}$$
$$t \approx 2.84 \text{ hours}$$

139. (c)
$$\text{Formula: } A = P\left(1 + \frac{r}{n}\right)^{nt}$$
$$6200 = 5000\left(1 + \frac{r}{12}\right)^{12(3)}$$
$$1.24 = \left(1 + \frac{r}{12}\right)^{36}$$
$$(1.24)^{1/36} = 1 + \frac{r}{12}$$
$$1.005993204 = 1 + \frac{r}{12}$$
$$0.005993204 = \frac{r}{12}$$
$$0.0719184469 = r$$
$$7.2\% \approx r$$

—CONTINUED—

139. —CONTINUED—

(d) Formula: $A = Pe^{rt}$

$$7500 = 5000e^{0.06t}$$

$$1.5 = e^{0.06t}$$

$$\ln 1.5 = \ln e^{0.06t}$$

$$\ln 1.5 = 0.06t$$

$$\frac{\ln 1.5}{0.06} = t$$

$$6\frac{3}{4} \text{ years} \approx t$$

(e) Formula: $A = P\left(1 + \frac{r}{n}\right)^{nt}$

$$A = 1000\left(1 + \frac{0.08}{4}\right)^{4(1)}$$

$$A = 1000(1.02)^4$$

$$A = \$1082.43$$

Effective yield $= \dfrac{82.43}{1000} = 0.08243 \approx 8.24\%$

(f) Formula: $A = Pe^{rt}$

$$10,000 = 5000e^{0.06t}$$

$$2 = e^{0.06t}$$

$$\ln 2 = \ln e^{0.06t}$$

$$\ln 2 = 0.06t$$

$$\frac{\ln 2}{0.06} = t$$

$$11.6 \text{ years} \approx t$$

$$20,000 = 5000e^{0.06t}$$

$$4 = e^{0.06t}$$

$$\ln 4 = \ln e^{0.06t}$$

$$\ln 4 = 0.06t$$

$$\frac{\ln 4}{0.06} = t$$

$$23.1 \text{ years} \approx t$$

141. $2^{x-1} = 30$ requires logarithms because $2^{x-1} = 32$ can be rewritten as $2^{x-1} = 2^5$ and the exponents set equal.

143. To solve an exponential equation, first isolate the exponential expression, then take the logarithms of both sides of the equation, and solve for the variable.

To solve a logarithmic equation, first isolate the logarithmic expression, then exponentiate both sides of the equation, and solve for the variable.

Section 11.6   Applications

1.
$$A = P\left(1 + \frac{r}{n}\right)^{nt}$$

$$1004.83 = 500\left(1 + \frac{r}{12}\right)^{12(10)}$$

$$2.00966 = \left(1 + \frac{r}{12}\right)^{120}$$

$$(2.00966)^{1/120} = 1 + \frac{r}{12}$$

$$1.0058333 = 1 + \frac{r}{12}$$

$$0.0058333 = \frac{r}{12}$$

$$0.07 \approx r$$

$$7\% \approx r$$

3.
$$A = P\left(1 + \frac{r}{n}\right)^{nt}$$

$$36,581.00 = 1000\left(1 + \frac{r}{365}\right)^{365(40)}$$

$$36.581 = \left(1 + \frac{r}{365}\right)^{14,600}$$

$$(36.581)^{1/14,600} = 1 + \frac{r}{365}$$

$$1.0002466 = 1 + \frac{r}{365}$$

$$0.0002466 = \frac{r}{365}$$

$$0.0899981 = r$$

$$9\% \approx r$$

5.
$$A = Pe^{rt}$$
$$8267.38 = 750e^{r(30)}$$
$$11.023173 = e^{r(30)}$$
$$\ln 11.023173 = \ln e$$
$$\ln 11.023173 = 30r$$
$$\frac{\ln 11.023173}{30} = r$$
$$0.08 \approx r$$
$$8\% \approx r$$

7.
$$A = P\left(1 + \frac{r}{n}\right)^{nt}$$
$$22{,}405.68 = 5000\left(1 + \frac{r}{365}\right)^{365(25)}$$
$$4.481136 = \left(1 + \frac{r}{365}\right)^{9125}$$
$$(4.481136)^{1/9125} = 1 + \frac{r}{365}$$
$$1.000164384 = 1 + \frac{r}{365}$$
$$0.00164384 = \frac{r}{365}$$
$$0.059 \approx r$$
$$6\% \approx r$$

9.
$$A = P\left(1 + \frac{r}{n}\right)^{nt}$$
$$12{,}000 = 6000\left(1 + \frac{0.08}{4}\right)^{4t}$$
$$2 = (1.02)^{4t}$$
$$\log_{1.02} 2 = \log_{1.02} 1.02^{4t}$$
$$\frac{\log 2}{\log 1.02} = 4t$$
$$\frac{\log 2}{\log 1.02} \div 4 = t$$
$$8.75 \text{ years} \approx t$$

11.
$$A = P\left(1 + \frac{r}{n}\right)^{nt}$$
$$4000 = 2000\left(1 + \frac{0.105}{365}\right)^{365t}$$
$$2 = (1.0002877)^{365t}$$
$$\log_{1.0002877} 2 = \log_{1.0002877} 1.0002877^{365t}$$
$$\frac{\log 2}{\log 1.0002877} = 365t$$
$$\frac{\log 2}{\log 1.0002877} \div 365 = t$$
$$6.60 \text{ years} \approx t$$

13.
$$A = Pe^{rt}$$
$$3000 = 1500e^{0.075t}$$
$$2 = e^{0.075t}$$
$$\ln 2 = \ln e^{0.075t}$$
$$\ln 2 = 0.075t$$
$$\frac{\ln 2}{0.075} = t$$
$$9.24 \text{ years} \approx t$$

15.
$$A = P\left(1 + \frac{r}{n}\right)^{nt}$$
$$600 = 300\left(1 + \frac{0.05}{1}\right)^{1(t)}$$
$$2 = 1.05^t$$
$$\log_{1.05} 2 = \log_{1.05} 1.05^t$$
$$\log_{1.05} 2 = t$$
$$14.21 \text{ years} \approx t$$

17.
$$1587.75 = 750\left(1 + \frac{0.075}{n}\right)^{n(10)}$$
$$1587.75 = 750e^{0.075(10)}$$
$$1587.75 = 1587.75$$
Continuous compounding

19.
$$141.48 = 100\left(1 + \frac{0.07}{n}\right)^{n(5)}$$
$$141.48 = 100\left(1 + \frac{0.07}{4}\right)^{4(5)}$$
$$141.48 = 141.48$$
Quarterly compounding

21.
$$A = Pe^{rt}$$
$$A = 1000e^{0.08(1)}$$
$$A = \$1083.29$$
$$\text{Effective yield} = \frac{83.29}{1000}$$
$$= 0.08329 \approx 8.33\%$$

23.
$$A = P\left(1 + \frac{r}{n}\right)^{nt}$$
$$A = 1000\left(1 + \frac{0.07}{12}\right)^{12(1)}$$
$$A = \$1072.29$$
$$\text{Effective yield} = \frac{72.29}{1000}$$
$$= 0.07229 \approx 7.23\%$$

25. $A = P\left(1 + \dfrac{r}{n}\right)^{nt}$

$A = 1000\left(1 + \dfrac{0.06}{4}\right)^{4(1)}$

$A = \$1061.36$

Effective yield $= \dfrac{61.36}{1000}$

$= 0.06136$

$\approx 6.136\%$

27. $A = P\left(1 + \dfrac{r}{n}\right)^{nt}$

$A = 1000\left(1 + \dfrac{0.08}{12}\right)^{12(1)}$

$A = \$1083.00$

Effective yield $= \dfrac{83.00}{1000}$

$= 0.083 = 8.300\%$

29. No. Each time the amount is divided by the principal, the result is always 2.

31. $A = Pe^{rt}$

$10,000 = Pe^{0.09(20)}$

$\dfrac{10,000}{e^{1.8}} = P$

$\$1652.99 \approx P$

33. $A = P\left(1 + \dfrac{r}{n}\right)^{nt}$

$750 = P\left(1 + \dfrac{0.06}{365}\right)^{365(3)}$

$\dfrac{750}{(1.0001644)^{1095}} = P$

$\$626.46 \approx P$

35. $A = P\left(1 + \dfrac{r}{n}\right)^{nt}$

$25,000 = P\left(1 + \dfrac{0.07}{12}\right)^{12(30)}$

$\dfrac{25,000}{(1.005833)^{360}} = P$

$\$3080.15 \approx P$

37. $A = P\left(1 + \dfrac{r}{n}\right)^{nt}$

$1000 = P\left(1 + \dfrac{0.05}{365}\right)^{365(1)}$

$\dfrac{1000}{(1.000136986)^{365}} = P$

$\$951.23 \approx P$

39. $A = \dfrac{P(e^{rt} - 1)}{e^{r/12} - 1}$

$A = \dfrac{30(e^{0.08(10)} - 1)}{e^{0.08/12} - 1}$

$A \approx \$5496.57$

41. $A = \dfrac{P(e^{rt} - 1)}{e^{r/12} - 1}$

$A = \dfrac{50(e^{0.10(40)} - 1)}{e^{0.10/12} - 1}$

$A \approx \$320,250.81$

43. $A = \dfrac{P(e^{rt} - 1)}{e^{r/12} - 1}$

$A = \dfrac{30(e^{0.08(20)} - 1)}{e^{0.08(20)} - 1}$

$A \approx \$17,729.42$

Total interest $= \$17,729.42 - 7200 \approx \$10,529.42$

45. $y = Ce^{kt}$

$3 = Ce^{k(0)}$

$3 = C$

$8 = 3e^{k(2)}$

$\dfrac{8}{3} = e^{2k}$

$\ln\dfrac{8}{3} = \ln e^{2k}$

$\ln\dfrac{8}{3} = 2k$

$\dfrac{\ln\frac{8}{3}}{2} = k \approx 0.4904$

47. $y = Ce^{kt}$

$400 = Ce^{k(0)}$

$400 = C$

$200 = 400e^{k(3)}$

$\dfrac{1}{2} = e^{3k}$

$\ln\dfrac{1}{2} = \ln e^{3k}$

$\ln\dfrac{1}{2} = 3k$

$\dfrac{\ln\frac{1}{2}}{3} = k \approx -0.2310$

49.

$y = Ce^{kt}$ $14.3 = 12.2e^{k(21)}$ $y = 12.2e^{0.0076t}$

$12.2 = Ce^{k(0)}$ $\dfrac{14.3}{12.2} = e^{21k}$ $y = 12.2e^{0.0076(26)}$

$12.2 = C$ $y \approx 14.9$ million

$\ln\dfrac{143}{122} = \ln e^{21k}$

$\ln\dfrac{143}{122} = 21k$

$\dfrac{1}{21}\ln\dfrac{143}{122} = k$

$0.0076 \approx k$

51.

$y = Ce^{kt}$ $23.4 = 14.7e^{k(21)}$ $y = 14.7e^{0.0221t}$

$14.7 = Ce^{k(0)}$ $\dfrac{23.4}{14.7} = e^{21k}$ $y = 14.7e^{0.0221(26)}$

$14.7 = C$ $y \approx 26.1$ million

$\ln\dfrac{234}{147} = \ln e^{21k}$

$\ln\dfrac{234}{147} = 21k$

$\dfrac{1}{21}\ln\dfrac{234}{147} = k$

$0.0221 \approx k$

53.

$y = Ce^{kt}$ $10.6 = 10.5e^{k(21)}$ $y = 10.5e^{0.0005t}$

$10.5 = Ce^{k(0)}$ $\dfrac{10.6}{10.5} = e^{21k}$ $y = 10.5e^{0.0005(26)}$

$10.5 = C$ $y \approx 10.6$ million

$\ln\dfrac{106}{105} = \ln e^{21k}$

$\dfrac{\ln\left(\frac{106}{105}\right)}{21} = k$

$0.0005 \approx k$

55.

$y = Ce^{kt}$ $18.8 = 15.5^{k(21)}$ $y = 15.5e^{0.0092t}$

$15.5 = Ce^{k(0)}$ $\dfrac{18.8}{15.5} = e^{21k}$ $y = 15.5e^{0.0092(26)}$

$15.5 = C$ $y \approx 19.7$ million

$\ln\left(\dfrac{188}{155}\right) = \ln e^{21k}$

$\dfrac{\ln\left(\frac{188}{155}\right)}{21} = k$

$0.0092 \approx k$

57. (a) k is larger in Exercise 51, because the population of Shanghai is increasing faster than the population of Osaka.

(b) k corresponds to r; k gives the annual percentage rate of growth.

59.

$y = Ce^{kt}$ $3 = 6e^{k(1620)}$ $y = 6e^{-0.00043(1000)}$

$6 = Ce^{k(0)}$ $0.5 = e^{1620k}$ $y \approx 3.91$ grams

$6 = C$ $\ln 0.5 = \ln e^{1620k}$

$\dfrac{\ln 0.5}{1620} = k$

$-0.00043 \approx k$

61.

$y = Ce^{kt}$ $4 = Ce^{-0.00012(1000)}$

$0.5C = Ce^{k(5730)}$ $4 = Ce^{-0.12}$

$0.5 = e^{5730k}$ $\dfrac{4}{e^{-0.12}} = C$

$\ln 0.5 = \ln e^{5730k}$ 4.51 grams $\approx C$

$\ln 0.5 = 5730k$

$\dfrac{\ln 0.5}{5730} = k$

$-0.00012 \approx k$

63.

$y = Ce^{kt}$ $2.1 = 4.2e^{k(24,360)}$ $y = 4.2e^{-0.00003(1000)}$

$4.2 = Ce^{k(0)}$ $0.5 = e^{24,360k}$ $y \approx 4.08$ grams

$4.2 = C$ $\ln 0.5 = \ln e^{24,360k}$

$\dfrac{\ln 0.5}{24,360} = k$

$-0.00003 \approx k$

65. $y = Ce^{kt}$

$5 = Ce^{k(0)}$

$5 = C$

$2.5 = 5e^{k(1620)}$

$0.5 = e^{1620k}$

$\ln 0.5 = \ln e^{1620k}$

$\ln 0.5 = 1620k$

$\dfrac{\ln 0.5}{1620} = k$

$-0.00043 \approx k$

$y = 5e^{-0.00043(1000)}$

$y \approx 3.25$ grams

67. $y = Ce^{kt}$

$5 = Ce^{k(0)}$

$5 = C$

$2.5 = 5e^{k(5730)}$

$0.5 = e^{5730k}$

$\ln 0.5 = \ln e^{5730k}$

$\ln 0.5 = 5730k$

$\dfrac{\ln 0.5}{5730} = k$

$-0.00012 \approx k$

$y = 5e^{-0.00012(1000)}$

$y \approx 4.43$ grams

69. $\quad 16{,}500 = 22{,}000e^{k(1)}$

$\dfrac{16{,}500}{22{,}000} = e^k$

$\ln \dfrac{16{,}500}{22{,}000} = \ln e^k$

$\ln \dfrac{16{,}500}{22{,}000} = k$

$-0.2876821 = k$

$y = 22{,}000e^{-0.2876821(3)} \approx \$9281.25$

71. $R = \log_{10} I$

Alaska:

$8.4 = \log_{10} I$

$10^{8.4} = 10^{\log_{10} I}$

$10^{8.4} = I$

San Fernando Valley:

$6.6 = \log_{10} I$

$10^{6.6} = 10^{\log_{10} I}$

$10^{6.6} = I$

Ratio of two intensitiies:

$\dfrac{I \text{ for Alaska}}{I \text{ for San Fernando Valley}} = \dfrac{10^{8.4}}{10^{6.6}}$

$= 10^{8.4-6.6} = 10^{1.8} \approx 63$

The earthquake in Alaska was 63 times as great.

73. $R = \log_{10} I$

Mexico City:

$8.1 = \log_{10} I$

$10^{8.1} = 10^{\log_{10} I}$

$10^{8.1} = I$

Nepal:

$6.5 = \log_{10} I$

$10^{6.5} = 10^{\log_{10} I}$

$10^{6.5} = I$

Ratio of two intensities:

$\dfrac{I \text{ for Mexico City}}{I \text{ for Nepal}} = \dfrac{10^{8.1}}{10^{6.5}}$

$= 10^{8.1-6.5} = 10^{1.6} \approx 40$

$= 10^{1.6} \approx 40$

The earthquake in Mexico City was 40 times as great.

75. $\text{pH} = -\log_{10}[\text{H}^+]$

$\text{pH} = -\log_{10}(9.2 \times 10^{-8}) \approx 7.04$

77. $\text{pH} = -\log_{10}[\text{H}^+]$

fruit:

$2.5 = -\log_{10}[\text{H}^+]$

$-2.5 = \log_{10}[\text{H}^+]$

$10^{-2.5} = 10^{\log_{10}[\text{H}^+]}$

$0.0031623 = \text{H}^+$

tablet:

$9.5 = -\log_{10}[\text{H}^+]$

$-9.5 = \log_{10}[\text{H}^+]$

$10^{-9.5} = 10^{\log_{10}[\text{H}^+]}$

$3.1623 \times 10^{-10} = \text{H}^+$

$\dfrac{\text{H}^+ \text{ of fruit}}{\text{H}^+ \text{ of tablet}} = \dfrac{0.0031623}{3.1623 \times 10^{-10}}$

$= 10{,}000{,}071$

The H^+ of fruit is 10^7 times as great.

79. (a) *Keystrokes:*

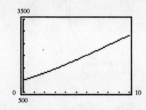

$\boxed{\text{Y=}}$ 5000 $\boxed{\div}$ $\boxed{(}$ $\boxed{1}$ $\boxed{+}$ 4 $\boxed{e^x}$ $\boxed{(}$ $\boxed{(-)}$ $\boxed{\text{X,T,}\theta}$ $\boxed{\div}$ 6 $\boxed{)}$ $\boxed{)}$ $\boxed{\text{GRAPH}}$

(b) $p(0) = \dfrac{5000}{1 + 4e^{-0/6}} = \dfrac{5000}{5} = 1000$

(c) $p(9) = \dfrac{5000}{1 + 4e^{-9/6}} \approx 2642$

(d)
$$2000 = \frac{5000}{1 + 4e^{-t/6}}$$
$$1 + 4e^{-t/6} = 2.5$$
$$4e^{-t/6} = 1.5$$
$$e^{-t/6} = 0.375$$
$$\ln e^{-t/6} = \ln 0.375$$
$$-\frac{t}{6} = \ln 0.375$$
$$t = (\ln 0.375)(-6)$$
$$t \approx 5.88 \text{ years}$$

81. (a)
$$S = 10(1 - e^{kx})$$
$$2.5 = 10(1 - e^{k(5)})$$
$$0.25 = 1 - e^{5k}$$
$$-0.75 = -e^{5k}$$
$$0.75 = e^{5k}$$
$$\ln 0.75 = \ln e^{5k}$$
$$\ln 0.75 = 5k$$
$$\frac{\ln 0.75}{5} = k$$
$$-0.0575 \approx k$$
$$S = 10(1 - e^{-0.0575x})$$

(b) $S = 10(1 - e^{-0.0575(7)})$
$\quad = 10(1 - e^{-0.4025})$
$\quad = 10(0.3313536611)$
$\quad \approx 3.314$

Thus, 3314 units must be sold.

83. If the equation $y = Ce^{kt}$ models exponential decay, $k < 0$ because decay is decreasing so k must be negative.

85. The effective yield of an investment collecting compound interest is the simple interest rate that would yield the same balance at the end of 1 year. To compute the effective yield, divide the interest earned in 1 year by the amount invested.

87. If the reading on the Richter scale is increased by 1, the intensity of the earthquake is increased by a factor of 10.

Review Exercises for Chapter 11

1. (a) $f(-3) = 2^{-3} = \frac{1}{8}$

(b) $f(1) = 2^1 = 2$

(c) $f(2) = 2^2 = 4$

3. (a) $g(-3) = e^{-(-3)/3} = e^1 \approx 2.718$

(b) $g(\pi) = e^{-\pi/3} \approx 0.351$

(c) $g(6) = e^{-6/3} = e^{-2} \approx 0.135$

5. (c) Basic graph

7. (a) Basic graph reflected in the x-axis

9.

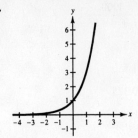

Table of values:

x	-1	0	1
y	$\frac{1}{3}$	1	3

11.

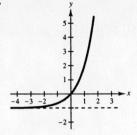

Table of values:

x	-1	0	1
y	$-\frac{2}{3}$	0	2

13.

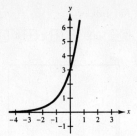

Table of values:

x	-1	0	1
y	1	3	9

15.

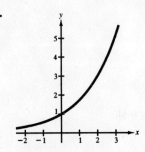

Table of values:

x	0	2	-2
y	1	3	$\frac{1}{3}$

17.

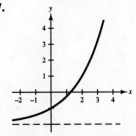

Table of values:

x	-2	0	2
y	$-\frac{7}{3}$	-1	1

19. *Keystrokes:*

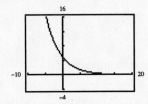

21. *Keystrokes:*

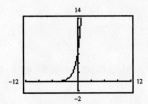

23. (a) $(f \circ g)(x) = x^2 + 2$

so $(f \circ g)(2) = 2^2 + 2 = 6$

(b) $(g \circ f)(x) = (x + 2)^2$

$= x^2 + 4x + 4$

so $(g \circ f)(-1) = (-1)^2 + 4(-1) + 4$

$= 1 - 4 + 4$

$= 1$

25. (a) $(f \circ g)(x) = \sqrt{x^2 - 1 + 1}$

$= \sqrt{x^2}$

$= |x|$

so $(f \circ g)(5) = |5| = 5$

(b) $(g \circ f)(x) = \left(\sqrt{x + 1}\right)^2 - 1$

$= x + 1 - 1$

$= x$

so $(g \circ f)(-1) = -1$

27. (a) $(f \circ g) = \sqrt{2x - 4}$

Domain: $[2, \infty)$

(b) $g \circ f = 2\sqrt{x - 4}$

Domain: $[4, \infty)$

29. No, $f(x)$ does not have an inverse. f is not one-to-one.

31. Yes, $h(x)$ does have an inverse. f is one-to-one.

33.
$$f(x) = 3x + 4$$
$$y = 3x + 4$$
$$x = 3y + 4$$
$$x - 4 = 3y$$
$$\frac{x - 4}{3} = y$$
$$\frac{x - 4}{3} = f^{-1}(x) = \frac{1}{3}(x - 4)$$

35.
$$y = \sqrt{x}$$
$$x = \sqrt{y}$$
$$x^2 = y$$
$$x^2 = f^{-1}(x)$$
$$(x \geq 0)$$

37.
$$f(t) = t^3 + 4$$
$$y = t^3 + 4$$
$$t = y^3 + 4$$
$$t - 4 = y^3$$
$$\sqrt[3]{t - 4} = y$$
$$\sqrt[3]{t - 4} = f^{-1}(t)$$

39. $\log_4 64 = 3$

41. $e^1 = e$

43. $\log_{10} 1000 = 3$ because $10^3 = 1000$.

45. $\log_3 \frac{1}{9} = -2$ because $3^{-2} = \frac{1}{9}$.

47. $\ln e^7 = 7 \ln e = 7$

49. $\ln 1 = 0$

51. (a) $f(1) = \log 31 = 0$

(b) $f(27) = \log_3 27 = 3$

(c) $f(0.5) = \log_3 0.5 = \dfrac{\log 0.5}{\log 3} \approx -0.631$

53. (a) $f(e) = \ln 3 = 1$

(b) $f\left(\dfrac{1}{3}\right) = \ln \dfrac{1}{3} \approx -1.099$

(c) $f(10) = \ln 10 \approx 2.303$

55. (a) $g(-2) = \ln e^{3(-2)} = -6$

(b) $g(0) = \ln e^{3(0)} = 0$

(c) $g(7.5) = \ln e^{3(7.5)} = \ln e^{22.5} = 22.5$

57.

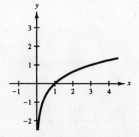

Table of values:

x	1	3
y	0	1

59.

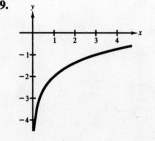

Table of values:

x	1	3
y	-2	-1

61.

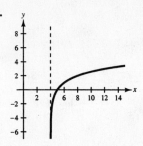

Table of values:

x	5	6
y	0	1

63.

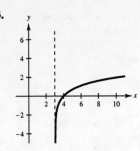

Table of values:

x	4	5
y	0	0.7

65.

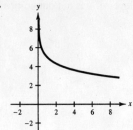

Table of values:

x	1	e
y	5	4

67. $\log_4 9 = \dfrac{\log 9}{\log 4} \approx 1.585$

69. $\log_{12} 200 = \dfrac{\log 200}{\log 12} \approx 2.132$

71. $\log_5 18 = \log_5 3^2 + \log_5 2$

$\qquad = 2\log_5 3 + \log 2$

$\qquad \approx 2(0.6826) + 0.43068$

$\qquad \approx 1.79588$

73. $\log_5 \frac{1}{2} = \log_5 1 - \log_5 2$

$\qquad \approx 0 - (0.43068)$

$\qquad \approx -0.43068$

75. $\log_5 (12)^{2/3} = \frac{2}{3}[2\log_5 2 + \log_5 3]$

$\qquad \approx \frac{2}{3}[2(0.43068) + 0.6826]$

$\qquad \approx 1.02931$

77. $\log_4 6x^4 = \log_4 6 + 4\log_4 x$

79. $\log_5 \sqrt{x+2} = \frac{1}{2}\log_5(x+2)$

81. $\ln\dfrac{x+2}{x-2} = \ln(x+2) - \ln(x-2)$

83. $\ln\left[\sqrt{2x}(x+3)^5\right] = \ln\sqrt{2x} + \ln(x+3)^5$

$\qquad = \ln(2x)^{1/2} + 5\ln(x+3)$

$\qquad = \frac{1}{2}[\ln 2 + \ln x] + 5\ln(x+3)$

85. $-\dfrac{2}{3}\ln 3y = \ln(3y)^{-2/3} = \ln\left(\dfrac{1}{3y}\right)^{2/3}$

87. $\log_8 16x + \log_8 2x^2 = \log_8(16x \cdot 2x^2)$

$\qquad = \log_8(32x^3)$

89. $-2(\ln 2x - \ln 3) = \ln\left(\dfrac{2x}{3}\right)^{-2}$

$\qquad = \ln\left(\dfrac{3}{2x}\right)^2 = \ln\dfrac{9}{4x^2}, \; x > 0$

91. $4[\log_2 k - \log_2(k-t)] = 4\left[\log_2\left(\dfrac{k}{k-t}\right)\right]$

$\qquad = \log_2\left(\dfrac{k}{k-t}\right)^4, \; t < k$

93. $3\ln x + 4\ln y + \ln z = \ln x^3 + \ln y^4 + \ln z$

$\qquad = \ln x^3 y^4 z, \; x > 0, \; y > 0, \; z > 0$

95. False

$\qquad \log_2 4x = \log_2 4 + \log_2 x$

$\qquad\qquad = 2 + \log_2 x$

97. True

$\qquad \log_{10} 10^{2x} = 2x\log_{10} 10 = 2x$

99. True

$\qquad \log_4 \dfrac{16}{x} = \log_4 16 - \log_4 x$

$\qquad\qquad = 2 - \log_4 x$

101. $2^x = 64$

$2^x = 2^6$

$x = 6$

103. $4^{x-3} = \frac{1}{16}$

$4^{x-3} = 4^{-2}$

$x - 3 = -2$

$x = 1$

105. $\log_3 x = 5$

$3^{\log_3 x} = 3^5$

$x = 243$

107. $\log_2 2x = \log_2 100$

$2x = 100$

$x = 50$

109. $\log_3(2x + 1) = 2$

$3^{\log_3(2x+1)} = 3^2$

$2x + 1 = 9$

$2x = 8$

$x = 4$

111. $3^x = 500$

$\log_3 3^x = \log_3 500$

$x = \dfrac{\log 500}{\log 3}$

$x \approx 5.66$

113. $\ln x = 7.25$

$e^{\ln x} = e^{7.25}$

$x = e^{7.25}$

$x \approx 1408.10$

115. $2e^{0.5x} = 45$

$e^{0.5x} = 22.5$

$\ln e^{0.5x} = \ln 22.5$

$0.5x = \ln 22.5$

$x = 2 \ln 22.5$

$x \approx 6.23$

117. $12(1 - 4^x) = 18$

$1 - 4^x = \frac{18}{12}$

$-4^x = \frac{3}{2} - 1$

$-4^x = \frac{1}{2}$

$4^x = -\frac{1}{2}$

No solution; there is no power
that will raise 4 to $-\frac{1}{2}$.

119. $\log_{10} 2x = 1.5$

$2x = 10^{1.5}$

$x = \dfrac{10^{1.5}}{2} \approx 15.81$

121. $\frac{1}{3} \log_2 x + 5 = 7$

$\frac{1}{3} \log_2 x = 2$

$\log_2 x = 6$

$2^{\log_2 x} = 2^6$

$x = 2^6 = 64$

123. $\log_2 x + \log_2 3 = 3$

$\log_2 x(3) = 3$

$2^{\log_2 3x} = 2^3$

$3x = 8$

$x = \dfrac{8}{3} \approx 2.67$

125.

$A = P\left(1 + \dfrac{r}{n}\right)^{nt}$

$410.90 = 250\left(1 + \dfrac{r}{4}\right)^{4(10)}$

$1.6436 = \left(1 + \dfrac{r}{4}\right)^{40}$

$(1.6436)^{1/40} = 1 + \dfrac{r}{4}$

$1.0124997 = 1 + \dfrac{r}{4}$

$0.0124997 = \dfrac{r}{4}$

$0.0499 = r$

$5\% \approx r$

127.

$A = P\left(1 + \dfrac{r}{n}\right)^{nt}$

$15399.30 = 5000\left(1 + \dfrac{r}{365}\right)^{365(15)}$

$3.07986 = \left(1 + \dfrac{r}{365}\right)^{5475}$

$(3.07986)^{1/5475} = 1 + \dfrac{r}{365}$

$1.000205479 = 1 + \dfrac{r}{365}$

$0.000205479 = \dfrac{r}{365}$

$0.074999 = r$

$7.5\% \approx r$

129.
$$A = Pe^{rt}$$
$$24{,}666.97 = 1500e^{r(40)}$$
$$16.44464667 = e^{40r}$$
$$\ln 16.4464667 = \ln e^{40r}$$
$$\ln 16.4464667 = 40r$$
$$\frac{\ln 16.4464667}{40} = r \approx 7\%$$

131.
$$A = P\left(1 + \frac{r}{n}\right)^{nt}$$
$$A = 1000\left(1 + \frac{0.055}{365}\right)^{365(1)}$$
$$A = \$1056.54$$
$$\text{Effective yield} = \frac{56.54}{1000} = 0.0565 \approx 5.65\%$$

133.
$$A = P\left(1 + \frac{r}{n}\right)^{nt}$$
$$A = 1000\left(1 + \frac{0.075}{4}\right)^{4(1)}$$
$$A = \$1077.14$$
$$\text{Effective yield} = \frac{77.14}{1000} - 0.07714 \approx 7.71\%$$

135.
$$A = Pe^{rt}$$
$$A = 1000e^{0.075(1)}$$
$$A = \$1077.88$$
$$\text{Effective yield} = \frac{77.88}{1000} = 0.07788 \approx 7.79\%$$

137.

$y = Ce^{kt}$

$3.5 = Ce^{k(0)}$

$3.5 = C$

$1.75 = 3.5e^{k(1620)}$

$0.5 = e^{1620k}$

$\ln 0.5 = \ln e^{1620k}$

$\ln 0.5 = 1620k$

$\dfrac{\ln 0.5}{1620} = k$

$-0.00043 \approx k$

$y = 3.5e^{-0.00043(1000)}$

$y \approx 2.282 \text{ grams}$

139.

$y = Ce^{kt}$

$0.5C = Ce^{k(5730)}$

$0.5 = e^{k(5730)}$

$\ln 0.5 = \ln e^{5730k}$

$\ln 0.5 = 5730k$

$\dfrac{\ln 0.5}{5730} = k$

$-0.00012 \approx k$

$2.6 = Ce^{-0.00012(1000)}$

$2.6 = Ce^{-0.12}$

$\dfrac{2.6}{e^{-0.12}} = C$

$2.934 \text{ grams} \approx C$

141.

$y = Ce^{kt}$

$5 = Ce^{k(0)}$

$5 = C$

$2.5 = 5e^{k(24{,}360)}$

$0.5 = e^{24{,}360k}$

$\ln 0.5 = \ln e^{24{,}360k}$

$\ln 0.5 = 24{,}360k$

$\dfrac{\ln 0.5}{24{,}360} = k$

$-0.000028 \approx k$

$y = 5e^{-0.000028(1000)}$

$y \approx 4.860 \text{ grams}$

143.
$$30.00 = 24.95(1.05)^t$$

$$\frac{30.00}{24.95} = 1.05^t$$

$$\log_{1.05} \frac{30.00}{24.95} = \log_{1.05} 1.05^t$$

$$\frac{\log \dfrac{30.00}{24.95}}{\log 1.05} = t$$

$$3.8 \text{ years} \approx t$$

145.
$$A = Pe^{rt}$$

$$1500 = 750e^{0.055t}$$

$$2 = e^{0.055t}$$

$$\ln 2 = \ln e^{0.055t}$$

$$\ln 2 = 0.055t$$

$$\frac{\ln 2}{0.055} = t \approx 12.6 \text{ years}$$

147.
$$B = 10 \log_{10}\left(\frac{I}{10^{-16}}\right)$$

$$125 = 10 \log_{10}\left(\frac{I}{10^{-16}}\right)$$

$$12.5 = \log_{10}\left(\frac{I}{10^{-16}}\right)$$

$$10^{12.5} = 10^{\log_{10}(I/10^{-16})}$$

$$10^{12.5} = \frac{I}{10^{-16}}$$

$$10^{12.5}(10^{-16}) = I$$

$$10^{-3.5} = I$$

$$= 3.16 \times 10^{-4} \text{ watts per square centimeter}$$

149. *Keystrokes:*

$\boxed{Y=}$ 600 $\boxed{\div}$ $\boxed{(}$ 1 $\boxed{+}$ 2 $\boxed{e^x}$ $\boxed{(}$ $\boxed{(-)}$ .2 $\boxed{X,T,\theta}$ $\boxed{)}$ $\boxed{)}$ $\boxed{GRAPH}$

The limiting size of the population in this habitat is 600.

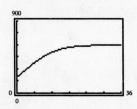

151. (a) *Keystrokes:*

$\boxed{Y=}$ 78.56 $\boxed{-}$ 11.6314 $\boxed{LN}$ $\boxed{X,T,\theta}$ $\boxed{GRAPH}$

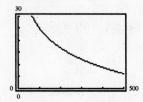

(b) $V = 14.3$ cubic feet per minute per person

Trace to $x = 250$

Chapter Test for Chapter 11

1. (a) $f(-1) = 54\left(\frac{2}{3}\right)^{-1}$
$= 54\left(\frac{3}{2}\right)$
$= 81$

(b) $f(0) = 54\left(\frac{2}{3}\right)^0$
$= 54$

(c) $f\left(\frac{1}{2}\right) = 54\left(\frac{2}{3}\right)^{1/2}$
≈ 44.09

(d) $f(2) = 54\left(\frac{2}{3}\right)^2$
$= 54\left(\frac{4}{7}\right)$
$= 24$

2.

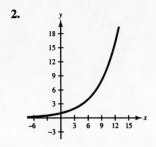

3. (a) $f \circ g = f(g(x)) = f(x^2 + 1) = 3(x^2 + 1) - 4$
$$= 3x^2 + 3 - 4$$
$$= 3x^2 - 1$$

Domain: $(-\infty, \infty)$

(b) $g \circ f = g(f(x)) = g(3x - 4) = (3x - 4)^2 + 1$
$$= 9x^2 - 24x + 16 + 1$$
$$= 9x^2 - 24x + 17$$

Domain: $(-\infty, \infty)$

4. $f(x) = 5x + 6$ $\qquad$ $x = 5y + 6$ $\qquad$ $\dfrac{x-6}{5} = y$ $\qquad$ $f^{-1}(x) = \dfrac{1}{5}(x-6)$

$\qquad$ $y = 5x + 6$ $\qquad$ $x - 6 = 5y$

5. $f(g(x)) = f(-2x + 6)$

$\qquad$ $= -\dfrac{1}{2}(-2x + 6) + 3$

$\qquad$ $= x - 3 + 3$

$\qquad$ $= x$

$\quad g(f(x)) = g\left(-\dfrac{1}{2}x + 3\right)$

$\qquad$ $= -2\left(-\dfrac{1}{2}x + 3\right) + 6$

$\qquad$ $= x - 6 + 6$

$\qquad$ $= x$

6. *f* and *g* are inverse functions.

7. $\log_4 \dfrac{5x^2}{\sqrt{y}} = \log_4 5 + 2\log_4 x - \dfrac{1}{2}\log_4 y$

8. $\ln x - \ln y = \ln \dfrac{x}{y^4},\ y > 0$

9. $\log_5(5^3 \cdot 6) = 3\log_5 5 + \log_5 6$

$\qquad$ $= 3 + \log_5 6$

10. $\log_4 x = 3$

$\quad 4^{\log_4 x} = 4^3$

$\qquad$ $x = 64$

11. $\qquad 10^{3y} = 832$

$\qquad \log 10^{3y} = \log 832$

$\qquad\qquad 3y = \log 832$

$\qquad\qquad y = \dfrac{\log 832}{3}$

$\qquad\qquad y \approx 0.973$

12. $400e^{0.08t} = 1200$

$\qquad e^{0.08t} = 3$

$\quad \ln e^{0.08t} = \ln 3$

$\qquad 0.08t = \ln 3$

$\qquad\quad t = \dfrac{\ln 3}{0.08}$

$\qquad\quad t \approx 13.733$

13. $3\ln(2x - 3) = 10$

$\qquad \ln(2x - 3) = \dfrac{10}{3}$

$\qquad e^{\ln(2x-3)} = e^{10/3}$

$\qquad 2x - 3 = e^{10/3}$

$\qquad\quad x = \dfrac{e^{10/3} + 3}{2}$

$\qquad\quad x \approx 15.516$

14. $8(2 - 3^x) = -56$

$\qquad 2 - 3^x = -7$

$\qquad -3^x = -9$

$\qquad 3^x = 9$

$\qquad 3^x = 3^2$

$\qquad$ so $x = 2$

15. $\log_2 x + \log_2 4 = 5$

$\qquad \log_2 x(4) = 5$

$\qquad 2^{\log_2 4x} = 2^5$

$\qquad 4x = 32$

$\qquad x = 8$

16. $\ln x - \ln 2 = 4$

$\qquad \ln \dfrac{x}{2} = 4$

$\qquad e^{\ln(x/2)} = e^4$

$\qquad \dfrac{x}{2} = e^4$

$\qquad x = 2e^4$

$\qquad x \approx 109.196$

17. $30(e^x + 9) = 300$

$\qquad e^x + 9 = 10$

$\qquad e^x = 1$

$\qquad e^x = e^0$

$\qquad$ so $x = 0$

18. (a) $A = 2000\left(1 + \dfrac{0.07}{4}\right)^{4(20)}$

$\qquad = \$8012.78$

(b) $A = 2000e^{0.07(20)}$

$\qquad = \$8110.40$

19. $\quad 100,000 = P\left(1 + \dfrac{0.09}{4}\right)^{4(25)}$

$\qquad \dfrac{100,000}{(1.0225)^{100}} = P$

$\qquad \$10,806.08 = P$

20. $\quad 1006.88 = 500e^{r(10)}$

$\qquad 2.01376 = e^{10r}$

$\qquad \ln 2.01376 = \ln e^{10r}$

$\qquad \ln 2.01376 = 10r$

$\qquad \dfrac{\ln 2.01376}{10} = r$

$\qquad 0.07 \approx r$

$\qquad 7\% \approx r$

21. $\quad y = Ce^{kt}$

$\qquad 18,000 = Ce^{k(0)}$

$\qquad 18,000 = C$

$14,000 = 18,000e^{k(1)}$

$\dfrac{14,000}{18,000} = e^k$

$\ln \dfrac{14}{18} = \ln e^k$

$\ln \dfrac{14}{18} = k$

$-0.2513144 = k$

$y = 18,000^{-0.2513144(3)}$

$\quad = \$8469.14$

22. $p(0) = \dfrac{2400}{1 + 3e^{-0/4}} = 600$

23. $p(4) = \dfrac{2400}{1 + 3e^{-4/4}} \approx 1141$

24. $\quad 1200 = \dfrac{2400}{1 + 3e^{-t/4}}$

$\qquad 1 + 3e^{-t/4} = \dfrac{2400}{1200}$

$\qquad 3e^{-t/4} = 1$

$\qquad e^{-t/4} = \dfrac{1}{3}$

$\qquad \ln e^{-t/4} = \ln \dfrac{1}{3}$

$\qquad -\dfrac{t}{4} = \ln \dfrac{1}{3}$

$\qquad t = -4 \ln \dfrac{1}{3} \approx 4.4 \text{ years}$

CHAPTER 12
Sequences, Series, and Probability

CHAPTER 12
Sequences, Series, and Probability

Section 12.1   Sequences and Series

Solutions to Odd-Numbered Exercises

1. $a_1 = 2(1) = 2$

$a_2 = 2(2) = 4$

$a_3 = 2(3) = 6$

$a_4 = 2(4) = 8$

$a_5 = 2(5) = 10$

$2, 4, 6, 8, 10, \ldots, 2n, \ldots$

3. $a_1 = (-1)^1 \cdot 2(1) = -2$

$a_2 = (-1)^2 \cdot 2(2) = 4$

$a_3 = (-1)^3 \cdot 2(3) = -6$

$a_4 = (-1)^4 \cdot 2(4) = 8$

$a_5 = (-1)^5 \cdot 2(5) = -10$

$-2, 4, -6, 8, -10, \ldots, (-1)^n 2n, \ldots$

5. $a_1 = \left(\dfrac{1}{2}\right)^1 = \dfrac{1}{2}$

$a_2 = \left(\dfrac{1}{2}\right)^2 = \dfrac{1}{4}$

$a_3 = \left(\dfrac{1}{2}\right)^3 = \dfrac{1}{8}$

$a_4 = \left(\dfrac{1}{2}\right)^4 = \dfrac{1}{16}$

$a_1 = \left(\dfrac{1}{2}\right)^5 = \dfrac{1}{32}$

$\dfrac{1}{2}, \dfrac{1}{4}, \dfrac{1}{8}, \dfrac{1}{16}, \dfrac{1}{32}, \ldots, \left(\dfrac{1}{2}\right)^n, \ldots$

7. $a_1 = \left(-\dfrac{1}{2}\right)^2 = \dfrac{1}{4}$

$a_2 = \left(-\dfrac{1}{2}\right)^3 = -\dfrac{1}{8}$

$a_3 = \left(-\dfrac{1}{2}\right)^4 = \dfrac{1}{16}$

$a_4 = \left(-\dfrac{1}{2}\right)^5 = -\dfrac{1}{32}$

$a_5 = \left(-\dfrac{1}{2}\right)^6 = \dfrac{1}{64}$

$\dfrac{1}{4}, \dfrac{1}{8}, -\dfrac{1}{16}, \dfrac{1}{32}, -\dfrac{1}{64}, \ldots, \left(-\dfrac{1}{2}\right)^{n+1}, \ldots$

9. $a_1 = (-0.2)^{1-1} = (-0.2)^0 = 1$

$a_2 = (-0.2)^{2-1} = (-0.2)^1 = -0.2$

$a_3 = (-0.2)^{3-1} = (-0.2)^2 = 0.04$

$a_4 = (-0.2)^{4-1} = (-0.2)^3 = -0.008$

$a_5 = (-0.2)^{5-1} = (-0.2)^4 = 0.0016$

11. $a_1 = \dfrac{1}{1+1} = \dfrac{1}{2}$

$a_2 = \dfrac{1}{2+1} = \dfrac{1}{3}$

$a_3 = \dfrac{1}{3+1} = \dfrac{1}{4}$

$a_4 = \dfrac{1}{4+1} = \dfrac{1}{5}$

$a_5 = \dfrac{1}{5+1} = \dfrac{1}{6}$

$\dfrac{1}{2}, \dfrac{1}{3}, \dfrac{1}{4}, \dfrac{1}{5}, \dfrac{1}{6}, \ldots, \dfrac{1}{n+1}, \ldots$

13. $a_1 = \dfrac{2(1)}{3(1) + 2} = \dfrac{2}{5}$

$a_2 = \dfrac{2(2)}{3(2) + 2} = \dfrac{4}{8} = \dfrac{1}{2}$

$a_3 = \dfrac{2(3)}{3(3) + 2} = \dfrac{6}{11}$

$a_4 = \dfrac{2(4)}{3(4) + 2} = \dfrac{8}{14} = \dfrac{4}{7}$

$a_5 = \dfrac{2(5)}{3(5) + 2} = \dfrac{10}{17}$

$\dfrac{2}{5}, \dfrac{1}{2}, \dfrac{6}{11}, \dfrac{10}{17}, \cdots, \dfrac{2n}{3n + 2}, \cdots$

15. $a_1 = \dfrac{(-1)^1}{1^2} = -1$

$a_2 = \dfrac{(-1)^2}{2^2} = \dfrac{1}{4}$

$a_3 = \dfrac{(-1)^3}{3^2} = -\dfrac{1}{9}$

$a_4 = \dfrac{(-1)^4}{4^2} = \dfrac{1}{16}$

$a_5 = \dfrac{(-1)^5}{5^2} = -\dfrac{1}{25}$

$-1, \dfrac{1}{4}, -\dfrac{1}{9}, \dfrac{1}{16}, -\dfrac{1}{25}, \cdots, \dfrac{(-1)^n}{n^2}, \cdots$

17. $a_1 = 5 - \dfrac{1}{2^1} = \dfrac{9}{2}$

$a_2 = 5 - \dfrac{1}{2^2} = \dfrac{19}{4}$

$a_3 = 5 - \dfrac{1}{2^3} = \dfrac{39}{8}$

$a_4 = 5 - \dfrac{1}{2^4} = \dfrac{79}{16}$

$a_5 = 5 - \dfrac{1}{2^5} = \dfrac{159}{32}$

$\dfrac{9}{2}, \dfrac{19}{4}, \dfrac{39}{8}, \dfrac{79}{16}, \dfrac{159}{32}, \cdots, 5 - \dfrac{1}{2^n}, \cdots$

19. $a_1 = \dfrac{(1 + 1)!}{1!} = \dfrac{2!}{1!} = \dfrac{2 \cdot 1}{1} = 2$

$a_2 = \dfrac{(2 + 1)!}{2!} = \dfrac{3!}{2!} = \dfrac{3 \cdot 2!}{2!} = 3$

$a_3 = \dfrac{(3 + 1)!}{3!} = \dfrac{4!}{3!} = \dfrac{4 \cdot 3!}{3!} = 4$

$a_4 = \dfrac{(4 + 1)!}{4!} = \dfrac{5!}{4!} = \dfrac{5 \cdot 4!}{4!} = 5$

$a_5 = \dfrac{(5 + 1)!}{5!} = \dfrac{6!}{5!} = \dfrac{6 \cdot 5!}{5!} = 6$

21. $a_1 = \dfrac{2 + (-2)^1}{1!} = 0$

$a_2 = \dfrac{2 + (-2)^2}{2!} = \dfrac{6}{2 \cdot 1} = 3$

$a_3 = \dfrac{2 + (-2)^3}{3!} = \dfrac{-6}{3 \cdot 2 \cdot 1} = -1$

$a_4 = \dfrac{2 + (-2)^4}{4!} = \dfrac{18}{4 \cdot 3 \cdot 2 \cdot 1} = \dfrac{3}{4}$

$a_5 = \dfrac{2 + (-2)^5}{5!} = \dfrac{-30}{5 \cdot 4 \cdot 3 \cdot 2 \cdot 1} = \dfrac{-1}{4}$

23. $a_{15} = (-1)^{15}[5(15) - 3]$

$= -1[72]$

$= -72$

25. $a_8 = \dfrac{8^2 - 2}{(8 - 1)!} = \dfrac{62}{7!} = \dfrac{62}{7 \cdot 6 \cdot 5 \cdot 4 \cdot 3 \cdot 2 \cdot 1} = \dfrac{31}{2520}$

27. $\dfrac{5!}{4!} = \dfrac{5 \cdot 4 \cdot 3 \cdot 2 \cdot 1}{4 \cdot 3 \cdot 2 \cdot 1} = 5$

29. $\dfrac{10!}{12!} = \dfrac{10!}{12 \cdot 11 \cdot 10!} = \dfrac{1}{132}$

31. $\dfrac{25!}{20! \, 5!} = \dfrac{25 \cdot 24 \cdot 23 \cdot 22 \cdot 21 \cdot 20!}{20! \, 5!}$

$= \dfrac{25 \cdot 24 \cdot 23 \cdot 22 \cdot 21}{5 \cdot 4 \cdot 3 \cdot 2 \cdot 1} = 5 \cdot 6 \cdot 23 \cdot 11 \cdot 7$

$= 53130$

33. $\dfrac{n!}{(n+1)!} = \dfrac{n \cdot 1}{(n+1)n \cdot 1} = \dfrac{1}{n+1}$ **35.** $\dfrac{(n+1)!}{(n-1)!} = \dfrac{(n+1)n(n-1)!}{(n-1)!}$
$= (n+1)n$

37. $\dfrac{(2n)!}{(2n-1)!} = \dfrac{(2n)(2n-1)!}{(2n-1)!} = 2n$

39. (c)

41. (b)

43. *Keystrokes* (calculator in sequence and dot mode):

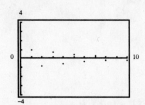

45. *Keystrokes* (calculator in sequence and dot mode):

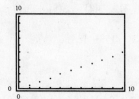

47. *Keystrokes* (calculator in sequence and dot mode):

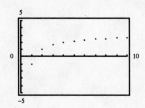

49.

n:	1	2	3	4	5
Terms:	3	6	9	12	15

Apparent pattern: Each term is three times n.

$a_n = 3n$

51.

n:	1	2	3	4	5
Terms:	1	4	7	10	13

Apparent pattern: Each term is three times n minus two.

$a_n = 3n - 2$

53.

n:	1	2	3	4	5
Terms:	0	3	8	15	24

Apparent pattern: Each term is the square of n minus one.

$a_n = n^2 - 1$

55.

n:	1	2	3	4	5
Terms:	2	-4	6	-8	10

Apparent pattern: The terms have alternating signs with those in the even position being negative. Each term is double n.

$a_n = (-1)^{n+1}2n$

57.

n:	1	2	3	4	5
Terms:	$\dfrac{2}{3}$	$\dfrac{3}{4}$	$\dfrac{4}{5}$	$\dfrac{5}{6}$	$\dfrac{6}{7}$

Apparent pattern: The numerator is 1 more than n and the denominator is 2 more than n.

$a_n = \dfrac{n+1}{n+2}$

59.

n:	1	2	3	4
Terms:	$\dfrac{1}{2}$	$-\dfrac{1}{4}$	$\dfrac{1}{8}$	$-\dfrac{1}{16}$

Apparent pattern: The numerator is 1 and each denominator is two to the nth power. The terms have alternating signs with those in the even position being negative.

$a_n = \dfrac{(-1)^{n+1}}{2^n}$

61.

n:	1	2	3	4
Terms:	1	$\dfrac{1}{2}$	$\dfrac{1}{4}$	$\dfrac{1}{8}$

Apparent pattern: The numerator is 1 and the denominator is two to the $n - 1$ power.

$a_n = \dfrac{1}{2^{n-1}}$

63.

n:	1	2	3	4	5
Terms:	$1 + \dfrac{1}{1}$	$1 + \dfrac{1}{2}$	$1 + \dfrac{1}{3}$	$1 + \dfrac{1}{4}$	$1 + \dfrac{1}{5}$

Apparent pattern: The sum of one and $\dfrac{1}{n}$.

$$a_n = 1 + \frac{1}{n}$$

65.

n:	1	2	3	4	5
Terms:	1	$\dfrac{1}{2}$	$\dfrac{1}{6}$	$\dfrac{1}{24}$	$\dfrac{1}{120}$

Apparent pattern: The numerator is one and the denominator is n factorial.

$$a_n = \frac{1}{n!}$$

67. $\displaystyle\sum_{k=1}^{6} 3k = 3(1) + 3(2) + 3(3) + 3(4) + 3(5) + 3(6)$

$\qquad = 3 + 6 + 9 + 12 + 15 + 18$

$\qquad = 63$

69. $\displaystyle\sum_{i=0}^{6} (2i + 5) = [2(0) + 5] + [2(1) + 5] + [2(2) + 5] + [2(3) + 5] + [2(4) + 5] + [2(5) + 5] + [2(6) + 5]$

$\qquad = 5 + 7 + 9 + 11 + 13 + 15 + 17$

$\qquad = 77$

71. $\displaystyle\sum_{j=3}^{7} (6j - 10) = (6 \cdot 3 - 10) + (6 \cdot 4 - 10) + (6 \cdot 5 - 10) + (6 \cdot 6 - 10) + (6 \cdot 7 - 10)$

$\qquad = (18 - 10) + (24 - 10) + (30 - 10) + (36 - 10) + (42 - 10)$

$\qquad = 8 + 14 + 20 + 26 + 32$

$\qquad = 100$

73. $\displaystyle\sum_{j=1}^{5} \frac{(-1)^{j+1}}{j^2} = \frac{(-1)^{1+1}}{1^2} + \frac{(-1)^{2+1}}{2^2} + \frac{(-1)^{3+1}}{3^2} + \frac{(-1)^{4+1}}{4^2} + \frac{(-1)^{5+1}}{5^2}$

$\qquad = 1 - \dfrac{1}{4} + \dfrac{1}{9} - \dfrac{1}{16} + \dfrac{1}{25}$

$\qquad = \dfrac{3600}{3600} - \dfrac{900}{3600} + \dfrac{400}{3600} - \dfrac{225}{3600} + \dfrac{144}{3600}$

$\qquad = \dfrac{3019}{3600}$

75. $\displaystyle\sum_{m=2}^{6} \frac{2m}{2(m-1)} = \frac{2(2)}{2(2-1)} + \frac{2(3)}{2(3-1)} + \frac{2(4)}{2(4-1)} + \frac{2(5)}{2(5-1)} + \frac{2(6)}{2(6-1)}$

$\qquad = \dfrac{4}{2} + \dfrac{6}{4} + \dfrac{8}{6} + \dfrac{10}{8} + \dfrac{12}{10}$

$\qquad = 2 + \dfrac{3}{2} + \dfrac{4}{3} + \dfrac{5}{4} + \dfrac{6}{5}$

$\qquad = \dfrac{437}{60} \approx 7.283$

77. $\displaystyle\sum_{k=1}^{6} (-8) = (-8) + (-8) + (-8) + (-8) + (-8) + (-8) = -48$

79.

$$\sum_{i=1}^{8}\left(\frac{1}{i}-\frac{1}{i+1}\right)=\left[\frac{1}{1}-\frac{1}{1+1}\right]+\left[\frac{1}{2}-\frac{1}{2+1}\right]+\left[\frac{1}{3}-\frac{1}{3+1}\right]+\left[\frac{1}{4}-\frac{1}{4+1}\right]+\left[\frac{1}{5}-\frac{1}{5+1}\right]+\left[\frac{1}{6}-\frac{1}{6+1}\right]+$$

$$\left[\frac{1}{7}-\frac{1}{7+1}\right]+\left[\frac{1}{8}-\frac{1}{8+1}\right]$$

$$=1+\left(-\frac{1}{2}+\frac{1}{2}\right)+\left(-\frac{1}{3}+\frac{1}{3}\right)+\left(-\frac{1}{4}+\frac{1}{4}\right)+\left(-\frac{1}{5}+\frac{1}{5}\right)+\left(-\frac{1}{6}+\frac{1}{6}\right)+\left(-\frac{1}{7}+\frac{1}{7}\right)+\left(-\frac{1}{8}+\frac{1}{8}\right)-\frac{1}{9}$$

$$=1-\frac{1}{9}=\frac{8}{9}$$

81.

$$\sum_{n=0}^{5}\left(-\frac{1}{3}\right)^{n}=\left(-\frac{1}{3}\right)^{0}+\left(-\frac{1}{3}\right)^{1}+\left(-\frac{1}{3}\right)^{2}+\left(-\frac{1}{3}\right)^{3}+\left(-\frac{1}{3}\right)^{4}+\left(-\frac{1}{3}\right)^{5}$$

$$=1+\left(-\frac{1}{3}\right)+\frac{1}{9}+\left(-\frac{1}{27}\right)+\frac{1}{81}+\left(-\frac{1}{243}\right)$$

$$=\frac{243-81+27-9+3-1}{243}$$

$$=\frac{182}{243}$$

83. *Keystrokes*:

$\boxed{\text{LIST}}$ $\boxed{\text{MATH 5}}$ $\boxed{\text{LIST}}$ $\boxed{\text{OPS 5}}$ 3 $\boxed{\text{X,T,}\theta}$ $\boxed{x^2}$ $\boxed{,}$ $\boxed{\text{X,T,}\theta}$ $\boxed{,}$ 1 $\boxed{,}$ 6 $\boxed{,}$ 1 $\boxed{,}$ $\boxed{)}$ $\boxed{\text{ENTER}}$

$$\sum_{n=1}^{6}3n^{2}=273$$

85. *Keystrokes*:

$\boxed{\text{LIST}}$ $\boxed{\text{MATH 5}}$ $\boxed{\text{LIST}}$ $\boxed{\text{OPS 5}}$ $\boxed{\text{X,T,}\theta}$ $\boxed{\text{MATH}}$ $\boxed{\text{PRB 4}}$ $\boxed{-}$ $\boxed{\text{X,T,}\theta}$ $\boxed{,}$ $\boxed{\text{X,T,}\theta}$ $\boxed{,}$ 2 $\boxed{,}$ 6 $\boxed{,}$ 1 $\boxed{)}$ $\boxed{\text{ENTER}}$

$$\sum_{j=2}^{6}(j!-j)=852$$

87. *Keystrokes*:

$\boxed{\text{LIST}}$ $\boxed{\text{MATH 5}}$ $\boxed{\text{LIST}}$ $\boxed{\text{OPS 5}}$ $\boxed{\text{X,T,}\theta}$ 6 $\boxed{\div}$ $\boxed{\text{MATH}}$ $\boxed{\text{PRB 4}}$ $\boxed{,}$ $\boxed{\text{X,T,}\theta}$ $\boxed{,}$ 0 $\boxed{,}$ 4 $\boxed{,}$ 1 $\boxed{)}$ $\boxed{\text{ENTER}}$

$$\sum_{j=0}^{4}\frac{6}{j!}=16.25$$

89. *Keystrokes*:

$\boxed{\text{LIST}}$ $\boxed{\text{MATH 5}}$ $\boxed{\text{LIST}}$ $\boxed{\text{OPS 5}}$ $\boxed{\text{LN}}$ $\boxed{\text{X,T,}\theta}$ $\boxed{,}$ $\boxed{\text{X,T,}\theta}$ $\boxed{,}$ 0 $\boxed{,}$ 6 $\boxed{,}$ 1 $\boxed{)}$ $\boxed{\text{ENTER}}$

$$\sum_{k=1}^{6}\ln k=6.5793$$

91. $\displaystyle\sum_{k=1}^{5}k$

93. $\displaystyle\sum_{k=1}^{5}2k$

95. $\displaystyle\sum_{k=1}^{10}\frac{1}{2k}$

97. $\displaystyle\sum_{k=1}^{20}\frac{1}{k^{2}}$

99. $\displaystyle\sum_{k=0}^{9}\frac{1}{(-3)^{k}}$

101. $\displaystyle\sum_{k=1}^{20}\frac{4}{k+3}$

103. $\displaystyle\sum_{k=1}^{11}\frac{k}{k+1}$

105. $\displaystyle\sum_{k=1}^{20}\frac{2k}{k+3}$

107. $\displaystyle\sum_{k=0}^{6} k!$

109. $\bar{x} = \dfrac{3 + 7 + 2 + 1 + 5}{5}$

$= \dfrac{18}{5} = 3.6$

111. $\bar{x} = \dfrac{0.5 + 0.8 + 1.1 + 0.8 + 0.7 + 0.7 + 1.0}{7}$

$= \dfrac{5.6}{7} = 0.8$

113. (a) $A_1 = 500(1 + 0.07)^1 = \$535.00$

$A_2 = 500(1 + 0.07)^2 = \$572.45$

$A_3 = 500(1 + 0.07)^3 = \$612.52$

$A_4 = 500(1 + 0.07)^4 = \$655.40$

$A_5 = 500(1 + 0.07)^5 = \$701.28$

$A_6 = 500(1 + 0.07)^6 = \$750.37$

$A_7 = 500(1 + 0.07)^7 = \$802.89$

$A_8 = 500(1 + 0.07)^8 = \$859.09$

(b) $A_{40} = 500(1 + 0.07)^{40} = \$7487.23$

(c) *Keystrokes* (calculator in sequence and dot mode):

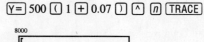

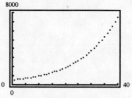

(d) Yes. Investment earning compound interest increases at an increasing rate.

115. $a_5 = \dfrac{180(5 - 2)}{5} = 108°$

$a_6 = \dfrac{180(6 - 2)}{6} = 120°$

$a_5 + 2a_6 = 108° + 240° = 348° < 360°$

117. $d_7 = \dfrac{180(7 - 6)}{7} = 25.7°$

$d_8 = \dfrac{180(8 - 6)}{8} = 45°$

$d_9 = \dfrac{180(9 - 6)}{9} = 60°$

$d_{10} = \dfrac{180(10 - 6)}{10} = 72°$

$d_{11} = \dfrac{180(11 - 6)}{11} = 81.8°$

119. An infinite sequence has an unlimited number of terms such as $a_n = 3n$.

121. The nth term of a sequence is $a_n = (-1)^n n$. When n is odd, the terms are negative.

123. True. $\displaystyle\sum_{k=1}^{4} 3k = 3 \sum_{k=1}^{4} k$.

$\displaystyle\sum_{k=1}^{4} 3k = 30 = 3 \sum_{k=1}^{4} k = 3(10)$

Section 12.2   Arithmetic Sequences

1. $d = 3$

$5 - 2 = 3, \ 8 - 5 = 3, \ 11 - 8 = 3$

3. $d = -6$

$94 - 100 = -6, \ 88 - 94 = -6, \ 82 - 88 = -6$

5. $d = -12$

$-2 - 10 = -12, \; -14 - {}^-2 = -12,$

$-26 - {}^-14 = -12, \; -38 - {}^-26 = -12$

7. $d = \frac{2}{3}$

$\frac{5}{3} - 1 = \frac{2}{3}, \; \frac{7}{3} - \frac{5}{3} = \frac{2}{3}, \; 3 - \frac{7}{3} = \frac{2}{3}$

9. $d = -\frac{5}{4}$

$\frac{9}{4} - \frac{7}{2} = -\frac{5}{4}, \; 1 - \frac{9}{4} = -\frac{5}{4}, \; -\frac{1}{4} - 1 = -\frac{5}{4}$

11. The sequence is arithmetic.

$d = 2$

$4 - 2 = 2, \; 6 - 4 = 2, \; 8 - 6 = 2$

13. arithmetic; $d = -2$

$8 - 10 = -2, \; 6 - 8 = -2, \; 4 - 6 = -2, \; 2 - 4 = -2$

15. The sequence is arithmetic.

$d = -16$

$16 - 32 = -16, \; 0 - 16 = -16, \; -16 - 0 = -16$

17. The sequence is arithmetic.

$d = 0.8$

$4 - 3.2 = 0.8, \; 4.8 - 4 = 0.8, \; 5.6 - 4.8 = 0.8$

19. The sequence is arithmetic.

$d = \frac{3}{2}$

$\frac{7}{2} - 2 = \frac{3}{2}, \; 5 - \frac{7}{2} = \frac{3}{2}, \; \frac{13}{2} - 5 = \frac{3}{2}$

21. The sequence is not arithmetic.

$\frac{2}{3} - \frac{1}{3} = \frac{1}{3}$

$\frac{4}{3} - \frac{2}{3} = \frac{2}{3}$

The difference is NOT the same.

23. The sequence is not arithmetic.

$\sqrt{2} - 1 = \sqrt{2} - 1 \approx .41$

$\sqrt{3} - \sqrt{2} = \sqrt{3} - \sqrt{2} \approx .31$

The difference is NOT the same.

25. The sequence is not arithmetic.

$\ln 8 - \ln 4 = \ln 8 - \ln 4 \approx .69$

$\ln 12 - \ln 8 = \ln 12 - \ln 8 \approx .41$

The difference is NOT the same.

27. $a_1 = 3(1) + 4 = 7$

$a_2 = 3(2) + 4 = 10$

$a_3 = 3(3) + 4 = 13$

$a_4 = 3(4) + 4 = 16$

$a_5 = 3(5) + 4 = 19$

29. $a_1 = -2(1) + 8 = 6$

$a_2 = -2(2) + 8 = 4$

$a_3 = -2(3) + 8 = 2$

$a_4 = -2(4) + 8 = 0$

$a_5 = -2(5) + 8 = -2$

31. $a_1 = \frac{5}{2}(1) - 1 = \frac{3}{2}$

$a_2 = \frac{5}{2}(2) - 1 = 4$

$a_3 = \frac{5}{2}(3) - 1 = \frac{13}{2}$

$a_4 = \frac{5}{2}(4) - 1 = 9$

$a_5 = \frac{5}{2}(5) - 1 = \frac{23}{2}$

33. $a_1 = \frac{3}{5}(1) + 1 = \frac{8}{5}$

$a_2 = \frac{3}{5}(2) + 1 = \frac{11}{5}$

$a_3 = \frac{3}{5}(3) + 1 = \frac{14}{5}$

$a_4 = \frac{3}{5}(4) + 1 = \frac{17}{5}$

$a_5 = \frac{3}{5}(5) + 1 = \frac{20}{5} = 4$

35. $a_1 = -\frac{1}{4}(1 - 1) + 4 = 4$

$a_2 = -\frac{1}{4}(2 - 1) + 4 = \frac{15}{4}$

$a_3 = -\frac{1}{4}(3 - 1) + 4 = \frac{7}{2}$

$a_4 = -\frac{1}{4}(4 - 1) + 4 = \frac{13}{4}$

$a_5 = -\frac{1}{4}(5 - 1) + 4 = 3$

37. $a_n = a_1 + (n - 1)d$

$a_n = 3 + (n - 1)\frac{1}{2}$

$a_n = 3 + \frac{1}{2}n - \frac{1}{2}$

$a_n = \frac{1}{2}n + \frac{5}{2}$

39. $a_n = a_1 + (n - 1)d$

$a_n = 1000 + (n - 1)(-25)$

$a_n = 1000 - 25n + 25$

$a_n = -25n + 1025$

41. $a_n = a_1 + (n - 1)d$

$20 = a_1 + (3 - 1)(-4)$ so $a_n = 28 + (n - 1)(-4)$

$20 = a_1 - 8$ $a_n = 28 - 4n + 4$

$28 = a_1$ $a_n = -4n + 32$

43. $a_n = a_1 + (n - 1)d$

$a_n = 3 + (n - 1)\frac{3}{2}$

$a_n = 3 + \frac{3}{2}n - \frac{3}{2}$

$a_n = \frac{3}{2}n + \frac{3}{2}$

45. $a_n = a_1 + (n - 1)d$

$15 = 5 + (5 - 1)d$ so $a_n = 5 + (n - 1)\frac{5}{2}$

$15 = 5 + 4d$ $a_n = 5 + \frac{5}{2}n - \frac{5}{2}$

$10 = 4d$ $a_n = \frac{5}{2}n + \frac{5}{2}$

$\frac{5}{2} = \frac{10}{4} = d$

47. $a_n = a_1 + (n - 1)d$

$16 = a_1 + (3 - 1)4$

$8 = a_1$

$a_n = 8 + (n - 1)(4)$

$a_n = 4n + 4$

49. $a_n = a_1 + (n - 1)d$

$30 = 50 + (3 - 1)d$

$-20 = 2d$

$-10 = d$

$a_n = 50 + (n - 1)(-10)$

$a_n = -10n + 60$

51. $d = \dfrac{8 - 10}{4} = -\dfrac{1}{2}$

$a_n = a_1 + (n - 1)d$

$10 = a_1 + (2 - 1)\left(-\dfrac{1}{2}\right)$

$10 = a_1 - \dfrac{1}{2}$

$\dfrac{21}{2} = a_1$

$a_n = \dfrac{21}{2} + (n - 1)\left(-\dfrac{1}{2}\right)$

$a_n = \dfrac{21}{2} - \dfrac{1}{2}n + \dfrac{1}{2}$

$a_n = -\dfrac{1}{2}n + 11$

53. $d = \dfrac{0.30 - 0.35}{1} = -0.05$

$a_n = a_1 + (n - 1)d$

$a_n = 0.35 + (n - 1)(-0.05)$

$a_n = 0.35 - 0.05n + 0.05$

$a_n = -0.05n + 0.40$

55. $a_n = a_1 + (n - 1)d$

$a_1 = 25$ and $d = 3$

$a_2 = 25 + (2 - 1)(3) = 28$

$a_2 = 25 + (3 - 1)(3) = 31$

$a_2 = 25 + (4 - 1)(3) = 34$

$a_2 = 25 + (5 - 1)(3) = 37$

57. $a_n = a_1 + (n - 1)d$

$a_1 = 9$

$a_2 = a_{1+1} = a_1 - 3 = 9 - 3 = 6$

$a_3 = a_{2+1} = a_2 - 3 = 6 - 3 = 3$

$a_4 = a_{3+1} = a_3 - 3 = 3 - 3 = 0$

$a_5 = a_{4+1} = a_4 - 3 = 0 - 3 = -3$

59. $a_n = a_1 + (n - 1)d$

$a_1 = -10$

$a_2 = a_{1+1} = a_1 + 6 = -10 + 6 = -4$

$a_3 = a_{2+1} = a_2 + 6 = -4 + 6 = 2$

$a_4 = a_{3+1} = a_3 + 6 = 2 + 6 = 8$

$a_5 = a_{4+1} = a_4 + 6 = 8 + 6 = 14$

61. $a_n = a_1 + (n - 1)d$

$a_1 = 100$ and $d = -20$

$a_2 = 100 + (2 - 1)(-20) = 80$

$a_3 = 100 + (3 - 1)(-20) = 60$

$a_4 = 100 + (4 - 1)(-20) = 40$

$a_5 = 100 + (5 - 1)(-20) = 20$

63. $\displaystyle\sum_{k=1}^{20} k = 20\left(\dfrac{1 + 20}{2}\right)$

$= 210$

65. $\displaystyle\sum_{k=1}^{50} (k + 3) = 50\left(\dfrac{4 + 53}{2}\right)$

$= 1425$

67. $\displaystyle\sum_{k=1}^{10} (5k - 2) = 10\left(\dfrac{3 + 48}{2}\right)$

$= 255$

69. $\displaystyle\sum_{n=1}^{500} \frac{n}{2} = 500\left(\frac{\frac{1}{2} + 250}{2}\right)$

$\qquad = 62,625$

71. $\displaystyle\sum_{n=1}^{30} \left(\frac{1}{3}n - 4\right) = 30\left(\frac{-\frac{11}{3} + 6}{2}\right)$

$\qquad = 35$

73. $\displaystyle\sum_{n=1}^{12} (7n - 2) = 12\left(\frac{5 + 82}{2}\right)$

$\qquad = 522$

75. $\displaystyle\sum_{n=1}^{25} (6n - 4) = 25\left(\frac{2 + 146}{2}\right)$

$\qquad = 1850$

77. $\displaystyle\sum_{n=1}^{8} (225 - 25n) = 8\left(\frac{200 + 25}{2}\right)$

$\qquad = 900$

79. $\displaystyle\sum_{n=1}^{50} (12n - 62) = 50\left(\frac{-50 + 538}{2}\right)$

$\qquad = 12,200$

81. $\displaystyle\sum_{n=1}^{12} (3.5n - 2.5) = 12\left(\frac{1 + 39.5}{2}\right)$

$\qquad = 243$

83. $\displaystyle\sum_{n=1}^{10} (0.4n + 0.1) = 10\left(\frac{0.5 + 4.1}{2}\right)$

$\qquad = 23$

85. (b)   **87.** (e)   **89.** (c)

91. *Keystrokes* (calculator in sequence and dot mode):

Y= (−) 2 n + 21 TRACE

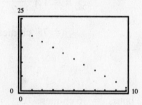

93. *Keystrokes* (calculator in sequence and dot mode):

Y= .6 n + 1.5 TRACE

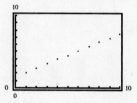

95. *Keystrokes* (calculator in sequence and dot mode):

Y= 2.5 n − 8 TRACE

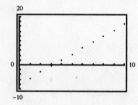

97. *Keystrokes:*

LIST MATH 5 LIST OPS 5 750 − 30 X,T,θ , X,T,θ , 1 , 25 , 1) ENTER

$\displaystyle\sum_{j=1}^{25} (750 - 30j) = 9000$

99. *Keystrokes:*

LIST MATH 5 LIST OPS 5 300 − 8 X,T,θ ÷ 3 , X,T,θ , 1 , 60 , 1) ENTER

$\displaystyle\sum_{i=1}^{60} \left(300 - \frac{8}{3}i\right) = 13,120$

101. *Keystrokes:*

LIST MATH 5 LIST OPS 5 2.15 X,T,θ + 5.4 , X,T,θ , 1 , 50 , 1) ENTER

$\displaystyle\sum_{n=1}^{50} (2.15n + 5.4) = 3011.25$

103. $\displaystyle\sum_{n=1}^{75} = 75\left(\frac{1+75}{2}\right) = 2850$

105. $\displaystyle\sum_{n=1}^{50} 2n = 50\left(\frac{2+100}{2}\right) = 2550$

107. 36,000, 38,000, 40,000, 42,000, 44,000, 46,000

Total salary $= 6\left(\dfrac{36,000+46,000}{2}\right) = \$246,000$

109. Sequence $= 20, 21, 22, \ldots \quad n = 20 \quad d = 1$

$a_n = a_1 + (n-1)d \qquad \displaystyle\sum_{n=1}^{20}(19+n) = 20\left(\frac{20+39}{2}\right)$

$a_n = 20 + (n-1)1 \qquad\qquad\qquad = 590$ seats

$a_n = 19 + n$

$\dfrac{\text{Total cost}}{\text{Total seats}} = \text{Cost per ticket}$

$\dfrac{15,000}{590} = 25.43$

Charge \$25.43 to make at least \$15,000

111. Sequence $= 93, 89, 85, 81, \ldots$

$\displaystyle\sum_{n=1}^{8}(97-4n) = 8\left(\frac{93+65}{2}\right)$

$\qquad\qquad\qquad = 632$ bales

113. Sequence $= 1, 2, 3, 4, \ldots$

$a_n = a_1 + (n-1)d \qquad \displaystyle\sum_{n=1}^{12} n = 12\left(\frac{1+12}{2}\right)$

$a_n = 1 + (n-1)(1) \qquad\qquad = 78$ chimes

$a_n = 1 + n - 1$

$a_n = n$

3 chimes each hour $\times$ 12 hours $= 36$ chimes

Total chimes $= 78 + 36 = 114$ chimes

115. Sequence $= 16, 48, 80, \ldots \quad n = 8 \quad d = 32$

$a_n = a_1 + (n-1)d \qquad a_n = 16 + (n-1)32$

$a_n = 16 + (8-1)32 \qquad\quad = 16 + 32n - 32$

$a_n = 16 + 224 \qquad\quad a_n = 32n - 16$

$a_n = 240$

$\displaystyle\sum_{n=1}^{8}(32n-16) = 8\left(\frac{16+240}{2}\right)$

$\qquad\qquad\qquad = 1024$ feet

117. (a) $1 + 3 = 4$

$\quad 1 + 3 + 5 = 9$

$\quad 1 + 3 + 5 + 7 = 16$

$\quad 1 + 3 + 5 + 7 + 9 = 25$

$\quad 1 + 3 + 5 + 7 + 9 + 11 = 36$

(b) The sums of positive odd integers yield perfect squares.

$\quad 1 + 3 + 5 + 7 + 9 + 11 + 13 = 49$

(c) $\displaystyle\sum_{k=1}^{n}[1+(k-1)2] = n\left(\frac{1+(2n-1)}{2}\right)$

$\qquad\qquad\qquad\qquad = n\left(\frac{2n}{2}\right) = n^2$

119. $a_n = a_1 + (n-1)d$

$d = 15 - 12 = 3$

$12 = a_1 + (2-1)3$

$9 = a_1$

121. A recursion formula gives the relationship between the terms a_{n+1} and a_n.

123. Sequence $= 100, 101, 102, \ldots, 200$

$\displaystyle\sum_{n=100}^{200} n = 101\left(\frac{100+200}{2}\right)$

$\qquad\quad = 15,150$

(*Note:* $a_n = a_1 + (n-1)d$

$\qquad a_n = 100 + (n-1)1$

$\qquad a_n = n + 99$ if n begins at 1.

To start at 100, use n.)

Section 12.3   Geometric Sequences and Series

1. $r = 3$ since

$$\frac{6}{2} = 3, \quad \frac{18}{6} = 3, \quad \frac{54}{18} = 3$$

3. $r = -3$ since

$$\frac{-3}{1} = -3, \quad \frac{9}{-3} = -3, \quad \frac{-27}{9} = -3$$

5. $r = -\frac{1}{2}$ since

$$\frac{-6}{12} = -\frac{1}{2}, \quad \frac{3}{-6} = -\frac{1}{2}, \quad \frac{-\frac{3}{2}}{3} = -\frac{1}{2}$$

7. $r = -\frac{3}{2}$ since

$$\frac{-\frac{3}{2}}{1} = -\frac{3}{2}, \quad \frac{\frac{9}{4}}{-\frac{3}{2}} = -\frac{3}{2}, \quad \frac{-\frac{27}{8}}{\frac{9}{4}} = -\frac{3}{2}$$

9. $r = \pi$ since

$$\frac{\pi}{1} = \pi, \quad \frac{\pi^2}{\pi} = \pi, \quad \frac{\pi^3}{\pi} = \pi$$

11. $r = 1.06$ since

$$\frac{500(1.06)^2}{500(1.06)} = 1.06, \quad \frac{500(1.06)^3}{500(1.06)^2} = 1.06$$

13. The sequence is geometric.

$r = \frac{1}{2}$ since

$\frac{32}{64} = \frac{1}{2}, \quad \frac{16}{32} = \frac{1}{2}, \quad \frac{8}{16} = \frac{1}{2}$

15. The sequence is not geometric, because

$\frac{15}{10} = \frac{3}{2}$ and $\frac{20}{15} = \frac{4}{3}$

17. The sequence is geometric.

$r = 2$ since $\frac{10}{5} = 2, \quad \frac{20}{10} = 2, \quad \frac{40}{20} = 2$

19. The sequence is not geometric, because

$\frac{8}{1} = 8$ and $\frac{27}{8} = \frac{27}{8}$

21. The sequence is geometric.

$r = -\frac{2}{3}$ since

$\frac{-\frac{2}{3}}{1} = -\frac{2}{3}, \quad \frac{\frac{4}{9}}{-\frac{2}{3}} = -\frac{2}{3}, \quad \frac{-\frac{8}{27}}{\frac{4}{9}} = -\frac{2}{3}$

23. The sequence is geometric.

$r = (1 + 0.02)$ since

$\frac{10(1 + 0.02)^2}{10(1 + 0.02)} = (1 + 0.02), \frac{10(1 + 0.02)^3}{10(1 + 0.02)^2} = (1 + 0.02)$

25. $a_n = a_1 r^{n-1}$

$a_n = 4(2)^{n-1}$

$a_1 = 4(2)^{1-1} = 4$

$a_2 = 4(2)^{2-1} = 8$

$a_3 = 4(2)^{3-1} = 16$

$a_4 = 4(2)^{4-1} = 32$

$a_5 = 4(2)^{5-1} = 64$

27. $a_n = a_1 r^{n-1}$

$a_n = 6\left(\frac{1}{3}\right)^{n-1}$

$a_1 = 6\left(\frac{1}{3}\right)^{1-1} = 6$

$a_2 = 6\left(\frac{1}{3}\right)^{2-1} = 2$

$a_3 = 6\left(\frac{1}{3}\right)^{3-1} = \frac{2}{3}$

$a_4 = 6\left(\frac{1}{3}\right)^{4-1} = \frac{2}{9}$

$a_5 = 6\left(\frac{1}{3}\right)^{5-1} = \frac{2}{27}$

29. $a_n = a_1 r^{n-1}$

$a_n = 1\left(-\frac{1}{2}\right)^{n-1}$

$a_1 = 1\left(-\frac{1}{2}\right)^{1-1} = 1$

$a_2 = 1\left(-\frac{1}{2}\right)^{2-1} = -\frac{1}{2}$

$a_3 = 1\left(-\frac{1}{2}\right)^{3-1} = \frac{1}{4}$

$a_4 = 1\left(-\frac{1}{2}\right)^{4-1} = -\frac{1}{8}$

$a_5 = 1\left(-\frac{1}{2}\right)^{5-1} = \frac{1}{16}$

31. $a_n = a_1 r^{n-1}$

$a_n = 4\left(-\dfrac{1}{2}\right)^{n-1}$

$a_1 = 4\left(-\dfrac{1}{2}\right)^{1-1} = 4$

$a_2 = 4\left(-\dfrac{1}{2}\right)^{2-1} = -2$

$a_3 = 4\left(-\dfrac{1}{2}\right)^{3-1} = 1$

$a_4 = 4\left(-\dfrac{1}{2}\right)^{4-1} = -\dfrac{1}{2}$

$a_5 = 4\left(-\dfrac{1}{2}\right)^{5-1} = \dfrac{1}{4}$

33. $a_n = a_1 r^{n-1}$

$a_n = 1000(1.01)^{n-1}$

$a_1 = 1000(1.01)^{1-1} = 1000$

$a_2 = 1000(1.01)^{2-1} = 1010$

$a_3 = 1000(1.01)^{3-1} = 1020.1$

$a_4 = 1000(1.01)^{4-1} = 1030.301$

$a_5 = 1000(1.01)^{5-1} = 1040.604$

35. $a_n = a_1 r^{n-1}$

$a_n = 4000\left(\dfrac{1}{1.01}\right)^{n-1}$

$a_1 = 4000\left(\dfrac{1}{1.01}\right)^{1-1} = 4000\left(\dfrac{1}{1.01}\right)^{0} = 4000(1) = 4000$

$a_2 = 4000\left(\dfrac{1}{1.01}\right)^{2-1} = 4000\left(\dfrac{1}{1.01}\right)^{1} \approx 3960.40$

$a_3 = 4000\left(\dfrac{1}{1.01}\right)^{3-1} = 4000\left(\dfrac{1}{1.01}\right)^{2} \approx 3921.18$

$a_4 = 4000\left(\dfrac{1}{1.01}\right)^{4-1} = 4000\left(\dfrac{1}{1.01}\right)^{3} \approx 3882.36$

$a_5 = 4000\left(\dfrac{1}{1.01}\right)^{5-1} = 4000\left(\dfrac{1}{1.01}\right)^{4} \approx 3843.92$

37. $a_n = a_1 r^{n-1}$

$a_n = 10\left(\dfrac{3}{5}\right)^{n-1}$

$a_1 = 10\left(\dfrac{3}{5}\right)^{1-1} = 10\left(\dfrac{3}{5}\right)^{0} = 10(1) = 10$

$a_2 = 10\left(\dfrac{3}{5}\right)^{2-1} = 10\left(\dfrac{3}{5}\right)^{1} = 6$

$a_3 = 10\left(\dfrac{3}{5}\right)^{3-1} = 10\left(\dfrac{3}{5}\right)^{2} = 10\left(\dfrac{9}{25}\right) = \dfrac{18}{5}$

$a_4 = 10\left(\dfrac{3}{5}\right)^{4-1} = 10\left(\dfrac{3}{5}\right)^{3} = 10\left(\dfrac{27}{125}\right) = \dfrac{54}{25}$

$a_5 = 10\left(\dfrac{3}{5}\right)^{5-1} = 10\left(\dfrac{3}{5}\right)^{4} = 10\left(\dfrac{81}{625}\right) = \dfrac{162}{125}$

39. $a_n = a_1 r^{n-1}$

$a_{10} = 6\left(\dfrac{1}{2}\right)^{10-1} = \dfrac{3}{256}$

41. $a_n = a_1 r^{n-1}$

$a_{10} = 3(\sqrt{2})^{10-1} = 48\sqrt{2}$

43. $a_n = a_1 r^{n-1}$

$a_{12} = 200(1.2)^{12-1}$

$a_{12} \approx 1486.02$

45. $a_n = a_1 r^{n-1}$

$a_{10} \approx 120\left(-\dfrac{1}{3}\right)^{10-1}$

$a_{10} \approx -0.0061$

47. $a_n = a_1 r^{n-1}$

$a_5 = 4\left(\dfrac{3}{4}\right)^{5-1} = \dfrac{81}{64}$

49. $a_n = a_1 r^{n-1}$

$a_6 = 1\left(\pm\dfrac{3}{2}\right)^{6-1} = \pm\dfrac{243}{32}$

51. $r = \dfrac{a_3}{a_2} = \dfrac{16}{12} = \dfrac{4}{3}$

$a_n = a_1 r^{n-1}$

$12 = a_1\left(\dfrac{4}{3}\right)^{2-1}$

$12 = a_1\left(\dfrac{4}{3}\right)$

$9 = a_1$

$a_n = 9\left(\dfrac{4}{3}\right)^{n-1}$

$a_4 = 9\left(\dfrac{4}{3}\right)^{4-1}$

$a_4 = 9\left(\dfrac{4}{3}\right)^{3} = 9\left(\dfrac{64}{27}\right) = \dfrac{64}{3}$

53. $a_n = a_1 r^{n-1}$

$a_n = 2(3)^{n-1}$

55. $a_n = a_1 r^{n-1}$

$a_n = 1(2)^{n-1}$

57. $a_n = a_1 r^{n-1}$

$a_n = 1\left(-\dfrac{1}{5}\right)^{n-1}$

59. $a_n = a_1 r^{n-1}$

$a_n = 4\left(-\dfrac{1}{2}\right)^{n-1}$

61. $a_n = a_1 r^{n-1}$

$a_n = 8\left(\dfrac{1}{4}\right)^{n-1}$

63. $r = \dfrac{a_2}{a_1} = \dfrac{\frac{21}{2}}{14} = \dfrac{3}{4}$

$a_n = ar^{n-1}$

$a_n = 14\left(\dfrac{3}{4}\right)^{n-1}$

65. $4r = -6$

$r = -\dfrac{6}{4} = -\dfrac{3}{2}$

$a_n = a_1 r^{n-1}$

$a_n = 4\left(-\dfrac{3}{2}\right)^{n-1}$

67. (b)

69. (a)

71. $\displaystyle\sum_{i=1}^{10} 2^{i-1} = 1\left(\dfrac{2^{10}-1}{2-1}\right) = \dfrac{1024-1}{1} = 1023$

73. $\displaystyle\sum_{i=1}^{12} 3\left(\dfrac{3}{2}\right)^{i-1} = 3\left(\dfrac{\left(\frac{3}{2}\right)^{12}-1}{\frac{3}{2}-1}\right) = 3\left(\dfrac{128.74634}{0.5}\right) \approx 772.48$

75. $\displaystyle\sum_{i=1}^{15} 3\left(-\dfrac{1}{3}\right)^{i-1} = 3\left(\dfrac{\left(-\frac{1}{3}\right)^{15}-1}{-\frac{1}{3}-1}\right)$

$= 3\left(\dfrac{-1.0000001}{-1.3333333}\right)$

≈ 2.25

77. $\displaystyle\sum_{i=1}^{12} 4(-2)^{i-1} = 4\left(\dfrac{(-2)^{12}-1}{-2-1}\right)$

$= 4\left(\dfrac{4095}{-3}\right)$

$= -5460$

79. $\displaystyle\sum_{i=1}^{8} 6(0.1)^{i-1} = 6\left(\dfrac{(0.1)^8-1}{0.1-1}\right)$

$= 6\left(\dfrac{-.99\overline{9}}{-.9}\right)$

$= 6(1.\overline{1})$

≈ 6.67

81. $\displaystyle\sum_{i=1}^{10} 1(-3)^{i-1} = \left(\dfrac{(-3)^{10}-1}{-3-1}\right)$

$= -14,762$

83. $\displaystyle\sum_{i=1}^{15} 8\left(\dfrac{1}{2}\right)^{i-1} = 8\left(\dfrac{\left(\frac{1}{2}\right)^{15}-1}{\frac{1}{2}-1}\right)$

≈ 16

85. $\displaystyle\sum_{i=1}^{8} 4(3)^{i-1} = 4\left(\dfrac{3^8-1}{3-1}\right)$

$= 4\left(\dfrac{6560}{2}\right)$

$= 13,120$

87. $\displaystyle\sum_{i=1}^{12} 60\left(-\dfrac{1}{4}\right)^{i-1} = 60\left(\dfrac{\left(-\frac{1}{4}\right)^{12}-1}{\left(-\frac{1}{4}\right)-1}\right)$

$\approx 60\left(\dfrac{-1.000}{-1.25}\right)$

≈ 48

89. $\displaystyle\sum_{i=1}^{20} 30(1.06)^{i-1} = 30\left(\dfrac{1.06^{20}-1}{1.06-1}\right)$

≈ 1103.57

91. $\displaystyle\sum_{i=1}^{18} 500(1.04)^{i-1} = 500\left(\dfrac{1.04^{18}-1}{1.04-1}\right)$

$= 500\left(\dfrac{1.025816515}{0.04}\right)$

$\approx 12,822.71$

93. $\displaystyle\sum_{n=0}^{\infty} \left(\dfrac{1}{2}\right)^n = \dfrac{1}{1-\frac{1}{2}} = \dfrac{1}{\frac{1}{2}} = 2$

95. $\displaystyle\sum_{n=0}^{\infty} \left(-\dfrac{1}{2}\right)^n = \dfrac{1}{1-\left(\frac{1}{2}\right)}$

$= \dfrac{1}{\frac{3}{2}}$

$= \dfrac{2}{3}$

97. $\displaystyle\sum_{n=0}^{\infty} 2\left(-\dfrac{2}{3}\right)^n = \dfrac{2}{1-\left(-\frac{2}{3}\right)}$

$= \dfrac{2}{\frac{5}{3}}$

$= \dfrac{6}{5}$

99. $\displaystyle\sum_{n=0}^{\infty} 8\left(\dfrac{3}{4}\right)^n = \dfrac{8}{1-\frac{3}{4}}$

$= \dfrac{8}{\frac{1}{4}}$

$= 32$

101. $a_n = 20(-0.6)^{n-1}$

Keystrokes (calculator in sequence and dot mode):

$\boxed{Y=}$ 20 $\boxed{(}$ $\boxed{(-)}$ 0.6 $\boxed{)}$ $\boxed{\wedge}$ $\boxed{(}$ $\boxed{n}$ $\boxed{-}$ 1 $\boxed{)}$ $\boxed{\text{TRACE}}$

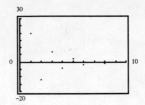

103. $a_n = 15(0.6)^{n-1}$

Keystrokes (calculator in sequence and dot mode):

$\boxed{Y=}$ 15 $\boxed{(}$ 0.6 $\boxed{)}$ $\boxed{\wedge}$ $\boxed{(}$ $\boxed{n}$ $\boxed{-}$ 1 $\boxed{)}$ $\boxed{\text{TRACE}}$

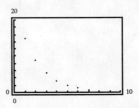

105. $a_0 = 250{,}000$

$a_1 = 250{,}000(0.75)$

$a_2 = 250{,}000(0.75)^2$

$a_3 = 250{,}000(0.75)^3$

$a_4 = 250{,}000(0.75)^4$

(a) $a_n = 250{,}000(0.75)^n$

(b) $a_5 = 250{,}000(0.75)^5 = \$59{,}326.17$

(c) the first year

107. Total salary $= \displaystyle\sum_{n=1}^{40} 30{,}000(1.05)^n$

$= 30{,}000\left(\dfrac{1.05^{40} - 1}{1.05 - 1}\right)$

$= \$3{,}623{,}993.23$

109. $\qquad A = P\left(1 + \dfrac{r}{n}\right)^{nt}$

$a_{120} = 100\left(1 + \dfrac{0.09}{12}\right)^{12(10)} = 100(1.0075)^{120}$

$a_1 = 100(1.0075)^1$

balance $= \left[100(1.0075)\right]\left[\dfrac{1.0075^{120} - 1}{1.0075 - 1}\right] \approx \$19{,}496.56$

111. $\qquad A = P\left(1 + \dfrac{r}{n}\right)^{nt}$

$a_{480} = 30\left(1 + \dfrac{0.08}{12}\right)^{12(40)} = 30\left(\dfrac{151}{150}\right)^{480}$

$a_1 = 30\left(\dfrac{151}{150}\right)^1$

balance $= \left[30\left(\dfrac{151}{150}\right)\right]\left[\dfrac{\left(\frac{151}{150}\right)^{480} - 1}{\left(\frac{151}{150}\right) - 1}\right] \approx \$105{,}428.44$

113. $\qquad A = P\left(1 + \dfrac{r}{n}\right)^{nt}$

$a_{360} = 75\left(1 + \dfrac{0.06}{12}\right)^{12(30)} = 75(1.005)^{360}$

$a_1 = 75(1.005)^1$

balance $= \left[75(1.005)\right]\left[\dfrac{1.005^{360} - 1}{1.005 - 1}\right] \approx \$75{,}715.32$

115. $a_n = 0.01(2)^{n-1}$

(a) Total income $= \displaystyle\sum_{n=1}^{29} 0.01(2)^{n-1}$

$= 0.01\left[\dfrac{2^{29} - 1}{2 - 1}\right]$

$\approx \$5{,}368{,}709.11$

(b) Total income $= \displaystyle\sum_{n=1}^{30} 0.01(2)^{n-1}$

$= 0.01\left[\dfrac{2^{30} - 1}{2 - 1}\right]$

$\approx \$10{,}737{,}418.23$

117. (a) $P = (0.999)^n$

(b) $P = (0.999)^{365}$
$= .694069887$
$\approx 69.4\%$

(c) *Keystrokes* (calculator in sequence and dot mode):

$\boxed{Y=}$ .999 $\boxed{\wedge}$ $\boxed{n}$ $\boxed{\text{TRACE}}$

700 days

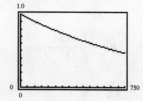

119. $a_1 = 6^2 = 36$

$a_2 = (3\sqrt{2})^2 = 18$

$r = \dfrac{a_2}{a_1} = \dfrac{18}{36} = \dfrac{1}{2}$

$a_n = 36\left(\dfrac{1}{2}\right)^{n-1}$

Total area $= \displaystyle\sum_{n=1}^{6} 36\left(\dfrac{1}{2}\right)^{n-1} = 36\left(\dfrac{\left(\frac{1}{2}\right)^6 - 1}{\frac{1}{2} - 1}\right)$

$= 70.875$ square inches

121. $\displaystyle\sum_{i=1}^{10} 2(100)(0.75)^n = 2(100)(0.75)\left[\dfrac{0.75^{10} - 1}{0.75 - 1}\right]$

$= 150(3.774745941)$

≈ 566.21

Total distance $= 100 + \displaystyle\sum_{i=1}^{10} 2(100)(0.75)^n$

$= 100 + 566.21$

$= 666.21$ feet

123. (a) Sequence $= 2, 4, 8, 16, \ldots$

$a_n = 2^n$

(b) Total ancestors $= \displaystyle\sum_{i=1}^{66} 2^n$

$= 2\left(\dfrac{2^{66} - 1}{2 - 1}\right)$

$= 1.4757 \times 10^{20}$

(c) It is likely that you have had no common ancestors in the last 2000 years.

125. The general formula for the nth term of a geometric sequence is $a_n = a_1 r^{n-1}$.

127. An example of a geometric sequence whose terms alternate in sign is $a_n = \left(-\frac{2}{3}\right)^{n-1}$.

129. An increasing annuity is an investment plan where equal deposits are made in an account at equal time intervals.

Mid-Chapter Quiz for Chapter 12

1. $a_1 = 32\left(\dfrac{1}{4}\right)^{1-1} = 32$

$a_2 = 32\left(\dfrac{1}{4}\right)^{2-1} = 8$

$a_3 = 32\left(\dfrac{1}{4}\right)^{3-1} = 2$

$a_4 = 32\left(\dfrac{1}{4}\right)^{4-1} = \dfrac{1}{2}$

$a_5 = 32\left(\dfrac{1}{4}\right)^{5-1} = \dfrac{1}{8}$

2. $a_1 = \dfrac{(-3)^1 \cdot 1}{1 + 4} = -\dfrac{3}{5}$

$a_2 = \dfrac{(-3)^2 \cdot 2}{2 + 4} = 3$

$a_3 = \dfrac{(-3)^3 \cdot 3}{3 + 4} = -\dfrac{81}{7}$

$a_4 = \dfrac{(-3)^4 \cdot 4}{4 + 4} = \dfrac{81}{2}$

$a_5 = \dfrac{(-3)^5 \cdot 5}{5 + 4} = -135$

3. $\displaystyle\sum_{k=1}^{4} 10k = 4\left(\dfrac{10 + 40}{2}\right) = 100$

4. $\displaystyle\sum_{i=1}^{10} 4 = 10\left(\dfrac{4 + 4}{2}\right) = 40$

5. $\displaystyle\sum_{j=1}^{5}\frac{60}{j+1} = \frac{60}{2} + \frac{60}{3} + \frac{60}{4} + \frac{60}{5} + \frac{60}{6}$

$\qquad\qquad = 30 + 20 + 15 + 12 + 10$

$\qquad\qquad = 87$

6. $\displaystyle\sum_{n=1}^{8} 8\left(-\frac{1}{2}\right) = 8(-4) = -32$

7. $\displaystyle\sum_{k=1}^{20}\frac{2}{3k}$

8. $\displaystyle\sum_{k=1}^{25}\frac{(-1)^{k-1}}{k^3}$

9. $d = \frac{1}{2}$

10. $d = -6$

11. $r = \frac{6}{2} = 3$

12. $r = \frac{1}{2}$

13. $a_n = a_1 + (n-1)d \qquad a_n = 20 + (n-1)(-3)$

$\qquad 11 = 20 + (4-1)d \qquad a_n = 20 - 3n + 3$

$\qquad -9 = 3d \qquad\qquad\qquad a_n = -3n + 23$

$\qquad -3 = d$

14. $a_n = a_1 r^{n-1}$

$\qquad a_n = 32\left(-\frac{1}{4}\right)^{n-1}$

15. $\displaystyle\sum_{n=1}^{50}(3n+5) = 50\left(\frac{8+155}{2}\right)$

$\qquad\qquad\qquad = 4075$

16. $\displaystyle\sum_{n=1}^{300}\frac{n}{5} = 300\left(\frac{\frac{1}{5}+60}{2}\right)$

$\qquad\qquad\quad = 9030$

17. $\displaystyle\sum_{i=1}^{8} 9\left(\frac{2}{3}\right)^{i-1} = 9\left(\frac{\left(\frac{2}{3}\right)^8 - 1}{\frac{2}{3} - 1}\right)$

$\qquad\qquad\qquad = 9\left(\frac{\frac{256}{6561} - 1}{-\frac{1}{3}}\right)$

$\qquad\qquad\qquad = 9\left(\frac{-.96098}{-.33\overline{3}}\right)$

$\qquad\qquad\qquad \approx 25.947$

18. $\displaystyle\sum_{j=1}^{20} 500(1.06)^{j-1} = 500\left(\frac{1.06^{20}-1}{1.06-1}\right)$

$\qquad\qquad\qquad\qquad = 500\left(\frac{2.2071}{.06}\right)$

$\qquad\qquad\qquad\qquad \approx 18,392.796$

19. $\displaystyle\sum_{i=0}^{\infty} 3\left(\frac{2}{3}\right)^i = 3\left(\frac{1}{1-\frac{2}{3}}\right)$

$\qquad\qquad\qquad = 3(3) = 9$

20. $\displaystyle\sum_{i=0}^{\infty}\frac{4}{5}\left(\frac{1}{4}\right)^i$

$\qquad = \frac{4}{5}\left(\frac{1}{1-\frac{1}{4}}\right)$

$\qquad = \frac{4}{5}\left(\frac{4}{3}\right) = \frac{16}{15}$

21. Geometric sequence with $a_1 = 625$ and $r = -.4$.

$\qquad a_n = a_1 r^{n-1}$

$\qquad a_n = 625(-.4)^{n-1}$

$\qquad a_{12} = 625(-.4)^{12-1}$

$\qquad\quad \approx -0.026$

22. $a_n = 10\left(\frac{1}{2}\right)^{n-1} \Rightarrow$ upper graph

$\qquad b_n = 10\left(-\frac{1}{2}\right)^{n-1} \Rightarrow$ lower graph

23. Sequence $= 25.75,\ 23.5,\ 21.25,\ 19, \ldots$

arithmetic with $a_1 = 25.75,\ d = -2.25$

$\qquad a_n = 25.75 + (n-1)(-2.25)$

$\qquad a_n = 25.75 + (10-1)(-2.25)$

$\qquad a_{10} = 5.5°$

24. $b_n = \ln a_n$ is arithmetic.

Section 12.4   The Binomial Theorem

1. $_6C_4 = {}_6C_2 = \dfrac{6 \cdot 5}{2 \cdot 1} = 15$ **3.** $_{10}C_5 = \dfrac{10 \cdot 9 \cdot 8 \cdot 7 \cdot 6}{5 \cdot 4 \cdot 3 \cdot 2 \cdot 1} = 252$ **5.** $_{20}C_{20} = 1$

7. $_{18}C_{18} = 1$ **9.** $_{50}C_{48} = {}_{50}C_2 = \dfrac{50 \cdot 49}{2 \cdot 1} = 1225$ **11.** $_{25}C_4 = \dfrac{25 \cdot 24 \cdot 23 \cdot 22}{4 \cdot 3 \cdot 2 \cdot 1} = 12{,}650$

13. *Keystrokes:*

30 [MATH] [PRB 3] 6 [ENTER] $_{30}C_6 = 593{,}775$

15. *Keystrokes:*

12 [MATH] [PRB 3] 7 [ENTER] $_{12}C_7 = 792$

17. *Keystrokes:*

52 [MATH] [PRB 3] 5 [ENTER] $_{52}C_5 = 2{,}598{,}960$

19. *Keystrokes:*

200 [MATH] [PRB 3] 195 [ENTER] $_{200}C_{195} = 2{,}535{,}650{,}040$

21. *Keystrokes:*

25 [MATH] [PRB 3] 12 [ENTER] $_{25}C_{12} = 5{,}200{,}300$

23. $_6C_2 = 15$

Row 6: 1 6 15 20 15 6 1

entry 2

25. $_7C_3 = 35$

Row 7: 1 7 21 35 35 21 7 1

entry 3

27. $_8C_4 = 70$

Row 8: 1 8 28 56 70 56 28 8 1

entry 4

29. $(a + 2)^3 = (1)a^3 + (3)a^2(2) + (3)a(2^2) + 1(2^3)$

$\qquad = a^3 + 6a^2 + 12a + 8$

31. $(x + y)^8 = 1x^8 + 8x^7y + 28x^6y^2 + 56x^5y^3 + 70x^4y^4 + 56x^3y^5 + 28x^2y^6 + 8xy^7 + 1y^8$

33. $(2x - 1)^5 = 1(2x)^5 + 5(2x)^4(-1) + 10(2x)^3(-1)^2 + 10(2x)^2(-1)^3 + 5(2x)(-1)^4 + (-1)^5$

$\qquad = 32x^5 - 80x^4 + 80x^3 - 40x^2 + 10x - 1$

35. $(2y + z)^6 = (1)(2y)^6 + 6(2y)^5z + 15(2y)^4z^2 + 20(2y)^3z^3 + 15(2y)^2z^4 + 6(2y)z^5 + 1z^6$

$\qquad = 64y^6 + 192y^5z + 240y^4z^2 + 160y^3z^3 + 60y^2z^4 + 12yz^5 + z^6$

37. $(x^2 + 2)^4 = 1(x^2)^4 + 4(x^2)^3(2) + 6(x^2)^2(2)^2 + 4(x^2)(2)^3 + 1(2)^4$

$\qquad = x^8 + 8x^6 + 24x^4 + 32x^2 + 16$

39. $(x + 3)^6 = 1x^6 + 6x^5(3) + 15x^4(3)^2 + 20x^3(3)^3 + 15x^2(3)^4 + 6x(3)^5 + 1(3)^6$

$\qquad = x^6 + 18x^5 + 135x^4 + 540x^3 + 1215x^2 + 1458x + 729$

41. $(x - 4)^6 = (1)x^6 - (6)x^5(4) + (15)x^4(4^2) - (20)x^3(4^3) + (15)x^2(4^4) - (6)x(4^5) + (1)4^6$

$\qquad = x^6 - 24x^5 + 240x^4 - 1280x^3 + 3840x^2 - 6144x + 4096$

43. $(x + y)^4 = 1x^4 + 4x^3y + 6x^2y^2 + 4xy^3 + 1y^4$

45. $(u - 2v)^3 = 1u^3 - 3u^2(2v) + 3u(2v)^2 - 1(2v)^3$

$\qquad = u^3 - 6u^2v + 12uv - 8v^3$

47. $(3a + 2b)^4 = 1(3a)^4 + 4(3a)^3(2b) + 6(3a)^2(2b)^2 + 4(3a)(2b)^3 + 1(2b)^4$

$= 81a^4 + 216a^3b + 216a^2b^2 + 96ab^3 + 16b^4$

49. $(2x^2 - y)^5 = 1(2x^2)^5 + 5(2x^2)^4(-y) + 10(2x^2)^3(-y)^2 + 10(2x^2)^2(-y)^3 + 5(2x^2)(-y)^4 + 1(-y)^5$

$= 32x^{10} - 80x^8y + 80x^6y^2 - 40x^4y^3 + 10x^2y^4 - y^5$

51. $_nC_r x^{n-r}y^r$

$n = 10, \quad n - r = 7, \quad r = 3, \quad x = x, \quad y = 1$

$_{10}C_3 x^7 1^3$

$_{10}C_3 = \dfrac{10 \cdot 9 \cdot 8}{3 \cdot 2 \cdot 1} = 120$

53. $_nC_r x^{n-r}y^r$

$n = 15, \quad n - r = 4, \quad r = 11, \quad x = x, \quad y = (-y)$

$_{15}C_{11} x^4(-y)^{11} = -_{15}C_{11} x^4 y^{11}$

$-_{15}C_{11} = -_{15}C_4 = -\dfrac{15 \cdot 14 \cdot 13 \cdot 12}{4 \cdot 3 \cdot 2 \cdot 1} = -1365$

55. $_nC_r x^{n-r}y^r$

$n = 12, \quad n - r = 3, \quad r = 9, \quad x = 2x, \quad y = y$

$_{12}C_9 (2x)^3 y^9$

$_{12}C_9 = _{12}C_3 = \dfrac{12 \cdot 11 \cdot 10}{3 \cdot 2 \cdot 1} = 220$

$(2)^3 {}_{12}C_9 = 8(220) = 1760$

57. $_nC_r x^{n-r}y^r$

$n = 4, \quad n - r = 2, \quad r = 2, \quad x = x^2, \quad y = (-3)$

$_4C_2 (x^2)^2(-3)^2$

$_4C_2 = \dfrac{4 \cdot 3}{(2 \cdot 1)} = 6$

$(-3)^2 {}_4C_2 = 9(6) = 54$

59. $_nC_r x^{n-r}y^r$

$n = 8, \quad n - r = 4, \quad r = 4, \quad x = \sqrt{x}, \quad y = 1$

$_8C_4 (\sqrt{x})^4(1)$

$_8C_4 = \dfrac{8 \cdot 7 \cdot 6 \cdot 5}{4 \cdot 3 \cdot 2 \cdot 1} = 70$

61. $(1.02)^8 = (1 + 0.02)^8$

$= (1)^8 + 8(1)^7(0.02) + 28(1)^6(0.02)^2 + 56(1)^5(0.02)^3 + \cdots$

$\approx 1 + 0.16 + 0.0112 + 0.000448$

≈ 1.172

63. $(2.99)^{12} = (3 - 0.01)^{12}$

$= 1(3)^{12} - 12(3)^{11}(0.01) + 66(3)^{10}(0.01)^2 - 220(3)^9(0.01)^3 + 495(3)^8(0.01)^4 - 792(3)^7(0.01)^5 + \cdots$

$\approx 531{,}441 - 21{,}257.64 + 389.7234 - 4.33026 + 0.03247695 - 0.0001732104$

$\approx 510{,}568.785$

65. $\left(\frac{1}{2} + \frac{1}{2}\right)^5 = 1\left(\frac{1}{2}\right)^5 + 5\left(\frac{1}{2}\right)^4\left(\frac{1}{2}\right) + 10\left(\frac{1}{2}\right)^3\left(\frac{1}{2}\right)^2 + 10\left(\frac{1}{2}\right)^2\left(\frac{1}{2}\right)^3 + 5\left(\frac{1}{2}\right)\left(\frac{1}{2}\right)^4 + 1\left(\frac{1}{2}\right)^5$

$= \frac{1}{32} + \frac{5}{32} + \frac{10}{32} + \frac{10}{32} + \frac{5}{32} + \frac{1}{32}$

67. $\left(\frac{1}{4} + \frac{3}{4}\right)^4 = 1\left(\frac{1}{4}\right)^4 + 4\left(\frac{1}{4}\right)^3\left(\frac{3}{4}\right) + 6\left(\frac{1}{4}\right)^2\left(\frac{3}{4}\right)^2 + 4\left(\frac{1}{4}\right)\left(\frac{3}{4}\right)^3 + 1\left(\frac{3}{4}\right)^4$

$= \frac{1}{256} + \frac{12}{256} + \frac{54}{256} + \frac{108}{256} + \frac{81}{256}$

69. The difference between consecutive entries increases by 1.
2, 3, 4, 5

71. There are $n + 1$ terms in the expansion of $(x + y)^n$.

73. The signs in the expansion of $(x + y)^n$ are all positive.
The signs in the expansion of $(x - y)^n$ alternate.

75. $_nC_r = {}_nC_{n-r}$

Section 12.5   Counting Principles

1. $\{0, 2, 4, 6, 8\}$ 5 ways

3.

First number	Second number
1	9
2	8
3	7
4	6
5	5
6	4
7	3
8	2
9	1

9 ways

5.

First number	Second number
1	9
2	8
3	7
4	6
6	4
7	3
8	2
9	1

8 ways

7. $\{1, 3, 5, 7, 9, 11, 13, 15, 17, 19\}$ 10 ways

9. $\{2, 3, 5, 7, 11, 13, 17, 19\}$ 8 ways

11. $\{3, 6, 9, 12, 15, 18\}$ 6 ways

13.

First number	Second number
1	7
2	6
3	5
4	4
5	3
6	2
7	1

7 ways

15.

First number	Second number
1	7
2	6
3	5
5	3
6	2
7	1

6 ways

17. $3 \cdot 2 = 6$ ways

19. label = letter   number
26   ·   10   = 260 labels

21. plate = digit   digit   digit   digit   letter   letter
10   · 10   · 10   · 10   · 26   · 26   = 6,760,000 plates

23. (a) $9 \cdot 10 \cdot 10 = 900$ numbers   (b) $10 \cdot 9 \cdot 8 = 720$ numbers   (c) $4 \cdot 10 \cdot 10 = 400$ numbers

25. $3 \cdot 3 \cdot 2 \cdot 1 = 18$ ways   **27.** $3 \cdot 2 \cdot 1 \cdot 5 \cdot 4 \cdot 3 \cdot 2 \cdot 1 = 720$ ways

29. A, B, C, D; A, B, D, C; A, C, B, D; A, C, D, B; A, D, B, C; A, D, C, B;

B, A, C, D; B, A, D, C; B, C, A, D; B, C, D, A; B, D, A, C; B, D, C, A;

C, A, B, D; C, A, D, B; C, B, A, D; C, B, D, A; C, D, A, B; C, D, B, A;

D, A, B, C; D, A, C, B; D, B, A, C; D, B, C, A; D, C, A, B; D, C, B, A

31. | | |
|---|---|
| AB | BA |
| AC | CA |
| AD | DA |
| BC | CB |
| BD | DB |
| CD | DC |

33. $6! = 6 \cdot 5 \cdot 4 \cdot 3 \cdot 2 \cdot 1 = 720$ ways

35. $40 \cdot 40 \cdot 40 = 64{,}000$ ways   **37.** $8! = 40{,}320$ ways   **39.** $_{10}P_4 = 10 \cdot 9 \cdot 8 \cdot 7 = 5040$

41. $_6C_2 = \dfrac{6!}{4!\,2!} = \dfrac{6 \cdot 5}{2 \cdot 1} = 15$ subsets

{A, B}, {A, C}, {A, D}, {A, E}, {A, F}, {B, C},
{B, D}, {B, E}, {B, F}, {C, D}, {C, E}, {C, F},
{D, E}, {D, F}, {E, F}

43. $_{20}C_3 = \dfrac{20!}{17!\,3!} = \dfrac{20 \cdot 19 \cdot 18}{3 \cdot 2 \cdot 1} = 1140$ ways

45. $_9C_4 = \dfrac{9!}{5!\,4!} = \dfrac{9 \cdot 8 \cdot 7 \cdot 6}{4 \cdot 3 \cdot 2 \cdot 1}$
$= 126$ ways

47. $_{12}C_9 = \dfrac{12!}{3!\,9!} = \dfrac{12 \cdot 11 \cdot 10}{3 \cdot 2 \cdot 1}$
$= 220$ ways

49. $_{15}C_5 = \dfrac{15!}{5!\,10!}$
$= 3003$ ways

51. (a) $_6C_4 = \dfrac{6!}{2!\,4!} = \dfrac{6 \cdot 5}{2 \cdot 1}$
$= 15$ ways

(b) $_4C_2 = \dfrac{4!}{2!\,2!} = \dfrac{4 \cdot 3}{2 \cdot 1} = 6$

$_2C_2 = \dfrac{2!}{0!\,2!} = 1$

$_4C_2 \cdot {_2C_2} = 6 \cdot 1$
$= 6$ ways

53. (a) $_8C_4 = \dfrac{8!}{4!\,4!} = \dfrac{8 \cdot 7 \cdot 6 \cdot 5}{4 \cdot 3 \cdot 2 \cdot 1} = 70$

(b) $_2C_1 \cdot {_2C_1} \cdot {_2C_1} \cdot {_2C_1} = 2 \cdot 2 \cdot 2 \cdot 2 = 16$

55. $_7C_2 = \dfrac{7!}{5!\,2!} = \dfrac{7 \cdot 6}{2 \cdot 1} = 21$

57. Diagonals of Hexagon $= {_6C_4} - {_6C_1} = 9$

59. Diagonals of Decagon $= {_{10}C_8} - {_{10}C_1} = 35$

61. The Fundamental Counting Principle: Let E_1 and E_2 be two events that can occur in m_1 ways and m_2 ways, respectively. The number of ways the two events can occur is $m_1 \cdot m_2$.

63. Permutation: The ordering of five students for a picture.

Combination: The selection of three students from a group of five students for a class project.

Section 12.6   Probability

1. {a, b, c, d, e, f, g, h, i, j, k, l, m, n, o, p, q, r, s, t, u, v, w, x, y, z}

number of outcomes = 26

3. {AB, AC, AD, AE, BC, BD, BE, CD, CE, DE}

number of outcomes = 10

5. {ABC, ACB, BAC, BCA, CAB, CBA}

7. {WWW, WWL, WLW, WLL, LWW, LWL, LLW, LLL}

9. $1 - 0.35 = 0.65$

11. $P(E) = 1 - p = 1 - 0.82 = 0.18$

13. $P(E) = \dfrac{n(E)}{n(S)} = \dfrac{3}{8}$

15. $P(E) = \dfrac{n(E)}{n(S)} = \dfrac{7}{8}$

17. $P(E) = \dfrac{n(E)}{n(S)} = \dfrac{26}{52} = \dfrac{1}{2}$

19. $P(E) = \dfrac{n(E)}{n(S)} = \dfrac{12}{52} = \dfrac{3}{13}$

21. $P(E) = \dfrac{n(E)}{n(S)} = \dfrac{1}{6}$

23. $P(E) = \dfrac{n(E)}{n(S)} = \dfrac{5}{6}$

25. $P(E) = 1 - \dfrac{n(F)}{n(S)} = 1 - \dfrac{1}{10} = \dfrac{9}{10}$

(*F* is event that person does have type *B*.)

27. $P(E) = \dfrac{n(E)}{n(S)} = \dfrac{24.3}{100} = 0.243$

29. $P(E) = \dfrac{n(E)}{n(S)} = \dfrac{60.9}{100} = 0.609$

31. (a) $P(E) = \dfrac{n(E)}{n(S)} = \dfrac{1}{5}$

(b) $P(E) = \dfrac{n(E)}{n(S)} = \dfrac{1}{3}$

(c) $P(E) = \dfrac{n(E)}{n(S)} = 1$

33. (a) $P(\text{candidate } A \text{ or candidate } B) = 0.5 + 0.3 = 0.8$

(b) $P(\text{Candidate } 3) = 1 - 0.5 - 0.3 = 0.2$

35. $P(E) = \dfrac{n(E)}{n(S)} = \dfrac{70}{325} = \dfrac{14}{65}$

37. (a) $P(E) = \dfrac{n(E)}{n(S)} = \dfrac{57{,}510{,}000}{196{,}950{,}000} = \dfrac{1917}{6565}$

(b) $P(E) = \dfrac{n(E)}{n(S)} = \dfrac{139{,}440{,}000}{196{,}950{,}000} = \dfrac{4648}{6565}$

39. $P(E) = \dfrac{n(E)}{n(S)} = \dfrac{45^2}{60^2} = \dfrac{2025}{3600} = 0.5625$

(*E* is the probability that they do not meet.)

$1 - P(E) = 1 - 0.5625 = 0.4375$

41. (a)

Female

	X	X
X	XX	XX
Y	XY	XY

Male

Probability of a girl $= \frac{2}{4} = \frac{1}{2}$

Probability of a boy $= \frac{2}{4} = \frac{1}{2}$

(b) Because the probabilities are the same, it is equally likely that a newborn will be a boy or a girl.

43. $P(E) = \dfrac{n(E)}{n(S)} = \dfrac{1}{1 \cdot 4 \cdot 3 \cdot 2 \cdot 1} = \dfrac{1}{24}$

45. $P(E) = \dfrac{n(E)}{n(S)} = \dfrac{1}{10 \cdot 10 \cdot 10 \cdot 10 \cdot 10} = \dfrac{1}{100{,}000}$

47. $P(E) = \dfrac{n(E)}{n(S)} = \dfrac{1}{{}_{10}C_8} = \dfrac{1}{45}$

49. $P(E) = \dfrac{n(E)}{n(S)} = \dfrac{1}{{}_{10}C_2} = \dfrac{1}{\dfrac{10}{8!\,2!}} = \dfrac{1}{\dfrac{10 \cdot 9}{2 \cdot 1}} = \dfrac{1}{45}$

51. $P(E) = \dfrac{n(E)}{n(S)} = \dfrac{{}_4C_4}{{}_{10}C_4} = \dfrac{1}{\dfrac{10!}{6!\,4!}} = \dfrac{1}{\dfrac{10 \cdot 9 \cdot 8 \cdot 7}{4 \cdot 3 \cdot 2 \cdot 1}} = \dfrac{1}{210}$

53. $P(E) = \dfrac{n(E)}{n(S)} = \dfrac{{}_{13}C_5}{{}_{52}C_5} = \dfrac{\dfrac{13!}{8!\,5!}}{\dfrac{52!}{47!\,5!}} = \dfrac{\dfrac{13 \cdot 12 \cdot 11 \cdot 10 \cdot 9}{5 \cdot 4 \cdot 3 \cdot 2 \cdot 1}}{\dfrac{52 \cdot 51 \cdot 50 \cdot 49 \cdot 48}{5 \cdot 4 \cdot 3 \cdot 2 \cdot 1}} = \dfrac{13 \cdot 11 \cdot 9}{52 \cdot 51 \cdot 5 \cdot 49 \cdot 4} = \dfrac{11 \cdot 3}{4 \cdot 17 \cdot 5 \cdot 49 \cdot 4} = \dfrac{33}{66{,}640}$

55. (d) $8 \cdot 5 \cdot 3 = 120$

(e) The drawing will be done without replacement since each person receives only one gift.

(f) (a) $P(E) = \dfrac{n(E)}{n(S)} = \dfrac{1}{150}$

(b) $P(E) = \dfrac{n(E)}{n(S)} = \dfrac{1}{105}$

57. The probability that the event does not occur is $1 - \frac{3}{4} = \frac{1}{4}$.

59. Over an extended period, it will rain 40% of the time under the given weather conditions.

Review Exercises for Chapter 12

1. $a_1 = 3(1) + 5 = 8$

$a_2 = 3(2) + 5 = 11$

$a_3 = 3(3) + 5 = 14$

$a_4 = 3(4) + 5 = 17$

$a_5 = 3(5) + 5 = 20$

3. $a_1 = \dfrac{1}{2^1} + \dfrac{1}{2} = 1$

$a_2 = \dfrac{1}{2^2} + \dfrac{1}{2} = \dfrac{3}{4}$

$a_3 = \dfrac{1}{2^3} + \dfrac{1}{2} = \dfrac{1}{8} + \dfrac{4}{8} = \dfrac{5}{8}$

$a_4 = \dfrac{1}{2^4} + \dfrac{1}{2} = \dfrac{1}{16} + \dfrac{8}{16} = \dfrac{9}{16}$

$a_5 = \dfrac{1}{2^5} + \dfrac{1}{2} = \dfrac{1}{32} + \dfrac{16}{32} = \dfrac{17}{32}$

5. $a_n = 2n - 1$

7. $a_n = \dfrac{n}{(n + 1)^2}$

9. (a)   **11.** (b)   **13.** (d)

15. $\displaystyle\sum_{k=1}^{4} 7 = 7 + 7 + 7 + 7 = 28$

17. $\displaystyle\sum_{n=1}^{4}\left(\frac{1}{n} - \frac{1}{n+1}\right) = \frac{1}{2} + \frac{1}{6} + \frac{1}{12} + \frac{1}{20} = \frac{30 + 10 + 5 + 3}{60} = \frac{48}{60} = \frac{4}{5}$

19. $\displaystyle\sum_{n=1}^{4}(5n - 3)$ **21.** $\displaystyle\sum_{n=1}^{6}\frac{1}{3n}$ **23.** $d = -2.5$

25. $a_1 = 132 - 5(1) = 127$
$\quad a_2 = 132 - 5(2) = 122$
$\quad a_3 = 132 - 5(3) = 117$
$\quad a_4 = 132 - 5(4) = 112$
$\quad a_5 = 132 - 5(5) = 107$

27. $a_1 = \frac{3}{4}(1) + \frac{1}{2} = \frac{5}{4}$
$\quad a_2 = \frac{3}{4}(2) + \frac{1}{2} = 2$
$\quad a_3 = \frac{3}{4}(3) + \frac{1}{2} = \frac{11}{4}$
$\quad a_4 = \frac{3}{4}(4) + \frac{1}{2} = \frac{7}{2}$
$\quad a_5 = \frac{3}{4}(5) + \frac{1}{2} = \frac{17}{4}$

29. $a_1 = 5$
$\quad a_2 = 5 + 3 = 8$
$\quad a_3 = 8 + 3 = 11$
$\quad a_4 = 11 + 3 = 14$
$\quad a_5 = 14 + 3 = 17$

31. $a_1 = 80$
$\quad a_2 = 80 - \frac{5}{2} = \frac{160}{2} - \frac{5}{2} = \frac{155}{2}$
$\quad a_3 = \frac{155}{2} - \frac{5}{2} = \frac{150}{2} = 75$
$\quad a_4 = \frac{150}{2} - \frac{5}{2} = \frac{145}{2}$
$\quad a_5 = \frac{145}{2} - \frac{5}{2} = \frac{140}{2} = 70$

33. $a_n = dn + c$
$\quad 10 = 4(1) + c$
$\quad 6 = c$
$\quad a_n = 4n + 6$

35. $a_n = dn + c$
$\quad 1000 = -50(1) + c$
$\quad 1050 = c$
$\quad a_n = -50n + 1050$

37. $\displaystyle\sum_{k=1}^{12}(7k - 5) = 12\left(\frac{2 + 79}{2}\right) = 486$

39. $\displaystyle\sum_{j=1}^{100}\frac{j}{4} = 100\left(\frac{\frac{1}{4} + 25}{2}\right) = 1262.5$

41. *Keystrokes:*

[LIST] [MATH 5] [LIST] [OPS 5] 1.25 [X,T,θ] [+] 4 [,] [X,T,θ] [,] 1 [,] 60 [,] 1 [)] [ENTER]

$\displaystyle\sum_{i=1}^{60}(125i + 4) = 2527.5$

43. $r = \frac{3}{2}$

45. $a_n = a_1 r^{n-1}$
$\quad a_n = 10(3)^{n-1}$
$\quad a_1 = 10(3)^{1-1} = 10$
$\quad a_2 = 10(3)^{2-1} = 30$
$\quad a_3 = 10(3)^{3-1} = 90$
$\quad a_4 = 10(3)^{4-1} = 270$
$\quad a_5 = 10(3)^{5-1} = 810$

47. $a_n = a_1 r^{n-1}$

$a_n = 100\left(-\frac{1}{2}\right)^{n-1}$

$a_1 = 100\left(-\frac{1}{2}\right)^{1-1} = 100$

$a_2 = 100\left(-\frac{1}{2}\right)^{2-1} = -50$

$a_3 = 100\left(-\frac{1}{2}\right)^{3-1} = 25$

$a_4 = 100\left(-\frac{1}{2}\right)^{4-1} = -12.5$

$a_5 = 100\left(-\frac{1}{2}\right)^{5-1} = 6.25$

49. $a_1 = 3$

$a_2 = 2(3) = 6$

$a_3 = 2(6) = 12$

$a_4 = 2(12) = 24$

$a_5 = 2(24) = 48$

51. $a_n = a_1 r^{n-1}$

$a_n = 1\left(-\frac{2}{3}\right)^{n-1}$

53. $a_n = a_1 r^{n-1}$

$a_n = 24(2)^{n-1}$

55. $a_n = a_1 r^{n-1}$

$a_n = 12\left(-\frac{1}{2}\right)^{n-1}$

57. $\displaystyle\sum_{n=1}^{12} 2^n = 2\left(\frac{2^{12} - 1}{2 - 1}\right) = 8190$

59. $\displaystyle\sum_{k=1}^{8} 5\left(-\frac{3}{4}\right)^k = -\frac{15}{4}\left(\frac{\left(-\frac{3}{4}\right)^8 - 1}{-\frac{3}{4} - 1}\right) \approx -1.928$

61. $\displaystyle\sum_{i=1}^{8} (1.25)^{i-1} = 1\left(\frac{1.25^8 - 1}{1.25 - 1}\right) \approx 19.842$

63. $\displaystyle\sum_{n=1}^{120} 500(1.01)^n = 505\left(\frac{1.01^{120} - 1}{1.01 - 1}\right) \approx 116,169.54$

65. $\displaystyle\sum_{i=1}^{\infty} \left(\frac{7}{8}\right)^{i-1} = \frac{1}{1 - \frac{7}{8}} = \frac{1}{\frac{1}{8}} = 8$

67. $\displaystyle\sum_{k=1}^{\infty} 4\left(\frac{2}{3}\right)^{k-1} = \frac{4}{1 - \frac{2}{3}} = \frac{4}{\frac{1}{3}} = 12$

69. *Keystrokes:*

$\boxed{\text{LIST}}$ $\boxed{\text{MATH 5}}$ $\boxed{\text{LIST}}$ $\boxed{\text{OPS 5}}$ 50 $\boxed{(}$ 1.2 $\boxed{)}$ $\boxed{\wedge}$ $\boxed{(}$ $\boxed{\text{X,T,}\theta}$ $\boxed{-}$ 1 $\boxed{)}$ $\boxed{,}$ $\boxed{\text{X,T,}\theta}$ $\boxed{,}$ 1 $\boxed{,}$ 50 $\boxed{,}$ 1 $\boxed{)}$ $\boxed{\text{ENTER}}$

$\displaystyle\sum_{k=1}^{50} 50(1.2)^{k-1} \approx 2.275 \times 10^6$

71. $_8C_3 = \dfrac{8!}{3!\,5!} = \dfrac{8 \cdot 7 \cdot 6 \cdot 5!}{3 \cdot 2 \cdot 5!} = 56$

73. $_{12}C_0 = 1$

75. *Keystrokes:*

40 $\boxed{\text{MATH}}$ $\boxed{\text{PRB 3}}$ 4 $\boxed{\text{ENTER}}$ $_{40}C_4 = 91,390$

77. *Keystrokes:*

25 $\boxed{\text{MATH}}$ $\boxed{\text{PRB 3}}$ 6 $\boxed{\text{ENTER}}$ $_{25}C_6 = 177,100$

79. $(x + 1)^{10} = 1x^{10} + 10x^9(1) + 45x^8(1)^2 + 120x^7(1)^3 + 210x^6(1)^4 + 252x^5(1)^5 + 210x^4(1)^6 + 120x^3(1)^7 + 45x^2(1)^8$
$+ 10x(1)^9 + 1(1)^{10}$

$= x^{10} + 10x^9 + 45x^8 + 120x^7 + 210x^6 + 252x^5 + 210x^4 + 120x^3 + 45x^2 + 10x + 1$

81. $(3x - 2y)^4 = 1(3x)^4 + 4(3x)^3(-2y) + 6(3x)^2(-2y)^2 + 4(3x)(-2y)^3 + (-2y)^4$

$= 81x^4 - 216x^3y + 216x^2y^2 - 96xy^3 + 16y^4$

83. $(u^2 + v^3)^9 = 1(u^2)^9 + 9(u^2)^8(v^3) + 36(u^2)^7(v^3)^2 + 84(u^2)^6(v^3)^3 + 126(u^2)^5(v^3)^4 + 126(u^2)^4(v^3)^5 + 84(u^2)^3(v^3)^6 + 36(u^2)^2(v^3)^7$

$+ 9(u^2)(v^3)^8 + (v^3)^9$

$= u^{18} + 9u^{16}v^3 + 36u^{14}v^6 + 84u^{12}v^9 + 126u^{10}v^{12} + 126u^8v^{15} + 84u^6v^{18} + 36u^4v^{21} + 9u^2v^{24} + v^{27}$

85. $_nC_r x^{n-r} y^r$

$n = 10, \quad n - r = 5, \quad r = 5, \quad x = 3, \quad y = (-3)$

$_{10}C_5 = 252 \cdot (-3)^5 = -61,236$

87. $_nC_r x^{n-r} y^r$

$n = 7, \quad r = 3, \quad n - r = 4, \quad x = x, \quad y = (2y)$

$_7C_3(2)^3 = 35 \cdot 8 = 280$

89. $\displaystyle\sum_{n=1}^{50} 4n = 50\left(\frac{4 + 200}{2}\right) = 5100$

91. $\displaystyle\sum_{n=1}^{12} (3n + 19) = 12\left(\frac{22 + 55}{2}\right) = 462$

93. (a) $a_n = 85,000(1.012)^n$

    (b) $a_{50} = 85,000(1.012)^{50}$

        $\approx 154,328$

95. $2 \cdot 2 \cdot 2 = 8$

97. $_{15}C_5 = \dfrac{15 \cdot 14 \cdot 13 \cdot 12 \cdot 11}{5 \cdot 4 \cdot 3 \cdot 2 \cdot 1}$

             $= 3003$

99. $P(E) = \dfrac{n(E)}{n(S)} = \dfrac{2}{6} = \dfrac{1}{3}$

101. $P(E) = \dfrac{n(E)}{n(S)} = \dfrac{1}{4 \cdot 3 \cdot 2 \cdot 1} = \dfrac{1}{24}$

103. $P(E) = \dfrac{n(E)}{n(S)} = \dfrac{_{74}C_8}{_{84}C_8} \approx 0.346$

Chapter Test for Chapter 12

1. $a_n = \left(-\frac{2}{3}\right)^{n-1}$

$a_1 = \left(-\frac{2}{3}\right)^{1-1} = 1$

$a_2 = \left(-\frac{2}{3}\right)^{2-1} = -\frac{2}{3}$

$a_3 = \left(-\frac{2}{3}\right)^{3-1} = \frac{4}{9}$

$a_4 = \left(-\frac{2}{3}\right)^{4-1} = -\frac{8}{27}$

$a_5 = \left(-\frac{2}{3}\right)^{5-1} = \frac{16}{81}$

2. $\displaystyle\sum_{j=0}^{4} (3j + 1) = 1 + 4 + 7 + 10 + 13 = 35$

3. $\displaystyle\sum_{n=1}^{5} (3 - 4n) = 5\left(\frac{-1 + -17}{2}\right) = -45$

4. $\displaystyle\sum_{n=1}^{12} \frac{2}{3n + 1}$

5. $a_n = a_1 + (n - 1)d$

$a_n = 12 + (n - 1)4$

   $= 12 + 4n - 4 = 4n + 8$

$a_1 = 4(1) + 8 = 12$

$a_2 = 4(2) + 8 = 16$

$a_3 = 4(3) + 8 = 20$

$a_4 = 4(4) + 8 = 24$

$a_5 = 4(5) + 8 = 28$

6. $a_n = a_1 + (n - 1)d$

$a_n = 5000 + (n - 1)(-100)$

$a_n = 5000 - 100n + 100$

$a_n = -100n + 5100$

7. $\displaystyle\sum_{n=1}^{50} = 50\left(\frac{3 + 150}{2}\right) = 3825$

8. $r = -\dfrac{3}{2}$

9. $a_n = a_1 r^{n-1}$

$a_n = 4\left(\dfrac{1}{2}\right)^{n-1}$

10. $\displaystyle\sum_{n=1}^{8} 2(2^n) = 4\left(\frac{2^8 - 1}{2 - 1}\right) = 1020$

11. $\displaystyle\sum_{n=1}^{10} 3\left(\frac{1}{2}\right)^n = \frac{3\left(\frac{1}{2}^{10} - 1\right)}{2\left(\frac{1}{2} - 1\right)} = \frac{3069}{1024}$

12. $\displaystyle\sum_{i=1}^{\infty}\left(\frac{1}{2}\right)^i = \frac{\frac{1}{2}}{1 - \frac{1}{2}} = \frac{\frac{1}{2}}{\frac{1}{2}} = 1$

13. $\displaystyle\sum_{i=1}^{\infty} 4\left(\frac{2}{3}\right)^{i-1} = \frac{4}{1 - \frac{2}{3}} = \frac{4}{\frac{1}{3}} = 12$

14. $A = P\left(1 + \dfrac{r}{n}\right)^{nt}$

$$a_{300} = \left(1 + \frac{0.08}{12}\right)^{12(25)} = 50(1.0066667)^{300}$$

$$a_1 = 50(1.0066667)^1$$

$$\text{balance} = [50(1.0066667)^1]\left[\frac{1.0066667^{300} - 1}{1.0066667 - 1}\right] = \$47,868.64$$

15. $\displaystyle {}_{20}C_3 = \frac{20 \cdot 19 \cdot 18}{3 \cdot 2 \cdot 1} = 1140$

16. $(x - 2)^5 = 1(x^5) - 5x^4(2) + 10x^3(2)^2 - 10x^2(2)^3 + 5x(2)^4 - 1(2)^5$
$\qquad = x^5 - 10x^4 + 40x^3 - 80x^2 + 80x - 32$

17. The coefficient of x^3y^5 in expansion of $(x + y)^8$ is 56, since ${}_8C_3 = 56$.

18. plates = letter digit digit digit
$\qquad\qquad = 26 \quad\cdot\quad 10 \quad\cdot\quad 10 \quad\cdot\quad 10 \quad = 26{,}000 \text{ plates}$

19. $\displaystyle {}_{25}C_4 = \frac{25!}{4! \, 21!} = \frac{25 \cdot 24 \cdot 23 \cdot 22}{4 \cdot 3 \cdot 2 \cdot 1} = 12{,}650$

20. $1 - 0.75 = 0.25$

21. $\displaystyle P(E) = \frac{n(E)}{n(S)} = \frac{6}{52} = \frac{3}{26}$

22. $\displaystyle P(E) = \frac{n(E)}{n(S)} = \frac{1}{{}_4C_2} = \frac{1}{\frac{4!}{2! \, 2!}} = \frac{1}{\frac{4 \cdot 3}{2 \cdot 1}} = \frac{1}{6}$

Cumulative Test for Chapters 10–12

1. $(x - 5)^2 + 50 = 0$
$\qquad (x - 5)^2 = -50$
$\qquad\quad x - 5 = \pm\sqrt{-50}$
$\qquad\qquad\quad x = 5 \pm 5i\sqrt{2}$

2. $3x^2 + 6x + 2 = 0$
$\qquad x^2 + 2x + 1 = -\dfrac{2}{3} + 1$
$\qquad\quad (x + 1)^2 = \dfrac{1}{3}$
$\qquad\quad\; x + 1 = \pm\sqrt{\dfrac{1}{3}}$
$\qquad\qquad\; x = -1 \pm \dfrac{\sqrt{3}}{3}$

3. $x + \dfrac{4}{x} = 4$
$\qquad x\left[x + \dfrac{4}{x}\right] = [4]x$
$\qquad\qquad x^2 + 4 = 4x$
$\qquad\; x^2 - 4x + 4 = 0$
$\qquad\qquad (x - 2)^2 = 0$
$\qquad\qquad\quad x - 2 = 0$
$\qquad\qquad\qquad\; x = 2$

4. $\sqrt{x + 10} = x - 2$
$\quad (\sqrt{x + 10})^2 = (x - 2)^2$
$\qquad\; x + 10 = x^2 - 4x + 4$
$\qquad\qquad\; 0 = x^2 - 5x - 6$
$\qquad\qquad\; 0 = (x - 6)(x + 1)$
$\quad 0 = x - 6 \qquad 0 = x + 1$
$\quad 6 = x \qquad\qquad -1 = x$
$\qquad\qquad\qquad \text{not a solution}$

Check: $\sqrt{6 + 10} \overset{?}{=} 6 - 2$
$\qquad\quad \sqrt{16} = 4$
$\qquad\qquad\; 4 = 4$
$\quad \sqrt{-1 + 10} \overset{?}{=} -1 - 2$
$\qquad\qquad \sqrt{9} = -3$
$\qquad\qquad\; 3 \neq -3$

5. $y = a(x - h)^2 + k$

$1 = a(1 - 2)^2 + 3$

$1 = a(1) + 3$

$-2 = a$

$y = -2(x - 2)^2 + 3 = -2(x^2 - 4x + 4) + 3$

$\qquad\qquad\qquad\quad = -2x^2 + 8x - 8 + 3$

$\qquad\qquad\qquad\quad = -2x^2 + 8x - 5$

6. *Critical numbers:* $x = 0, \frac{7}{2}$

Test intervals:

Positive: $(-\infty, \, 0]$

Negative: $\left[0, \frac{7}{2}\right]$

Positive: $\left[\frac{7}{2}, \infty\right)$

Solution: $(-\infty, 0] \cup \left[\frac{7}{2}, \infty\right)$

7.
$$\frac{1}{x + 1} + \frac{2x}{x - 1} - 1 < 0$$

$$\frac{x - 1 + 2x(x + 1) - (x^2 - 1)}{(x + 1)(x - 1)} < 0$$

$$\frac{x - 1 + 2x^2 + 2x - x^2 + 1}{(x + 1)(x - 1)} < 0$$

$$\frac{x^2 + 3x}{(x + 1)(x - 1)} < 0$$

$$\frac{x(x + 3)}{(x + 1)(x - 1)} < 0$$

Critical numbers: $x = -3, -1, 0, 1$

Test intervals:

Positive: $(-\infty, -3)$

Negative: $(-3, -1)$

Positive: $(-1, 0)$

Negative: $(0, 1)$

Positive: $(1, \infty)$

Solution: $(-3, -1) \cup (0, 1)$

8.

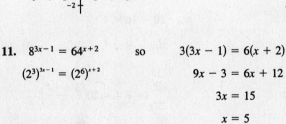

9.

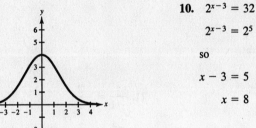

10. $2^{x-3} = 32$

$2^{x-3} = 2^5$

so

$x - 3 = 5$

$x = 8$

11. $8^{3x-1} = 64^{x+2}$ so $3(3x - 1) = 6(x + 2)$

$(2^3)^{3x-1} = (2^6)^{x+2}$

$9x - 3 = 6x + 12$

$3x = 15$

$x = 5$

12. $\ln 1 = 0$ because $e^0 = 1$

13. $\log_3 \frac{1}{81} = -4$ because $3^{-4} = \frac{1}{81}$

14.

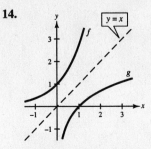

f and *g* are inverse functions, because graphs are symmetric to the line $y = x$

15. $3(\log_2 x + \log_2 y) - \log_2 z = \log_2(xy)^3 - \log_2 z = \log_2\dfrac{(xy)^3}{z}$

16. $4\log_2 x = 10$

$\log_2 x = \dfrac{10}{4}$

$\log_2 x = \dfrac{5}{2}$

$x = 2^{5/2}$

$x = \left(\sqrt{2}\right)^5$

$x = 4\sqrt{2}$

17. $3(1 + e^{2x}) = 20$

$1 + e^{2x} = \dfrac{20}{3}$

$e^{2x} = \dfrac{17}{3}$

$\ln e^{2x} = \ln\dfrac{17}{3}$

$2x = \ln\dfrac{17}{3}$

$x = \dfrac{\ln\dfrac{17}{3}}{2}$

$x \approx 0.8673$

18. $e^{3x+1} - 4 = 7$

$e^{3x+1} = 11$

$\ln e^{3x+1} = \ln 11$

$3x + 1 = \ln 11$

$x = \dfrac{\ln 11 - 1}{3}$

$x \approx 0.4660$

19. $2\log_3(x - 1) = \log_3(2x + 6)$

$2\log_3(x - 1)^2 = \log_3(2x + 6)$

$\log_3(x - 1)^2 = \log_3(2x + 6)$

$(x - 1)^2 = 2x + 6$

$x^2 - 2x + 1 = 2x + 6$

$x^2 - 4x - 5 = 0$

$(x - 5)(x + 1) = 0$

$x = 5 \qquad x = -1$

extraneous

not a solution

20. $6 - 8 = -2$

$4 - 6 = -2$

$2 - 4 = -2$

$d = -2$

$a_n = -2n + 10$

21. $\dfrac{\frac{2}{3}}{-1} = -\dfrac{2}{3}$

$\dfrac{-\frac{4}{9}}{\frac{2}{3}} = -\dfrac{2}{3}$

$r = -\dfrac{2}{3}$

$a_n = (-1)^n\left(-\dfrac{2}{3}\right)^{n-1}$

22. $12 - 8 = 4$

$16 - 12 = 4$

$20 - 16 = 4$

$d = 4$

$a_n = 4n + 4$

$\displaystyle\sum_{n=1}^{20} 4n + 4 = 20\left(\dfrac{8 + 84}{2}\right)$

$= 20\left(\dfrac{92}{2}\right) = 920$

23. $\displaystyle\sum_{i=0}^{\infty} 3\left(\dfrac{1}{2}\right)^i = \dfrac{3}{1 - \frac{1}{2}} = \dfrac{3}{\frac{1}{2}} = 6$

24. $(z - 3)^4 = 1(z^4) + 4(z^3)(-3) + 6(z^2)(-3)^2 + 4(z)(-3)^3 + 1(-3)^4$

$= z^4 - 12z^3 + 54z^2 - 108z + 81$

25. $25{,}000(0.8)^t = 15{,}000$

$$(0.8)^t = \frac{15{,}000}{25{,}000}$$

$$(0.8)^t = 0.6$$

$$\log_{0.8}(0.8)^t = \log_{0.8} 0.6$$

$$t = \frac{\log 0.6}{\log 0.8}$$

$$t \approx 2.3$$

Let $y_2 = 15{,}000$ and find the intersection, $t \approx 2.3$ years.

Keystrokes:

$\boxed{Y=}\ 25{,}000\ \boxed{(}\ 0.8\ \boxed{)}\ \boxed{\wedge}\ \boxed{X,T,\theta}\ \boxed{GRAPH}$

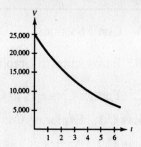

26.

$$A = P\left(1 + \frac{r}{n}\right)^{nt}$$

$$3750 = 3000\left(1 + \frac{r}{4}\right)^{4(3)}$$

$$1.25 = \left(1 + \frac{r}{4}\right)^{12}$$

$$1.25^{1/12} = 1 + \frac{r}{4}$$

$$1.25^{1/12} - 1 = \frac{r}{4}$$

$$4(1.25^{1/12} - 1) = r$$

$$0.075 = r$$

$$7.5\% = r$$

27. Total salary $= \displaystyle\sum_{n=1}^{10} 32{,}000(1.05)^n$

$$= 32{,}000\left(\frac{1.05^{10} - 1}{1.05 - 1}\right)$$

$$\approx \$402{,}493$$

28. $\displaystyle {}_{10}C_3 = \frac{10!}{7!\,3!} = \frac{10 \cdot 9 \cdot 8}{3 \cdot 2 \cdot 1} = 120$

29. $\displaystyle P(E) = \frac{n(E)}{n(S)} = \frac{1}{2 \cdot 2 \cdot 1} = \frac{1}{4}$

APPENDICES

APPENDICES
Appendix A   Conic Sections

Solutions to Odd-Numbered Exercises

1. $x^2 + y^2 = 25$

3. $(x - h)^2 + (y - k)^2 = r^2$
$(x + 4)^2 + (y - 1)^2 = 9$

5. $r = \sqrt{(5 - 0)^2 + (2 - 0)^2}$
$r = \sqrt{25 + 4}$
$r = \sqrt{29}$
$x^2 + y^2 = 29$

7. $r = \sqrt{(1 - 7)^2 + (3 + 5)^2}$
$r = \sqrt{36 + 64}$
$r = \sqrt{100}$
$r = 10$
$(x - 7)^2 + (y + 5)^2 = 100$

9. center $= (0, 0)$
radius $= 4$

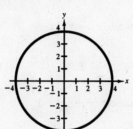

11. center $= (2, 3)$
radius $= 2$

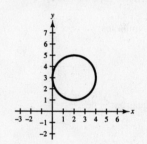

13. $25x^2 + 25y^2 - 144 = 0$
$x^2 + y^2 = \frac{144}{25}$
center $= (0, 0)$
radius $= \frac{12}{5}$

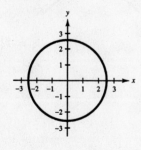

15. $\qquad x^2 + y^2 - 4x - 2y + 1 = 0$
$\qquad x^2 - 4x + y^2 - 2y = -1$
$(x^2 - 4x + 4) + (y^2 - 2y + 1) = -1 + 4 + 1$
$\qquad\qquad (x - 2)^2 + (y - 1)^2 = 4$
center $= (2, 1)$
radius $= 2$

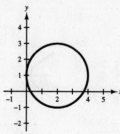

17. Equation: $\dfrac{x^2}{16} + \dfrac{y^2}{9} = 1$

19. Equation: $\dfrac{x^2}{25} + \dfrac{y^2}{4} = 1$

21. Equation: $\dfrac{x^2}{100} + \dfrac{y^2}{36} = 1$

23. Vertices: $(-4, 0)$, $(4, 0)$

Co-Vertices: $(0, 2)$, $(0, -2)$

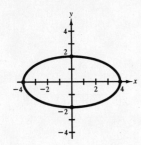

25. Vertices: $\left(-\dfrac{5}{3}, 0\right)$, $\left(\dfrac{5}{3}, 0\right)$

Co-Vertices: $\left(0, \dfrac{4}{3}\right)$, $\left(0, -\dfrac{4}{3}\right)$

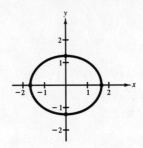

27. $4x^2 + y^2 - 4 = 0$

$$\dfrac{x^2}{1} + \dfrac{y^2}{4} = 1$$

Vertices: $(0, 2)$, $(0, -2)$

Co-Vertices: $(1, 0)$, $(-1, 0)$

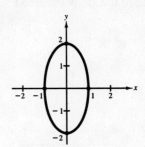

29. Vertices: $(3, 0)$, $(-3, 0)$

Asymptotes: $y = \dfrac{5}{3}x$

$$y = -\dfrac{5}{3}x$$

Equation: $\dfrac{x^2}{9} - \dfrac{y^2}{25} = 1$

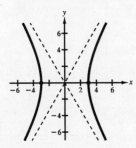

31. Vertices: $(0, 3)$, $(0, -3)$

Asymptotes: $y = \dfrac{3}{5}x$

$$y = -\dfrac{3}{5}x$$

Equation: $\dfrac{y^2}{9} - \dfrac{x^2}{25} = 1$

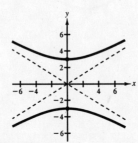

33. Vertices: $(0, 3)$, $(0, -3)$

Asymptotes: $y = \dfrac{9}{9}x$ $y = -\dfrac{9}{9}x$

$$y = x \quad\quad y = -x$$

Equation: $y^2 - x^2 = 9$

$$\dfrac{y^2}{9} - \dfrac{x^2}{9} = 1$$

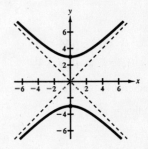

35. Vertices: $(4, 0)$, $(-4, 0)$

Asymptotes: $y = \dfrac{2}{4}x = \dfrac{1}{2}x$

$$y = -\dfrac{2}{4}x = -\dfrac{1}{2}x$$

Equation: $4y^2 - x^2 + 16 = 0$

$$\dfrac{4y^2}{-16} - \dfrac{x^2}{-16} = \dfrac{-16}{-16}$$

$$\dfrac{-y^2}{4} + \dfrac{x^2}{16} = 1$$

$$\dfrac{x^2}{16} - \dfrac{y^2}{4} = 1$$

37. $\dfrac{x^2}{16} - \dfrac{y^2}{64} = 1$

39. $\dfrac{x^2}{81} - \dfrac{y^2}{36} = 1$

41. $x^2 + y^2 = 9$ (c)

43. $\dfrac{x^2}{4} + \dfrac{y^2}{9} = 1$ (e)

45. $x^2 - y^2 = 4$ (a)

47. $x^2 + y^2 = 4500^2$

$x^2 + y^2 = 20{,}250{,}000$

49. Equation of ellipse $= \dfrac{x^2}{50^2} + \dfrac{y^2}{40^2} = 1$

or

$\dfrac{x^2}{2500} + \dfrac{y^2}{1600} = 1$

$\dfrac{45^2}{2500} + \dfrac{y^2}{1600} = 1$

$\dfrac{y^2}{1600} = 0.19$

$y^2 = 304$

$y = 17.435596 \approx 17.4$ feet

51. (a) $x^2 + y^2 = 625$ (equation of circle)

(x, y) of the rectangle is also the point on the circle, so y-coordinate equals:

$x^2 + y^2 = 625$

$y^2 = 625 - x^2$

$y = \sqrt{625 - x^2}$

Width $= 2\left(\sqrt{625 - x^2}\right)$

Area $= 2x \cdot 2\left(\sqrt{625 - x^2}\right)$

Area $= 4x\sqrt{625 - x^2}$

(b)

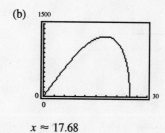

$x \approx 17.68$

53. $d_2 - d_1 = \sqrt{(12 - -12)^2 + (12 - 0)^2} - \sqrt{(12 - 12)^2 + (12 - 0)^2}$

$= \sqrt{576 + 144} - \sqrt{0 + 144}$

$= \sqrt{720} - \sqrt{144}$

$= 12\sqrt{5} - 12$

$d_4 - d_3 = (12 + a) - (12 - a)$

$= 12 + a - 12 + a$

$= 2a$

$2a = 12\sqrt{5} - 12$

$a = 6\sqrt{5} - 6$

vertex $= \left(6\sqrt{5} - 6, 0\right)$

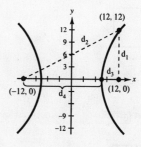

Appendix B   Introduction to Graphing Utilities

1. *Keystrokes:*

$\boxed{Y=}$ $\boxed{(-)}$ 3 $\boxed{X,T,\theta}$ $\boxed{GRAPH}$

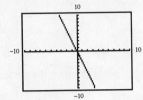

3. *Keystrokes:*

$\boxed{Y=}$ $\boxed{(}$ 3 $\boxed{\div}$ 4 $\boxed{)}$ $\boxed{X,T,\theta}$ $\boxed{-}$ 6 $\boxed{GRAPH}$

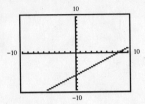

5. *Keystrokes:*

$\boxed{Y=}$ $\boxed{(}$ 1 $\boxed{\div}$ 2 $\boxed{)}$ $\boxed{X,T,\theta}$ $\boxed{x^2}$ $\boxed{GRAPH}$

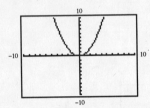

7. *Keystrokes:*

$\boxed{Y=}$ $\boxed{X,T,\theta}$ $\boxed{x^2}$ $\boxed{-}$ 4 $\boxed{X,T,\theta}$ $\boxed{+}$ 2 $\boxed{GRAPH}$

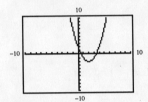

9. *Keystrokes:*

$\boxed{Y=}$ $\boxed{ABS}$ $\boxed{(}$ $\boxed{X,T,\theta}$ $\boxed{-}$ 3 $\boxed{)}$ $\boxed{GRAPH}$

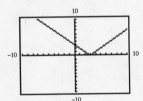

11. *Keystrokes:*

$\boxed{Y=}$ $\boxed{ABS}$ $\boxed{(}$ $\boxed{X,T,\theta}$ $\boxed{x^2}$ $\boxed{-}$ 4 $\boxed{)}$ $\boxed{GRAPH}$

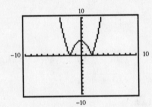

13. *Keystrokes:*

$\boxed{Y=}$ 27 $\boxed{X,T,\theta}$ $\boxed{+}$ 100 $\boxed{GRAPH}$

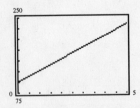

15. *Keystrokes:*

$\boxed{Y=}$ 0.001 $\boxed{X,T,\theta}$ $\boxed{x^2}$ $\boxed{+}$ 0.5 $\boxed{X,T,\theta}$ $\boxed{GRAPH}$

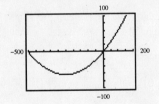

17. *Keystrokes:*

$\boxed{Y=}$ 15 $\boxed{+}$ $\boxed{ABS}$ $\boxed{(}$ $\boxed{X,T,\theta}$ $\boxed{-}$ 12 $\boxed{)}$ $\boxed{GRAPH}$

$$
\begin{array}{l}
\text{Xmin} = 4 \\
\text{Xmax} = 20 \\
\text{Xscl} = 1 \\
\text{Ymin} = 14 \\
\text{Ymax} = 22 \\
\text{Yscl} = 1
\end{array}
$$

19. *Keystrokes:*

$\boxed{Y=}$ $\boxed{(-)}$ 15 $\boxed{+}$ $\boxed{ABS}$ $\boxed{(}$ $\boxed{X,T,\theta}$ $\boxed{+}$ 12 $\boxed{)}$ $\boxed{GRAPH}$

Xmin = -20
Xmax = -4
Xscl = 1
Ymin = -16
Ymax = -8
Yscl = 1

21. *Keystrokes:*

y_1 $\boxed{Y=}$ 2 $\boxed{X,T,\theta}$ $\boxed{+}$ $\boxed{(}$ $\boxed{X,T,\theta}$ $\boxed{+}$ 1 $\boxed{)}$ $\boxed{ENTER}$

y_2 $\boxed{(}$ 2 $\boxed{X,T,\theta}$ $\boxed{+}$ $\boxed{X,T,\theta}$ $\boxed{)}$ $\boxed{+}$ 1 $\boxed{GRAPH}$

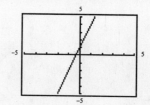

Associative Property of Addition

23. *Keystrokes:*

y_1 $\boxed{Y=}$ 2 $\boxed{(}$ 1 $\boxed{\div}$ 2 $\boxed{)}$ $\boxed{ENTER}$

y_2 1 $\boxed{GRAPH}$

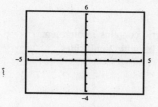

Multiplicative Inverse Property

25. *Keystrokes:*

$\boxed{Y=}$ 9 $\boxed{-}$ $\boxed{X,T,\theta}$ $\boxed{x^2}$ $\boxed{GRAPH}$

Trace to *x*-intercepts: $(-3, 0)$ and $(3, 0)$

Trace to *y*-intercept: $(0, 9)$

27. *Keystrokes:*

$\boxed{Y=}$ 6 $\boxed{-}$ $\boxed{ABS}$ $\boxed{(}$ $\boxed{X,T,\theta}$ $\boxed{+}$ 2 $\boxed{)}$ $\boxed{GRAPH}$

Trace to *x*-intercepts: $(-8, 0)$ and $(4, 0)$

Trace to *y*-intercept: $(0, 4)$

29. *Keystrokes:*

$\boxed{Y=}$ 2 $\boxed{X,T,\theta}$ $\boxed{-}$ 5 $\boxed{GRAPH}$

Trace to *x*-intercept: $\left(\frac{5}{2}, 0\right)$

Trace to *y*-intercept: $(0, -5)$

31. *Keystrokes:*

$\boxed{Y=}$ $\boxed{X,T,\theta}$ $\boxed{x^2}$ $\boxed{+}$ 1.5 $\boxed{X,T,\theta}$ $\boxed{-}$ 1 $\boxed{GRAPH}$

Trace to *x*-intercepts: $(-2, 0)$ and $\left(\frac{1}{2}, 0\right)$

Trace to *y*-intercept: $(0, -1)$

33. *Keystrokes:*

y_1 $\boxed{Y=}$ $\boxed{(-)}$ 4 $\boxed{ENTER}$

y_2 $\boxed{(-)}$ $\boxed{ABS}$ $\boxed{X,T,\theta}$ $\boxed{GRAPH}$

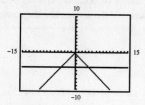

Triangle

35. *Keystrokes:*

y_1 $\boxed{Y=}$ $\boxed{ABS}$ $\boxed{X,T,\theta}$ $\boxed{-}$ 8 $\boxed{ENTER}$

y_2 $\boxed{(-)}$ $\boxed{ABS}$ $\boxed{X,T,\theta}$ $\boxed{+}$ 8 $\boxed{GRAPH}$

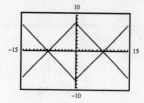

Square

37. *Keystrokes:*

y_1 $\boxed{Y=}$ 0.07 $\boxed{X,T,\theta}$ $\boxed{x^2}$ $\boxed{+}$ 1.06 $\boxed{X,T,\theta}$ $\boxed{+}$ 88.97 $\boxed{ENTER}$

y_2 0.02 $\boxed{X,T,\theta}$ $\boxed{x^2}$ $\boxed{-}$ 0.23 $\boxed{X,T,\theta}$ $\boxed{+}$ 10.70 $\boxed{GRAPH}$

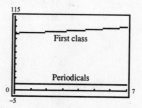

Appendix C   Further Concepts in Geometry

Appendix C.1   Exploring Congruence and Similarity

1. Answers will vary.

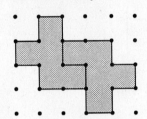

3.

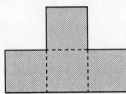

5. Two figures are similar if they have the same shape. Figures (a) and (b) are similar.

7.

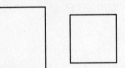

9. The grid contains 7 congruent triangles with 2-unit sides.

11. No. All the triangles in the grid are equilteral triangles, and all of these triangles have the same shape. Therefore all the triangles in the grid are similar to each other.

13. False. For example, the two squares shown below are similar, but they are not congruent.

15. True. Any two squares have the same shape, so any two squares are similar.

17. The ray from P through Q is matched with notation (d).

19. The length of the segment between P and Q is matched with notation (b).

21. $\angle ZXW$ and $\angle WXZ$ are names for the same angle.

$\angle ZXY$ and $\angle YXZ$ are names for the same angle.

$\angle YXW$ and $\angle WXY$ are names for the same angle.

$\angle ZXY$ and $\angle YXW$ are adjacent angles.

23. (b) $m\angle WXY \approx 30°$

25. (d) Equiangular

A triangle with angle measures of 60°, 60°, and 60° is an equiangular triangle because all the angles are the same size.

27. (f) Right

A triangle with angle measures of 30°, 60°, and 90° is a right triangle because it contains a right angle.

29. (c) Obtuse

A triangle with angle measures of 20°, 145°, and 15° is an obtuse triangle because it contains an obtuse angle.

31. The three points of congruent sides are

$\overline{LM} \cong \overline{NO}, \overline{MP} \cong \overline{NQ},$ and $\overline{LP} \cong \overline{OQ}.$

33. If $\triangle ABC \cong \triangle TUV$, then $m\angle C = m\angle V.$

35. If $\triangle LMN \cong \triangle TUV$, then $\overline{LN} \cong \overline{TV}.$

37.

	Scalene	Isosceles	Equilateral
Acute	Yes	Yes	Yes
Obtuse	Yes	Yes	No
Right	Yes	Yes	No

  (Not possible)

 (Not possible)

39.
$$AC = BC$$
$$2x + 6 = 12$$
$$2x = 6$$
$$x = 3$$
$$BC = 12$$
$$AC = 2(3) = 6 + 6 = 12$$
$$AB = 4x = 4(3) = 12.$$

Therefore, all three sides of the triangle are of length 12. Yes, the triangle is equilateral.

41.
$$AC = BC$$
$$2x = 4x - 6$$
$$-2x = -6$$
$$x = 3$$
$$AC = 2x = 2(3) = 6$$
$$BC = 4x - 6 = 4(3) - 6 = 12 - 6 = 6$$
$$AB = x + 3 = 3 + 3 = 6.$$

Therefore, all three sides of the triangles are of length 6. Yes, the triangle is equilateral.

43. $\dfrac{1/8 \text{ inch}}{1 \text{ foot}} = \dfrac{1/(8 \cdot 12) \text{ foot}}{1 \text{ foot}} = \dfrac{1/96 \text{ foot}}{1 \text{ foot}} = \dfrac{1}{96}$

$\dfrac{1}{96} \cdot 1200 = 12.5$

The scale drawing would be 12.5 feet by 12.5 feet. No, such a large drawing does not seem reasonable.

45. If V is located at either $(3, 1)$ or $(3, 5)$, then $\triangle PQR \cong \triangle TUV$.

47. Form a tetrahedron, a three-dimensional figure with four congruent triangular faces.

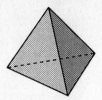

Appendix C.2 Angles

1. Answers will vary.

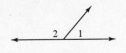

3. Answers will vary.

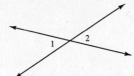

5. Answers will vary.

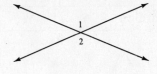

7. $\angle AOC$ and $\angle COD$ are adjacent, congruent, supplementary angles.

9. $\angle BOC$ and $\angle COE$ are adjacent, supplementary angles.

11. $\angle BOC$ and $\angle COF$ are adjacent, complementary angles.

13. False.

$m\angle 3 = 30° \implies m\angle 1 = 150° \implies m\angle 4 = 30°$.

15. False.

For example,
$m\angle 3 = 30° \implies m\angle 1 = 150° \implies m\angle 2 = 150°$, and thus $\angle 2 \neq \angle 3$.

17. True.

19. $x = 110°$ because vertical angles are congruent.

21. $(2x - 5°) + 75° = 180°$ because two angles that form a linear pair are supplementary.

$$2x - 5° + 75° = 180°$$
$$2x + 70° = 180°$$
$$2x = 110°$$
$$x = \frac{110°}{2}$$
$$x = 55°$$

23. $\qquad 3x + 20° = 5x - 50°$ because vertical angles are congruent.

$$3x - 5x + 20° = -50°$$
$$-2x + 20° = -50°$$
$$-2x = -70°$$
$$x = \frac{-70°}{-2}$$
$$x = 35°$$

25. Answer (c)

$m\angle PSQ + m\angle QST = 180°$ because two angles that form a linear pair are supplementary.

$$m\angle PSQ + 110° = 180°$$
$$m\angle PSQ = 70°$$

$m\angle PSQ + m\angle P + m\angle Q = 180°$ because the sum of the measures of the interior angles of a triangle is 180°.

$$70° + 40° + m\angle Q = 180°$$
$$110° + m\angle Q = 180°$$
$$m\angle Q = 70°$$

27. ∠3 and ∠5 are alternate interior angles because they lie between *l* and *m* and on opposite sides of *t*.

∠4 and ∠6 are alternate interior angles because they lie between *l* and *m* and on opposite sides of *t*.

29. ∠3 and ∠6 are corresponding interior angles because they lie between *l* and *m* and on the same side of *t*.

∠4 and ∠5 are corresponding interior angles because they lie between *l* and *m* and on the same side of *t*.

31. $m\angle 1 + 70° = 180°$ because two angles that form a linear pair are supplementary.

$$m\angle 1 = 110°$$

$m\angle 2 = m\angle 1$ by the Alternate Exterior Angle Theorem

$m\angle 2 = 110°$

33. $m\angle 2 + 110° = 180°$ because two angles that form a linear pair are supplementary.

$$m\angle 2 = 70°$$

$m\angle 1 = m\angle 2$ by the Alternate Interior Angles Theorem

$m\angle 1 = 70°$

Alternate approach for angle 1:

$m\angle 1 + 110° = 180°$ by Consecutive Interior Angles Theorem

$$m\angle 1 = 70°$$

35. $a = 4a - 90°$ because corresponding angles are congruent.

$-3a = 90°$

$a = \dfrac{-90°}{-3}$

$a = 30°$

$60° - b = 2b$ because corresponding angles are congruent.

$60° = 3b$

$\dfrac{60°}{3} = b$

$20° = b$

37. ∠2, ∠5, and ∠7 are the interior angles of the triangle.

(These are the original three angles of the triangle.)

39. *Step 1*: $m\angle 1 + 110° = 180°$ because two angles that form a linear pair are supplementary.

$$m\angle 1 = 70°$$

Step 2: $m\angle 3 + 110° = 180°$ because two angles that form a linear pair are supplementary.

$$m\angle 3 = 70°$$

Step 3: $m\angle 2 = 110°$ because vertical angles are congruent.

Step 4: $m\angle 7 + 155° = 180°$ because two angles that form a linear pair are supplementary.

$$m\angle 7 = 25°$$

Step 5: $m\angle 8 = 155°$ because vertical angles are congruent.

—CONTINUED—

39. **—CONTINUED—**

Step 6: $m\angle 5 + m\angle 7 + m\angle 2 = 180°$ because the sum of the measures of the interior angles of a triangle is 180°.

$$m\angle 5 + 25° + 110° = 180°$$
$$m\angle 5 + 135° = 180°$$
$$m\angle 5 = 45°$$

Step 7: $m\angle 4 + m\angle 5 = 180°$ because two angles that form a linear pair are supplementary.

$$m\angle 4 + 45° = 180°$$
$$m\angle 4 = 135°$$

Step 8: $m\angle 6 = 135°$ because vertical angles are congruent.

41. $m\angle B = 35°$ because corresponding angles of congruent triangles are congruent.

43. $m\angle D + m\angle E + m\angle F = 180°$
$$105° + 35° + m\angle F = 180°$$
$$140° + m\angle F = 180°$$
$$m\angle F = 40°$$

45. True.

The sum of the measures of three angles of a triangle is 180°.

The sum of the measures of the two 60° angles is 120°.

Therefore, the measure of the third angle is $180° - 120° = 60°$.

Thus, the triangle has three 60° angles, so the triangle is equiangular.

47. *Step 1:* $m\angle 1 + 60° + 90° = 180°$ because the sum of the measures of three angles of a triangle is 180°.
$$m\angle 1 + 150° = 180°$$
$$m\angle 1 = 30°$$

Step 2: $m\angle 2 = 60°$ because vertical angles are congruent.

Step 3: $m\angle 3 + 60° + 70° = 180°$ because the sum of the measurers of the three angles of a triangle is 180°.
$$m\angle 3 + 130° = 180°$$
$$m\angle 3 = 50°$$

Step 4: $m\angle 6 + 70° + 55° = 180°$ because the three angles combine to form a straight angle.
$$m\angle 6 + 125° = 180°$$
$$m\angle 6 = 55°$$

Step 5: $m\angle 7 = 55°$ because vertical angles are congruent.

Step 6: $m\angle 8 + 55° = 180°$ because two angles that form a linear pair are supplementary.
$$m\angle 8 = 125°$$

Step 7: $m\angle 4 + 55° + 90° = 180°$ because the sum of the measures of the three angles of a triangle is 180°.
$$m\angle 4 + 145° = 180°$$
$$m\angle 4 = 35°$$

Step 8: $m\angle 5 + 90° = 180°$ becaused two angles that form a linear pair are supplementary.
$$m\angle 5 = 90°$$

Step 9: $m\angle 9 + 90° + 55° = 180°$ because the sum of the measures of the three angles of a triangle is 180°.
$$m\angle 9 + 145° = 180°$$
$$m\angle 9 = 35°$$

49.

$m\angle A + m\angle B + m\angle C = 180°$ because the sum of the measures of the three interior angles of a triangle is 180°.

$$13° + m\angle B + 90° = 180°$$
$$m\angle B + 103° = 180°$$
$$m\angle B = 77°$$

51.

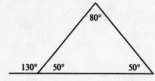

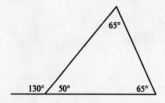

53. The sum of the measures of the three interior angles of a triangle is 180°.

$$x + (x + 30°) + (x + 60°) = 180°$$
$$x + x + 30° + x + 60° = 180°$$
$$3x + 90° = 180°$$
$$3x = 90°$$
$$x = 30°$$
$$x + 30° = 30° + 30° = 60°$$
$$x + 60° = 30° + 60° = 90°$$

The measures of the three interior angles are 30°, 60°, and 90°.

55. The sum of the measures of the three interior angles of a triangle is 180°.

$$(3x + 2°) + (5x - 1°) + (6x + 11°) = 180° \qquad 3x + 2° = 3(12°) + 2° = 38°$$
$$3x + 2° + 5x - 1° + 6x + 11° = 180° \qquad 5x - 1° = 5(12°) - 1° = 59°$$
$$14x + 12° = 180° \qquad 6x + 11° = 6(12°) + 11° = 83°$$
$$14x = 168°$$
$$x = 12°$$

The measures of the three interior angles are 38°, 59°, and 83°.

Appendix D   Further Concepts in Statistics

1. Organize scores by ordering the numbers. Let the leaves represent the units digits. Let the stems represent the tens digits.

Stems	Leaves
7	0 5 5 5 7 7 8 8 8
8	1 1 1 1 2 3 4 5 5 5 5 7 8 9 9 9
9	0 2 8
10	0 0

3. Organize scores by ordering the numbers. Let the leaves represent the units digits. Let the stems represent the tens digits.

Stems	Leaves
5	2 5 9
6	2 3 6 6 7
7	0 1 2 3 4 7 8 8 9
8	0 1 3 4 5 7 9
9	0 0 2 3 3 3 5 6 8 9
10	0 0

5. Frequency Distribution

Interval	Tally
[15, 22)	ЖЖ |||
[22, 29)	ЖЖ	
[29, 36)	ЖЖ
[36, 43)	||||
[43, 50)	ЖЖ ||

Histogram

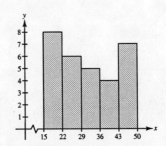

7.

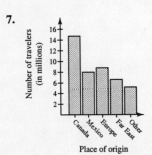

Place of origin

9. 1985: 165 million tons

1995: 210 million tons

11. Total waste and recycled waste increased every year.

13. Total waste equals the sum of the other three quantities.

15.

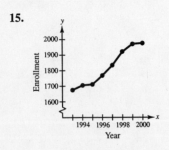

17.

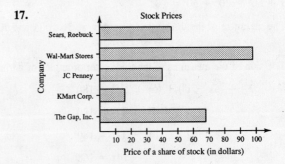

19.

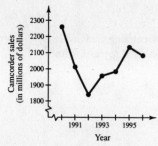

21. x and y have a positive correlation because as x increases y also increases.

23. Yes, it appears that players with more hits tend to have more runs batted in.

25. Negative correlation, because as the age of the car increases the value of the car decreases.

27. Positive correlation, because as the age of a tree increases the height also increases.

29.

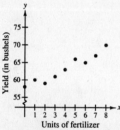

31. The air pressure at 42,500 feet is approximately 2.45 pounds per square inch.

33.

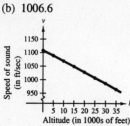

Use graphing utility by entering data in 2 lists with STAT PLOT graph.

35. Use graphing utility to find regression line STAT CALC4 [Lin Reg($ax + b$)].

(a) $y = 57.49 + 1.43x$

(b) 71.8

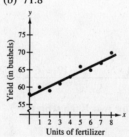

37.

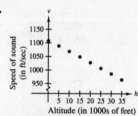

Use graphing utility by entering data in 2 lists with STAT PLOT graph.

39. Use graphing utility to find regression line STAT CALC4 [Lin Reg($ax + b$)].

(a) $v = 1117.3 - 4.1h$

(b) 1006.6

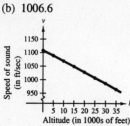

41. $y = -2.179x + 22.964$

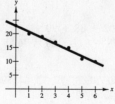

Use graphing utility by entering data in 2 lists with STAT PLOT graph. Find regression line with STAT CALC4.

43. $y = 2.378x + 23.546$

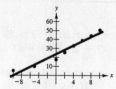

Use graphing utility by entering data in 2 lists with STAT PLOT graph. Find regression line with STAT CALC4.

45. (a) $y = 11.1 + 0.28t$ 12.78

(b)

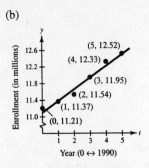

Year (0 ↔ 1990)

(c) $r \approx 0.987$

Use graphing utility by entering data in 2 lists with STAT PLOT graph. Find regression line with STAT CALC4 .

47. Mean: $\dfrac{5 + 12 + 7 + 14 + 8 + 9 + 7}{7} = 8.86$

Median: 14 12 9 8 7 7 5 = 8
$\qquad\qquad\uparrow$
Middle score

Mode: 7 occurs twice = 7

49. Mean: $\dfrac{5 + 12 + 7 + 24 + 8 + 9 + 7}{7} = 10.29$

Median: 24 12 9 8 7 7 5 = 8
$\qquad\qquad\uparrow$
Middle score

Mode: 7 occurs twice = 7

51. (a) $(67.92 + 59.84 + 52 + 52.50 + 57.99 + 65.35 + 81.76 + 74.98 + 87.82 + 83.18 + 65.35 + 57) \div 12 = \$67.14$

(b) Median: 87.82
83.18
81.76
74.98
67.92
65.35 ⎤
65.35 ⎦ average of 2 middle bills = $65.35
59.84
57.99
57.00
52.50
52.00

53. (a) Mean: $\dfrac{0 \cdot 1 + 1 \cdot 24 + 2 \cdot 45 + 3 \cdot 54 + 4 \cdot 50 + 5 \cdot 19 + 6 \cdot 7}{200} \approx 3.07$

(b) Median: List all the data. Find the average of the two 100th scores = 3

(c) Mode: 3 occurs 54 times

55. Answers vary. One possibility: {4, 4, 10}

57. (a) $(99 + 64 + 80 + 77 + 59 + 72 + 87 + 79 + 92 + 88 + 90 +$
$\quad 42 + 20 + 89 + 42 + 100 + 98 + 84 + 78 + 91) \div 20 = 76.55$

(b) Median: list the scores from highest to lowest
$\qquad$ 100, 99, 98, 92, 91, 90, 89, 88, 87, 84, 80, 79, 78, 77, 72, 64, 59, 42, 42, 20
Find the average of the two tenth scores: 84 and 80 = 82

(c) Mode: 42 occurs twice
The median gives the most representative description since more of the test scores are in the 80's.

Appendix E   Introduction to Logic

Appendix E.1   Statements and Truth Tables

1. Statement, because only one truth value can be assigned.

3. Open statement, because a specific figure is needed to assign a truth value.

5. Open statement, because a value of x is needed to assign a truth value.

7. Open statement, because values of x and y are needed to assign a truth value.

9. Nonstatement, because no truth value can be assigned.

11. Open statement, because a specific place is needed to assign a truth value.

13. (a) $2^2 - 5(2) + 6 \overset{?}{=} 0$

$\quad\quad 4 - 10 + 6 = 0$

$\quad\quad\quad\quad 0 = 0$ True

(b) $(-2)^2 - 5(-2) + 6 \overset{?}{=} 0$

$\quad\quad 4 + 10 + 6 = 0$

$\quad\quad\quad\quad 20 \neq 0$ False

15. (a) $(-2)^2 \overset{?}{\leq} 4$

$\quad\quad 4 \leq 4$ True

(b) $0^2 \overset{?}{\leq} 4$

$\quad\quad 0 \leq 4$ True

17. (a) $4 - |0| \overset{?}{=} 2$

$\quad\quad 4 \neq 2$ False

(b) $4 - |1| \overset{?}{=} 2$

$\quad\quad 3 \neq 2$ False

19. (a) $\dfrac{-4}{-4} \overset{?}{=} 1$

$\quad\quad 1 = 1$ True

(b) $\dfrac{0}{0} \overset{?}{=} 1$

Undefined $\neq 1$ False

21. (a) $\sim p$: The sun is not shining.

(b) $\sim q$: It is not hot.

(c) $p \wedge q$: The sun is shining and it is hot.

(d) $p \vee q$: The sun is shining or it is hot.

23. (a) $\sim p$: Lions are not mammals.

(b) $\sim q$: Lions are not carnivorous.

(c) $p \wedge q$: Lions are mammals and lions are carnivores.

(d) $p \vee q$: Lions are mammals or lions are carnivorous.

25. (a) $\sim p \wedge q$: The sun is not shining and it is hot.

(b) $\sim p \vee q$: The sun is not shining or it is hot.

(c) $p \wedge \sim q$: The sun is shining and it is not hot.

(d) $p \vee \sim q$: The sun is shining or it is not hot.

27. (a) $\sim p \wedge q$: Lions are not mammals and lions are carnivorous.

(b) $\sim p \vee q$: Lions are not mammals or lions are carnivorous.

(c) $p \wedge \sim q$: Lions are mammals and lions are not carnivorous.

(d) $p \vee \sim q$: Lions are mammals or lions are not carnivorous.

29. p: It is four o'clock.

q: It is time to go home.

$p \wedge \sim q$

31. p: It is four o'clock.

q: It is time to go home.

$\sim p \vee q$

33. p: The dog has fleas.

q: The dog is scratching.

$\sim p \vee \sim q$

35. p: The dog has fleas.

q: The dog is scratching.

$\sim p \wedge q$

37. The bus is blue.

39. x is not equal to 4.

41. The earth is flat.

43.

p	q	~p	~p ∧ q
T	T	F	F
T	F	F	F
F	T	T	T
F	F	T	F

45.

p	q	~p	~q	~p ∨ ~q
T	T	F	F	F
T	F	F	T	T
F	T	T	F	T
F	F	T	T	T

47.

p	q	~q	p ∨ ~q
T	T	F	T
T	F	T	T
F	T	F	F
F	F	T	T

49.

p	q	~p	~q	~p ∧ q	p ∨ ~q
T	T	F	F	F	T
T	F	F	T	F	T
F	T	T	F	T	F
F	F	T	T	F	T

not identical
not logically equivalent

51.

p	q	~p	~q	p ∨ ~q	~(p ∨ ~q)	~p ∧ q
T	T	F	F	T	F	F
T	F	F	T	T	F	F
F	T	T	F	F	T	T
F	F	T	T	T	F	F

identical
logically equivalent

53.

p	q	~p	~q	p ∧ ~q	~p ∨ q	~(~p ∨ q)
T	T	F	F	F	T	F
T	F	F	T	T	F	T
F	T	T	F	F	T	F
F	F	T	T	F	T	F

identical
logically equivalent

55. Let p = The house is red.

q = It is made of wood.

(a) = p ∧ ~q

(b) p ∨ ~q

p	q	~q	p ∧ ~q	p ∨ ~q
T	T	F	F	T
T	F	T	T	T
F	T	F	F	F
F	F	T	T	T

not identical
not logically equivalent

57. Let p = The house is white.

q = It is blue.

(a) ~(p ∨ q)

(b) ~p ∧ ~q

p	q	p ∨ q	~(p ∨ q)	~p	~q	~p ∧ ~q
T	T	T	F	F	F	F
T	F	T	F	F	T	F
F	T	T	F	T	F	F
F	F	F	T	T	T	T

identical
logically equivalent

59.

p	~p	~p ∧ p
T	F	F
T	F	F
F	T	F
F	T	F

not a tautology

61.

p	$\sim p$	$\sim(\sim p)$	$\sim(\sim p) \vee \sim p$
T	F	T	T
T	F	T	T
F	T	F	T
F	T	F	T

a tautology

63.

p	q	$\sim p$	$\sim q$	$p \wedge q$	$\sim(p \wedge q)$	$\sim p \vee \sim q$
T	T	F	F	T	F	F
T	F	F	T	F	T	T
F	T	T	F	F	T	T
F	F	T	T	F	T	T

identical
logically equivalent

Appendix E.2   Implications, Quantifiers, and Venn Diagrams

1. (a) $p \rightarrow q$: If the engine is running, then the engine is wasting gasoline.

(b) $q \rightarrow p$: If the engine is wasting gasoline, then the engine is running.

(c) $\sim q \rightarrow \sim p$: If the engine is not wasting gasoline, then the engine is not running.

(d) $p \rightarrow \sim q$: If the engine is running, then the engine is not wasting gasoline.

3. (a) $p \rightarrow q$: If the integer is even, then it is divisible by 2.

(b) $q \rightarrow p$: If it is divisible by 2, then the integer is even.

(c) $\sim q \rightarrow \sim p$: If if it is not divisible by 2, then the integer is not even.

(d) $p \rightarrow \sim q$: If the integer is even, then it is not divisible by 2.

5. Let p = The economy is expanding.

q = Interest rates are low.

$q \rightarrow p$

7. Let p = The economy is expanding.

q = Interest rates are low.

$p \rightarrow q$

9. Let p = The economy is expanding.

q = Interest rates are low.

$p \rightarrow q$

11.

Hypothesis	Conclusion	Implication
T	T	T

13.

Hypothesis	Conclusion	Implication
F	T	T

15.

Hypothesis	Conclusion	Implication
T	F	F

17.

Hypothesis	Conclusion	Implication
F	T	T

19.

Hypothesis	Conclusion	Implication
T	T	T

21. Converse:
If you can see the eclipse, then the sky is clear.

Inverse:
If the sky is not clear, then you cannot see the eclipse.

Contrapositive:
If you cannot see the eclipse, then the sky is not clear.

23. Converse:
If the deficit increases, then taxes were raised.

Inverse:
If taxes are not raised, then the deficit will not increase.

Contrapositive:
If the deficit does not increase, then taxes were not raised.

25. Converse:
It is necessary to apply for the
visa to have a birth certificate.

Inverse:
It is not necessary to have a birth
certificate to not apply for the visa.

Contrapositive:
It is not necessary to apply for the
visa to not have a birth certificate.

27. Negation:
Paul is not a junior
and not a senior.

29. Negation:
If the temperature increases,
then the metal rod will not expand.

31. Negation:
We will go to the ocean and the
weather forecast is not good.

33. Negation: No students are in extracurricular activities.

35. Negation: Some contact sports are not dangerous.

37. Negation: Some children are allowed at the concert.

39. Negation: None of the $20 bills are counterfeit.

41.

p	q	$\sim q$	$p \rightarrow \sim q$	$\sim(p \rightarrow \sim q)$
T	T	F	F	T
T	F	T	T	F
F	T	F	T	F
F	F	T	T	F

43.

p	q	$q \rightarrow p$	$\sim(q \rightarrow p)$	$\sim(q \rightarrow p) \wedge q$
T	T	T	F	F
T	F	T	F	F
F	T	F	T	T
F	F	T	F	F

45.

p	q	$(p \vee q)$	$\sim p$	$(p \vee q) \wedge (\sim p)$	$[(p \vee q) \wedge (\sim p)] \rightarrow q$
T	T	T	F	F	T
T	F	T	F	F	T
F	T	T	T	T	T
F	F	F	T	F	T

47.

p	q	$\sim p$	$\sim q$	$p \rightarrow \sim q$	$\sim q \rightarrow p$	$p \leftrightarrow \sim q$	$(p \leftrightarrow \sim q) \rightarrow \sim p$
T	T	F	F	F	T	F	T
T	F	F	T	T	T	T	F
F	T	T	F	T	T	T	T
F	F	T	T	T	F	F	T

49.

p	q	$q \rightarrow p$	$\sim p$	$\sim q$	$\sim p \rightarrow \sim q$
T	T	T	F	F	T
T	F	T	F	T	T
F	T	F	T	F	F
F	F	T	T	T	T

identical

51.

p	q	$p \rightarrow q$	$\sim(p \rightarrow q)$	$\sim q$	$p \wedge \sim q$
T	T	T	F	F	F
T	F	F	T	T	T
F	T	T	F	F	F
F	F	T	F	T	F

identical

53.

p	q	$p \to q$	$\sim q$	$(p \to q) \vee \sim q$	$\sim p$	$p \vee \sim p$
T	T	T	F	T	F	T
T	F	F	T	T	F	T
F	T	T	F	T	T	T
F	F	T	T	T	T	T

identical

55.

p	q	$\sim p$	$\sim p \wedge q$	$p \to (\sim p \wedge q)$
T	T	F	F	F
T	F	F	F	F
F	T	T	T	T
F	F	T	F	T

identical

57. Let p = A number is divisible by 6.

q = It is divisible by 2.

Statement is $p \to q$

p	q	$p \to q$	$\sim p$	$\sim q$	$\sim q \to \sim p$
T	T	T	F	F	T
T	F	F	F	T	F
F	T	T	T	F	T
F	F	T	T	T	T

identical

(c) If a number is not divisible by 2, then it is not divisible by 6.

$\sim q \to \sim p$

$p \to q \equiv \sim q \to \sim p$

59. (a) Some citizens over the age of 18 have the right to vote is *not* logically equivalent to above statement.

61. Let A = people who are happy

B = college students

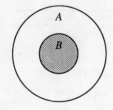

63. Let A = people who are happy

B = college students

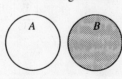

65. Let A = people who are happy

B = college students

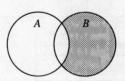

67. Let A = people who are happy

B = college students

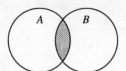

69. Let A = people who are happy

B = college students

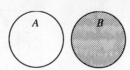

71. (a) Statement does not follow.

(b) Statement follows.

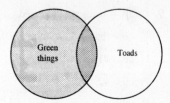

73. (a) Statement does not follow.

(b) Statement does not follow.

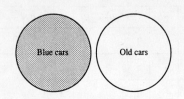

Appendix E.3   Logical Arguments

1.

p	q	~p	~q	p→~q	(p→~q) ∧ q
T	T	F	F	F	F
T	F	F	T	T	F
F	T	T	F	T	T
F	F	T	T	T	F

[(p→~q) ∧ q]→~p
T
T
T
T

3.

p	q	~p	p ∨ q	(p ∨ q) ∧ ~p
T	T	F	T	F
T	F	F	T	F
F	T	T	T	T
F	F	T	F	F

[(p ∨ q) ∧ ~p]→q
T
T
T
T

5.

p	q	~p	~q	~p→q	(~p→q) ∧ p
T	T	F	F	T	T
T	F	F	T	T	T
F	T	T	F	T	F
F	F	T	T	F	F

[(~p→q) ∧ p]→~q
F
T
T
T

7.

p	q	p ∨ q	(p ∨ q) ∧ q	[(p ∨ q) ∧ q]→p
T	T	T	T	T
T	F	T	F	T
F	T	T	T	F
F	F	F	F	T

9. Let: p = taxes are increased

   q = businesses will leave the state

Premise #1: p→q

Premise #2: p

Conclusion: q

p	q	p→q	(p→q) ∧ p	(p→q) ∧ p→q
T	T	T	T	T
T	F	F	F	T
F	T	T	F	T
F	F	T	F	T

Argument is valid.

11. Let: p = taxes are increased

q = businesses will leave the state

Premise #1: $p \rightarrow q$

Premise #2: q

Conclusion: p

p	q	$p \rightarrow q$	$(p \rightarrow q) \wedge q$	$(p \rightarrow q) \wedge q \rightarrow p$
T	T	T	T	T
T	F	F	F	T
F	T	T	T	F
F	F	T	F	T

Argument is invalid.

13. Let: p = doors are locked

q = car was not stolen

Premise #1: $p \rightarrow q$

Premise #2: $\sim q$

Conclusion: $\sim p$

p	q	$p \rightarrow q$	$\sim p$	$\sim q$	$(p \rightarrow q) \wedge \sim q$	$(p \rightarrow q) \wedge \sim q \rightarrow \sim p$
T	T	T	F	F	F	T
T	F	F	F	T	F	T
F	T	T	T	F	F	T
F	F	T	T	T	T	T

Argument is valid.

15.

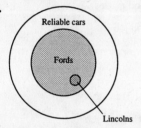

Argument is valid.

17.

Argument is invalid.

19. Let: p = Eric is at the store.

q = He is at the handball court.

Premise #1: $p \vee q$

Premise #2: $\sim p$

Conclusion: q

p	q	$\sim p$	$p \vee q$	$(p \vee q) \wedge \sim p$	$(p \vee q) \wedge \sim p \rightarrow q$
T	T	F	T	F	T
T	F	F	T	F	T
F	T	T	T	T	T
F	F	T	F	F	T

Argument is valid.

21. Let: p = It is a diamond.

q = It sparkles in the sunlight.

p	q	$p \wedge q$	$\sim(p \wedge q)$	$\sim(p \wedge q) \wedge q$	$\sim(p \wedge q) \wedge q \rightarrow p$
T	T	T	F	F	T
T	F	F	T	F	T
F	T	F	T	T	F
F	F	F	T	F	T

Argument is invalid.

Premise #1: $\sim(p \wedge q)$

Premise #2: q

Conclusion: p

23. Let: p = 7 is a prime number

q = 7 does not divide evenly into 21

Premise #1: $p \rightarrow q$

Premise #2: $\sim q$

So conclusion must be $\sim p$ or 7 is not a prime number which is (b).

25. Let: p = Economy improves

q = Interest rates lowered

Premise #1: $p \rightarrow q$

Premise #2: $\sim q$

So conclusion must be $\sim p$ or the economy does not improve which is (c).

27. Let: p = Smokestack emissions must be reduced

q = Acid rain will continue as an environmental problem

Premise #1: $p \vee q$

Premise #2: $\sim p$

So conclusion must be q or acid rain will continue as an environmental problem which is (b).

29. Let: p = Rodney studies

q = He will make good grades

r = He will get a good job

Premise #1: $p \rightarrow q$

Premise #2: $q \rightarrow r$

Conclusion: $p \rightarrow r$ Law of Transitivity

$\sim r$: If Rodney doesn't get a good job

$\sim p$: He didn't study

So by the Law of Contraposition the answer is (c).

31. Let A = All numbers divisible by 5

B = All numbers divisible by 10

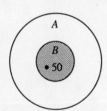

Argument is valid.

33. Let A = People eligible to vote

B = People under the age of 18

C = College students

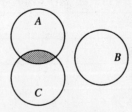

Argument is invalid.

35. Let p represent the statement "Sue drives to work," let q represent "Sue will stop at the grocery store," and let r represent "Sue will buy milk."

First write:

> Premise #1: $p \rightarrow q$
>
> Premise #2: $q \rightarrow r$
>
> Premise #3: p

Reorder the premises:

> Premise #3: p
>
> Premise #1: $p \rightarrow q$
>
> Premise #2: $q \rightarrow r$
>
> Conclusion: r

Then we can conclude r. That is, "Sue will buy milk."

37. Let p represent "This is a good product," let q represent "We will buy it," and let r represent "the product was made by XYZ Corporation."

First write:

> Premise #1: $p \rightarrow \sim q$
>
> Premise #2: $r \vee \sim q$
>
> Premise #3: $\sim r$

Note that $p \rightarrow q \equiv q \rightarrow \sim p$, and reorder the premises:

> Premise #2: $r \vee \sim q$
>
> Premise #3: $\sim r$
>
> (Conclusion from Premise #2, Premise #3: $\sim q$)
>
> Premise #1: $\sim q \rightarrow \sim p$
>
> Conclusion: $\sim p$

Then we can conclude $\sim p$. That is, "It is not a good product."